Informatik-Fachberichte 219

Herausgeber: W. Brauer
im Auftrag der Gesellschaft für Informatik (GI)

H. Burkhardt K. H. Höhne
B. Neumann (Hrsg.)

Mustererkennung 1989

11. DAGM-Symposium
Hamburg, 2.-4. Oktober 1989

Proceedings

Springer-Verlag
Berlin Heidelberg New York
London Paris Tokyo Hong Kong

Herausgeber

Hans Burkhardt
Technische Universität Hamburg-Harburg, Technische Informatik I
Harburger Schloßstraße 20, D-2100 Hamburg 90

Karl Heinz Höhne
Universität Hamburg, Institut für Mathematik und
Datenverarbeitung in der Medizin
Martinistraße 52, D-2000 Hamburg 20

Bernd Neumann
Universität Hamburg, Fachbereich Informatik
Bodenstedtstraße 16, D-2000 Hamburg 50

CR Subject Classification (1987): I.2, I.4, I.5

ISBN-13:978-3-540-51748-1   e-ISBN-13:978-3-642-75102-8
DOI: 10.1007/978-3-642-75102-8

2145/3140 – 543210 – Gedruckt auf säurefreiem Papier

Veranstalter

DAGM: Deutsche Arbeitsgemeinschaft für Mustererkennung

Tagungsleitung

H. Burkhardt, Technische Informatik I, Technische Universität Hamburg–Harburg

K.H. Höhne, Institut für Mathematik und Datenverarbeitung in der Medizin, Universität Hamburg

B. Neumann, FB Informatik, Universität Hamburg

Programmkomitee

H. Burkhardt	Hamburg
E. Dorrer	München
G. Hirzinger	Oberpfaffenhofen
K.H. Höhne	Hamburg
H. Kazmierczak	Ettlingen
W.G. Kropatsch	Graz
O. Kübler	Zürich
M. Kuhn	Hamburg
B. Neumann	Hamburg
H. Niemann	Erlangen
E. Paulus	Braunschweig
S. Pöppl	Neuherberg
D.P. Pretschner	Hildesheim
B. Schleifenbaum	Wetzlar
W. von Seelen	Mainz

DAGM   Deutsche Arbeitsgemeinschaft für Mustererkennung

Die DAGM veranstaltet seit 1978 jährlich an verschiedenen Orten ein wissenschaftliches Symposium mit dem Ziel, Aufgabenstellungen, Denkweisen und Forschungsergebnisse aus verschiedenen Gebieten der Mustererkennung vorzustellen, den Erfahrungs- und Ideenaustausch zwischen den Fachleuten anzuregen und den Nachwuchs zu fördern. Die DAGM wird durch folgende wissenschaftliche Trägergesellschaften gebildet:

DGaO	Deutsche Gesellschaft für angewandte Optik
GMDS	Deutsche Gesellschaft für medizinische Dokumentation, Informatik und Statistik
GI	Gesellschaft für Informatik
ITG	Informationstechnische Gesellschaft
DGNM	Deutsche Gesellschaft für Nuklearmedizin
IEEE	The Institute of Electrical and Electronic Engineers, Deutsche Sektion
DGPF	Deutsche Gesellschaft für Photogrammetrie und Fernerkundung

Die DAGM ist Mitglied der International Association for Pattern Recognition (IAPR).

Zum Geleit

Anläßlich der 10. Jahrestagung der DAGM in Zürich schied Herr Prof. Dr. H. H. Nagel als Vorsitzender der DAGM satzungsgemäß aus seinem Amte aus. Es würde den Rahmen dieses Geleitworts sicherlich sprengen, wollte man an dieser Stelle die Laudatio für Herrn Prof. Nagel anläßlich der DAGM-Tagung im September 1988 in Zürich wiederholen.

Als neuem Vorsitzenden der DAGM ist es mir aber eine sehr angenehme Pflicht und persönliche Freude, die Zusammenfassung an dieser Stelle zu wiederholen:

> Herr Prof. Nagel hat sich um die DAGM verdient gemacht.

In diesen Dank möchte ich auch den scheidenden stellvertretenden Vorsitzenden der DAGM, Herrn Prof. Dr.-Ing. H. Kazmierczak, mit einschließen.

Wesentliche, heute fast selbstverständlich erscheinende Merkmale der DAGM sind unter der Amtszeit von Herrn Kollegen Nagel entstanden:

- qualifizierte Proceedings mit entsprechender Tradition und Akzeptanz,

- Anerkennung in der International Association for Pattern Recognition (IAPR),

- Öffnung nach außen, z. B. Tagungen mit den österreichischen und schweizerischen Schwestergesellschaften,

- die nun schon beinahe traditionellen DAGM-Preise für herausragende Arbeiten.

Wie im vergangenen Jahr hat die Trägerversammlung beschlossen, die Träger der DAGM-Preise für das Jahr 1988 in den diesjährigen Tagungsband aufzunehmen, um dieser Ehrung die ihr angemessene Würdigung auch über den Kreis der Symposiumsteilnehmer hinaus zu verschaffen.

Für ihren Einsatz bei der Vorbereitung und Durchführung dieses Symposiums möchte ich den Kollegen Burkhardt, Höhne und Neumann sehr herzlich danken.

Die Tatsache, daß das DAGM-Symposium nun zum zweiten Male in Hamburg stattfindet, spricht für den Erfolg der Mustererkennungsaktivitäten im Hamburger Bereich.

Neuherberg, den 25. Juli 1989

Prof. Dr. Dr. S. J. Pöppl
Vorsitzender der DAGM

Der mit 1000 DM dotierte

DAGM–Preis 1988

wurde

D.Morgue und G.Gerig

Ecole Nationale Supérieure des Telecommunications
Paris
Institut für Kommunikationstechnik, ETH Zürich

für den folgenden Beitrag verliehen:

Recognition of Nonrigid Objects Using the Generalized Hough Transform

---

Der mit 1000 DM dotierte

DAGM–Preis 1988

wurde

C.K. Sung

Fraunhofer Institut für Informations- und Datenverarbeitung
Karlsruhe

für den folgenden Beitrag verliehen:

Extraktion von typischen und komplexen Vorgängen aus einer langen Bildfolge einer Verkehrsszene

Weitere Preise für das Jahr 1988 wurden verliehen an

M. Dresselhaus, G. Hartmann
B. Mertsching

Universität–Gesamthochschule
Paderborn
Fachbereich Elektrotechnik

Positionserfassung und Verfolgung
von Objekten in hierarchisch
codierten Bildern

E. Gmür, H. Bunke

Institut für Informatik und
angewandte Mathematik
Universität Bern

PHI–1: Ein CAD–basiertes
Roboter–Sichtsystem

E. Hiltebrand

Institut für Elektrotechnik
ETH–Zürich

Interaktive Bearbeitung und
Darstellung medizinischer
Volumen–Bilddaten

A. Luhn, A. Dengel

Siemens AG, Zentralbereich Forschung
und Technik, München
Institut für Informatik,
Universität Stuttgart

Modellgestützte Segmentierung
und Hypothesengenerierung für
die Analyse von Papierdokumenten

R. Mester, U.Franke, T. Aach

Institut für Elektrische
Nachrichtentechnik
RWTH Aachen

Segmentation of Image Pairs and
Sequences by Contour Relaxation

Vorwort

Mustererkennung heißt, ähnlich wie bei Sinneswahrnehmungen, mit Sensoren Signale aus der technischen Umwelt zu empfangen und mit Hilfe zuvor gelernter Situationen momentane Messungen zu interpretieren und dabei im Hinblick auf neue Eindrücke lernfähig zu sein. Es ist eine kluge Entscheidung gewesen, ein Symposium einer solch schwierigen Aufgabenstellung zu widmen, welche uns sicher noch viele Jahre beschäftigen wird. Methoden kommen und gehen, und es werden diejenigen Bestand haben, welche uns bei der Lösung dieser Aufgabe weiterhelfen. Gerade deshalb wird auch der Themenkatalog einem gewissen Wandel unterliegen und offen sein für neue Lösungsansätze und Anwendungsgebiete. So könnten zur Zeit neue methodische Anregungen aus dem Forschungsgebiet der neuronalen Netze und neue Anwendungsaufgaben aus dem Bereich der Robotik kommen.

Nahezu 100 eingereichte Arbeiten dokumentieren das große Interesse an diesem Fachgebiet. Die seit Zürich eingeführte Regel, vollständige Entwürfe zu verlangen, hat sich bewährt und die Arbeit des Programmkomitees im Sinne einer sorgfältigen Auswahl von Beiträgen unterstützt. Es wurden 42 Vorträge und 38 Plakatpräsentationen ausgewählt, welche zusammen mit 3 eingeladenen Arbeiten diesmal einen recht umfangreichen Tagungsband bilden.

Die Teamarbeit der letztjährigen Tagungsleitung in Zürich war uns ein Vorbild, und so wurde das 11. DAGM-Symposium gemeinsam von der Technischen Universität Hamburg-Harburg und der Universität Hamburg organisiert. Der 800. Hafengeburtstag schien uns ein willkommener Rahmen für diese Veranstaltung zu sein. Ganz herzlich bedanken möchten wir uns für die großzügige Unterstützung durch den Hamburger Senat und für die Spenden von seiten der Industrie. Unser Dank geht auch an das Programmkomitee für die fruchtbare Zusammenarbeit, an Herrn Dr. Thamer, Frau Löbkens sowie an alle Mitarbeiter, welche die Vorbereitung und die Durchführung des Symposiums tatkräftig unterstützt haben.

Verbleibt noch, uns allen einen erfolgreichen Verlauf der Veranstaltung zu wünschen mit lebhaften und fruchtbaren Diskussionen, so daß auch zukünftig das DAGM-Symposium ”Mustererkennung” eine Begegnungsstätte für einen lebendigen und ideenreichen Gedankenaustausch sein wird.

H. Burkhardt, K. H. Höhne, B. Neumann

Inhalt

Digitale Geometrie, Morphologie, Topologie

Plakate

Dreidimensionale Bildauswertung

Anwendungen

Plakate

Medizinische Anwendungen I

Plakate

Segmentierung

Plakate

Bildfolgen

Plakate

Objektmodelle

Plakate

Hardware und Systeme

Plakate

Medizinische Anwendungen II

Grundlagen, Objekterkennung

Plakate

Bewegungs- und Stereobildverarbeitung

Spracherkennung

Standardisierungen

Application of Mathematical Morphology
to Machine Vision

Hyonam Joo and Robert M Haralick

Department of Electrical Engineering, FT-10
University of Washington
Seattle, WA 98195

Abstract

This paper gives an analysis of a variety of morphological vision procedures. The analysis is designed to illustrate the power and flexibility of mathematical morphology for the extraction of shape information from gray tone and binary images.

1   Introduction

Usually, when a vision expert is given a task, he or she first analyzes the task, makes a plan, produces a procedure, tries it, evaluates the test results, and refines or updates the procedure. In doing so, the expert uses the knowledge of the given problem domain and his or her knowledge of the vision algorithms and is able to reason through using the knowledge to determine a reasonable vision procedure.

It is natural for us to ask how a human vision expert can solve the vision tasks and to investigate the mechanism involved in the process of developing such solutions. In particular, the problem we hope to solve is to reduce any machine vision task to a sequence of given morphological operations. We must use them at the right moment, and with the right parameters to accomplish what the vision procedure is designed for. Research on this topic can be found in [3, 7, 12, 14, 16, 20].

Considering that a vision procedure can be decomposed into a sequence of primitive operators, the following constitutes a first set of questions that can be raised toward the solution of the automatic compilation of vision procedures: What is a set of basic primitive operators that do not require finer grained decomposition relative to the task at hand and what does each primitive operator do when applied to the image ? How do primitive operators aggregate together to encode knowledge which we express using the terms of coarse grain system ? Are there any properties of an image that suggest to a human expert which operator to use on the image to solve a particular vision problem ? These are questions regarding the knowledge about the primitive operators and the discrimination information contained in a distorted noisy image they work on. Even though there exists a number of known image descriptions available, an image in general contains a huge amount of information. However, it appears that there exists a set of image descriptions relevant to the selection of each primitive operator. An application of an operator results in an output

"

image with a different set of descriptions that we can possibly predict. These operators can be described in terms of the relationship between the class of images on which the operator works and the parameters of the operator. We need to discover and explicitly describe this knowledge in order to utilize it in the automatic compilation of machine vision procedure.

The problem of the automatic compilation of vision procedures involves more subproblems to be solved than the ones listed above. For example, here we have not asked questions about the reasoning mechanism, the evaluation method, and the representational method. In this paper, as part of the solution to the automatic compilation problem, we only discuss the analysis performed on a set of morphological vision procedures developed by vision experts.

In section 2 we review the basic definitions for both binary and grayscale morphological operators. Haralick *et al* [6] has a more complete discussion. A set of example morphological algorithms developed by vision experts is described in section 3. A summary and some comments relative to the solution of automatic compilation of vision procedures are given in section 4.

2 Definitions

Let E denote the set of integers. A pixel in an image is a tuple $(r,c) \in E \times E$ where r and c represents the row and column coordinates of the pixel in an image. A binary image can be thought of as subset of $E \times E$ whose elements are the pixels of binary value one. The following are the definitions for the binary morphological operators used in this paper. For any set $A \subseteq E^N$ and $x \in E^N$, let A_x denote the translation of A by x which is defined by

$$A_x = \{y \mid \text{for some } a \in A, y = a + x\}$$

For any set $A \in E^N$, let $\check{A}$ denote the reflection of A about the origin which is defined by

$$\check{A} = \{x \mid \text{for some } a \in A, x = -a\}$$

The dilation of a set $A \subseteq E^N$ by a set $B \subseteq E^N$ is defined by

$$A \oplus B = \{x \mid \text{for some } a \in A \text{ and } b \in B, x = a + b\}$$
$$= \bigcup_{b \in B} A_b$$

The erosion of A by B is defined by

$$A \ominus B = \{x \mid \text{for every } b \in B, x + b \in A\}$$
$$= \bigcap_{b \in B} A_{-b}$$

The opening of A by B is defined by

$$A \circ B = (A \ominus B) \oplus B$$
$$= \{x \mid \text{for some } y. \, x \in B_y \subseteq A\}$$

The closing of A by B is defined by

$$A \bullet B = (A \oplus B) \ominus B$$
$$= \{x \mid x \in \check{B}_y \text{ implies } \check{B}_y \cap A \neq \phi\}$$

The hit–or–miss transformation of a set A by a structuring element $B = \{B^0, B^1\}$ is defined as:

$$A \circledast B = (A \ominus B^0) - (A \oplus \check{B}^1)$$

where the minus sign means a set subtraction. A point x belongs to $A \circledast B$ if and only if $B_x^0 \subset A$ and $B_x^1 \subset A^c$.

Let $M = \{1, \cdots, m\}$ and $N = \{1, \cdots, n\}$ be the sets of indices. Let $I^k, k \in M$ and $J^k, k \in N$ be the connected components in binary images I and J respectively. the conditional dilation of the image I with respect to the image J by a structuring element B is defined as:

$$I \oplus |_J B = \bigcup_i \{J^i \mid (J^i \oplus B) \cap I^k \neq \phi, \text{ for some } k \in M\}$$

This operation extracts only the connected components of J, when dilated by B intersect with I. The conditional dilation can be implemented as the following iterative procedure.

$$I_0 = I$$
$$I_n = (I_{n-1} \oplus B) \cap J$$

and

$$I \oplus |_J B = I_n$$

where n satisfies $I_n = I_m$, for all $m > n$.

The morphological operators can be defined for a real valued function $f : E^N \to R$, where R is a set of real numbers. For any function $f : F \to R$, define the reflection of f by $\check{f} : \check{F} \to R$ where

$$\check{f}(x) = f(-x)$$

Now, before we define the grayscale morphological operators, let us define the umbra and top. A set $A \subseteq E^N \times R$ is an umbra if and only if $(x, y) \in A$ implies that $(x, z) \in A$ for every $z \leq y$. The top of A is a function $T[A]$ mapping the spatial domain of A, $\{x \in E^N \mid \text{for some } y \in E, (x, y) \in A\}$, to the top surface of A,

$$T[A](x) = \max\{y \in R \mid (x, y) \in A\}$$

If $F \subseteq E^N$ and $f : F \to R$, the umbra of f is the set

$$U[f] = \{(x, y) \in (F \times R) \mid y \leq f(x)\}$$

Now, the definitions for the morphological operators follow: For $F, K \subseteq E^N, f : F \to R$, and $k : K \to R$, the dilation of f by k is the mapping $(f \oplus k) : (F \oplus K) \to R$ defined by

$$(f \oplus k)(x) = T[U[f] \oplus U[k]](x)$$
$$= \max_{z \in K, (x-z) \in F} \{f(x - z) + k(z)\}$$

The erosion of f by k is the mapping $(f \ominus k) : (F \ominus K) \to R$ defined by

$$(f \ominus k)(x) = T[U[f] \ominus U[k]](x)$$
$$= \min_{z \in K} \{f(x + z) - k(z)\}$$

The opening $(f \circ k)$ and closing $(f \bullet k)$ for functions are similarly defined as:

$$f \circ k = (f \ominus k) \oplus k$$
$$f \bullet k = (f \oplus k) \ominus k$$

We use the following special type of binary structuring elements in our discussion in this chapter.

- (c, r): A point at column c and row r.

- box(w, h): A box of width w and height h.

- diamond(r): A diamond whose diagonal is r pixels long.

- disk(r): A disk of radius r such that disk$(r) = \{x \mid \|x\| \leq r\}$.

- line$(((c_1, r_1), (c_2, r_2))$: A straight line segment connecting two points (c_1, r_1) and (c_2, r_2).

- ring(r): An annulus defined by disk$(r + 1)$ - disk(r).

The following structuring function are used for the grayscale operations. The function values are all zero while the domain of the structuring functions are different.

- rod(r): The domain of function is a disk of radius r.

- brick(w, h): The domain of function is a box of width w and height h.

3 Compendium of Morphological Image Processing Algorithms

This section presents a collection of morphological image processing algorithms developed by vision experts. It is hoped that by analyzing the algorithms in this collection we can find the knowledge and the reasoning mechanism employed by vision experts in their process of developing those algorithms. Each of the following subsections provides an analysis for each algorithm. The analysis is divided into four parts – input image description, goal, morphological procedure, and reasoning and knowledge. The description of the input image that the algorithm is expecting and the goal image that the algorithm is supposed to produce are given in plain English. The morphological procedure that accomplishes the task is presented. The procedure is written in two different fashions. It is first given in step by step fashion, where each step corresponds to a single meaningful primitive morphological operation. A short description of each step is also given. In each step i, G_i or B_i represents the grayscale or the binary image produced at the step i, where G_0 or B_0 designates the original input image. Next, the procedure is expressed in as compact an equation as possible using the smallest number of intermediate images. Finally, we present the reasoning and knowledge constituting the morphologic basis for the algorithm.

3.1 Identify leads coming from a SMD component

input image: [1]

- The image consists of one very dark component mounted on a medium bright background.

- The background can have both dark and bright stripes.

- The component is rectangular shaped and located in the middle of the image. It is the largest object in the image. The leads are horizontally oriented in a vertical column and connected to the two vertical boundaries of the component. The body of the rectangle may contain medium bright or bright objects which are smaller or thinner than the leads.

- The leads are very bright and small rectangular shaped objects. The distance between a pair of leads next to each other is longer than the height of one lead. The leads themselves may contain some small dark areas.

goal:

- Detect only the bright almost rectangular shaped blobs with known size, the leads, which are located inside the very dark large rectangular shaped component in the middle of the image. The small dark areas that may exist inside the lead should also be detected as being a part of the lead. Thus, each detected lead should be a convex shape with no holes in it.

morphological procedure:

1. $B_1 = G_0 < 200$:
   Binarization to extract dark region.

2. $B_2 = B_1 \oplus \mathrm{box}(5,5)$:
   Improve connectivity. It fills holes caused by the medium bright objects inside the component, but not the leads.

3. $B_3 = B_2 \ominus \text{box}(80,100)$:
The component is the only blob which is larger than a box which is 80 pixels in column and 100 pixels in row. An erosion with a large box removes all blobs smaller than the size of the box.

4. $G_4 = B_3 \oplus \text{box}(160,90)$:
Generate a mask that covers the component and its leads. The width of the box is bigger than the box used in the previous erosion.

5. $G_5 = G_0 \wedge G_4$:
Extract the part of the image which contains the area of interest, the component with leads.

6. $G_6 = G_5 \circ \text{rod}(4)$:
Removes small noisy relatively bright areas within the dark extracted area.

7. $B_7 = G_6 > 174$:
Binarization to extract the bright leads.

8. $B_8 = B_7 \bullet \text{box}(8,2)$:
Improve connectivity of the detected leads.

9. $B_9 = B_8 \circ \text{box}(20,3)$:
Removes noisy blobs smaller than the leads.

$$B_a = ((G_0 < 200) \oplus \text{box}(80,100) \ominus \text{box}(80,100) \oplus \text{box}(160,90)) \wedge G_0$$
$$B_b = ((B_a \circ \text{rod}(4)) > 174) \bullet \text{box}(8,2) \circ \text{box}(20,3)$$

reasoning and knowledge:

- We need to distinguish the bright blobs in the component from the ones in the background. Using the fact that the leads are *inside* the dark component, try to find the dark component including the leads. Use the detected component as a mask to select only the leads (steps 1 to 5).

- To distinguish the darker component from the leads, use a thresholding for absolute darkness (step 1). We can predict that the thresholding will produce an image consisting of the component with many holes because of the leads and the smaller bright objects inside the component. It could also have dark blobs in the background.

- Remove holes inside the component by a dilation operation (step 2).

- Remove the dark objects in the background by an opening operation (step 3 and 4). These two operations should be applied in the order mentioned above.

- We need to find only the connected convex blobs (leads) with certain size and shape (steps 6 and 9).

- A grayscale opening operation can be used to remove small noisy peaks (step 6).

- To detect bright leads, use a thresholding for absolute brightness (step 7).

- Fill in the holes inside the detected leads by a closing operation (step 8). It also connects small but clustered pixels. Step 6 helped to preclude this condition.

- Select only the leads by an opening operation by a proper structuring element that can be fit inside the leads but not inside other noisy blobs (step 9).

3.2 Hot spot detection in IR images

input image: [4]

- The input image consists of bright blobs with different sizes, shapes, and brightness in a generally relatively dark background.

- The background is quite arbitrary except the fact that it is relatively dark.

goal:

- To detect relatively bright blobs with certain size and shape.

- Each bright blob contains some points that are brighter than a given brightness value.

morphological procedure:

1. $B_1 = G_0 > 112$:
   Extract bright blobs with absolute brightness value greater than 112.

2. $G_2 = G_0 \bullet \mathrm{rod}(2)$:
   Removes small relatively dark areas.

3. $G_3 = G_2 - (G_2 \circ \mathrm{rod}(8))$:
   Detects relatively bright blobs whose size is smaller than a disk of radius 8.

4. $G_4 = G_3 \circ \mathrm{rod}(1)$:
   Removes small relatively bright peaks whose support is is less than a disk of radius 1.

5. $B_5 = G_4 > 13$:
   Binarization to detect bright areas.

6. $B_6 = B_1 \oplus |_{B_5} \mathrm{disk}(1)$:
   Combine information in B_1 and B_5. Detect relatively bright blobs which contain at least one pixel whose grayscale value is greater than 112.

$$G_a = G_0 \bullet \mathrm{rod}(2)$$
$$B_b = ((G_a - G_a \circ \mathrm{rod}(8)) \circ \mathrm{rod}(1)) > 13$$
$$B_c = (G_0 > 112) \oplus |_{B_b} \mathrm{disk}(1)$$

reasoning and knowledge:

- By thresholding the image at the specified level it is guaranteed that we get at least some part of the blobs to be detected (step 1).

- Relatively bright areas with certain sizes can be detected by a thresholding of an opening residue operation (step 2 to 5).

- The detected objects that are larger than the specified size can be filtered out by an opening residue operation (step 3).

- The detected objects which are smaller than the specified size can be removed by an opening operation (step 4).

- When one operator detects some part of the objects to be detected and the other operator detects all the objects to be detected together with some more unwanted objects, conditionally dilate the result of the first operator with respect to the result of the second operator (step 6).

3.3 Inspection of watch gears for missing or broken teeth

input image: [19]

- The input image is a binary image with the watch gears as foreground pixels. The background has no other objects. There can be at most two watch gears in the image.

- Each watch gear is a disk with teeth around the boundary of the disk. Some of its teeth may be broken or missing causing different shape than expected. The inside of the disk contains four holes with a known size, shape, and arrangement.

- If there are two watch gears in the image, they can touch each other at their boundaries. They never occlude each other.

goal

- Find the location of any missing or broken teeth. The broken tooth is the one whose height is shorter than the height of the normal tooth.

morphological procedure:

1. $B_1 = B_0 \ominus \mathrm{ring}(30, 16)$:
   Find the center of the gear holes inside the body.

2. $B_2 = B_1 \oplus \mathrm{octagon}(35)$:
   Construct an octagonal hole mask centered at the centers of the gear holes found in step 1.

3. $B_3 = B_0 \vee B_2$:
   Fill in the holes by combining the original image with the mask found in step 2.

4. $B_4 = B_3 \circ \mathrm{disk}(75)$:
   Select only the gear body without the teeth.

5. $B_5 = B_4 \oplus \text{disk}(3.5)$:
   Extend the gear body a little bit so that it covers the broken teeth completely but not the normal teeth.

6. $B_6 = (B_5 \oplus \text{disk}(7.5)) - B_5$:
   Construct a ring around the extended gear body that covers only the normal teeth.

7. $B_7 = B_0 \wedge B_6$:
   Mask out only the normal teeth.

8. $B_8 = B_7 \oplus \text{disk}(6.5)$:
   Connect the teeth by dilating them by a disk whose diameter is equal to the tip_to_tip gear tooth spacing.

9. $B_9 = B_6 > B_8$:
   Determine the place where the teeth can not be connected by the dilation operation in step 9 by taking the difference between the connected teeth and the teeth mask constructed in step 6.

$$B_a = (((B_0 \ominus \text{ring}(30, 16)) \oplus \text{octagon}(35)) \vee B_0)$$
$$= \quad \text{o } \text{disk}(75) \oplus \text{disk}(3.5)$$
$$B_b = B_a < (B_a \oplus \text{disk}(7.5))$$
$$B_c = B_b > ((B_b \wedge B_0) \oplus \text{disk}(6.5))$$

reasoning and knowledge

- Find the difference between the ones we want to detect and the others. Does the missing or broken teeth have unique characteristics ? If they are missing, the spacing between the teeth in the missing part of the teeth train is longer than the known regular spacing. If they are broken, the height of the broken tooth is shorter than the standard tooth.

- Isolate the normal teeth from the main gear body. Main body can be extracted by opening operation (step 1 to 4).

- If no proper structuring element can be found for the opening, fill in the holes in the body (step 1 to 3) and then open it with a proper disk (step 4).

- The normal tooth is taller than the broken tooth. To generate a mask that covers only the normal teeth, extend the gear body by the maximum height of broken tooth (step 5), and then get the annulus surrounding the extended gear body by a dilation residue operation (step 6).

- Since the spacing between the teeth where a tooth is missing is longer than normal tip–to–tip gear tooth spacing, if we dilate the teeth only image by a disk whose diameter is equal to the tip–to–tip gear tooth spacing, all teeth will be connected except at the location where a tooth is missing (step 8).

- Disconnected portion of the teeth can be detected by subtracting the result obtained in step 8 from the normal teeth mask (step 9).

3.4 Road detection in radar images

input image: [9]

- There are two images. One shows only a part of the desired roads, while the other image shows all the roads with many other noisy objects.

- The image contains dark valleys with all different width and length. Some are clustered and some are isolated. They are relatively dark within a relatively bright region of the image.

- The image consists of vertical stripes of large dark and bright regions with similar size. The dark regions contains small bright blobs.

goal

- Detect only the dark valleys with certain width and certain length within the bright region of the image. Detected valleys must be connected.

morphological procedure:

1. $G_1 = G_0 \bullet \mathrm{rod}(1)$:
   Removes small dark noisy spots. G_0 is the image that shows only parts of all the roads to be detected with small amount of noisy valleys.

2. $B_2 = ((G_1 \bullet \mathrm{rod}(5)) - G_1) > thr_1$:
   Detects relatively dark valleys whose width is smaller than 10 pixels.

3. $B_3 = \mathrm{sieve\_filter}(B_2, \mathrm{box\_frame}, \mathrm{box}, N)$:
   Removes small isolated objects and retains only the connected longer valleys.
   The sieve_filter(I, box_frame, box, N), which sifts the image through increasingly coarse sieves, is defined as follows:

$$I^1 = I$$
$$\textbf{for each } i = 1, \cdots, N$$
$$\quad A = I^i \oplus \mathrm{box\text{-}frame}_i$$
$$\quad B = A \ominus \mathrm{box}_i$$
$$\quad I^{i+1} = I^i \wedge B$$

   where the sizes of box_frame$_i$ and box$_i$ increase as i increases.

4. $B_4 = B_3 \bullet \mathrm{disk}(20)$:
   Connects a set of isolated blobs close to each other and makes a single cluster which is larger than the valleys we want to detect. This will also connect the disconnected valleys.

5. $B_5 = (B_4 \circ \text{disk}(5)) \oplus \text{disk}(8)$:
   Generates a mask that covers the larger cluster of noisy textured blobs.

6. $B_6 = B_4 - B_5$:
   Removes the noisy textured blobs from the valley image.

7. $B_7 = B_6 \oplus |_{B_0} \text{disk}(1)$:
   B_0 can be obtained using the same procedure (step 1 to 6) from the second image that shows all the valleys clearly including a lot of noisy valleys. Since B_6 contains pixels of valleys with certain amount of confidence, conditionally dilate B_6 with respect to B_0 to recover all the valleys that we want to detect.

reasoning and knowledge

- The small noisy peaks and pits can be removed by an opening and closing operations respectively (step 1). A closing operation will help remove smaller noisy valleys.

- The valleys can be detected by thresholding a closing residue operation (step 2).

- The narrow valleys can be smoothed out by a closing operation (step 1).

- The shorter valleys can be removed by the sieve filter (step 3).

- If a group of small isolated objects make a cluster, they cannot be removed easily by the sieve filter only. If the size of the cluster is bigger than the valleys and quite far apart from the valleys, they can be made to form a single larger set by a closing operation (step 4). Then, it can be removed by an opening residue operation (step 5 and 6).

- If we can mark part of the objects we want to extract, they can be extended to recover the original whole object by a conditional dilation operation (step 7).

3.5 Airplane detection

input image:

- The image contains a relatively bright airplane shaped objects with some horizontal and vertical bright stripes in a generally dark background. The object can be partially occluded by the bright stripes. The lengths of strips are longer than the airplane.

goal

- Detect only the airplane shaped object.

morphological procedure:

1. $G_1 = G_0 - (G_0 \circ \text{brick}(1, 49, 0))$:
   Removes vertical relatively bright strips that are longer than 49 pixels.

2. $G_2 = G_1 - (G_1 \circ \text{brick}(89, 1, 0))$:
   Removes horizontal relatively bright stripes that are longer than 89 pixels.

3. $G_3 = G_2 \bullet \text{brick}(5,5,0)$:
   Removes small dark pits that are smaller than the size of 5 pixels by 5 pixels square. These dark areas are the noise inside the bright airplane.

4. $B_4 = G_3 > 14$:
   Threshold to detect bright areas.

$$G_a = G_0 - G_0 \circ \text{brick}(1,49,0)$$
$$B_b = ((G_a - G_a \circ \text{brick}(89,1,0)) \bullet \text{brick}(5,5,1)) > 14$$

reasoning and knowledge

- The horizontal and vertical stripes can be removed by an opening residue operation by a proper structuring element.

3.6 IR ship recognition

input image:

- It contains a single relatively highly contrasted dark ship in the near middle of the image. The boundary of the ship is quite blurred. The ship shows up as a long horizontal shape with some vertical structures at the top of it. The bottom of the ship is almost straight horizontal parallel to the water level.

goal

- Detect significant features of the ship. This means finding the horizontal edges comming from the top boundary of the ship.

morphological procedure:

1. $G_1 = G_0 \bullet \text{rod}(3)$:
   Removes small noisy dark pits or ridges.

2. $B_2 = (G_1 \bullet \text{brick}(1,70,0) - G_1) > 40$:
   Finds relatively dark areas whose shape can not be covered by a long vertical stripe. In this example, ship is the only object that can be detected by this thresholding a closing residue operation.

3. $B_3 = B_2 \ominus \text{disk}(3)$:
   Removes any unwanted dark areas that can be detected by step 2. It also shrinks the area of the detected ship, thus making sure we only detect inside parts of the ship.

4. $B_4 = ((G_1 \oplus \text{brick}(12,2,0)) - G_1) > 20$:
   If we let the origin of the brick to be at the center of the top row of its support, a thresholding a closing residue operation will find the top boundary of the ship.

5. $B_5 = B_4 \wedge (G_1 < 85)$:
   The detected boundary should be inside the dark area (ship) whose brightness value is less than 85. The grayscale value of the ship area should be absolutely less than 85.

6. $B_6 = B_3 \vee B_5$:
   Combines information gathered through two mutually compensating operations.

7. $B_7 = B_6 \wedge (B_2 \oplus \text{disk}(8))$:
   Make sure the areas detected in B_6 come from B_2.

8. $B_8 = B_7 \bullet \text{box}(2, 10)$:
   Fills in small zig_zag boundaries.

9. $B_9 = (B_8 \bullet \text{disk}(4) \ominus \text{disk}(4)) \vee B_8$:
   Fills in possible holes inside the ship.

10. $B_{10} = (B_9 \oplus \text{points}\{(0,0),(0,-2)\}) > B_9$:
    Detects only the top edges.

reasoning and knowledge

- The uneven structures on the top of the objects are the significant features. Try to determine the characteristics of these features.

- The relatively dark region of horizontally long object can be detected by a thresholding a closing residue operation especially by a vertical line segment longer than the height of the object (step 2).

- The top boundary can be detected by a thresholding a dilation residue operation by a brick structuring element whose origin is at the top row instead of the center row (step 4).

- To have consistent features, we need to smooth the boundaries This can be done by a closing operation. The structuring element used in this operation should be small enough not to destroy the details of the shape but to remove small noisy jaggedness of the boundaries (step 8).

- To fill in holes inside the detected ship but not to destroy the details of the boundary, close the image and shrink it by an erosion operation. Then or the result with the original image (step 9).

- If one operation finds the core of the ship while another operation can find the top portion of the ship more easily but not the core, combine the result of both operation for the better recognition of the ship (step 6). We can find an object if different part of the object can be detected by different operations.

- The top boundaries can be detected by a dilation residue operation (step 10). The structuring element consists of two points, one at the origin and the other at vertically shifted point.

3.7 Barcode detection

input image:

- It is a binary image with barcodes as foreground pixels. The barcode is shown as a group of small boxes with the same height but with different widths. They are highly regular. The size of the group of barcode is known. However, they can be arbitrarily oriented and positioned.

- There are many noisy blobs with similar sizes but not regularly grouped as the barcodes. The background can also contain large and small objects allowing it to be quite arbitrary.

goal

- Find the barcodes, clusters of bars with known size, shape, and arrangement. The bars are parallel to each other. They can almost be included in a single rectangular shaped box. The distance between each bar is almost same.

morphological procedure:

1. $B_1 = B_0 - (B_0 \circ \mathrm{disk}(6))$:
   Removes objects larger than the size of barcode.

2. $B_2 = B_1 \circ \mathrm{disk}(2)$:
   Removes objects smaller than the size of barcode.

3. $B_3 = B_2 \ominus \mathrm{points}\{(0,3),(0,4),(0,-3),(0,-4),(3,0),(4,0),(-3,0),(-4,0)\}$:
   Detects regular patterns (barcodes) by an erosion operation with special set of points.

4. $B_4 = B_3 \oplus \mathrm{box}(15,15)$:
   Generates the barcode mask by dilating the seeds detected in step 3.

5. $B_5 = B_2 \wedge B_4$:
   Masks out only the barcodes.

6. $B_6 = B_5 \ominus \mathrm{disk}(1)$:
   Separates possibly connected bars in each barcode group.

$$B_a = (B_0 - B_0 \circ \mathrm{disk}(6)) \circ \mathrm{disk}(2)$$
$$B_b = (B_a \wedge (B_a \ominus \mathrm{pointset} \oplus \mathrm{box}(15,15))) \ominus \mathrm{disk}(1)$$

reasoning and knowledge

- Remove larger blobs by an opening residue operation (step 1).

- Remove smaller blobs by an opening operation (step 2).

- The clusters of blobs arranged regularly can be detected by an erosion operation by a set of regularly arranged points (step 3). The set of points should be center symmetric so that it can detect arbitrarily positioned barcodes.

3.8 Recognition of broken rice grains

input image: [16]

- It is a grayscale image where the grains show up as bright highly contrasted objects with respect to the dark background.

- The grains can touch each other.

goal

- Find nonconvex bright blobs with certain size which do not touch the boundary of the input image. The detected blobs should be isolated.

morphological procedure:

1. $B_1 = (G_0 - (G_0 \circ \mathrm{rod}_1)) > \mathrm{thr}_1$:
   Detects relatively bright blobs whose size are smaller than the rod_1 used in the opening operation.

2. $B_2 = \mathrm{watershed}(B_1)$:
   Separates touching blobs by a watershed operation. The watershed operation is described in [15].

3. $B_3 = (G_0 \geq 0) - ((G_0 \geq 0) \ominus \mathrm{disk}(1))$:
   Finds the boundary of the input image.

4. $B_4 = B_3 \oplus |_{B_2} \mathrm{disk}(1)$:
   Extracts blobs touching the boundary of the image.

5. $B_5 = B_2 - (B_2 \circ \mathrm{disk}(2))$:
   Marks concave blobs which are not open by a disk.

6. $B_6 = B_5 \oplus |_{B_2} \mathrm{disk}(1)$:
   Recover the concave blobs from the seeds marked by step 5.

7. $B_7 = B_6 > B_4$:
   Remove blobs touching the boundary of the image from the detected concave blobs.

reasoning and knowledge

- The convex shaped objects are open to a convex shaped disk which is smaller than the size of the objects (step 5).

- If operators are are independent, we can use them in any order in the procedure.

- Some operations presuppose special operations. For example the operation which extracts convex grains presupposes the segmentation by watershed which separates touching grains.

3.9 Particle marking

input image: [15]

- In a dark background, there are some distinguishably bright objects, cells with nuclei and the cytoplasm.

- The medium bright cells may or may not contain very bright nuclei. The shape of nuclei is convex round or sometimes smeared making them to appear as concave blobs. The cytoplasm is medium bright and does not contain nuclei.

goal

- Find medium bright cells with a convex shaped nuclei inside them. The size of nuclei is approximately known.

morphological procedure:

1. $B_1 = G_0 > \mathrm{thr}_1$:
   Detects only the very bright nuclei.

2. $B_2 = G_0 > \mathrm{thr}_2$:
   Detects all bright blobs. $B_2 \supset B_1$ since $\mathrm{thr}_1 > \mathrm{thr}_2$.

3. $B_3 = B_1 \oplus |_{B_2} \mathrm{disk}(1)$:
   Detects cells with nuclei in them.

4. $B_4 = (B_1 \bullet \mathrm{disk}(5)) - B_1$:
   Marks concave blobs.

5. $B_5 = B_4 \oplus |_{B_3} \mathrm{disk}(1)$:
   Recovers concave blobs from the image of cells with nuclei.

6. $B_6 = B_3 - B_5$:
   Genarates an image of convex shaped cells with nuclei.

7. $B_7 = B_1 \ominus \mathrm{disk}(10)$:
   Marks nuclei larger than the disk of radius 10 pixels.

8. $B_8 = B_7 \oplus |_{B_6} \mathrm{disk}(1)$:
   Recovers convex cells with nuclei. The nuclei are larger than the disk of radius 10 pixels.

reasoning and knowledge

- Use two thresholding operations with different threshold values to detect areas with different grayscale intensity values (step 1 and 2).

- To remove the cytoplasm without nuclei, mark the nuclei (step 1) and then conditionally dilate the marked image with respect to the image of all bright objects (step 2) to find only the ones with nuclei (step 3).

- Convex objects are closed to a convex disk.

- Mark the concave objects by a closing residue operation (step 4).

3.10 Prunings on the thinning of a daisy

input image: [15]

- It is a binary image which contains an arbitrary shaped object.

goal

- Determine the skeleton image without noisy small branches.

morphological procedure:

The following morphological operations use the structuring elements of the .Golay alphabet. When the Golay symbol appears inside a brace, for example $\{L\}$, the sequence $\{L\}$ has the successive rotations of L as its elements.

1. $B_1 = B_0 \bigcirc \{L\}(\text{iteration} = N_1)$:
   Thins the daisy to get the skeleton image. The symbol $\bigcirc$ represents a thinning operation defined as follows:
$$A \bigcirc L = A - (A \circledast L)$$
   where L_0 is a set of pixel positions with 0 in L, and L_1 is a set of pixel positions with 1 in L. The hit–or–miss operation ($\circledast$) can be used to detect a portion of image which matches the pattern defined by the structuring element used in the operation. The thinning operation is applied N_1 times where N_1 is the half of the maximum width of the object to be thinned.

2. $B_2 = B_1 \bigcirc \{E\}(\text{iteration} = N_2)$:
   Prunes the branches shorter than N_2 pixels long. This also shortens the main skeleton.

3. $B_3 = B_2 \circledast \{E\}$:
   Detects the end points of the main skeleton.

4. $B_4 = B_3 \oplus |_{B_1} \text{disk}(1)$:
   Recovers the shortened main skeleton by conditionally dilating the end points found in step 3 with respect to the original skeleton image. The conditional dilation operation is applied N_2 times since we know the main skeleton was shortened N_2 times in step 2.

5. $B_5 = B_2 \vee B_4$:
   Combines the pruned skeleton with the lengthened end points image.

reasoning and knowledge

- To shorten the branches in the skeleton image, perform a thinning operation with structuring elements which can detect only the end points of a line. It will also shorten the longer branches.

- If we conditionally dilate the above shortened image, we get the original image back. We want to lengthen only the long branches left in the shortened image.

- The end points of branches can be detected by a pattern matching using a hit–or–miss operation.

3.11 Neighbor Analysis

input image: [15]

- It is a grayscale image of polished section of polycrystalline ceramic (bright) seen in a rectangular mask. The bright region can have small dark noisy spots.

- The boundaries of bright grains are relatively dark and are of known thickness. The boundaries are quite homogeneously relatively dark. The grains can have some dark noise inside them.

goal

- Find grains whose boundaries are completely shown in the image. Make the boundaries one pixel thick.

morphological procedure:

1. $B_1 = (G_0 \geq 0) \ominus \text{disk}(2)$:
   Mask of whole image shrunk 2 pixels from the boundary of the image.

2. $B_2 = (B_1 \oplus \text{disk}(1)) - B_1$:
   Extracts the boundary of input image.

3. $B_3 = G_0 > \text{thr}_1$:
   Detects the bright grains. The detected grains can have holes in them because of dark noise.

4. $B_4 = (\text{skiz}(B_3^c))^c$:
   The skiz operation is defined as $\text{skiz}(A) = [A \bigcirc \{L\}] \bigcirc \{E\}$ and it detects skeletons without short branches. The skiz of dark areas (B_3^c) will give a single pixel wide grain boundaries. It does not contain the skeleton of dark noise because the noisy spot does not touch two different grains. Thus the complement of the skiz extracts zones of influence of each bright grains.

5. $B_5 = B_4 \wedge B_1$:
   Take out the boundary effect of the input image.

6. $B_6 = B_2 \oplus |_{B_5} \text{disk}(1)$:
   Extracts grains touching the image boundary (B_2).

7. $B_7 = B_5 - B_6$:
   Detects grains whose boundaries are completely shown in the image.

reasoning and knowledge

- The grains touching the image boundary are not completely shown in the image.

- The boundaries of grains can be thinned to a single pixel by a skiz operation.

- The skiz of an isolated blob is a null image.

3.12 Defect lines detection

input image: [15]

- It is a binary image with black stripes. The stripes have a specified width and jaggedness. Most of the stripes start from one side of the image and end at some other side of the image.

- Some of the stripes are broken inside the image. Consecutive broken stripes form an imaginary line in the image if we suppose connecting the end points of the broken stripes.

- There could be several such imaginary lines formed by different groups of consecutive broken stripes.

goal

- Find the regions where the stripes are disconnected. Connect the end points of those disconnected stripes.

morphological procedure:

1. $B_1 = B_0^c \bigcirc \{L\}(\text{iteration} = N_1)$:
   Homotopically thins the dark stripes to get the skeletons of them. Detects the end points of the main skeleton.

2. $B_2 = B_1 \bigcirc \{E\}(\text{iteration} = N_2)$:
   Prunes small branches of the detected skeletons.

3. $B_3 = B_2 \circledast \{E\}$:
   Detects the end points of pruned skeletons.

4. $B_4 = B_2 \vee (B_3 \oplus |_{B_1} \text{disk}(1))$:
   Recovers the shortened main skeletons by conditionally dilating the end points of the shortened main skeletons.

5. $B_5 = B_4 \circledast \{E\}$:
   Detects the end points of the main skeletons.

6. $B_6 = (B_4 - (B_5 \oplus \text{disk}(1))) \vee B_5$:
   Separates the end points from the main skeletons.

7. $B_7 = (\text{skiz}(B_6^c))^c$:
   Determines zones of influence of end points as well as zones of influence of main skeletons.

8. $B_8 = B_5 \oplus |_{B_7} \text{disk}(1)$:
   Extracts only the zones of influence of end points by conditionally dilating the end points with respect to the zones of influence image obtained in step 7.

9. $B_9 = B_8 \oplus \text{disk}(1)$:
   Connects the detected zones of influence of end points.

10. $B_{10} = B_9 \bigcirc \{L\}$(iteration $= N_3$):
    Thins the detected zones of influence to get the imaginary line connecting the end points.

reasoning and knowledge

- The overall plan is to detect the end points of each of the broken stripes and connect them (restatement of the goal).

- The end points of the broken stripes can be detected by finding the end points of the longer lines in the skeleton image of the stripes (step 5).

- We can predict that the skeleton of the stripes will have small branches because of the jaggedness of the stripes.

- Use the procedure described in section 3.10 to extract only the main skeletons (step 2 to 4).

- There could be several regions of broken stripes. Since they are not guaranteed to be separated father than the distance between the end points inside one such region, a dilation to connect the end points followed by a thinning may not work.

- The end points of one group of consecutive broken stripes are next to each other. Thus, if we compute the zones of influence image from the image of skeletons and separated end points zones of influence of a single group of end points will be next to each other (step 6 and 7).

- The end points of a line can be separated from the line by subtracting the dilated end points from the line and ORing with the end points (step 6).

- Fill in each zones of influence of end points by a conditional dilation operation (step 8).

- Connect zones of influence of end points next to each other by a dilation operation (step 9).

3.13 Discrimination between cells and artifacts

input image: [13, 15]

- It consists of relatively bright particles in a dark background. The particles are generally round. Some of them are nonconvex and relatively large.

- Some of the particles are overlapping each other. Some have holes.

goal

- Find only the large roundly shaped particles. If a particle is round, the curvature of its boundary is bounded.

morphological procedure:

1. $B_1 = G_0 > thr_1$:
   Detects bright objects.

2. $B_2 = \text{conditional\_bisector}(B_1, 2)$:

The conditional bisector is defined as follows:

$$\text{conditional\_bisector}(A, N) = \bigcup_i [(A \oplus \text{disk}(i)) - ((A \ominus \text{disk}(i+1))$$
$$\oplus \text{disk}(N))]$$

It can detect the local peaks in the distance transformed image. It does not detect the ridges. If a blob is not round, the conditional bisector detects more than one isolated point inside the blob.

3. $B_3 = B_1 - B_2$:

Make holes inside blobs detected in step 1.

4. $B_4 = \text{skiz}(B_3)$:

Gets the skiz of the bright blobs with holes. For a blob with single hole, the skiz is a single loop inside the blob. For a blob with more than one hole, the skiz consists of more than one loop connected each other by a line. Thus, if a blob is not round, the skiz inside the blob includes points where more than three lines meet.

5. $B_5 = (B_1 \oplus \text{disk}(2)) - B_1$:

Detects boundaries of the bright blobs.

6. $B_6 = B_5 \bigcirc \{L\}(\text{iteration} = N_1)$:

Thin the boundaries of the bright blobs. If a blob is not round, the thinned boundary includes small branches because of its jaggedness.

7. $B_7 = B_4 \vee B_6$:

Combines the skiz image with the boundary image.

8. $B_8 = B_7 \circledast \{F\}$:

Finds points where three lines meet by a pattern matching operation (hit–or–miss).

9. $B_9 = B_8 \oplus |_{B_1} \text{disk}(1)$:

Extract blobs marked by step 8.

10. $B_{10} = B_1 - B_9$:

Extract only the round blobs.

reasoning and knowledge

- Mark the overlapping particles, non-round shaped particles, and the particles with holes in them. We can accomplish the task if we can mark all the particles that we do not want to detect.

- If two round blobs overlap each other, a conditional bisector operation results in two points inside the overlapped particles (step 2).

- Thus, if we take the skiz of the overlapping blobs without its conditional bisector, it gives two loops around each points detected by the conditional bisector operation connected together by a line segment (step 4).

- If the boundary of a particle is not round, its boundary obtained by a dilation residue operation will have some jaggedness (step 5).

- If we thin a jagged object, the thinned image will contain small branches (step 6).

- The points where three lines meet can be detected by a hit–or–miss operation (step 8).

3.14 Inspection of hybrid circuits

input image: [19]

- It is a grayscale image of hybrid circuit boards. It shows relatively bright stripes of all different widths in a dark background. The image is quite noisy.

goal

- Detect only the bright thin stripes.

morphological procedure:

1. $G_1 = \text{isotropic\_filter}(G_0, N_1, N_2)$:
   Remove noise. The isotropic filtering can be done by an alternating application of opening and closing by balls of increasing size. Thus, isotropic_filter (I, N_1, N_2) is implemented as follows:

$$
\begin{aligned}
&I^{N_1} = I \\
&\textbf{for each } i = N_1, \cdots, N_2 \\
&\quad A = I^i \circ \text{ball}_i \\
&\quad I^{i+1} = A \bullet \text{ball}_i
\end{aligned}
$$

2. $B_2 = G_1 > \text{thr}$:
   Detects bright areas.

3. $B_3 = B_2 - (B_2 \bullet \text{disk}(6))$:
   Removes larger blobs which can be closed to the disk of radius 6 pixels.

4. $B_4 = B_3 \circ \text{disk}(2)$:
   Removes small blobs smaller than the disk of radius 2 pixels.

reasoning and knowledge

- Detect large blobs by a closing operation.

- Removes small objects by an opening operation.

- Combine the above two operations to detect blobs with certain size.

3.15   Feature detection in Dental Image

input image: [5]

- It is a grayscale image containing several teeth shown relatively bright in a dark background. It also contains relatively bright gums near to one end of each tooth and between a pair of neighboring teeth.

- Within each tooth the grey tone is approximately constant. Eah tooth can have fillings or metal caps resulting in a very bright region inside the tooth.

- A tooth can have enamel at its vertical boundary which is shown brighter than the intensity of tooth. The shape of enamel region is sharp thin triangle which is vertically oriented between the background and tooth.

goal

- Find the starting position of the enamel region close to the gum region in between the teeth.

morphological procedure:

1. $G_1 = G_0 \bullet \mathrm{rod}(r_1)$:
   Removes small grey tone pits.

2. $B_2 = (G_1 - G_1 \circ \mathrm{rod}(r_2)) > thr_1$:
   Detects relatively bright region which cannot contain the rod.

3. $B_3 = B_2 \circ \mathrm{box}(w_1, h_1)$:
   Removes artifacts which cannot contain the $\mathrm{box}(w_1, h_1)$. The box chosen here is a vertically long and thin that can be contained in the detected enamel region.

4. $B_4 = B_3 - B_3 \circ \mathrm{box}(w_2, h_2)$:
   Finds corners by an opening residue with a box structuring element of small width.

5. $B_5 = (G_0 < thr_2) \oplus \mathrm{disk}(r_3)$:
   Detects the dark background and expand it a little bit by dilating it with a small disk.

6. $B_6 = B_4 \wedge B_5$:
   Detects only the corner points touching the background.

$$G_a = G_0 \bullet \mathrm{rod}(r_1)$$

$$B_b = [(G_a - G_a \circ \mathrm{rod}(r_2)) > thr_1] \circ \mathrm{box}(w_1, h_1)$$

$$B_c = (B_b - B_b \circ \mathrm{box}(w_2, h_2)) \wedge [(G_0 < thr_2) \oplus \mathrm{disk}(r_3)]$$

reasoning and knowledge

- The plan is finding the grey tone hill oriented along the column axis and then extracting the corners close to the background or to the gum area.

- Remove small relatively dark pits by a closing operation (step 1). Since we are interested in the sharp corners of the relatively bright region, we do not open the image to remove the small relatively bright peaks.

- After finding the relatively bright areas (step 2), remove noisy small artifacts that does not conform to the shape we are interested in by an opening operation (step 3). Use thin vertical box for the opening to keep the corners intact but to remove the artifacts.

- The sharp corner of the vertically thin triangle can be detected by an opening residue with a horizontally thin but short line segment (step 4).

- The corners touching the background or near the background can be extracted by expanding the background a little bit and ANDing it with the corner image (step 5 and 6).

3.16 Detecting occluded object

input image: [2]

- It is a grayscale image of a single target which is relatively dark and roughly oval in shape. The dimension of the target and its aspect ratio are approximately known.

- Parts of the target can be obscured by brighter areas of approximately known size.

- The background is relatively brighter than the target but may not be of uniform brightness.

- The image may contain areas of bright glint which are relatively brighter than their neighboring pixels. The size of glint area is small and approximately known. Areas of this type may aggregate to form larger regions of bright glint but no more than about 25% of the target area..

- Very small dark areas are present in the image and commonly occur near glint.

goal

- Detect the target with occlusions removed.

morphological procedure:

1. $B_1 = G_0 > thr_1$:
   Detects absolutely bright glint.

2. $G_2 = G_0 \bullet \text{brick}(w_1, h_1)$:
   Eliminates small dark regions.

3. $B_3 = [G_2 - (G_2 \circ \text{brick}(w_2, h_2))] > thr_2$:
   Detects small relatively bright regions (glint areas).

4. $G_4 = [(B_1 \vee B_3) \oplus \text{box}(3,3)] > 0$:
   Forms mask of absolutely bright or locally relatively bright glint areas.

5. $G_5 = G_2 \wedge G_4^c$:
   Replaces the grayscale values of the pixels in the glint area with zero in the image free of small dark regions.

6. $G_6 = \text{fill}(G_5, G_4)$:
   Fills in glint areas with the grayscale values surrounding them. The morphological fill operation, $\text{fill}(f, g)$, can be defined iteratively as follows:

$$G^0 = g$$
$$F^0 = f$$
$$G^i = G^{i-1} \ominus \text{rod}(1)$$
$$F^i = F^{i-1} \vee \left[(F^{i-1} \oplus \text{rod}(1)) \wedge G^{i-1}\right]$$
$$\text{fill}(f, g) = F^n \quad \text{where} \quad G^{n+1} = 0$$

7. $G_7 = G_6 \circ \text{brick}(w_3, h_3)$:
   Removes the relatively bright areas occluding the target.

8. $B_8 = [G_7 - (G_7 \bullet \text{brick}(w_4, h_4))] > thr_3$:
   Detects the large relatively dark region, which constitutes the target.

reasoning and knowledge

- Overall plan is to remove unwanted noise and glints, to remove occlusion, and to detect the targets.

- Detect absolutely bright and larger glints and small relatively bright glints by the thresholding and the thresholding of the opening residue operation respectively (step 1, 3, and 4).

- Fill in the glint areas with the grayscale values of the pixels surrounding them (step 5 and 6). A simple opening will do the similar job, but it will also destroy the other areas.

4 Summary and comments

In this paper, we analyzed a small set of vision algorithms where the mathematical morphology is applied to the solution of machine vision tasks. However, it is still an open question to find and explicitly describe the functional descriptions of all the vision tasks that the morphological procedures can solve. One should first look for the solution from the theory side on the mathematical morphology such as [11, 15, 17, 6, 8, 10, 18]. One can also examine the vision algorithms already developed by vision experts such as the ones listed in this paper. This helps us understand the main role of the morphological operator in the algorithm sequence and the context in which the operator appears. On the experimental side, we are currently designing a system that can propose a morphological procedure and try to interpret them in a controlled search environment. This research will help us achieve our goal of automatically producing morphological vision algorithms.

References

[1] ARCHIBALD, C. Identify leads coming from a smd component. Tech. rep., Machine Vision International, Ann Arbor, MI. July 1985. Internal memo.

[2] BALLESTRASSE, C., KATZ. P., AND HARALICK, R. Detecting targets. Tech. rep., Applied Physics Lab, University of Washington, Seattle, WA, May 1989. Technical Report.

[3] GILLIES, A. *Machine learning procedures for generating image domain feature detectors.* Ph.D. dissertation, Department of Computer and Communication Science, University of Michigan, Ann Arbor, MI, 1985.

[4] HARALICK, R. Sub-pixel precision hot spot detection and recognition: A morphologic/facet approach. Tech. rep., Department of EE, University of Washington, Seattle, WA, May 1988.

[5] HARALICK, R., RAMESH, V., HAUSMANN, E., AND ALLEN, K. Computerized detection of cemento-enamel junctions in digitized dental radiographs. Tech. rep., Department of EE, University of Washington, Seattle, WA, May 1989.

[6] HARALICK, R., STERNBERG, S., AND ZHUANG, X. Image analysis using mathematical morphology. *IEEE Trans. on Pattern Analysis and Machine Intelligence PAMI-9*, 4 (1987), 532–550.

[7] HASEGAWA, J., KUBOTA, H., AND TORIWAKI, J. Automated construction of image processing procedures by sample-figure presentation. In *Proc. 8th ICPR* (Oct. 1986), pp. 586–588.

[8] HEIJNANS, H., AND RONSE, C. The algebraic basis of mathematical morphology, part 1: Dilations and erosions. Tech. rep., Center for mathematics and computer science, Dept. of Applied Mathematics, PO. Box 4079, 1009 AB, Amsterdam, The Netherlands, June 1988. Report AM-R8807.

[9] LEE, J. Multiple sensor fusion based on morphological processing. Tech. rep., Boeing electronics high tech center, P.O. Box 24969, MS 7J-24, Seattle, WA, 1988. Internal memo.

[10] MARAGOS, P. *A unified theory of translation-invarient systems with applications to morphological analysis and coding of images.* Ph.D. dissertation, School of Electrical Engineering, Georgia Institute of Technology, July 1985.

[11] MATHERON, G. *Random sets and integral geometry.* Wiley, New York, 1975.

[12] MATSUYAMA, T. Expert systems for image processing: Knowledge-based composition of image analysis processes. Tech. rep., Dept. of Information Engineering, Tohoku University, Sendai, Miyagi 980, Japan, 1988. submitted to CVGIP.

[13] MEYER, F. Iterative image transformation for an automatic screening of cervical smears. *J. of Histochemistry and Cytochemistry 27*, 1 (1978), 128–135.

[14] SAKAUE, K., AND TAMURA, H. Automatic generation of image processing programs by knowledge-based verification. *IEEE CH2145* (1985), 189–192.

[15] SERRA, J. *Image analysis and mathematical morphology.* Academic Press Inc., 1982.

[16] SERRA, J. From mathematical morphology to artificial intelligence. In *Proc. 8th International Conference on Pattern Recognition* (Paris, France, Oct 1986), pp. 1336–1343.

[17] SERRA, J., Ed. *Image analysis and mathematical morphology, Volume 2: Theoretical advances.* Academic Press Inc., 1988.

[18] STERNBERG, S. Cellular computers and biomedical image processing. In *Biomedical images and computers*, J. Sklansky and J. Bisconte, Eds., vol. 17. Springer-Verlag, Berlin, 1982, pp. 294–319. Lecture notes in medical informations.

[19] STERNBERG, S. An overview of image algebra and related architectures. In *Integrated technology for parallel image processing.* Academic Press Inc., 1985, pp. 79–100.

[20] VOGT, R. *Automatic generation of morphological set recognition algorithms.* Ph.D dissertation, Electrical Engineering and Computer Science Department, University of Michigan, 1988.

Objekterkennung durch Monomorphie von Anordnungsgraphen

Herbert Müller
Lehrstuhl für Allgemeine Nachrichtentechnik
Universität der Bundeswehr Hamburg

Zusammenfassung

Die Einführung nichtmetrischer Anordnungsrelationen ermöglicht eine graphentheoretische Beschreibung von Punktstrukturen ohne zusätzliches a–priori–Wissen. Durch die Definition geeigneter Isomorphie– bzw. Monomorphiekriterien kann das Korrespondenzproblem zwischen zwei beliebigen orthogonalen Abbildungen planarer Objekte durch einen effizienten Suchalgorithmus gelöst werden. Ferner lassen sich die Anordnungsrelationen zur Erkennung eines 3D–Objektes in einer Abbildung innerhalb eines vorgegebenen Aspektwinkelbereiches anwenden.

1. Einführung

Viele Verfahren zur Objekterkennung in (Kamera–)Bildern basieren auf der Repräsentation der Modell– und Bildstrukturen durch (attributierte) Graphen (z.B. */Gmu88/*). Der Erkennungsalgorithmus besteht dannn aus der Hypothesenbildung durch die Suche nach (Teil)graph-Isomorphien und anschließender Verifikationsphase anhand zusätzlicher Kriterien und Randbedingungen. Die Generierung eines Modellgraphen kann direkt aus der Wissensbasis erfolgen; die Extraktion von Kanten aus einem Grauwertbild bereitet hingegen häufig Schwierigkeiten, während die Knotenpunkte als Schnitt– oder Eckpunkte von Kantensegmenten noch detektiert werden können */Tho87/*. Ferner erscheint die Anwendung graphentheoretischer Ansätze auch in solchen Fällen wünschenswert, in denen (lokale) Merkmale **nur** durch ihren geometrischen Ort ohne Relationen untereinander definiert sind. Als Beispiel seien klassische Punktefinder–Operatoren */Mor79/* und morphologische Verfahren zur Detektion von Objektprimitiven */Sha88/* erwähnt. In */Mue88/* wurde eine Methode beschrieben, Kantenrelationen für Punktstrukturen durch Delaunay-Triangulation zu erzeugen, um das Korrespondenzproblem zu lösen.

In der vorliegenden Arbeit werden neue, metrikunabhängige Beziehungen zwischen zwei und drei Punkten vorgestellt. In den Kapiteln 2 und 3 werden zunächst diese Anordnungsrelationen und die zugehörigen Monomorphiekriterien definiert. Ein Algorithmus für eine effiziente Monomorphiesuche ist in Kap.4 beschrieben. Kap.5 stellt formal den Zusammenhang zwischen Modell und Anordnungsgraph her und skizziert die Anwendung der Anordnungsrelation zur Erkennung von 3D–Objekten in Abbildungen.

2. Drei–Punkt–Anordnungsrelationen

Seien P_i, P_j, P_k drei verschiedene Punkte einer ebenen Punktstruktur V aus N Punkten und Δ_{ijk} das hierdurch festgelegte Dreieck. Als **Drei–Punkt–Anordnungsrelation (3PAR)** a_{ijk} wird eine numerische Größe festgelegt, die den Zählsinn im Dreieck Δ_{ijk} beschreibt:

$$
\textbf{Definition 2.1:} \qquad a_{ijk} = \begin{cases} +1 & \text{falls } P_i, P_j, P_k \text{ im Uhrzeigersinn} \\ -1 & \text{falls } P_i, P_j, P_k \text{ im Gegenuhrzeigersinn angeordnet sind.} \\ 0 & \text{falls } P_i, P_j, P_k \text{ kollinear} \end{cases}
$$

Aus der Definition der (dreidimensionalen) **Drei–Punkt–Anordnungsmatrix** $A = (a_{ijk})$ folgt unmittelbar die Invarianz bzgl. Skalierung, Translation und Rotation in der Bildebene.

Äquivalent zur *Def. 2.1* kann die 3P–Anordnungsrelation auch durch die geometrische Lage des Punktes P_k zur Verbindungslinie $P_i P_j$ definiert werden. Stellt man die zweidimensionalen Punktkoordinaten eines Punktes P_i durch einen dreidimensionalen, homogenen Vektor $P_i = (w \cdot p_{ix}, w \cdot p_{iy}, w)$ mit $w \neq 0$ dar, so läßt sich die Verbindungslinie l_{ij} zwischen P_i und P_j durch das Vektorprodukt $l_{ij} = p_i \times p_j$ und der senkrechte Abstand d des Punktes P_k zur Linie l_{ij} durch das Skalarprodukt $d \propto p_k \cdot l_{ij}$ beschreiben. Da es sich bei l_{ij} um eine gerichtete Größe handelt, gibt das Vorzeichen von d an, auf welcher Seite der Punkt P_k bezüglich der Verbindungslinie l_{ij} liegt. Somit läßt sich die 3PAR auch definieren als:

$$
\textbf{Definition 2.2:} \qquad a_{ijk} = \begin{cases} \operatorname{sign}(p_k \cdot l_{ij}) & \text{falls } d \geq d_{min} \\ 0 & \text{sonst} \end{cases}
$$

Für $d_{min} = 0$ sind *Def. 2.1* und *Def. 2.2* zueinander äquivalent, aber *Def. 2.2* ermöglicht eine fehlertolerante Berechnung der 3P–Anordnungsmatrix: ein Matrixelement a_{ijk} wird gelöscht, d.h. zu null gesetzt, falls der Abstand des Punktes P_k zur Verbindungslinie l_{ij} innerhalb der möglichen Fehlergrenzen liegt, d.h. eine Links–/Rechtsentscheidung aufgrund möglicher Störungen oder Toleranzen bei der Berechnung der Punktkoordinaten nicht mehr getroffen werden kann.

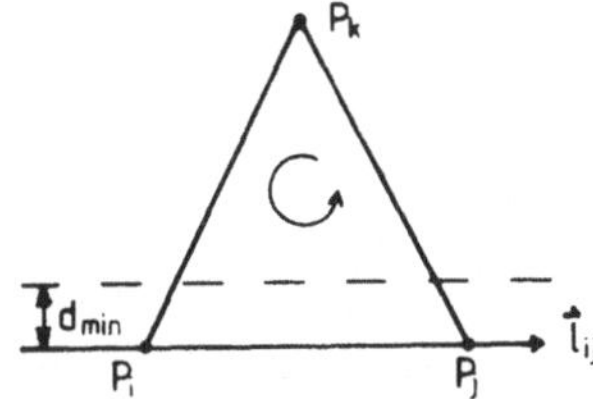

Abb.1: Schema zur Interpretation von Def.2.1 (Zählsinn im Dreieck) und Def.2.2

Anhand der 3PAR kann ein tolerantes Kriterium für die Isomorphie zwischen zwei Bildpunktstrukturen definiert werden:

Definition 2.3:   Eine planare Punktstruktur V aus N Punkten ist **3PA–monomorph** zu einer Punktstruktur V′, wenn es eine eineindeutige Zuordnung V→V′ gibt, so daß für die Elemente der zugehörigen 3P–Anordnungsmatrizen **A** und **A′** gilt:

$$(\forall i\ \forall j\ \forall k)\ ((a_{ijk} \neq 0)\ \Rightarrow\ (a_{ijk} = a'_{lmn}))$$

P'_l, P'_m, P'_n sind die zu P_i, P_j, P_k korrespondierenden Punkte.

Im Unterschied zu einem Isomorphismus V↔V′ weist ein Monomorphismus V→V′ eine Zuordnungsrichtung auf und ist i.a. nicht umkehrbar. Da nach *Def.2.3* nur die Elemente von **A** auf $a_{ijk}\neq 0$ geprüft werden, sei vorausgesetzt, daß ein Toleranzkriterium $d_{min}\neq 0$ nur bei der Ermittlung von **A** Anwendung findet, während **A′** mit $d_{min}=0$ zu berechnen ist. Eine Fehlerabschätzung für V′ muß daher bereits bei der Berechnung von **A** berücksichtigt werden. Der Richtungssinn der Zuordnung ermöglicht in späteren Definitionen einen Vergleich durch Kleiner–/Gleich– Beziehungen.

3. Zwei–Punkt–Anordnungsrelationen

Das Kriterium *(2.3)* erfordert die Überprüfung von bis zu $N(N-1)(N-2)/2$ Matrixelementen. Außerdem benötigt man für die direkte Anwendung gängiger Verfahren zur Isomorphiebestimmung Zwei–Punkt–Relationen ähnlich den Kantenrelationen von Adjazenzgraphen. Daher wird eine Dimensionsreduktion der 3P–Anordnungsmatrix zu einer zweidimensionalen Matrix **B** der Zwei–Punkt–Anordnungsrelationen (2PAR) b_{ij} durchgeführt:

Definition 3.1:   Die **2P–Anordnungsrelation** b_{ij} ist definiert durch:

$$b_{ij}=\begin{cases} \sum_{k=1}^{N} \frac{1}{2}\cdot(a_{ijk}+1) & \text{falls } i<j \\[2mm] \sum_{k=1}^{N} -\frac{1}{2}\cdot(a_{ijk}-1) & \text{falls } i>j \\[2mm] 0 & \text{falls } i=j \end{cases}$$

Diese formale Definition läßt sich anschaulich interpretieren, wenn man berücksichtigt, daß $a_{ijk}\in\{-1,0,1\}$ gilt: der Wert von b_{ij} entspricht der Anzahl der Punkte P_k, die links der gerichteten Verbindunglinie l_{ij} liegen. Aus der Definition folgt unmittelbar:

(3.2) $b_{ij} \leq N-2-b_{ji}$

Das Gleichheitszeichen gilt für den Fall, daß sich alle Punkte P_k außerhalb des Toleranzbereiches $\pm d_{min}$ um l_{ij} befinden. Unter dieser einschränkenden Bedingung kann die **B**–Matrix direkt zur Bestimmung der konvexen Hülle der Punktstruktur herangezogen werden:

Lemma 3.3:   Falls $b_{ij} = 0$ oder $b_{ij} = N-2$, dann gehört das Polygon P_iP_j zur konvexen Hülle der Punktanordnung V.

Als Beispiel ist für die Punktstruktur in *Abb.2* die zugehörige 2P–Anordnungsmatrix in *Abb.3* wiedergegeben. Aufgrund des Toleranzkriteriums ($d_{min}=5$ Pixel) ist die konvexe Hülle in diesem Fall nicht eindeutig definiert.

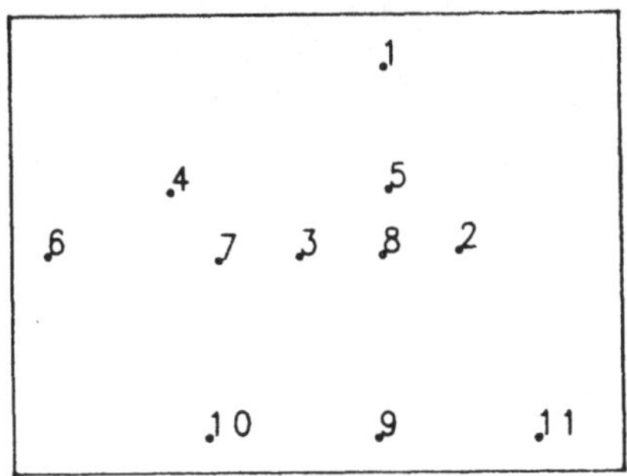

Abb.2: Punktstruktur

$$\begin{bmatrix} - & 0 & 5 & 8 & 2 & 8 & 5 & 2 & 4 & 5 & 0 \\ 0 & - & 6 & 7 & 8 & 5 & 4 & 1 & 1 & 3 & 0 \\ 2 & 3 & - & 6 & 2 & 6 & 6 & 3 & 5 & 5 & 4 \\ 0 & 2 & 3 & - & 1 & 8 & 6 & 3 & 6 & 8 & 5 \\ 6 & 1 & 7 & 8 & - & 6 & 5 & 2 & 3 & 4 & 2 \\ 0 & 4 & 2 & 0 & 2 & - & 6 & 5 & 8 & 9 & 7 \\ 3 & 5 & 3 & 2 & 4 & 3 & - & 6 & 6 & 5 & 6 \\ 6 & 8 & 5 & 5 & 7 & 3 & 3 & - & 1 & 2 & 3 \\ 5 & 7 & 4 & 2 & 6 & 1 & 2 & 8 & - & 0 & 8 \\ 2 & 6 & 2 & 1 & 5 & 0 & 3 & 7 & 8 & - & 8 \\ 8 & 8 & 5 & 4 & 7 & 2 & 3 & 6 & 0 & 0 & - \end{bmatrix}$$

Abb.3: 2PA–Matrix zu Abb.2

In Analogie zur *Def.2.3* läßt sich nun ein Kriterium für eine 2PA–Monomorphie formulieren:

Definition 3.4: Eine planare Punktstruktur V aus N Punkten ist **2PA–monomorph** zu einer Punktstruktur V′, wenn es eine eineindeutige Zuordnung V→V′ gibt, so daß für die Elemente der zugehörigen 2P–Anordnungsmatrizen B und B′ gilt:

$$(\forall i\ \forall j)\ ((b_{ij} \le b'_{kl}) \wedge (b_{ji} \le b'_{lk}))$$

P'_k, P'_l sind die zu P_i, P_j korrespondierenden Punkte.

Das Monomorphiekriterium *(3.4)* ist weniger streng als Kriterium *(2.3)*, da die Einzelzuordnung der Punkte P_k in der Summe in *(3.4)* nicht eingeht. Aus *(3.1)* läßt sich unmittelbar folgern:

Lemma 3.5: Alle 3PA–monomorphen Bildpunktstrukturen V,V′ sind auch 2PA–monomorph.

Die 2PAR b_{ij} können als gerichtete, bewertete Kanten zwischen zwei Knotenpunkten P_i und P_j eines **Anordnungsgraphen G** aufgefaßt werden. Diese "Kantenrelationen" b_{ij} ermöglichen gemäß *Lemma 3.5* eine Vorab–Selektion möglicher 3PA–Isomorphien durch den Test auf eine 2PA–Monomorphie nach dem Kriterium *(3.4)* und somit eine erhebliche Reduzierung des Suchraumes für 3PA–Monomorphien. Die Gesamtstrategie der Korrespondenzsuche ist in *Abb.4* dargestellt: Aus den Bilddaten (bzw. Modelldaten, s.Kap.6) werden Anordnungsgraphen generiert und zunächst 2PA–Monomorphien gesucht, die anschließend auf 3PA–Monomorphie getestet werden. Besteht sowohl 2PA– als auch 3PA–Monomorphie, wird eine Korrespondenzhypothese ausgegeben und kann aufgrund anderer Kriterien verifiziert werden. Die 2PAR erlauben die Anwendung von Algorithmen zur Bestimmung von Graphenisomorphien. Ein mögliches Verfahren wird im nächsten Kapitel beschrieben.

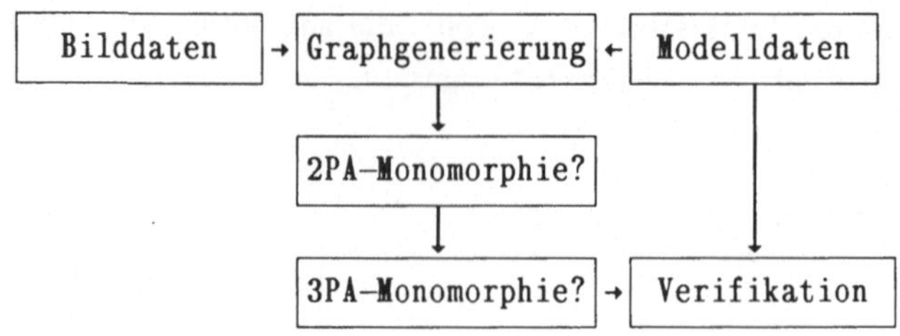

Abb.4: Schema der Modell/Bild –Bild–Korrespondenzlösung

4. Korrespondenzlösung durch Monomorphie von Anordnungsgraphen

Die Korrespondenzsuche zwischen zwei Abbildungen V,V′ entspricht graphentheoretisch der Suche nach Isomorphien (bzw. Monomorphien) der bildbeschreibenden Graphen G,G′. Sofern die Anzahl N der Punkte in V **kleiner** ist als die Anzahl N′ der Punkte in V′, lassen sich die Monomorphie-kriterien *(2.3)* und *(3.4)* aufgrund der Kleiner–/Gleichbeziehungen ebenso zur Suche aller Teilgraphen G⊂G′ anwenden. In diesem Fall existieren N′!/(N′−N)! mögliche Einbettungen von G in G′. Um eine extensive Suche und die Anwendung des (aufwendigeren) 3PA– Monomorphie-kriteriums auf **alle** kombinatorisch möglichen Korrespondenzen zu vermeiden, kann mit Hilfe der 2PAR eine effiziente Reduktion des Suchraumes erfolgen.

Hierzu wird die **Zuordnungsmatrix M** mit N Zeilen und N′ Spalten eingeführt, deren Elemente m_{ik} die Zulässigkeit einer Punktkorrespondenz $P_i{\rightarrow}P_k'$ angeben und zu $m_{ik}=1$ initialisiert werden. Ist eine Korrespondenz $P_i{\rightarrow}P_k'$ a priori ausgeschlossen, so wird $m_{ik}=0$ gesetzt. Eine Refinement–Prozedur vergleicht für alle möglichen Punktkorrespondenzen $P_i{\rightarrow}P_k'$ die zugehörigen Zeilen und Spalten der 2P–Anordnungsmatrizen B und B′ und überprüft, ob folgende Bedingung erfüllt ist:

Bedingung 4.1: $\quad \underset{1\leq x\leq N}{(\forall\ x)}\ ((m_{ik}=1) \Rightarrow \underset{1\leq y\leq N'}{(\exists\ y)}\ ((m_{xy}=1) \wedge (b_{ix} \leq b_{ky}') \wedge (b_{xi} \leq b_{yk}')))$

Für jedes Zeilen- (Spalten-)Element b_{ix} (b_{xi}) muß gem.*(4.1)* mindestens ein Zeilen- (Spalten-) Element b_{ky}' (b_{yk}') existieren, welches das 2PA-Monomorphiekriterium erfüllt, wobei die Korrespondenz $P_x{\rightarrow}P_y'$ erlaubt sein muß. Ist die Bedingung *(4.1)* nicht erfüllt, wird das Matrixelement m_{ik} gelöscht. Da die Refinement–Prozedur auch das Element m_{xy} verwendet, werden iterativ alle Indizes i und j durchlaufen, bis keine Änderung der **M**–Matrix mehr erfolgt.

Die Refinement–Prozedur kann in einen Baumsuche–Algorithmus integriert werden */Ull76/*, der das Refinement jeweils bei Erreichen einer neuen Suchtiefe (d.h. Hinzunahme einer weiteren Punktkorrespondenz–Hypothese) erneut aufruft und nur solche Korrespondenzen zuläßt, die mit **M** verträglich sind. Dies führt selbst in Fällen, in denen die Matrix **M** zu Beginn (Suchtiefe d=0) noch stark besetzt ist, zu einer erheblichen Einschränkung der Suchraumes.

Prüft man die Punktstruktur aus *Abb.2* auf Monomorphien mit sich selbst, so ergibt bereits die erste Anwendung des Refinements eine Diagonalmatrix **M**, so daß in diesem Beispiel die Korrespondenzlösung ohne Suchalgorithmus direkt gefunden wird.

5. Korrespondenzlösung zwischen einem 3D–Modell und einer Abbildung

Während bisher die Korrespondenzlösung zwischen zwei Bildpunktstrukturen V,V′ beschrieben wurde, soll nun die Korrespondenz zwischen einer räumlichen Punktstruktur W und einer orthogonalen Abbildung V′ von W′ (W unter verändertem Aspektwinkel) mit Hilfe der Anordnungsrelationen gefunden werden. Die 3D–Koordinaten von W im Beobachtersystem seien bekannt und es sei W′=T(W), V=P(W), V′=P(W′), wobei T die Transformation des Modells in eine neue Raumlage und P die orthogonale Projektion auf die (x,y)–Ebene symbolisiert.

Eine weitere Definition der 3P–AR stellt zunächst den Zusammenhang zwischen der (Parallel–) Projektion auf die (x,y)–Ebene und der dreidimensionalen Modellstruktur her. Beschreibt man die Bildpunktkoordinaten formal durch dreidimensionale Vektoren $\mathbf{p}_i=(p_{ix},p_{iy},0)^t$, können die 3PAR durch die z–Komponente eines Vektorproduktes dieser Größen ausgedrückt werden:

Definition 5.1:
$$a_{ijk} = \mathrm{sign}((\mathbf{p}_j-\mathbf{p}_i)\times(\mathbf{p}_k-\mathbf{p}_i)_z)$$
$$= -\mathrm{sign}((0,0,-1)^t\cdot(\mathbf{p}_j-\mathbf{p}_i)\times(\mathbf{p}_k-\mathbf{p}_i))$$

In *Def.5.1* wird die 3P–AR durch die Flächennormale des Dreieckes Δ_{ijk} bestimmt. Das Vorzeichen von a_{ijk} gem. *Def.5.1* gibt an, ob die beiden Verbindungsvektoren $(\mathbf{p}_j-\mathbf{p}_i)$ und $(\mathbf{p}_k-\mathbf{p}_i)$ zusammen mit dem Einheitsvektor in z–Richtung ein rechts– oder linkshändiges System bilden. Dies stellt eine Möglichkeit der Berechnung des Zählsinnes des Dreieckes Δ_{ijk} dar und ist somit der *Def.2.1* äquivalent.

Die Relationen a_{ijk} können nach *Def.5.1* auch direkt aus den z–Komponenten der Flächennormalen der räumlichen Dreiecke berechnet werden, da diese von den z–Koordinaten der Ortsvektoren unabhängig sind. Interpretiert man den Vektor (0,0,–1) als Beobachtungsvektor, so stimmt *(5.1)* mit dem Sichtbarkeitskriterium für die (Dreiecks–)Fläche eines konvexen 3D–Körpers bei Parallel- projektion überein */New79/*. Hieraus folgt, daß sich das Vorzeichen von a_{ijk} genau dann ändert, wenn die (virtuelle) Dreiecksfläche von der jeweils anderen Seite (Vorder– oder Rückseite) betrachtet wird. Hieraus ergibt sich eine Schlußfolgerung für den Fall planarer Modellstrukturen:

Lemma 5.2:   Für planare Modellpunktstrukturen W gibt es genau zwei 3P–Anordnungs- matrizen $A^+=(a_{ijk})$ und $A^-=(-a_{ijk})$, die für die Abbildungen der Vorder– und Rückseite des Modells unter beliebigem Aspektwinkel gelten.

Im Gegensatz zur Definition der Orientierungskonsistenz eines starren Körpers /MB83/, bei der an die Stelle des Beobachtungsvektors in *(5.1)* der Ortsvektor zu einem weiteren Modellpunkt tritt und die eine Lösung des Korrespondenzproblems zwischen zwei 3D–Punktobjekten ermöglicht /Chen88/, gilt die Invarianz der AR bei Abbildung von **dreidimensionalen** Modellen nur für Skalierung, Translation und Drehung um die z–Achse, nicht jedoch für für Drehungen um die x– oder y–Achse.

Gibt man aber eine Modell–Lage W und die maximal zu tolerierenden Drehwinkel γ und β um diese Achsen vor, so kann die aus W berechnetete Matrix **A** in eine für diesen Drehbereich $D(\gamma,\beta)$ **charakteristische Matrix A_d** konvertiert werden, indem alle Relationen a_{ijk}, die sich innerhalb des Drehbereiches ändern, zu null gesetzt werden. Die resultierende Matrix A_d ist gem. den Kriterien *(2.3)* und *(3.4)* monomorph zu allen Matrizen **A′** von Modellansichten V′ innerhalb des Drehbereiches D, da nur Relationen $a_{ijk}\neq0$ berücksichtigt werden. Die zu löschenden Relationen a_{ijk} kann man nach *(5.1)* leicht anhand des Winkel der Flächennormalen der räumlichen Dreiecke zur (x,y)–Ebene selektieren.

Als Beispiel ist in *Abb.5* eine charakteristische Ansicht eines Pyramiden–(Draht–)Modells mit 11 ausgezeichneten Punkten wiedergegeben. Als Drehwinkelbereich werden Drehungen um jeweils ±20 Grad um die x– und y– Achse zugelassen. Bei der Suche nach Monomorphie zu der Punktstruktur

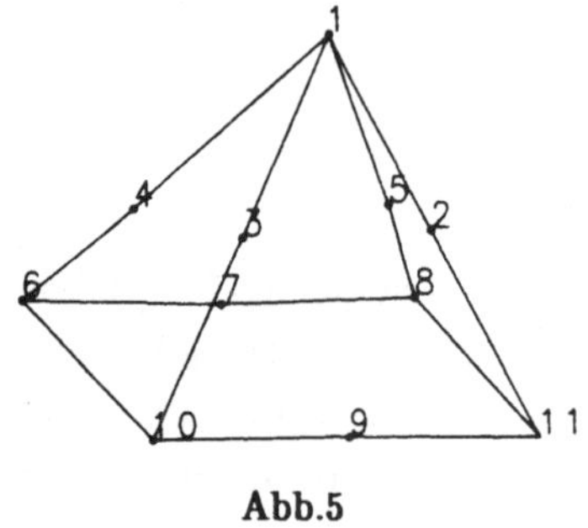

Abb.5

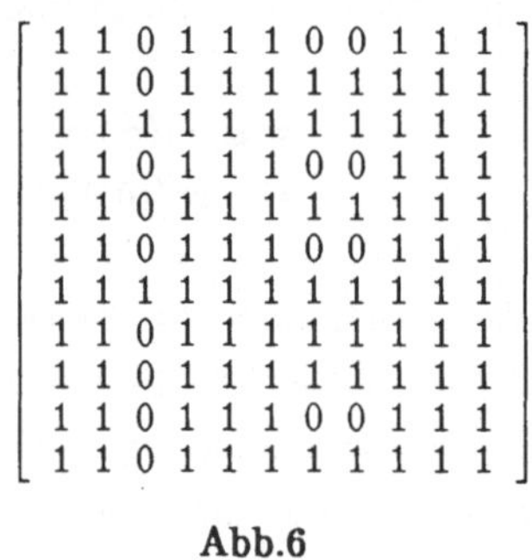

$$\begin{bmatrix}
1 & 1 & 0 & 1 & 1 & 1 & 0 & 0 & 1 & 1 & 1 \\
1 & 1 & 0 & 1 & 1 & 1 & 1 & 1 & 1 & 1 & 1 \\
1 & 1 & 1 & 1 & 1 & 1 & 1 & 1 & 1 & 1 & 1 \\
1 & 1 & 0 & 1 & 1 & 1 & 0 & 0 & 1 & 1 & 1 \\
1 & 1 & 0 & 1 & 1 & 1 & 1 & 1 & 1 & 1 & 1 \\
1 & 1 & 0 & 1 & 1 & 1 & 0 & 0 & 1 & 1 & 1 \\
1 & 1 & 1 & 1 & 1 & 1 & 1 & 1 & 1 & 1 & 1 \\
1 & 1 & 0 & 1 & 1 & 1 & 1 & 1 & 1 & 1 & 1 \\
1 & 1 & 0 & 1 & 1 & 1 & 1 & 1 & 1 & 1 & 1 \\
1 & 1 & 0 & 1 & 1 & 1 & 0 & 0 & 1 & 1 & 1 \\
1 & 1 & 0 & 1 & 1 & 1 & 1 & 1 & 1 & 1 & 1
\end{bmatrix}$$

Abb.6

aus *Abb.2* (einer Ansicht des Modells aus diesem Drehbereich) erhält man bei erster Anwendung der Refinement–Prozedur die Matrix **M** in *Abb.6*. Der vollständige Algorithmus ergibt zwei (von 11! $\simeq$ 4×10⁷ theoretisch möglichen !) 2PA–Monomorphien, von denen eine (Punkt 2 und 8 vertauscht) durch den nachfolgenden Test auf 3PA–Monomorphie verworfen wird, so daß nur die richtige Korrespondenz als Hypothese ausgegeben wird.

6. Schlußfolgerungen

Die Anordnungsrelationen lassen sich sowohl zur Korrespondenzlösung zwischen zwei beliebigen, orthogonalen Abbildungen planarer Punktstrukturen als auch zur Objekterkennung von 3D–

Modellen anhand der Projektionen in einem eingeschränkten Drehbereich einsetzen. Ein großer Drehwinkel- Toleranzbereich wird hierbei zweckmäßigerwise in mehrere Teilbereiche Di mit den zugehörigen charakteristischen Matrizen $\mathbf{A_{di}}$ zerlegt, die in einer geeigneten Datenstruktur (vgl./Sha82/,/Cha82/) gespeichert und einzeln auf Isomorphie mit $\mathbf{A}'$ geprüft werden. Aufgrund ihrer Invarianzeigenschaften eignen sich die AR insbesondere zur Lösung von Korrespondenzen in Bildfolgen eines langsam bewegten 3D–Objektes bei bekannter Ausgangsposition, aber unbekannter Änderung der Kameraausrichtung. Außerdem können die AR auch als robustes Kriterium zur Verifikation von Korrespondenzhypothesen, insbesondere bei planaren Teilstrukturen eines Modells, in andere Erkennungsalgorithmen integriert werden.

Literatur:

/Cha82/ I.Chakravarty, H.Freeman: Characteristic Views as a Basis for Three–dimensional Object Recognition, SPIE Vol.336: Robot Vision (1982)

/Che88/ H.H.Chen, T.S.Huang: Maximal Matching of 3D–Points for Multiple Object Motion Estimation, Pattern Recognition Vol.21, No.2, pp.75–90 (1988)

/Gmu88/ E.Gmür, H.Bunke,: PHI–1: Ein CAD–basiertes Roboter Sichtsystem, in: Informatik–Fachberichte 180, Springer, Berlin, pp.240–247 (1988)

/Mit85/ A.Mitiche, P.Bouthemy: Tracking Modelled Objects Using Binocular Images, Computer Vision, Graphics and Image Processing 32, pp.384–396 (1985)

/Mor79/ H.P.Moravec, Visual Mappping by a Robot Rover, 6th Int.Joint Conf. on Artificial Intelligence, pp. 598–600 (1979)

/Mue88/ H.Müller: Korrespondenzlösung zwischen zwei Abbildungen durch relationale Isomorphie, in: Informatik–Fachberichte 180, Springer, Berlin, pp.135–141 (1988)

/New79/ W.Newman, R.Sproull: Principles of Interactiv Computer Graphics, McGraw–Hill, New York (1979)

/Sha82/ L.G.Shapiro, R.M.Haralick: Organization of Relational Models for Scene Analysis, IEEE Trans.on PAMI –4, No.6 (1982)

/Sha88/ L.G.Shapiro, R.MacDonald, S.R.Sternberg: Ordered Structural Shape Matching with Primitive Extraction by Mathematical Morphology, Pattern Recognition Jan.1987

/Tho87/ D.W.Thomson, J.L.Mundy: Three–Dimensional Model Matching from an Unconstrained Viewpoint, IEEE Robotics and Automation Conf. 1987,1, pp.208–220

/Ull76/ J.R.Ullmann: An Algorithm for Subgraph Isomorphism, J.ACM,17, No.1, pp.31–42 (1976)

GEOMETRISCHE TRANSFORMATIONEN IN DER DISKRETEN EBENE

Albrecht Hübler

Friedrich-Schiller-Universität Jena

Sektion Mathematik

UHH, 17.OG., Jena, DDR-6900

1. Motivation

In der digitalen Bildverarbeitung und Computergrafik ist für viele Anwendungsfälle die Beherrschung verschiedener geometrischer Transformationen, insbesondere von Bewegungstransformationen wie Verschiebungen, Drehungen und Spiegelungen wichtig.

Während die Realisierung von Verschiebungen auf den üblicherweise als Trägerstrukturen digitaler Bilder verwendeten Gittern/Rastern keine Probleme bereitet, ist die Situation bezüglich der Rotationen und Spiegelungen wesentlich komplizierter.

Obwohl beispielsweise für die Ausführung von Drehungen in der digitalen Ebene eine ganze Reihe,zum Teil auch technisch realisierter Lösungen existiert (vgl. z.B. [1,2]), ist das Problem aus mathematisch-theoretischer Sicht noch unbefriedigend geklärt.

Heuristische, auf ganz spezielle praktische Situationen zugeschnittene Herangehensweisen sind typisch.

Einigen Aufschluß über die Kompliziertheit dieser Fragestellung geben in der entsprechenden Literatur die auf das Studium digitaler Kreise gerichteten Arbeiten (z.B. [8,9,10]).

Viele Forderungen, die man an derartige geometrische Abbildungen stellen möchte, lassen sich nicht oder nur in sehr eingeschränktem Maße erfüllen, beispielsweise die Invarianz von topologischen Eigenschaften der transformierten Menge.

Alle bisherigen Erfahrungen legen die Vermutung nahe, daß in der digitalen Ebene die Klasse wohldefinierter, auch theoretisch beherrschbarer geometrischer Transformationen relativ arm strukturiert ist.

Um Aussagen hierzu zu erhalten, ist es sinnvoll, sich bewußt von der unmittelbaren Bezugnahme auf die euklidische Geometrie zu lösen und ausschließlich die Struktur der diskreten Ebene und die ihr inhärenten Eigenschaften zu nutzen.

Ein möglicher Weg, Einsichten über eine in diesem Sinne abstrakte Geometrie der diskreten Ebene zu gewinnen, ist der Aufbau einer axiomatischen Theorie für eine Geometrie diskreter Punkträume wie er in [5] demonstriert wird. Dieser in [6] weitergeführte Ansatz fixiert im Rahmen einer in sich abgeschlossenen axiomatischen Theorie die wesentlichen Merkmale diskreter Räume, die als Trägerstrukturen digitaler Bilder geeignet sind.

Hervorzuheben sind dabei die im Vergleich zur euklidischen Geometrie unterschiedliche Festlegung der Parallelität, die es zuläßt, daß nichtparallele Geraden der diskreten Ebene keinen gemeinsamen Punkt besitzen, sowie die axiomatische Fixierung der Diskretheit, die alle kontinuierlichen Geometrien als Modelle ausschließt.

Die Diskretheit wird erzwungen durch die Forderungen, daß zwischen je zwei Punkten höchstens endlich viele Punkte liegen dürfen, und daß ebenso die Anzahl aller zwischen zwei gegebenen parallelen Geraden liegenden und zu diesen beiden Geraden parallelen Geraden endlich ist.

Das Axiomensystem ist so angelegt, daß es die Verhältnisse in den für die digitale Bildverarbeitung geeigneten Gittern genau beschreibt und das übliche Orthogonalgitter als Standardmodell verwendet werden kann. Da es nicht möglich ist, die gesamte Theorie hier einzuführen, beschränken wir unsere Betrachtungen auf dieses Modell. Im nächsten Abschnitt werden dafür in aller Kürze einige wichtige Begriffe bereitgestellt.

2. Die Gitterpunktebene

Wie weiter vorn bereits erwähnt, kann das ebene quadratische Gitter als ein Standardmodell für die axiomatisch beschriebene diskrete Ebene angesehen werden.

Dabei identifiziert man die *Punktmenge* der Gitterpunktebene mit der Menge $\mathbb{Z}^2$ aller geordneten Paare ganzer Zahlen.

Mit $\mathcal{T}$ sei die Menge aller *Verschiebungen* auf $\mathbb{Z}^2$ bezeichnet, wobei die Verschiebungen $\tau \in \mathcal{T}$ eineindeutig mit den ganzzahligen Vektoren $\mathfrak{v} \in \mathbb{Z}^2$ korrespondieren: Für die durch einen Vektor $\mathfrak{v} \in \mathbb{Z}^2$ erklärte Verschiebung τ und alle Punkte $p \in \mathbb{Z}^2$ gilt $\tau(p):=p+\mathfrak{v}$. Eine Verschiebung τ heißt genau dann eine *einfache* Verschiebung, wenn sie nicht durch mehrfache Hintereinanderausführung einer anderen Verschiebung erzeugt werden kann, d.h. wenn für alle Verschiebungen $\tau'\neq\tau$ und für alle $i \in \mathbb{Z}$ mit $|i|>1$ stets $(\tau')^i\neq\tau$ ist.

Die *Geraden* auf $\mathbb{Z}^2$ sind genau die durch wiederholte Anwendung einer einfachen Verschiebung τ auf einen Punkt $p \in \mathbb{Z}^2$ erzeugten Punktmengen. Für eine Gerade g gibt es also eine bis auf Inversenbildung eindeutig bestimmte einfache Verschiebung τ, so daß für jeden Punkt $p \in g$ die Darstellung $g=\{\tau^i(p): i \in \mathbb{Z}^2\}$ richtig ist. Die Verschiebung τ wird *Generator* von g genannt.

Für drei Punkte p, q und r sagen wir, der Punkt q *liegt zwischen* p und r, falls es eine Verschiebung τ und positive ganze Zahlen i und j gibt, für die $\tau^i(p)=q$ und $\tau^j(q)=r$ gelten. Abkürzend verwenden wir für die Zwischenrelation die Schreibweise $Z(p,q,r)$.

Zwei Geraden g und g' sind genau dann *parallel*, wenn es eine Verschiebung τ mit $\tau(g)=g'$ gibt.

Ein Paar $\mathcal{K}=[g_x,g_y]$ von Geraden mit den Generatoren τ_x und τ_y heißt genau dann ein *Koordinatensystem*, wenn es zu jedem Punkt $p \in \mathbb{Z}^2$ ganze Zahlen i und j gibt, für die $\{p\} = \tau_x^i(g_y) \cap \tau_y^j(g_x)$ ist. Die Zahlen i und j, die für ein gegebenes Koordinatensystem $\mathcal{K}$ stets eindeutig bestimmt sind, heißen die x- bzw. die y-Koordinaten von p bezüglich $\mathcal{K}$, die kurz auch durch $x_{\mathcal{K}}(p)$ und $y_{\mathcal{K}}(p)$ bezeichnet werden. Der eindeutig bestimmte Schnittpunkt $\mathcal{O}$ von g_x und g_y ist dann der Ursprung des Koordinatensystems $\mathcal{K}$.

Offensichtlich ist für die einfachen Verschiebungen τ_0 und τ_1 mit $\tau_0(p) := p+(1,0)$ und $\tau_1(p) := p+(0,1)$ und für $\mathcal{O}:=(0,0)$ das Paar $\mathcal{K}_0:=[g_0,g_1]$ mit $g_0:=\{\tau_0^i(\mathcal{O}): i \in \mathbb{Z}\}$ und $g_1:=\{\tau_1^i(\mathcal{O}): i \in \mathbb{Z}\}$ ein Koordinatensystem. Aus der Konstruktion von $\mathcal{K}_0$ folgt, daß für einen beliebigen Punkt $p=(a,b)$ stets $x_{\mathcal{K}_0}(p)=a$ und $y_{\mathcal{K}_0}(p)=b$ sind.

Für alle weiteren Betrachtungen sei das Koordinatensystem $\mathcal{K}_0$ als festes Bezugskoordinatensystem vorausgesetzt.

Es ist klar, daß für einfache Verschiebungen τ_x mit $\tau_x(p)=p+(a,b)$ und τ_y mit $\tau_y(p)=p+(c,d)$ die Geraden $g_x:=\{\tau_x^i(\mathcal{O}): i \in \mathbb{Z}\}$ und $g_y:=\{\tau_y^j(\mathcal{O}): j \in \mathbb{Z}\}$ genau dann ein Koordinatensystem bilden, wenn die Determinante der Matrix $M(\tau_x,\tau_y) := \begin{bmatrix} a & c \\ b & d \end{bmatrix}$ den Wert 1 oder -1 hat.

Gleichzeitig beschreibt diese Matrix M eine lineare Abbildung φ_M, die das Koordinatensystem $\mathcal{K}_0$ auf das Koordinatensystem $\mathcal{K}:=[g_x,g_y]$ abbildet, d.h. für jeden Punkt $p=(u,v)$ gilt $x_{\mathcal{K}}(\varphi(p))=u$ und $y_{\mathcal{K}}(\varphi(P))=v$.

Ein geordnetes Tripel dreier paarweise verschiedener Punkte (p,q,r) mit $p=(u_1,v_1)$, $q=(u_2,v_2)$ und $r=(u_3,v_3)$ nennen wir genau dann *positiv (negativ) orientiert*, wenn die Determinante

$$\det \begin{bmatrix} 1 & u_1 & v_1 \\ 1 & u_2 & v_2 \\ 1 & u_3 & v_3 \end{bmatrix} \qquad \text{größer (kleiner) als Null ist.}$$

Für ein durch die Matrix $M = \begin{bmatrix} a & c \\ b & d \end{bmatrix}$ gegebenes und folglich durch das Punkttripel $(\mathcal{O},(a,b),(c,d))$ eindeutig beschriebenes Koordinatensystem $\mathcal{K}$ ist die Orientierung von $\mathcal{K}$ gleich der des Tripels $(\mathcal{O},(a,b),(c,d))$, also genau dann positiv, wenn $\det(M) = 1$ ist.

3. Geometrische Abbildungen in der Gitterpunktebene

Ausgehend von der anschaulichen Vorstellung, daß eine geometrische Abbildung die geometrische Struktur einer Punktmenge, die sich in der Anordnung, d.h. der relativen Lage der Punkte zueinander manifestiert, nicht verändern soll, definieren wir:

Eine bijektive Abbildung $\varphi : \mathbb{Z}^2 \longrightarrow \mathbb{Z}^2$ heißt genau dann *geometrische Abbildung*, wenn die Zwischenrelation Z bezüglich φ invariant ist, d.h. wenn für je drei Punkt p, q und r genau dann $Z(p,q,r)$ gilt, wenn $Z(\varphi(p),\varphi(q),\varphi(r))$ ist.

Man sieht sofort, daß alle Verschiebungen geometrische Abbildungen sind.

Mit $\mathcal{G}$ sei die Klasse aller geometrischen Abbildungen bezeichnet, mit $\mathcal{G}_0$ die Klasse derjenigen Abbildungen aus $\mathcal{G}$, die den Punkt $\mathcal{O}$ fest lassen. Es gilt [6]:

SATZ: Für eine Bijektion $\varphi : \mathbb{Z}^2 \longrightarrow \mathbb{Z}^2$ sind folgende Aussagen äquivalent:

    1) φ ist eine geometrische Abbildung.

    2) φ bildet parallele Geraden auf parallele Geraden ab.

    3) φ bildet das Koordinatensystem $\mathcal{K}_0$ auf ein Koordinatensystem $\mathcal{K}=[\varphi(g_0),\varphi(g_1)]$ ab.

Damit ist offensichtlich jede geometrische Abbildung $\varphi \in \mathcal{G}_0$ eindeutig durch eine Matrix $M_\varphi = \begin{bmatrix} a & c \\ b & d \end{bmatrix}$ mit $|\det(M_\varphi)|=1$ beschrieben, d.h. es gilt für alle Punkte p: $\begin{bmatrix} x(\varphi(p)) \\ y(\varphi(p)) \end{bmatrix} = M_\varphi \begin{bmatrix} x(p) \\ y(p) \end{bmatrix}$.

Im weiteren interessieren wir uns speziell für die Klasse $\mathcal{G}_0$.

Eine als *Bewegung* zu bezeichnende Abbildung $\varphi \in \mathcal{G}_0$ sollte folgenden, noch zu präzisierenden Kriterien genügen:

(B1) Bei mehrfacher Anwendung von φ darf sich das Bild eines Punktes p nicht "beliebig weit" von $\mathcal{O}$ entfernen.

(B2) φ darf das Koordinatensystem nicht unzulässig "deformieren".

Aufgrund der Diskretheit der Gitterpunktebene Z^2 kann die Bedingung (B1) offensichtlich durch die Forderung formal präzisiert werden, daß es eine natürliche Zahl i geben muß, für die die i-fache Hintereinanderausführung φ^i von φ die identische Abbildung *id* ist.

Ein Resultat aus [4] löst dieses Problem: Es sei $E:=\begin{pmatrix} 1 & 0 \\ 0 & 1 \end{pmatrix}$.

SATZ: Es sei $\varphi \in \mathcal{G}_0$ mit $M_\varphi=\begin{pmatrix} a & c \\ b & d \end{pmatrix}$ und $M_\varphi \neq \pm E$.

Die durch φ erzeugte zyklische Gruppe ist genau dann endlich, wenn $0 \leq |a+d| \cdot \det(M_\varphi) \leq 1$ gilt.

Die Ordnung einer durch φ mit $M_\varphi \neq \pm E$ erzeugten endlichen Gruppe ist entweder

- gleich 2, wenn $\det(M_\varphi)=-1$ und a+d=0 sind   oder
- gleich 3, wenn $\det(M_\varphi)=1$ und a+d=-1 sind   oder
- gleich 4, wenn $\det(M_\varphi)=1$ und a+d=0 sind   oder
- gleich 6, wenn $\det(M_\varphi)=1$ und a+d=1 sind.

Eine Präzisierung der Bedingung (B2) erfolgt durch die Festlegung, daß die φ-Bilder der "Einheitsvektoren" des Bezugskoordinatensystems, $e_x=(1,0)$, $e_y=(0,1)$, $-e_x$, $-e_y$, auf die Quadranten des Bezugskoordinatensystems verteilt sind, d.h. es soll in jedem der vier Quadranten eines der Elemente $\varphi(e_x)$, $\varphi(e_y)$, $\varphi(-e_x)$, $\varphi(-e_y)$ liegen.

Das führt zu folgenden Bedingungen an eine Abbildung $\varphi \in \mathcal{G}_0$ mit $M_\varphi=\begin{pmatrix} a & c \\ b & d \end{pmatrix}$, eine *Rotation* zu sein:

(R1) $\det(M_\varphi) = 1$.

(R2) Die durch φ erzeugte zyklische Gruppe ist endlich.

(R3) $\varphi=id$ oder O ist der einzige Fixpunkt von φ.

(R4) $a \cdot d \geq 0$ und $b \cdot c \leq 0$.

Für *Spiegelungen* φ aus $\mathcal{G}_0$ mit $M_\varphi=\begin{pmatrix} a & c \\ b & d \end{pmatrix}$ sind die adäquaten Bedingungen:

(S1) $\det(M_\varphi)=-1$.

(S2) Die durch φ erzeugte zyklische Gruppe ist endlich.

(S3) Es gibt genau eine Fixpunktgerade bezüglich φ, die alle Fixpunkte von φ enthält.

(S4) $a \cdot d \leq 0$ und $b \cdot c \geq 0$.

Es zeigt sich, daß genau 12 Abbildungen aus $\mathcal{G}_0$ existieren, die den Forderungen (R1)-(R4) genügen:

SATZ: Die Klasse $\mathcal{R}$ aller geometrischen Abbildungen aus $\mathcal{G}_0$, die die Bedingungen (R1)-(R4) erfüllen, ist genau durch die folgende Menge von Abbildungsmatrizen beschrieben:

$$\mathfrak{M}_\mathcal{R}=\left\{ \begin{pmatrix} 0 & -1 \\ 1 & a \end{pmatrix}, \begin{pmatrix} 0 & 1 \\ -1 & a \end{pmatrix}, \begin{pmatrix} a & -1 \\ 1 & 0 \end{pmatrix}, \begin{pmatrix} a & 1 \\ -1 & 0 \end{pmatrix}, \begin{pmatrix} b & 0 \\ 0 & b \end{pmatrix} : a\in\{0,1,-1\}, b\in\{1,-1\} \right\}.$$

Als Spiegelungen kommen unendlich viele Abbildugen in Frage.

SATZ: Die Klasse $\mathscr{S}$ aller geometrischen Abbildungen aus $\mathscr{G}_o$, die die
Bedingungen (S1)–(S4) erfüllen, ist genau durch die
Abbildungsmatrizen aus

$$\mathfrak{M}_{\mathscr{S}}=\{\begin{pmatrix}0 & 1\\ 1 & a\end{pmatrix},\begin{pmatrix}0 & -1\\ -1 & 0\end{pmatrix},\begin{pmatrix}-1 & a\\ 0 & 1\end{pmatrix},\begin{pmatrix}1 & a\\ 0 & -1\end{pmatrix},\begin{pmatrix}-1 & 0\\ a & 1\end{pmatrix},\begin{pmatrix}1 & 0\\ a & -1\end{pmatrix} : a\in\mathbb{Z}, a\neq 0\}$$

beschrieben.

Die sinnvolle Forderung, daß sowohl die Rotationen als auch
Rotationen und Spiegelungen zusammen, jeweils eine Gruppe bilden,
führt zu weiteren Einschränkungen. Es gibt genau drei maximale
Gruppen, die aus Elementen von $\mathscr{R}$ bestehen:

$$\mathscr{R}_1 = \{\varphi: M_\varphi=\begin{pmatrix}0 & -1\\ 1 & 0\end{pmatrix}^i, \ i \in \{1,2,3,4\}\},$$

$$\mathscr{R}_2 = \{\varphi: M_\varphi=\begin{pmatrix}1 & -1\\ 1 & 0\end{pmatrix}^i, \ i \in \{1,2,3,4,5,6\}\},$$

$$\mathscr{R}_3 = \{\varphi: M_\varphi=\begin{pmatrix}0 & -1\\ 1 & 1\end{pmatrix}^i, \ i \in \{1,2,3,4,5,6\}\}.$$

Diese drei möglichen Rotationsgruppen der Gitterpunktebene können
durch Hinzunahme jeweils eindeutig bestimmter, mit $\mathscr{R}_1$, $\mathscr{R}_2$, $\mathscr{R}_3$
verträglichen Teilmegen von $\mathscr{S}$ zu Bewegungsgruppen ergänzt werden:

$$\text{Für } \mathscr{S}_1 := \{ \varphi: M_\varphi\in \{\begin{pmatrix}0 & a\\ a & 1\end{pmatrix}, \begin{pmatrix}a & 0\\ 0 & -a\end{pmatrix}\}, \ a \in \{1,-1\} \},$$

$$\mathscr{S}_2 := \{ \varphi: M_\varphi\in \{\begin{pmatrix}0 & a\\ a & 0\end{pmatrix}, \begin{pmatrix}a & 0\\ a & -a\end{pmatrix}, \begin{pmatrix}a & -a\\ 0 & -a\end{pmatrix}\}, \ a \in \{1,-1\} \} \quad \text{und}$$

$$\mathscr{S}_3 := \{ \varphi: M_\varphi\in \{\begin{pmatrix}0 & a\\ a & 0\end{pmatrix}, \begin{pmatrix}-a & 0\\ a & a\end{pmatrix}, \begin{pmatrix}a & a\\ 0 & -a\end{pmatrix}\}, \ a \in \{1,-1\} \}$$

sind $\mathscr{B}_i:=\mathscr{R}_i\cup\mathscr{S}_i$, $i \in \{1,2,3\}$, die maximalen Bewegungsgruppen des
Gitterpunktraumes $\mathbb{Z}^2$.

Je nach Wahl von $\mathscr{B}_1$, $\mathscr{B}_2$ oder $\mathscr{B}_3$ als Bewegungsgruppe ist also der
Gitterpunktraum mit genau 4 Rotationen und 4 Spiegelungen (im Falle
von $\mathscr{B}_1$) bzw. mit genau 6 Rotationen und 6 Spiegelungen (für $\mathscr{B}_2$ und
$\mathscr{B}_3$) versehen. Geeignete Interpretationen der Gitterpunktebene führen
zu sinnvollen Interpretationen dieser Gruppen:
Die Wahl von $\mathscr{B}_1$ als Bewegungsgruppe entspricht der Interpretation der
diskreten Ebene als 4-Nachbarschaftsgraph mit der $90°$-Drehung und
deren Vielfachen als Rotationen sowie den Spiegelungen mit den
Spiegelachsen x=0, y=0, x=y und y=−x.
Wählt man $\mathscr{B}_2$ oder $\mathscr{B}_3$ als Gruppe der Bewegungen, ist die
Interpretation der diskreten Ebene als ebener 6-Nachbarschaftsgraph
adäquat. Die Drehgruppe entspricht dann der $60°$-Drehung und deren
Vielfachen, die Spiegelungen besitzen als Spiegelachsen gerade die
drei Hauptachsen und die Halbierenden der Sektoren zwischen jeweils
zwei benachbarten Hauptachsen. Der einzige Unterschied zwischen der

Wahl von $\mathcal{B}_2$ und $\mathcal{B}_3$ besteht bei dieser Interpretation in der unterschiedlichen Wahl der x- und y-Achse des Koordinatensystems unter den Hauptrichtungen.

4. Schlußbemerkungen

Die vorgestellten Ergebnisse bestätigen die eingangs formulierte Vermutung, daß der Versuch, wohldefinierte geometrische Bewegungsgruppen zu erklären, zu sehr arme Klassen führt, die bei der Lösung praktischer Probleme kaum weiterhelfen. Dies wird durch die Tatsache unterstrichen, daß alle endlichen Gruppen von Elementen aus $\mathcal{G}_0$ maximal 12 Elemente haben können [3].
Auch andere Herangehensweisen, z.B. Drehungen als spezielle isometrische Abbildungen aufzufassen, führen zu ähnlichen Resultaten. In [7] wurde gezeigt, daß die "Einheitskreise" der dafür geeigneten digitalen Metriken höchstens 6 "Eckpunkte" besitzen, was gleibedeutend damit ist, daß die zugehörigen Drehgruppen aus nicht mehr als 6 Elementen bestehen können.
Man wird also für die Bearbeitung praktischer Probleme auch weiter nach anderen Lösungen fragen müssen, die stets einen Verzicht auf gewünschte Eigenschaften bedeuten. Es kommt aber immer, auch bei der Suche nach praktikablen Lösungen darauf an, genau aufzuklären, auf welche der möglichen Forderungen man eigentlich verzichtet, welche Kompromisse man eingeht. Es kann dies der Verzicht auf Eineindeutigkeit oder gar auf die Eindeutigkeit der Abbildungen sein, man kann die Forderung der Gruppeneigenschaften fallen lassen usw.; stets sollte man aber genau wissen und beschreiben können, was man wirklich tut. Nur so ist es möglich, neue Impulse für die weitere theoretische Aufarbeitung – nicht zuletzt eine wichtige Bedingung für die Lehre – zu geben bzw. zu erhalten.

5. Literatur

[1] Anderson, K.L., F.C.Mintzer, G.Goertzel, L.Mitchell: Method for rotating a binary image. United States Patent 4,627,020, Dec.1986

[2] Arabnia, H.R., M.A.Oliver: Arbitrary rotation of raster images with SIMD machine architectures. Computer Graphics Forum 6 (1987), 3-11

[3] Baum, V.: Geometrische Transformationen in der diskreten Geometrie der Bildverarbeitung. Diplomarbeit, Sektion Mathematik der Friedrich-Schiller-Universität Jena, Juni 1989

[4] Bitter, M.: Some properties of the unimodular group in $\mathbb{R}^2$. In:
     A. Hübler, W. Nagel, B.D. Ripley, G. Werner (eds.): Geobild'89,
     Akademie-Verlag Berlin, Mathematical Research 51, 1989, 37-41

[5] Hübler, A.: An axiomatic approach to discrete geometry and its
     relation to usual digital geometry for image processing. In:
     L.P. Yaroslavskij, A. Rosenfeld, W. Wilhelmi (eds):   Computer
     Analysis of Images and Patterns. Akademie-Verlag Berlin, Mathem.
     Research 40, 1987, 174-186

[6] Hübler, A.: Diskrete Geometrie für die digitale Bildverarbeitung
     Diss. B, Math.-Naturwiss.-Technische Fakultät der Friedrich-
     Schiller-Universität Jena, 1989

[7] Kotowski, O.: Der digitale metrische Raum $[\mathbb{Z}^2,\delta]$ und Eichfiguren
     Diplomarbeit an der Sektion Mathematik der Friedrich-Schiller-
     Universität Jena, Juni 1988

[8] Kulpa, Z.: On the properties of discrete circles, rings, and
     discs. Computer Graphics and Image Processing 10 (1979), 348-365

[9] Kulpa, Z., B.Kruse: Algorithms for circular propagation in
     discrete images. Computer Vision, Graphics, and Image Processing
     24 (1983), 305-328

[10] Nakamura, A., K.Aizawa: Digital circles.   Computer Vision,
     Graphics, and Image Processing 26 (1984), 242-255

Lage- und skalierungsinvariante Skelette zur robusten Beschreibung und Erkennung binärer Formen

Robert Ogniewicz, Olaf Kübler,
Fernand Klein und Ulrich Kienholz

Institut für Kommunikationstechnik
Fachgruppe Bildwissenschaft
ETH–Zentrum, CH–8092 Zürich

Zusammenfassung

Das in der kontinuierlichen Ebene sehr einfache Konzept der *Mittelachsentransformation* zur Beschreibung binärer Formen konnte bisher nur in Teilen auf die diskrete Ebene übertragen werden, da topologische Aspekte und euklidische Metrik sich nicht zwanglos vereinigen lassen, solange am diskreten Raster festgehalten wird. Die Evaluation des *Voronoi-Diagramms* einer Objektberandung fördert die frühzeitige Loslösung vom diskreten Bildräster. Im Sinne einer alternativen Definition des *euklidischen Skeletts* wird ein einfaches Kriterium zur Extraktion von *signifikanten Diagrammkomponenten* postuliert, einschliesslich der Garantie *zusammenhängender* Skelette und ihrer *Invarianz* gegenüber Ähnlichkeitsabbildungen. Die sog. *Voronoi-* resp. *V-Skelette* dienen als Ausgangsbasis für einen nachfolgenden Erkennungsprozess, wobei *Isolation* der Objekte ohne Rückgriff auf vorangehende *Identifikation* stattfinden kann. Die Identifikation ist sowohl für geometrisch starre als auch "elastische" Objekte durchführbar.
Die Methode als Ganzes präsentiert sich als erfolgreiche diskrete Umsetzung des Blumschen Skelettkonzepts.

1 Einführung

Ein zentrales Anliegen der Bildverarbeitung ist der Schritt von der enormen Datenmenge zu einer kompakteren und handlicheren Beschreibung. In diesem Zusammenhang begegnet uns wiederholt das Problem der *Skelettierung*, d.h. der Reduktion eines flächigen binären Objekts auf linienhafte Strukturen. Das Konzept der *Mittelachsentranformation* (MAT) mit ihrer bekannten "Präriefeueranalogie" nach Blum [BLU67] steht Pate für die Formulierung des Begriffs *Skelett*. Leider ist die Umsetzung der Blumschen Ideen nicht trivial, da seine Definitionen auf der Geometrie im Kontinuum basieren, was im Konflikt zur naturgemäss diskreten Darstellung der Bildinformation in technisch realisierbaren Bildverarbeitungssystemen steht.

Die topologische Ausprägung der MAT als *Verdünnung* bedeutet in der algorithmischen Umsetzung in der Regel, dass alle Objektpixel hinsichtlich ihrer topologischen Relevanz iterativ getestet und getilgt werden. Charakteristikum dieses Verfahrens sind die extrem langen Laufzeiten, solange nicht spezielle parallel arbeitende Hardware zum Einsatz kommt.

Die geometrisch strukturellen Ausprägungen der MAT gehen direkt oder indirekt von der Existenz einer *Distanzkarte* für jedes Objekt aus. Montanari [MON68] definiert "sein" Skelett als die Menge aller Endpunkte der *optimalen Pfade*, die ins Objektinnere führen. In der alternativen Definition nach Rosenfeld und Pfaltz [ROS66] erscheint das Skelett als Ort der Zentren *grösster "Kreisscheiben"* innerhalb des Objekts, die sinngemäss nicht von anderen Scheiben überdeckt werden. Arbeiten von Arcelli [ARC85] oder Dorst [DOR86] beruhen auf *Gratverfolgung* (entlang der *Sattelpunkte*) innerhalb der Distanzkarte. Durch eine Reihe zusätzlicher Regeln kann dabei erreicht werden, dass topologisch korrekte, zusammenhängende Skelette resultieren.

Montanari's Skelett enthält leider weitaus mehr Punkte als unbedingt nötig zur Rekonstruktion des Objekts. Die zweite Definition liefert uns einen minimalen Satz von Skelettpunkten, aber: beide Definitionen leiden unter dem Fehlen von Information über den lokalen Zusammenhalt der Skelettkomponenten.
Separat dazu stellt uns die Berechnung der *Distanztransformation* vor einige Schwierigkeiten. Die schnellsten Algorithmen profitieren von den Eigenschaften regulärer Metriken, doch bedeuten die damit verbundenen Abweichungen zur euklidischen Distanz erhebliche Hindernisse für die Skelettextraktion. Eine korrekte euklidische DT lässt sich durch eine sequentielle DT nach Danielson [DAN80] mit nachgeschaltetem

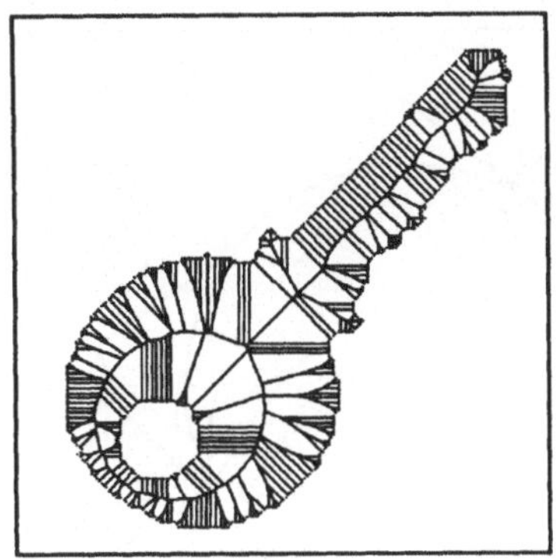 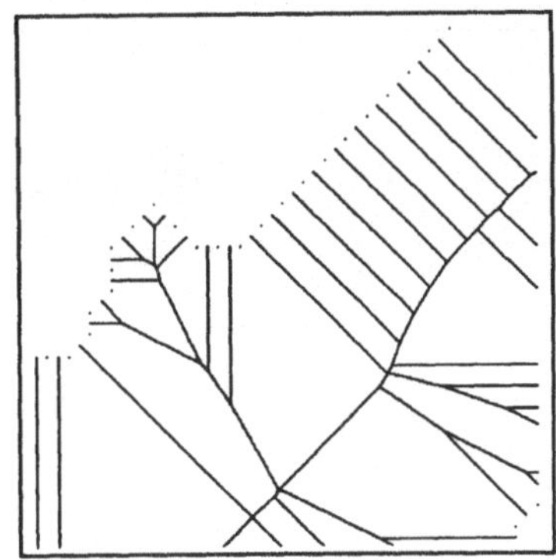

Figur 1: Inneres Voronoi-Diagramm, rechts Ausschnitt

Korrekturlauf (Klein [KLE87]) bestimmen resp. mittels paralleler Implementation nach Yamada [YAM84], allerdings um den Preis erhöhten Rechenaufwands.

Alle Verfahren weisen die gemeinsamen Nachteile auf, *dass sich die entstehenden Skelette nur bei grossherziger Betrachtung als lage- und skalierungsinvariant ansehen lassen und dass eine immense Anzahl Pixel bearbeitet und auf ihre Zugehörigkeit zum Skelett geprüft werden muss.*

Die vorliegende Arbeit zeigt, wie durch eine alternative Definition des Skeletts robustere Strukturen erzeugt und gleichsam eine "Versöhnung" des Diskreten und des Kontinuums ganz im Sinne von Blum erzielt werden kann.

2 Vom diskreten Raster zum Kontinuum

Montanari [MON69] schlägt eine direkte analytische Erfassung der kollidierenden Wellenfronten unter Verzicht auf die Distanztransformation vor. Ausgehend von der Objektbeschreibung mittels eines Polygonzuges ist eine mathematisch exakte Verfolgung der Wellenausbreitung ins Objektinnere im Kontinuum möglich. Die Effizienz der Skelettgenerierung steigt damit erheblich, da lediglich die für das Skelett essentiellen Punkte herangezogen werden. Hingegen bleibt das Problem ungelöst, wie eine *Regularisierung*, d.h. die Extraktion der signifikanten Skelettkomponenten zu realisieren ist, die gleichzeitig die Topologie des Skeletts bewahrt.

Dennoch erkennen wir die Vorteile, die ein frühzeitiges Lösen vom diskreten Bildraster bietet. Zum einen ist die nötige Rechenzeit zur Skelettierung weniger mit den geometrischen Dimensionen der Szene gekoppelt als vielmehr mit der Komplexität der Bildinformation. Zum anderen entfällt die Frage nach der Form der Wellenfronten im Subpixelbereich; in der Folge sind die Relationen zwischen den Skelettkomponenten leichter zu eruieren.

Im Gegensatz zu Montanari, der Polygonzüge als Initiatoren der Wellenfronten heranzieht, beschränken wir uns auf die Randpunkte des Objekts. Dementsprechend erscheinen die Wellenfronten als konzentrische Kreise; der geometrische Ort der Begegnung zweier Fronten ist die Mittelsenkrechte auf die Verbindungsstrecke der erzeugenden Punkte. Diese Formulierung entspricht weitgehend der Definition des *Voronoi-Diagramms* (VD).

3 Voronoi-Diagramm

Dem Voronoi-Diagramm liegt folgende Problemstellung zugrunde:

> *Gegeben sei eine Menge S von N Punkten in der Ebene. Gesucht ist für jeden Punkt $P_i \in S$ der geometrische Ort aller Punkte U, so dass die (hier: euklidische) Distanz von U zu P_i geringer ist als zu jedem anderen Punkt $P_{j,j \neq i} \in S$.*

Die obige Bedingung segmentiert die Ebene in konvexe *Voronoi-Polygone* um jedes $P_i \in S$, die Seiten der Polygone bilden die Mittelsenkrechten auf die Verbindungsstrecken je zweier benachbarten Punkte aus S.

45

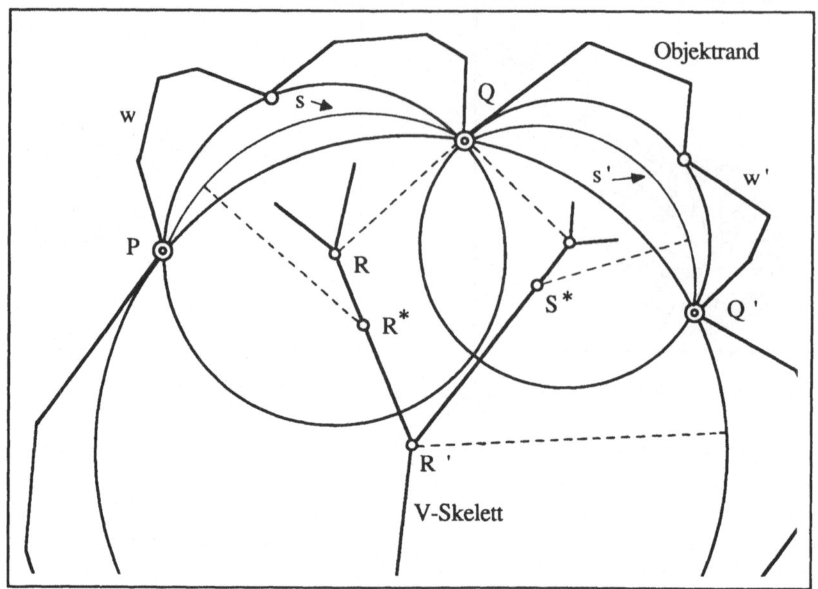

Figur 2: *Skelettextraktion durch Kreisfehlerrechnung*

Im Unterschied zur allgemeinen Definition beschränken wir uns in dieser Arbeit auf das Voronoi-Diagramm im Inneren eines Objekts (Figur 1). Shamos und Preparata [PRE86] führen an, dass die praktische Bestimmung des Voronoi-Diagramms relativ effizient durchgeführt werden kann, nämlich bei $O(N \log N)$ Zeit- und $O(N)$ Speicherbedarf. Darüberhinaus lässt sich die Berechnung des VD in parallel durchführbare Schritte zergliedern, sodass Echtzeitimplementierungen im Rahmen des Möglichen liegen dürften. Die Rekonstruierbarkeit der ursprünglichen Punktwolke resp. des Objekts ergibt sich unmittelbar aus der Tatsache, dass das Voronoi-Diagramm ein duales Pendant zur Triangulation (sog. *Delaunay-Triangulation*) darstellt. Diese Interpretation des VD erlaubt die Abschätzung der maximal zu erwartenden Daten mit $3N - 6$ Kanten und $2N - 5$ Schnittpunkten (im folgenden meist *V-Punkte* genannt).

4 Regularisierung und Kreiskriterium

Klein [KLE87] postuliert ein einfaches und effizientes Verfahren zur Reduktion des oft immensen Datensatzes unter gleichzeitiger Topologieerhaltung des Skeletts (Figur 2). Definitionsgemäss bildet jeder Punkt R^* des Voronoi-Diagramms das Zentrum der maximal eingepassten Kreisscheibe. Als diskriminierendes Merkmal definiert Klein die Genauigkeit der Approximation zwischen Kreisscheibe und dem Verlauf der Objektberandung. Die Kante RR' ist Teil der Mittelsenkrechten auf die Strecke PQ. Für jeden Punkt R^* innerhalb der Kante RR' ist mit s die Länge des Kreisbogens zwischen P und Q mit Zentrum in R^*, mit w der Weg längs des Randes von P nach Q gegeben. Das *Kreiskriterium* fordert für die Zulassung eines Punktes R^* zum Skelett eine Mindestabweichung (*Kreisfehler*)

$$\Delta K = w - s > T$$

mit T als Schwellwert.

Der Kreisfehler ist somit für jeden Punkt des Skeletts definiert, unter Berücksichtigung der Rechenzeiten ist jedoch die Beschränkung auf die Schnittpunkte (V-Punkte) des Voronoi-Diagramms von Vorteil.

Die topologiebewahrende Eigenschaft des Kreiskriteriums lässt sich direkt aus der Definition herleiten. Auf dem Weg in das Innere des Objekts wächst ΔK monoton, da w konstant bleibt, jedoch s monoton abnimmt. Hingegen ist der Verlauf von ΔK nicht stetig: erreichen wir auf dem Weg über die Kante RR' ihren Endpunkt R', vereinigen sich die entsprechenden Wege PQ und QQ' und an die Stelle zweier

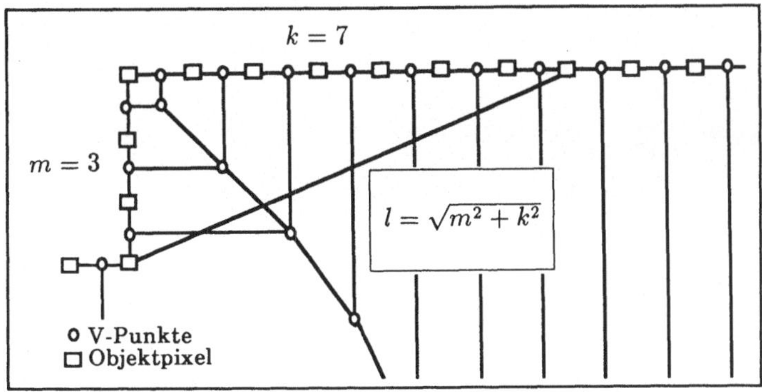

Figur 3: *Abschätzung des Schwellwertes T zur Skelettierung*

approximierender Kreisbögen s und s' tritt ein einzelner. ΔK an der Stelle R' entspricht folglich der Summe der Kreisfehler aller einlaufender Kanten.

Im Falle "löchriger" Objekte gehören u.U. die erzeugenden Punkte P und Q zu verschiedenen Konturen, dann wird per definitionem ΔK auf "unendlich" gesetzt, sodass diese Skeletteile für jeden Wert von T erhalten bleiben. Erwähnenswert ist die elegante Möglichkeit, den Konturverlauf als Potentialfunktion $W(P)$ zu interpretieren, somit ist die Integration der Kontur nur einmal durchzuführen. Folglich erhalten wir alle Wege w aus

$$w = \min(|W(P) - W(Q)|, L_0 - |W(P) - W(Q)|)$$

wobei L_0 die Gesamtlänge der Kontur bedeutet.

Während kein Supremum für ΔK existiert, kann der Kreisfehler nicht beliebig tiefe Werte annehmen. Weil das Voronoi-Diagramm nur auf das Objektinnere beschränkt bleibt, enden die äussersten Kanten in die Mitte zwischen zwei Randpunkten. Die dort auftretenden minimalen Kreisscheiben überlappen die Objektberandung. Daraus resultiert ein (negatives) ΔK_{min} von $1 - \pi/2$ für die 4-er Metrik resp. $\Delta K_{min} = \sqrt{2}(1 - \pi/2)$ für 8-er Metrik.

Bei der Abschätzung des notwendigen Schwellwertes T greifen wir auf die Überlegung zurück, dass eine Stufe der Höhe m, d.h. eine Abweichung vom geradlinigen Randverlauf, zur Ausbildung eines Skelettastes mit monoton wachsendem Schwellwert ins Objektinnere führen muss (Figur3). Der analytische Verlauf des Astes ist durch sukzessives Schneiden der entsprechenden Mittelsenkrechten berechenbar; einer ersten Abschätzung liegt eine wesentlich trivialere Betrachtung zugrunde: für sehr weit im Inneren liegende Skelettpunkte lässt sich der approximierende Kreisbogen durch eine Strecke nähern und wir erhalten

$$\Delta K <\approx m + k - \sqrt{m^2 + k^2} <\approx m, \qquad k \gg m.$$

Die praktisch zu erwartenden Abweichungen durch Binarisierung eines gutartigen Grauwertbildes liegen in der Grössenordnung von einem Pixel, ähnliche Fehler kann man ebenfalls von einer nachfolgenden Ähnlichkeitsabbildung annehmen. Die Wahl von $2.0 \leq T \leq 3.0$ dürfte demnach ausreichend invariante Skelette garantieren (Figur 4).

5 Anwendungen

Alternatives Datenformat

Obwohl selbst das regularisierte Voronoi-Skelett eine beträchtliche Datenmenge bedeutet, lässt sich die darin enthaltene Information zur einfachen Durchführung geometrischer Operationen mit anschliessender Rekonstruktion heranziehen. Klein [KLE87] zeigt, wie ausgehend von Skelettdaten die Darstellung eines Buchstabensatzes in diversen Skalierungen und Lagen erfolgen kann, was besonders dann von Interesse erscheint, wenn Hardware zur schnellen Generierung von Kreisscheiben zur Verfügung steht.

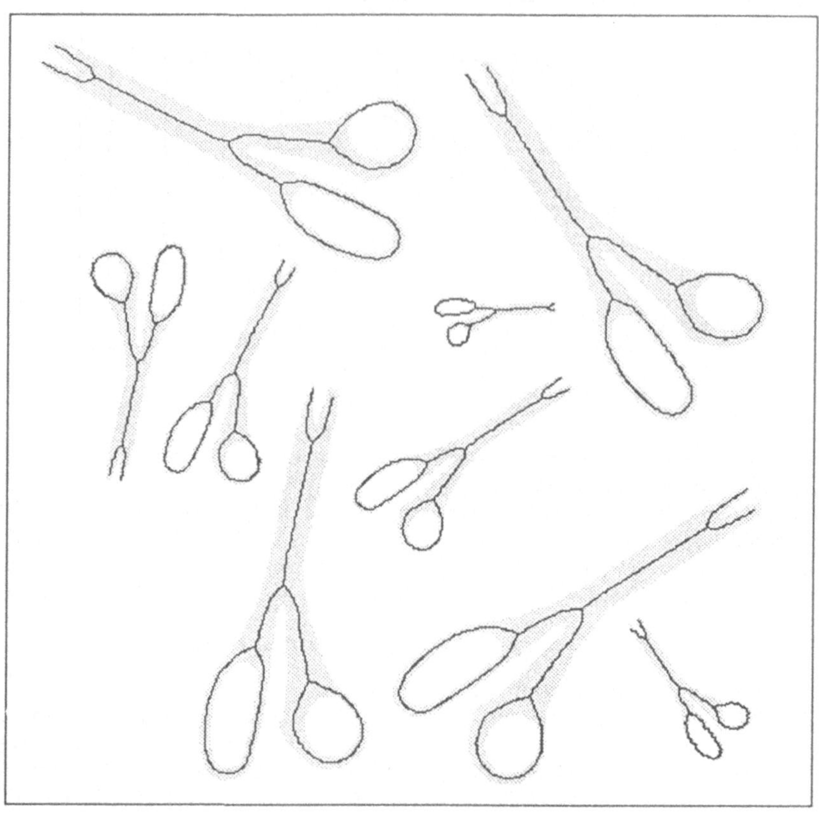

Figur 4: *Voronoi-Skelette*

*Voronoi-Skelette in variabler Lage und Skalierung bei Schwellwert $T = 2.0$. Die Skelette
zeigen weitgehende Invarianz gegenüber Ähnlichkeitsabbildungen.*
Rechenzeiten VAX 11/780: VD 34 s, Kreisfehler 12 s
Komplexität VD: 6633 Punkte, 12668 Kanten, 12655 V-Punkte

Isolation

I.a. erfolgt die Isolation *parallel* zur Identifikation; das erfolgreiche Erkennen eines Gegenstandes unterstützt
das Herauslösen diesselben aus der Umgebung. In vielen Fällen kann aber bereits aufgrund der Konturen in
einer Szene eine Aussage getroffen werden über die wahrscheinlichen Objektgrenzen. Insbesondere können
Einschnürungen dazu beitragen, den Bildinhalt in leichter zu verarbeitende Teile zu segmentieren und
im folgenden Identifikationsschritt die kombinatorischen Möglichkeiten drastisch zu reduzieren. Kienholz
und Ogniewicz [KIO87] haben eine Implementation der obigen Skelettdefinition erarbeitet, die in der Lage
ist, die Isolation vor der Identifikation durchzuführen. Als Auftrennungskriterium gilt das Auftreten von
signifikanten Engpässen innerhalb der untersuchten Szene, die sich als lokale Minima des Radienverlaufs
längs der Skelettäste manifestieren. Es muss an dieser Stelle darauf hingewiesen werden, dass die Suche
nach den lokalen Minima sich nur über die V-Punkte des Voronoi-Skeletts erstreckt, somit ist eine besonders
präzise Lokalisierung der Einschnürung nicht möglich. Allerdings dürfte die Geometrie innerhalb einer
Überlappungsstelle von sekundärer Bedeutung sein für die Identifikation.

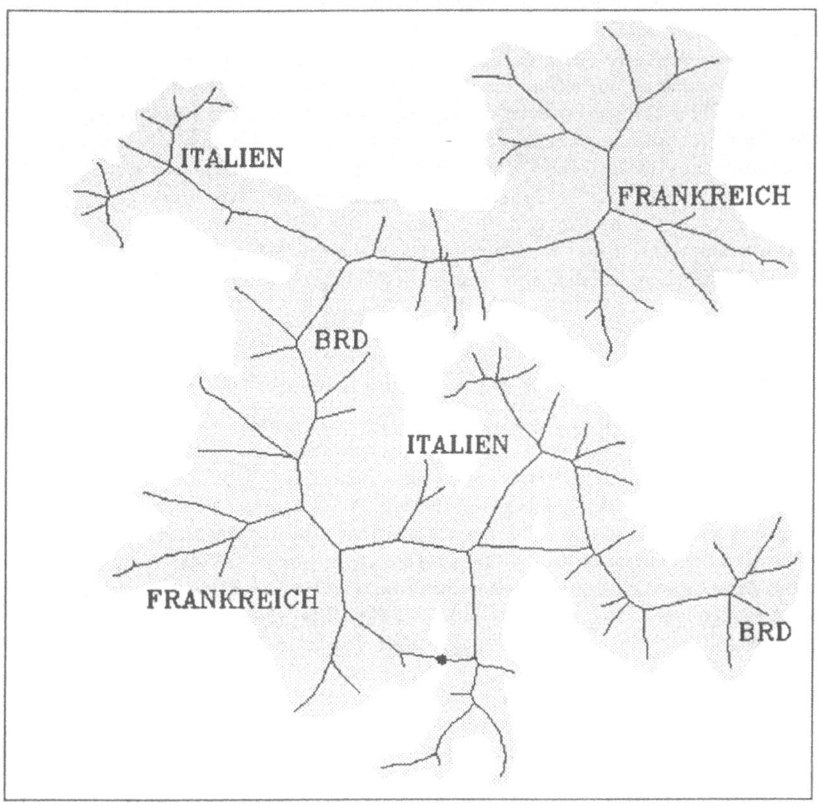

Figur 5: *Knotenbasierte Identifikation*
Die Identifikation mittels Skelettknoten zeigt gute Resultate in Objekten mit komplexen Konturen. Diese Methode würde im nächsten Beispiel (Figur 6) versagen, da sich die Skelette von "Herz" und "Klee" nicht hinreichend voneinander unterscheiden.
Rechenzeiten VAX 11/780: VD 24 s, Kreisfehler 6 s, Isolation/Identifikation 93 s
Komplexität VD: 3333 Punkte, 6457 Kanten, 6456 V-Punkte

Identifikation

Die Identifikation von sich berührenden resp. überlappenden Gegenständen ist realisierbar, sofern die gegenseitigen Einwirkungen zweier ineinanderfliessender Objekte die Form der Skelette nicht entscheidend modifizieren (Figur 5). Ein V-Skelett repräsentiert immer einen Mittelwert zwischen lokalen und globalen Einflüssen seiner erzeugenden Umgebung, während Teile der Kontur als ausgeprägt lokale Merkmale in Erscheinung treten.

In ihrer Implementierung gehen Kienholz und Ogniewicz von den Knoten des Skeletts aus, d.h. von V-Punkten mit mehr als zwei anstossenden Ästen. Knoten entstehen als Treffpunkte mehrerer signifikanter Einflusszonen, ihre Heranziehung als Ausgangsbasis für die Identifikation erscheint deshalb legitim. Übereinstimmung mehrerer Objektknoten mit einem vorher gelernten Modell erlaubt die Postulierung einer Ähnlichkeitstransformation, die Hypothesen zur Lage der anderen Modellknoten ermöglicht. Im Falle fortschreitender Verifikation der Hypothesen erfolgt parallel die Optimierung der Abbildungsparameter.
Diese Methode ist nicht brauchbar für Objekte, deren Skelette keine oder nur sehr wenige Knoten aufweisen. Ausserdem reagieren V-Skelette empfindlich auf Konvexitäten, jedoch nur beschränkt auf Konkavitäten, es ist mit dem Auftreten fast gleichartiger Skelette bei zwei in der Konturcharakteristik verschiedenen Objekten zu rechnen.

In einer zweiten Variante wird der Knotenbegriff relativiert und prinzipiell alle V-Punkte zum Vergleich herangezogen, wobei das Kreiskriterium die massgebliche Grösse für die "Wichtigkeit" eines V-Punktes im Erkennungsprozess beisteuert. Der Ausschnitt des betrachteten Voronoi-Diagramms hängt von der Komplexität der Erkennung ab, in schwierigen Fällen geht das Verfahren nahtlos in einen Konturvergleich über. Dieser Ansatz führte zu längeren Laufzeiten, erbrachte jedoch weitaus bessere Resultate (Figur 6).

Das V-Skelett lässt sich auch zur Erkennung von "elastischen" Objekten einsetzen. Hierbei wird zuerst die Skelettinformation in eine Liste von Prolog-Fakten übersetzt. Darin sind nicht nur rigide Grössen wie absolute Lage und geometrischer Verlauf sondern auch globale Werte wie gegenseitige Relationen, Gesamtlänge oder Tendenz des Radienverlaufs enthalten. Der anschliessende Vergleich zwischen Modellen und Szene fällt dementsprechend weit weniger streng aus, die Einführung von sogenannten "Joker"-Ästen, die in jede nicht identifizierbare Konfiguration passen, ermöglicht auch die Identifikation teilweise verdeckter Gegenstände.

6 Diskussion

Das Voronoi-Diagramm der Randpunkte erlaubt die Definition eines *euklidischen Skeletts im Kontinuum*. Die Extraktion des Skeletts aus dem Voronoi-Diagramm des Objektinneren ist einfach zu bewerkstelligen, der sogenannte *Kreisfehler* ΔK als Kriterium stellt eine monotone, wenn auch nicht stetige Funktion dar, somit ist die Erhaltung der Skelettopologie gewährleistet. Die so erhaltene Datenstruktur lässt sich sowohl für die *Identifikation* geometrisch starrer als auch elastischer Objekte einsetzen. In vielen Fällen ist eine *Isolation vor der Identifikation* möglich, allerdings setzt das Auftreten sehr stark überlappender oder durch starke Einschnürungen charakterisierter Objekte hier eine Grenze. Sowohl die Berechnung des VD als auch die Identifikation erfolgreich isolierter Objekte kann mit spezieller Hardware *parallel* bewältigt werden. Die in unserer Implementation charakteristischen langen Rechenzeiten dürften demnach kein Hinderungsgrund für einen praktischen Einsatz sein.

Ein gewichtiger Nachteil ist die *Rauschempfindlichkeit* von V-Skeletten, da selbst winzigste Löcher die Struktur des Voronoi-Diagramms entscheidend verändern. Schliesslich muss für rundliche Objekte explizit die Erhaltung des Skeletts garantiert werden, weil das Kreiskriterium runde Formen übermässig unterdrückt. In der Bilanz sehen wir das V-Skelett als vielversprechenden Ansatz für weiterführende Arbeiten. Als Thema käme beispielsweise die *Elimination von Störstellen* (Löchern) in bereits vorhandenen Voronoi-Diagrammen durch schnelle lokale Operationen in Frage.

Referenzen

[ARC85] Arcelli C., Sanniti di Baja G., *A width-independent fast thinning algorithm*, IEEE-PAMI-7/4, pp.463-474, 1985

[BLU67] Blum H., *A Transformation for Extracting New Descriptors of Shape*. from: Models for the Perception of Speech and Visual Form, W. Wathen-Dunn, Ed., 1967. MIT Press, 1967

[DAN80] Danielson P., *Euclidean distance mapping*, CGIP-14, pp.227-248, 1980

[DOR86] Dorst L., *Pseudo-Euclidean skeletons*, Proc. 8.ICPR, pp.286-288, 1986

[KLE87] Klein F., *Vollständige Mittelachsenbeschreibung binärer Objekte mit euklidischer Metrik und korrekter Geometrie*. Dissertation ETH-Z 1987

[KIO87] Kienholz U., Ogniewicz R., *Isolation und Identifikation einander berührender und/oder überlappender Teile in Binärbildern*. Diplomarbeit ETH-Z 1987

[MON68] Montanari U., *A method for obtaining skeletons using a quasi-Euclidean distance*, JACM-15/4, pp.600-624, 1968

[MON69] Montanari U., *Continuous skeletons from digitized images*, JACM-16/4, pp.534-549, 1969

[PRE86] Preparata F.P., Shamos M.I., *Computational Geometry*, Texts and Monographs in Computer Science, Springer Verlag, 1986

[ROS66] Rosenfeld A., Pfaltz J.L., *Sequential operations in digital picture processing*, JACM-13/4, pp.471-494, 1966

[YAM84] Yamada H., *Complete Euclidean distance transformation by parallel operation*, Proc. 7.ICPR, pp.69-71, 1984

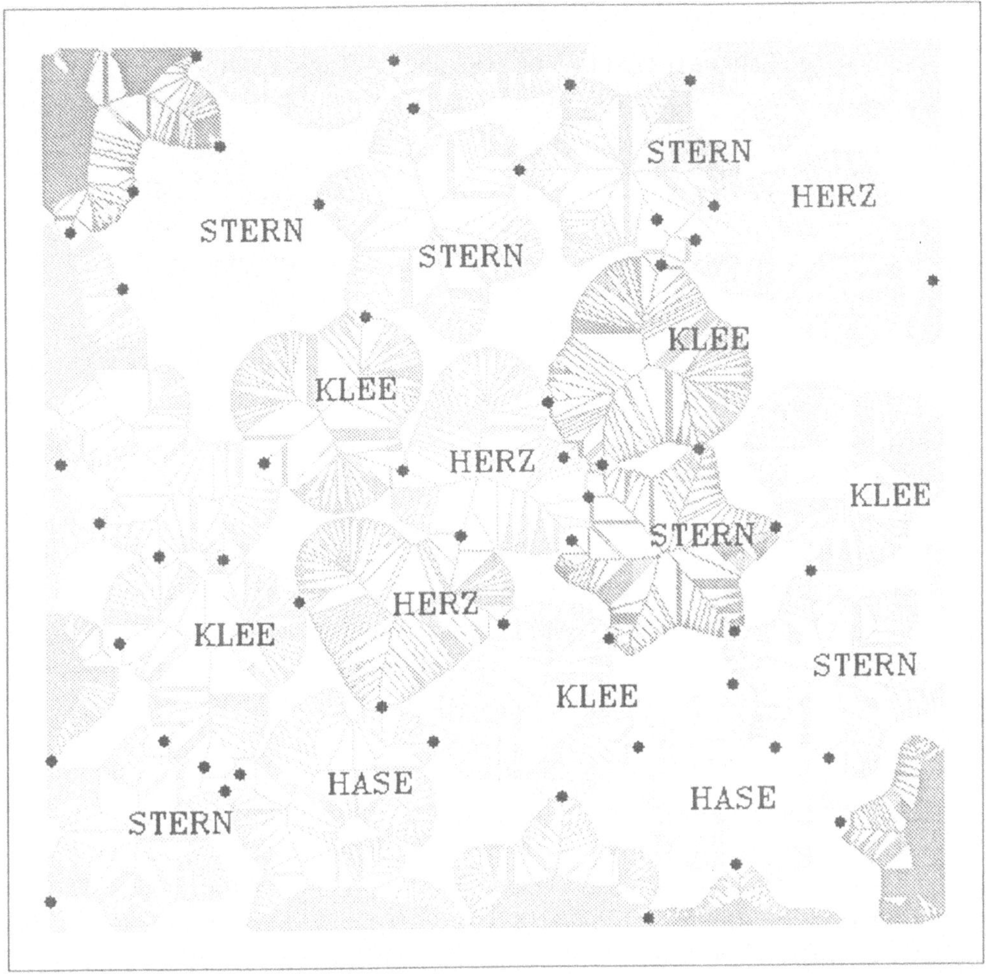

Figur 6: *Generalisierte Identifikationsmethode*

Identifikation von Weihnachtsgebäck. Die dunklen Punkte markieren die Einschnürungen, die verschiedenen Grautöne die Separierung in Einzelobjekte. Die Auftrennung wäre für eine korrekte Identifikation nicht notwendig, man beachte den richtig erkannten "Hasen" am unteren Bildrand (im Zweifelsfall steht der Verzicht auf Isolation, denn falsche Segmentierung führt unweigerlich zu Fehlern).
Rechenzeiten VAX 11/780: VD 54 s, Kreisfehler 14 s, Isolation/Identifikation 541 s
Komplexität VD: 7262 Punkte, 14190 Kanten, 14166 V-Punkte

Parametrization of the Hough Transform

Ulrich Eckhardt
Institut für Ange-
wandte Mathematik
Universität Hamburg
Bundesstraße 55

D-2000 Hamburg 13

Gerd Maderlechner
Zentrale Aufgaben
Informationstechnik
Siemens AG
Otto-Hahn-Ring 6

D-8000 München 83

<u>0. Introduction</u>. The Hough transform is used in picture processing for detecting linear structures in binary images. It associates to each line running through the image the integral along this line of the gray value function. There are many different possibilities for parametrizing lines in the plane and, depending on the applications under consideration, there exist many approaches for parametrization having different properties. The aim of this paper is to give a unified theoretical framework for the parametrization of the Hough transform. On the basis of this theoretical work it becomes possible to compare different parametrizations proposed so far and to construct parametrizations having desirable properties.

<u>1. Homogeneous Coordinates in the Plane</u>. In projective geometry one associates to a point in the plane with Cartesian coordinates (x, y) the *homogeneous coordinates* $(x, y, 1)$. Analogously, to each set of homogeneous coordinates (x, y, z) with $z \neq 0$ there belongs the point $(x/z, y/z)$ in the plane. In case $z = 0$ there is no corresponding (proper) point.

A *line* in the plane is given by a triple $\Xi = (\xi, \eta, \zeta)$ of numbers. The point $P = (x, y, z)$ belongs to the line given by Ξ if and only if

$$\langle P, \Xi \rangle := x \cdot \xi + y \cdot \eta + z \cdot \zeta = 0. \tag{1}$$

Note that for $\alpha \neq 0$ the parameter vectors Ξ and $\alpha \cdot \Xi$ represent the same line. Generally we assume $\xi^2 + \eta^2 \neq 0$ and $z \neq 0$.

The set of all points $P = (x, y, 1)$ is the *domain* of the Hough transform. It contains the image to be transformed. The set of all line parameters Ξ is termed the *Hough space*.

The main advantage in using homogeneous coordinates is that in equation (1) the point coordinates and the line parameters appear completely symmetrical. This is a manifestation of the well-known point-line duality in projective geometry (see e.g. SPERNER, 1959, § 16): to each fixed Ξ in the Hough space there corresponds via (1) a point set which is a line in the domain of the Hough transform. Conversely, for each fixed P in the domain equation (1) yields the set of all lines running through P. Obviously, the set of all parameters for such lines is itself a line in Hough space.

2. **Linear Transformations**. The homogeneous coordinates are extremely well suited for investigating geometrical questions. We give some examples. Given two vectors $\Xi_1 = (\xi_1, \eta_1, \zeta_1)$ and $\Xi_2 = (\xi_2, \eta_2, \zeta_2)$ in Hough space. The intersection point of the lines represented by these two vectors has the coordinates

$$x = \frac{\eta_2\zeta_1 - \eta_1\zeta_2}{\xi_2\eta_1 - \xi_1\eta_2} \qquad \text{and} \qquad y = \frac{\xi_1\zeta_2 - \xi_2\zeta_1}{\xi_2\eta_1 - \xi_1\eta_2}. \tag{2}$$

The lines are parallel if and only if

$$\xi_2\eta_1 = \xi_1\eta_2 \tag{3}$$

and their distance is in this case

$$h = \left| \frac{\zeta_1}{\sqrt{\xi_1^2 + \eta_1^2}} - \frac{\zeta_2}{\sqrt{\xi_2^2 + \eta_2^2}} \right|. \tag{4}$$

If

$$\xi_1\xi_2 = -\eta_1\eta_2 \tag{5}$$

then the lines are orthogonal.

Given a (3,3)-matrix A and denote by A^T it transpose. Then

$$\langle \Xi, A \cdot P \rangle = \langle A^T\Xi, P \rangle. \tag{6}$$

This means: the effect of a linear transformation in the domain is equivalent to the effect of the adjoint transformation in Hough space.

Let Q be a *translation vector*. Then

$$\langle \Xi, P + Q \rangle = \langle \Xi, P \rangle + \langle \Xi, Q \rangle, \tag{7}$$

hence a translation in the domain can be represented in Hough space by the transformation

$$(\xi, \eta, \zeta) \rightarrow (\xi, \eta, \zeta + \langle \Xi, Q \rangle), \tag{8}$$

which is a rotation around an axis through the origin which is perpendicular to Q.

Although equations (6) and (7) hold generally, in the present context they make only sense if the transformation matrix A keeps z fixed and if Q has the form $(x_o, y_o, 0)$.

3. The Hough Transform. Given an image by means of its gray value function $\beta(P)$ which is defined for all $P = (x, y, 1)$ in the plane. For a line with parameter vector Ξ let σ be the line element. Then the *Hough transform* or *Radon transform* of the function β is defined as follows (see e.g. DEANS, 1983, NATTERER, 1986)

$$H(\beta)(\Xi) = \int_{\langle P, \Xi \rangle = 0} \beta(P)\, d\sigma. \tag{9}$$

The line belonging to Ξ has the form

$$\{P \mid \langle P, \Xi \rangle = 0\} = \{(x, y) \mid x \cdot \xi + y \cdot \eta + \zeta = 0\} =$$

$$= \{(x, y) \mid x = x_o - t \cdot \eta,\ y = y_o + t \cdot \xi,\ t\ \text{real}\}, \tag{10}$$

where (x_o, y_o) is a fixed point on the line, i.e. $x_o \cdot \xi + y_o \cdot \eta + \zeta = 0$. As a consequence, we get

$$H(\beta)(\Xi) = \frac{1}{\sqrt{\xi^2 + \eta^2}} \cdot \int_{-\infty}^{\infty} \beta \begin{pmatrix} x_o - t \cdot \eta \\ y_o + t \cdot \xi \end{pmatrix} dt. \tag{11}$$

The Hough transform has the following properties:

$$H(\beta)(\alpha \cdot \Xi) = H(\beta)(\Xi) \quad \text{for all real numbers } \alpha \neq 0 \tag{12}$$

(homogeneity with respect to Ξ). Furthermore,

$$H(\beta_1 + \beta_2)(\Xi) = H(\beta_1)(\Xi) + H(\beta_2)(\Xi),$$
$$H(\lambda \cdot \beta)(\Xi) = \lambda \cdot H(\beta)(\Xi) \tag{13}$$

(linearity with respect to β). Finally, by (6) with $\beta'(P) = \beta(A \cdot P)$, $\det A \neq 0$:

$$H(\beta')(\Xi) = \frac{1}{\det A} \cdot H(\beta)((A^{-1})^T \Xi) \tag{14}$$

and by (8) with $\beta'(P) = \beta(P + Q)$ with fixed translation vector Q:

$$H(\beta')(\Xi) = H(\beta)(\xi, \eta, \zeta - \langle Q, \Xi \rangle). \tag{15}$$

If in the latter relation Q is orthogonal to the line determined by Ξ, then $\langle Q, \Xi \rangle = 0$, hence $H(\beta')(\Xi) = H(\beta)(\Xi)$. This means that the value of the Hough transform at Ξ is invariant with respect to a translation perpendicular to Ξ.

4. Normalization. The Hough space as defined here is three-dimensional. By the homogeneity (12) it is sufficient to know the function value of the transform at one point on the line determined by Ξ. Therefore we introduce a *normalization functional* $\Phi : \mathbb{R}^3 \to \mathbb{R}$. Among all possible representations of a line those with $\Phi(\Xi) = 0$ is chosen. Φ should have the following properties:

(1) $\Phi(\Theta_3) \neq 0$ (Θ_3 is the zero vector of $\mathbb{R}^3$),

(2) For each $\Xi \neq \Theta_3$ there exists at most one λ with $\Phi(\lambda \cdot \Xi) = 0$.
If in addition the following condition holds:

(3) For each $\Xi \neq \Theta_3$ there exists exactly one λ with $\Phi(\lambda \cdot \Xi) = 0$,
then the normalizing functional Φ is said to be *complete*.

When Φ is an affine-linear functional, i.e. $\Phi(\Xi) = <C, \Xi> + d$ with
$C = (a, b, c)$ and $d \neq 0$ (by property (1) above), then there always exists a *singular line* $\Xi_o \neq \Theta_3$ with $\Phi(\lambda \cdot \Xi_o) \neq 0$ for all λ. Obviously each Ξ_o with $<C, \Xi_o> = 0$ leads to a singular line. This means, that only with a nonlinear nomalizing functional a complete normalization can be obtained.

The following four normalizations are used by most authors:

Linear normalization functionals $\Phi(\Xi) = a \cdot \xi + b \cdot \eta + c \cdot \zeta + d$. Here always a singular line is present. The advantage of this normalization is that the point-line duality of projective geometry is preserved. This is of importance when dealing with polygonal sets, i.e. sets whose boundaries are closed polygons. The image of a polygonal set in the plane under the duality map is a polygonal set in the Hough space. The image polygons in Hough space, however, can intersect themselves und thus not form boundary lines of sets (see e.g. WAHL, 1988 or STAHS and WAHL, 1989 for examples). Another favourable property of linear normalizations is that the extremely transparent behaviour of the transformation with respect to linear transformations (see § 2) is essentially preserved.

Piecewise linear normalization. In order to avoid singular lines and yet retain the favourable properties of linear normalization, the image space can be subdivided in appropriately chosen subsets and a linear normalization functional can be chosen in each of these subsets. We cite for example the "muff" transformation of WALLACE (1985) and the "twin Hough space" of Wahl (BILAND and WAHL 1988, WAHL 1988, STAHS and WAHL 1989). Also here, polygons are transformed into polygons.

Almost all authors use a nonlinear normalization functional which leads to *Hesse's normal form* of a line (see e.g. SPERNER, 1959)

$$\Phi(\xi, \eta, \zeta) = \xi^2 + \eta^2 - 1. \tag{16}$$

There are, of course, many other nonlinear normalizations possible. QUAN and MOHR (1989) use the *Gaussian sphere* normalization:

$$\Phi(\xi, \eta, \zeta) = \xi^2 + \eta^2 + \zeta^2 - 1. \tag{17}$$

5. Parametrization. A *parametrization* is a biunique mapping of the components of the normalized vector Ξ onto two parameters describing the line corresponding to Ξ in a geometrically transparent way. We give

some examples:

The most well-known linear parametrization is the *slope-intercept parametrization* of a line. Here, the normalization functional is

$$\Phi(\xi, \eta, \zeta) = \eta + 1. \tag{18}$$

Parameters of a line are its *slope* with respect to the x-axis $\xi = \tan \phi$ and its *intercept* with the y-axis ζ. This parametrization was used in the original paper of HOUGH (1962). Singular lines are all parallels to the y-axis.

Some authors used piecewise linear normalization functionals in order to avoid singular lines and simultaneously to retain as much as possible of the favourable behaviour with respect to linear transformations. BILAND and WAHL (1988; see also WAHL 1988 and STAHS and WAHL 1989) used the piecewise linear functional

$$\Phi(\xi, \eta, \zeta) = \begin{cases} \eta + 1 & \text{for } |\xi| \leq |\eta| \\ \xi + 1 & \text{for } |\eta| \leq |\xi| \end{cases} \tag{19}$$

For $|\xi| \leq |\eta|$ this coincides with the slope-intercept parametrization (18), for $|\eta| \leq |\xi|$ with the same parametrization, but rotated by 90° with normalization functional $\Phi(\Xi) = \xi + 1$.

Another piecewise linear parametrization was proposed by WALLACE (1985; so-called "muff" transformation) and by RISSE (1988). For this parametrization it is assumed that all relevant details of the image are contained in the rectangle

$$R = \{P = (x, y, 1) \mid |x| \leq X, |y| \leq Y\}. \tag{20}$$

$\Phi(\Xi)$ is defined as proposed by Biland and Wahl above. For parametrization the following numbers are calculated

$$\left. \begin{aligned} Y_X &= X \cdot \xi + \zeta \\ Y_{-X} &= -X \cdot \xi + \zeta \end{aligned} \right\} \quad \text{if } |\xi| \leq |\eta|, \text{ i.e. } \eta = -1$$

$$\left. \begin{aligned} X_Y &= Y \cdot \eta + \zeta \\ X_{-Y} &= -Y \cdot \eta + \zeta \end{aligned} \right\} \quad \text{if } |\eta| \leq |\xi|, \text{ i.e. } \xi = -1 \tag{21}$$

Y_X and Y_{-X} are the intersection points of the line determined by Ξ with the vertical lines $x = X$ and $x = -X$, analogously for X_Y and X_{-Y}. If the line belonging to Ξ meets the interior of R, then exactly two of the following inequalities are true:

$$\begin{aligned} |Y_X| &\leq Y, & |Y_{-X}| &\leq Y, \\ |X_Y| &\leq X, & |X_{-Y}| &\leq X. \end{aligned} \tag{22}$$

The parameters of a line are those two among the numbers X_Y, X_{-Y}, Y_X and Y_{-X} fulfilling the inequalities together with an indication as to which inequality is meant. Geometrically speaking, the parameters of a line are the both intersection points of it with the rectangle R.

A very important property of this parametrization is that the parameters belong to the domain of the Hough transformation. Therefore they can be chosen consistently to the discretization of the image under consideration in a quite natural way. By means of this approach automatically all problems caused by the arbitrary choice of the discretization of the Hough space are eliminated. For details of this aspect see the paper of RISSE (1988).

The most widely used normalization functional is (16). An obvious parametrization is

$$\xi = \quad \sin \phi,$$
$$\eta = - \cos \phi, \tag{23}$$
$$\zeta = - p.$$

This parametrization is not quite unique, since parameters ϕ, p and $\phi + \pi$, $- p$ describe the same line. Therefore, either $p \geq 0$ or $0 \leq \phi < \pi$ can be required. p is the distance of the line given by Ξ from the origin and ϕ is the angle of the line with the x-axis. The image of a point in the domain is a sinusoidal curve in p,ϕ Hough space given by

$$x \sin \phi - y \cos \phi = p. \tag{24}$$

A very closely related parametrization is the *fan-beam parametrization* (see NATTERER, 1986). Given a number S which is assumed to be so large that all relevant details of the image under consideration are within the set $\{x^2 + y^2 \leq S^2\}$. ϕ is defined as in (23), instead of p, however, a parameter α is introduced by

$$\zeta = - S \cdot \sin(\phi - \alpha). \tag{25}$$

Each (relevant) line crosses the circle with radius S centered at the origin in exactly two points. α is the angle between the direction from the origin to one of these points and the x-axis. This parametrization was investigated by ECKHARDT and MADERLECHNER (1988).

The parameters in (23) can be interpreted in an obvious way as polar coordinates of a point in a parameter space (whenver p is chosen to be nonnegative). If we put $x = \rho \cdot \sin \psi$, $y = \rho \cdot \cos \psi$, then we get from (1) as an equation relating points of the domain and of the Hough space

$$p = \rho \cdot \sin(\phi - \psi). \tag{26}$$

For given ρ and ψ this is the equation of a circle with radius ρ and center ($- \rho \cdot \sin \psi, \rho \cdot \cos \psi$) in Hough space.

YU (1989) proposed to parametrize a line by the parameters $q = \frac{1}{\rho}$ and ϕ. Then (26) becomes

$$q \cdot \rho \cdot \sin(\phi - \psi) = 1, \tag{27}$$

and this relation is again completely symmetrical with respect to parameters in the domain and the Hough space, hence a duality relation is valid. By means of this duality Yu was able to derive a stability result for the line oriented Hough transform (ECKHARDT, SCHERL, YU, 1987). Parametrization (27) is closely related to polarity for convex sets. There is, however, a singular point (0, 0, 1). For details se YU (1989).

6. <u>Applications</u>. The normalizations and parametrizations of the Hough transform presented here have different advantages and disadvantages. Therefore it is necessary to decide in each concrete application which parametrization will be optimal. It is easily possible to transform one parametrization to any other by means of a simple nonlinear point transformation of the Hough space (at least if both normalizations are complete). The following requirements can act as criteria:

- A transparent relation between the geometry of the original plane and the geometry of the Hough space (linear and piecewise linear normalization, parametrization of Yu (27)),
- A bounded set of parameters if the image under consideration is a bounded set in the domain (Hesse's normal form (16), Gaussian sphere normalization (17), most piecewise linear normalizations, polar coordinate parametrization (26)),
- The discretization of the Hough space can be made consistent to the discretization of the domain. This is very important if aliasing effects are to be suppressed (RISSE 1988).

If only a part of the Hough space is of interest, a linear normalization can be introduced which is focused onto this part and maps the singular lines into the complement. This approach was used by HOUGH (1962). Such a situation arises when a document is to be adjusted under the condition that it is only slightly misaligned with respect to the ideal position.

Many researchers proposed to project the Hough space onto a subspace of smaller dimension in order to reduce the search effort. The properties of such projections depend heavily on the normalization and parametrization used (ECKHARDT and MADERLECHNER, 1988).

References

BILAND HP, WAHL FM (1988) Hough-space decomposition for polyhedral scenes. Computer and Systems Sciences 45:197-216. Berlin: Springer

DEANS SR (1983) The Radon Transform and Some of Its Applications. New York: John Wiley and Sons

ECKHARDT U, MADERLECHNER G (1988) Application of the projected Hough transform in picture processing. Lecture Notes in Computer Science 301:370-379. Berlin: Springer

ECKHARDT U, SCHERL W, YU Z (1987) Representation of plane curves by means of descriptors in Hough space. Universität Hamburg, Regionales Rechenzentrum, Berichte

HOUGH PVC (1962) Method and means for recognizing complex patterns. U.S. Patent 3,069,654

NATTERER F (1986) The Mathematics of Computerized Tomography. Stuttgart: B.G. Teubner

QUAN L, MOHR R (1989) Determining perspective structures using hierarchical Hough transform. Pattern Recognition Letters 9:279-286

RISSE T (1988) Yet another parametrization for Hough transformation. Informatik-Fachberichte 180:142-150. Berlin: Springer

SPERNER E (1959) Einführung in die Analytische Geometrie und Algebra. 1. Teil, 4. durchgesehene Auflage, 2. Teil, 3. durchgesehene Auflage. Göttingen: Vandenhoeck & Ruprecht

STAHS TG, WAHL FM (1989) Polyhedral object recognition by Hough space analysis. Mathematical Research 51:165-172. Berlin: Akademie-Verlag

WAHL FM (1988) Analysing Hough nets for recognition of polyheder-like objects. 9th ICPR Rom, Proceedings 200-206

WALLACE RS (1985) A modified Hough transform for lines. IEEE Computer Science Conf. on CVPR, San Francisco, Proceedings 665-667

YU Z (1989) Stabile Analyse von Binärbildern. Thesis, Universität Hamburg

TEXTURANALYSE, FRAKTALE UND SCALE SPACE FILTERING

Uwe Müssigmann

Fraunhofer Institut für Produktionstechnik und Automatisierung (IPA),
Nobelstr. 12, 7000 Stuttgart 80

Das Verfahren des Scale Space Filtering wurde bisher in der Bildverarbeitung zur Beschreibung und Erkennung von ebenen Kurven sowie zweidimensionalen Formen benutzt. In diesem Beitrag wird eine darauf basierende neue Methode zur Berechnung der fraktalen Dimension von digitalen Grauwertbildern vorgestellt. Die fraktale Dimension wird als ein quantitatives Maß zur Texturanalyse, speziell der Textursegmentation, eingesetzt.

1. Einführung

Seit den ersten Arbeiten von Benoit B. Mandelbrot über Strukturen und physikalische Prozesse, die Skalenverhalten zeigen, hat der neue Forschungszweig, Fraktale Geometrie, in vielen Wissenschaftsbereichen Beachtung gefunden. Das Skalenverhalten der untersuchten Mengen ist dadurch gekennzeichnet, daß bei jeder Auflösung das Objekt, die Menge, als extrem irregulär oder bruchstückhaft, aber niemals glatt erscheint [1]. Dadurch sind die Fraktale, so der Name dieser Objekte, eindeutig gegen die Elemente der Euklidschen Geometrie wie beispielsweise Punkt, Gerade, Ebene abgegrenzt.

Die Idee der fraktalen Geometrie beschränkt sich nicht nur auf synthetische, mathematische Muster, auch die meisten in der Natur auftretenden Formen lassen sich der Klasse der Fraktale zuordnen. Beispiele solcher Formen sind Wolken, Gebirge und Küstenlinien. Diese Objekte sind auch Gegenstand der automatischen Bildanalyse. Es ist also nur natürlich, wenn das Konzept der Fraktale auch im Bereich der Bildverarbeitung verstärkt zum Einsatz gelangt. Erste Anwendungen finden sich in Arbeiten von Quinqueton [2], Pentland [3], Peleg et al. [4] sowie Medioni et al. [5].

Mandelbrot schlägt zur Beschreibung fraktaler Mengen u.a. die fraktalen Dimensionen vor. Ein Beispiel dafür ist die Hausdorff Dimension. In einer exakten mathematischen Formulierung sind Fraktale dadurch gekennzeichnet, daß ihre Hausdorff Dimension größer ist als ihre topologische Dimension. Sie kann auch nichtganzzahlige Werte annehmen und wird deshalb als gebrochene Dimension bezeichnet. In dem vorliegenden Beitrag soll gezeigt werden, daß die Hausdorff Dimension bzw. allgemeiner die fraktalen Dimensionen in der Texturanalyse als ein quantitatives Maß zur Klassifikation und Segmentation genutzt werden können. Diese Betrachtung basiert auf zwei wesentliche Aussagen: nach Rosenfeld ist Texturrauheit eine meßbare charakteristische Eigenschaft, die in der Texturklassifikation verwendet werden kann [6]. Die zweite Feststellung stammt von Pentland und besagt, daß die Hausdorff Dimension einer Oberfläche eng mit unserem intuitiven Begriff von Rauheit zusammenhängt [3]. Die Hausdorff Dimension ist durch mehrere Grenzwert- und Supremumbildungen definiert und dadurch für die meisten fraktalen Mengen numerisch nur mit erheblichem Aufwand zu berechnen. Aus diesem Grund werden üblicherweise andere fraktale Dimensionen zur numerischen Charakterisierung eines Fraktals herangezogen.

2. Fraktalanalyse

Ein Standardverfahren in der Mathematik zur "Zähmung" von irregulären Kurven besteht darin, das Objekt durch möglichst glatte Kurven zu approximieren. Mit zunehmendem Approximationsgrad werden immer feinere Strukturen des Ausgangssignals berücksichtigt. Anders ausgedrückt bedeutet diese Vorgehensweise, daß das Objekt auf verschiedenen Maßstäben (Skalen) bzw. bei verschiedener Auflösung untersucht wird. Aus dem Verhalten von Merkmalen der glatten Kurven kann dann auf die Merkmale des ursprünglichen Signals geschloßen werden. Ein solches Merkmal kann die Länge des Signals sein. Um die Länge zu berechnen, wird das Ausgangssignal beispielsweise durch Polygonzüge mit konstanter, aber bei jedem feineren Näherungsschritt kleinerwerdender Kantenlänge approximiert (Bild 1).

Bild 1. Polygonapproximation einer Kochkurve

Die Länge des Signals sollte sich als Grenzwert der Längen der Polygonzüge ergeben. Kurven, deren Länge so ermittelt werden kann, für die also ein Grenzwert existiert, heißen rektifizierbar. Fraktalen Kurven läßt sich auf diese Weise keine Länge zuordnen; sie wächst gegen unendlich. Eine solche Divergenz entdeckte auch Lewis F. Richardson bei der Bestimmung der Länge von Küstenlinien mit dem eben genannten Polygonzugverfahren. Bei seinen Untersuchungen stellte er für die Länge L der Polygonzüge für immer kleinerwerdende Kantenlänge ϵ folgende Gesetzmäßigkeit fest:

$$L(\epsilon) \sim \epsilon^{1-D} \quad ,$$

wobei D von der jeweiligen Küste abhängt.

Richardsons Untersuchungen wurden von Mandelbrot als "Coastline of Britain Analysis" aufgegriffen. Mandelbrot erkannte, daß nicht die Länge sondern der im Potenzgesetz auftretende Parameter D als fraktale Dimension zur Charakterisierung eines fraktalen Signals interpretiert werden kann. Die Berechnung der Dimension erfolgt mittels Regressionsanalyse von ln $L(\epsilon)$ über ln ϵ. Bild 2 zeigt die doppellogarithmische Darstellung des obigen Potenzgesetzes für eine fraktale, streng selbstähnliche Kurve, die Kochkurve. Die fraktale Dimension ergibt sich aus der Steigung der Regressionsgeraden durch die experimentell bestimmte Punktwolke.

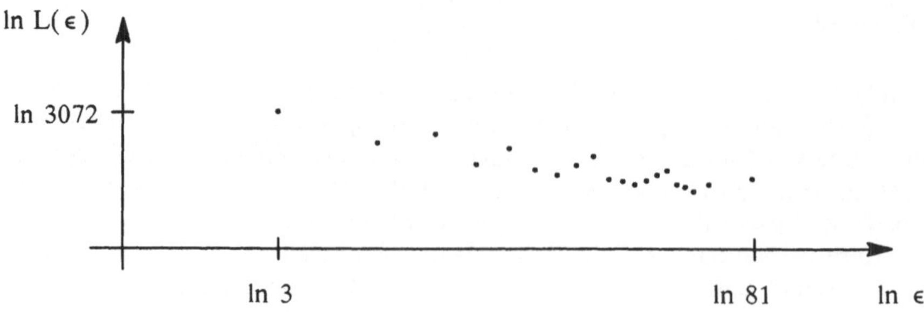

Bild 2. "Coastline of Britain Analysis" für eine Kochkurve

Wie aus Bild 2 ersichtlich ist, ist die Auswahl von geeigneten Maßstäben bzw. Approximationsgrad der Polygonzüge (Wahl der ε), die zur Bestimmung der fraktalen Dimension dienen, aufgrund der großen Schwankungen innerhalb der Punktwolke keine triviale Aufgabe. Diese Wahl kann leicht zu Mißinterpretationen der fraktalen Dimension führen.

Bei allen Verfahren zur Bestimmung einer fraktalen Dimension realer Strukturen tritt im allgemeinen ein Hauptproblem auf, das auch bei der Analyse von Texturen von Bedeutung ist: im Gegensatz zu mathematischen Fraktalen können reale Objekte nur in einem begrenzten aber unbekannten Skalenbereich als fraktale Strukturen interpretiert werden. Nur innerhalb dieses Bereichs wird die Struktur ein Skalenverhalten zeigen. An den Bereichsgrenzen tritt eine mehr oder weniger drastische Änderung des Skalenverhaltens auf. Bei der "Coastline of Britain Analysis" ist aber aufgrund der großen Schwankungen innerhalb der Punktwolke eine solche Änderung nur schwer festzustellen und damit eine Bestimmung der Bereichsgrenzen nahezu unmöglich.

Das Problem der Bestimmung geeigneter objektangepaßter Skalen existiert auch in anderen Bereichen der Bildverarbeitung. Die Beschreibung eines Signals anhand bestimmter Merkmale hängt nicht nur vom Signal selber ab, sondern auch von der gewählten Auflösung, bei der die Untersuchung erfolgt. Sie ist eindeutig, wenn die Analyse auf allen Skalen vorgenommen und zu einem Ergebnis zusammengefaßt wird. Ein Verfahren hierzu wurde von Witkin vorgeschlagen, das Scale Space Filtering (SSF) [7]: das gegebene Signal wird durch Faltung mit einem Faltungskern geglättet, wobei der Kern von einem stetig veränderbaren Skalierungsparameter abhängt. Nach der Glättung werden die Merkmale, beispielsweise die Extrema, der durch die Faltung erhaltenen Folge von Signalen in Abhängigkeit vom Skalierungsparameter analysiert. Durch die Analyse gelangt man zu einem "Fingerabdruck" der ausgewerteten Funktion. In Bild 3 ist der Fingerabdruck einer Realisierung einer gebrochenen Brownschen Bewegung dargestellt.

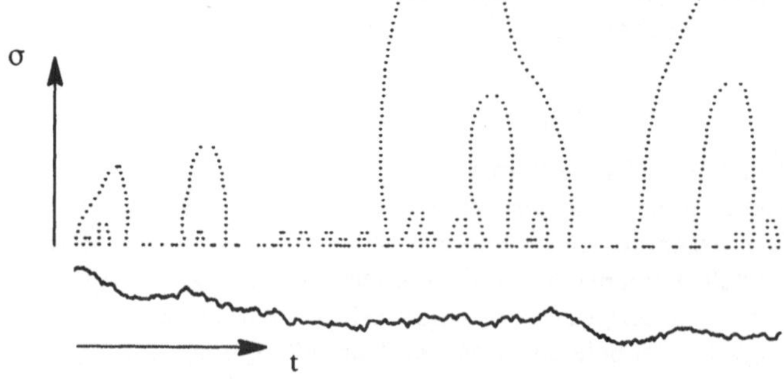

Bild 3. "Fingerabdruck" einer gebrochenen Brownschen Bewegung

Jeder Punkt des Fingerabdrucks symbolisiert ein Extremum, das zur Zeit t und Skalierungsparameter σ erscheint.

Die mögliche Klasse von Faltungskernen, die für eine solche Analyse in Frage kommen, wird durch Forderungen eingeschränkt, die an das Ergebnis der Glättung gestellt werden. Die wichtigsten Bedingungen sind, daß das geglättete Signal für die Grenzübergänge – Skalierungsparameter gegen null bzw. unendlich – bestimmte wohldefinierte Werte annimmt und daß bei einer Auflösung auftretende Extremas bei feinerer Skala nicht mehr verschwinden. Witkin hat in seiner Arbeit gezeigt, daß die einzige Funktion, die alle diese Voraussetzungen erfüllt, die Gaußfunktion ist. Im eindimensionalen Fall ist die Faltung eines Signals I(x) mit der Gaußfunktion g(x,σ) gegeben durch

$$C(x, \sigma) = I(x) * g(x, \sigma) = \int_{-\infty}^{\infty} I(u)\ \frac{1}{\sigma\sqrt{2\pi}}\ e^{\frac{-(x-u)^2}{2\sigma^2}}\ du\ .$$

Durch diese Faltung wird eine kontinuierliche Folge von geglätteten stetigen Signalen definiert, wobei jedes Folgenglied von der jeweiligen Standardabweichung σ der Gaußfunktion abhängt (Bild 4). Eine Verallgemeinerung auf höhere Dimensionen ist möglich [8].

Bild 4. Eine Folge von Gauß–geglätteten Signalen mit von oben nach
unten abnehmender Standardabweichung

Die von Witkin vorgeschlagene Methode wurde bisher in der Bildverarbeitung zur Beschreibung und Erkennung von ebenen Kurven sowie zweidimensionalen Formen eingesetzt [9]. Offensichtlich ist die Analyse einer Struktur durch Betrachtung der Folge von geglätteten Signalen, die man durch Faltung mit Gaußfunktionen mit unterschiedlicher Standardabweichung erhält, auch ein geeignetes Mittel, um das Skalenverhalten einer fraktalen Menge zu studieren [10]. Der bereits erwähnte Algorithmus von Richardson suggeriert einen möglichen Weg, um die SSF zur Fraktalanalyse zu nutzen: die Berechnung der Länge des geglätteten Signals in Abhängigkeit von der Auflösung σ. Tatsächlich zeigen erste experimentelle Ergebnisse folgende Gesetzmäßigkeit (s.z.B. [11]):

$$L(\sigma) \sim \sigma^p\ .$$

In Bild 5 ist die doppellogarithmische Darstellung dieses Potenzverhaltens wiederum für die Kochkurve dargestellt.

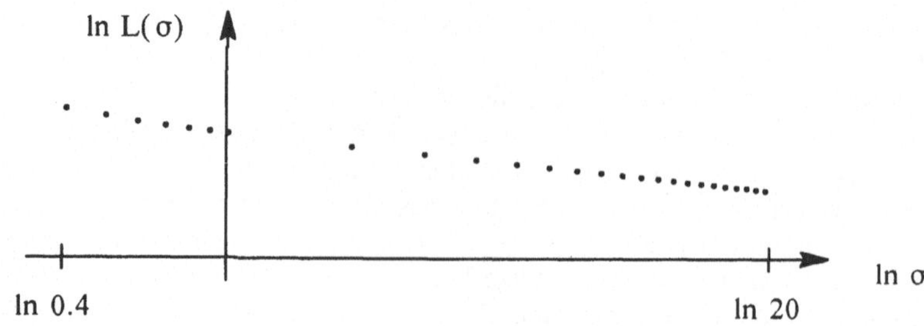

Bild 5. Analyse der Signallänge L in Abhängigkeit von der Varianz σ des
Faltungskerns für eine Kochkurve.

Im Gegensatz zur "Coastline of Britain Analysis" tritt hier das Problem der Wahl geeigneter Maßstäbe (bzw. Standardabweichung σ) nicht auf; die Länge L stellt sich als monoton fallende Funktion in Abhängigkeit von σ dar [12].

Mit Hilfe des Scale Space Filtering ist es auch möglich, das Skalenverhalten eines Grauwertbildes G(x,y) als charakteristisches Maß zur Texturanalyse zu benutzen. Hierzu wird das Originalbild analog zum eindimensionalen Fall durch eine Folge von geglätteten, stetigen Signalen C(x,y;σ) approximiert, die man durch Faltung des Originals mit einer zweidimensionalen Gaußfunktion mit variabler Standardabweichung σ erhält. Als charakteristisches Merkmal betrachten wir im zweidimensionalen Fall die Oberfläche des Objekts unter verschiedenen Auflösungen. Sie wird wie folgt berechnet:

$$F(\sigma) = \int \sqrt{1 + C_x(x,y;\sigma)^2 + C_y(x,y;\sigma)^2} \; dx \, dy$$

wobei C_x und C_y die partiellen Ableitungen des geglätteten Signals C sind.

Auch hier hat sich ein Potenzgesetz für eine Vielzahl verschiedener Texturen als gültig erwiesen:

$$F(\sigma) \sim \sigma^p \qquad (*)$$

wobei p als charakteristischer Skalenexponent für die jeweilige Textur interpretiert werden kann. Bild 6 zeigt die doppellogarithmische Darstellung von (*) für verschiedene Texturen, zusammen mit einem für die Signale gültigen Skalenbereich.

Ein bemerkenswertes Resultat unserer hier dokumentierten Methode zur Analyse von fraktalen Mengen durch Faltung mit einem Gaußkern und anschließender Berechnung der Oberfläche der Struktur ist, daß alle von uns untersuchten Strukturen innerhalb eines festen Bereichs Skalenverhalten zeigen. Die ermittelte fraktale Dimension ist unabhängig von einer Änderung der Folge von Auflösungen {σ} innerhalb des Skalenbereichs. Im Gegensatz zu anderen Auswertemethoden aus dem Bereich der fraktalen Geometrie wie zum Beispiel die "Coastline of Britain Analysis" führt der hier vorgeschlagene Weg zu einer monotonen Funktion F(σ). Aufgrund der Monotonie ist es ausreichend, nur eine sehr kleine Zahl von geglätteten Signalen des Originalsignals zu untersuchen, um das Skalenverhalten exakt wiederzugeben. Diese Eigenschaft favorisiert die SSF nicht nur zur Texturanalyse, sondern auch zur Charakterisierung von seltsamen Attraktoren von nichtlinearen physikalischen Prozessen.

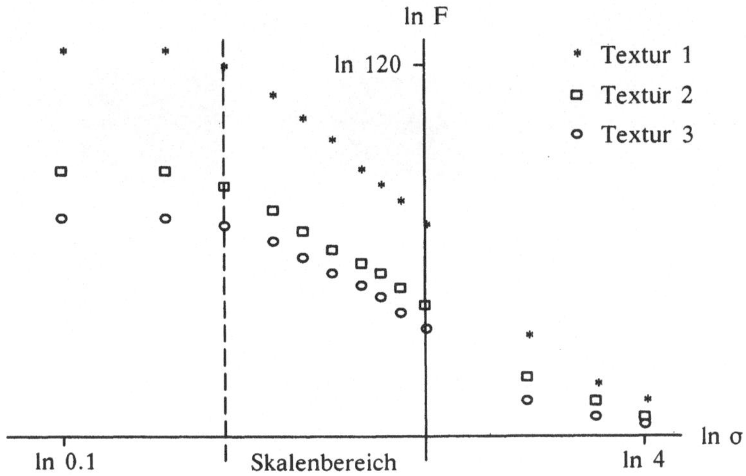

Bild 6. Doppellogarithmische Darstellung der Oberfläche eines Signals
in Abhängigkeit von der Auflösung

3. Textursegmentation

Wir haben die eben beschriebene Methode der Fraktalanalyse zur Textursegmentation von Grauwertbildern mit einem Format von 256*256 Bildpunkten und einer Punktinformation von 256 Grauwerten eingesetzt; jedes Bild setzt sich aus zwei verschiedenen Texturen zusammen. Um das charakteristische Skalenverhalten der jeweiligen Textur mit Hilfe der SSF zu ermitteln, wird für jeden Bildpunkt der Skalenexponent durch Analyse einer gewissen Punktumgebung berechnet, d.h. Faltung der Punktumgebung mit einer Folge von Gaußfunktionen mit unterschiedlichen Standardabweichungen und Berechnung der Oberfläche der geglätteten Signale innerhalb der Punktumgebung in Abhängigkeit von der Auflösung. Eine detailierte Abhandlung zur numerischen Berechnung einer Gauß-Faltung auf einem diskreten Gitter findet sich z.B. in [13,14]. Der mittels dem obengenannten Potenzgesetz bestimmte Skalenexponent wird dem jeweiligen Bildpunkt in Form eines Grauwertes zugeordnet. Das Ergebnis dieses Zwischenschritts zur Textursegmentation ist das "Dimensionsbild". Der letzte Schritt besteht darin, alle Punkte mit ungefähr gleichem Skalenexponent zu einer Textur zusammenzufassen. Da sich die hier gezeigten Strukturen nur jeweils aus zwei verschiedenen Texturen zusammensetzen, ist diese letzte Aufgabenstellung mit dem in der Bildverarbeitung bekannten Problem der automatischen Binarisierung gleichzusetzen; hierfür gibt es in der Literatur zahlreiche Lösungsvorschläge, genannt sei nur das Verfahren von Johannsen und Bille [15].

Im folgenden zeigen wir das Ergebnis der Textursegmentation für einige Testbilder. Für alle vorgestellten Bilder wurden die Werte σ = 0.3, 0.6 und 0.9 als Standardabweichung der Glättungsfunktionen gewählt. In Bild 7 ist die erste Textur dargestellt. Sie setzt sich zusammen aus Rechnersimulationen verschiedener selbstaffiner fraktaler Strukturen [16]. Das Bild zeigt das Originalbild (links), das Dimensionsbild (Mitte) und das Ergebnis der Segmentation.

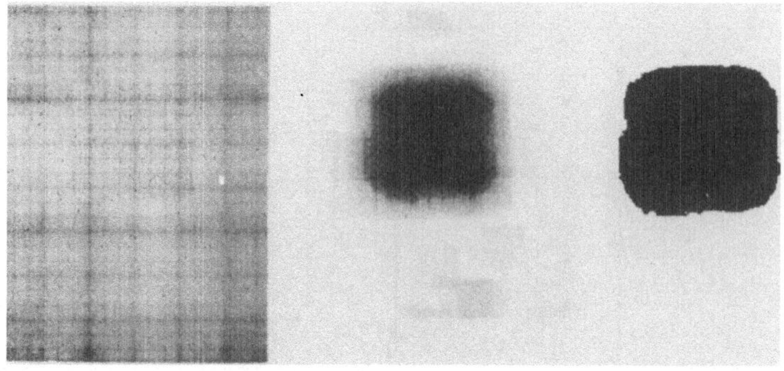

Bild 7. Textursegmentation für ein rechnererzeugtes Testbild
(selbstaffine Struktur)

Die zwei folgenden Beispiele (Bild 8, 9) zeigen Kompositionen verschiedener natürlicher Texturen (Brodatz [17]).

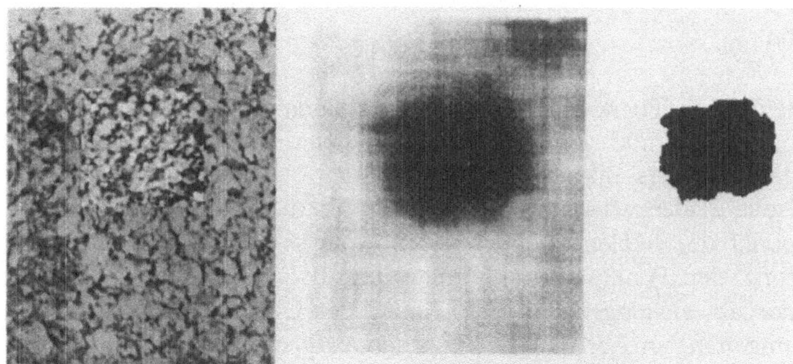

Bild 8. Textursegmentation von verschiedenen Brodatz-Texturen (D33, D29)

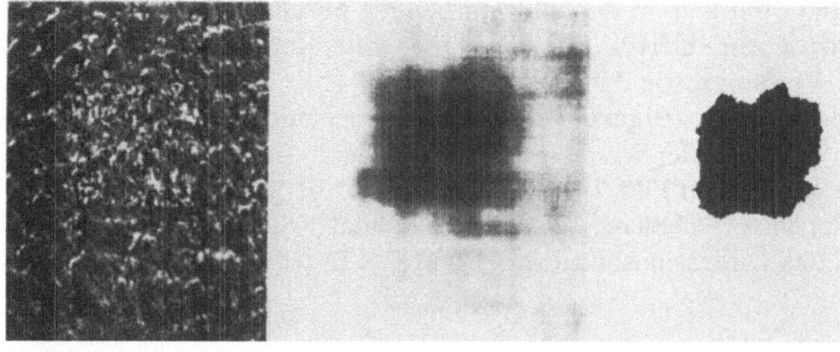

Bild 9. Textursegmentation von verschiedenen Brodatz-Texturen (D92, D9)

4. Zusammenfassung

Wir haben einige wichtige Eigenschaften des Scale Space Filtering zur Berechnung der fraktalen Dimension von Texturen genutzt. Scale Space Filtering stellt ein Werkzeug dar, mit dessen Hilfe einige wesentliche Probleme bei der numerischen Berechnung der fraktalen Dimension in einer zufriedenstellenden Weise gelöst werden können: die Festlegung eines Skalenbereichs und Auswahl einer geeigneten Folge von Skalenparametern. Die Ergebnisse, die wir bei der Textursegmentation erhalten haben, sind sehr vielversprechend und können dazu beitragen, die Fraktale Geometrie im Bereich der Texturanalyse weiter zu etablieren.

5. Literatur

[1]   Mandelbrot, B.B., "The Fractal Geometry of Nature", W.H.Freeman, San Francisco, 1983.

[2]   Nguyen, P.T. and Quinqueton, J., "Space Filling Curves and Texture Analysis", Proc. 6th Int. Conf. Patt. Recog., Munich 1982.

[3]   Pentland, A., "Fractal-Based Description Of Natural Scenes", IEEE Patt. Anal. Mach. Intel., Vol. 6, No. 6, 1984.

[4]   Peleg, S., Naor, J., Hartley, R. and Avnir, D., "Multiple Resolution Texture Analysis and Classification", 4th Jerusalem Conf. Info. Techniques, IEEE Comp. Soc. Press 1984.

[5]   Medioni, G. and Yasumoto, Y., "A Note on Using the Fractal Dimension for Segmentation", IEEE Comp. Vision Workshop, Annapolis MD, 1984.

[6]   Rosenfeld, A. and Troy, E., "Visual Texture Analysis", Tecn. Report 70-116, Comp. Science Center, Univ. Maryland 1970.

[7]   Witkin, A.P., "Scale Space Filtering", Proc. Int. Joint Conf. Art. Intel., Karlsruhe 1983.

[8]   Yuille, A.L. and Poggio, T.A., "Scaling Theorems for Zero Crossings", IEEE Patt. Anal. Mach. Intel., Vol.8, No.1, 1986.

[9]   Mokhtarian, F. and Mackworth, A., "Scale-Based Description and Recognition of Planar Curves and Two-Dimensional Shapes", IEEE Patt. Anal. Mach. Intel., Vol.8, No.1, 1986.

[10] Schmutz, M., private Mitteilung, 1988.

[11] Rueff, M., "Scale Space Filtering and the Scaling Regions of Fractals", in From the Pixels to the Features, ed. J.C. Simon, North Holland, 1989.

[12] Müssigmann, U., in Vorbereitung.

[13] Hummel, R. and Lowe, D., "Computational Considerations in Convolution and Feature-extraction in Images", in From the Pixels to the Features, ed. J.C. Simon, North Holland, 1989.

[14] Lindeberg, T., "Scale-Space for Discrete Images", Proc. 6.SCIA, Oulu, Finland, June 1989.

[15] Johannsen, G. and Bille, J., "A Threshold Selection Method using Information Measure", 6th Int. Conf. Patt. Recog., Munich 1982.

[16] Barnsley, M. and Demko, S., "Iterated function systems and the global construction of fractals", Proc. R. Soc. London A399, 243-275, 1985.

[17] Brodatz, P., "Texture - A Photographic Album for Artists and Designers", Dover, New York, 1966.

R.Bollhorst
Siemens AG, Werk für Textendgeräte
Rohrdamm 7, 1000 Berlin 13

F.Leschonski
TU-Berlin, Institut für Regelungstechnik
Fachgebiet Regelungstechnik und Systemdynamik (Prof.I.Hartmannn)
Einsteinufer, 1000 Berlin 10

Zusammenfassung:

Es wird ein Verfahren zur Kodierung/Konvertierung von Binär,- Quasibinär- und Grauwertbildern mit relativ großen homogenen Flächen in eine vektorähnliche Struktur vorgestellt. Der Code basiert auf einer nach topologischen Gesichtspunkten vorgenommenen Segmentierung des Rasterbildes und enthält eine vollständige Beschreibung des zu analysierenden Dokumentes.
Die generierten Basis-Listen, die im wesentlichen Koordinaten von Polygonzügen enthalten, finden nach einer mehrstufigen Erweiterung bei der Interpretation von ICONS (Sinnbildern) Verwendung.
ICONS sind Sinnbilder, die mehr oder weniger komplexe Sachverhalte graphisch illustrieren und veranschaulichen. ICONS können die unterschiedlichsten Ausprägungen und Ausdehnungen annehmen. Daher sind ICONS besonders geeignete Untersuchungsobjekte für Interpretationsaufgaben.

Grundlagen:

Das zunächst als Rasterbild vorliegende Element wird im folgenden nicht als eine Matrix von Grauwerten beschrieben, sondern unter Verwendung des topologischen Begriffes des *Zellenkomplexes* /1/ segmentiert. Die Elemente eines Rasterbildes werden in 0-d, 1-d und 2-d-Elemente unterteilt (s. *Abb.1*). Die Pixel erscheinen hier als 2-d-Elemente (mittlere Helligkeit einer Elementarfläche), die "Begrenzungen" als 1-d-

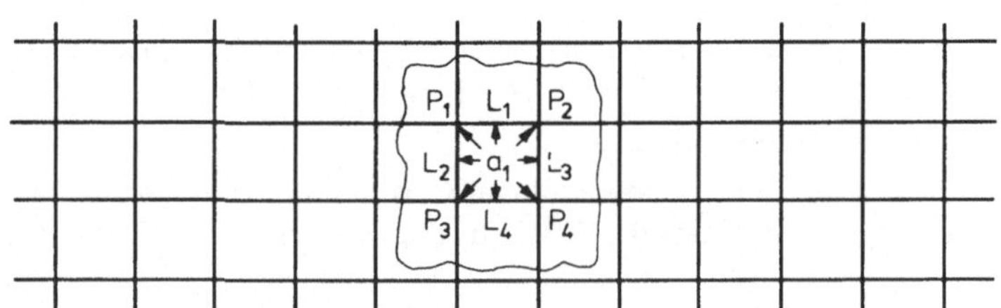

Abb.1: Der 2-d-Zellenkomplex

Als *2-d-Zellenkomplex* (s. *Abb.1*) wird ein zusammenhängendes Gebiet bezeichnet, dessen Pixel ein gemeinsames Merkmal, z.B. den Grauwert haben. Ein solches Gebiet wird auch *2-Zelle* genannt. Die Begrenzungen einer *2-Zelle* sind Polygonzüge. Die Gabelungspunkte der Begrenzungen bestehen aus 0-d-Elementen, die Verbindungen zwischen ihnen aus 1-d-Elementen.

Die Zuordnung der 0-d und 1-d-Elemente zu den 2-d-Elementen kann nicht durch die häufig in der Bildverarbeitung verwendete 4-er - oder 8-er-Nachbarschaft erfolgen. Es treten topologische Widersprüche auf. Beispielsweise ist der Jordan'sche Kurvensatz nicht mehr erfüllt. Dieser Satz besagt, daß eine einfache geschlossene Kurve in einer Ebene die Ebene in zwei Komponenten unterteilt. Eine widerspruchsfreie Zugehörigkeit wird definiert, in dem jedem 0-d-Element und 1-d-Element ein einziges 2-d-Element nach folgender Vorschrift zugewiesen wird:
Jedes 0-d-Element wird dem zu ihm südöstlich liegenden 2-d-Element zugewiesen. Jedes 1-d-Element wird dem zu ihm südlich oder östlich liegenden 2-d-Element zugewiesen.
Diese Zuordnungsvorschrift entspricht einer 6-er-Nachbarschaft der Pixel.
Jedem 2-d-Element wird also ein Speicherfeld zugewiesen, das den Grauwert selbst und die Werte der zugeordneten 0-d- und 1-d-Elemente enthält..

Unter Verwendung geeigneter Vorverarbeitungsmethoden (Kantenerkennung, Konturverfolgung) wird das Dokument in *2-Zellen* zerlegt. Die Begrenzungen dieser *2-Zellen* werden als Polygonzüge in Listen abgelegt - mit einer zusätzlichen Markierung, die den Grauwert kennzeichnet.
Die zusätzliche Verknüpfung der Datenliste durch Zeiger ermöglicht eine Kodierung, auf die sehr schnell und gezielt zugegriffen werden kann. Die weitere Bearbeitung der Bildinhalte wird dadurch wesentlich effektiver.

Die Listen liefern eine vollständige Beschreibung des Dokumentes. Es ist zweckmäßig, binäre oder besser quasi-binäre Vorlagen zu verwenden.
Als quasi-binäres Bild ist ein Bild mit einer Auflösung von maximal 3 bit zu verstehen. Die Konvertierung einer höher aufgelösten Grauwertvorlage in ein quasi-binäres Bild verhindert das bei einer einfachen Binärisierung auftretende "Ausfransen" von Linien und Konturen.
Als Nebeneffekt der Listenrepräsentation tritt in der Regel eine erhebliche Datenreduktion ohne Informationsverlust auf. Die Datenreduktion kann durch die Einführung von Kreissegmenten und durch eine Linienapproximation (Linienglättung) weiter erhöht werden. Es läßt sich zeigen, daß die Binarisierungsfehler in Relation zu den Approximationfehlern auf dem quasi-binärem Bild größer sind.

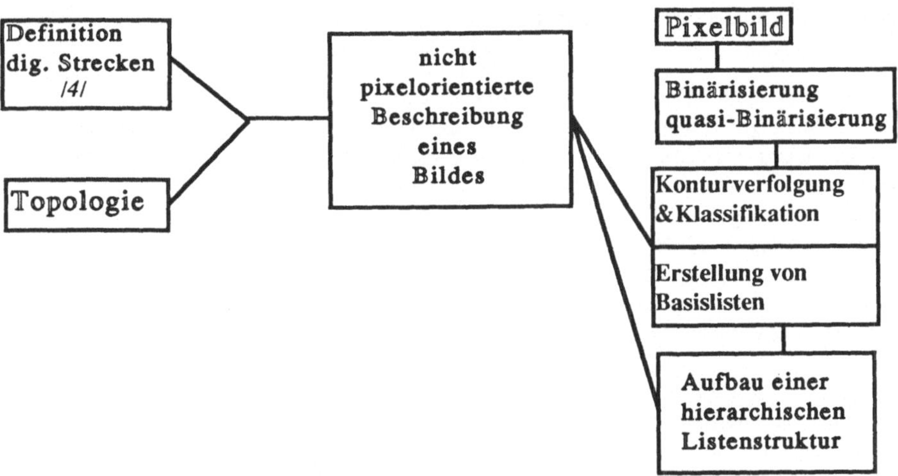

Abb.2: Gesamtschema zur Erstellung topologischer Listen

Realisierung / Zielsetzungen

Die in parametrisierter Form vorliegenden Polygone werden in nächsten Schritten zum Aufbau hierarchischer Listenstrukturen verwendet. Diese Listenstrukturen ermöglichen eine schrittweise und gezielte Überprüfung einzelner Merkmale und eine anschließende Klassifizierung der bearbeiteten ICONS /2/, /3/ .ICONS sind häufig aus elementaren graphischen Objekten mit einfachen Symmetrieeigenschaften zusammengesetzt. Ziel der Auswertung ist es, diese konstruktiven Grundelemente zu klassifizieren.
In einem ersten Schritt werden die Informationen der digitalen Strecken der Polygonzüge ausgewertet. Diese Auswertung ermöglicht folgende Aussagen:

- Länge der Strecke
- Steigung
- Charakter der Endpunkte
- Linientyp

In einem weiteren Schritt werden Listen erstellt mit den folgenden Eigenschaften und Beziehungen zwischen Strecken:

- Parallelität von Strecken
- Winkel
- Bildung von Mittellinien
- Zusammenfassung von Endpunkten

Für den Anwendungsfall der Interpretation von ICONS müssen Informationen abgelegt werden über:

- nicht klassifizierte Linien
- nicht klassifizierte Regionen
- Rechtecke / Quadrate
- Dreiecke (gleichschenklig / gleichseitig)
- Kreisbögen
- Pfeile
- sonstige Umrandungslinien
- Regionen mit symmetrischer Umrandung

Die Implementierung und Untersuchungen erfolgen auf einem *Apple Mac II* unter der graphischen Benutzeroberfläche *QuickDraw*. Es hat sich gezeigt, daß diese Entwicklungsumgbung für die Untersuchungen besonders geeignet ist. Die Ergebnisse können in vorhandenen Bildbearbeitungsprogrammen und objektorientierten graphischen Editoren verifiziert und modifiziert werden.

Zunächst ist es möglich, ICONS aus dem gescannten Bild herauszulösen, und in einem Objektfenster darzustellen und interaktiv zu bearbeiten.

Die folgenden Operationen sind:

- Erstellung der Basis-Listen
- Erstellung einer Hierarchie
- Verknüpfung einzelner Regionen

Bei der Erstellung der Hierarchie findet die Tatsache Berücksichtigung, daß ICONS auch aus mehreren *2-Zellen* zusammengesetzt sein können.

Literatur:

[1] V.A. Kovalevsky; Finite Topolgy as Applied to Image Analysis; Computer Vision, Graphics, and Image Processing 45; 1989

[2] M.Baretta, P.Mussio, M.Protti; ICONS: Interpretation and Use; 1986 IEEE Computer Society; Workshop on Visual Languages; Dallas, Texas, USA

[3] P.Mussio, M.Padula, M.Protti; Description Based ICON Design; 1987 IEEE Workshop on Visual Languages; Linköping, Sweden

[4] A.Rosenfeld; Digital Straight Line Segments; IEEE Transactions on Computers, No.12; December 1974

Ein neues Verfahren zur Verdünnung von Binärbildern

Yu Ji

Institut für Informatik der Technischen Universität München

Orleansstr. 34

8000 München 8o

Zusammenfassung:

In dieser Arbeit wird ein neues Verfahren eingeführt, das "der Rapid Thinning (RT) Algorithmus" genannt wird. Dieses Verfahren basiert auf dem einfachen und gültigen Skelettmodell digitaler Bilder. Dabei wird im Vergleich zum Algorithmus von Pavlidis (Classical Thinning (CT) Algorithms)/(vgl. Pavlidis 1982:199) eine wesentlich kürzere Laufzeit erreicht.

1. Einleitung

In den letzten Jahren erschienen viele Verfahren zur Verdünnung von Binärbildern. In der Regel hat jeder Ansatz zwei Ziele: die Qualität der Verdünnung zu erhöhen und die Laufzeit zu verkürzen. Die Erhöhung der Qualität hängt vom Skelettmodell und die Verkürzung der Laufzeit von der algorithmischen Umsetzung ab. Ein Verfahren zur Verdünnung einer digitalen Menge wird realisiert, indem iterativ-sequentielle oder parallele einfache Punkte von der Menge eliminiert werden. Die idealen Skelettpunkte sind als die Mittellinie des Objets definiert.

1982 stellte Pavlidis einen klassischen Verdünnungs- (CT)Algorithmus vor, der zu den parallelen Verfahren gehört. Dabei benutzte er ein Skelettmodell, um die Regionen des Bildes zu verdünnen (Abb.1). Dieses Skelettmodell ist folgendermaßen definiert: "Neighborhood patterns of multiple pixels . At least one of each group of pixels marked with A or B must be nonzero. " (vgl. Pavlidis 1982: 198). Dieses Skelettmodell ist mehrdeutig, d.h. es werden mehrere Gitterpunkte P und N(P) nicht nur bei Modell (1), sondern auch bei Modell (2) besetzt.

In dem hier vorgestellten Algorithmus wird ein Skelettmodell (Abb. 3) eingeführt, das von dem klassischen abgeleitet ist. Trotz seiner Ableitung paßt es immer noch auf das ursprüngliche Skelettmodell. Eine genaue Darstellung des abgeleiteten Skeletts ist im dritten Teil angeführt.

Ziel des neuen Algorithmus ist es, durch Veränderung des Skelettmodells und der Abarbeitung die Laufzeit zu verkürzen. Die grundlegende Idee bei diesem Algorithmus ist die folgende: zuerst wird das ganze Bild abgetastet, wobei alle Konturpunkte des Objektes in einem Puffer abgespeichert werden. Dann werden nur noch Punkte aus dem Puffer verarbeitet. Wenn ein Punkt kein Skelettpunkt ist, wird er eliminiert. In der Nachbarschaft eines Konturpunktes werden alle neu entstandenen Konturpunkte gefunden und im Puffer wieder abgespeichert. Dies wird wiederholt, bis nur noch Skelettpunkte übrig sind.

Im folgenden werden die Definition des Bildes, das Skelettmodell und der Algorithmus dargestellt, sowie das Skelettmodell und die Laufzeit der beiden Algorithmen verglichen.

A A A	A A A
O P O	B P O
B B B	B O C
(1)	(2)

Abb.1: "Klassisches Skelettmodell", bei dem mindestens eines der beiden mit A und B gekennzeichneten Pixel eine Nichtnull sein muß. C muß Nichtnull sein.

2. Definition

Es sei R die Menge aller Punkte der zu verarbeitenden Region. Die Punkte aus der Menge R nennen wir die "schwarzen Punkte", das Komplement die "weißen Punkte". Sei P Punkt aus R. N(P) ist 3 x 3 Nachbarschaft von P (Abb.2). Ein Punkt aus N(P) wird abkürzend mit L, OL, O, RO, R, UR, U, bzw. UL bezeichnet. Dabei steht L, O, R., U für Links, Oben, Rechts, Unten. L, O, R, U sind direkte Nachbarpunkte von P. OL, OR, UR, UL sind indirekte Nachbarpunkte von P.

Wenn der Punkt von N(P) aus schwarzen und weißen Punkten besteht, dann wird P Konturpunkt genannt. Wenn alle Punkte von N(P) aus schwarzen Punkten bestehen, dann wird P als Innenpunkt bezeichnet. Wenn N(P) das Skelettmodell (Abb. 3) erfüllt, heißt P Skelettpunkt.

C(R) ist die Menge aller Konturpunkte von P, I(R) die Menge der Innenpunkte, T(R) die Menge aller Skelettpunkte. Es werden zwei Puffer verwendet. In Puffer 1 werden alle Koordinaten der Konturpunkte, in Puffer 2 die Koordinaten der Konturpunkte bei jeder Iteration zwischengespeichert. R, C(R), I(R), und T(R) haben folgende Beziehung: Am Anfang ist R die Summe von C(R) und I(R). T(R) ist leer. Während des Prozesses wird ein Punkt P aus C(R),wenn seine Nachbarschaft N(P) das Skelettmodell erfüllt, in T(R) abgespeichert. R ist die Summe von C(R), I(R) und T(R). Zum Schluß sind C(R) und I(R) leer. Ist R gleich T(R), so ist R eine Menge von Skelettpunkten des Objektes.

OR	O	OL
R	P	L
UR	U	UL

Abb. 2: Der Gitterpunkt P und seine Nachbarschaft

73

3. Skelettmodell

In dem Verfahren zur parallelen Verdünnung wird ein invariantes Kriterium verwendet. Wenn ein Gitter P das Kriterium nicht erfüllt, wird er eliminiert. Dieses Kriterium wird als Skelettmodell bezeichnet.

Die Skelettpunkte im Bild sind die Punkte entlang der Mittelachsen. Ihre Umgebung weist eine der folgenden Konfigurationen auf, wobei Modell (1) eine Rotation von 90 Grad und Modell (2) eine viermalige Rotation von je 90 Grad durchläuft. Hierbei soll P der zu untersuchende Punkt sein, "A" muß ein Nichtnull-Wert sein, "0" gehört zu S (Abb. 3). Dieses Skelettmodell ist eindeutig, d.h. ein Skelettpunkt P und N(P) erfüllt entweder Modell (1) oder Modell (2).

```
        A
   0  P  0                P  0
        A                 0  A

      (1)                 (2)
```

Abb. 3: *Skelettmodell*
"A" muß ein Nichtnullwert sein

4. RT-Algorithmus

Der RT-Algorithmus unterscheidet sich vom CT-Algorithmus. Im CT-Algorithmus werden die Skelettpunkte durch die mehrmalige Abtastung des Bildes gefunden. Die Skelettpunkte des RT-Algorithmus werden so ermittelt, daß die Konturpunkte in zwei Puffern abwechselnd bearbeitet werden. Alle (neuen) Konturpunkte werden in Puffer 1 gespeichert. Anhand der Richtung von L, O, R, U werden die Punkte P, die als Koordinaten in Puffer 1 gespeichert werden, mit dem Skelettmodell verglichen. Wenn P und N(P) das Skelettmodell erfüllen, ist der Gitterpunkt P Skelettpunkt, der mit "SP" markiert wird, ansonsten wird der Gitterpunkt P mit "HP" gekennzeichnet. Die Koordinaten dieser beschriebenen Gitterpunkte P werden in Puffer 2 gespeichert. Mit den Koordinaten in Puffer 2 werden neue Konturpunkte gefunden, die zu den direkten Nachbarpunkten von P gehören. Die Koordinaten dieser neuen Punkte werden wieder im Puffer 1 gespeichert. Weiter werden die Punkte, die durch "HP" im Puffer 2 markiert wurden, eliminiert, d.h. die schwarzen Punkte verwandeln sich in weiße Punkte. Der RT-Algorithmus läßt sich folgendermaßen darstellen:

1)   Markiere alle Punkte im Bild gemäß Gitter P (Abb.2)
2)   Speichere die Koordinaten aller Konturpunkte im Puffer 1.
3)   Während Puffer 1 nicht leer ist, werden die nachfolgenden Schritte 4)-10) wiederholt.
4)   Für alle Markierungen M = L, O, R, U. führe aus Schritte 5)-9)
5)   Für alle Punkte M in Puffer 1 führe aus Schritte 6)-8).
6)   Prüfe, ob Punkt ein Skelettpunkt ist.
7)   Im Falle eines Skelettpunktes markiere "SP", sonst "HP".

8)   Speichere den Punkt nach Puffer 2.

9)   Markiere alle mit "HP" gekennzeichneten Punkte in Puffer 2 als weiße Punkte.

10) Für alle Punkte in Puffer 2 untersuche die Nachbarschaft und speichere sie im Puffer 1, falls Konturpunkte entstanden sind.

5. Vergleich

Zum Vergleich der Laufzeit wurden Messsungen auf einer Mikro-VAX II durchgeführt.

In Abb. 4(a) erhalten wir das Ergebnis vom RT-Algorithmus, ebenso wie das Ergebnis des CT-Algorithmus (Abb.4(b)). Die beiden Ergebnisse sind sich ähnlich. Im Vergleich zum CT-Algorithmus hat der RT-Algorithmus folgende Besonderheiten:

1. Das Skelettmodell wurde vereinfacht und dadurch die Mehrdeutigkeit des CT-Algorithmus vermieden. 2. Es wird nicht das ganze Bild, sondern nur die zu verarbeitenden Regionen abgetastet. Im CT-Algorithmus wird jeder Punkt mehrmals abgetastet, im RT-Algorithmus nur einmal.

Weil das Skelettmodell und die Art und Weise der Verarbeitung im RT-Algorithmus verändert wurden, ergeben sich kürzere Laufzeiten. Abb. 5 zeigt den Unterschied der Laufzeit zwischen dem RT-Algorithmus und dem CT-Algorithmus. Die dargestellten Bilder sind Ausschnitte aus den Originalbildern, die eine Größe von 512 x 512 haben.

RT	2.566 s (Abb. 4a (1))	16.499 s (Abb.4a(2))
CT	255.258 s (Abb. 4b (1))	1318.82 s (Abb.4b(2))

Abb. 5: Laufzeit der beiden Algorithmen, gemessen auf einer MIKRO-VAX-II

6. Literatur

T. Pavlidis, (1982) Algorithms for Graphics and Image Processing. Washington/DC: Computer Science Press, 1982

Tamura, H. (1978) A Comparison of Line Thinning Algorithms from Digital Geometry Viewpoint. Proc. of the 4th Int. Conf. on Pattern Recognition, Kyoto/Japan, Nov. 7-10, 1978, P. 715-719

Xia, X. (1986) A New Thinning Algorithm for Binary Images. Proc. of the 8th Int. Conf. on Pattern Recognition, Paris/France, Oct. 27-31, 1986, P. 995-997

Rosenfeld, A. (1979) Digital Topology. American Mathematical Monthly 86: 621-630

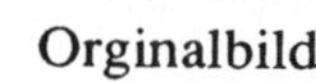

Orginalbild   Orginalbild

Abb.4a(1) Ergebnis von RT Algorithmus   Abb.4a(2) Ergebnis von RT Algorithmus

Abb.4b(1) Ergebnis von CT Algorithmus   Abb.4b(2) Ergebnis von CT Algorithmus

Intrinsic Topology of Medial Axis

Zhangzheng Yu
Institut für Angewandte Mathematik
Universität Hamburg
Bundesstraße 55
2000 Hamburg 13
West Germany

1. Introduction

The medial axis transform is used frequently in the segmentation of binary images, where it is necessary, to seperate unconnected objects into different classes. The instable behavior of the medial axis transform, as known in the literatur, makes the application of this technique more difficult. A metrical regularization of the transform was proposed in [3], in which two parameters were contained. The suitable determination of the both parameters, as in the regularization of many ill-posed problems, depends on the context, so that a general theory on the choice of the parameters is impossible.

In this paper, we give a topological criterion, through which the medial axis can be divided into two parts: The one part is sensitive to the changes of the set to be transformed, whereas the other part has a stable behavior. The two parts are called the instable part and the stable part of the medial axis, respectively. It is proved in this paper that the stable part bears essential property of the set to be transformed, which is exactly the information needed for segmentation of binary images.

The topological criterion: The stable part of the medial axis of a plane set G consists of the centers of such discs whose interiors lie entirely in the set G and whose boundaries have common points with at least two different connected components of the complement of the set G, respectively.

2. A theorem on the behavior of the stable part of a medial axis

For a given set G in the plane and a positive number δ, the δ-parallel set of G, called $G(\delta)$, is the set of such points y for which there exists another point x so, that the distance between x and y is not greater than δ. For two sets G_1 and G_2 in the plane, the Hausdorff-distance between G_1 and G_2 is defined as

$$h(G_1, G_2) = \inf\{\delta : G_1 \subset G_2(\delta), G_2 \subset G_1(\delta)\}$$

Theorem 1: Let G_1 and G_2 be two compact sets in the plane whose diameters are not greater than a positive number r. Let $K_1^{(1)}, ..., K_m^{(1)}$ and $K_1^{(2)}, ..., K_m^{(2)}$ be the connected components of the complements G_1^c und G_2^c of the sets G_1 und G_2, respectively. The Euclidean distances between different $K_j^{(1)}$ and between different $K_j^{(2)}$ are assumed to be not smaller than a positive number d. The Hausdorff-distances between $K_j^{(1)}$ and $K_j^{(2)}$ for j from 1 to m are assumed to be limited from above by a positive number δ. Unter these conditions, the Hausdorff-distance between the stable part, called $T(G_1)$, of the medial axis of the set G_1 and the stable part, called $T(G_2)$, of the medial axis of the set G_2 is bounded from above by $\frac{2r^2\delta}{d^2 - 4\delta^2}$.

Proof: Without loss of generality, the number m of components is assumed to be two. All symbols appearing in the proof can be found in figure 1.

For one point x_0 of $T(G_1)$, there is a disc $K(x_0, r_1)$ with x_0 as its center and r_1 as its radius which is contained entirely in the set G_1. According to the topological criterion, the boundary of the disc $K(x_0, r_1)$ has one common point with $K_1^{(1)}$ and another one with $K_2^{(1)}$. Because of $d(K_1^{(1)}, K_2^{(1)}) \geq d$ we have $r_1 \geq \frac{d}{2}$. Because of $h(K_1^{(1)}, K_1^{(2)}) \leq \delta$, the interior of the disc

$K(x_0, r_1 - \delta)$ has commen points neither with $K_1^{(2)}$ nor with $K_2^{(2)}$. There is a point x_1 in the set $K_1^{(2)}$ so, that the distance of x_0 and x_1 equals the distance of x_0 and $K_1^{(2)}$. There is also a point x_2 in the set $K_2^{(2)}$ so, that the distance of x_0 and x_2 equals the distance of x_0 and $K_2^{(2)}$. If the distance of x_0 and x_1 equals the distance of x_0 and x_2, then x_0 belongs to $T(G_2)$.

Since the investigation of the two cases, $d(x_0, x_1) < d(x_0, x_2)$ or $d(x_0, x_2) < d(x_0, x_1)$, is symmetric, we consider in the following only the first case: $d(x_0, x_1) < d(x_0, x_2)$. For an arbitrary point x lying in $[x_0, x_2]$, the segment joining the two points x_0 and x_2, we have the inequalities

$$d(x, x^*) \le d(x, K_1^{(2)}) \le d(x, x_1)$$

and the equality

$$d(x, K_2^{(2)}) = d(x, x_2).$$

The proof consists in finding a point $\bar{x}$ in the segment $[x_0, x_2]$ so, that this point belongs to $T(G_2)$, i.e., the stable part of the medial axis of the set G_2.

For this purpose, we consider at first the solution of the equation $d(x, x^*) = d(x, x_2)$. From

$$d^2(x, x^*)$$
$$= \|x - x_0\|^2 + \|x_0 - x^*\|^2 - 2\|x - x_0\|\|x_0 - x^*\|\cos\theta$$
$$= \|x - x_0\|^2 + \|x_0 - x_1\|^2 - 2\|x - x_0\|\|x_0 - x_1\|\cos\theta$$

and

$$d^2(x, x_2) = \|x_0 - x_2\|^2 + \|x - x_0\|^2 - 2\|x_0 - x_2\|\|x - x_0\|$$

we have

$$\|x - x_0\|$$
$$= \frac{\|x_2 - x_0\|^2 - \|x_1 - x_0\|^2}{\|x_2 - x_0\| - \|x_1 - x_0\|\cos\theta}$$
$$= \frac{\|x_2 - x_0\| + \|x_1 - x_0\|}{\|x_2 - x_0\| - \|x_1 - x_0\|\cos\theta}\left(\|x_2 - x_0\| - \|x_1 - x_0\|\right)$$
$$\le \frac{4\delta}{1 - \frac{\|x_1 - x_0\|}{\|x_2 - x_0\|}\cos\theta}$$
$$\le \frac{4\delta}{1 - \cos\theta}.$$

The equation

$$d^2(x^*, x_2) = d^2(x_0, x_2) + d^2(x_0, x_1) - 2d(x_0, x_2)d(x_0, x_1)\cos\theta$$

delivers

$$\cos\theta = \frac{d^2(x_0, x_2) + d^2(x_0, x_1) - d^2(x^*, x_2)}{2d(x_0, x_2)d(x_0, x_1)}$$

and

$$1 - \cos\theta = \frac{d^2(x^*, x_2) - [d(x_0, x_2) - d(x_0, x_1)]^2}{2d(x_0, x_2)d(x_0, x_1)}.$$

Because the diameters of G_1 and G_2 are bounded from above by a number r, we obtain

$$1 - \cos\theta \ge \frac{d^2 - d\delta^2}{2(\frac{r}{2})^2} = \frac{2(d^2 - 4\delta^2)}{r^2},$$

from which we get an estimate for the distance between the point x_0 and a solution $\bar{x}$ of the equation $d(x, x^*) = d(x, x_2)$

$$\|\bar{x} - x_0\| \le \frac{4\delta r^2}{2(d^2 - 4\delta^2)} = \frac{2\delta r^2}{d^2 - 4\delta^2}.$$

Because of

$$d(x_0, K_1^{(2)}) < d(x_0, K_2^{(2)})$$

and

$$d(\bar{x}, K_1^{(2)}) \geq d(\bar{x}, x^*) = d(\bar{x}, x_2) = d(\bar{x}, K_2^{(2)})$$

there exists a point $\bar{x}_0$ in the segment $[x_0, \bar{x}]$ so, that

$$d(\bar{x}_0, K_1^{(2)}) = d(\bar{x}_0, K_2^{(2)})$$

and

$$d(\bar{x}_0, x_0) \leq d(\bar{x}, x_0) \leq \frac{2\delta r^2}{d^2 - 4\delta^2}.$$

Therefore, we have proved, for every point x_0 from $T(G_1)$ there is another point $\bar{x}_0$ from $T(G_2)$ so, that the inequality

$$d(x_0, \bar{x}_0) \leq \frac{2\delta r^2}{d^2 - 4\delta^2}$$

is satisfied. That means, the stable part of the medial axis of the set G_1 lies in the $\frac{2\delta r^2}{d^2-4\delta^2}$-parallel set of the stable part of the medial axis of the set G_2. With the same arguments we can prove, that $T(G_2)$ lies in the $\frac{2\delta r^2}{d^2-4\delta^2}$-parallel set of $T(G_1)$.

3. Discretization of Metrics

Before forwarding to an algorithm to calculate the stable part of the medial axis of a plane set, we consider the discretization of the metric used in the algorithm. A metric d defined in the plane induces a metric d_0 on the plane lattice, i.e., the set of all plane points with integer coordinates. The induced metric is often required to take integer values on the lattice, which is not always the case. For instance, the Euclidean metric on the plane induces no integer metric on the lattice. To overcome this discrepancy, the metric is often rounded to an integer function, which is no metric in the lattice any more. We have found a new rounding method yielding a metric in the lattice.

Rounding Method: A real number x is rounded to an integer $\{x\}$ which is the smallest of all the integers not smaller than x. For example, $\{5\} = 5$ and $\{3.5\} = 4$.

Theorem 2: According to the method above, the rounding $\{d\}$ of every metric d defined on a set G is an integer metric.

Proof: We need only to verify the validity of the triangle inequality. Because d is a metric on the set G, the inequality

$$d(x, y) \leq d(x, z) + d(y, z),$$

is valid for arbitrary three points x, y and z of the set G, which delivers

$$\begin{aligned}
&\{d(x, y)\} \\
&\leq \{d(x, z) + d(y, z)\} \\
&\leq \{d(x, z) + \{d(y, z)\}\} \\
&\leq \{\{d(x, z)\} + \{d(y, z)\}\} \\
&= \{d(x, z)\} + \{d(y, z)\},
\end{aligned}$$

That means $\{d\}$ is a metric.

As an example, some circles in the plane lattice in sense of this rounded metric of the Euclidean metric are drawn in figure 2.

4. The Algorithm

The following algorithm is used to calculate the stable part of the medial axis of a plane set. After little changes, it can also applied for calculation of the whole medial axis.

The algorithm is implemented in four steps:
1. colouring of all connected components of the complement of the set to be transformed;
2. According to this rounding method, all points of the discretized circles issuing from the origin are stored in a look-up table;
3. Calculation of the distances between every point of the set to be transformed and the complement of this set by application of theorem 2;
4. Simultaneously to step 3, it is investigated if the actual point is qualified to be a member of the stable part of the medial axis.

The efficiency of the approach can be greatly enhanced by clever use of the information obtained during execution. So it is not necessary in step 3 to take all pairs of points into account, a rather small subset of them will be sufficient. Details of the implementation can be found in my dissertation.

An example is displayed in figure 3.

5. Conclusion

The results in this paper have practical significance. Applying theorem 1, we can estimate, for instance, the changes of segmentation of disturbed binary images, whereas such a prediction is not possible in other known algorithm. The algorithm is implemented on the computer *Siemens 7.882* in Rechenzentrum der Universität Hamburg. It has proved to be reliable and efficient.

My special thanks belong to Prof. Dr. Eckhardt who has introduced me into the application of mathematics in the picture processing.

Literature

[1] Eckhardt U.: Verdünnung mit perfekten Punkten. Informatik-Fachberichte 108, 204-210. Springer-Verlag. 1988
[2] Pavlidis T.: Structural Pattern Recognition. Springer-Verlag. 1977
[3] Yu Z.: Regularisierung der Mittelachsentransformation. Informatik-Fachberichte 180, 211-218. Springer-Verlag. 1988
[4] Yu Z.: Stabile Analyse von Binärbildern. Dissertation. Universität Hamburg. 1989
[5] Yu Z., Maderlechner G.: Determination of Global Properties in Document Processing. IAPR Workshop on Computer Vision - Special Hardware and Industrial Applications, Oct. 12-14, 1988, Tokyo, Proceedings, pp. 312-314

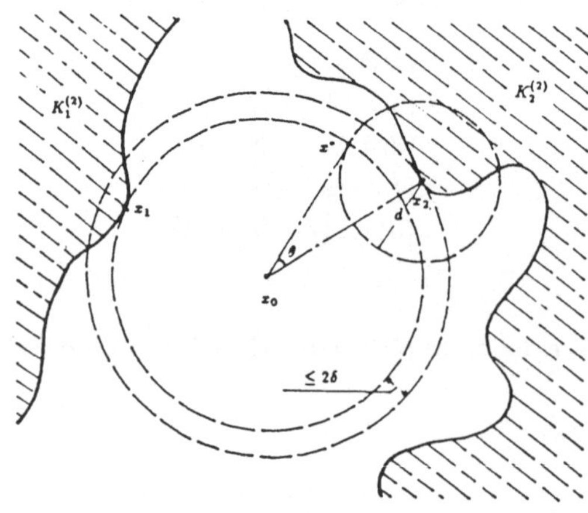

figure 1

figure 2

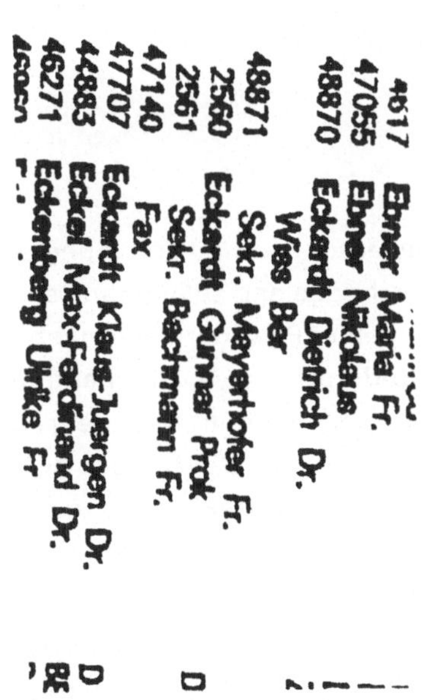

4617	Ebner Maria Fr.	
47055	Ebner Nikolaus	
48870	Eckardt Dietrich Dr.	
	Wiss Ber	
48871	Sekr. Mayerhofer Fr.	
2560	Eckardt Gunnar Prok	D
2561	Sekr. Bachmann Fr.	
47140	Fax	
47707	Eckardt Klaus-Juergen Dr.	D
44883	Eckel Max-Ferdinand Dr.	BE
46271	Eckenberg Ulrike Fr	

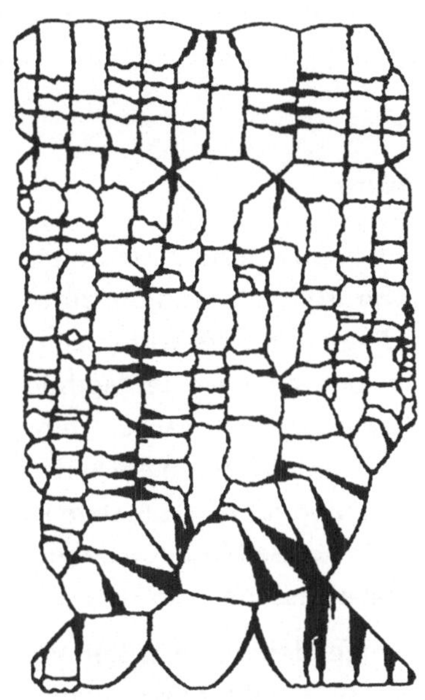

figure 3

81

Verwendung einer Bildauswertungsmethode für polyedrische Szenen zur Erkennung von Werkstücken aus gebogenem Blech

Claus Müller, Susanne Rössle und Hans-Hellmut Nagel
Deutsch-Französisches Institut für Automation und Robotik (I.A.R.),
Universität Karlsruhe,
c/o Fraunhofer-Institut für Informations- und Datenverarbeitung (IITB),
Fraunhoferstr. 1, 7500 Karlsruhe 1

Zusammenfassung:

Auf dem Bildauswertungssystem VISTA (Visuelles Interpretationssystem für Technische Anwendungen) wurde der von Sugihara /6/ vorgeschlagene Algorithmus zur Interpretation von Strichzeichnungen implementiert. Seine Methode erlaubt es, aus zweidimensionalen Strichzeichnungen polyedrischer Objekte eine qualitative, topologische Beschreibung abzuleiten. Die Methode soll benutzt werden, um geeignete Werkstücke in ihrer Fertigungsumgebung erkennen zu können. Blechbiegeteile (siehe Abb. 1a-1b) sind Werkstücke, die dafür in Frage kommen, da man sie unter der Annahme einer örtlich eng begrenzten Biegezone als Polyeder betrachten und deren berechnete Konturbilder als Eingang für den Interpretations-Algorithmus verwenden kann.

1. Einleitung

Nach einer sich über rund 25 Jahre erstreckenden wissenschaftlichen Forschung zur Interpretation von Strichzeichnungen durch zahlreiche Forscher gelang es Sugihara, eine hinreichende Bedingung für die korrekte Interpretierbarkeit einer Polyederszene zu entwickeln. Ausgehend von einer nach der Methode von Clowes und Huffman /1/, /3/ erhaltenen konsistenten Markierung der „Kanten" einer Strichzeichnung, stellt er mit Hilfe der räumlichen Hinweise der Kantenmarken ein System von Ungleichungen und Gleichungen auf. Die Existenz einer Lösung für dieses System ist das Kriterium für eine korrekte Interpretierbarkeit. Sugihara wollte diese Methode benutzen, um z. B. vom Konstrukteur handgefertigte Skizzen einer Polyederszene in korrekte Zeichnungen umsetzen zu können.

In dem im folgenden dargestellten Verfahren soll diese Methode benutzt werden, um gebogene Blechteile in ihrer Fertigungsumgebung erkennen zu können. Die Methode bietet den Vorteil, daß man für den Erkennungsprozeß keine expliziten Modelle der Werkstücke

Abb. 1a-1b: Für die Interpretation geeignete Werkstücke aus gebogenem Blech

vorgeben muß. Das Modell wird vielmehr durch die Erzeugung der mathematischen Beschreibung automatisch „gelernt".

Das Verfahren wird in drei Schritten durchlaufen. Erster Schritt ist die Erzeugung einer Eingangsstruktur, die den Randbedingungen der weiteren Schritte genügen muß. Im zweiten Abschnitt wird versucht, der erzeugten Struktur eine konsistente Markierung zuzuordnen. Diese Markierung erfolgt durch Zuweisung von charakterisierenden Marken an Kanten, so daß die sich ergebenden Kantenverbindungstypen in der von Huffman, Clowes und Waltz /7/ ausgearbeiteten Tabelle enthalten sind. Im dritten Schritt wird aus der markierten Eingangsstruktur das System von Gleichungen und Ungleichungen abgeleitet, das es erlaubt, die Interpretierbarkeit der Szene zu überprüfen. Abschließend soll auch ein Verfahren zur Nutzung von zusätzlichen Informationen aus dem Grauwertverlauf der sichtbaren Flächen im Bild kurz beschrieben werden, mit dem es möglich ist, die nach dem Ansatz von Sugihara noch verbleibenden Freiheitsgrade der Beschreibung weiter einzuschränken.

2. Erzeugung einer Eingangsstruktur

Kern des ersten Verfahrensabschnittes ist das Gradientenverfahren nach Korn /4/. Hier soll eine Strichzeichnung aus einem Videobild einer Polyederszene erzeugt werden, die dann Eingang finden kann zum zweiten Teil des Verfahrens. Das Gradientenverfahren liefert Gradientenbetrag und -richtung für jeden Grauwert im Grauwertgebirge des Eingangsvideobildes. Durch Vergleichsoperationen von Betrag und Richtung des jeweiligen Pixels mit den Werten der acht Pixel in der unmittelbaren Nachbarschaft kann dann entschieden werden, ob ein lokales Maximum des Gradientenbetrages vorliegt. Man erhält am Ausgang zwei synthetische Bilder, in denen sich nur noch dort ein von Null verschiedener Wert befindet, an denen ein lokales Maximum des Betrages vorliegt. Die so erhaltenen Konturbilder der Polyederszene weisen im allgemeinen noch erhebliche Unterschiede auf zu einer von Hand angefertigten Skizze derselben Szene.

Erste Ursache dafür ist, daß ein Algorithmus zur Kantendetektion bei entsprechender Maskengröße auch dann noch Maxima detektiert, wenn die Betragsunterschiede zu den Nachbarpixeln sehr klein werden, d.h. nur noch durch ins Bild eingestreutes Rauschen verursacht

werden. In dem verwendeten Verfahren ist dem Rechnung getragen durch eine Schwelle
für den Gradientenbetrag. Außerdem hat man für den anschließenden Konturverkettungs-
schritt die Konturen nach deren Gradientenbetragsgröße sortiert vorliegen. Durch die Angabe, wieviele der betragsgrößten Konturen für die weitere Verarbeitung verwendet werden sollen, kann man die Anzahl der weiter zu verarbeitenden Konturen verringern. Die im
Algorithmus enthaltenen Maßnahmen reichen im allgemeinen aus, um zufällig erzeugte
Konturen zu eliminieren.

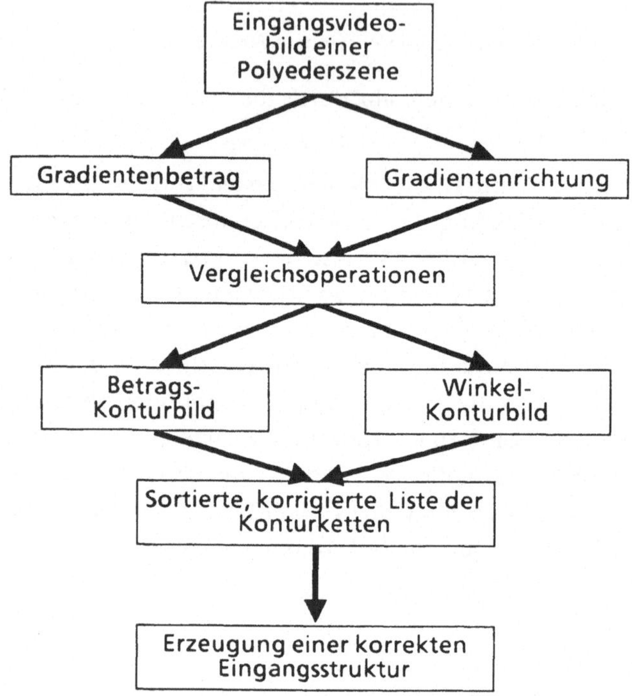

Abb. 2: Die schematische Darstellung für die Erzeugung einer Eingangsstruktur
aus einem Videobild.

Zweite Ursache für die Unterschiede zwischen einer handgefertigten Skizze und einem
vom Algorithmus erzeugten Konturbild sind durch die Beleuchtung hervorgerufene physikalische Effekte auf den Objekten, wie Schattenwurf oder Glanzlicht, die bei durchschnittlichen Beleuchtungsverhältnissen im Bild unweigerlich zu sehen sind. Gebogene Blechteile
weisen an den Zonen der äußeren Biegekanten solche Glanzlichter auf, wodurch die eigentliche Kante teilweise in zwei Konturlinien aufgespalten wird, die sich entlang des
Glanzlichtes ziehen und so die Lokalisierung der wahren Kante erschweren. Effekte dieser
Art werden bei dem jetzigen Stand der Arbeiten noch nicht korrigiert. Es wurden bis jetzt
die Beleuchtung und die Form der Körper so gewählt, daß Abweichungen dieser Art gering
blieben.

Dritte Ursache für Differenzen sind schließlich fehlende Pixel in den Konturketten. Für die
Wiedergabe eines polyedrischen Objektes ist es notwendig, von geraden durchgehenden

Linienstücken mit eindeutigen Knotenpunkten in den Ecker auszugehen. Man muß daher bei der Verkettung aufgebrochene Konturketten wieder verschmelzen, um sie anschließend an Stellen starker Krümmung, wo räumliche Objektecken vermutet werden, wieder aufbrechen zu können. Durch diesen Vorgang ist es möglich, eine zweidimensionale Struktur festzulegen, bei der nur noch Geradenstücke vorhanden sind, die an beiden Enden auf einen eindeutigen Knotenpunkt treffen. In Abb. 2 ist der erläuterte Vorgang zur Erzeugung einer Eingangsstruktur für die anschließenden Schritte noch einmal schematisch dargestellt.

3. Konsistente Markierung

Der Markierungsvorgang, der aus der Literatur auch als „Waltz-Filterung" /7/ bekannt ist, weist für den Fall einer geeigneten Eingangsstruktur jeder Kante eine sie räumlich charakterisierende Marke zu. Eine Linie kann eine konvexe, konkave oder verdeckende Kante im Bild darstellen, je nachdem, wie die Flächen, deren gemeinsame Schnittlinie der Kante entspricht, im Raum zum Betrachter orientiert und sichtbar sind. Eine Eingangsstruktur gilt als geeignet, wenn jedes Linienelement eine Verbindung zu genau zwei Knotenpunkten herstellt. Die Methode geht auf Clowes und Huffman zurück, die für den Fall von maximal trihedralen Kantenverbindungen einen Katalog möglicher Verbindungstypen ausarbeiteten. Die Aufgabe dieses Verfahrensabschnittes besteht nun darin, allen Kanten gleichzeitig eine der sie räumlich charakterisierenden Marken so zuzuordnen, daß die sich ergebenden Verbindungen an den Knotenpunkten jeweils einem Verbindungstyp aus der Tabelle entsprechen. So ergeben sich konsistente Markierungen der im ersten Abschnitt erhaltenen zweidimensionalen Struktur, die erste Hinweise auf die räumliche Interpretierbarkeit der abgebildeten Objekte ergeben. Der Begriff „Hinweise" soll zum Ausdruck bringen, daß es nicht immer möglich ist, aus einer zweidimensionalen, konsistent markierten Struktur die Beschreibung für einen räumlichen Körper abzuleiten. Die anomalen Bilder von Draper /2/ und Huffman /3/ (Abb. 3) sind Beispiele für korrekt markierbare Eingangsstrukturen, die aber im dritten Teil des Verfahrens zu Widersprüchen führen. Es ist jedoch unwahrscheinlich, daß das aus einer realen Aufnahme extrahierte Konturbild und damit die erzeugte Eingangsstruktur eine solche anomale Struktur aufweist.

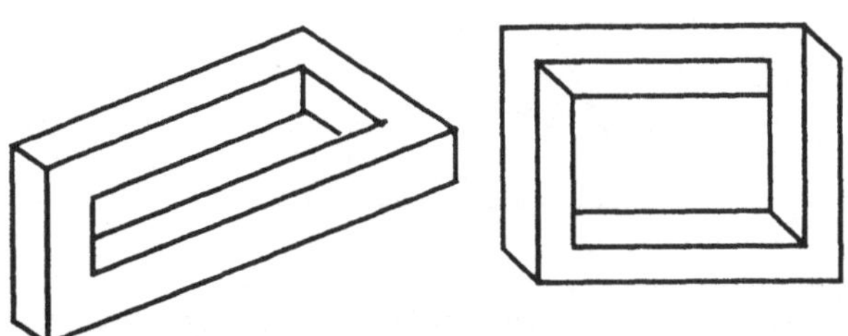

Abb. 3:   Anomale, korrekt markierbare Linienzeichnungen nach Draper /2/ und Huffman /3/.

Die Existenz einer konsistenten Markierung für die Struktur ist Voraussetzung dafür, daß die nachfolgende Interpretation durchgeführt werden kann.

4. Interpretation einer markierten Eingangsstruktur

Im dritten Abschnitt des Verfahrens wird der Versuch unternommen, die markierte Eingangsstruktur räumlich zu interpretieren. Basis dieses Vorganges ist die Definition einer räumlichen Struktur durch die im ersten Abschnitt erzeugte Eingangsstruktur mit einer der konsistenten Markierungen aus dem zweiten Verfahrensabschnitt. Zur Definition gehören:

- Festlegung der Knotenpunkte durch deren Bildkoordinaten.

- Definition von Flächen im Raum durch einen Parametersatz mit Hilfe von geschlossenen Linienzügen im Bild.

- Definition von Vertex*-Flächenpaaren.

- Erzeugung von Tiefenaussagen und zusätzlichen Vertizes* auf der Grundlage der Linienmarkierungen.

* Sugihara bezeichnet die zu einem Knotenpunkt gehörigen räumlichen Punkte als Vertizes

Die Methode von Sugihara erlaubt es nun, aus dieser definierten, räumlichen Struktur ein System von Ungleichungen und Gleichungen abzuleiten.

Ungleichungen werden aus den vom zweiten Verfahrensabschnitt erzeugten Markierungen der Linienelemente abgeleitet. Eine markierte Eingangsstruktur, wie sie bei den in Abb. 3 dargestellten Linienzeichnungen möglich ist, ergibt widersprüchliche Tiefenaussagen und dadurch sich gegenseitig ausschließende Bedingungen in den Ungleichungen.

Das schon erwähnte System von Gleichungen kann man aus den Aussagen der räumlichen Struktur - welche Vertizes auf welchen Flächen liegen - ableiten. Dabei kann es auch vorkommen, daß redundante Gleichungen erzeugt werden. Die Lösbarkeit wird beeinflußt, wenn einzelne Knotenpunkte im Bild auf positionsabhängige Bedingungen zwischen solchen redundanten Gleichungen führen. Abb. 4 zeigt dies für den Fall eines Pyramiden-

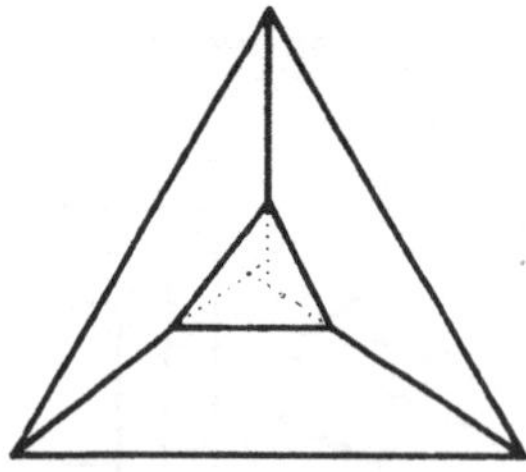

Abb. 4: Die abgeschnittene Pyramide stellt nur dann eine korrekte Abbildung eines Polyeders dar, wenn sich die drei Seitenlinien in der Mitte in einem Punkt treffen.

stumpfes. Hier ist einer der drei Knotenpunkte, die sich auf der oberen Schnittfläche befinden, positionsabhängig. Die Abbildung der Pyramide ist nur dann korrekt, wenn dieser Punkt so positioniert ist, daß jede der unterbrochen eingezeichneten Linienverlängerungen durch den Schnittpunkt der anderen beiden Linienverlängerungen geht.

Sugihara konnte zeigen, daß es mittels eines Netzwerkes möglich ist, aus der Konstellation von Flächen und den auf ihnen liegenden Vertizes redundante Aussagen ausfindig zu machen und die durch sie erzeugten Gleichungen aus dem aufgestellten Gleichungssystem zu eliminieren. Dadurch kann man die exakte Position der Knotenpunkte im Bild berechnen und die Gleichungen der entsprechenden Vertex-Flächenpaare mit der geforderten Redundanz in das Gleichungssystem wieder aufnehmen. Auf diese Weise ist gewährleistet, daß auch Positionierfehler durch die Digitalisierung des Videobildes oder durch darauf beruhende Verarbeitungsfehler die Lösbarkeit des Ungleichungs- und Gleichungssystems nicht beeinflussen können. Daher kann so ein Polyeder mit korrekt erzeugter Eingangsstruktur in jedem Fall erkannt werden.

5. Zwischenergebnisse und Ausblick

Der beschriebene Algorithmus wurde mit Videoaufnahmen von Körpern getestet, die den Verfahrensvoraussetzungen weitestgehend entsprachen. Die Abb. 5a-5h zeigen das schrittweise Vorgehen des Verfahrens. In Abb. 5a ist das Eingangsvideobild zu sehen. Es handelt sich um einen Würfel, dessen vordere Ecke abgeschnitten ist. In Abb. 5b sind die vom Gradientenverfahren gelieferten Betragsmaxima dargestellt. Die in die Datenlisten übernommenen, nach der Betragsgröße sortierten ersten zehn Konturen sind in Abb. 5c zu sehen. Abb. 5d zeigt die in gerade Linienstücke unterteilten Konturen, die dann so verschmolzen werden, daß eine korrekte Eingangsstruktur (Abb. 5e) entsteht. Das Beispiel des Würfels wurde gewählt, weil durch die abgeschnittene Ecke in der Eingangsstruktur genau wie bei der Pyramide in Abb. 4 ein positionsabhängiger Knotenpunkt vorhanden ist und dadurch ein Durchlaufen des Korrekturmoduls eingeleitet wird. Das Ergebnis des Netzwerkes zur Suche von positionsabhängigen Knotenpunkten kennzeichnet dann den gefundenen Punkt, in dem es einen Rahmen in das Bild um den Punkt einblendet (Abb. 5f). Es muß hier erwähnt werden, daß es auch möglich wäre, einen der beiden anderen an dem kleinen, inneren Dreieck sich befindenden Knotenpunkte als positionsabhängig zu bezeichnen. In Abb. 5g ist die Korrektur der Struktur zu sehen, die sich dadurch ergibt, daß man die Flächen, wie sie die Lösung des Verfahrens liefert, räumlich schneidet, und den berechneten Vertex als Knotenpunkt wieder ins Bild einblendet.

Wie in Abschnitt 1 bereits erwähnt, wurde das Verfahren erweitert /5/, in dem zusätzliche Informationen, die man mit Hilfe eines Referenzbildes (Bild einer Kugel mit gleichen Oberflächenreflexionseigenschaften) aus dem Grauwertverlauf der Objektflächen extrahiert, zur Erzeugung einer Szenenbeschreibung mit auswertet. In Abb. 5h ist das Ergebnis dieses Verfahrens zu sehen. Es liefert an sogenannten Y-förmigen Knotenpunkten die Normalen-

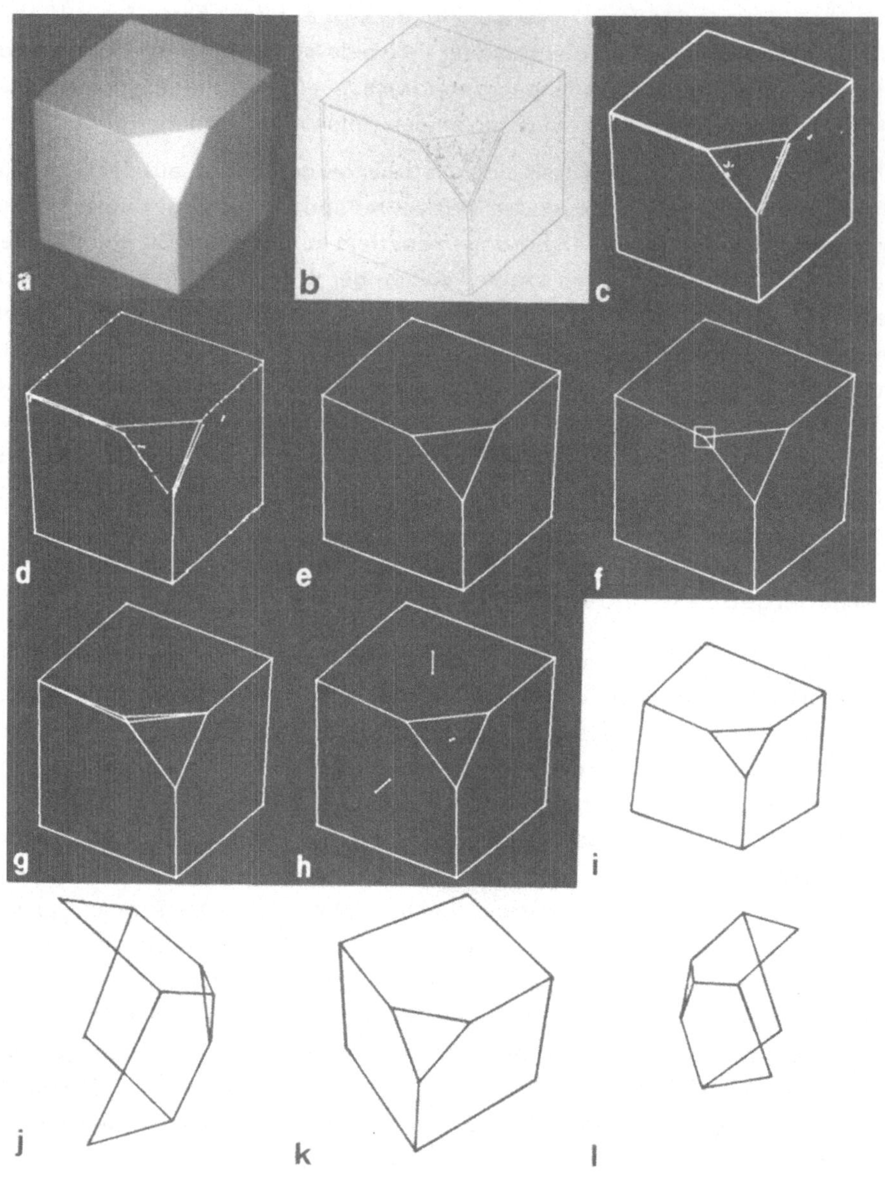

Abb. 5a-5l: Zwischenergebnisse wie sie beim Durchlaufen des Verfahrens anfallen.
(Erläuterungen sind in Abschnitt 5 des Textes zu finden)

vektoren der angrenzenden Flächen, die diese Ecke räumlich erzeugen. Diese zusätzlichen Informationen schränken die noch verbleibenden Freiheitsgrade weiter ein. In den Abb. 5i-5l wurde versucht, das sich mit diesen zusätzlichen Informationen ergebende Ergebnis der Interpretation zu illustrieren. Die Abb. 5i zeigt die Projektion der sich aus den räumlichen Koordinaten der Lösungen ergebenden Figur auf die Bildebene. In Abb. 5j wurde die Figur um 90° gedreht. Die Figur in Abb. 5k entspricht einer Drehung um 180°, was einen Blick von hinten in das Drahtmodell gestattet. Abb. 5l ist die entsprechende Ansicht der um 270° gedrehten Figur.

Im weiteren Verlauf der Untersuchungen soll der Algorithmus für die Erkennung von Blechbiegeteilen optimiert werden.

Literatur

/1/ **Clowes, M. B. (1971);** „On seeing things", Artificial Intelligence 2 (1971) 79-116.

/2/ **Draper, S. W. (1978);** „The Penrose triangle and a family of related figures", Perception 7 (1978) 283-296.

/3/ **Huffman, D. A. (1978);** „Impossible objects as nonsense sentences", in Meltzer, B. and Michie, D.(eds.), Machine Intelligence 6, Edinburgh Univ. Press., Edinburgh/UK, pp. 295-323.

/4/ **Korn, A. (1988);** „Toward a symbolic representation of intensity changes in images", IEEE Transactions on Pattern Analysis and Machine Intelligence, PAMI-10 (1988) 610-625.

/5/ **Rössle, S. (1989);** „Berechnung der Orientierung sichtbarer Oberflächen einer Polyederszene anhand des Grauwertverlaufs einer Videoaufnahme", Studienarbeit an der Universität Karlsruhe, Fakultät für Informatik, Institut für Algorithmen und Kognitive Systeme.

/6/ **Sugihara, K. (1986);** „Machine interpretation of line drawings", MIT Press, Cambridge/Mass.

/7/ **Waltz, D. (1975);** „Understanding line drawings of scenes with shadows", in Winston, P. H.(ed.), The Psychology of Computer Vision, McGraw-Hill New York, pp. 19-91.

EIN ATTRIBUTIERTES RELAXATIONSVERFAHREN
ZUR 3D-LAGEERKENNUNG VON OBJEKTEN

M. Heuser, C.-E.Liedtke,
Institut für Theoretische Nachrichtentechnik
und Informationsverarbeitung
Universität Hannover
Appelstr. 9a, 3000 Hannover 1

Zusammenfassung

Es wird ein recheneffizientes und robustes Verfahren zur Lagebestimmung dreidimensionaler Objekte aus einer monokularen Ansicht vorgestellt. Das Verfahren basiert auf einer attributierten Relaxation. Es ist in der Lage auch bei komplexen Szenen mit mehreren sich überlappenden Objekten gute Resultate zu liefern.

1. Einleitung

Eine der Schlüsselaufgaben im Bereich des industriellen Einsatzes von Bilddeutungssystemen ist die Lageerkennung von Objekten. Sie ist Voraussetzung für die Planung von Greif- und Bewegungsvorgängen sowie die räumliche Orientierung bei der Beurteilung oder Überprüfung von Objekteigenschaften, beispielsweise bei Qualitätsüberwachungs- oder Vermessungsaufgaben. Die Anforderungen an die Zuverlässigkeit der Lageerkennung in industriellen Anwendungen sind hoch. Insbesondere müssen Qualitätsverluste des Kamerabildes bedingt durch Beleuchtungseffekte, Kamerarauschen, leichte Defokussierung, teilweise Verdeckung durch andere Objekte usw. toleriert werden können. Darüberhinaus muß von Verfahren, die in der Praxis Einsatz finden sollen, eine hohe Recheneffizienz gefordert werden. Um diesen Anforderungen weitgehend gerecht zu werden wurde ein robustes Lageerkennungsverfahren entwickelt, das auf der Analyse einer monokularen Objektansicht basiert und von der Voraussetzung ausgeht, daß ein rechnerinternes Modell des in seiner unbekannten Lage zu erkennenden Objektes verfügbar ist. Das Prinzip des Erkennungsprozesses beruht darauf, daß die rechnerinterne Modellbildung mit dem Bild im Hinblick auf für die Lageerkennung relevante Eigenschaften des Objektes, wie die Relativlagen von Ecken, Bohrungen, Achsen, Kanten usw. zueinander verglichen wird. Derartige Ansätze sind schon an anderer Stelle in der Literatur dargestellt worden (/1,2/). Im Hinblick auf eine besondere Berücksichtigung der Robustheit in Verbindung mit einer hohen Recheneffizienz hat sich ein attributiertes Relaxationsverfahren als sehr vorteilhaft herausgestellt, das im folgenden dargestellt werden soll.

2. Bild- und Prototypenbeschreibung

Das Prinzip einer modellgestützten Bildanalyse kann anhand von Abb.1 dargestellt werden. Aus dem zu analysierenden ikonischen Bild **G** der industriellen Szene wird unter Verwendung eines Zwischenresultates in Form eines segmentierten Bildes eine symbolische Beschreibung gewonnen. Die symbolische Beschreibung besteht aus der Menge **S** der Bildbereichshinweise, die eine relevante Beschreibung des zweidimensionalen Bildes darstellen. Aus der symbolischen Beschreibung des zweidimensionalen Bildes können Rückschlüsse auf die symbolische Beschreibung der dreidimensionalen Szene gezogen werden, aus der das Bild durch eine projektive Abbildung ursprünglich gewonnen wurde. Die 3D-Szene wird durch die Menge der dreidimensionalen Szenenbereichshinweise **U** beschrieben.

Für das Objekt, dessen Lage aus dem Bild bestimmt werden soll, liegt die Modellbeschreibung eines Prototypen vor. Das Modell beinhaltet sowohl eine symbolische Beschreibung im dreidimensionalen Raum, die als Szenenskizze **B** bezeichnet wird, als auch eine daraus hergeleitete zweidimensionale Abbildung in Form der sog. Bildskizze **A**. Falls erforderlich, kann darüberhinaus die bildliche Ausprägung des synthetischen segmentierten Bildes bzw. die eines synthetischen Bildes gewonnen werden.

Elemente der symbolischen Beschreibung **U** bzw. **B**, die bei industriellen Objekten zur Lagebeschreibung besonders geeignet erscheinen, sind charakteristische Kanten des Objektes, wie kreisförmige Bohrungs- oder Achsenkanten oder orthogonale Mehrfachecken. Die Elemente von **S** und **A** sind mögliche 2D-Ausprägungen davon, also Mehrschenkelecken und Ellipsen. Die hier im wesentlichen verwendeten 2D- und 3D-Beschreibungselemente und deren Zusammenhang ist in Abb.2 dargestellt. Die automatische Extraktion der Bildbereichshinweise, die einen wesentlichen Bestandteil des Lageerkennungsverfahrens darstellt, wurde an anderer Stelle (/3/) im Detail beschrieben.

Die Erkennung eines 3D-Objektes und dessen Lage aus einer monokularen Ansicht beruht darauf, daß das reale Bild und die daraus hergeleiteten Elemente der bildlichen und symbolischen Beschreibung mit den für verschiedene Lagehypothesen berechneten entsprechenden Elementen des Prototypen verglichen werden. Zu besonders effizienten Verfahren kommt man, wenn man sich auf den Vergleich der symbolischen Beschreibungen beschränken kann.

3. Lagehypothesen

Die Zuordnung zwischen einem Element des Prototypen (**B** oder **A**) und einem Element des realen Bildes (**U** oder **S**) wird als Hypothese bezeichnet. Unterschieden wird zwischen Primärhypothesen und Sekundärhypothesen. Die Primärhypothese stellt die Zuweisung eines Szenenbereichshinweises U_l zu einem Element der Szenenskizze B_j dar. Sie ist eine Zuweisung im 3D-Raum, während die Sekundärhypothesen auf der Basis der 2D-Ausprägungen aufgestellt werden und die Zuweisung eines Elementes der Bildskizze A_k zu einem Bildbereichshinweis S_h beinhalten.

Nach Aufstellen einer Primärhypothese werden mehrere Sekundärhypothesen herangezogen, um die Verträglichkeit der mit der Primärhypothese getroffenen Zuordnung eines Szenenbereichshinweises zu einem Element der Szenenskizze zu bestimmen. Die Wichtung der Primärhypothese B–U mit Hilfe einer Sekundärzuordnung A–S geschieht über ein Verträglichkeitsmaß $c_{BU}(A,S)$, das angibt, wie verträglich die paarweisen Zuordnungen miteinander sind. In die Berechnung dieses Verträglichkeitsmaßes geht das Wissen über die Struktur des Prototypen, insbesondere im Hinblick auf die räumliche Orientierung einzelner Elemente der Szenenskizze ein.

Bei der Aufstellung der Primärhypothesen wird zunächst eine orthogonale Projektion der Elemente der Szenenskizze in die Bildebene angenommen. Mit dieser Annahme kann eine Bestimmung der z-Komponente des Translationsvektors aus der Projektion nicht erfolgen. Statt dessen sei angenommen, daß die mit einer Variation des Objektabstandes von der Kamera auftretende Änderung der Projektionsgröße durch eine Variation eines linearen Größenfaktors σ des Objektes hervorgerufen würden. Als unbestimmte Parameter der Objektlage kann somit der 3-Tupel $x=(\mathbf{R}, \vec{t}, \sigma)$ mit der räumlichen Rotationsmatrix $\mathbf{R}$, dem Translationsvektor $\vec{t}$ und dem Größenfaktor der Abbildung σ gelten. Dieser 3-Tupel kann auf letztlich sechs voneinander unabhängige Unbekannte zurückgeführt werden.

Die Primärhypothese erlaubt es, *einen Teil* der Unbekannten der Objektlage zu spezifizieren. Die Anzahl m der verbleibenden unbestimmten Parameter der Objektlage ist vom Typ
der Elemente der Primärhypothese abhängig. Sie beträgt m=1 bei Dreischenkelecken, da
der Skalierungsfaktor als unbekannte Größe verbleibt, ebenfalls m=1 bei Ellipsen, da ein
Drehwinkel im Raum nicht bestimmt werden kann und m=2 bei Zweischenkelecken, bei
denen zwei Drehwinkel im Raum nicht eindeutig bestimmbar sind.

Das Lageerkennnungsverfahren beruht im Prinzip auf der Aufstellung und Bewertung der
Primärhypothesen. Die letztlich ermittelte Lage des Objektes ergibt sich aus der Primärhypothese, deren Bewertung den höchsten Wert annimmt. Bei der Bewertung der Hypothesen ist man bestrebt zunächst diejenigen zu verfolgen, bei denen die Anzahl der unbestimmten Parameter niedrig ist, also beispielsweise Hypothesen basierend auf Drei-
Schenkel-Ecken.

4. Verträglichkeitsmaße zur Hypothesenbewertung

Die Berechnungsvorschrift für die Bewertungskoeffizienten ist abhängig vom Typ der betrachteten Beschreibungselemente. Sie soll anhand von Abb.3 erläutert werden. Betrach-
tet sei hier nur der Fall, daß die Primärhypothese auf einer Drei-Schenkel-Ecke basiert.
Wenn die Sekundärhypothesen Mehrschenkelecken darstellen, müssen die vorhandenen
Bildbereichshinweise daraufhin untersucht werden, inwieweit sich S_k finden lassen, die
ausgehend von einer Projektion von U_l wie das Element A_h der Bildskizze unter dem
Winkel Θ erscheinen (siehe Abb.3a). Die in der Realität auftretende Abweichung des
Winkels unter dem ein S_k tatsächlich erscheint, wird als Sichtwinkelabweichung $d\Theta$ be-
zeichnet. Neben dieser Sichtwinkelabweichung kann im Fall von Mehrschenkelecken als
Sekundärhypothese weiterhin überprüft werden, ob die vermutete Drehlage von A_h mit
der Drehlage von S_k übereinstimmt. Im Einzelnen können dazu die Drehlagen aller Schen-
kel der Ecke betrachtet werden (Abb.3b). Für jeden Schenkel l ergibt sich dabei eine
Drehwinkelabweichung $d\Phi_l$.

Ein räumlicher Kreis in der Szenenskizze wird zu einer Ellipse, deren Mittelpunkt auf einem
Strahl zu liegen kommt, der unter dem Winkel Θ von der Position des projizierten Elementes der Szenenskizze B_l ausgeht (Abb.3c). Die Abweichung des Strahls durch den Mittelpunkt einer tatsächlich detektierten Ellipse S_k von dem Strahl durch das projizierte Element der Szenenskizze A_h wird mit dem Maß $d\Theta$ erfaßt. Die Drehlage einer Ellipse ergibt
sich aus der Lage der Haupt- und Nebenachse. Der Unterschied zwischen der Soll- und
Istlage, das heißt der Drehlage von A_h und S_k, wird mit dem Maß $d\Phi$ erfaßt (Abb.3d).
Darüberhinaus kann bei Ellipsen zur Bewertung der Abweichung auch die Größendifferenz herangezogen werden (siehe Abb.3e). Die Größendifferenz kann dabei als die Differenz der großen bzw. kleinen Hauptachsen von A_h und S_k berechnet werden. Insgesamt
wird das Verträglichkeitsmaß $c_{lj}(h,k)$ aus drei Komponenten gebildet, die multiplikativ entsprechend Abb.3f miteinander verknüpft werden.

5. Attributierte Relaxation

Als Verfahren zur Ermittlung der Primärhypothese mit der höchsten Bewertung wird ein
attributiertes Relaxationsverfahren eingesetzt. Das Prinzip der Relaxation besteht dabei in
der iterativen Berechnung einer Matrix von numerischen Maßen $p_i(j)$, die angeben, wie
wahrscheinlich es ist, daß der Szenenbereichshinweis U_i dem Element B_j der Szenenskizze des Prototypen entspricht. Die Relaxationsvorschrift ist in Abb.4 dargestellt. Das Verfahren der Relaxation weist aufgrund des iterativen Prozesses einen erheblichen Rechen

zeitbedarf auf. Wenn *ein* unbekannter Lageparameter der Primärhypothesen vorliegt, kann man ohne Iteration auskommen, wenn man die Verträglichkeit der Sekundärhypothesen bzgl. der mit einem Attribut bewerteten Primärhypothese betrachtet, wobei als Attribut hier der o.g. Skalierungsfaktor s gewählt wird. Abb.5 zeigt den Einfluß des Skalierungsfaktors bei während des Analyseprozesses hypothetisch erzeugten Szenenskizzen. Akzeptiert wird *der* Skalierungsfaktor, für den am häufigsten Unterstützungen bei der Betrachtung verschiedener Primär- und Sekundärhypothesenpaare gefunden wird.

Mit Hilfe des Skalierungsfaktors können die sechs räumlichen Lageparameter für die wahrscheinlichste Lage des Objektes im 3D-Raum ermittelt werden. Über die aus dem Prototypenmodell bekannte Kanteninformation kann eine Verfikation der ermittelten Lage durchgeführt werden.

Abb.6 zeigt die einzelnen Zwischenresultate bei der Verarbeitung ausgehend vom ikonischen Bild (Abb.6a) über das Segmentierte Bild (Abb.6b), zu den extrahierten Bildbereichshinweisen(Abb.6c). In Abb.6d bzw. Abb. 6e wurde die Szenenskizze erzeugt, die der mit dem o.g. Verfahren berechneten Lage entspricht und dem ikonischen Bild überlagert. Abb.6f demonstriert die Verifikation über die Kanteninformation.

/1/ *T.M.Silberberg, L.Davis, D.Harwood, "An iterative Hough procedure for three-dimensional Object recognition", Pattern recognition, Vol. 17, No. 6, 621-629 (1984)*

/2/ *R.Horaud, "New Methods for Matching 3D Objects with single perspective Views", IEEE PAMI, Silver Springs, 401-412 (1987)*

/3/ *Heuser, M.; Liedtke, C.-E.: Detection of the Position of 3D Industrial Objects under Consideration of Reduced Image Quality, The 6th Scandinavian Conference on Image Analysis, University of Oulu, Finnland, Juni 1989.*

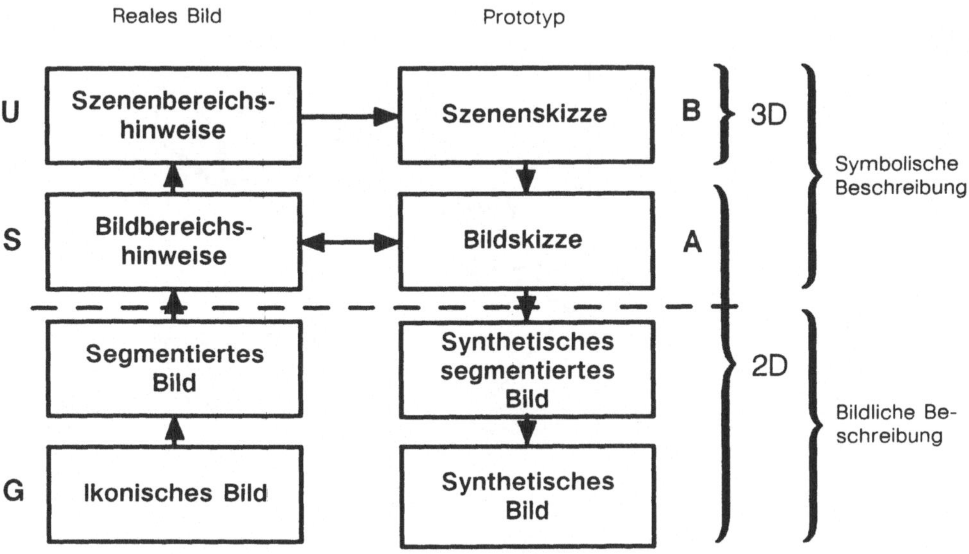

Abb. 1 : Formen der Bild- und Prototypenbeschreibung

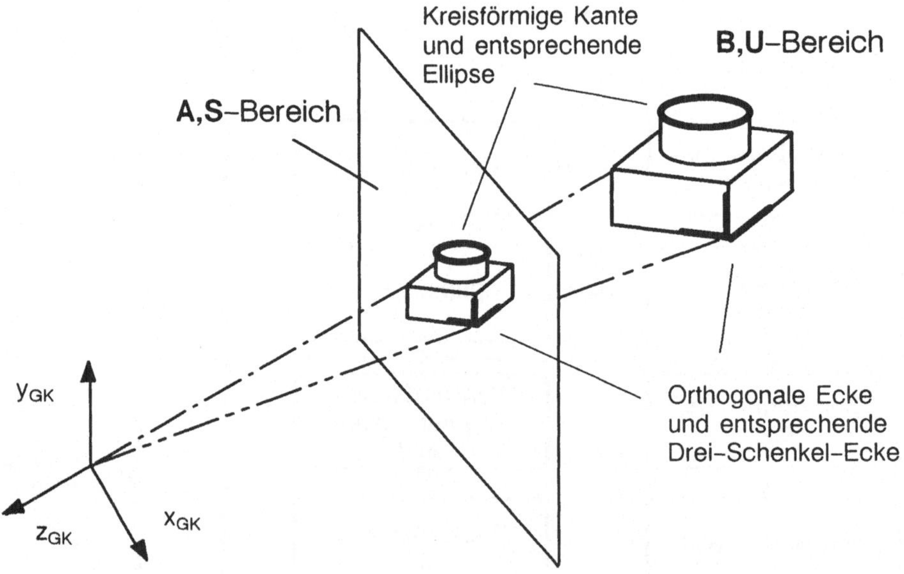

Abb. 2 :   a) Bildbereichs-/Szenenbereichshinweise
          b) Relationen zwischen korrespondierenden Elementen

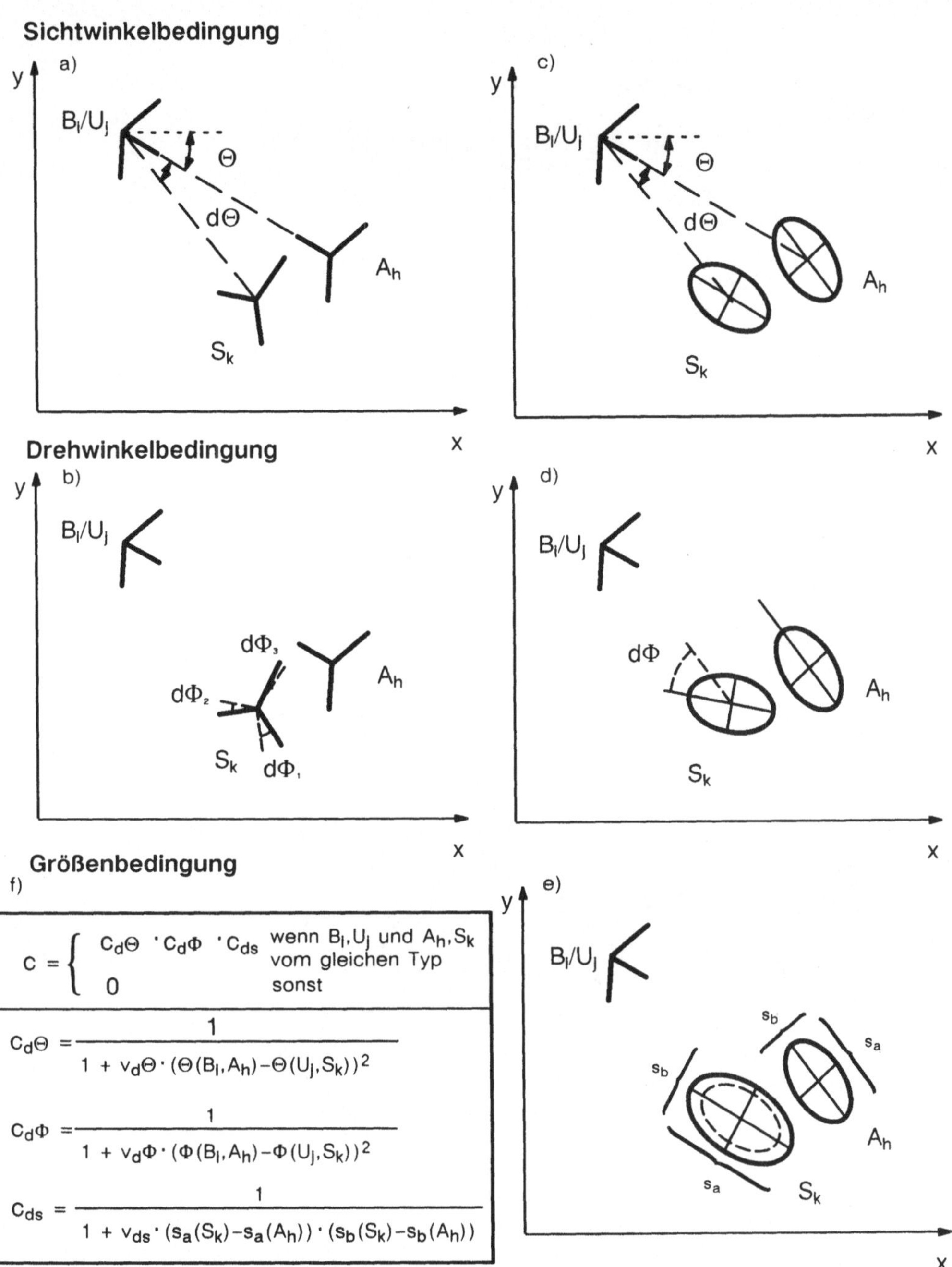

f)

$$C = \begin{cases} C_{d\Theta} \cdot C_{d\Phi} \cdot C_{ds} & \text{wenn } B_l, U_j \text{ und } A_h, S_k \\ & \text{vom gleichen Typ} \\ 0 & \text{sonst} \end{cases}$$

$$C_{d\Theta} = \frac{1}{1 + v_{d\Theta} \cdot (\Theta(B_l, A_h) - \Theta(U_j, S_k))^2}$$

$$C_{d\Phi} = \frac{1}{1 + v_{d\Phi} \cdot (\Phi(B_l, A_h) - \Phi(U_j, S_k))^2}$$

$$C_{ds} = \frac{1}{1 + v_{ds} \cdot (s_a(S_k) - s_a(A_h)) \cdot (s_b(S_k) - s_b(A_h))}$$

Abb. 3: Verträglichkeitsmaße

For r **from** 1 **to** *Zahl der Iterationen*
  For i **from** 0 **to** $|B|-1$
    For j **from** 0 **to** $|U|-1$

$$q_i^r(j) = \sum \underset{h^\circ \neq i \; k^\circ \neq j}{\text{MAX}} [\; c_{ij}(h,k) \cdot p_{h^\circ}^{r-1}(k^\circ)\;]$$

    End
  End

  For i **from** 0 **to** $|B|-1$
    For j **from** 0 **to** $|U|-1$

$$p_i^r(j) = \frac{q_i^r(j)}{\underset{i^*,j^*}{\text{MAX}}\; q_i^*(j^*)}$$

    End
  End
End

Abb. 4 : Algorithmus der Relaxation

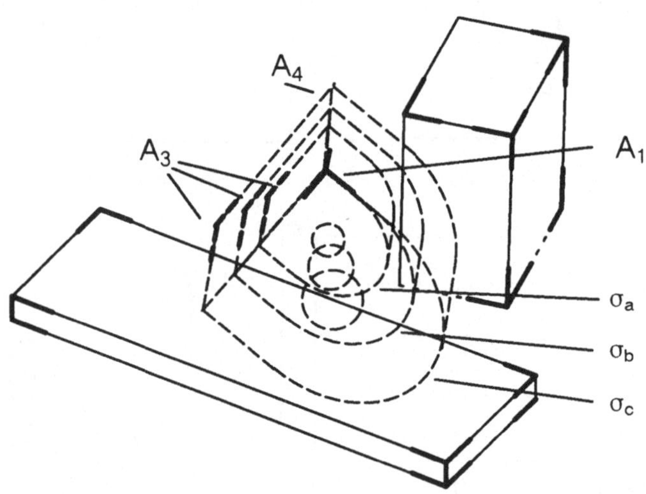

Abb. 5 : Hypothesen mit unterschiedlichen Größenfaktoren

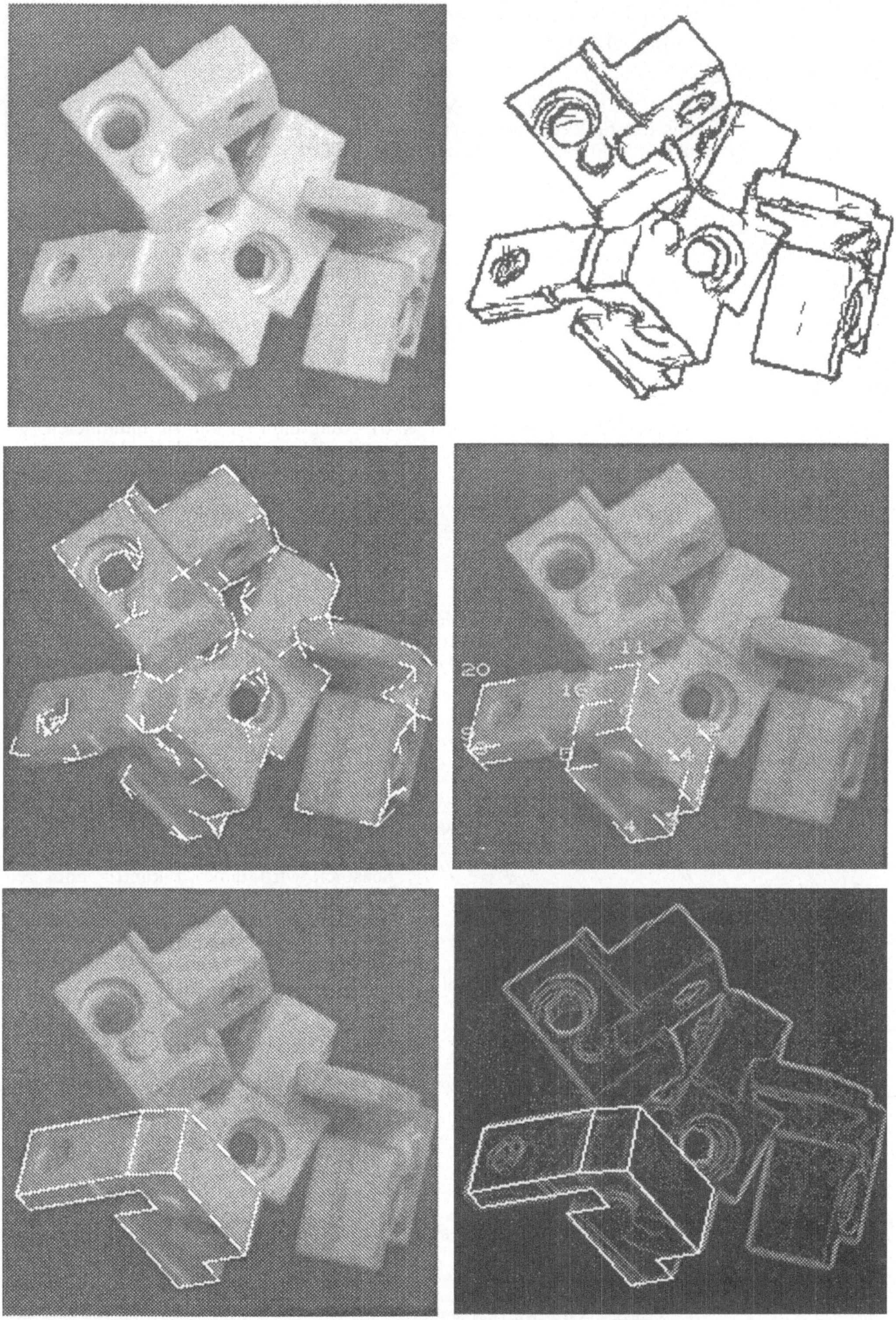

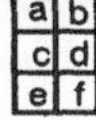

Abb. 6 : Ergebnisse der Lagebestimmung

Adaptive Light Encoding for 3–D–Sensing
with Maximum Measurement Efficiency

R. Malz
Institut für Technische Optik, Prof. Dr. H. J. Tiziani
Universität Stuttgart, Pfaffenwaldring 9, 7000 Stuttgart 80

Kurzfassung

Für das 3–D–Maschinensehen mit einer Triangulationskamera wird eine neue Klasse hybrider Codes zur flächenhaften Objektmarkierung eingeführt. Diese Codes vereinigen die Genauigkeit analoger Markierungsfunktionen mit der Robustheit digitaler Codes. Maximale Meßgeschwindigkeit wird erzielt, indem die Codes die verfügbare Qualität des Videosignals ausnutzen.
Eine aktive 3–D–Kamera, bestehend aus Laser–Codeprojektor, CCD–Sensor und Codierungssystem kann sowohl für Messungen im Submillimeter–Bereich, als auch für die Objekterkennung im Meterbereich eingesetzt werden. Durch automatische Analyse des Nachrichtenkanals "Projektor–Objekt–Kamera–Bildspeicher" und variable Codegenerierung kann die jeweils verfügbare Kanalkapazität ausgenutzt werden, um die Übertragungszeit der Markierungsinformation zu minimieren. Die Projektion und Bildsequenzaufnahme für eine topographische Karte mit absoluten Distanzwerten erfordert zwischen 160 ms bis 320 ms.

Keywords: coding; three–dimensional (3–D) imaging; topographic mapping; triangulation; photogrammetry; stereo vision; laser ranging; machine vision; artificial intelligence.

1 Einleitung

Optische 3–D–Sensoren werden zunehmend auch für zeitkritische Anwendungen in der industriellen Produktion benötigt. Hier sind es vor allem geometrische Triangulationsverfahren in unterschiedlicher technischer Realisierung, die in der Fahrzeugnavigation, Robotik und Qualitätskontrolle Anwendung finden /1–9/.

Die Verarbeitungsleistungen von Bildverarbeitungsrechnern konnten seit der Einführung der Mikrocomputer um viele Größenordnungen gesteigert werden, während die Videofrequenzbandbreite aus physikalischen Gründen nahezu unverändert geblieben ist (10 MByte/s). Die Aufnahmezeit der erforderlichen Videodaten bei 3–D–Sensoren wird daher zukünftig zum zeitbestimmenden Faktor, der die Verweildauer der Objekte bestimmt oder die Reaktionszeit der Roboter begrenzt.

Diese Akquisitionszeit kann gegenüber herkömmlichen Verfahren wesentlich reduziert werden, wenn mit adaptiven Codes die durch Reflexionsschwankungen des Meßobjektes stark beeinträchtigte Informationsübertragungskapazität des Videokanals mit möglichst niedriger Redundanz genutzt wird.

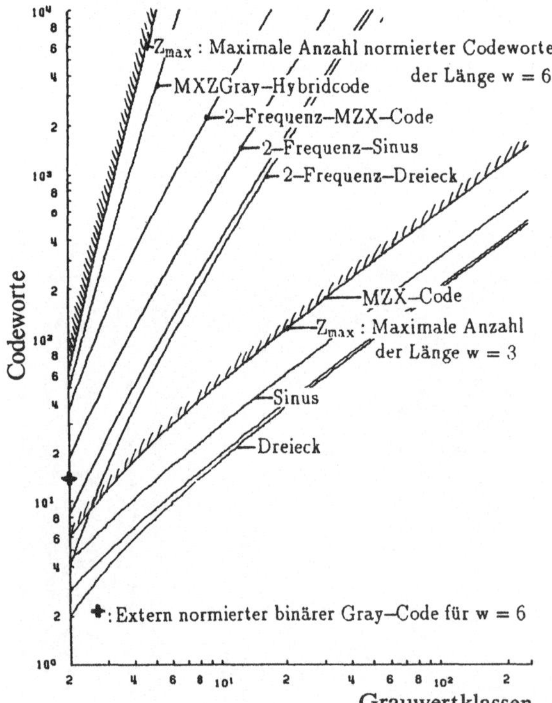

Abb. 1: Vergleich der Codierungseffizienz verschiedener Markierungsfunktionen. Nach oben ist die Zahl der Codeworte bzw. der unterscheidbaren Distanzwerte aufgetragen, nach rechts die für die Codierung frei verfügbare Dynamik (Kamera– minus Objektdynamik), gemessen in unterscheidbaren Grauwertklassen. Der MZXGray–Code erreicht nahezu die theoretische Grenze.

2 Markierungsfunktionen für triangulationsbasierte 3–D–Sensoren

Triangulationsbasierte Verfahren werden bevorzugt aktiv eingesetzt, d.h. die Objekte werden optisch markiert. Die Geometrie entspricht dabei der beim passiven Stereosehen, wobei ein Auge bzw. eine Kamera durch einen Projektor ersetzt wird. Durch die Markierung kann die suchende Korrespondenzanalyse entfallen, da jeder Objektpunkt eine zeitlich codierte geometrische Information aussendet. Als Markierungen werden bisher eingesetzt:

- singuläre Markierungen (Punkte bei der Lasertriangulation, Linien beim Lichtschnitt),
- analoge periodische Markierungen (Moiré, Sinus–, Dreieck–, Rechteck–Streifen, Farbspektren),
- digitale Streifencodierungen mit binären Gray–Codes.

Analoge Markierungsfunktionen

Analoge Verfahren verwenden meist periodische Beleuchtungsstrukturen mit sinus– oder dreieckförmiger örtlicher Intensitätsmodulation. Durch drei– oder mehrfache Projektion der Streifenstrukturen mit unterschiedlichen Phasenverschiebungen können additive und multiplikative Störungen eliminiert werden /3,9/.

Die Markierungsgröße, d.i. die relative Position oder Phase, ist nur innerhalb einer Periode eindeutig. Mehrfachmessungen mit unterschiedlichen Periodenlängen können die Unbestimmtheit auflösen. Bei geringem Signal–Rausch–Abstand wird die Zahl der erforderlichen Bilder relativ groß, da mehrere Phasenmessungen mit geringer Relativauflösung verknüpft werden müssen.

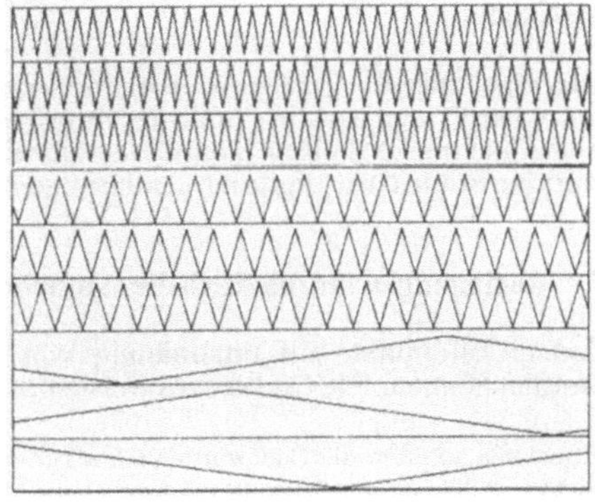

Abb. 2: Analoges Markierungsverfahren mit Dreiecksfunktionen und drei verschiedenen Periodenlängen.

Digitale Binärcodes

Codesequenzen, die bei der Überlappung benachbarter Codeworte keine Fehlzuweisung verursachen und damit für eine optische Längencodierung geeignet sind, wurden erstmals von Gray vorgeschlagen /Gray, F., US Patent 1953/. Graycodes werden seit langem in der elektrischen oder optischen Längen– und Winkelmeßtechnik verwendet. Es liegt daher nahe, sie auch für die triangulationsbasierte 3–D–Meßtechnik einzusetzen /6,7,8/.

Abb. 3: Binärer Graycode mit $Z = 2^9 = 512$ Worten. Nach rechts ist der Ort bzw. die Nummer des Codewortes aufgetragen, nach unten die Zeit bzw. die Stellen des Codewortes ($c_0...c_{11}$). Die Projektion von $c_0 = 0$ und $c_1 = 1$ liefert die lokale Entscheidungsschwelle durch einfache Mittelung.
In dieser Form ist der Code redundant. Die Pfeile deuten auf die beiden einzigen aus $c_2...c_{11}$ gebildeten Codeworte, die keine interne Normierung enthalten (000000000 und 111111111).
Alle übrigen $Z = 2^w - 2$ Codeworte tragen bereits die Information über die lokale Entscheidungsschwelle.

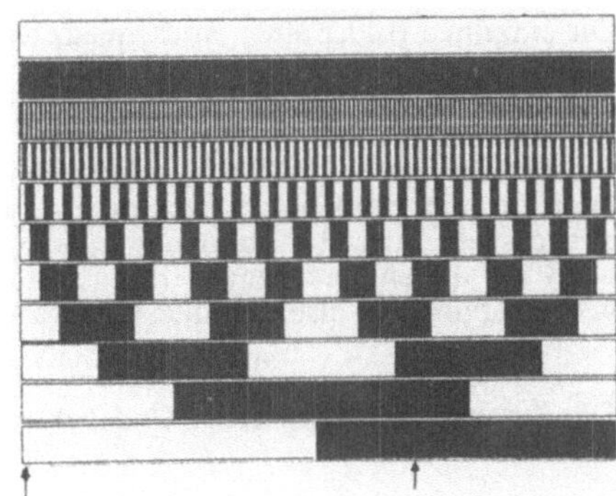

Vergleicht man analoge und digitale Markierungen, so ergeben sich interessante Unterschiede:

Analoge Markierungsfunktionen haben die prinzipielle Eigenschaft, daß sich ihre Meßauflösung mit dem Signal–Rauschabstand verändert. Ihre begrenzten Ortsfrequenzspektren wirken sich günstig im Hinblick auf die Abtastung durch den ortsdiskreten Sensor aus. Periodische Grundfunktionen mit mehreren Periodenlängen nutzen aber die kombinatorische Vielfalt nicht aus, die sich durch eine vollständige Permutation aller zeitlich nacheinander gesendeten Intensitätswerte ergeben könnte.

Digitale Binärcodes sind am wenigsten störanfällig, erfordern aber die größten Wortlängen w zur Darstellung einer bestimmten Information bzw. zur Erzeugung einer Anzahl Z von Worten. Sie sind daher nur bei niedrigster Dynamik zweckmäßig. Höherer Signal–Rausch–Abstand kann für die Auflösung nicht genutzt werden; lediglich die Erkennungssicherheit geht asymptotisch gegen 100%. Die relative Ortsauflösung ist durch die gewählte Zahl der Codeworte festgelegt. Hohe Ortsauflösungen sind auch mit größeren Codewortlängen wegen der breitbandigen Ortsfrequenzspektren nicht erreichbar.

Im folgenden soll gezeigt werden, wie sich dieses Dilemma auflösen läßt. Es werden hybride Codes entwickelt, die sowohl analoge, interpolierbare Funktionen als auch Digitalcodes verschiedener Wertigkeiten enthalten. Damit lassen sich die Vorteile der genauen analog–optischen Meßtechnik mit der Robustheit digitaler Codierungen verbinden.

3 Bedingungen für die Realisierung hybrider und mehrwertiger Codes

Jedem Bildpunkt soll unabhängig von seiner örtlichen Umgebung ein Distanzwert zugeordnet werden können. Die Codeworte werden daher als zeitliche Sequenz von Intensitätswerten projiziert.

Prinzipiell könnten die Codeworte auch als räumlich parallele Intensitätsmuster projiziert werden. Dies würde aber nicht nur die laterale Ortsauflösung des Distanzbildes in einer Richtung um den Faktor w reduzieren, sondern auch einschränkende Anforderungen an die Homogenität und Stetigkeit der Objektoberfläche stellen. Auch die Verwendung zusätzlicher informationstragender Größen, wie z.B. Polarisation (2 Teilbilder) oder Farbe (3 Teilbilder) kann in Betracht kommen; für die Codierung ergeben sich dabei keine grundsätzlich neuen Aspekte.

Codes konstanter Wortlänge

Betrachtet werden hier ausschließlich Codes konstanter Wortlänge. Sie sind nicht nur aus technischen Gründen zweckmäßig, sondern auch aus informationstheoretischer Sicht optimal, wenn man unterstellt, daß es sich bei Distanzmeßwerten um gleichwahrscheinliche Ereignisse handelt /10/. Ein Codewort oder –vektor $\underline{c}$ ist dann eine geordnete Liste mit w Buchstaben, die einem Alphabet A mit n Elementen entnommen sind:

$$\underline{c} = (c_0, c_1, c_2, ..., c_{w-1}) \qquad \text{mit } c_i \in A, \ A := \{a_0, a_1, a_2, ..., a_{n-1}\}.$$

Die einzelnen Bilder einer Bildsequenz entsprechen den w Stellen eines Codewortes, die Wertigkeit n des Codes entspricht der Zahl aktuell verfügbarer Grauwertklassen (die bei einer bestimmten Dynamik als noch unterscheidbar gelten können).

Bei der Projektion der Codes muß jede der gesendeten Intensitätsfolgen $\underline{l} = (l_0, l_1, l_2, ..., l_{w-1})$ eindeutig einem gültigen Codevektor $\underline{c} = (c_0, c_1, c_2, ..., c_{w-1})$ entsprechen. Im Modulator wird daher zunächst die bijektive Abbildung $a_j \rightarrow l_j$ der n möglichen Buchstaben aus dem Alphabet $A = \{a_0 ... a_{n-1}\}$ auf die entsprechenden Intensitätswerte $l_0 ... l_{n-1}$ hergestellt. Mit äquidistanten Intensitätswerten ergibt sich also eine quasilineare Modulatorkennlinie der Steilheit Δl:

$$l_j = l_{min} + \Delta l * a_j \quad \text{mit } a_j = j \text{ und } \Delta l = \frac{l_{max}-l_{min}}{n-1}.$$

Der Zusammenhang zwischen Codevektor $\underline{c}$ und Sendevektor $\underline{l}$ ist dann $\underline{l} = l_{min} + \Delta l * \underline{c}$.

Nachrichtenkanal

Die Übertragungstrecke vom Projektor über Objekt, Kamera, AD–Wandler bis zum Bildspeicher kann als Bündel von Nachrichtenkanälen aufgefaßt werden. Der digitale Bildspeicher ist das Empfängerarray. Jeder dieser Kanäle ist gekennzeichnet durch eine Reihe unbestimmter und nichtidealer Eigenschaften:

- Übertragungsfaktor   (Objektreflektanz, Vignettierung, CCD–Empfindlichkeit, Verstärker)
- Gleichanteil   (Fremdlicht, Schwarzwertklemmung des Videosignals, A/D–Wandler)
- Nichtlinearität   (CCD, Verstärker, A/D–Wandler, Numerische Begrenzung)
- Kanaltrennung   (Defokussierung, Ladungsdiffusion, Verstärkerbandbreite)
- Rauschen   (Quanten–$\sim$, Thermisches $\sim$, Digitalisierungs–$\sim$, Numerisches $\sim$)

Die theoretische Bestimmung der Übertragungskapazität solchermaßen gestörter Nachrichtenkanäle ist sehr aufwendig /10,11/. Für die folgenden Betrachtungen wird daher ein vereinfachtes Modell gewählt, das die wesentlichen Eigenschaften der verschiedenen Markierungen und Codierungen deutlich macht: sämtliche zeitabhängige Störungen werden in einer Größe, dem effektiven Quantisierungsrauschen, zusammengefaßt. Der Quotient aus der minimal noch akzeptierten Signaldynamik und dem Spitzenwert des Quantisierungsrauschen wird als Zahl n unterscheidbarer Grauwertklassen eingeführt.

Weiter wird davon ausgegangen, daß der nichtlineare Anteil der Intensitätsübertragungsfunktion innerhalb des zulässigen Signalbereiches gegenüber anderen Störungen vernachlässigt werden kann (Übertragungsexponent $\gamma = 1.0$), was durch sorgfältige technische Realisierung erreichbar ist.

Die registrierte Bildfolge $p = 0, ..., w-1$ liefert für jede Pixelkoordinate x,y eine Sequenz von Intensitätswerten, die den gesendeten Stellen eines bestimmten Codewortes entsprechen:

$$\mathbf{g}\,(x,y) = (g_0, g_1, g_2, ..., g_{w-1}).$$

Die Komponenten eines solchen Empfangsvektors lassen sich durch die Gleichungen

$$g_P = T_P\, l_P + f_P + r_P$$

beschreiben, für die folgendes gelten soll:

- Der Übertragungsfaktor $\underline{T} = (\,T_0, T_1, T_2, ..., T_{w-1})$ sei unbestimmt, aber während der Bildaufnahme konstant, d.h. $T_0 = T_1 = T_2 = ... = T_{w-1}$.
- Die Beleuchtungssequenz $\underline{l} = (l_0, l_1, l_2, ..., l_{w-1})$ entspreche dem Sendevektor (bzw. einem gewichteten Integral über benachbarte Sendevektoren) auf dem Oberflächenelement, daß dem Pixel zugeordnet ist.
- Der Gleichanteil $\underline{f} = (\,f_0, f_1, f_2, ..., f_{w-1}\,)$ sei unbestimmt, aber während der Bildaufnahme konstant, d.h. $f_0 = f_1 = f_2 = ... = f_{w-1}$.
- Die statistischen Eigenschaften der Störterme $\underline{r} = (\,r_0, r_1. r_2, ..., r_{w-1}\,)$ seien ebenfalls invariant gegenüber dem Parameter p der Bildfolge.

Anforderungen an Markierungscodes

- Die Wertigkeit, d.h. die Zahl der unterscheidbaren Grauwertklassen n der Codierung sollte in weiten Grenzen variabel sein, um eine Anpassung an unterschiedliche Signal–Rausch–Abstände (inkonstantes Reflexionsvermögen der Objektoberflächen) zu ermöglichen.
- Objekt– und systembedingte multiplikative Intensitätsstörungen müssen eliminiert werden können. Daher sind nur Codeworte zulässig, die mindestens einmal den Buchstaben l_{n-1} für maximale Sendeintensität enthält.
- Additive Intensitätsstörungen (Umgebungslicht) müssen eliminiert werden können. Daher sind nur Codeworte zulässig, die mindestens einmal den Buchstaben l_0 enthalten, der die Intensität 0 repräsentiert.
- Die Überlappung benachbarte Codeworte darf keine großen Fehler verursachen. Die räumliche Codesequenz muß gewährleisten, daß bei Überlappung ein Codewort entsteht, das entweder als fehlerhaft erkannt und zurückgewiesen werden kann, oder das einem der beiden beteiligten Codeworte entspricht.
- Die Länge Z der räumlichen Codesequenz, d.h. die Zahl der unterscheidbaren Codeworte, richtet sich nach der geforderten relativen Ortsauflösung. Eine periodische Codewiederholung kann in Betracht gezogen werden, wenn die durch Distanzsprünge hervorgerufenen Unstetigkeiten in der empfangenen Codesequenz kleiner als Z/2 sind.

4 Maximale Anzahl intensitätsnormierter Codeworte für n Grauwertklassen

Gegeben sei ein bestimmter Signal–Rausch–Abstand des Gesamtsystems. Für die Codeerzeugung wird eine dementsprechende Zahl n unterscheidbarer Grauwertklassen festgelegt. Aus dem Alphabet $A := \{ a_{n-1}, \ldots a_2, a_1, a_0 \}$ können zunächst $Z = n^w$ Codeworte gebildet werden. Alle $(n-1)^w$ Codeworte, die den Wert a_{n-1} nicht enthalten, müssen entfallen, d.h. $Z = n^w - (n-1)^w$. Weiter entfallen alle Codeworte, die den Wert a_0 nicht enthalten: $Z = n^w - 2(n-1)^w$. Alle $(n-2)^w$ Worte, die a_0 *und* a_{n-1} nicht enthalten und zweimal subtrahiert wurden, müssen wieder addiert werden. Damit ergibt sich für den allgemeinen Fall mit $n \geq 2$ und $w \geq 3$ eine maximale Anzahl von

$$Z_{max} = n^w - 2(n-1)^w + (n-2)^w$$

selbstnormierenden Codeworten. Alle weiteren Betrachtungen zur Effizienz von Codes beziehen sich auf diese obere Grenze (s.a. Abb.1). Mit dem Verbot nicht implizit normierter Codeworte tritt ein Problem auf: Stetigkeit und innere Symmetrie der Graycodes werden gestört (Abb. 4 oben). Es stellt sich die Frage, ob Codesequenzen konstruiert werden können, die alle verbleibenden Codeworte nutzen und dabei den obengenannten Anforderungen genügen. Ein allgemeines Prinzip zur Lösung dieses Problems ist nicht bekannt.

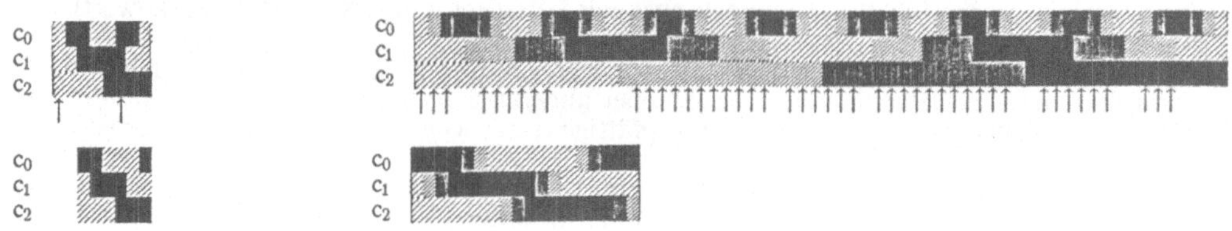

Abb. 4: Die markierten Worte der Graycodes erlauben keine implizite Intensitätsnormierung. Die Sequenzierung der Restmenge (hier für 3–stellige Codes) ist für höherstellige Codes ein Problem.

5 Binäre und höherwertige MZX–Codesequenzen mit w = 3 Stellen

Zunächst kann gezeigt werden, daß sich bei einer Wortlänge von w = 3 vollständige Codesequenzen aufbauen lassen (Abb. 4 unten und Abb. 5). Für solche Codes wird hier die Abkürzung **MZX**–Code eingeführt, da jedes Codewort einen Buchstaben für die volle Amplitude (**Maximum**) und einen für verschwindende Amplitude (**Zero**) enthält, während der dritte Buchstabe den (für größere n quasikontinuierlichen) aufsteigenden oder absteigenden Übergang (**X**) kennzeichnet. Der 3–stellige MZX–Code nutzt 100% der Codeworte, mit denen eine implizite Normierung möglich ist. Aus der o.g. Gleichung ergibt sich für w=3 :

$$Z = 6\,(n-1)\,.$$

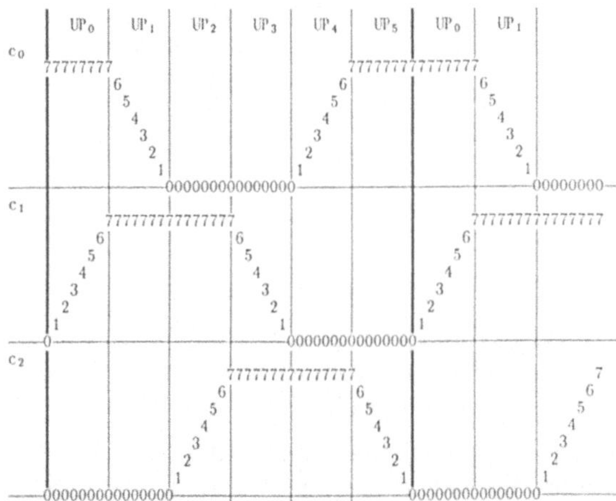

Abb. 5: Darstellung eines MZX–Codes der Wertigkeit n=8. Der quasianaloge Charakter ist deutlich erkennbar.

Der trapezförmige MZX–Code ist aufgrund seiner inneren Symmetrie einfach zu decodieren. Jede Codierungsperiode der Länge $Z = 6(n-1)$ ist in 6 gleichartige Bereiche der Länge n–1 unterteilt, die aus einer Permutation der Stellen c_0, c_1 und c_2 hervorgehen. Diese Eigenschaft ist unabhängig von der Zahl n der Graustufen und gilt auch für den analogen Fall. Durch den Grenzübergang $n \rightarrow \infty$ unter Beibehaltung der geometrischen Periodenlänge wird dieser Code zu einer analogen Markierungsfunktion mit der Möglichkeit zur Interpolation, bestmöglicher Dynamikausnutzung, niederfrequentem Ortsfrequenzspektrum, vernachlässigbaren Moiré–Effekten bei voller Nutzung der Kamera–Auflösung.

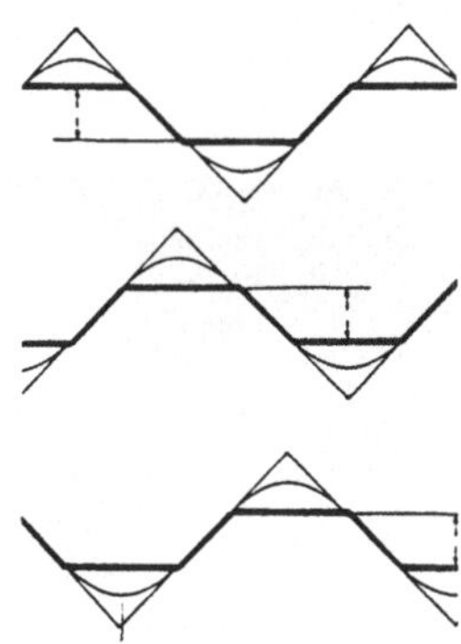 **Abb. 6:** Die MZX–Codesequenz im Vergleich zu häufig verwendeten analogen Markierungsfunktionen. Die Skizze zeigt die örtlichen Intensitätsverläufe für

• Sinusfunktionen mit Phasenschiebungen zur Intensitätsnormierung, wie sie z.B. beim Drei–Interferogramm–Verfahren eingesetzt werden /3/,

• Dreieckfunktionen mit Phasenschiebungen, wie sie bei einer Moiré–Projektion mit zwei gekreuzten Rechteckgittern oder bei zeitlicher Integration linear verschobener Rechteckgitter auftreten /9/,

• und den MZX–Code, der auch als Tripel von Trapezfunktionen mit Phasenschiebungen von jeweils $2\pi/3$ aufgefaßt werden kann. Dabei entspricht die Codierungssequenz Z der Periodenlänge 2π.

Aus den in Abb.6 gezeigten Grundfunktionen $I_1(x)$, $I_2(x)$, $I_3(x)$ lassen sich für jeden Code die zugehörigen Modulationshüllkurven

$$E_{max}(x) - E_{min}(x) = Max(I_1(x), I_2(x), I_3(x)) - Min(I_1(x), I_2(x), I_3(x))$$

bestimmen. Normiert auf den erforderlichen Signalbereich $I_{max} - I_{min}$ ergibt sich daraus ein Maß für die Ausnutzung des verfügbaren Signalbereichs:

$$\eta = (E_{max} - E_{min}) \, / \, (I_{max} - I_{min}).$$

Für den MZX–Code ergibt sich $\eta = 1$, für die Sinusfunktionen $\eta = 3/4$, und für die dreieckförmige Modulation $\eta = 2/3$. Bei gleichem (absolutem) Quantisierungsrauschen erfordert die Sinusfunktion gegenüber dem MZX–Code eine um den Faktor $4/3$ höhere maximale Signalamplitude und damit einen um $4/3$ höheren Signal–Rausch–Abstand. Bei der Dreiecksfunktion ist das Verhältnis $3/2$. Geht man umgekehrt von einem gegebenen Signal–Rausch–Abstand bzw. von einer bestimmten Zahl n unterscheidbarer Grauwertklassen aus und betrachtet die analogen Markierungsfunktionen als Codes, so erhält man z.B. beim Dreieck–Code nur $2(n-1)$ Codeworte, also ein Drittel des gesamten Codevorrates (vgl. Abb.1).

6 Vier– und mehrstellige Codes

Für w=3 lassen sich alle verwertbaren Codeworte in einer regelmäßigen Codesequenz unterbringen. Dieser Fall ist deshalb einfach zu behandeln, weil jeweils 2 von 3 Codebuchstaben strengen Restriktionen unterliegen (M und Z); der Rest ist Permutation. Aus dem gleichen Grund ist auch der Vorteil gegenüber Sinus– oder Dreieckfunktionen nicht allzu groß. Flächendeckende Objektcodierungen bei voller Nutzung der Ortsauflösung der Kamera erfordern meist wesentlich längere Codesequenzen. Eine dem MZX–Code vergleichbare Lösung läßt sich für w>3 jedoch nicht angeben. Es stellt sich hier wiederum die Frage, wie alle möglichen (normierten) Codeworte, die sich mit w > 3 Stellen bilden lassen, in eine quasianaloge Codesequenz gepackt werden können. Ein Beispiel für eine Erweiterung des 3–stelligen MZX–Codes ist die Multiplikation mit einem n–wertigen Graycode, der aus den verbleibenden w–3 Stellen erzeugt wird. Dabei wird der MZX–Code periodisch wiederholt, und mit einem Wort des Gray–Codes ergänzt. Die damit erzielte Codesequenz ist mit

$$Z = 6 \, (n{-}1) \, n^{w-3} \qquad n \geq 2, \, w \geq 3.$$

schon erheblich länger als bei allen anderen bisher betrachteten Verfahren (Abb. 1). Die vorteilhaften Eigenschaften des MZX–Codes bleiben bei dieser Form der Erweiterung erhalten. Für höchste Auflösung wird der MZX–Codeanteil wertekontinuierlich projiziert und mit einem Interpolationsdecoder ausgewertet.

MZXGray–Codes liefern bereits ausgezeichnete Ergebnisse. Allerdings tritt noch ein geringer Verlust an Codeworten auf: jedes Mal, wenn eine der Stellen des Graycodes mit dem Buchstaben für M oder für Z besetzt ist, entstehen im ursprünglichen MZX–Code Freiheitsgrade, die ungenutzt bleiben. Eine weitere Optimierung der Codes erfordert einen größeren Aufwand, und zwar sowohl bei der Generierung als auch bei der Decodierung. Es soll hier nur erwähnt werden, daß es durchaus quasianaloge, normierte Codesequenzen gibt, die (bis auf geringste vernachlässigbare Verluste) den gesamten Wortvorrat ausschöpfen.

7 Technische Realisierung der Code–Projektion

LCD–Codeprojektor
Projektoren mit Flüssigkristallanzeigen (LCD's) und anderen elektrisch adressierbaren Lichtventilen eignen sich derzeit nur für binäre /7/, ternäre oder quaternäre Codes mit diskreter niedriger Ortsauflösung. Eigene Versuche mit zwei verschiedenen LCD–Projektoren (100x128 und 640x400 Elemente) lieferten bei quaternärer Codierung (n=4) keine befriedigenden und temperaturstabilen Ergebnisse.

Laser–Codeprojektor
Für die Projektion beliebiger Codes wurde ein Laserprojektor mit quasianalogen Eigenschaften entwickelt. Er erlaubt die Projektion von Codes mit maximal 3000 Worten und 2 bis 4096 Graustufen. Die maximale Orts- und Intensitätsauflösung ist im Vergleich zur Orts– und Wertequantisierung einer Matrixkamera bzw. eines digitalen Bildspeichers so hoch, daß auch Signale gesendet werden können, die von optisch erzeugten Analogsignalen nicht zu unterscheiden sind. Andererseits lassen sich auch beliebige orts– und/oder wertdiskrete Codes projizieren. Die Projektion erfolgt mit einer maximalen Frequenz von 50 Hz. Damit ist eine zeitlich lückenlose Bildaufnahme möglich (6 Vollbilder in 240 Millisekunden).

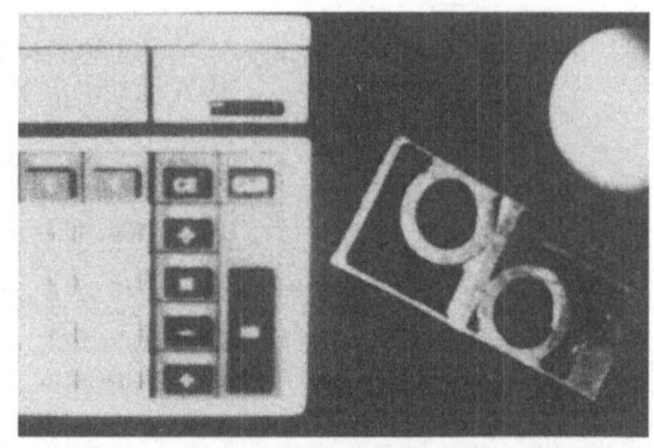

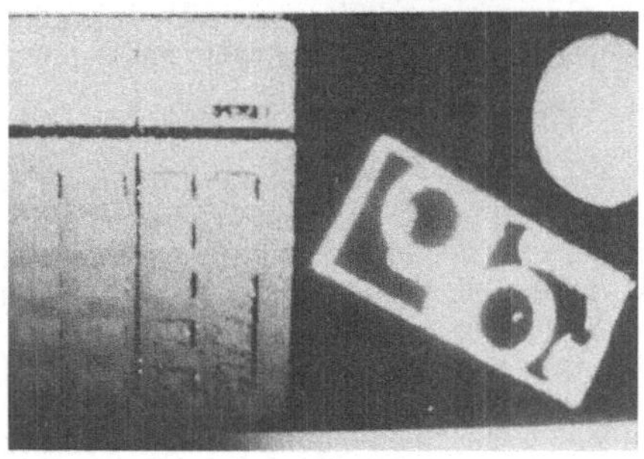

Abb. 7: Das Graubild oben zeigt drei Objekte mit stark unterschiedlichen Reflexionseigenschaften: einen schräggestellten Taschenrechner, einen weißen Tischtennisball und ein korrodiertes Metallobjekt. Das Bild in der Mitte zeigt eine der Beleuchtungsstrukturen eines hybriden MZXGray–Codes mit w=6 Stellen bzw. Bildern. Ein Binärcode mit zusätzlicher Normierung würde hier 11 Bilder erfordern, ein Verfahren mit Dreiecksfunktionen verschiedener Periodenlängen 12 oder 15 Bilder. Das Bild rechts unten zeigt einen Ausschnitt aus dem 512 x 512 –Distanzbild in Pseudo–Grau–Darstellung. Weiß entspricht der kameranächsten Höhenschicht, Referenzfläche und zurückgewiesene Meßwerte werden schwarz dargestellt.

8 Automatische System– und Objektanalyse und Codeoptimierung

Die geforderte Flexibilität des Codierungssystems kann unter wechselnden Anforderungen nur durch eine automatische Optimierung genutzt werden. Durch eine Analyse der Systemeigenschaften, der optischen und topologischen Eigenschaften von Meßobjekt–Typ und Umgebung sowie der aufgabenspezifischen Randbedingungen sollen alle erforderlichen Kriterien für eine automatische Generierung geeigneter Codes gesammelt und gegeneinander abgewogen werden. Wegen der Nichtlinearität der Zusammenhänge und Vielfalt der Bedingungen wird eine wissensbasierte Verarbeitung angestrebt. Die Entwicklung dieses Systems ist in Teilen fertiggestellt.

9 Diskussion und Ausblick

Die eindeutige und effiziente optische Markierung der Objektoberfläche ist eine wesentliche Voraussetzung für die triangulationsbasierte 3–D–Meßtechnik. Das vorgestellte Codierungssystem hat folgende Eigenschaften:

- Unempfindlichkeit gegenüber Fremdlichteinflüssen und Lichtreflexen durch binäre Codeanteile. Einsetzbar bei Objekten mit komplexer Topologie und variierenden Reflexionseigenschaften (3–D–Robot Vision).
- Genauigkeit durch analoge Codeanteile. Die erreichbare Auflösung liegt im Submillimeterbereich (Qualitätsprüfung in Fertigung und Montage).
- Geschwindigkeit durch effiziente Codes. Die Erzeugung und Aufnahme der optischen Information zur Berechnung von 260.000 3–D–Koordinatenpunkten erfolgt in 0.240 Sekunden. Die weitere Verarbeitung hängt von der Rechnerstruktur ab und kann unter 1 Sekunde liegen.
- Wählbare Prioritäten für maximale Geschwindigkeit, maximale Genauigkeit oder maximale Störsicherheit. Die jeweils zweit– und drittrangigen Eigenschaften werden automatisch stärker in Anspruch genommen.
- Automatische Anpassung des Sensors an System, Objekt und optische Umgebungsbedingungen, d.h. der verfügbare Signal–Rausch–Abstand wird durch die Codierung ausgeschöpft.

Literatur:

1 Levi, P.: Planen für autonome Montageroboter.
Informatik–Fachberichte 191. Springer–Verlag 1988.
2 Tiziani, H.J.: Optische Abstandsmessung mit hoher örtlicher Auflösung.
Microwaves and Optronics '89, Conference Proceedings. Network GmbH 1989.
3 Tiziani, H. J.: Rechnerunterstützte Laser–Meßtechnik. Technisches Messen **tm 54** (1987).
4 Strand, T.C. : Optical Three–Dimensional Sensing for Machine Vision.
Optical Engineering 24 (1985), S. 33–40.
5 Seitz, G.; Tiziani, H.J.; Litschel, R.: 3–D–Koordinatenmessung durch optische Triangulation.
Feinwerktechnik & Meßtechnik 94 (1986) 7, S. 423–425.
6 Altschuler, M. D., Altschuler, B. R. Taboada, J.: Laser Electro–Optic System for Rapid Three–Dimensional Topographic Mapping of Surfaces. Opt. Eng., 20 (6) (1981).
7 Wahl, F. M.: A Coded Light Approach for Depth Map Acquisition.
DAGM 1986, Informatik Fachberichte 125. Springer 1986.
8 Sato, K.; Inokuchi, S.: Range–Imaging System Utilizing Nematic Liquid Crystal Mask.
Proc. 1st Intern. Conf. on Computer Vision, London, 1987.
9 Zumbrunn, R.: Automatic Fast Shape Determination of Diffuse Reflecting Objekts at Close Range, by Means of Structured Light and Digital Phase Measurement. ISPRS Intercommission Conf. on Fast Processing of Photogrammetric Data, Interlaken, Switzerland, June 2–4, 1987.
10 Berlekamp, E.R.: Algebraic Coding Theory. McGraw–Hill Book Company (1968).
11 Cattermole, K. W.: Statistische Analyse und Struktur von Information, VCH, 1988.

Gewinnung von Oberflächenformen aus einem Grauwertbild durch Shape from Shading

Heinz Brünig   Reinhard Prechtel

Lehrstuhl für Informatik 5 (Mustererkennung)
Universität Erlangen-Nürnberg
Martensstraße 3, 8520 Erlangen

Übersicht

Es werden verschiedene Verfahren des *Shape from Shading* (Form aus Schattierung) vorgestellt und Ergebnisse präsentiert, die mit diesen Methoden erzielt wurden.

Bei Shape from Shading werden aus den Daten eines einzigen Grauwertbildes die in der Szene vorliegenden Oberflächenformen berechnet. Das erste hier gezeigte Verfahren liefert eine schnelle Grobklassifikation der Oberflächen als *eben, zylindrisch* oder *sphärisch*. Bei zwei weiteren Methoden wird durch eine lokale Untersuchung der Grauwertänderung in der Umgebung eines jeden Bildpunktes die Oberflächennormale in jedem Punkt bestimmt. Aus diesen Daten kann dann die Oberflächenform rekonstruiert werden. Das vierte Verfahren beruht auf einer biologisch motivierten Approximation der Reflexionsfunktion der Oberfläche. Unter dieser Voraussetzung kann direkt aus der Grauwertfunktion auf die Oberflächenform geschlossen werden.

Bei der Erprobung der Methoden hob sich speziell ein Verfahren durch gute Ergebnisse hervor.

1 Einleitung

Bei der Bildanalyse dreidimensionaler Szenen ist es zunehmend von Interesse, Informationen über die Form von Oberflächen aus dem Bildmaterial zu gewinnen. Hierzu wurden verschiedene Methoden für Shape from Shading implementiert. Diese Verfahren erlauben es, aus den lokalen Grauwertänderungen in einem Grauwertbild auf die Form der Oberfläche zurückzuschließen. Im Gegensatz zu Stereo-Verfahren, welche zwei Bilder für die Analyse benötigen (wobei das damit verbundene Korrespondenzproblem zu bewältigen ist), wird bei *Form aus Schattierung* nur ein einziges Grauwertbild benötigt. Allerdings kann dann auch kein echtes Tiefenbild (mit Koordinaten im dreidimensionalen Raum) berechnet werden, sondern nur ein relatives Tiefenbild — das Oberflächenrelief. Diese Information ist aber für weite Bereiche der Bildanalyse vollkommen ausreichend.

2 Grundlagen für Shape from Shading

Bevor die einzelnen Verfahren genauer betrachtet werden, muß noch der Bilderzeugungsprozeß, der die Grundlage für die Berechnungen bildet, analysiert werden. Nach Horn [Hor 75,Hor 77] sind folgende Komponenten für die Bilderzeugung bestimmend:

- Die Beleuchtung (z.B. Art, Anzahl und Richtung der Lichtquellen)

- Das Reflexionsverhalten der Oberfläche (z.B. matt, spiegelnd)

- Die Oberflächengeometrie (d.h. die Oberflächenkrümmung in jedem Punkt)

- Der Blickpunkt (z.B. Betrachtungsabstand und -richtung, Perspektive)

Der Zusammenhang zwischen diesen Komponenten ist durch Differentialgleichungen gegeben. Um die Berechnung der Oberflächenform durchführen zu können, sind einige Annahmen notwendig. Als wichtigster Punkt wird bei allen Verfahren ein *Lambert'sches Reflexionsverhalten* der Oberfläche vorausgesetzt. Ein Körper mit Lambert'schem Reflexionsverhalten besitzt eine ideal matte Oberfläche, bei der die Intensität des reflektierten Lichtes nur vom Winkel zwischen der Lichteinfallsrichtung und der Oberflächennormalen abhängt:

$$I(x,y) = \rho\lambda\, \mathbf{N}(x,y) \cdot \mathbf{L} = \rho\lambda\, cos(i_{x,y}) \tag{1}$$

Dabei ist I(x,y) die im Bildpunkt (x,y) vom Betrachter wahrgenommene Intensität, $\mathbf{N}$(x,y) die Oberflächennormale im Punkt (x,y) und $\mathbf{L}$ die Richtung des einfallendes Lichtes. Die entsprechenden Winkel sind der Abbildung 1 zu entnehmen. Die Beleuchtungsstärke λ und das Reflexionsvermögen (Albedo) der Oberfläche ρ werden als konstant angenommen.

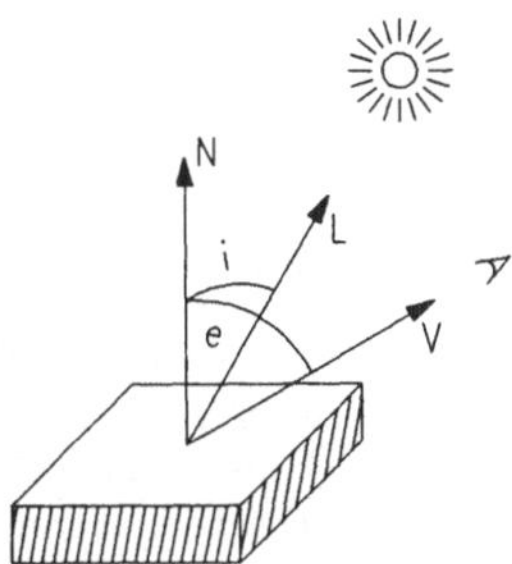

Abbildung 1: Veranschaulichung der für die Reflexionseigenschaften von Oberflächen
wichtigen Vektoren und Winkel

Die Annahme einer Lambert'schen Oberfläche mag zunächst als sehr einschränkend empfunden werden. Es sei jedoch vorweggenommen, daß es sich für die Berechnungen als nicht gravierend erweist, wenn diese Voraussetzung etwas verletzt wird. Zudem gilt allgemein, daß Spiegelungen nur in relativ kleinen Bildbereichen einen relevanten Beitrag zum Bildgenerierungsprozeß liefern. Maximale Spiegelung liegt dort vor, wo $\mathbf{V}$, $\mathbf{N}$ und $\mathbf{L}$ koplanar sind und Einfallswinkel i und Ausfallswinkel e gleich sind. Der Beitrag der spiegelnden Reflexion nimmt dann exponentiell um diese Richtung maximaler Spiegelung ab. Außerhalb dieser Bereiche spiegelnder Reflexion kann in sehr guter Näherung ein Lambert'sches Reflexionsverhalten der Oberfläche angenommen werden. Dadurch wird zwar die Oberfläche in spiegelnden Bereichen falsch berechnet, außerhalb dieser Bereiche sind die Ergebnisse jedoch gut.

Eine weitere Voraussetzung für die Berechnungen betrifft die Beleuchtung. Es wird in den Theorien eine Punktlichtquelle im Unendlichen angenommen. Bei der Erprobung der Verfahren zeigte sich, daß die Verletzung dieser Voraussetzung keine Schwierigkeiten bereitet. So wurden häufig Szenen bei normalem Umgebungslicht in einem Raum verwendet, wobei jeweils mehrere Lichtquellen vorhanden waren. Solange die Szene nicht mit mehreren stark gebündelten Lichtquellen aus verschiedenen Richtungen beleuchtet wird, ergibt sich in den meisten Fällen eine mittlere Beleuchtungsrichtung, die für die Verfahren auszureichen scheint.

3 Verfahren zur Berechnung der Oberflächenform aus der Schattierung

In diesem Abschnitt werden die Vorgehensweisen mehrerer Verfahren zur Berechnung von Oberflächenformen aus der Schattierung vorgestellt. Die erste Methode liefert eine schnelle Grobklassifikation der vorliegenden Oberflächen als *eben, zylindrisch* oder *sphärisch*. Bei den zwei folgenden Verfahren wird die Oberflächennormale in jedem Bildpunkt berechnet. Daraus kann dann in einem nachfolgenden Schritt die Oberfläche rekonstruiert werden. Der letzte Ansatz verwendet einen direkten Zusammenhang zwischen Grauwertfunktion und Tiefenrelief, um die Oberflächen zu berechnen.

3.1 Grobklassifikation von Oberflächenformen

Dieses auf [Coh 84] basierende und in [Brä 89] realisierte Verfahren verwendet eine wichtige Eigenschaft Lambert'scher Oberflächen für die Klassifikation von Oberflächentypen. Bei einer Lambert'schen Oberfläche hängt, wie aus Gleichung 1 ersichtlich, der Grauwert im einem Bildpunkt nur vom Winkel i zwischen der Beleuchtungsrichtung und der in diesem Punkt vorliegenden Oberflächennormalen ab. Daraus ist direkt ersichtlich, daß die Grauwerte in einem Bildbereich, je nach Oberflächenform, charakteristisch angeordnet sind. Bei einer ebenen Oberfläche ist in jedem Punkt der Winkel i und damit der Grauwert gleich. Bei einer zylindrischen Oberfläche liegt auf Linien parallel zur Zylinderachse jeweils der gleiche Winkel i, und damit der gleiche Grauwert, vor. Bei sphärischen Oberflächen treten die Linien konstanten Grauwertes als Ellipsen auf.

Dieses Verhalten der Grauwerte wird nun zur Bestimmung der Oberflächenform herangezogen. Da bei realen Bildern zum einen das Lambert'sche Gesetz nie exakt erfüllt sein wird und zudem z.B. durch Rauschen bedingte Störungen vorhanden sind, werden hier auf den „Linien gleichen Grauwertes" die Grauwerte sicherlich nicht genau gleich sein. Für die Berechnungen werden daher statistische Verfahren verwendet. Dazu wird das Bild zunächst in quadratische Fenster zerlegt. In jedem Fenster wird dann die statistische Verteilung der Grauwerte ermittelt. Eine elliptische Verteilung der Grauwerte — wie sie bei sphärischen Oberflächen vorliegt — ist nur schwer direkt zu ermitteln und zudem müssen dafür große Fenster betrachtet werden. Deshalb wird zunächst in jedem Fenster nur untersucht, ob die Grauwerte gleichmässig verteilt sind, oder eine ausgezeichnete Richtung bei der Grauwertverteilung vorliegt. Danach wird jedes Fenster als *eben* oder *nicht-eben* klassifiziert, wobei in letzterem Fall aus der Grauwertverteilung auf die Richtung der Oberflächenkrümmung geschlossen werden kann. In einem weiteren Schritt werden dann die Fenster in der Umgebung eines jeden Fensters untersucht. Für eine zylindrische Oberfläche wird die Richtung in allen Fenstern gleich sein und auf die Ausrichtung der Zylinderachse schließen lassen. Ändert sich die Richtung der Grauwertverteilungen in charakteristischer Weise, so kann auf eine sphärische Oberfläche geschlossen werden und die beteiligten Fenster werden entsprechend klassifiziert. Ist keine eindeutige Klassifikation als *zylindrisch* oder *sphärisch* möglich, so wird das Fenster als *undefiniert* gekennzeichnt.

Durch seinen schnellen Ablauf, der zudem hochgradig parallelisierbar ist, eignet sich dieses Verfahren gut für eine schnelle Klassifikation einfacher Oberflächen. Zudem ist dies das einzige hier vorgestellte Verfahren, das die Oberflächen direkt klassifiziert. Bei den anderen Methoden wird zwar eine exaktere Rekonstruktion der Oberflächen durchgeführt, für die weitere Verwendung dieses Ergebnisses in einem Bildanalysesystem ist aber anschließend noch eine Approximation und Klassifikation der Oberflächen erforderlich.

3.2 Gewinnung der Oberflächenform aus den Oberflächennormalen

In diesem Abschnitt werden zwei Verfahren vorgestellt, die jeweils durch lokale Untersuchung der Grauwertänderungen in der Umgebung eines Bildpunktes, die Berechnung der Oberflächennormale in jedem Bildpunkt durchführen. Die Normale wird dabei durch die beiden Winkel *Tilt* τ und *Slant* σ festgelegt. Wie aus Abbildung 2 ersichtlich, ist τ der Winkel zwischen der x-Achse und der Projektion der Normalen in die x-y-Ebene. σ ist der Winkel zwischen der Normalen und der z-Achse.

Zum weiteren Verständnis der Verfahren halten wir uns das Bild einer Kugel vor Augen, auf der ja alle möglichen Orientierungen von Flächennormalen vorliegen. Beleuchtet man diese aus der Betrachtungsrichtung, so nimmt die Helligkeit radial vom Mittelpunkt der sichtbaren Kugelscheibe nach außen hin ab. Die Abnahme des Grauwertes ist dabei kosinus-förmig, wenn man eine Lambert'sche Oberfläche annimmt (siehe Gleichung 1). Konstanter Grauwert liegt auf konzentrischen Kreisen um den Mittelpunkt vor. Damit wird der Zusammenhang zu den gesuchten Bestimmungswinkeln τ und σ der Oberflächennormalen in einem Bildpunkt deutlich. Der Tilt τ,

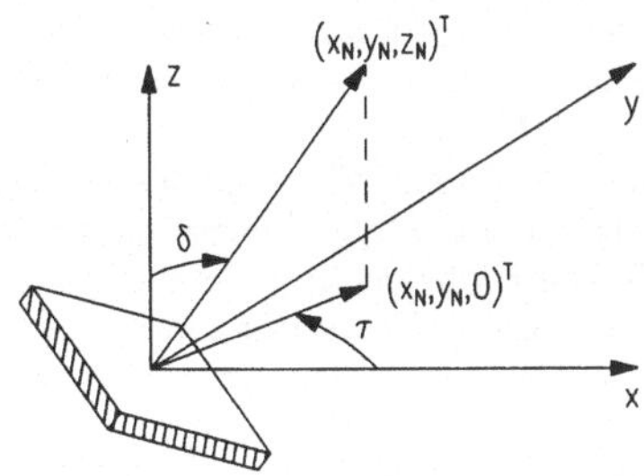

Abbildung 2: Definition von Tilt τ und Slant σ

also die Projektion der Normalen in die Bildebene, stimmt mit der Richtung des Grauwertgradienten überein. Der Slant σ läßt sich aus der Stärke des Grauwertgradienten ableiten, der bei der Kugel zum Rand hin abnimmt (Kosinus !). Bei der Betrachtung beliebiger Oberflächen wird der gleiche Schluß durchgeführt. Dazu wird vorausgesetzt, daß die Oberfläche in der lokalen Umgebung des betrachteten Punktes durch ein Segment einer Kugeloberfläche approximiert werden kann. Diese Voraussetzung erweist sich allerdings in der Praxis als nicht problematisch, solange die Oberfläche glatt ist, also keine Knickstellen und Unstetigkeiten aufweist.

Das erste hier realisierte Verfahren beruht auf der von A. Pentland in [Pen 84] vorgestellten Methode. Hierbei werden die ersten und zweiten Ableitungen der Grauwertintensität in der Umgebung eines Punktes für die Berechnungen herangezogen. Dies erweist sich jedoch bei der Anwendung auf reale Bilder, durch die dort vorhandene Quantisierung und die geringe Grauwertauflösung von maximal 256 Stufen, als äußert problematisch. Ergebnisse hierzu sind in Kapitel 4 zu finden.

Diese Problematik konnte von Lee und Rosenfeld [Lee 85] unter der Voraussetzung einer bekannten Beleuchtungsrichtung umgangen werden. Es werden dann nur noch die ersten Ableitungen der Grauwertintensität für die Berechnungen benötigt. Das Verfahren wurde in [Pre 88] weiter verbessert und liefert die zur Zeit besten Ergebnisse. Die entscheidende Verbesserung ist die Berechnung der Oberflächennormalen bei vier verschiedenen Auflösungsstufen. Die drei gröberen Auflösungsstufen werden durch Medianfilterungen aus dem Ausgangsbild erzeugt. Damit erweist sich das Verfahren auch bei realen Bildern als sehr robust.

Beide Verfahren liefern als Ergebnis die Oberflächennormale in jedem Bildpunkt. Aus diesen Daten kann theoretisch direkt die Oberflächenform rekonstruiert werden. In der Praxis sind die berechneten Oberflächennormalen jedoch mit einem Fehler behaftet. Ordnet man der Oberflächennormalen in jedem Punkt eine Einheitsfläche zu, so würde sich in diesem Fall keine geschlossene Oberfläche ergeben. In einem nachfolgenden Schritt wird deshalb eine *Integrabilitätsbedingung* berücksichtigt. Dazu wurden zwei Methoden erprobt, die hier kurz charakterisiert werden sollen. Zunächst kann jeder Oberflächennormalen ein entsprechender Oberflächengradient zugeordnet werden. Aus diesem Vektorfeld von Gradienten soll nun ein skalares Feld, welches die Tiefeninformation beinhaltet, bestimmt werden. Diese Tiefeninformation z(x,y) stellt das Potential des Vektorfeldes dar. Ist die Integrabilitätsbedingung für die berechneten Gradienten nicht erfüllt, was man i.A. annehmen muß, so ist das Kurvenintegral über das Gradientenfeld zwischen zwei Punkten von der Wahl des Weges abhängig. Es werden nun, von einem Bezugspunkt ausgehend, verschiedene Kurvenintegrale berechnet und iterativ die Integrabilität hergestellt. Das zweite Verfahren, basierend auf [Fra 87], beruht auf der Annahme, daß die ideale Oberflächenfunktion z(x,y) durch eine Linearkombination von Basisfunktionen darstellbar ist, deren Ableitungen nach x und y jeweils orthogonal seien. In diesem Funktionenraum werden nun die berechneten Oberflächengradienten auf neue Gradienten abgebildet, die der Integrabilitätsbedingung genügen und deren Abstand zu den alten Gradienten minimal ist. Das zweite Verfahren liefert dabei die besseren Resultate. Daher wurde es bei der Berechnung der hier gezeigten Ergebnisse verwendet.

Abschließend soll noch auf eine grundsätzliche Problematik der Shape from Shading Verfahren dieses Abschnittes hingewiesen werden. Durch die rein lokale Untersuchung der Grauwertände-

109

rung kann nicht entschieden werden, ob eine konkave oder eine konvexe Krümmung der Oberfläche vorliegt. So werden hier alle Oberflächennormalen als zu konvexen Krümmungen gehörend interpretiert. Wie aus den Ergebnissen ersichtlich ist, werden zwar auch leicht konkav gekrümmte Flächen richtig rekonstruiert, liegen jedoch in einem Bild sehr starke 'Berge' und 'Täler' vor, so werden die 'Täler' als 'Berge' rekonstruiert. Eine richtige Unterscheidung zwischen konkaven und konvexen Oberflächenteilen kann nur durch eine Einbeziehung globaler Informationen erreicht werden. Dieses Problem kann leicht durch die sogenannte *Kraterillusion* veranschaulicht werden. In Abbildung 3 ist links deutlich ein Krater und rechts eine vulkanartige Erhebung zu erkennen. Es handelt sich jedoch um das gleiche Bild, das nur einmal auf dem Kopf stehend wiedergegeben ist. Dreht man die Abbildung herum, so sieht man im Bild, das vorher den Krater darstellte, den Vulkan und umgekehrt. Diese Interpretationen beruhen auf der intuitiven Annahme, daß die Szene von oben beleuchtet wird und damit ein bestimmter Schattenwurf entsteht. Die Unterscheidung erfolgt somit nicht durch lokale, sondern vielmehr durch globale Informationen.

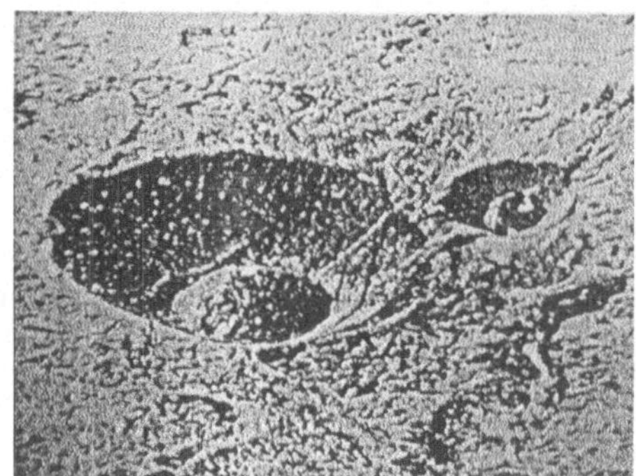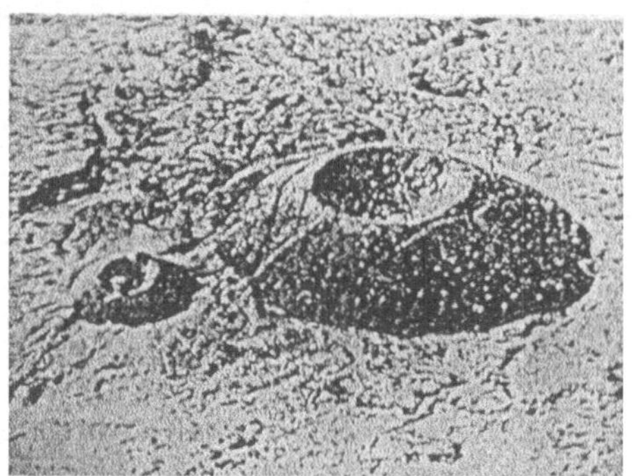

Abbildung 3: Veranschaulichung der Kraterillusion

3.3 Berechnung der Oberflächenform bei Approximation der Reflexionsfunktion

Dieses in [Pen 88a,Pen 88b] vorgestellte Verfahren basiert auf einer Approximation der Reflexionsfunktion der Oberfläche. Wie bei den anderen Verfahren wird auch hier eine Lambert'sche Reflexionsfunktion der Oberfläche und eine Beleuchtung durch entfernte Punktlichtquellen vorausgesetzt. Zudem darf im Bild keine Selbstabschattung auftreten. Pentland entwickelt die Bildfunktion I(x,y), welche von der Beleuchtungsrichtung und dem in jedem Bildpunkt vorliegenden Oberflächengradienten (p,q) abhängt, in einer Taylorreihe um p,q = 0. Für kleine Oberflächengradienten, bzw. bei flacher Beleuchtung der Szene, dominieren dann die linearen Terme dieser Entwicklung. Vernachlässigt man die Terme quadratischer und höherer Ordnung, so erhält man einen linearen Zusammenhang zwischen der Bildfunktion I(x,y) und der Oberfläche z(x,y). Die Annahme, daß die Reflexionsfunktion der Oberfläche eine lineare Funktion der Oberflächenorientierung ist, stellt eine gute Approximation der Lambert'schen Reflexionsfunktion dar. Pentland kann durch Experimente zeigen, daß auch der menschliche Betrachter eine solche lineare Approximation der Reflexionsfunktion bei der Interpretation von Bildern verwendet.

4 Ergebnisse

In diesem Abschnitt werden Ergebnisse, die mit den oben beschriebenen Verfahren erzielt wurden, gezeigt und kurz diskutiert. Die Verfahren wurden dabei jeweils auf den gleichen Satz von zwei Bildern angewandt. Diese Grauwertbilder sind in Abbildung 4 dargestellt. Es handelt sich dabei um das synthetisch erzeugte Bild einer aus Betrachtungsrichtung beleuchteten Kugel, sowie das unter normalen Umgebungbedingungen aufgenommene Bild eines Bleistiftes an dem sich eine Verzierung mit komplexer Oberflächenstruktur befindet.

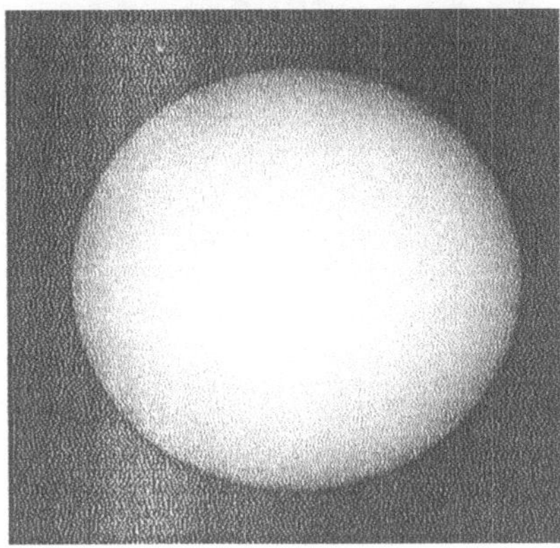
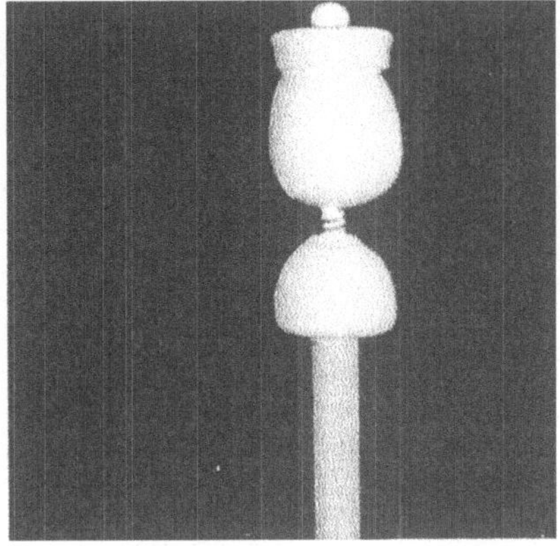

Abbildung 4: Grauwertbilder von Kugel und Stift

Die Resultate der Grobklassifikation aus Abschnitt 3.1 sind in Abbildung 5 dargestellt. Die Klassifikation der Kugeloberfläche ist vollkommen richtig. Beim Stift wird der zylindrische Schaft auch richtig klassifiziert. Die restlichen Oberflächen werden auch gut erkannt, wobei natürlich, wegen der komplexen Struktur der Oberfläche, die Klassifikation einiger Fenster als *undefiniert* durchaus gerechtfertigt ist.

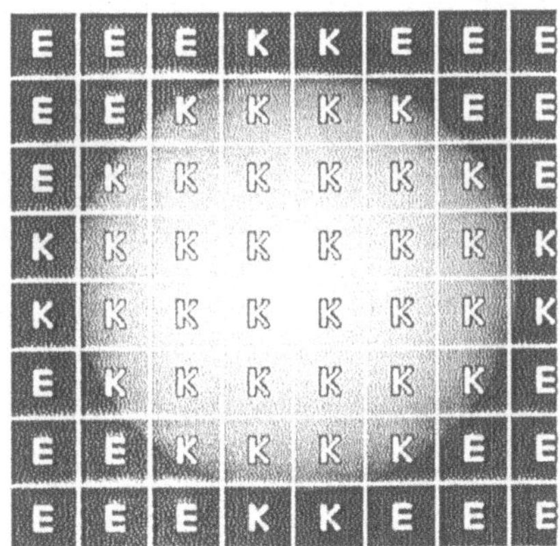
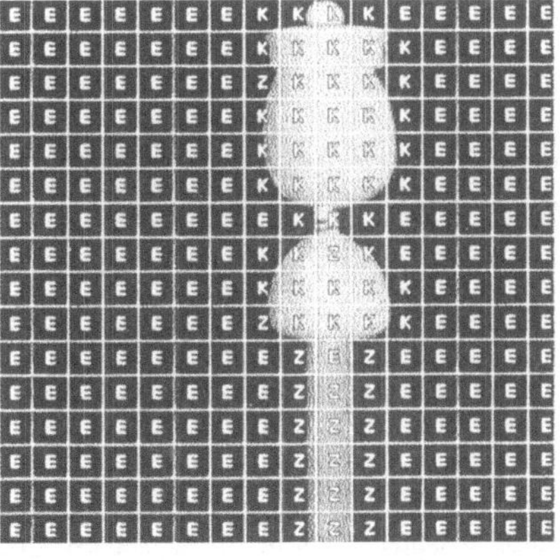

Abbildung 5: Ergebnisse der Grobklassifikation aus Abschnitt 3.1

Bei den Ergebnissen des auf [Pen 84] beruhenden ersten Verfahrens aus Abschnitt 3.2 werden die dort angedeuteten Probleme deutlich. Wie in Abbildung 6 zu sehen, treten schon beim synthetisch erzeugten Bild der Kugel große Fehler auf. Diese sind auf die Quantisierung der Grauwerte des Eingabebildes zurückzuführen. Bei der Verwendung von synthetisch erzeugten Bildern mit reellwertigen Grauwerten liefert auch dieses Verfahren brauchbare Ergebnisse. Da jedoch eine Anwendung auf Bilder in natürlicher Umgebung im Vordergrund steht, erweist sich dieses Verfahren als kaum geeignet. Die Oberfläche des Bleistiftes wird dann auch stark fehlerhaft berechnet.

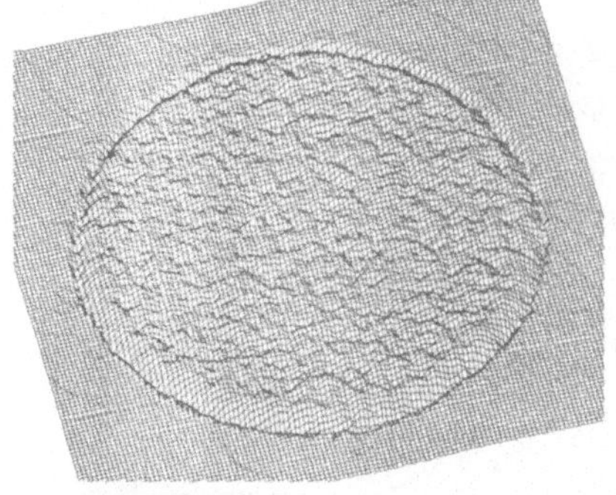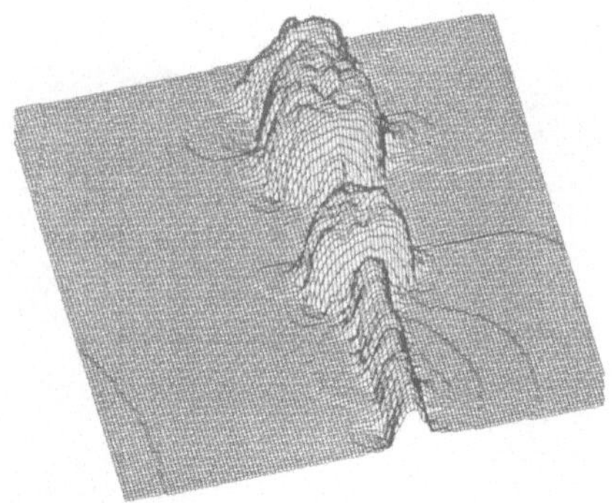

Abbildung 6: Ergebnisse des Verfahrens nach [Pen 84] aus Abschnitt 3.2

In Abbildung 7 sind die Resultate des Verfahrens aus [Pre 88] zu finden. Wie schon in Kapitel 3.2 angedeutet, liefert dieses die besten Ergebnisse. Die Kugeloberfläche wird sehr gut rekonstruiert. Auch die berechnete Oberfläche des Stiftes stimmt sehr gut mit dem wirklichen Aussehen überein. Dieses Verfahren erwies sich auch bei der Anwendung auf verschiedenste andere reale Szenen als sehr robust.

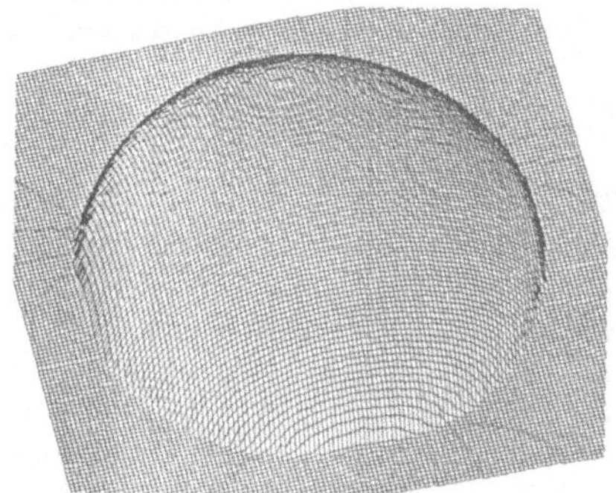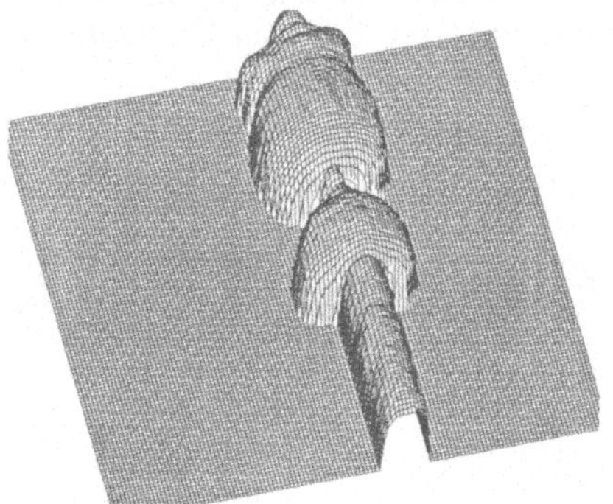

Abbildung 7: Ergebnisse des Verfahrens nach [Pre 88] aus Abschnitt 3.2

Abschließend sind die Ergebnisse des Verfahrens aus Abschnitt 3.3 in Abbildung 8 zu finden. Die Resultate sind noch gut, es zeigt sich jedoch eine wesentlich geringere Detailtreue als bei dem Verfahren aus [Pre 88]. Bei starken Änderungen der Oberflächenkrümmung treten immer Verschleifungen auf, was sich hier besonders an den Objektgrenzen sehr unangenehm auswirkt. Häufig ist der Oberflächenfunktion auch noch eine sehr tiefe Frequenz überlagert — ein Effekt, der auch in [Pen 88b] angesprochen wird.

5 Ausblick

Die hier vorgestellten Verfahren zur Gewinnung von Oberflächenformen aus einem Grauwertbild liefern — mit einer Ausnahme — auch auf Bildern realer Szenen gute Ergebnisse. Das Verfahren zur Grobklassifikation von Oberflächen scheint für einfache Klassifikationsaufgaben durchaus geeignet. Dabei kann die problemspezifische Adaption des Klassifikationsschrittes in Richtung komplexerer

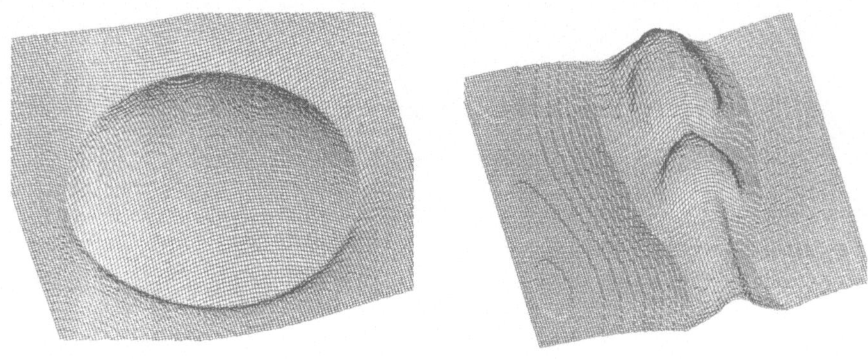

Abbildung 8: Ergebnisse des Verfahrens nach [Pen 88b] aus Abschnitt 3.3

Oberflächen sinnvoll sein. Bei den Verfahren zur genauen Rekonstruktion des Oberflächenreliefs erwies sich die auf [Pen 84] beruhende Methode als ungeeignet. Das in [Pre 88] realisierte Vorgehen bringt die besten Ergebnisse und erweist sich bei Verwendung verschiedensten Bildmaterials als sehr robust. Der in [Pen 88b] vorgestellte Algorithmus liefert dagegen nicht ganz so gute Resultate, zeichnet sich dafür aber durch eine kürzere Rechenzeit aus.

Auf Grund der erzielten Ergebnisse zeigt sich, daß Shape from Shading Verfahren eine sinnvolle und hilfreiche Ergänzung bei der Segmentierung und Analyse dreidimensionaler Szenen darstellen.

Literatur

[Brä 89] Brähler B.: Extraktion dreidimensionaler Information aus Grauwertbildern durch "Contours of constant image intensity"(Arbeitstitel), Studienarbeit am Lehrstuhl für Informatik 5 (Mustererkennung), Universität Erlangen-Nürnberg, 1989.

[Coh 84] Cohen F.S., Cayula J.-F. P.: 3-D object recognition from a single image. SPIE Vol. 521, Intelligent Robots and Computer Vision, 1984, 7-15.

[Fra 87] Frankot R.T., Chellappa R.: A Method for Enforcing Integrability in Shape from Shading Algorithms. Proc. First Int. Conf. on Computer Vision (ICCV), London 1987, 118-127.

[Hor 75] Horn B.K.P.: Obtaining Shape from Shading Information. In Winston P.H.(ed.): The Psychology of Computer Vision. Mc Graw-Hill, New York 1975, 115-155.

[Hor 77] Horn B.K.P.: Understanding Image Intensities. Artificial Intelligence 8 (1977), 201-231.

[Lee 85] Lee C.-H., Rosenfeld A.: Improved Method of Estimating Shape from Shading using the Light Source Coordinate System. Artificial Intelligence 26 (1985), 125-143.

[Pen 84] Pentland A.: Shape From Shading. IEEE Transactions on Pattern Analysis and Machine Intelligence, Vol. PAMI-6 (1984), 170-187.

[Pen 88b] Pentland A.: Shape Information from Shading: A Theory About Human Perception. Proc. Second Int. Conf. on Computer Vision (ICCV), Tampa, Florida, 1988, 404-413.

[Pen 88a] Pentland A.: On the Extraction Of Shape Information From Shading". Proc. of the American Assoc. for Artificial Intelligence, St. Paul, MN., USA, Aug. 22-27, 1988

[Pre 88] Prechtel R.: Extraktion dreidimensionaler Information aus Grauwertbildern durch "Local Shading Analysis". Diplomarbeit am Lehrstuhl für Informatik 5 (Mustererkennung), Universität Erlangen-Nürnberg, 1988.

A SYSTEM THAT LEARNS TO RECOGNIZE 3-D OBJECTS

G. Gabrielides

Department of Computer Studies
Loughborough University of Technology
LE11 3TU ENGLAND

ABSTRACT

A system that uses induction in order to learn to recognize simulated scenes of 3-D
objects is presented. A simulator generates line drawings representing polyhedra which
are used as input to a learner. It also composes polyhedral scenes which are given as
input to a recognizer. The learner builds up a knowledge base of definitions that describe these polyhedra from single or multiple views presented to it by the user. The
definitions use a structural representation and a minimum of input primitives, based on
a number of simple assumptions about the polyhedra. The recognizer attempts to interpret the scenes according to its experience in the learnt polyhedra. The system achie-
ves to recognize all the objects that it has learnt within a reasonable amount of time.

INTRODUCTION

Inductive learning is the process of acquiring knowledge by drawing inductive infer-
ences from facts provided by a teacher or the environment [MI'83], and it appears in
two major forms: learning from examples and learning from observation. The problem of
learning from examples [WI'75, BU'78, DI'81, SI'83, NE'85] is to find plausible general
descriptions explaining a given database, using a set of positive (examples) and negative (antiexamples) training instances. It starts with a set of specific descriptions,
in an appropriate description language, and gradually produces a more general set by a
repeated application of generalization rules.
   Visual understanding relates an input and its implicit structure to an explicit
structure that already exists in the internal representation of the world. High-level
vision systems require rich knowledge representations called knowledge bases, that reflect the system's previous experience. Line drawings are 2-D representations of 3-D
objects as groups of points connected with one another by straight lines and they be-
came an natural target from the early days of computer vision [GU'69, CL'71, HU'71, WA'
75, KA'79]. This project employs the method of learning from examples (near-miss case),
in order to create a knowledge base that is ·capable of interpreting polyhedral scenes
represented by line drawings.

THE SYSTEM

1. SYSTEM OVERVIEW

The system consists of three main parts: a simulator of 3-D figures, a learner and
a recognizer.
   The 3-D figure simulator generates and plots line drawings corresponding to different views of certain polyhedra. It allows the user to compose her/his own scene and
produces the primitives that are used as input to the next two parts.
   The learner works in three stages. In the first stage, an Elementary Concept Learner (ECL) learns the basic entities that make up a line drawing. In the second stage a
Multiple View Learner (MVL) learns the definitions of the polyherda that are to be recognized from more than one views. In the third stage a Single View Learner (SVL) learns
how to recognize the same polyhedra from a single view.
   The recognizer is presented with scenes of line drawings. A Single View Recognizer

(SVR) segments the input into faces of possible polyhedra and attempts to match the
segmented scene with a set of Single View Definitions (SVD) of polyhedra, which may
lead to several alternative interpretations. A unique interpretation can be obtained by
a Multiple View Recognizer (MVR) that makes assupmtions about hidden elements using a
set of ad-hoc rules. The MVR uses a set of Multiple View Definitions (MVD) of polyhedra.
   The system uses predicate logic to represent knowledge and is mainly implemented
in PROLOG. Therefore the PROLOG notation is adopted in the following sections.

2. THE 3-D FIGURE SIMULATOR

The main purpose of the simulator is to get round the difficult problem of low-level
vision, so that the system can concentrate only on the high-level part. It receives an
input of two arrays: the Vertex Array (N x 3) representing the (N) vertices of a polyhedron in Cartesian coordinates, and the Face Array (M x 3) representing the faces (M polygons) of the above polyhedron in form of closed cyclic lists of the vertices ($<$ R-2)
that make up each face. The input polyhedron is subsequently subjected to a number of
transformations i. e. rotation, scaling, translation, hidden-line removal and projection (orthographic or central) [HA'83]. If the polyhedron is to be used by the learner,
the system produces five more views of it (back, top, bottom, left and right). If the
input is to be given to the recognizer, the user is allowed to insert more than one polyhedra and thus compose her/his own scene.
   The output of the simulator consists of a set of PROLOG predicates that represent
the visible connections between the vertices of the polyhedra in the scene. A simple
connection between two vertices a and b is given by the primitive: conn(a,b,N). N
is an integer ($0=<N=<2$) called the Face Counter (FC), and its use is discussed in sections 3, 4 and 5. The simulator is written in C.

3. THE FIGURE DEFINITIONS

The definitions are PROLOG clauses of the form:
$$C :- H1 , H2 , ... , Hn. \tag{1}$$

where C is the conclusion (rule-head) and Hi a condition of the hypothesis (rule-body).
In this case C is the name of the figure and Hi are the components that make it up.
The sipmlest 1-D figure is a straight line segment between two points A and B :

    line(A,B):-(conn(A,B,N);conn(B,A,N)),atom(A),atom(B),integer(N),A\=B,N>0,N=<2. (2)

N is the FC that indicates tne no. of faces (polygons) that share the side AB. Every
time that a visible face of a polyhedron is recognized as such, the FC's of its sides
are decremented by 1. Thus, lines with FC = 1 in the data are important cues indicating the existence of hidden faces. This is interpreted as a need for another view by
the MVL (sec. 4), or an assumption (sec.5) by the recognizer.
   The 2-D definitions contain convex quadrilaterals and triangles defined as:

    c_quadril(A,B,C,D):-line(A,B),line(B,C),line(C,D),line(D,A),
                 A\=C,B\=D,not(line(A,C)),not(line(B,D)),
                 not(poin_qul(A,B,C,D)),not(nocnvx(A,B,C,D)).   (3)
    trian(A,B,C):-line(A,B),line(B,C),line(C,A),not(poin_trn(A,B,C)).

which represent visible faces of a polyhedron. The predicates 'poin_trn', 'poin_qul'
and 'nocnvx' are called Special Features and are added to the data at a certain stage
(sec. 5). The definitions (3) are based on the following simple assumptions:
   a) the containment of a point (vertex) by a polygon implies that it is not a vis-
      ible face (or even a plane at all)
   b) the existence of a diagonal in a polygon (with > 3 sides) implies two planar
      faces with a common edge, that are not co-planar.
These give rise to the definition of a possible face, which for a triangle is:

    p_trian(A,B,C):-line(A,B),line(B,C),line(C,A),poin_trn(A,B,C).   (4)

Similar definitions are used for possible quadrilaterals.

Another group of definitions contains the 'set_' definitions used by the SVR. Every time that a visible (or possible) face is detected, it is marked by the system by asserting a new predicate of the form: 'a2D_figure' (or 'p2D_figure'). A triangular face is defined for example as:

$$set\_atr(A,B,C):-atrian(A,B,C);atrian(A,C,B);atrian(B,A,C);$$
$$atrian(B,C,A);atrian(C,A,B);atrian(C,B,A). \qquad (5)$$

There are also definitions for possible faces (e.g. set_ptr etc) according to the same principle. The substitution of low-level primitives ('conn's) by high-level ones ('a2D_figure's) is used to speed up the recognition.

The 3-D figure definitions are divided into MVD and SVD representing four classes of convex polyhedra: tetrahedron, pyramid, prism and truncated pyramid (or box) (Fig. 1). The MVD are unique for every polyhedron, make use of the 2-D definitions in (3) and require more than one views or an assumption about their hidden elements in order to succeed. The SVD have a one (figure) to many (views) correspondence for each polyhedron,

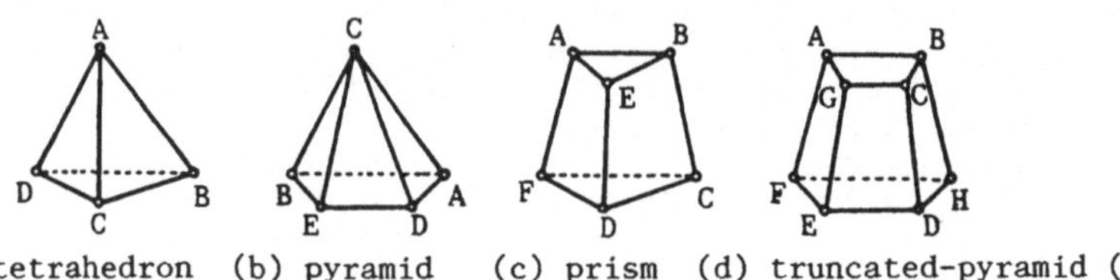

(a) tetrahedron   (b) pyramid   (c) prism   (d) truncated-pyramid (box)

Figure 1.

$$box(A,B,C,D,E,F,G,H):-c\_quadril(A,B,C,G),c\_quadril(G,C,D,E),c\_quadril(A,G,E,F),$$
$$c\_quadril(B,A,F,H),c\_quadril(C,B,H,D),c\_quadril(H,F,E,D).$$
$$prism(A,B,C,D,E,F):-c\_quadril(E,A,F,D),c\_quadril(A,B,C,F),c\_quadril(E,B,C,D),$$
$$trian(A,B,E),trian(F,C,D). \qquad (6)$$
$$pyram(A,B,C,D,E):-c\_quadril(A,D,E,B),trian(A,B,C),trian(A,D,C),trian(C,D,E),$$
$$trian(B,E,C).$$
$$tetra(A,B,C,D):-trian(A,B,C),trian(A,C,D),trian(A,B,D),trian(B,C,D).$$

use the 2-D 'set_' definitions (e.g. (5)) and are able to succeed with only one view. Fig. 2 illustrates the SVD of a tetrahedron. The 'non_convex_contour_angle' predicates are also Special Features and prevent cases like those in Fig. 2b.

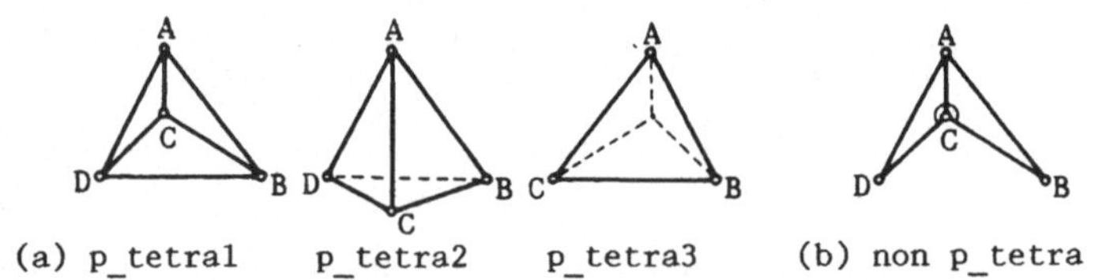

(a) p_tetra1   p_tetra2   p_tetra3   (b) non p_tetra

Figure 2.

$$p\_tetra1(A,B,C,D):-set\_atr(A,B,C),set\_atr(A,C,D),set\_atr(C,B,D),set\_ptr(A,B,D).$$
$$p\_tetra2(A,B,C,D):-set\_atr(A,B,C),set\_atr(A,C,D),B\backslash=D, \qquad (7)$$
$$not(non\_convex\_contour\_angle(A)),not(non\_convex\_contour\_angle(C)).$$
$$P\_tetra3(A,B,C):-set\_atr(A,B,C).$$

4. THE LEARNER

The learner consists of the ECL that learns the 1-D and 2-D definitions of (2), (3) and (4), the MVL and the SVL that learn the 3-D definitions of (6) and (7) from mul-

tiple and single views respectively. The basic strategy of the learners is to:
  a) create a rule (like (1)) by generalizing a positive training instance with the
     method of turning constants into variables
  b) test the rule using a series of training instances, and identify any existing
     faults by a procedure called critic
  c) modify the rule so that it behaves correctly by a procedure called modifier, us-
     ing the technique of the discriminating algorithm [LA'81]. This looks for the
     difference between a selection and a rejection context called the discriminating
     literal, from a fixed set of primitives supplied by the user called the descrip-
     tion space. Steps a) and b) are repeated until no more faults can be identified
     by the critic.

	training instance	truth value of new rule	type of error
1.	positive	true	no error
2.	positive	false	omission
3.	negative	true	commission
4.	negative	false	no error

Table 1.

The critic detects two kinds of faults (Table 1), commission errors caused by an
insufficiently constraind rule-body, and omission errors caused by an incorrectly constrained rule-body. Commission errors arise from the comparison of a selection and a
rejection context, and are corrected by conjuncting a new condition to the old rule-
body. If the discriminating element is a extra factor its negation becomes the new
condition. Otherwise, the new condition is a relation between the two different lit-
erals (near-miss assumption). Omission errors are detected by comparing two selection
contexts and are corrected by disjuncting the new condition from the old-body [GA'88].
   Although the basic mechanism of the three learning procedures is the same, there
are some worth mentioning differences. The ECL has an internal system (r_rules) that
tests the behaviour of the new rules by itself. The head of the new rule (i.e. the fig-
ure-name) is supplied by the user. The MVL uses only positive examples (legal views).
One of its most important features is its ability to determine the next optimal view
and to minimize the no. of required views. It begins with a first view, finds all the
visible faces and decrements the FC of their sides by 1. This leads to a database of
'conn's with FC = 0 or 1. The 'conn's with FC = 1 determine which of the 5 views (sec.
2) is the next one according to a criterion (view with most invisible 'conn's first).
The SVL uses the MVD in order to generate its own rule-head. Finally all three learners
make constructive generalizations.

5. THE RECOGNIZER

The recognizer accomplishes the interpetation of polyhedral scenes in two stages using
the procedures SEE and REC. SEE calls the simulator to create a scene and produces the
initinal input of 'conn's to the system. The system makes a raw segmentation of the
scene by extracting all the polygons that could be possible faces. At this point a special procedure called LINK looks for Special Features (i.e. 'poin_', 'nocnvx_' etc)
and adds them to the initial database. REC segments the scene again based on the new
augmented database and asserts 'a2D_figure' and 'p2D_figure' predicates (i.e. atrian,
pquadril etc) to the system. At the same time decrements by 1 the FC of the 'conn's involved in the scene.
   The recognition is performed by the SVR that uses the SVD and has as many rules as
there are possible interpretations of line drawings. Rules responsible for more complex polyhedra with higher Face Visibility (no. of invisible faces / no. of visible
faces) are tried first. If a certain view can be interpreted as more than one polyhedra, then there there are as many entries sharing the same definition. The general format of these rules is:

    p_figure((3D_FigureName,FaceVisibility,AssumptionCode)):-
                    a3D_figure;SVD_3D_figure,mark_3D_figure.   (8)

117

Assumption Code is the key to one of several assumptions that should be made if a unique
interpretation is sought. Predicate 'a3-Dfigure' (e.g. aprism) is similar to the 'a2D_-
figure' (sec. 3), and is asserted by the predicate 'mark_3D_figure' when a SVD succeeds.
At the same time the 'a2D_figure' predicates are removed from the database, in order to
speed up the process that looks for alternative interpretations. The system allows the
user to look for certain polyhedra with a given Face Visibility in a scene.

If a unique interpretation is required, the system makes an assumption about the
hidden part of the line drawing (or needs another view). Assumptions are ad-hoc rules
used by the system, in order to add the missing lines to the database. There are 7 different assumptions [GA'88] divided into four main categories: no hidden elements, hidden point, hidden line and hidden face, based again on the cues given by the 'conn's
with FC = 1. Assumptions can be made only for one polyhedron at a time and are followed
by a MVR attempt. The MVR contains only four rules of the following format:

    m_figure(3D_FigureName):-MVD_3D_figure,mark_3D_figure.   (9)

The 'mark_3D_figure' predicate is used to update the FC of the 'conn's every time that
a polyhedron is recognized.

CONCLUSION

A system that learns to recognize simulated polyhedral scenes represented by line drawings has been developed. A simulator generates the input scenes. A learner uses induction in order to learn the definitions of four classes of polyhedra from single or multiple views. A recognizer attempts to interpret the scenes using the above acquired
knowledge base. The introduction of a structural representation for line drawings is a
useful tool and shows the close interdependence of the two proceses. The system was
tested with several scenes consisting of combinations among the recognizable polyhedra
and two "illegal" ones. The lengthiest process is the raw scene-segmentation (up to 3
min.), while all other processes are a lot faster (few sec.). The system is extensible
and can be equipped with rules that cope with more complex polyhedra. Some suggestions
for further development (occlusion, sophisticated rule-learning) are made in [GA'88].

REFERENCES

1. BUCHANAN, B.G. & FEIGENBAUM, E.A. 1978: 'DENDRAL and Meta-DENDRAL: Their Applications Dimension', AI, Vol. 11, pp. 5-24.
2. CLOWES, M.B. 1971: 'On Seeing Things', AI 2, 1, Spring, pp. 79-116.
3. DIETTERICH, T. & MICHALSKI, R.S. 1981: 'Inductive Learning of Structural Descriptions, AI Vol. 16.
4. GABRIELIDES, G. 1988: 'A System that Learns to Recognize 3-D Objects', PhD Thesis,
   Deprt. of Comp. Studies, LUT, Loughborough, England.
5. GUZMAN, A. 1969: 'Decomposition of a Visual Scene into 3-D Bodies', Automatic Interpretation and Classification of Images, Ed.: Grasseli, PhD Thesis, Academic Press.
6. HARRINGTON, S. 1983: 'Computer Graphics: A Programming Approach', McGraw-Hill.
7. HUFFMAN, D.A. 1971: 'Impossible Objects as Nonsense Sentenses', M 16.
8. KANADE, T. 1979: 'Recovery of the 3-D Shape of an Object from a Single View', CMU-CS
   Comp. Scien. Deprt., Carnegie-Mellon Univ., pp. 79-153.
9. LANGLEY, P. 1981: 'Language Acquisition through Error Recovery', CIP, Working Paper
   432, Carnegie-Mellon Univ., June.
10. MICHALSKI, R.S. 1983: 'A Theory and Methodology of Inductive Learning' in Machine
    Learning: An AI Approach, Ed.: Michalski, Carbonell, Mitchell, Tioga, pp. 83-134.
11. NEVES, D. 1985: 'Learning from Examples and Doing', Proc. 9th JICAI, pp. 624-630.
12. SILVER, B. 1983: 'Learning Equation Solving Methods from Examples', Proc. 8th JICAI,
    pp. 429-431.
13. WALTZ, D.I. 1975: 'Generating Semantic Descriptions from Drawing of Scenes with Sha-
    dows, PVC.
14. WINSTON, P.H. 1975: 'Learning Structural Descriptions from Examples': The Psycho-
    logy of Computer Vision, McGraw-Hill, pp. 157-209.

A TWO-STEP MODELLING ALGORITHM FOR TOMOGRAPHIC SCENES

WŁODZIMIERZ KASPRZAK

Instytut Podstaw Informatyki PAN
Laboratorium Architektury Systemow Komputerowych
ul. Polna 18/20, PL– 00–625 Warszawa, Poland

Abstract

The boundary of 3–D tomographic objects is modelled by polygons for the purpose of visualization. At first boundary faces of 3-D regions are detected and transformed into a list of small edges. The detection step is solved by the boundary tracking algorithm of Artzy et al.[1] Next in a polygon growing step adjacent coplanar edges are merged into planar loops. The advantages of independent subscene analysis in relation to the implementation of Frieder et al. [2] are also discussed.

1. INTRODUCTION

In opposit to a direct visualization technique, called *voxel rendering* [3], there exist also *indirect* methods, which assume an intermediate representation of the discrete scene – in terms of discrete volumes [4] or surfaces [5]. This paper reports the modelling by surfaces- technique. In order to derive a surface model from the 3-D scene a three steps strategy is usually performed: scene segmentation, boundary detection and surface formation. It is proposed here to analyse subscenes independently on each other. Then the solution of the three steps strategy requires only one input transmission of each subscene and this is performed without sending the temporary results to the outer memory.

2. THE ALGORITHM

The scene is traversed slice by slice, row by row, and column after column. The *boundary detection* procedure is called when a new unmarked boundary face has been found. The boundary of a homogeneous 3-D region is tracked, the outer boundary voxels are marked by current region label and simultaneously the edge elements are formed. Next the *edge growing* procedure starts - the list of edges is examined and subsequent, coplanar edges are merged into polygonal loops. The scene traversion is continued in the main cycle from the next unmarked voxel.

2.1 Boundary detection

A *face* is represented by an ordered pair (b, w) of voxels which share it. An *edge* of face (b, w) is each pair of voxels (b, v), such that b and v share it and w and v are face adjacent.

Let (v, v_2) be an edge of face (v, v_1) perpendicular to axis $i(i = 1, 2, 3)$(v, v_1 are face adjacent along axis i). It is said that (v, v_2) is a *right edge* of face (v, v_1) if v_1, v_2 are adjacent along axis $j = i \bmod 3 + 1$; or it is a *left edge* if v_1, v_2 are adjacent along axis $l = (i + 1) \bmod 3 + 1$. In other words if the face is perpendicular to axis $i = 1, 2, 3$, then right edges are also perpendicular to axis $j = 2, 3, 1$ and left edges – to axis $l = 3, 1, 2$.

Let a face be perpendicular to axis i. If k is the component along axis i of face's direction vector, then the distinction between *in-* and *out-edges* of given face is as follow:

- for left edges (perpendicular to axis j): $out_j = i_j + k_i$
- for right edges (perpendicular to axis l): $out_l = i_l - k_i$

Let Q be a region, (b, w) be a boundary face and (b, v) be a left (right) edge of face (b, w). Let further $u(u \neq w)$ be face adjacent with b and v. The face p **left (right) adjacent** to (b, w) is given as follow (Fig. 1):

$$p = \left\{ \begin{array}{ll} (b, u) \text{ , if } u \notin Q & /\text{CASE 1}/ \\ (u, v) \text{ , if } u \in Q, v \notin Q & /\text{CASE 2}/ \\ (v, w) \text{ , if } u, v \in Q & /\text{CASE 3}/ \end{array} \right.$$

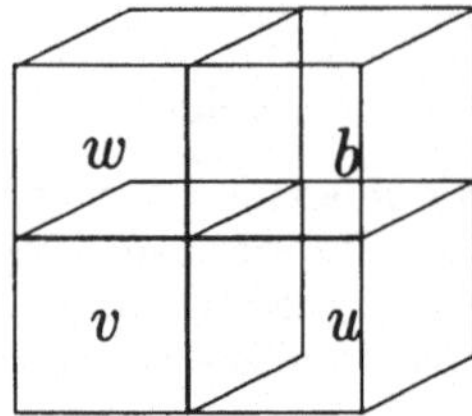

Figure 1: Face adjacency

The applied boundary tracking technique presented by Artzy et al. [1] is based on the traversion from actual boundary face, first to a left adjacent- and out- face and next to a right adjacent- and out- face.

Due to the efficiency of visualization the number of detected faces should be as small as possible. Hence it could be worthy to grow coplanar adjacent faces into polygons. The boundary edges are formed on the base of detected faces and they are grouped into six data vectors, one for each face direction. Two edges are formed if there is a face adjacency described above by CASE 1 or CASE 3. In CASE 2 two faces are coplanar and their common edge is inside of the approximating polygon. At any time when an out- edge from, say A to B, is located in the edge-vector indexed by the direction of given face F, the in-edge from B to A is located in the edge-vector indexed by the direction of the face adjacent to F.

2.2 Polygon formation

It is searched for all planar loops after the edge lists are formed. This step is performed by an edge growing procedure.

During edge growing some specific problems are solved:
- proper loop or hole classification
- a reducement of polygon when the number of vertices exceeds the limit
- smoothing of elementar step edges

In order to check the appearance of *holes* we store the sum of angles between subsequent polygon edges. Let a polygon plane be specified by a pair (i, k_i), where $i = 0, 1, 2$ is the normal vector axis, and $k_i = -1, 1$ is the normal vector length along axis i. Let j_R and j_l denote the right and left axes for plane $(i, 1)$. The angles between two subsequent edges Act_E and $Next\_E$ are specified in Table 1. For $k_i = -1$ the coordinates along axis j_L are taken with negative sign. This is equivalent with multiplication of the values in Table 1 by -1.

For a proper polygon the sum of all angles between subsequent two edges is equal to 2Π and for

	$j_L\ j_R$	Next_E 0 1	1 0	0 -1	-1 0
A	0 1	–	90°	err	−90°
c	1 0	90°	–	90°	err
t	0 -1	err	−90°	–	90°
E	-1 0	90°	err	−90°	–

Table 1: Loop vs. hole distinction (for $k_i = 1$)

a hole this sum is equal to -2Π.

If the size of an actual polygon exceeds a prespecified limit it is necessary to cut it in at least two parts. The convexity of derived polygon must be carefully checked. The new edge between the last and first vertex can be included in the polygon if all detected edges and all remained edges of the second part do not cross this edge. The direction between the last and first edge must be checked and eventually one should remove the first vertex.

We can reduce the number of vertices of a polygon by replacing small consecutive "step edges" by one straight diagonal edge. Consider the situation in Fig. 2. If two *edges* $12, 23$ have unit length each and the sum of *angles* $1, 2$ is null, then the *vertex* 2 will be eliminated from the polygon. If the third *angle* 3 is the same as the first one and the length of the third *edge* 34 has also unit length, then the *vertex* 3 will be eliminated.

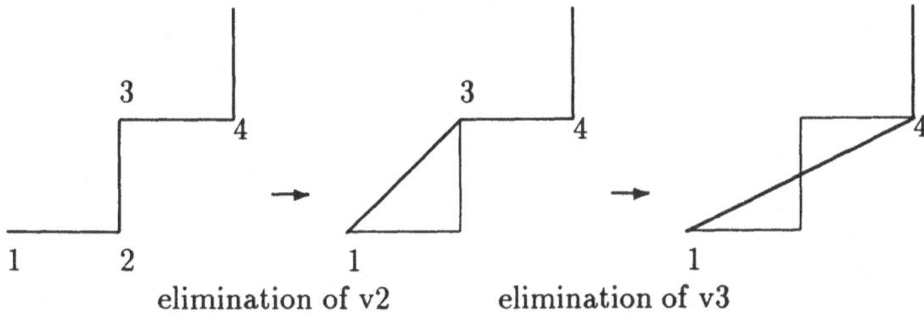

Figure 2: Smoothing of edges

3. SUBSCENE PROCESSING MODE

The **subscene mode** of discrete scene processing is proposed here. Particularly the "grey-level" subscenes of dimension 32 x 32 x 32 voxels with 4 bits/voxel are processed.

Originally the applied boundary tracking algorithm uses three lists of faces – the queue X, the mark list M and the output list [1]. Frieder et. al.[2] have stated five effectivity criteria for an implementation of the tracking algorithm: 1) handling input data, 2) data structure for list M, 3) queuing discipline, 4) output file organization, 5) encoding the faces.

Due to the mode of subscene-modelling the problems 1)-3) can be solved immediately. It is not required to provide a virtual memory for the whole scene, because the processed subscene is kept fully in the program memory and there is only one input transmission per subscene. The small size of the subscene allows us to implement the list M in a direct accessible way – by a 3-D bit array. Then the application of special hashing tables for efficient removing and relocating of marked faces is not required. In our particular implementation 16 kbyte of memory for the actual subscene and 24 kbyte for the face list M ($M/S = 150\%$) are allocated. The proportion M/S has clearly a maximum for binary scenes – 600% , but our ultimate goal is to process "grey-level" scenes with at least 16 density values per voxel. In the implementation [2] a 10 kbyte window for subscenes and a 12 kbyte mark list has been used. Hence the proportion between the mark list and direct available portion of scene was 120%.

For an array M the queuing discipline (LIFO or FIFO) is not important, because the number of faces on M (which is related to the number of faces on the queue list) is inrelevant and the memory requirements for the queue list are not crucial. Although the best results for two competitive performances of the que X were obtained by a FIFO technique [2], nevertheless in the subscene mode the LIFO technique can be used as well.

4. RESULTS

Two versions of the modelling algorithm were implemented in C language and tested on a PC AT microcomputer. The executable files need about 50 kbyte or 75 kbyte memory respectively. The first version has a continuous list M. The new added faces are always placed at the end of the list. The removing is only marking the places as "empty" and no relocation of remaining elements is done. In the second implementation a direct addressable array BD (for marking visited faces) instead of the list M is used. The output list can be omitted, as it is replaced by six edge lists for the polygon growing step (in both versions).

Sub-scene	Regions Faces	I. version		II.version		Polygons
		Track.	Grow.	Track.	Grow.	
S1	2 regions					
	542	25 s	2 s	5 s	2 s	11
	1182	105 s	22 s	9,5 s	29 s	286
S2	1 region					
	1200	106 s	51 s	11 s	57 s	305
S3	3 regions					
	1172	105 s	48 s	9,5 s	54 s	544
	900	60 s	30 s	6,5 s	32 s	421
	684	37 s	20 s	5 s	22 s	324

S1 – two cube- and one ball-object; S2 – one central ellipsoid
S3 - three rotated co-centered cubes

Table 2: Processing three subscenes

It was observed that the processing speed of the tracking step, while acting on simple, artificial subscenes, has increased 5–10 times (Table 2), if the array BD was used instead of a continuous list M. The time ratio was: at least $(5s/25s = 20\%)$ for the simplest region with small number of faces and maximal $(9,5s/105s \simeq 9\%)$ - for the most complex region with greatest number of faces.

In worst case the growing of polygons has reduced the number of surfaces to 50%. Maximal reduction was achieved for a simple two-cube object, well located in scene (its axes were parallel to the scene coordinate system axes) – $11polygons/542faces \simeq 2\%$. A middle reduction (for a ball) was: $286polygons/1182faces \simeq 24\%$. The minimal reduction occured for a small cube rotated by 45 degrees along each axis - $324polygons/684faces \simeq 47\%$.

The execution time of the polygon growing was in the worst case about 50% of the time for boundary detection with continuous list M. This took approximately 5 times longer than the fast version of the boundary detection,, however it should be related to the saved time during the later visualization. It is known that the time for viewing a set of surfaces is in proportion to the square number of surfaces. Nevertheless the polygon growing could be also improved by the application of a hashing technique or direct accessible arrays for the edge lists. If the same subscene is assumed, 96 kbytes of memory for an edge array will be required.

5. CONCLUSIONS

A discrete scene modelling approach with independently processed subscenes was proposed and implemented. The importance of efficient performance of the list of visited faces was shown (in extreme cases of list M performance the proportion of tracking time was at least 5 to 1). By the polygon growing the number of polygons needed for visualization has been decreased to 10–45%. The numbers of vertices of these polygons were relatively small due to application of the smoothing technique. The artificial surfaces between sub-scenes are not critical for visualization or can be easy detected and removed.

Although merely voxel rendering approaches will work in real-time, the indirect visualization techniques can be both useful and efficient. They are useful if images and image descriptions from various sources ought to be combined, like in image understanding and computer animation. Random accessible arrays for the lists of visited faces and detected edges should be used in order to profit from the technology improvement and to achieve enormous efficiency gain.

Acknowledgment: This work was partly performed during the author s stay at the Universität Erlangen–Nürnberg, IMMD 5 (Mustererkennung), Martensstr. 3, 8520 Erlangen (supported by the Alexander von Humboldt–Foundation, Bonn, F.R.G.).

References

[1] Artzy E.,Frieder G.,Herman G.T.: *The theory, design, implementation and evaluation of a three-dimensional surface detection algorithm*, **CGiP, 15** (1981), No.1, 11–24.

[2] Frieder G., Herman G.T., Meyer C., Udupa J.: *Large Software Problems for Small Computers. An Example from Medical Imaging*, **IEEE Software, 2** (1985), No.5, 37–47.

[3] Jaman K.A.,Gordon R.,Rangayyan R.M.: *Display of 3D Anisotropic Images from Limited-View Computed Tomograms*, **CVGIP, 30** (1985), 345–361.

[4] Meagher D.: *Geometric modeling using octree encoding*, **CGIP, 19** (1982), No.2, 129–147.

[5] Herman G.T.,Udupa I.K.: *Display of 3-D digital images: computational foundations and medical applications*, **IEEE CG&A, 3** (1983), No.5, 39–46.

[6] Gordon D., Udupa J.K.: *Fast Surface Tracking in Three– Dimensional Binary Images*, **CVGIP, 45** (1989), 196–214.

On the Capacity of Quantitative Computational Shape Recovery from Local Shading Information

Heiko Neumann

Universität Hamburg, Fachbereich Informatik
Bodenstedtstr.16, D-2000 Hamburg 50

Abstract. In the past decade of computational vision research considerable attention has been paid to shape estimation from shading information. Computational approaches incorporate numerous assumptions in order to constrain the solution space. Out of the set of techniques defined so far, Pentland's local shape estimation algorithm claims to be the one of most general applicability. With respect to the question what role this estimator can play in a general purpose vision system, a thorough evaluation has been carried out to assess the influence of inherent free parameters on the solution.

1. Introduction and Overview

One of the paramount tasks of biological and artificial vision systems is the reconstruction of depth that is lost due to retinal projection of a scene. Based on the known existence of monocular depth cues (see e.g. [7]) hierarchical computational models were proposed by [1] and [9] to estimate relative depth information from a single image. Computer Graphics models of light reflection contributed to formalization of intensity image formation which is a many-to-one mapping $a : (\rho, I_0, \phi_n, \theta_n, \phi_s, \theta_s) \rightarrow I$ typically used within local models for synthetic image generation ([3]). The variables explicate the surface reflectance (ρ), incident illumination (I_0), surface orientation parameters (ϕ_n, θ_n), and parameters defining the illuminant direction (ϕ_s, θ_s). Inferring local surface orientation from local intensity information thus is an "inverse graphics" problem involving a computational solution to a one-to-many mapping problem. Since this is an ill-posed problem, the inclusion of constraints derived from additional assumptions is necessary for achieving unambiguous results. Ikeuchi/Horn and Brooks/Horn ([6], [2]) proposed regularization approaches based on Horn's reflectance map ([5]). These authors incorporate knowledge about the light source position (in [6]) and knowledge about surface orientations (e.g. at occluding contours). Pentland was the first author who analyzed quantitative recovery of shape from shading without requiring a reflectance map. His argumentation is based upon known facts from the psychophysical (and in some part physiological) literature as well as on mathematical studies of the local differential geometry of illuminated surface patches. Modifications of this local shape from shading approach were published by Lee/Rosenfeld ([8]) and Ferrie/Levine ([4]). A detailed analysis of the above mentioned approaches (see [11] for details) exhibited that Pentland's approach involves the mildest requirements in order to constrain the inverse problem of shape recovery. This theoretical finding motivated further investigation in order to evaluate accuracy, efficiency, robustness, etc. as well as possible usability as a module in a general purpose vision system.

Assuming a Lambertian surface reflectance in the case of prior unknown reflection characteristics (e.g. outside specular highlights), the image irradiance equation is defined by $I = \rho I_0 \cdot (\vec{n} \cdot \vec{s})$. Assuming a point light source at infinity, the 1st and 2nd order differentials of I only depend on the 1st and 2nd order differentials of the surface normal $\vec{n}$, $d\vec{n}$ and $d^2\vec{n}$, respectively. Assuming a local (Monge) surface patch $\mathbf{r}(u, v)$, these differentials yield $d\vec{n} = -\kappa_1 \mathbf{r}_u\, du - \kappa_2 \mathbf{r}_v\, dv$ (first differential), where κ_1, κ_2 denote the principal curvatures of the surface patch, and, furthermore assuming a second order surface, $d^2\vec{n} = -\kappa_1^2 \vec{n}\, du^2 - \kappa_2^2 \vec{n}\, dv^2$ (second differential) (see [14] for details). Relating

124

these derivatives to those of the image intensity function requires a rotational/translational transformation, because derivatives of a surface normal are considered in the local tangential plane of the surface only. For a locally *spherical* surface with radius R, the possibility of pointwise mathematical evaluation of the system of equations, which allows for determination of all the parameters required including the local surface normal, has been proven in [15]. However, a shape estimator using local intensity information requires robust estimates based upon regions, rather than single points, for the purpose of estimating the tilt and slant components (ϕ_n, θ_n) of the surface normals. Based on both the image irradiance equation for a local sphere and its partial derivatives (and by using the normalized Laplacian and neglecting a factor that depends on the illuminant direction), a robust approach to estimation of the depth component of a surface normal $(z_n = \cos^{-1}\theta_n)$ has been derived by Pentland:

$$z_n = c \cdot (|\frac{\nabla^2 I}{I}| - c^2)^{-\frac{1}{2}} \tag{1}$$

with $c = 1/R$. Following a proof in [15] for correspondence of tilt direction and the direction of maximum 2nd order directional derivative, one can define a tilt estimator that incorporates weighted sums of the Laplacian output taken in different directions. In order to get more robust tilt estimates, one can compute the gradient of the slant field which has been determined by using eqn.(1) (see [15]). Thus, Pentland's approach is capable of independently computing the polar components of surface normals from local shading without knowing the illumination direction. However, the local shape estimation cannot resolve the convex/concave ambiguity without knowing the illuminant direction. This ambiguity can only be resolved by considering the sign of the derivative of the intensity function in the direction of the illuminant tilt. Pentland ([13], see [8] for an improved method) has proposed a robust maximum likelihood technique for estimation of the illumination direction from the distribution of image intensities, which requires the assumption of isotropic distribution of surface normals. This clearly is an improper assumption for images taken from short-range distances which possibly contain natural as well as man-made objects. Therefore, the estimation of the illumination direction was excluded in the following analysis and the considered objects were all assumed to be convex with respect to the view point.

2. Experimental Evaluation

Issues. The first issue in the evaluation of free parameters is the estimation of the sphere's curvature radius in the z_n-equation. Three approaches have been taken into account: Estimating the *main* value (minimum method) and the *mean* value (variance method) for the radius R as well as the direct mapping of estimated z_n-values (z_n^*) to the range of possible z_n-values of a hemisphere. For the minimum method, the value for R is estimated by using the quotient $2/\min[\|\nabla^2 I/I\|]$. The variance method is based on the observation that a measure of surface curvature can be derived from the variance of the intensity distribution within a considered region. For that reason, depth can be estimated up to $z_n \approx \sigma_\kappa \cdot (|\nabla^2 I/I| - \sigma_\kappa^2)^{1/2}$. In order to derive useful results, the measure σ_κ must be scaled by a constant k which has to be set empirically. In order to be independent of this constant, the third alternative for z_n-estimation is the direct mapping of the estimated z_n-values onto the interval $[0..1]$ according to their previously estimated min/max-values of z_n^* (see [11] for details).

Once a satisfactory estimate has been found, the shape estimator has been evaluated with respect to the following aspects:

- parametrization of the algorithm (i.e. a scaled filter function) with regard to a discrete scale space,

- influence of *i)* the region sizes and *ii)* the size of a local neighborhood for suppressing operator responses (which otherwise enter into the procedure for estimating surface slant and therefore produce erroneous results) in the vicinity of zero-crossings in the Laplacian on the local z_n-estimates,

- influence of changes of both illumination direction and viewer position,

- influence of specular reflection (e.g. local violation of Lambert's law),

- quality of reconstruction for spheres, ellipsoids, cylinders, etc., and

- usability and validity of the method for the case of real (e.g. CCD) camera images.

Brief Description of Experiments and Results. For the evaluation of the algorithm a number of experiments have been carried out within a carefully defined testbed ([12]). The first experiment was concerned with the influence of parametrization of the filter base function, an isotropic 2D Gaussian ($\sigma_x = \sigma_y$), on z_n-estimates. The image data used for this experiment was a synthetic Lambertian sphere of radius 70 pixels illuminated by a point source placed at the view point position. The correspondence of view and light source position was used in order to exclude effects that may influence the result. The results from this startup experiment are summarized below:

- since z_n-estimates in low slant areas derived from image data filtered with a small scale operator are unreliable, reliable estimates have been only obtained for parametrizations $\sigma \in \{6.01, 12.37\}$ ($\equiv \omega \in \{17,35\}$, $\omega = 2\sqrt{2}\sigma$) from a discrete scale space sampled logarithmically,

- the region size (R_s) for estimation of z_n from local intensity values directly influences the result,

- since depth estimation was performed regardless of image discontinuities, reconstruction is erroneous in the local neighborhood of LoG zero-crossings, and

- the variance method is more robust than the minimum method; the multiplication factor k can only be empirically determined ($k = 1000.0$ has been found to be a useful value)

The second experiment analyzed the influence of the region size R_s for estimation of z_n-values from image data and the region size R_{zc} for suppression of erroneous estimates that originate from results of preprocessing in the neighborhood of image discontinuities. The image data used was the same as in experiment 1. The results lead to the conclusions that

- the volume of the reconstructed object shrinks in correspondence with increasing values for the region size,

- for appropriate σ-parametrizations of the low pass filter function (see experiment 1), an increasing region size R_s smoothes the result of reconstruction a bit but has no significant influence (a useful value obtained is $R_s = \omega$),

- the variance method ($k = 1000.0$) works more robust than the minimum method, and

- based on the model of an ideal step edge profile filtered with a LoG, $R_{zc} = 1.4 \cdot \omega$ is a good compromize between error minimization and volume maximization.

Within the third experiment the influence of illuminant slant was tested. The image data used here contained the same synthetic sphere as in the previous experiments, but illuminated from positions of constant azimuth $\phi_s = 270^\circ$ and varying polar angle $\theta_s \in \{10^\circ, 30^\circ, 50^\circ, 70^\circ\}$. Due to the neglection of the illuminant component within the z_n-estimator, the results show asymmetric z_n-profiles in the $<\vec{v}, \vec{s}>$ plane. The maximum slant error was determined to be $\Delta\theta_s \approx 5^\circ$.

In the fourth experiment the z_n-estimator was applied to shiny surfaces. Synthetic images of spheres illuminated from $\theta_s \in \{0^\circ, 30^\circ\}$ with specular components of 30% and 70%, respectively, have been used (synthetic image generation with Cook/Torrance illumination model ([3])). The estimator was parametrized according to the results from experiments 1 and 2. The results can be shortly summarized: Since specular highlights introduce irradiance discontinuities, the results of reconstruction are extremely distorted. Although the region sizes R_s and R_{zc} have been varied, no significant improvements have been achieved (e.g. for illuminant slant $\theta_s = 0^\circ$ and 30% specular reflectance).

The z_n-estimation for elliptic surfaces with varying principal curvatures was analyzed in the fifth experiment. The image data used of synthetic images of cylinders with axis length ratios of 1:2:1, 1:4:1, 1:10:1 and 1:1:2, in a $< x,y,z >$-coordinate system ($< x,y >$ parallel to the image plane). Illumination direction was set to $(\phi_s, \theta_s) = (0°, 0°)$ and $(315°, 30°)$, respectively. The results for surfaces with 1:n:1 axis ratios show robust estimates even in the case of 1:10:1. For the obliquely illuminated volume minor asymmetries occur as already observed in experiment 3. The results of processing the 1:1:2-surface were not so promising. For both simulated illumination directions the sign of curvature of the intensity surfaces change, i.e. LoG-z.c.s of minimal slope are present. The fluctuations in the reconstruction produce distorted results in the z_n-estimation.

In the sixth experiment the viewer position was varied with respect to the major principal axis of an ellipsoid with axis ratios of 2:1:1. The Lambertian ellipsoid was slanted (rotation about x-axis) by angles of $\beta \in \{30°, 60°\}$ using illuminant slants of $0°$ and $30°$ for each surface slant. The results show that differently slanted surfaces cannot be reconstructed in such a way that corresponding z_n-values are affected by a factor of $\cos\beta$.

Within the seventh experiment real camera images of single illuminated objects have been processed. The objects used were a sphere with diameter $d = 8.0[\text{cm}]$ as well as a cylinder with diameter $d = 9.5[\text{cm}]$ and height $h = 8.5[\text{cm}]$. Both objects were viewed obliquely ($\beta = 25°$) and the surface material was of a matt silken-sheen finish lacquer (see [12]). The images were acquired in a laboratory setup using a CCD camera (see [12] for details). The object-to-camera distance was $d = 1.24[\text{m}]$ and the illumination direction was set to $(\phi_s, \theta_s) = (65°, 50°)$ at a distance $d = 4.20[\text{m}]$ with a divergence of light rays of $0.1283°$ per centimeter measured in the object plane. The background material used was a matte dark green cloth, thus semi-thresholding for visual background suppression was rendered possible. The reconstructed z_n-fields verified the robustness of estimates. Another series of experiments were performed for the reconstruction of an ellipsoid (main axis length: $d_1 = 5.4[\text{cm}]$, orthogonal axis length: $d_2 = 4.1[\text{cm}]$) illuminated from a distance of $d = 0.85[\text{m}]$ and viewed from positions of $d_s = 0.59[\text{m}]$ and $0.82[\text{m}]$, respectively. Different parametrizations of the algorithm were tested. The results were distorted because of only small viewer-to-object distance and missing graduation in image irradiance due to the minimum light source distance and high surface reflectivity.

Within the eighth experiment the third method for the z_n-estimation involving the mapping of the z_n^*-estimations onto the z_n-range $[0..1]$ ($\min[z_n^*] \equiv z_n = 0$ and $\max[z_n^*] \equiv z_n = 1$) was tested. This z_n-estimator was applied to a variety of images that had been used throughout the previous experiments: Synthetic images of the sphere and ellipses used in experiment 1, 2, 5 and 6; real camera images of the sphere and ellipsoid from experiment 7. The results can be characterized as useful. Estimates for the ellipsoid of 1:1:2 axis ratio viewed along the major axis, result in a z_n-distribution that looks like a sombrero. Therefore, the results were not useful as judged in experiment 5. Furthermore, the fidelity of the results has been found to depend critically on the parametrization of the diameter of R_s (for further details, see [11]).

3. Conclusion

The experiments demonstrated the robustness of the estimation of surface orientation for curved Lambertian surfaces. Regarding local directional derivatives by summing up the normalized Laplacian, only flat, cylindrical, and doubly curved surfaces can be distinguished. Saddle surfaces cannot be approximated by second order surface patches (as can be seen from the eqn. for $d^2\vec{n}$ in Ch.(1)). A more detailed surface classification is possible by considering the quotient I_{xx}/I_{yy} ([15]). A negative quotient always denotes a saddle, but, unfortunately, a positive result cannot always be associated with elliptic surfaces only. Counterexamples can be shown where a saddle also results in positive I_{xx}/I_{yy} values. The non-linear mapping of estimated z_n^*-values to the range $[0..1]$ always requires the presence of all possibly visible surface orientations. Precise reconstruction of object surfaces is not possible using this approach for local surface estimation due to the underestimated volume. Variations of the image irradiance function distort the result of shape reconstruction just in cases where *a)* specular highlights are present, or *b)* areas of high slant values and foreshortening (e.g. ellipsoids viewed from the tip point) produce inflections in the intensity surface and, therefore, sign changes in $\nabla^2 I$ operator outputs. The inability of reconstructing object slants based on shading information alone seems to correspond with human visual performance as reported in [10].

In summary, the experiments showed that the local shape estimator, relying upon minimal assumptions about the imaging process, works well for matte surfaces. Quantitative shape information can be determined without requiring information on the light source position however leaving open the convex/concave ambiguity. However, in its present formulation the approach is not an appropriate model for human visual shape inference as claimed by Pentland. Its significance within a theory of robust and reliable general shape reconstruction must be further explored.

Literature

[1] Barrow, H.G.; Tenenbaum, J.M.: Recovering Intrinsic Scene Characteristics from Images. in: Hanson, A.R.; Riseman, E.M. (Eds.): Computer Vision Systems. New York: Academic Press (1978).

[2] Brooks, M.J.; Horn, B.K.P.: Shape and Source from Shading. Proc. 9th IJCAI, Los Angeles, USA, Aug. 18-23 (1985) 932-936.

[3] Cook, R.L.; Torrance, K.E.: A Reflectance Model for Computer Graphics. Computer Graphics 15 (3) (1981) 307-317.

[4] Ferrie, F.P.; Levine, M.D.: Where and Why Local Shading Analysis Works. IEEE Trans. PAMI 11 (2) (1989) 198-206.

[5] Horn, B.K.P.: Understanding Image Intensities. AI 8 (2) (1977) 201-231.

[6] Ikeuchi, K.; Horn, B.K.P.: Numerical Shape from Shading and Occluding Boundaries. AI 17 (1-3) (1981) 141-184.

[7] Kaufman, L.: Sight and Mind - An Introd. to Visual Percept. New York: Oxford Univ. Press (1974).

[8] Lee, C.H.; Rosenfeld, A.: Improved Methods of Estimating Shape from Shading Using the Light Source Coordinate System. University of Maryland, Computer Science Center: TR-1277 (1983).

[9] Marr, D.: Vision. San Francisco: W.H.Freeman & Co (1982).

[10] Mingolla, E.; Todd, J.T.: Perception of Solid Shape from Shading. Biol. Cybern. 53 (1986) 137-151.

[11] Neumann, H.: Shape from Local Shading: A Quantitative Evaluation. Universität Hamburg, Fachbereich Informatik: Technical Report, in preparation (1989).

[12] Neumann, H.; Stiehl, H.S.: Toward a Testbed for Evaluation of Early Visual Processes. Proc. of the 2nd Int. Conf. on CAIP '87, Wismar, GDR, Sept. 2-4 (1987) 202-208.

[13] Pentland, A.P.: Finding the Illuminant Direction. J. Opt. Soc. of Am. 72 (4) (1982) 448-455.

[14] Pentland, A.P.: Local Analysis of the Image: Limitations and Uses of Shading. Proc. IEEE Workshop on Computer Vision: Representation and Control, Ringe, USA, Aug. 23-25 (1982) 153-161.

[15] Pentland, A.P.: Local Shading Analysis. IEEE Trans. PAMI 6 (2) (1984) 170-187.

Modellgestütztes Bildverstehen von Dokumenten

Joachim Kreich
Siemens AG, München, Zentralabteilung Forschung und Entwicklung
Otto-Hahn-Ring 6, 8000 München 83

Zusammenfassung

Will man Bilder einer erweiterten Nutzung durch Computer zuführen, ist es sinnvoll, sie tiefer zu strukturieren und gesuchten Klassen und Bedeutungsinhalten zuzuordnen, d.h. "verstehen" zu lassen. Wir stellen ein System zum Verstehen von Dokumenten nach Bildvorlagen vor, in welchem alles Wissen konsequent in semantisch eindeutigen Konzepten repräsentiert ist. Aus diesen Konzepten leitet ein flexibel steuerbarer Schlußfolgerungsprozeß logische Aussagen bzw. Hypothesen über das Bild her und versucht, diese im Bild zu bestätigen. Zur Beherrschung der mächtigen Suchräume in den Konzept-, Bildobjekt- und vor allem Hypothesennetzen verfügt das Analysesystem über reichhaltig Metawissen und Möglichkeiten, um die Schlußfolgerungsstrategie effizient zu gestalten. Stichwörter sind hier: Heuristiken, Focus of Attention *und* Truth Maintenance. *Anwendungszweck des als Prototyp entwickelten Systems ist die Überführung von Papierdokumenten in Textverarbeitungssysteme oder Datenbanken. Außerdem stellen Dokumente einen relativ einfachen Objektbereich dar, in welchem diese für die Bildanalyse allgemein interessanten Verfahren systematisch erprobt werden können.*

1. Überblick

Das bildverstehende System [Krei88] beginnt ein Dokument zu analysieren, indem das Dokument von einer Kamera oder einem Scanner eingelesen und in ein Binärbild umgewandelt wird. Dieses Bild wird in Text- und Grafikanteile segmentiert [Sche86], von denen im weiteren Verlauf derzeit nur Textbereiche weiter analysiert werden. Textanteile werden mit einem Schriftzeichen-Erkennungsverfahren [Bern84] als ASCII-Code erkannt und gespeichert. Die Umrahmungen der einzelnen Schriftzeichen werden zu Wort-, Zeilen- und Textblockrahmen zusammengefaßt [Luhn88] und als Ergebnisse der Vorverarbeitung als Objekte gespeichert.

Die weitere Analyse übernimmt der wissenbasierte Teil des Systems, welcher traditionelle KI-Ansätze in der Mustererkennung wie HEARSAY-II [Erma80], VISION [Hans78] und das dokumentverstehende System von Srihari [Srih86] aufgreift und versucht, diese über neue Verfahren der Künstlichen Intelligenz weiterzuentwickeln. Wissen über Dokumentobjekte wird hauptsächlich mittels einer grafisch orientierten Wissenserwerbskomponente erfaßt und in hierarchisch organisierten [Nagy84], frameähnlichen Konzepten analog [Higa86] und [Ejir89] modelliert. Zur Zeit wird fast nur Wissen über die Textgeometrie (Layout) und kaum Wissen über Textinhalte (Logik) gespeichert und angewendet. Zur Analyse werden die Definitionen in den Konzepten benutzt, um z.B. mittels konzeptspezifischen Prozeduren Position und Form von Textrahmen oder Teile- und Nachbarschaftsrelationen im Dokument zu überprüfen [Higa86].

Dabei erlaubt die heterarchische Organisation des Konzeptwissens wie in [Srih87] sowohl *Top-down-* als auch *Bottom-up-*Verarbeitung bzw. eine zustandsabhängige flexible Steuerung des Schlußfolgerungsprozesses. Auf diese Weise entsteht ein großes Netz von Hypothesen, dessen Komplexität maßgeblich Komponenten und Strategie der modellgestützten Analyse bestimmt.

2. Wissensrepräsentation in Konzepten

Zu jedem relevanten Begriff über Dokumente (Brief, Adresse, Signatur usw.) wird ein *Konzept* angelegt, in welchem alle Aussagen über diesen Begriff gesammelt werden. *Konzepte* werden in der bekannten, frameähnlichen Form [Brac85] repräsentiert, wobei zu jedem Dokumentbegriff je ein Layout- und ein Logikkonzept angelegt wird. In einem Konzept wird zu einem Begriff sowohl das deklarative Wissen in Form von relationalen Aussagen als auch das prozedurale Wissen dieses Begriffs gespeichert.

Adresse
    Teile: Name, Straße, Wohnort

Firmenadresse
    Oberklasse: Adresse
    Teile: Firma   ->   kumuliert zu
                             Teile: Name, Straße, Wohnort, Firma

Bild 1: Kumulative Vererbung von Rollen

Neben der Definition von Defaultwerten für Instanzen werden in den Konzepten die Klassen (Rollen) der möglichen Werte angegeben. Diese werden kumulativ vererbt, d.h. konkretere Konzepte (Unterklassen) summieren unter den einzelnen Relationen alle Rollen der eigenen Klasse und der Oberklassen (Bild 1). Diese Vererbungsart erspart Redundanzen bei der Definition von Konzepten und fordert bei konkreteren Konzepten nur die Angabe von spezifischen Aussagen.

Regelhafte Aussagen werden in relationaler Form gespeichert, so daß nicht nur Regeln, sondern auch Regelteile vererbt werden können. Solche regelhafte Aussagen können vorwärts und rückwärts angewendet werden, was für die Navigationsmöglichkeiten der Analyse sehr wichtig ist. Die meisten Relationen wie Klasse, Element, Teile (Bild 2), Teil-von, Nachbar-links usw. werden in eigenen Konzepten beschrieben, um dort deklarative und prozedurale Eigenschaften der Relationen zu definieren, und um so der Analysesteuerung Metawissen zur Verfügung zu stellen.

Teile
    Klasse: Relation
    invers: Teil-von
    relationale-Eigenschaft: transitiv
    notwendige-Relationen: Klasse
    Prüffunktion: Teile-p

Bild 2: Konzept einer Relation

Als konzeptspezifische Prozeduren sind in den Konzepten die Algorithmen der Bilderkenung in semantisch möglichst eindeutiger Weise enthalten. Dies sind in unserem System z. Zt. vor allem Prüfprozeduren, welche Positions-, Form- und Nachbarschaftsbestimmungen von Dokumentteilen durchführen. Später sollen hier auch syntaktische und semantische Prozeduren zur Analyse von Texten aufgenommen werden. Prozeduren werden in objektorientierter Weise vererbt.

3. Flexible Steuerung der Schlußfolgerungen

Analysieren von Bildern bedeutet in unserem System, daß das in Konzepten repräsentierte Wissen auf Bilder angewendet wird, indem Beziehungen zwischen einzelnen Konzepten und Bildteilen hergestellt und untersucht werden. Der Analysevorgang stellt also einen Schlußfolgerungsprozeß ("Reasoning") dar, der solche hypothetischen Beziehungen entlang unterschiedlicher Relationen z. T. über viele Hypothesenstufen hinweg vorwärts und rückwärts verfolgt und an Hand von bestimmten Bedingungen überprüft.

Wegen der großen Anzahl von Bildobjekten und Konzepten und der Komplexität der Entscheidungen kann der Schlußfolgerungsprozeß nicht schematisch gesteuert werden. Daher scheiden Steuerungsverfahren wie *Depth-first search* [Wins84] und blindes *Backtracking* oder die Beschränkung nur auf Rückwärts-verfolgung von regelhaften Beziehungen aus. Vielmehr benötigt die Analysesteuerung viele Freiheitsgrade, um sich an dem jeweiligen Analysezustand und an dem Wissen, das in den Konzepten enthalten ist, zu orientieren und um möglichst zielgerichtet und effizient die einzelnen Aktionen steuern zu können.

Dokumentanalyseheuristik
    Klasse: Heuristik
    Start-mit-Layout-oder-Logik: Layout
    Start-Konzepte: Adresse, Brief
    notwendige-Relationen: Teile
    Kontext-Relationen: Teil-von, Nachbar

Bild 3: Allgemeines Steuerungskonzept

Der Analyseprozeß kann als ein Zyklus aufgefaßt werden, in dem Hypothesen generiert, ausgewählt, verifiziert und ausgewertet werden. Die Analyse beginnt solche Anfangshypothesen zu generieren bzw. nach solchen Konzepten im Dokument zu suchen, wie sie in dem allgemeinen Heuristikkonzept *Dokument-analyseheuristik* (Bild 3) aufgabenabhängig festgelegt sind. Nach unserem Beispiel soll zunächst eine Adresse gesucht werden, weil Adressen eventuell leichter zu finden und maßgeblich für die weitere Analyse sind. Solch eine Hypothese in der Form

$\exists$ (x) Klasse (x, Adresse), x $\in$ {Dokumentobjekt1, Dokumentobjekt2 ...}   *h1*

wird ebenfalls als Objekt gespeichert (Bild 4).

h1
    Klasse: Hypothese
    Relation: Klasse
    Variable-1:   Dokumentobjekt1, Dokumentobjekt2 ...
    Variable-2:   Adresse
    Quantor: $\exists$
    Teile: h2, h3, h4
    Wahrheitswert: Nil

Bild 4: Repräsentation einer Hypothese

Hier können Hinweise auf abhängige Hypothesen, Zwischenergebnisse und viele andere Informationen aus der Bearbeitung von Hypothesen bzw. einzelner Schlußfolgerungen hinterlegt werden.

3.1 Verifikation von Hypothesen

Hypothesen gelten als bestätigt, wenn sie sowohl relational als auch prozedural wahr sind. Die relationa-le Verifikation einer Hypothese (*h1* in Bild 4) benutzt den deklarativen Teil der beteiligten Konzepte (Bild 5), um dort aufgeführte Forderungen im Bild zu überprüfen. Dies geschieht mit der Generierung von wei-teren Hypothesen, um z.B. die im Konzept einer Adresse geforderten Teile *Name*, *Straße* und *Wohnort* (*h2, h3 und h4* in Bild 5) zu finden. Falls alle diese Teilhypothesen wahr sind, ist auch die Ausgangshypothese relational wahr. Diese Verifikationsart entspricht der Rückwärtsverfolgung von Regeln.

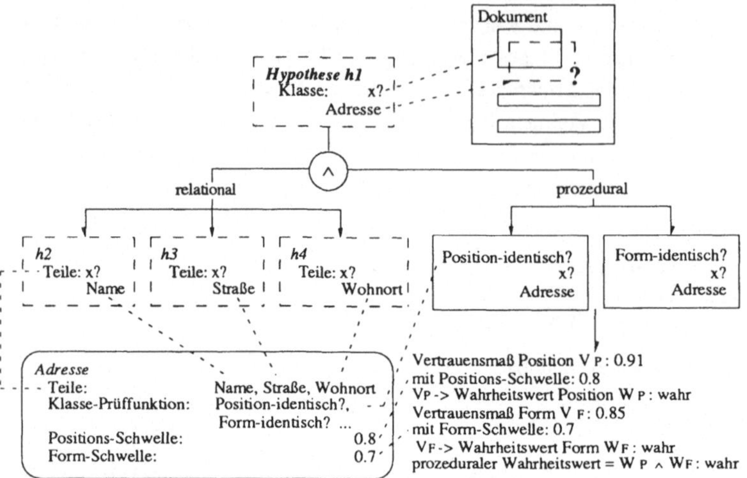

Bild 5: Verifikation einer Hypothese

Prozedural wird *h1* verifiziert, indem die im Konzept von *Adresse* angegebenen Prüffunktionen, hier *Position-identisch?* und *Form-identisch?*, auf Dokumentobjekte angewendet werden (Bild 5). *Position-identisch?* vergleicht die klassenmäßig definierte Position einer Adresse mit der Position der jeweiligen Dokumentobjekte und ermittelt ein Vertrauensmaß für die Positionstreue *VP* (0.91) nach

$$V_P = 1 - \sqrt{\left(x_{Klasse} - x_{Objekt}\right)^2 + \left(y_{Klasse} - y_{Objekt}\right)^2} \;/\, d_{max}$$

Hierin bedeuten *x* und *y* Positionsangaben und *d$_{max}$* die Diagonale einer Seite. *VP* wird an der konzeptspezifischen *Positions-Schwelle* (0.8) gemessen und der Wahrheitswert der Position *Wp* auf *wahr* (oder *falsch*) gesetzt. Solche Schwellen werden vom Entwickler als Erfahrungswerte eingegeben und defaultmäßig vererbt. Ebenso ermittelt *Form-identisch?* für die Form von Adressenkandidaten ein Vertrauensmaß für *VF* (0.85) nach

$$V_F = \left(\frac{Min\left(H\ddot{o}he_{Klasse}, H\ddot{o}he_{Objekt}\right)}{Max\left(H\ddot{o}he_{Klasse}, H\ddot{o}he_{Objekt}\right)} + \frac{Min\left(Breite_{Klasse}, Breite_{Objekt}\right)}{Max\left(Breite_{Klasse}, Breite_{Objekt}\right)}\right) / 2$$

und setzt mittels der *Form-Schwelle* (0.7) den Wahrheitswert der Formtreue *WF* auf *wahr*. Alle (in diesem Fall zwei) prozeduralen Wahrheitswerte werden über eine logische Konjunktion verbunden, so daß hier der gesamte prozedurale Wahrweitswert mit *wahr* bewertet wird.

3.2 Auswertung von Hypothesen

Nach ihrer endgültigen Verifikation werden Hypothesen ausgewertet. Ist eine Hypothese wahr, wird die damit bestätigte Relation in den beteiligten Objekten vermerkt, so daß nun z.B. Dokumentobjekt X tatsächlich als *Teil-von* Dokumentobjekt Y anerkannt ist. Solche Dokumentobjekte werden in den Objektfokus eingetragen. Bei falschen oder irrelevanten Hypothesen wird in die Wahrheitswerte der Teilhypothesen *irrelevant* eingetragen, so daß diese nicht tiefer verifiziert werden. Alle von einer verifizierten Hypothese direkt tangierten Hypothesen werden in den Hypothesenfokus eingereiht (siehe 3.4: Auswahl von Hypothesen). Diese Hypothesen können mit einer gewissen Aussicht auf Erfolg weiterbearbeitet werden.

Alle verifizierten Hypothesen bleiben mit ihren Ergebnissen als Begründungen [Dres89] der Schlußfolgerungen zugreifbar. Alle in den Dokumentobjekten gespeicherten Ergebnisse bleiben erhalten, auch wenn Hypothesen im größeren Kontext als falsch bewertet werden. Hiermit wird eine *Dependency-Directed-Backtracking*-Strategie [Wins84] analog einem *Truth-Maintenance*-System [Klee86] realisiert, indem nur solche Hypothesen zurückgenommen werden, die auch tatsächlich falsch sind, und indem wahre Zwischenergebnisse auf jeden Fall bestehen bleiben, damit sie eventuell in anderem Kontext weiterverarbeitet werden können.

3.3 Generierung von Hypothesen

Die Generierung von Hypothesen orientiert sich an Objekt- und Hypothesenfokus. Auf diese Weise können interessante Objekte und Hypothesen gezielt weiter untersucht werden. Dabei wird immer auf das in den Konzepten hinterlegte relationale Dokumentwissen zurückgegriffen. Die grobe Reihenfolge der Generierung von Hypothesen orientiert sich an bestimmten Typen von Relationen, die in *Dokumentanalyseheuristik* (Bild 3) definiert sind. Mit höchster Priorität (① in Bild 6) werden Hypothesen zu notwendigen Relationen generiert, wenn sie - wie oben beschrieben - zur Verifikation von anderen Hypothesen notwendig werden (Rückwärtsverkettung, Bild 5).

Mit niedrigerer Priorität werden Hypothesen zu Kontext-Relationen erzeugt, die zwar im Konzept enthalten sind, die aber nicht notwendigerweise erfüllt sein müssen, wie die *Teil-von*-Relation oder Nachbarschaften (Vorwärtsverkettung, ② in Bild 6). Die Analyse versucht also, ein untersuchtes Objekt mit seiner Umgebung zu verbinden und so vielleicht von einem lokalen Ergebnis gezielt auf einen größeren Zusammenhang zu schließen.

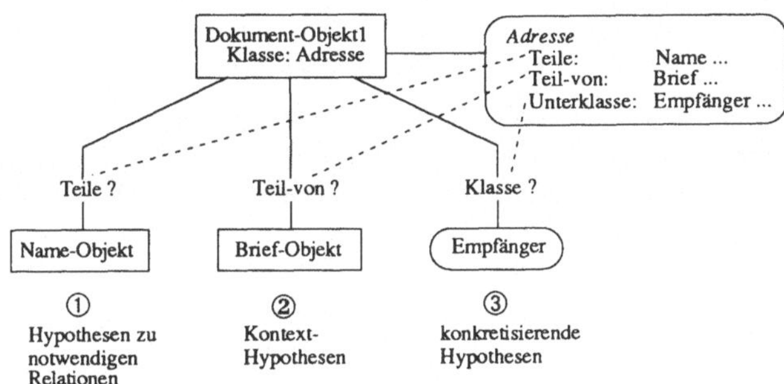

Bild 6: Generieren von Hypothesen

Mit niedrigster Priorität werden schließlich Hypothesen generiert, die versuchen, Objekte konkreter zu bestimmen (③ in Bild 6). Hier werden die Hinweise auf Unterklassen aus den für die Objekte bestätigten Konzepten benutzt, um zu versuchen, solche Objekte in Richtung dieser spezifischeren Konzepte genauer zu bestimmen (Adresse -> Empfänger).

3.4 Auswahl von Hypothesen

Obwohl das Hypothesennetz - wie beschrieben - vorsichtig und gezielt aufgebaut wird, ist die Anzahl der Hypothesen bei komplexeren Dokumenten bzw. Bildern zu groß, als daß die Auswahl der als nächstes zu bearbeitenden Hypothese nach einem starren Schema durchgeführt werden könnte. Aus der Menge der Hypothesen werden solche Hypothesen in den Hypothesenfokus übernommen, welche Aussagen über Objekte machen, die sich im Objektfokus befinden oder die von der Verifikation (wahr oder falsch) anderer Hypothesen betroffen sind.

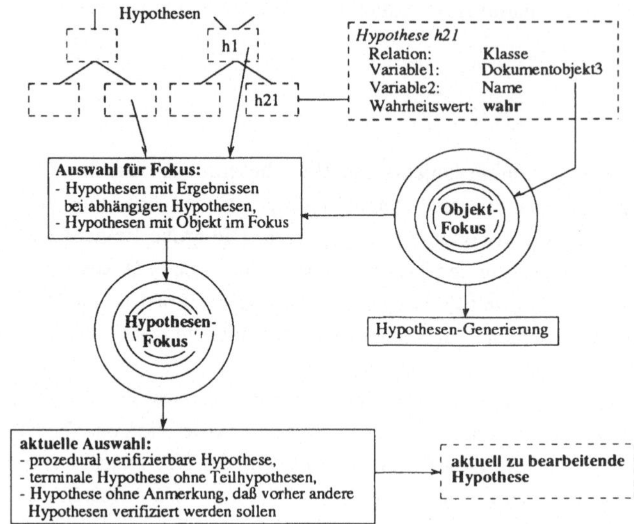

Bild 7: Auswahl von Hypothesen

Ist beispielsweise Hypothese *h21* in Bild 7 als wahre Hypothese bewertet worden, wird die abhängige Hypothese *h1* in den Hypothesenfokus übernommen. Gleichzeitig wird *Dokumentobjekt3* als betroffenes Objekt einer als wahr bestätigten Hypothese in den Objektfokus übernommen, so daß weitere Hypothesen, die sich auf dieses Objekt beziehen, ihrerseits in den Hypothesenfokus übernommen werden können. Aus dem Hypothesenfokus wird schließlich diejenige Hypothese zur Bearbeitung ausgewählt, deren Verifikationskosten minimal sind.

4. Stand der Arbeiten

Wir haben bisher etwa 180 Konzepte modelliert. Alle geschilderten Verfahren von der Wissensrepräsentation über die Wissenserwerbskomponente bis zur eigentlichen modellgestützten Dokumentanalyse sind auf Symbolics-Lispmaschinen implementiert und in der Lage, eine eingeschränkte Zahl einfacher Dokumente zu analysieren. Außerdem steht eine Erklärungskomponente zur Verfügung, welche in grafischer und texthafter Form den fortschreitenden Analyseprozeß protokolliert und auf gezielte Fragen des Benutzers hin Schlußfolgerungen (Hypothesen), Konzepte und Dokumentobjekte erklärt.

Um einen Analyseablauf zu beschreiben, sind in Bild 9 zwei Bildschirmdarstellungen eines Analyseprozesses gezeigt, der auf der Anwendung der in Bild 8 skizzierten Konzepte beruht. Die Analyse hatte zunächst mit Hilfe des Konzeptes *Dokument* ganz allgemein geprüft, ob die zu untersuchende Vorlage überhaupt als Dokument zu betrachten ist und mindestens einen größeren Textbereich (Paragraph) enthält.

Dokument
    Teile: Paragraph
    Unterklassen: Brief, Bericht

Brief
    Teile: Paragraph, Empfänger, Datum, Begrüßung
    Unterklassen: CCITT-Brief, Siemens-Brief

Empfänger
    Teile: Name, Straße, Wohnort

CCITT-Brief
    Teile: Paragraph, Empfänger, Datum, Begrüßung,
                          Firmen-Überschrift, Firmen-Info, Firmen-Unterschrift

Bild 8: Konzepte für Analysebeispiel

Nachdem dies zutraf, wurde eines der konkreteren Konzepte - hier *Brief* - vom System gewählt, um die Untersuchung weiter zu spezialisieren. Die Brief-Teile *Empfänger, Datum* und *Begrüßung* wurden gesucht und gefunden und somit das Dokument als Brief erkannt. Als weitere Konkretisierung wurde aus dem Briefkonzept die Unterklasse *CCITT-Brief* gewählt und als Gesamthypothese über das Dokument aufgestellt. Dieser Analyseschritt ist in Bild 9 so dargestellt, wie er während der Analyse auf dem Bildschirm gezeigt wurde.

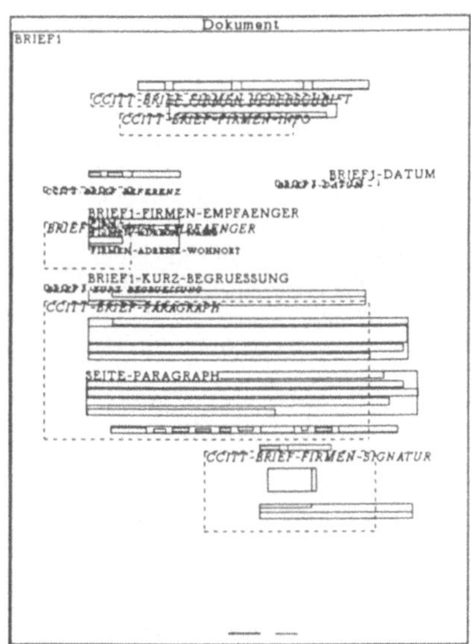
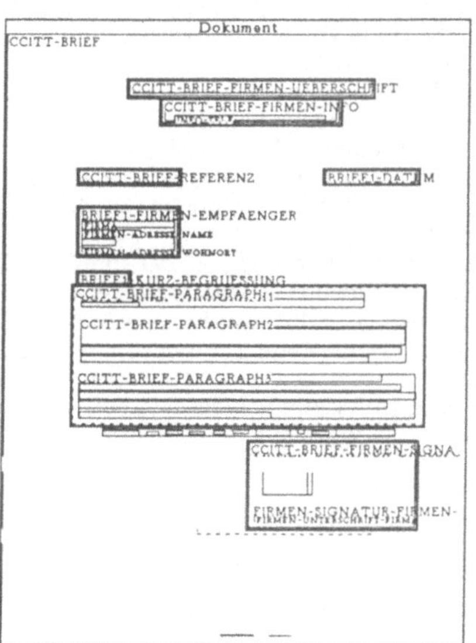

Bild 9 : Analysebeispiel: Links: Hypothesen, rechts: Analysierter Brief

Im linken Teil des Bildes sind die Ergebnisse der Vorverarbeitung als durchgezogene Rahmen und das Modell des *CCITT-Briefs* in Form seiner hypothetisierten Teile wie *CCITT-Brief-Referenz* und *CCITT-*

Brief-Firmen-Unterschrift gestrichelt dargestellt. Dokumentteile dieser Klassen und rekursiv deren Teile usw. wurden an den skizzierten Positionen gesucht und gefunden. Schließlich wurde, wie im rechten Teil von Bild 9 gezeigt, das gesamte Dokument als *CCITT-Brief* klassifiziert.

5. Ausblick

Wir wollen in die Dokumentanalyse wesentlich mehr Wissen vor allem bezüglich der inhaltlichen Analyse von Texten einbeziehen. Zu diesem Zweck wollen wir die schon eingerichteten Dokumentlogikkonzepte mit syntaktischem und semantischem Wissen über Dokumentinhalte versehen. Die Wissenserwerbskomponente wird erweitert werden müssen, so daß z.B. auch Lernen aus Beispielen und Widerspruchsüberprüfung unterstützt werden. Unser frameartige Repräsentationschema soll in Richtung auf *KL-ONE* [Brac85] bzw. *CLOS* erweitert und die Unifikation von logischen Formeln einbezogen werden.

Zur Verwaltung der Hypothesen sollen Standardmechanismen aus dem Bereich *Truth Maintenance* benutzt werden. Probabilistisches Schlußfolgern [Tani87] soll ermöglicht werden mit Verfahren wie *Fuzzy Reasoning* oder nach *Dempster-Shafer*. Die Vorverarbeitung der Dokumente und die wissensbasierte Analyse sollen weitestgehend integriert werden. Dies gilt einerseits für die Repräsentationsanforderung vom Pixelbereich bis zum Dokumentkonzept. Dies gilt jedoch auch für den Analyseprozeß, wo Entscheidungen über Linien und Schriftzeichen vom Kontext wesentlich höherer Ebenen (z.B. Adresskonzept) abhängen können.

Literaturverzeichnis

[Bern84] Bernhardt, L.: Three Classical Character Recognition Problems, Three New Solutions. Siemens Forschungs- und Entwicklungsberichte 13; 1984; Nr. 3; pp. 114-117.

[Brac85] Brachman, J.R.; Schmolze, J.R.: An Overview of the KL-ONE Knowledge Representation System. Cognitive Science 9; 1985; pp. 171-216.

[Dres89] Dressler, O.; Freitag, H.: Truth Maintenance Systeme. KI; 2/1989; pp. 13-19.

[Ejir89] Ejiri, M.: Knowledge-based Approaches to Practical Image Processing. International Workshop on Industrial Applications of Machine Intelligence and Vision (MIV-89); Tokyo 1989; pp. 1-8.

[Erma80] Erman, L.D.; Hayes-Roth, F.; Lesser, V.R., and Reddy, D.R.: The HEARSAY-II speech understanding system: Integrating knowledge to resolve uncertainty. Computing Surveys 12(2); 1980; pp. 213-253.

[Hans78] Hanson, A.; Riseman, E.: VISION: A Computer System for Interpreting Scenes. Hanson, Riseman (Ed.): Computer Vision Systems; Academic Press, Orlando; 1978; pp. 303-333.

[Higa86] Higashino, J.; Fujisawa, H.; Nakano, Y.; Ejiri, M.: A Knowledge-based Segmentation Method for Document Understanding. Proceedings of the 8th Int. Conf. Pattern Recognition, Paris 1986; pp. 745-748.

[Klee86] deKleer, J.: Assumption-Based TMS. Artificial Intelligence 28; 1986; pp. 127-162.

[Krei88] Kreich, J.: Wissensbasierte Dokumentanalyse. Proceedings 10. DAGM-Symposium, Zürich 1988; pp. 326-332.

[Luhn88] Luhn, A.; Dengel, A.: Modellgestütztes Segmentieren und Hypothesengenerierung für die Analyse von Papierdokumenten. Proceedings 10. DAGM-Symposium, Zürich 1988; pp. 226-232.

[Nagy84] Nagy, G.; Seth, S.: Hierarchical Representation of Optically Scanned Documents. Proceedings of the 7th Int. Conf. Pattern Recognition, Montreal 1984; pp. 347-349.

[Sche86] Scherl W.: Bildgraph, stochastische Grammatiken und Erkennungskriterien zur Analyse allgemeiner Druckvorlagen. Dissertation Techn. Fakultät Universität Erlangen-Nürnberg; 1986.

[Srih86] Srihari, S.N.: Document Image Understanding. IEEE, Proceedings of the Fall Joint Computer Conference (Cat. No. 86CH2345-7), Dallas, TX; 1986; pp. 87-96.

[Srih87] Srihari, S.N.; Bozinovic, R.: A Multi-Level Perception Approach to Reading Cursiv Script. Artificial Intelligence 33; 1987; pp. 217-255.

[Tani87] Tanimoto S.: The Elements of Artificial Intelligence. Computer Science Press, Rockville, Maryland; 1987.

[Wins84] Winston, P.: Artificial Intelligence. Addison-Wesley, Reading, MA; 1984.

Lageerkennung von Werkstücken innerhalb einer Blechbearbeitungszelle[*]

R.Lange   S.Schröder[†]   U.Geißler[‡]

Lehrstuhl für Informatik 5 (Mustererkennung)

und

Lehrstuhl für Fertigungstechnologie

Universität Erlangen Nürnberg

Übersicht

Flexible und automatisierte Produktionsanlagen sind eine wichtige Voraussetzung für die Verbesserung der Wettbewerbsfähigkeit der produzierenden Unternehmen. Gerade bei der automatischen Fertigung in kleinen Losgrößen mit möglichst geringen Rüstzeiten ist es notwendig, den Materialfluß flexibel zu gestalten. Dies bedingt jedoch die Identifikation und Lageerkennung der Werkstücke an den Bearbeitungseinheiten. Dazu bieten sich insbesondere Sensorlösungen an. Der vorliegende Artikel beschreibt ein kameragestütztes Lageerkennungssystem, daß unter Verwendung kostengünstiger Hard- und Software in eine Bearbeitungszelle integriert und dort mit Erfolg eingesetzt wurde.

1   Einleitung

Die Leistungsfähigkeit der Produktionstechnik bestimmt in der modernen Industriegesellschaft entscheidend die Entwicklung von Wohlstand, Lebensqualität und Sicherheit. Für die produzierenden Unternehmen ergeben sich aus den bestehenden Marktbedingungen, insbesondere durch den zunehmenden internationalen Wettbewerb, Forderungen nach Verbesserung der eigenen Wettbewerbsfähigkeit. Dies kann auch durch höhere Flexibilität in der Planung, Gestaltung und Herstellung der Produkte erreicht werden[Spu87].

Vor diesem Hintergrund wurde 1985 zwischen der Friedrich-Alexander-Universität und der Siemens AG mit Unterstützung durch den Freistaat Bayern die Durchführung des PAP-Projekts vereinbart. Dabei steht PAP für Projekt flexibel automatisierte Produktionssysteme. Mit dem Wissen, daß der industrielle Produktionsprozeß auf einem Zusammenwirken von Energietechnik, Fertigungstechnik, Materialtechnik und Informationstechnik beruht, haben sich im PAP-Projekt 6 Lehrstühle der Fertigungstechnik und der Informatik zusammengeschlossen, um kooperativ neue Konzepte für die Fabrik der Zukunft zu erarbeiten. Dabei bildet eine Modellfabrik die Basis für eine Reihe von gemeinsamen Forschungsprojekten bei denen ein besonderer Schwerpunkt auf den Aufgaben der Elektro- und Elektronikindustrie liegt [Fel87].

Zur Unterstützung des Automatisierungsprozesses sind insbesondere Sensorlösungen geeignet. Während bei der Fertigung von Massengütern mit hohen Loszahlen die Position von Werkstücken durch Paletten fest vorgegeben war, benötigt die flexible Fertigung mit kleinen Loszahlen einen flexiblen Materialfluß. Dies erfordert aber die Identifikation und Lageerkennung der Werkstücke an den Bearbeitungseinheiten.

Der hier vorliegende Artikel beschreibt ein solches System zur Lageerkennung, das am Lehrstuhl für Fertigungstechnik mit Unterstützung des Lehrstuhls für Mustererkennung aufgebaut wurde. Für

---

[*]Die hier dargestellte Arbeit entstand mit Unterstützung der Siemens AG Erlangen

[†]Lehrstuhl für Informatik 5, Martensstr. 3, 8520 Erlangen

[‡]Lehrstuhl für Fertigungstechnologie, Egerlandstr. 11, 8520 Erlangen

Abbildung 1: Fertigungsbeispiele des Blechbearbeitungssystems [Gei87]

eine existierende Blechbearbeitungszelle war die Notwendigkeit entstanden, die Lagekoordinaten beliebiger ebener Blechteile zu bestimmen. Der Schwerpunkt bei der Entwicklung des Systems bestand dabei nicht in der Realisierung eines allgemeinen Bildverarbeitungssystems, sondern in der Implementierung eines speziell auf eine Bearbeitungszelle ausgerichteten Systems. Wesentliche Voraussetzung für den flexiblen Einsatz ist die Möglichkeit, die für die Lageerkennung notwendigen Merkmale aus den CAD-Daten herzuleiten.

Obwohl dabei nur kostengünstige Hard- und Software verwendet wurde, sind die industriellen Anforderungen an die Qualität der Ergebnisse und an die Durchlaufzeit eingehalten worden. Dies ist durch die Einbettung des Systems in die Blechbearbeitungszelle der Modellfabrik gezeigt worden.

Blechbearbeitung gewinnt in heutiger Zeit in zunehmenden Maße an Bedeutung. Die Gründe dafür liegen in modernen Fertigungsstrategieen, die sinkende Losgrößen, hohe Innovationsgeschwindigkeit und schonenden Einsatz der Resourcen erfordern. Diesem Anspruch nach Flexibilität bei hoher Qualität kann bei der Blechbearbeitung in großem Maße Rechnung getragen werden.

Die Entwicklung, der Aufbau und die Erprobung eines derartigen flexiblen Systems stehen im Mittelpunkt des Teilprojkets Blechbearbeitung innerhalb des PAP-Projekts. Das Blechbearbeitungssystem soll der Herstellung abgekanteter Blechteile mit beliebigen Außenkonturen und Lochmustern dienen, wie sie vielfach als Gehäuse und Trägerelemente für elektrische und elektronische Geräte in Losgrößen bis zu wenigen tausend Stück benötigt werden [Gei87,Gei88]. Beispiele für solche Teile zeigt Abbildung 1

Die Struktur des Bearbeitungssystems zeigt Abbildung 2. In einem ersten Bearbeitungsschritt erfolgt das Zuschneiden der Blechteile durch Laser. Die Werkstücke verlassen danach die Schneidemaschine über eine Kleinteilerutsche und werden durch ein Transportsystem zur Biegemaschine gebracht. Das Handhabungssystem der Biegemaschine legt das zugeführte Teil ein und richtet es anhand der Lageinformation aus. Danach wird der Biegevorgang durchgeführt.

Die Verwendung einer Kleinteilerutsche ist eine für diese Maschinenkombination übliche Entsorgung. Andere Möglichkeiten, wie z.B. die Entnahme der zugeschnittenen Teile nach oben mit einem Sauggreifer oder aber eine separate Entnahmestation, sind wegen des erheblichen Aufwands nicht rentabel. Außerdem müßte aufgrund der unterschiedlichen Taktzeiten der beteiligten Maschinen auf jeden Fall ein Zwischenlager geschaffen werden. Soll die Lageposition des zugeschnittenen Teils erhalten bleiben, so müßte auch das Lager in diese Aufgabe einbezogen werden. Auch das würde einen erheblichen zusätzlichen Aufwand erfordern. Bei der aufgezeigten Lösung mit der Kleinteilerutsche geht zwar die Lageinformation verloren, als einfaches Zwischenlager dient dabei das Förderband des

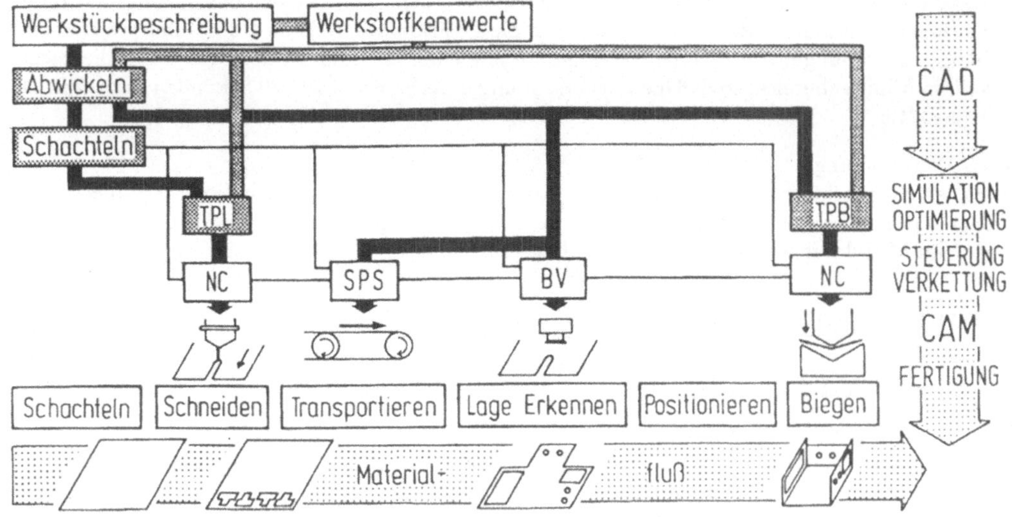

Abbildung 2: Struktur des Blechbearbeitungssystems [Gei87]

Transportsystems. Da für ein definiertes Einlegen die Lage des Werkstücks jedoch bekannt sein muß, ist es notwendig, vor der Handhabung die Lageinformation wiederzugewinnen.

Der hier dargestellte Anwendungsfall zeigt Charakteristika, die die Lagebestimmung offensichtlich vereinfachen. Es ist bekannt, welches Werkstück aufgenommen wurde. Damit entfällt die Notwendigkeit einer Klassifikation. Außerdem liegt das Teil vollständig und vereinzelt im Aufnahmebereich der Kamera. Desweiteren ergibt sich aus der Art der Werkstücke, daß eine Binärbildverarbeitung zur Bestimmung der äußeren und inneren Konturen für die Berechnung der Lage ausreicht. Andererseits enthalten die Werkstücke Details wie dünne Stege, die aufgrund der Auflösung nur bruchstückhaft oder überhaupt nicht im Bild wiedergegeben werden. Dies kann zu Problemen bei der Lagebestimmung führen.

Der Artikel gibt eine Übersicht über das realisierte Lageerkennungssystem. Die Details sind in [Lan88] beschrieben.

2 Anforderungen

Durch den Einsatz innerhalb der Blechzelle ergeben sich Rahmenbedingungen, die beim Entwurf der Lageerkennung berücksichtigt werden müssen:

- Einsatz in einer Fertigungsumgebung
  Das industrielle Umfeld beeinflußt vor allem die Aufnahmeeinrichtung. Sie muß mechanisch robust sein, bei verschiedenen Werkstoffen wie beschichteten und unbeschichteten Blechen und Aluminium hohen Kontrast liefern und Fremdlichteinflüsse tolerieren.

- Geringe Auflösung
  Die Teile sollen vollständig im erfaßten Bildbereich liegen. Durch das große Teilespektrum (Werkstücke von 50 – 400 Millimeter Kantenlänge) werden damit aber Details nur ungenügend wiedergegeben. Die minimale erkennbare Stegbreite beträgt beim installierten System 2 Millimeter. Bei der Analyse muß dies berücksichtigt werden.

- Zeitliche Beschränkung
  Um einen kontinuierlichen Fertigungsfluß sicherzustellen, soll bereits während des Biegens die

Handhabung des nächsten Werkstücks vorbereitet werden. Dazu müssen die notwendigen Informationen vom Zellenleitrechner übermittelt, die Position des Teils bestimmt und ein Handhabungsprogramm generiert werden. Bei der vorhandenen Maschinenausstattung liegt die Biegezeit im Minutenbereich, so daß die Lageerkennung innerhalb von 10 - 20 Sekunden durchführbar sein sollte.

- CAD-Anbindung
  Es ist zu beachten, daß die zur Lagebestimmung notwendige Information aus einer CAD-Datenbasis gewonnen werden soll. Die bei der Analyse eingesetzten Merkmale sind so zu wählen, daß im Mittel nur wenige Daten an das Lageerkennungssystem übermittelt werden müssen.

3 Bildverarbeitung im industriellen Umfeld

3.1 Hard- und Software

Für die Realisierung der Lageerkennung stand ein Personalcomputer (IBM AT) mit mathematischem Koprozessor zur Verfügung. Er wurde durch eine Bildverarbeitungskarte (PCVISION Frame Grabber) erweitert. Sie enthält im wesentlichen den Bildspeicher, dessen Inhalt über einen externen Monitor angezeigt wird. Als Aufnahmesensor wurde eine CCD-Kamera (Sony AVC-D5CE) verwendet. Die Konfiguration zeigt Abbildung 3.

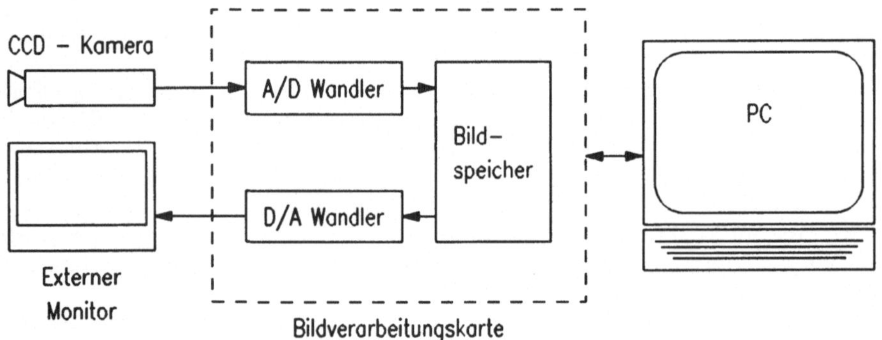

Abbildung 3: Konfiguration des Bildverarbeitungssystems

Die einzelnen Schritte der Lageerkennung und der Modellbildung verdeutlicht Abbildung 4. Man erkennt die für ein Binärbildsystem typische Bearbeitungsreihenfolge Binärbilderzeugung, Konturverfolgung und Merkmalextraktion. Auf der Modellseite werden aus den CAD-Daten des Werkstücks eine Reihe von für die Lageerkennung geeigneten Merkmalen bestimmt, von denen eine Untermenge im Analyseschritt mit Merkmalen des aufgenommenen Werkstücks verglichen wird.

3.2 Das System zur Lageerkennung

In jedem Bildverarbeitungssystem spielt die Aufnahme eine wichtige Rolle [Foi82]. Da im vorliegenden Fall die Aufnahmeeinrichtung weitgehend frei gewählt werden konnte, wurden verschiedene Anordnungen experimentell untersucht.

Bei der Verwendung von Durchlicht war zwar die Binärbilderzeugung durch einen festen Schwellwert besonders einfach und der Bildkontrast und die Detailabbildung waren sehr gut, die Oberfläche des Leuchtkastens erwies sich aber für den Dauerbetrieb als nicht robust genug. Ein daraufhin angebrachter Schutz der Milchglasoberfläche in Form eines dünnen Gitters erzwang eine große Kamerablende, damit das Gitter im Bild unsichtbar blieb. Hierbei erwies sich die automatische Verstärkungsregel der Kamera als gravierender Nachteil, da dadurch eine Binärisierung mit festem Schwellwert

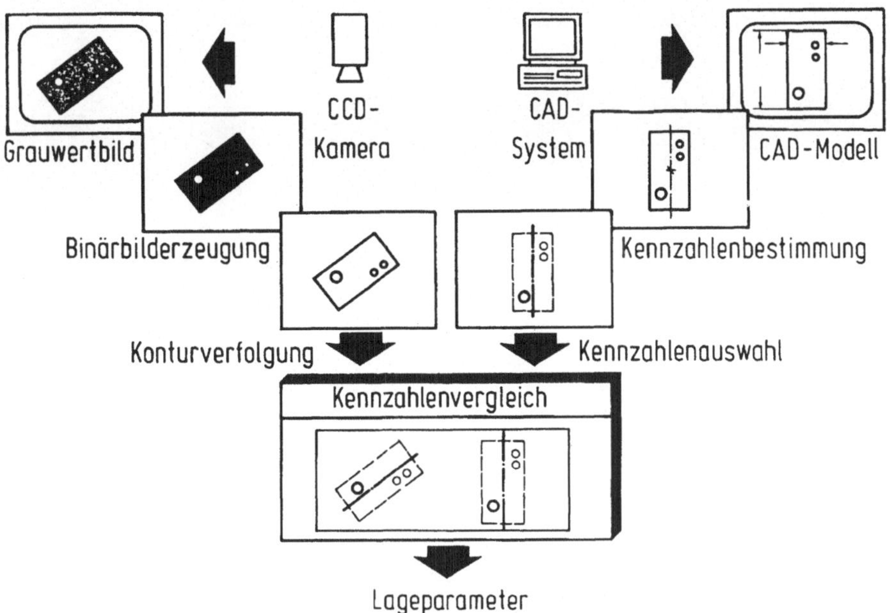

Abbildung 4: Bearbeitungsfolge der Lageerkennung und der Modellbildung

nicht mehr möglich war. Aufgrund dieser Voruntersuchungen entschied man sich, in der Zelle Auflicht einzusetzen. Die Teile werden auf der schwarzen, nur schwach reflektierenden Metalloberfläche eines Schrägförderers abgelegt, wobei alle im Bildbereich liegenden Objekte wie die Einfassung des Förderbandes und der Hallenfußboden geschwärzt wurden. Die mit dieser Konfiguration erzielten Ergebnisse waren so gut, daß der hohe Aufwand für die Verwendung von Durchlicht nicht gerechtfertigt schien.

Die Kamera wurde rund 2,5 Meter über der Werkstückablage an der Hallendecke montiert. Als Objektiv wurde ein Weitwinkel-Zoom (28–70mm Brennweite) verwendet. Die Beleuchtung der Szene kann durch die vorhandene Hallenbeleuchtung realisiert werden, da sich mehrere Leuchtstoffröhren in unmittelbarer Nähe zur Aufnahmeeinrichtung befinden. Diese Lichtquellen mußten jedoch seitlich abgedeckt werden, um direkten Lichteinfall in die Kamera zu vermeiden.

Bei der Aufnahme wird das Bild auf einen nichtquadratischen CCD-Sensor mit 512 x 512 Pixeln abgebildet. Diese Bildverzerrung muß bei den folgenden Bearbeitungsschritten berücksichtigt werden. Weitere Bildtransformationen, wie beispielsweise radiale Bildverzerrung durch Linsenabbildung, wurden aufgrund ihres geringen Einflusses vernachlässigt. Experimentelle Untersuchungen und die guten Ergebnisse des Gesamtsystems motivierten diese Entscheidung.

Die Transformation des Grauwertbildes in ein Binärbild wird durch ein Schwellwertverfahren vorgenommen. Die Verwendung eines globalen Schwellwerts ist dabei ausreichend, da die Szene gleichmäßig beleuchtet wird. Außerdem kann die Transformation dann direkt durch die Hardware der Bildverarbeitungskarte mit Look-Up-Tabellen durchgeführt werden. Dieser Schwellwert muß jedoch, vor allem aufgrund der Tageslichteinflüsse, automatisch auf die Aufnahme abgestimmt werden. Dazu wird das Grauwerthistogramm berechnet und der Schwellwert wie in [Hum86] beschrieben in das Tal zwischen den beiden Maxima gelegt, die zu Werkstück und Hintergrund gehören.

Das Binärbild wird nun nach Objektkonturen durchsucht. Der verwendete Algorithmus [Pav82] betrachtet dabei nicht das ganze Bild, sondern bricht ab, sobald ein dem Werkstück in der Fläche ähnliches Objekt gefunden wurde. Außerdem kann die Kontursuche an einer vorgegebenen Bildschirmposition gestartet werden. Dies ist günstig, da durch eine mechanische Zentriereinrichtung die Werkstücke immer an ähnlicher Stelle zu liegen kommen. Die Zahl zeitintensiver Bildspeicherzugriffe

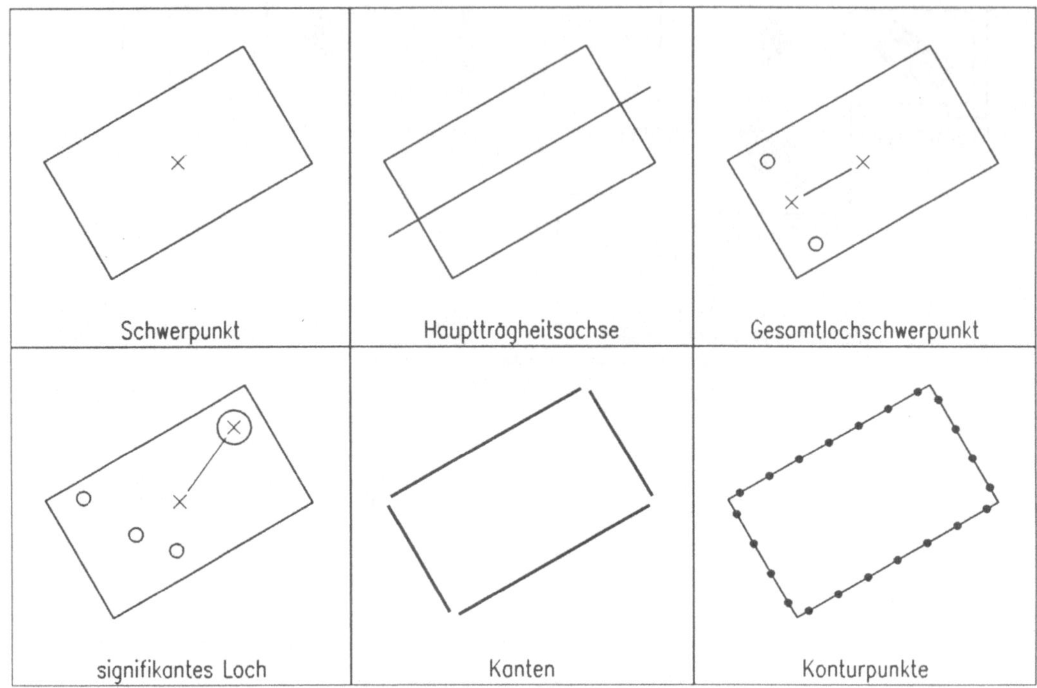

Abbildung 5: Merkmale

wird dadurch reduziert.

Aus den Außen- und Innenkonturen werden für die Lageerkennung geeignete Merkmale berechnet. Dabei sollen - wie in Abschnitt 2 angedeutet - meist schon wenige der in Abbildung 5 dargestellten Merkmale genügen, um eine Positionsbestimmung durchzuführen. Es wurden auch weitere Merkmale wie Eckpositionen untersucht. Diese brachten aber keine nennenswerte Leistungsverbesserung des Systems.

Zur Translationsbestimmung des Werkstücks wird immer der Objektschwerpunkt eingesetzt, da die Teile vollständig im Bild liegen. Dessen Berechnung erweist sich als sehr robust gegen kleine Aufnahmestörungen. Jedoch ist wichtig, daß bei der dazu nötigen Momentenberechnung nicht nur die Außenkontur alleine, sondern auch die Innenkonturen betrachtet werden. Durch die geringe Bildauflösung kann beispielsweise ein Loch, das nahe am Werkstückrand liegt, durchaus in die Außenkontur einbezogen werden. Die umschlossene Fläche und weitere Momente ändern sich dadurch natürlich.

Die Rotationsbestimmung kann oft schon durch Angabe der Achse des kleinsten zweiten Moments (Hauptträgheitsachse) erfolgen [Nie83]. Sie gibt bei langgestreckten, achssymmetrischen Teilen die Richtung dieser Ausdehnung an. Des weiteren können auch Objektlöcher (Gesamtlochschwerpunkt, Signifikantes Loch) als Merkmale eingesetzt werden. Da Geradenabschnitte in der Kontur das häufigste konstruktive Element sind, kann die Werkstückposition auch daraus ermittelt werden. Für Problemfälle kann man schließlich — ähnlich einer Schablone — Konturpunkte angeben.

Im Analyseschritt werden die Merkmale mit den entsprechenden Merkmalen des CAD-Modells verglichen. Meist genügt hier schon eine einfache Differenzbildung (Verschiebung des Schwerpunkts, Winkeldifferenz der Hauptträgheitsachsen), um die Position des Werkstücks festzustellen. Nur beim Vergleich von Geraden und von Konturpunkten wird die von der Hough-Transformation bekannte Akkumulatortechnik [Bal82] eingesetzt.

Es ist geplant, aus den möglichen Merkmalen aufgrund heuristischer Gütemaße eine für die Lageerkennung eines bestimmten Teils geeignete Untermenge auszuwählen. Vorschläge dafür wurden bereits in [Lan88] gemacht, ihre Anwendbarkeit bei einem größerem Teilespektrum muß jedoch erst untersucht werden.

4 Experimentelle Ergebnisse

Das System zur Lageerkennung wurde mit verschiedenen Blechzuschnitten getestet. Die Modelle wurden dazu durch Werkstückprototypen erzeugt und die Merkmale, die zur Lageerkennung verwendet werden sollen, manuell ausgewählt. Die Testwerkstücke mit den zugehörigen Merkmalen zeigt Abbildung 6

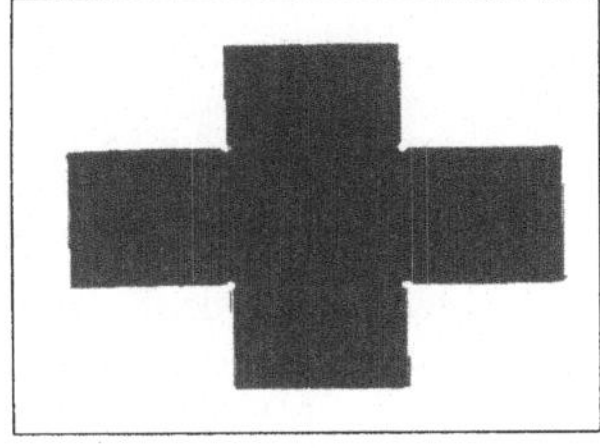

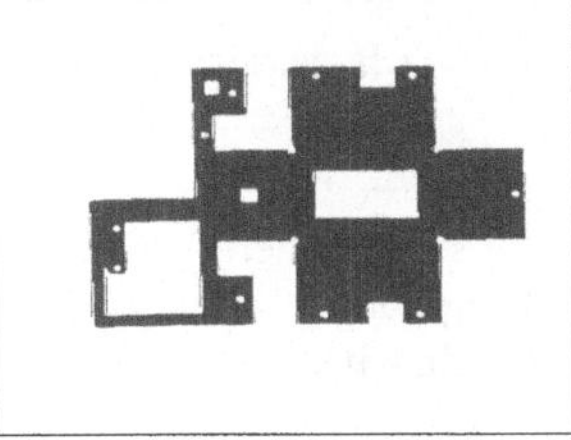

Schwerpunkt
Hauptträgheitsachse

Schwerpunkt
Hauptträgheitsachse
Kanten

Schwerpunkt
Signifikantes Loch

Abbildung 6: Testwerkstücke mit den verwendeten Merkmalen

Die Ergebnisse der Programmodule Binärbilderzeugung, Konturextraktion und Merkmalgewinnung, die subjektiv am Monitor beurteilt wurden, waren sehr zufriedenstellend.

Um auch ohne einen mechanisch aufwendigen drehbaren XY-Tisch quantitative Aussagen über die Genauigkeit der Lageerkennung zu erhalten, wurden mit einem CAD-System Werkstückkonturen erstellt. Davon wurde eine definiert verschobene und gedrehte Kopie erzeugt. Ein derartiges „virtuelles Werkstück" wurde als Prototyp angelernt, dann abgedeckt und die Kopie zur Lageerkennung angeboten. Die Translationsparameter konnten dabei sehr genau bestimmt werden (Abweichungen im Subpixelbereich). Die Resultate bei der Bestimmung der Rotationsparameter zeigt Tabelle 1. Die Abweichung liegt jeweils deutlich unter einem Grad.

Tabelle 1: Genauigkeit der Lageerkennung bei der Bestimmung des Rotationswinkels in Grad

Teil	Hauptr.	Ges.-Loch-Sp.	Sig. Loch	Kanten	Konturpkt.
1	0.09	0.30	0.29	0.22	0.40
2	0.16	0.18	0.10	0.31	0.57
3	0.33	–	–	0.35	0.24

Die Laufzeit der Lageerkennung hängt von den eingesetzten Merkmalen ab, wobei allerdings nur Kanten und Konturpunkte eine merkliche Verlängerung bewirken (siehe Tabelle 2). Bei einfachen Teilen wird die Analyse schon innerhalb von 3 Sekunden durchgeführt, im Durchschnitt erfolgt die Lageerkennung in rund 6 Sekunden. Die in Abschnitt 2 genannte Zeitschranke wird also in ausreichendem Maße eingehalten.

5 Zusammenfassung

Ein flexibler Materialfluß in der Fabrik der Zukunft erfordert die Lageerkennung und Identifikation der Werkstücke an den Bearbeitungseinheiten. Dazu wurde das in diesem Artikel vorgestellte Lageerkennungssystem kooperativ von Fertigungstechnikern und Informatikern entwickelt. Die Berücksich-

Tabelle 2: Laufzeiten der einzelnen Schritte der Lageerkennung

Schritt	Laufzeit
Binärbild erzeugen (einschließlich Schwellwertbestimmung)	1.6 sec
Konturextraktion	2.2 sec
Kantenbestimmung	3.8 sec (24 Kanten)
Konturpunktvergleich	4.0 sec (171 Punkte)
weitere Schritte (z.B. Vergleich signifikanter Löcher oder Kantenvergleich)	< 0.2 sec

tigung der spezifischen Anforderungen und Gegebenheiten des geplanten Einsatzgebiets — die Blechbearbeitungszelle — motivierten ein einfaches Bildanalysesystem zur Problemlösung. Dabei wurden bekannte Algorithmen der Mustererkennung und Standard-Hardware zu einem kostengünstigen System kombiniert, das die Anforderungen an Genauigkeit und Durchlaufzeit erfüllt. Restriktivere Zeitanforderungen können durch Einsatz eines leistungsfähigeren PC's (80386-Prozessor) ausgeglichen werden, ohne die Kosten des Systems stark anwachsen zu lassen.

Für eine vollständige Einbettung in den Informationsfluß der Bearbeitungszelle muß die Bestimmung der Merkmale aus den CAD-Daten des Werkstücks durch den Rechner erfolgen. Die zur Zeit implementierte interaktive Merkmalauswahl für das Lageerkennungssystem wird daher durch eine automatische Modellbildung ersetzt werden. Die Arbeiten dazu sind bereits begonnen worden.

Literatur

[Bal82] D. Ballard, C. Brown: *Computer Vision*. Prentice Hall, Englewood Cliffs N.J., 1982.

[Fel87] K. Feldmann: „Ziele, Aufbau und Arbeitsweise des Koorperationsprojektes PAP". In K. Feldmann, M. Geiger, U. Herzog, H. Niemann, B. Schmidt, H. Wedekind (Editoren): *Proc. Fachtagung Rechnerintegrierte Produktionssysteme*, S. 29–41, Erlangen, 1987.

[Foi82] J. Foith: *Intelligente Bildsensoren zum Sichten, Handhaben, Steuern und Regeln*. Springer Verlag, Berlin, 1982.

[Gei87] M. Geiger: „Flexibles Blechbearbeitungssystem im interdisziplinären Forschungsprojekt PAP". In K. Feldmann, M. Geiger, U. Herzog, H. Niemann, B. Schmidt, H. Wedekind (Editoren): *Proc. Fachtagung Rechnerintegrierte Produktionssysteme*, S. 199–218, Erlangen, 1987.

[Gei88] M. Geiger, U. Geißler: „Development of Software Modules for a Flexible Manufacturing System for Sheet Metal Parts". In *Proc. 16th North American Manufacturing Research Conference (NAMRC)*, S. 316–319, Urbana,Illinois, 1988.

[Hum86] H. Hümmer: *Konturdetektion in einem System zur Werkstückerkennung auf Binärbildbasis*. Diplomarbeit, Lehrstuhl für Informatik 5 (Mustererkennung) Universität Erlangen-Nürnberg, 1986.

[Lan88] R. Lange: *Lagebestimmung von ebenen Werkstücken in einer Blechbearbeitungszelle mit Hilfe eines Bildverarbeitungssystems*. Studienarbeit, Lehrstuhl für Informatik 5 (Mustererkennung) und Lehrstuhl für Fertigungstechnologie Universität Erlangen-Nürnberg, 1988.

[Nie83] H. Niemann: *Klassifikation von Mustern*. Springer Verlag, Berlin, 1983.

[Pav82] T. Pavlidis: *Algorithms for Graphics and Image Processing*. Springer Verlag, Berlin, 1982.

[Spu87] G. Spur: „Kooperation von Fertigungstechnik und Informatik für zukünftige Produktionssysteme". In K. Feldmann, M. Geiger, U. Herzog, H. Niemann, B. Schmidt, H. Wedekind (Editoren): *Proc. Fachtagung Rechnerintegrierte Produktionssysteme*, S. 1–21, Erlangen, 1987.

Temperaturbestimmung in Flammen mittels multispektraler Aufnahmen und tomographischer Bildverarbeitung *

Wolfgang Fischer

Technische Informatik I, Technische Universität Hamburg-Harburg
Harburger Schloßstr. 20, D-2100 Hamburg 90

Zusammenfassung

In dem vorliegenden Bericht werden Methoden zur Auswertung von multispektralen Aufnahmen leuchtender Prozesse (Flammen von z.B. Kerze, Bunsenbrenner, Satelliten-Steuertriebwerk) vorgestellt. Bei den Prozessen handelt es sich um Volumenstrahler, das heißt, die Aufnahmen sind Projektionen der dreidimensionalen Strahlungsdichte in die zweidimensionale Kameraebene. Aus diesen projektiven Aufnahmen der Prozesse muß zunächst die räumliche Verteilung der Meßgrößen gewonnen werden. Dies geschieht mit Hilfe von Methoden der Computer-Tomographie. Mit Verfahren der Pyrometrie wird dann die räumliche Temperaturverteilung bestimmt.

1   Einleitung

Zur Beurteilung und Klassifikation eines chemisch/physikalischen Prozeß können aus dem optischen Erscheinungsbild Merkmale gewonnen werden. Strukturelle Merkmale wie Laminarität und Verwirbelungsgrad von Strömungsprozessen werden häufig manuell ausgewertet. Hier bietet es sich an, Methoden der digitalen Bildverarbeitung und Mustererkennung zu nutzen. Physikalische Kenngrößen (z.B. Temperatur und Stoffkonzentrationen) , die für die Charakterisierung eines Prozesses wichtig sind, lassen sich mit Bildverarbeitungsmethoden ermitteln. Diese Kenngrößen dienen dann zur Klassifikation von Prozeßzuständen und zur optimalen Prozeßführung. An dieser Stelle soll die Errechnung der räumlichen Verteilung der Kenngröße Temperatur in Verbrennungsprozessen vorgestellt werden.

Die Methode der Temperaturbestimmung aufgrund der emittierten Strahlung basiert auf dem Planckschen Strahlungsgesetz (siehe (1)). Technisch wird die Anwendung des Planckschen Gesetzes in Strahlungspyrometern realisiert. In einem Pyrometer wird die Strahlungsintensität in bestimmten Wellenlängenbereichen gemessen und durch Vergleich mit einem geeichten Temperaturstrahler (schwarzer Körper) die Temperatur ermittelt [1]. Für hohe Temperaturen (über 1300^oC) werden Strahlungspyrometer wegen der großen Genauigkeit sogar für die Kalibrierung eingesetzt. Statt eines Punktsensors, wie in den Pyrometern üblich, kann man auch infrarotempfindliche, flächige Sensoren benutzen. Man erhält so eine Kamera, die das Bild der Temperaturverteilung wiedergibt (Wärmekamera oder Thermovision).

*Diese Arbeit wird im Rahmen des Sonderforschungsbereichs 238 "Prozeßnahe Meßtechnik und systemdynamische Modellbildung für mehrphasige Systeme" an der TU Hamburg-Harburg von der DFG unterstützt.

Für die Bestimmung von Oberflächentemperaturen haben sich die Pyrometer als geeignete Meßinstrumente erwiesen. So ist es naheliegend, auch in Flammen die Temperaturen mit Pyrometern zu bestimmen [2]. Jedoch ist die Lokalisierung eines Meßwertes nicht eindeutig. Die vom Sensor empfangende Strahlung stammt nicht von einem bestimmten Oberflächenpunkt, sondern ist die kumulierte Strahlung entlang einer Linie durch die Flamme (Volumenstrahler). Ein Strahlungspyrometer für die Temperaturbestimmung in Flammen muß so konstruiert sein, daß heiße Gase aus dem Prozeß abgesaugt und in einer Kammer unter kontrollierten Bedingungen auf die Temperatur untersucht werden können. Damit ist der Vorteil der berührungslosen Temperaturmessung, der die Pyrometer gegenüber den Thermoelementen auszeichnet, verloren gegangen. Es ist leicht einzusehen, daß das punktweise Durchmessen einer Flamme auf diese Weise recht mühsam und nur unvollständig möglich ist.

Mit einer Wärmekamera kann man einen Flammenprozeß als Ganzes erfassen. Aber auch hier besteht wie beim Pyrometer die Schwierigkeit der Zuordnung der Meßwerte zu Punkten des Ortsraumes im Prozeß. An dieser Stelle kann man Methoden der Computer-Tomographie einsetzen. Statt zunächst Temperaturen zu errechnen, wird ein räumliches Bild der Flamme erzeugt. Aus den dreidimensionalen Meßdaten wird dann die pyrometrisch Temperatur ermittelt.

Eine andere Methode der Temperaturbestimmung in Flammen wird in [3] vorgeschlagen. Diese Methode basiert auf Schlierenaufnahmen der Flammen und Verfahren der Computer-Tomographie.

2 Strahlung und Temperatur, physikalische Grundlagen

Die sichtbare Strahlung, die von einem Verbrennungsprozeß ausgesendet wird, ist in der Regel ein Gemisch von Strahlungskomponenten, die verschiedene Entstehungsursachen haben [4],[5]:

- Rotationen und Schwingungen innerhalb von Molekülen,

- Energiezustandsänderungen (Quantensprünge) von Elektronen in Atomen.

Aus der Strahlungsdichte kann auf den Energiezustand der Materie und damit auch auf die Temperatur geschlossen werden. Nimmt man kontinuierliche Energiezustände an, wird diese Strahlung durch das Plancksche Gesetz

$$J_s(\lambda, T) = \frac{2\pi hc^2}{\lambda^5} \frac{1}{e^{\frac{ch}{k\lambda T}} - 1} \tag{1}$$

beschrieben. Dabei ist $J_s(\lambda, T)$ die Strahlungsintensität, die von einer Oberfläche mit der Temperatur T bei der Wellenlänge λ emittiert wird (c: Lichtgeschwindigkeit, h: Plancksches Wirkungsquantum, k: Boltzmann-Konstante). Ein Körper, der exakt nach dem Planckschen Gesetz strahlt, wird schwarzer Körper genannt (geschwärzte Oberflächen kommen diesem Ideal sehr nahe). Ein Körper, der eine Strahlung $J(\lambda, T) = \varepsilon \cdot J_s(\lambda, T)$ mit einem Faktor $\varepsilon \leq 1$ aussendet, heißt grau und ε der Emissionskoeffizient.

Die Strahlung von Molekülschwingungen und angeregten Elektronen tritt nur in scharf abgegrenzten Wellenlängenbereichen auf (Banden). Stoffe mit einer solchen Strahlungscharakteristik werden selektive Strahler genannt. In den Banden gehorcht jedoch die Strahlung wieder dem Planckschen Gesetz. In der Praxis überlagern sich selektive und graue Strahlung (gemischte Strahlung).

Pyrometer sind am besten für schwarze bzw. graue Strahler geeignet. Gemischte Strahler werden durch geeignete Wahl von Emissionskoeffizienten grauen Strahlern angenähert. Es gibt verschiedene Arten von Pyrometern [1]:

- das Gesamtstrahlungspyrometer, bei dem der gesamte Wellenlängenbereich erfaßt und die Temperatur aus dem Stefan-Boltzmann-Gesetz

$$q_s(T) = \int_0^\infty J_s(\lambda, T)\, d\lambda = \sigma T^4 \tag{2}$$

(σ: Stefan-Boltzmann-Konstante) ermittelt wird,

- das Teilstrahlungspyrometer, das die Strahlung bei einer Wellenlänge λ mißt und die Temperatur durch die Wiensche Formel

$$J_s(\lambda, T) = 2\pi hc^2 \lambda^{-5} e^{-\frac{ch}{k\lambda T}} \tag{3}$$

(einer Annäherung an das Plancksche Gesetz für Temperaturen bis ca. $3500^\circ C$) bestimmt,

- das Farbpyrometer, mit dem nicht nur die Strahlungsintensität, sondern durch einen (manuellen) Farbenvergleich auch die Farbtemperatur bestimmt wird,

- das Quotientenpyrometer, bei dem aus dem Verhältnis der Strahlungen bei zwei Wellenlängen und der Planckschen bzw. Wienschen Formel die Temperatur errechnet wird.

Beim Gesamt- als auch beim Teilstrahlungspyrometer muß der Emissionskoeffizient des Strahlers bekannt sein, was in der Praxis häufig nicht gegeben ist. Der Emissionskoeffizient wird nicht nur von der Stoffart, sondern auch von der Stoffbeschaffenheit (z.B. Oberflächenrauhheit, Reinheitsgrad des Materials) beeinflußt. Mit dem Farbpyrometer kann man bei selektiven Strahlern einen Bereich einschränken, in dem die wahre Temperatur liegt. Bei der Temperaturermittelung mit dem Quotientenpyrometer kürzt sich der Emissionskoeffizient heraus, sobald er für die betrachteten Wellenlängen gleich ist. Bei nicht konstantem Emissionskoeffizient kann dieser durch mehrere Quotienten bei verschiedenen Wellenlängen (multispektrale Aufnahmen) annäherungsweise geschätzt werden.

Besonders bei Flammen ist es schwierig, den Emissionskoeffizienten zu bestimmen, so daß hier die Quotientenpyrometrie eine geeignete Methode ist, um Temperaturen zu bestimmen. Da außerdem der Emissionskoeffizient ortsabhängig sein kann, ist ein Verfahren notwendig, bei dem die Temperatur lokal bestimmt wird. Die für die Quotientenpyrometrie benötigten multispektralen Aufnahmen lassen sich durch eine Videokamera mit einem speziellen Objektivvorsatz gewinnen. Der Vorsatz besteht aus einem Strahlteiler und verschiedenen spektralen optischen Filtern.

3 Gewinnung räumlicher Daten, tomographische Grundlagen

Zunächst stellt sich die Frage, wie man die dreidimensionale Struktur eines leuchtenden, optisch nicht dichten Prozesses errechnen kann, obwohl mit Hilfe der Kamera nur seitliche, zweidimensional projektive Ansichten zu gewinnen sind. Man kann annehmen, daß die Strahlungsintensität, die man aufnimmt, das Integral der Intensitäten längs einer Geraden ist (Projektion, siehe Abbildung 1). Im Gegensatz zur Transmissionstomomographie, bei der Strahlung (z.B. Röntgenstrahlung) durch das zu untersuchende Objekt geschickt wird, nutzt man hier die Eigenstrahlung der Flamme aus (Emissionstomographie).

Im weiteren werden die Projektionen und Rekonstruktionen in Schichten (*gr. Tomos*) senkrecht zur Bildebene aufgeteilt, so daß für jede Schicht das Rekonstruktionsproblem unabhängig gelöst wird. Statt einer dreidimensionalen Funktion wird für jede Schicht eine zweidimensionale Funktion gesucht. Eine eindeutige Rekonstruktion der Intensitätsfunktion ist möglich, wenn die Projektionen für jeden Winkel $\phi \in [0, 2\pi]$ bekannt sind [6]. Diese Lösung des Problems hat schon 1917 J. Radon gegeben, weshalb das obige Problem auch Radonsches Problem genannt und durch die Radon-Transformation $\mathcal{R}$

$$\check{f}(p, \phi) = [\mathcal{R}f](p, \phi) = \int_{-\infty}^{\infty} f(p\cos\phi - t\sin\phi, p\sin\phi + t\cos\phi)\, dt \tag{4}$$

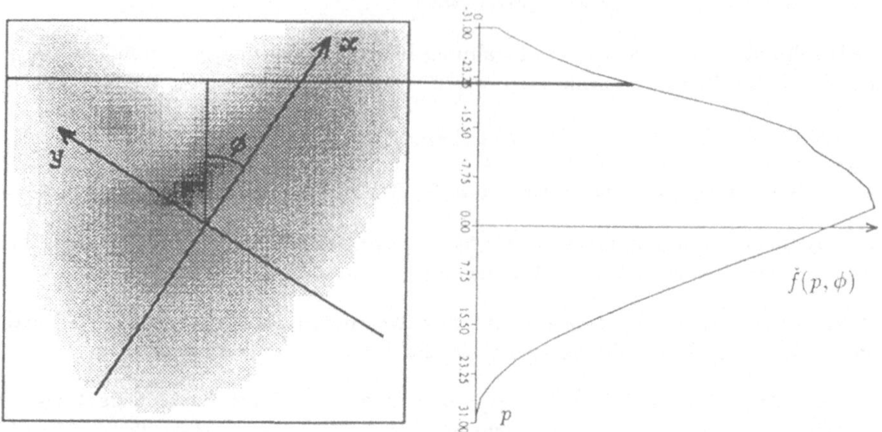

Abbildung 1: Projektion bei dem Winkel ϕ entlang der Senkrechten zu p

beschrieben wird [7]. Praktisch sind jedoch nie alle Projektionen bekannt. Man muß sich also mit einer unvollständigen Menge von Meßdaten begnügen. Für spezielle Probleme reicht jedoch auch eine geringe, endliche Anzahl von Projektionen aus, um das Rekonstruktionsproblem zu lösen. Falls die zu bestimmende Intensitätsfunktion rotationssymmetrisch ist, genügt genau eine Projektion. Im diesem Fall wird die Radon- durch die Abelsche Transformation $\mathcal{A}$ ersetzt:

$$f_{\mathcal{A}}(p) = [\mathcal{A}f](p) = \int_p^\infty r\, f(r)\,(r^2 - p^2)^{-\frac{1}{2}}\, dr \quad . \tag{5}$$

Eine Verallgemeinerung bildet eine elliptische Symmetrie. Das heißt, die gesuchte Funktion ist auf Ellipsen um einen Koordinatenursprung konstant:

$$f(x,y) = const. \qquad \text{mit} \quad \frac{x^2}{a^2} + \frac{y^2}{b^2} = s \quad (s \leq 1). \tag{6}$$

Mit der Kenntnis des Hauptachsenradius a und Nebenachsenradius b sowie der Lage der Ellipse reicht auch hier eine Projektion zur Bestimmung der gesuchten Funktion f. Die Lage der Ellipse wird durch den Mittelpunkt und dem Winkel ψ zwischen Hauptachse und X-Achse bestimmt. Dabei liegt der Koordinatenursprung im Mittelpunkt der Ellipse. Falls die Parameter a, b und ψ nicht a priori bekannt sind, lassen sich diese aus drei Projektionen ermitteln. Für die Radontransformierte gilt dann:

$$[\mathcal{R}f](p,\phi) = d_{\phi-\psi} \cdot [\mathcal{R}f](d_{\phi-\psi} \cdot p, \psi) \tag{7}$$

mit

$$d_\alpha = \left(\cos^2\alpha + \frac{b^2}{a^2}\sin^2\alpha\right)^{-\frac{1}{2}} \tag{8}$$

Die numerische Lösung des Rekonstruktionsproblems kann auf verschiedene Arten erfolgen:

- inverse Abelsche Transformation

- algebraische Rekonstruktionstechniken

- direkte Fouriermethode

- gefilterte Rückprojektionen

Die inverse Abelsche Transformation

$$f(r) = -\frac{1}{\pi} \int_r^\infty \frac{\partial f_A(p)}{\partial p} (p^2 - r^2)^{-\frac{1}{2}} \, dp \tag{9}$$

liefert theoretisch die gesuchte Funktion exakt zurück. Es zeigte sich aber, daß diese Rekonstruktionsmethode schlechter als die folgenden Verfahren ist.

Für die algebraischen Rekonstruktionstechniken wird das Objekt so diskretisiert, daß die Integrale als gewichtete Zeilensummen aufzufassen sind. Im Prinzip hat man dann ein lineares Gleichungssystem zu lösen. Da eine Rotationssymmetrie vorausgesetzt wird, erhält man eine invertierbare obere Dreiecksmatrix. Es wurden zwei Ansätze zur algebraischen Rekonstruktionstechnik gemacht, die beide zu vergleichbaren Ergebnissen führen. Der erste Ansatz diskretisiert die Linienintegrale, der zweite Ansatz berücksichtigt zusätzlich die flächige Ausdehnung der Bildpunkte. Bei dieser Rekonstruktionsmethode lassen sich leicht Nebenbedingungen (z.B. Hintergrund $= 0$) berücksichtigen.

Die Rekonstruktion mit der direkten Fouriermethode basiert auf dem Projektion-Slice-Theorem. Das Projektion-Slice-Theorem besagt, daß die zweidimensionale Fourier-Transformation $\mathcal{F}_2$ einer zweidimensionalen Funktion dasselbe Ergebnis liefert wie die Hintereinanderausführung einer Radon- und einer eindimensionalen Fourier-Transformation $\mathcal{F}_{RAD}$ in radialer Richtung:

$$\mathcal{F}_2(f) = \mathcal{F}_{RAD}(\mathcal{R}(f)) \quad . \tag{10}$$

Im Falle einer Rotationssymmetrie wird demnach die Abelsche Transformation durch eine eindimensionale Fourier-Transformation und eine daran anschließenden inverse Hankel-Transformation invertiert. Für rotationssymmetrische Funktionen ist die zweidimensionale Fourier-Transformation zu einer Hankel-Transformation äquivalent. Diese Beziehung ist als Abel-Fourier-Hankel-Zyklus bekannt [8].

Die populärste Methode ist die Rekonstruktion mittels gefilterter Rücktransformation. Auch diese Methode läßt sich auf das Projektion-Slice-Theorem zurückführen. Eine Umformung der inversen zweidimensionalen Fouriertransformation ergibt jedoch eine günstigere Rechenvorschrift, die wie folgt interpretiert werden kann. Zunächst wird die radontransformierte Funktion $\mathcal{R}f$ mit einer Filterfunktion gefaltet. Nach dieser Filterung wird eine Rückprojektion durchgeführt. Das Ergebnis ist eine tiefpaßgefilterte Version der ursprünglichen Funktion. Diese Rekonstruktionen sind in der Regel gut, weil hochfrequente Anteile, die den Großteil der Fehler ausmachen, herausgefiltert werden.

4 Anwendungen und Ergebnisse

Die Algorithmen zur tomographischen Rekonstruktion und zur pyrometrischen Temperaturbestimmung wurden auf einer VAX 8250 mit einem GOULD IP 8500 Bildverarbeitungssystem implementiert. Die Aufnahmen entstanden mit einer S/W-CCD-Kamera, die direkt an das GOULD-System angeschlossen war. Zur Gewinnung der multispekralen Aufnahmen wurden verschiedene optische Filter sowie ein spezieller Objektivvorsatz (Tricklinse) zur simultanen Verdreifachung des Bildes benutzt.

Durch dieses Aufnahmeverfahren hat man mehrere unterschiedlich spektral gefilterte Aufnahmen derselben Flamme auf einem Bild. Diese werden separiert und anschließend die tomographische Rekonstruktion getrennt durchgeführt. Die rekonstruierten räumlichen Intensitätsverteilungen dienen als Eingabe für die quotientenpyrometrische Temperaturbestimmung [9].

4.1 Bunsenbrennerflamme

Aufgrund des symmetrischen Aufbaus des Bunsenbrenners gilt hier die Annahme der Rotationssymmetrie. Der Brenner läßt sich so einstellen, daß die Flamme ruhig und stabil brennt. Zur
Überprüfung der Rekonstruktionen wurden die beiden Seiten der Gleichung

$$\int_{-\infty}^{\infty} f_A(x)dx = 2\pi \int_{0}^{\infty} f(r)r\,dr \tag{11}$$

ausgewertet. Es ergaben sich keine gravierende Unterschiede.

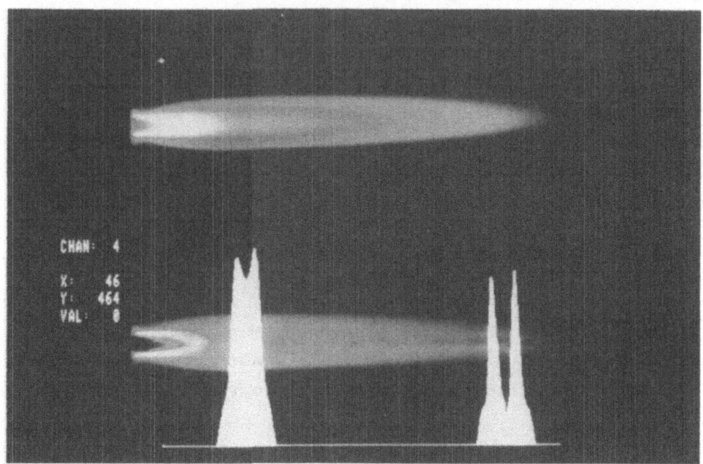

Abbildung 2: Bunsenbrenner, Projektion (oben) und rekonstruierter Längsschnitt (unten).

Die Flamme leuchtete nur wenig und hauptsächlich im blauen Spektralbereich. Dies ist ein
Zeichen dafür, daß die selektiven Strahler wie CO und CH überwogen. Dadurch waren aber
die Temperaturalgorithmen für graue Strahler nicht anwendbar. Man konnte am Brenner durch
Drosselung der Sauerstoffzufuhr auch eine leuchtende Flamme einstellen, so daß die Strahlung von
Rußpartikeln dominiert. Rußpartikel sind in der Regel schwarze bzw. graue Strahler. Hier würden
die pyrometrischen Methoden Anwendung finden, jedoch flackerte die leuchtende Flamme, und die
Annahme der Rotationssymmetrie konnte nicht aufrecht erhalten werden.

4.2 Kerzenflamme

Bei den Kerzenflammen ist der Anteil der selektiven Strahler gegenüber den leuchtenden Rußpartikeln relativ gering, so daß diese zunächst vernachlässigt werden können. Dadurch ist die Strahlung
im wesentlichen die eines schwarzen Strahlers. Die Kerzenflamme entsprach dem symmetrischen
Modell hinreichend. Mit Hilfe des Planckschen Gesetzes für schwarze Strahler konnten somit aus
multispektralen Aufnahmen Temperaturfelder ermittelt werden (siehe Abbildung 3).

4.3 Satellitentriebwerksabgase

Es wurden auch Aufnahmen von Satellitentriebwerken untersucht. Die Rekonstruktionen ergaben Strahlungsdichteverteilungen, die anders nicht zu beoachten waren (siehe Abbildung 4). Bei
den Triebwerksabgasen ist eine einfache Beziehung zu physikalischen Größen mit dem Plancksche

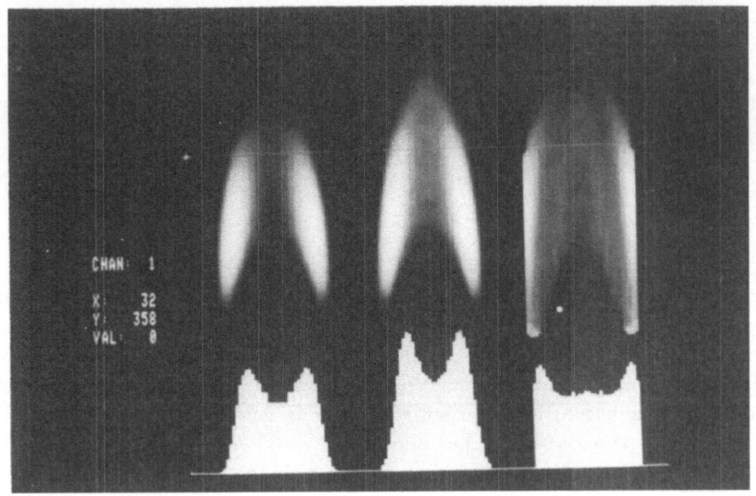

Abbildung 3: Verteilung der Intensitäten für multispektrale Aufnahmen (links: mit blauem Filter, Mitte: mit rotem Filter) und der relativen Temperatur (rechts) in einer Kerzenflamme, Längsschnitte.

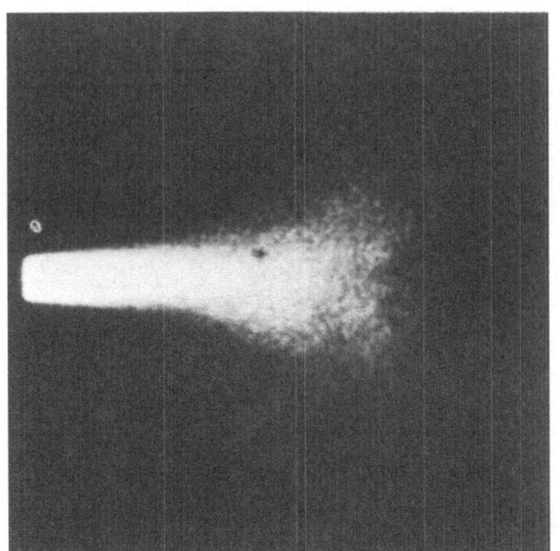
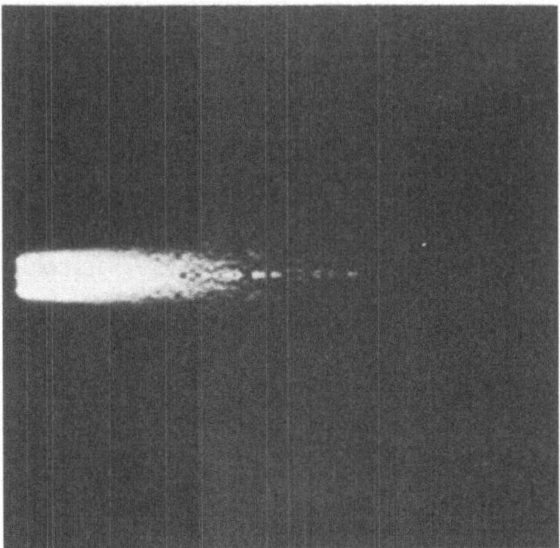

Abbildung 4: Satellitentriebwerk im Hochvakuum, Projektion (links) und rekonstruierter Längsschnitt (rechts).

Strahlungsgesetz nicht herleitbar. Die Abgase sind keine schwarzen oder grauen Strahler und die Anzahl der kurzlebigen, reaktiven Stoffe (Radikale) ist recht groß. Über das Gesetz zur Erhaltung der Massen wurde eine Beziehung zwischen der Verteilung der Triebwerksabgase und der Strahlungsdichte hergestellt. Insbesondere ergab die Untersuchung von Streamlines Erkenntnisse über das Verhalten der Abgase nach Verlassen der Triebwerksdüse ins Vakuum [10].

5 Ausblick

Die ersten Ergebnisse mit den pyrometrischen und tomographischen Algorithmen sind recht zufriedenstellend. Für die Zukunft ist die exakte Kalibrierung der Aufnahmeeinrichtung wichtig, um quantitativ genaue Temperaturfelder zu erstellen. Um die räumlichen Strahlungsdichten genauer zu errechnen, braucht man Projektionen bei vielen unterschiedlichen Winkeln, die aber häufig in der Praxis nicht gegeben sind. Ein anderer Weg zur Verbesserung der tomographischen Rekonstruktionen, der hier beschritten werden soll, ist eine flexible Anpassung der Algorithmen an die Geometrie der Prozesse, so daß eine minimale Anzahl von Projektionen genügt. Für eine Beurteilung der Methoden und Ergebnisse ist die Entwicklung von Gütekriterien zur Temperaturbestimmung und räumlichen Rekonstruktion notwendig.

Es zeigt sich, daß die Verbindung vom Kamera und Rechner neue Anwendungsgebiete erschließt. Auch ohne großen apparativen Aufwand kann durch geschickte Benutzung eines Bildverarbeitungssystems und einer multispektralen Aufnahmetechnik die räumliche Verteilung der Temperatur in Flammen ermittelt werden.

Literatur

[1] Euler, J.; Ludwig, R.: Arbeitsmethoden der optischen Pyrometrie. Verlag G. Braun, Karlsruhe, 1960.

[2] Chedaille, J.; Braud, Y.: Measurements in Flames. Edward Arnold Publishers, London, 1973.

[3] Braun, H.: Tomographie von Vektorfeldern. Interne Veröffentlichung des Instituts für Meß- und Regeltechnik der Universität Karlsruhe, 1986.

[4] Gaydon, A.G.; Wolfhard, H.G.: Flames, Their structure, radiation and temperature. Chapman and Hall, London , 1970.

[5] Pepperhoff, Werner: Temperaturstrahlung. Steinkopff, Düsseldorf, 1956.

[6] Herman, Gabor T.: Image Reconstructions from Projections, The Fundamentals of Computerized Tomography. Academic Press, New York, 1980.

[7] Deans, Stanley R.: The Radon Transform and Some of Its Applications. John Wiley and Sons, New York, 1983.

[8] Bracewell, Ronald N.: The Fourier Transform and its Applications. McGraw-Hill, Singapore, 1978.

[9] Fischer, W. ; Burkhardt H.: Anwendung von Bildverarbeitungs- und automatische Bildauswertemethoden auf multispektrale Bildaufnahmen von Mehrphasenströmungen. Arbeits- und Ergebnisbericht 1988 des DFG-Sonderforschungsbereich 238, TU Hamburg-Harburg 1988.

[10] Brekenfeld, A.; Trinks H.: Flammendiagnostik durch Videobildverarbeitung. Frühjahrstagung der Deutschen Physikalischen Gesellschaft, Essen, 1989.

Erfassung von optisch beobachtbarem Tierverhalten
mittels Bildverarbeitung

P. Herrmann, V. Schmitt, Dr. J. de Kramer

BASF AG, Abt. ZXT/T
D-6700 Ludwigshafen

Zusammenfassung

Es wird ein Bildverarbeitungssystem beschrieben, das Daten über die Wirkung von Schädlingsbekämpfungsmitteln gewinnt. Aus Bildsequenzen werden Gesamt- und Einzelaktivität sowie Populationsdynamik ermittelt. Ein spezielles Skelettierungsverfahren bestimmt Form und Bewegungsart einzelner, länglich geformter Objekte.

1. Einführung

Zur Beurteilung der Wirksamkeit von Schädlingsbekämpfungsmitteln sind eine Vielzahl von Methoden bekannt, bei kleinen Schädlingen bleibt jedoch oft nur die Beobachtung, zum Teil unter einem Mikroskop. Dabei werden Anzahl (Vermehrungsrate) und Aktivität der Tiere in einer mit dem zu prüfenden Wirkstoff versetzten Nährlösung ermittelt.

Die große Anzahl von neuen Wirkstoffen erlaubt nur seltene und kurze Kontrollen der einzelnen Probe, die Auswertung mit dem Auge ermöglicht nur grobe Beurteilungen. So muß man sich mit relativ wenigen und undifferenzierten Aussagen zur Wirksamkeit einer Testsubstanz begnügen.

Die Erkenntnis, daß die Entwicklung grundlegend neuer Wirkstoffe stagniert und die Schädlinge gegen die bekannten Mittel zunehmend resistent werden, begünstigt die Entwicklung genauerer Testverfahren.

Unser Ansatz ist die Verwendung eines Bildverarbeitungssystems, um in geringeren Zeitabständen genauere Meßdaten zu erhalten. Das Verfahren ist für längliche Objekte wie Larven, Raupen, Würmer und Fadenwürmer (Nematoden) vorgesehen, ist aber zumindest teilweise auch für andere Anwendungen interessant.

2. Material

Die Versuchstiere, in unserem Fall etwa 1 mm lange Nematoden, befinden sich in Petrischalen mit maximal 96 Kammern unter einem Mikroskop mit angeschlossener CCD-Kamera (Frame-Transfer-Verfahren). Durch Verwendung einer Dunkelfeldbeleuchtung erhält man eine kontrastreiche Darstellung. Für den entgültigen Einsatz der Anlage ist die Verwendung eines 25mm-Objektives mit Zwischenringen anstatt des Mikroskopes vorgesehen.

Den Kern des Auswertesystems bildet die VME-Bus-Karte IPC (Image Processing Computer) der Firma ELTEC (Abb. 1). Sie enthält einen 512 kBytes großen Bildspeicher mit 8-Bit A/D- und D/A-Wandlern. Zusätzlich befindet sich auf der Karte ein kompletter, lokaler 68020-Rechner mit 2 seriellen Schnittstellen.

Während der Programmentwicklung steckt die IPC in einem 68020-VME-Bus-Rechner (Host) mit Massenspeicher und umfangreicher Peripherie. Die lokale CPU wird stillgelegt und der Bildverarbeitungsteil der IPC läßt sich direkt ansprechen.

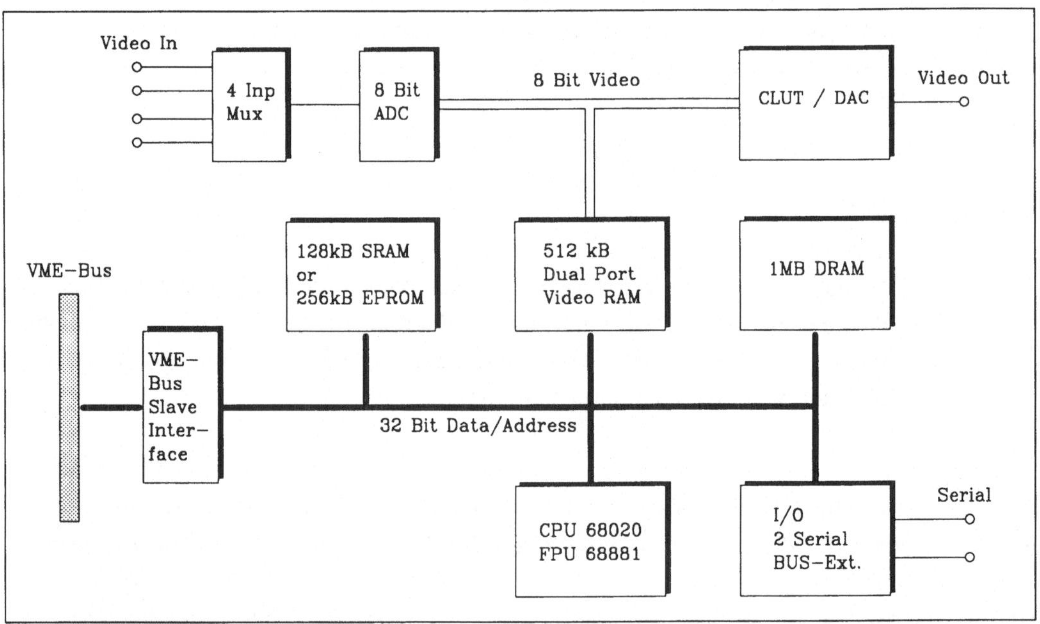

Abb. 1:   **Blockschaltbild der IPC-Karte**

Als Betriebssystem bietet sich OS-9 an, da es sowohl auf dem Host als auch in einer Minimalversion auf der IPC-CPU installierbar ist. Damit sind die gleichen Programme problemlos auf beiden Rechnern lauffähig.

Das Konzept einer lokalen CPU auf der Videokarte bietet mehrere Vorteile : Bei der Entwicklung von Bildverarbeitungsmethoden ist man frei von Vorgaben, die sich durch Spezialhardware (LUT's, ALU's usw, z.B. für Bildsubtraktionen) ergeben. Im anschließenden Einsatz kann die IPC-Karte für einfache Aufgaben die gesamte Auswertung übernehmen, ansonsten bietet sich die Aufgabenteilung mit einem Hostrechner an, wobei Parallelbetrieb möglich ist. Grundsätzlich entspricht dieser Ansatz der zunehmend praktizierten Methodik der dezentralen Vorverarbeitung ('Intelligente Sensoren') in der Meßtechnik.

Wir führen (auch aus Kostengründen) die gesamte Auswertung auf der IPC durch, die Ergebnisse gelangen über eine serielle Schnittstelle an einen PC/AT-Rechner, der die Daten auf Diskette archiviert und zur Präsentation aufbereitet.

Ein von der IPC gesteuerter Probenwechsler (X/Y-Tisch) erlaubt die vollautomatische Durchführung der Beobachtungen für bis zu 1000 Proben.

3. Methoden

Die folgenden Schritte werden in bestimmten Zeitabständen für jede Probe wiederholt. Sie sind in chronologischer Folge aufgeführt, die beiden ersten Schritte sind nur versuchsweise verwirklicht, da der Probenwechsler noch nicht zur Verfügung steht.

3.1 Probenwechsel
Nachdem eine Probe grob unter der Kamera positioniert ist, muß eventuell noch eine Feinjustierung vorgenommen werden. Ziel ist es, bei möglichst vollständiger Erfassung einer Probe nur wenig vom zylindrischen Rand der Probenkammer im Bildausschnitt haben, da dieser als Lichtleiter wirkt und die empfindliche CCD-Kamera übersteuert (Blooming-Effekt).

3.2 Fokussierung
Abhängig von der Blendenöffnung des Kameraobjektives und damit der Tiefenschärfe kann es nach dem Probenwechsel notwendig sein, die Bildschärfe nachzustellen. Dazu wird die Kamera in Z-Richtung so bewegt, daß die Summe S der Differenzenbeträge benachbarter Bildpunkte $P_{x,y}$ maximal wird (Gradientenbildung) :

$$S = \sum_{x,y} |P_{x,y} - P_{x+1,y}| + |P_{x,y} - P_{x,y+1}|$$

Um Fehler durch hochfrequente Rauschanteile zu vermeiden, kann man zuvor eine Mittelwertbildung (siehe auch 3.4) über n Pixel durchführen, so daß im Endeffekt nur mittlere Frequenzen berücksichtigt werden. Dem entspricht (in x-Richtung, y analog) :

$$S = \sum_{x,y} |(\sum_{i=1..n} P_{x+i,y})/n - (\sum_{i=1..n} P_{x+i+n,y})/n| + \ldots$$

$$= 1/n \sum_{x,y} |\sum_{i=1..n} P_{x+i,y} - P_{x+i+n,y}| + \ldots$$

Für die Maximalwertsuche kann die Division durch n entfallen.

3.3 Skalierung

Über 2 D/A-Wandler läßt sich der Nullpegel und die Verstärkung des Video-Eingangsverstärkers programmieren. Sie werden schrittweise verändert, bis in dem nach jedem Schritt neu ermittelten Histogramm sinnvolle Häufungen für minimale und maximale Helligkeiten auftreten (lineare Skalierung per Hardware).

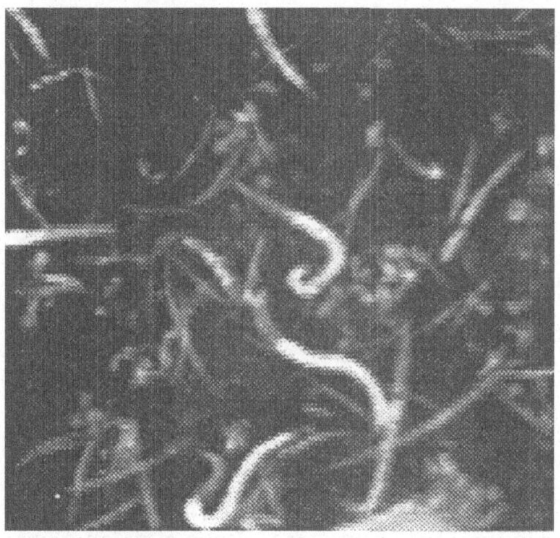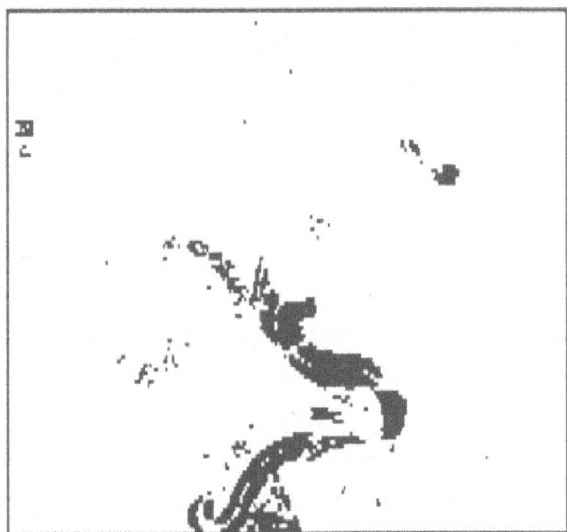

Abb.2 : **Originalbild und Binärbild dazu, Aktivitätszonen erscheinen schwarz**

3.4 Messung der Gesamtaktivität

Wir nehmen je 2 Einzelbilder mit 512 x 256 Punkten im zeitlichen Abstand von 2 bis 5 Bildern (40 bis 100 ms) auf. Durch Zusammenfassung von je 4 quadratisch angeordneten Pixeln zu einem neuen mittleren Wert vermindert man Rauschen und insbesondere Einflüsse aufgrund unterschiedlicher Synchronisation. Die beiden so erhaltenen, nur noch ein viertel so großen Bilder werden betragsmäßig voneinander subtrahiert (Ableitung nach der Zeit) und mit kleiner Schwelle S binarisiert. Im Binärbild finden sich dort 1-wertige Punkte, wo Änderungen bzw. Bewegungen stattgefunden haben (Abb. 2). Somit ist die Summe A aller Binärpunkte $B_{x,y}$ ein Maß für die allgemeine Änderung :

$$A_t = \sum_{x,y} B_{x,y,t} \quad \text{mit } B_{x,y,t} = \begin{cases} 1 \text{ für } |P_{x,y,t} - P_{x,y,t-dt}| > S \\ 0 \text{ sonst} \end{cases}$$

Diese Aktivitätsmessung kann nur relative Werte liefern, da Einflüsse wie Größe und Anzahl der Tiere sowie Bewegungen durch Stöße (insbesondere in Flüssigkeiten) unberücksichtigt bleiben.

3.5 Messung von Einzelaktivitäten

Es bietet sich an, das Aktivitäts-Binärbiid weiter auszuwerten. Wir betrachten eine beliebig zusammenhängende Ansammlung von 1-wertigen Punkten als eine Aktivitätszone, wobei Lücken von maximal einem Punkt in x- oder y-Richtung erlaubt sind. Die mit üblichen Methoden ([BB 82],[Ha 87]) ermittelte Anzahl und Flächen aller Zonen erlauben die Feststellung (Abb. 3), ob die Gesamtaktivität durch starke Bewegungen einiger weniger oder durch geringe Bewegung von vielen Tieren hervorgerufen wird.

```
    Summe Aktivität : 1830        Summe Zonen : 40


    Fläche 1      2       20      100     400     131072
    Anzahl |  19  |  16  |  3  |  0  |  2  |
```

Abb. 3 :   **Aktivitätszonen zu Abb. 2, als Summe und nach Pixelfläche gruppiert**

Bei dieser Messung werden tote Tiere nicht erfaßt und oft erhält man zuwenig (durch Überlappungen) oder zuviele Objekte (aufgrund mehrerer Bewegungszonen pro Tier). Die Ergebnisse sind jedoch zufriedenstellend, insbesondere unter Berücksichtigung der Schwierigkeiten einer Populationsbestimmung aus einem Standbild (starke Helligkeits- und Kontrastunterschiede, Überlappungen, Fremdobjekte wie Nährlösung, Bakterien, Schmutz).

3.6 Die Skelettierung

In Erweiterung der eben beschriebenen Aktivitätsmessungen (3.4 und 3.5), mit denen die bisherige Auswertung per Auge abgelöst und verbessert werden soll, interessiert das genaue Verhalten zumindest einiger Tiere. Dazu bilden wir ausgewählte Exemplare auf einen Polygonzug aus äquidistanten Längenstücken ab. Er ist durch die Angabe eines Startpunktes, der Länge eines Längensegments und einer Folge von Richtungen für jedes Längenstück vollständig beschrieben.

Unser Algorithmus zur Verdünnung von länglichen Objekten (LO) weicht von den gängigen Verfahren zur Skelettierung (z.B. [PD 81], [Br 87]), die meistens auf dem Prinzip fortgesetzter Erosion und praktisch immer mit Binärbildern arbeiten, ab. Wie verwenden das unveränderte Grauwertbild, die folgenden Schritte (Abb. 4) gehen von hellen LO's vor dunklem Grund aus :

a)   Berechne (einmal) eine Liste der Abstände dx,dy aller Pixel auf einem Kreis mit Radius r. Dabei ist r gleich der Länge l eines Polygonstückes.

b)   Wähle einen Startpunkt $P_{x,y}$ auf dem LO.

c)   Ermittle die Helligkeiten aller Pixel $P_{x+dx,y+dy}$ auf dem Kreisumfang um P.

d)   Anwendung der üblichen Methoden zur (Bild-) Signalverbesserung und -auswertung auf die Folge der Helligkeitswerte :

   - Tiefpaß gegen Rauschen

- Kontrastverbesserung durch Gradientenbildung
- Schwellenbildung mit gleitendem Mittelwert
- Bestimmung aller Maxima oberhalb des Schwellwertes

Man erhält eine Liste aller möglichen (hellen) Fortsetzungen des LO's im Abstand l von P. Mittels einfacher Annahmen (Heuristiken) wird die wahrscheinlichste Fortsetzung als neuer Punkt P ausgewählt, dessen Richtung gespeichert und mit Schritt c fortgefahren.

e)   Findet sich keine sinnvolle Fortsetzung, so wird abgebrochen. Die Folge der gespeicherten Richtungen beschreibt den Polygonzug.

Anmerkungen :

zu a) Dieser Schritt dient zur Einsparung von Rechenzeit. Natürlich muß r in sinnvollem Verhältnis zu der maximal erlaubten LO-Breite stehen.

zu b) Die Auswahl des Startpunktes ist nicht trivial. Eine Möglichkeit ist die Anwendung des Verfahrens auf alle Punkte eines Rasters unter Zurückweisung zu kurzer Polygone. Für bewegte Objekte bieten sich Punkte hoher Aktivität (aus 3.4) an.

zu d) Typische Heuristiken sind z.B., daß das LO seine bisherige Richtung, Richtungsänderung, Dicke, Helligkeit, Helligkeitsprofil usw. innerhalb gewisser Toleranzen beibehält. Sind Gabelungen oder Kreuzungen erlaubt, so können mehrere Fortsetzungspunkte akzeptiert und (rekursiv) untersucht werden. Der LO-Mittelpunkt muß nicht unbedingt mit dem Helligkeitsmaximum übereinstimmen. Sind die Ränder konstant heller als das Innere, bietet sich eine Randverfolgung an.

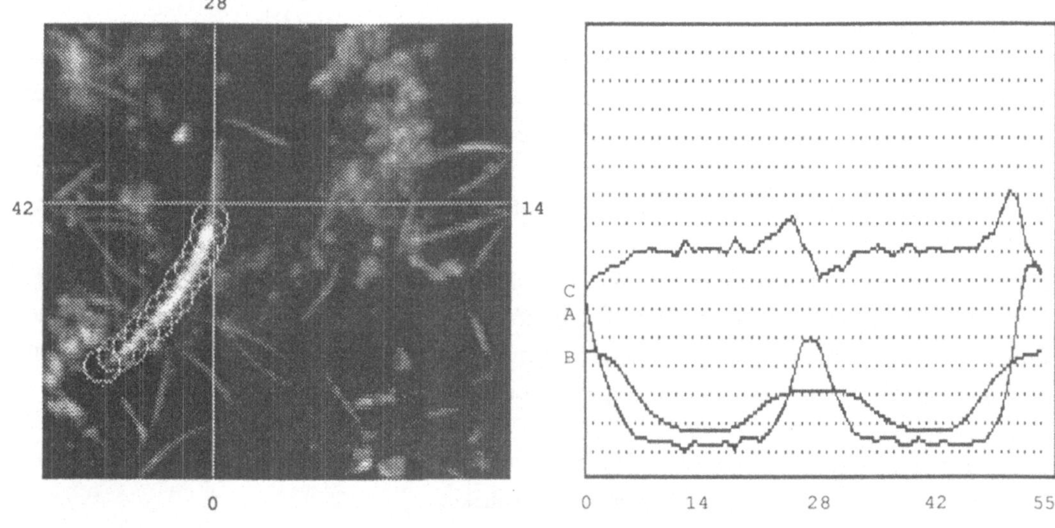

Abb 4 : **Bearbeitung eines Objekts, zur Verdeutlichung erscheinen abgetastete Punkte in Schritt c) hell. Rechts die Helligkeitsfolge A, Mittelwert B und Steigung C nach Schritt d) für die Punkte 0..55 des Kreisumfangs um das Fadenkreuz (im Gegenuhrzeigersinn, bei 6 Uhr beginnend).**

Das Verfahren, welches wir aufgrund gewisser Ähnlichkeiten auch als "Radar"-Methode bezeichen, zeigt einige erfreuliche Eigenschaften. Es arbeitet schnell, da die zeitaufwendigen Berechnungen zur Bildverbesserung nur für wenige Punkte durchgeführt werden. Die lokal begrenzte Auswertung erhöht die Empfindlichkeit und die Berücksichtigung von Objektmerkmalen vermeidet Fehler z.B. durch kreuzende Objekte oder Schmutzeffekte.

4. Ergebnisse

Versuchsweise durchgeführte Aktivitätsmessungen (3.4 und 3.5) speichern für jede Probe den Mittelwert aus 40 Einzelmessungen (Abb. 5). Insgesamt dauert dies etwa 30 Sekunden, somit lassen sich bis zu 100 Proben pro Stunde (inklusive Positionierung) bearbeiten.

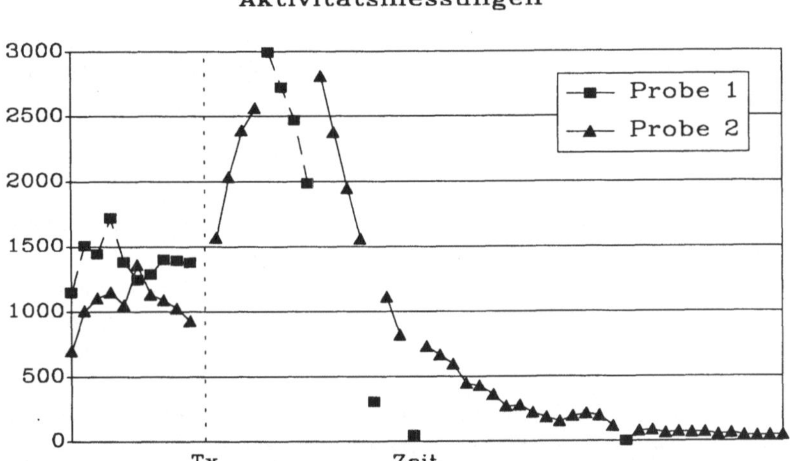

Abb. 5: **Aktivitätsverlauf, Wirkstoffzugabe bei Tx, Hyperaktivität und positive Wirkung**

Während die Aktivitätsmeßwerte (in Tabellen- oder Kurvenform) direkt vom Biologen oder Chemiker verwertbar sind, müssen die Daten zu den Polygonzügen (3.6) erst aufbereitet werden. In einem Parallelprojekt suchen wir daher nach geeigneten Auswerteverfahren. Das Optimum wäre die Entwicklung einer allgemein anwendbaren Lösung zur intelligenten Interpretationen von Meßreihen, die (Tier-) Verhalten beschreiben.

Literatur

[BB 82]   Ballard, Brown : *Computer Vision*, Prentice Hall, 1982
[Ha 87]   P. Haberäcker : *Digitale Bildverarbeitung*, Carl Hanser Verlag, 1987
[PD 81]   Davies, E.R. and A.P.N. Plummer (1981) : Thinning algorithms : A critique and a new methology, in: *Pattern Recognition* 14, S. 53-63
[Br 87]   O. Bruch (1987) : Line thinning by line following, in: *Pattern recognition Letters* 8 (1988), S. 271-276

Progressive Bildübertragung mit der $3 \times 3/2$ Pyramide

Harald F. Mayer und Walter G. Kropatsch
Institut für Digitale Bildverarbeitung und Graphik
Wastiangasse 6, A-8010 Graz, Austria

Kurzfassung

Digitale Bilder stellen eine größere Datenmenge (ab 256 kByte) dar, als sie bei konventioneller Datenübertragung auftritt. In dieser Arbeit wird eine Methode präsentiert, welche sowohl progressive Übertragung als auch komprimierte Speicherung von digitalen Bildern ermöglicht. Diese Methode wurde von Burt und Adelson für die $5 \times 5/4$ Pyramide entwickelt. Wir verwenden als Datenstruktur eine $3 \times 3/2$ Pyramide. Dadurch ist ein glatterer Bildaufbau bei der Übertragung möglich. Die Komprimierungsraten, die sich bei Versuchen ergaben, liegen zwischen 1:10 und 1:20.

1 Einleitung

Ein digitales Bild wird als zweidimensionales Feld (z.B.: 512×512 bzw. 1024×1024) von (256) Grauwerten gespeichert. Mehrere digitale Bilder stellen daher eine größere Datenmenge (ab 256 kByte) dar, als sie bei konventioneller Datenübertragung auftritt. Die Idee der *progressiven Bildübertragung* geht davon aus, daß zunächst mit einer geringen Datenmenge ein Bild mit den groben Umrissen erzeugt wird. Mit zunehmender Zeit werden jene Daten übertragen, die für eine qualitative Verbesserung (z.B.: Kontrastverstärkung) des Bildes sorgen.

Die Datenstruktur *Pyramide* eignet sich für diese Aufgabe [15]. Eine Pyramide ist in der Bildverarbeitung eine hierarchische Struktur von Bildern unterschiedlicher Auflösung aber gleichen Bildinhalts. Die einzelnen Bilder in einer Pyramide werden auch *Ebenen* genannt. Ein Element ($=$ *Pixel*) einer Ebene A_i wird dabei mittels einer *Reduktionsfunktion* aus einem zusammenhängenden Block von Elementen aus der darunterliegenden Ebene A_{i-1} berechnet. Diesen Block nennt man *Reduktionsfenster*. Der Faktor, um den sich die Auflösung pro Ebene verringert, wird *Reduktionsfaktor* genannt.

Folgende Schreibweise hat sich für Pyramiden eingebürgert: $m_1 \times m_2/r$; r ist der Reduktionsfaktor, $m_1 \times m_2$ die Größe des Reduktionsfensters. Beispiele für Pyramiden sind $2 \times 2/4$ (äquivalent zur Struktur eines Quadtree [14]), $5 \times 5/4$ [4], $2 \times 2/2$ [8] und $3 \times 3/2$ [7] (Abb. 1).

Pyramiden werden auf einem hochauflösenden Eingabebild durch lokale Rechenoperationen aufgebaut. Dadurch ist mit entsprechender Hardware [16] ein Pyramidenaufbau in Echtzeit möglich.

Eine Grauwertpyramide, bestehend aus den Ebenen $G_0, G_1, \ldots, G_n$, wird durch eine Funktion REDUCE auf dem Eingabebild G_0 aufgebaut.

$$G_i = \text{REDUCE}(G_{i-1}) \qquad i = 1, \ldots, n \qquad (1)$$

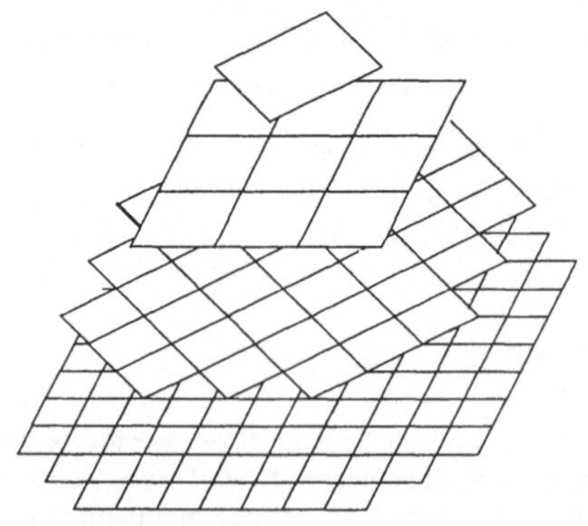

Abbildung 1: $3 \times 3/2$ **Pyramide**

Jeder Pixel der Ebene G_i wird als gewichtetes Mittel von Pixeln der Ebene G_{i-1} berechnet. Diese Operation wird *Faltung* genannt,

$$G_k(i_k, j_k) = \sum_{m=-1}^{1} \sum_{n=-1}^{1} w(m,n) G_{k-1}(i_{k-1} + m, j_{k-1} + n) \qquad (2)$$

wobei $w(m,n)$ ein (3×3) Gewichtsfeld ist, das als *Faltungskern* oder kurz *Kern* bezeichnet wird.

Jede Ebene G_i entspricht einer tiefpaß gefilterten Version von G_{i-1}. Dadurch werden hohe Frequenzen herausgefiltert. Die tiefen Frequenzen sind also in mehreren Ebenen $G_0, G_1, \ldots, G_k$ der Grauwertpyramide repräsentiert.

Diese Redundanz wird in der Laplace Pyramide beseitigt. Durch Interpolation berechnet man von einer Ebene G_{i+1} die Ebene G_i. Dazu definieren Burt und Adelson [4] eine Funktion EXPAND, welche die Umkehrung der Funktion REDUCE darstellen soll. Diese Umkehrfunktion ist nicht exakt, aber unter Verwendung von EXPAND kann jene Bildinformation bestimmt werden, die bei der Reduktion verloren ging.

Eine Laplace Pyramide ist definiert durch die Ebenen $L_0, \ldots, L_n$.

$$\begin{aligned} L_i &= G_i - \text{EXPAND}(G_{i+1}) \qquad i = 0, \ldots, n-1 \\ L_n &= G_n \end{aligned} \qquad (3)$$

Jede Ebene G_i kann durch die Ebenen $L_i, \ldots, L_n$ mit der Funktion EXPAND exakt rekonstruiert werden.

So wie die Grauwertpyramide als eine Menge von tiefpaß gefilterten Versionen des Originalbildes gesehen werden kann, stellt die Laplace Pyramide eine Menge von bandpaß gefilterten Versionen dar (Abb. 2).

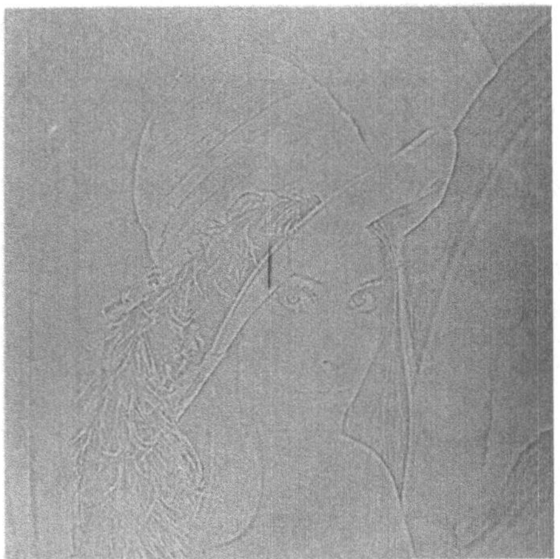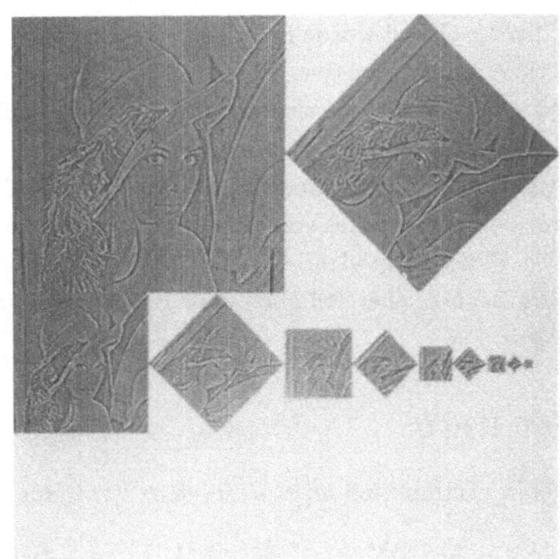

Abbildung 2: (a) die unterste Ebene L_0, (b) ie Ebenen $L_2, \ldots, L_{12}$ der $3 \times 3/2$ **Laplace Pyramide** (kontrastverstärkt)

2 Die $m \times m/2$ Pyramide

Der Aufbau erfolgt von unten nach oben, wobei jeder zweite Pixel als Zentrum des Reduktionsfensters genommen wird. Die Verteilung dieser Zentren ist schachbrettartig. In dieser Arbeit werden nur einander überlappende 3×3 Fenster verwendet.

Bei einer $m \times m/2$ Pyramide, bestehend aus den Ebenen $A_0, A_1, \ldots A_n$, ist jede Ebene A_i flächenmäßig um den Faktor 2 kleiner als die darunterliegende Ebene A_{i-1}. Aus der dazugehörigen geometrischen Reihe sieht man, daß eine $m \times m/2$ Pyramide höchstens doppelt soviel Speicherplatz benötigt wie das Ausgangsbild ($= A_0$), aus dem sie aufgebaut wurde.

Die Struktur der $3 \times 3/2$ Pyramide erfordert eine Rotation der Ebenen um 45 Grad [7] (siehe Abb. 3). Dabei tritt das Problem auf, daß zur Speicherung einer verdrehten Ebene A_i gleich viel Speicherplatz benötigt wird wie für die Ausgangsebene A_{i-1}. Alle üblichen Bildverarbeitungssysteme erlauben nämlich nur die Abspeicherung eines achsparallelen, rechteckigen Bildes. Will man eine um 45 Grad verdrehte Ebene abspeichern, so muß man ihr umschreibendes Rechteck abspeichern und die vier Ecken mit Ersatzwerten (0) füllen. Zu beachten ist weiters, daß nur jede zweite Ebene A_{2k+1} nicht achsparallel gegenüber dem Ausgangskoordinatensystem erscheint. Alle übrigen Ebenen A_{2k} sind entweder um 0, 90, 180 oder 270 Grad gegenüber dem Ausgangsbild verdreht. Das heißt, daß diese Ebenen wieder ohne zusätzlichen Speicherbedarf abgespeichert werden können. Daraus ergibt sich ein maximaler Speicherbedarf für eine $m \times m/2$ Pyramide, bei dem um 2/3 mehr Pixel gespeichert werden als notwendig.

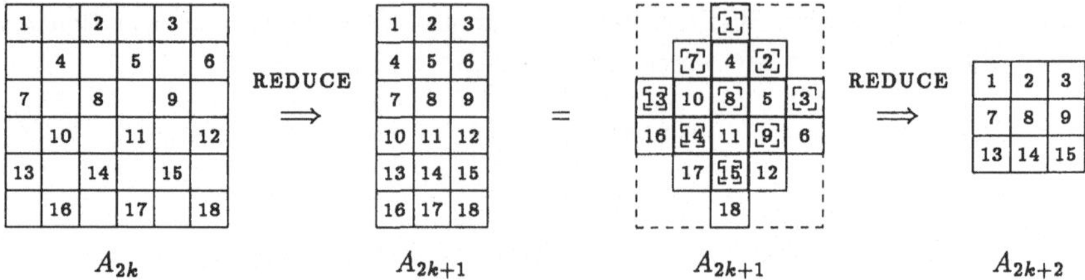

Abbildung 3: 2 Reduktionsschritte in der $3 \times 3/2$ Pyramide

Aus diesem Grund wurde in [10] eine *kompakte Speicherung* entwickelt, bei der keine überflüssigen Pixel abgespeichert werden müssen. Die Pixel einer um 45 Grad verdrehten Ebene A_{2k+1} (Abb. 3) werden dabei zeilenweise komprimiert in einem Rechteck abgespeichert. Dieses Rechteck besitzt gleich viele Zeilen wie A_{2k}, aber nur halb soviele Spalten (Abb. 4 a).

Bei einer $3 \times 3/2$ Pyramide enthält das Ergebnis G_i' der Funktion EXPAND doppelt soviele Pixel wie das Argument G_{i+1}. Die Pixel von G_i' werden durch Interpolation berechnet. Bei der Interpolation berechnet man jeden Pixel als gewichtetes Mittel seiner entsprechenden 3×3 Umgebung in G_i. Aus dieser 3×3 Umgebung können aber lediglich jene Pixel r verwendet werden, welche in der reduzierten Ebene G_{i+1} aufscheinen.

3 Der Kern

Für die Gewichtsfunktion $w(m,n)$ in einer Pyramide stellt Burt [3] folgende Bedingungen auf:

- Die Gewichtsfunktion ist *normiert*: $\sum_m \sum_n w(m,n) = 1$. Damit ist gewährleistet, daß sich der Grauwert einer homogenen Fläche in ihrem Inneren nicht ändert.

- *Symmetrie*: $w(m,n) = w(-m,-n)$ für alle m,n; keine Bevorzugung einer Richtung.

- *gleicher Einfluß* jedes Pixel: jeder Pixel einer Ebene muß mit dem insgesamt gleichen Gewicht bei der Berechnung der nächsten Ebene beteiligt sein. Jeder Pixel ist bei der Berechnung der nächsten Ebene gleich wichtig.

- *Unimodalität*: $w(i,j) \geq w(k,l) \geq 0$ für $|i| \leq |k|$ und $|j| \leq |l|$

Aus der Symmetrie folgt die Gestalt eines 3×3 Gewichtfeldes $w = \begin{pmatrix} a & b & a \\ b & c & b \\ a & b & a \end{pmatrix}$. Aus der Unimodalität folgt $0 \leq a \leq b \leq c$. Durch die beiden anderen Bedingungen erhält man zwei Gleichungen für die drei

Unbekannten a, b, c.

$$4b = c + 4a \qquad \text{(gleicher Einfluß)} \tag{4}$$

$$4a + 4b + c = 1 \qquad \text{(normiert)} \tag{5}$$

Die Gleichung (4) ist abhängig von der Struktur der $3 \times 3/2$ Pyramide. Der Wert des Zentrumspixel wird mit dem Gewicht c seiner Vaterzelle und mit Gewicht a an dessen vier Nachbarn weitergegeben. Jede Zelle ohne direkten Vater gibt ihren Wert mit Gewicht b an die vier über ihr liegenden Zellen weiter. Gleichen Einfluß haben die beiden Zellen daher dann, wenn die Summen der Gewichte, mit denen sie ihren Wert weitergeben, gleich sind. Durch Auflösen dieser Gleichungen erhält man Gewichtsfelder mit einem frei wählbaren Parameter.

$$a = \frac{1}{4}\left(\frac{1}{2} - c\right), \qquad b = \frac{1}{8}, \qquad \frac{1}{8} \le c \le \frac{1}{2} \tag{6}$$

Für den Parameter $c = 0.25$ erhält man beispielsweise ein Gewichtsfeld, das einer zweidimensionalen Gaußverteilung mit einer Streuung $\sigma = 0.85$ entspricht [9].

3.1 Wahl des Parameters c für den Kern

Unter der Annahme, daß die Grauwerte der Pixel in einem Bild statistisch unabhängig verteilt sind, kann die minimale Anzahl von Bits, die notwendig sind, um ein Pixel zu repräsentieren, durch die Entropie H angegeben werden [13]. Die Entropie H ist definiert durch $H = -\sum_{i=1}^{M} p(i) \log_2 p(i)$, wobei M die Anzahl der verschiedenen Graustufen, $p(i)$ die relative Häufigkeit der Graustufe i darstellt.

Bei der Anwendung der Laplace Pyramide zur Komprimierung von Bildern gilt es, die Entropie der einzelnen Ebene L_i möglichst niedrig zu halten.

Bei verschiedenen Versuchen hat sich herausgestellt, daß $c = 0.6$ in Gleichung (6) sowohl für den Reduktions- als auch Expansionskern die Entropie der einzelnen Ebenen am niedrigsten hält (Tabelle 1). Dieser Kern betont den Zentrumspixel sehr stark und hat dafür an den Ecken negative Gewichte ($a = -1/40$, $b = 5/40$, $c = 24/40$). Der Kern zeigt eine ähnliche Form wie jener 5×5 Kern, der sich bei Burt [4] in der $5 \times 5/4$ Pyramide als optimal herausstellte.

Die Ebenen der Grauwertpyramide erscheinen bei dieser Wahl des Kerns „schärfer" als etwa bei einer Gauß-ähnlichen Gewichtung ($c = 0.25$). Dadurch ist die Funktion EXPAND imstande, ein genaueres Interpolationsergebnis zu liefern.

		$c = 0.25$	$c = 0.4$	$c = 0.5$	$c = 0.6$	$c = 0.75$
	H	4.026	3.755	3.515	3.243	3.567
Ebene 0	σ	5.120	4.240	3.672	3.292	3.614
	H	3.623	3.476	3.362	3.233	3.791
Ebene 1	σ	4.548	4.079	3.813	3.751	4.688
	H	3.723	3.546	3.404	3.243	4.000
Ebene 2	σ	4.857	4.453	4.251	4.325	5.846
	H	3.967	3.776	3.624	3.443	4.275
Ebene 3	σ	5.437	5.018	4.849	5.053	7.241

Tabelle 1: Statistik der $3 \times 3/2$ Laplace Pyramide des „girl" - Bildes: vier Ebenen $L_0, \ldots, L_3$.

Die Unimodalitätsbedingung wird nur für Parameter c aus dem Intervall $[0.125, 0.5]$ erfüllt (vgl. [11]). Bei einer Wahl von $c = 0.6$ wird wie bei Burt diese Bedingung verletzt. Dadurch treten negative Gewichte auf, die nach der Faltung Pixel mit negativen Werten erzeugen können. Besonders wirkt sich diese Tatsache dann aus, wenn man durch mehrmaliges Anwenden der Funktion EXPAND eine höhere Ebene auf die Größe des Originalbildes vergrößern will. Dies ist im Rahmen einer progressiven Übertragung sinnvoll, um dem Betrachter immer ein Bild in Originalgröße zu zeigen. Bei Verwendung von EXPAND-Kernen mit negativen

Gewichten werden künstlich hohe Frequenzen erzeugt, die im Bild als Artefakte sichtbar werden. Die Form dieser Artefakte ist bei der $5 \times 5/4$ Pyramide stärker ausgeprägt als bei der $3 \times 3/2$ Pyramide (Abb. 4 b).

a

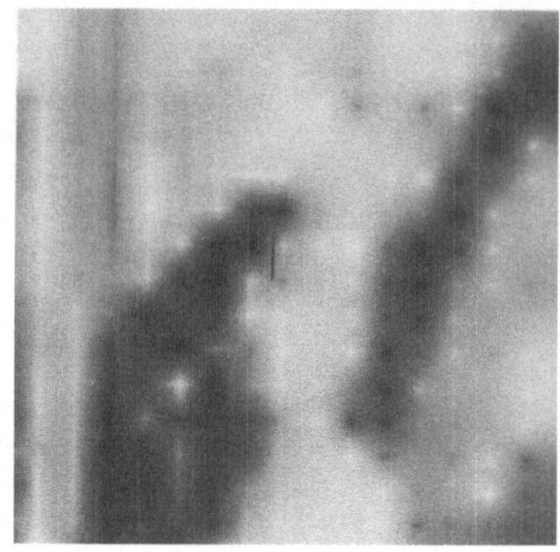
b

Abbildung 4: (a) die Ebenen $A_1, \ldots, A_{12}$ in kompakter Speicherung, (b) Vergleich der Artefakte in der $5 \times 5/4$ (links) und der $3 \times 3/2$ (rechts) Pyramide.

Diese Artefakte treten nicht auf, wenn man bei der EXPAND Funktion einen Kern wählt, der nur positive Gewichte enthält. Allerdings haben unsere Versuche gezeigt, daß dadurch die Entropie und damit der Speicherbedarf (= Übertragungszeit) erhöht wird.

Um sowohl die Artefakte zu beheben und trotzdem den Speicherbedarf minimal zu halten, bietet sich folgende Lösung an: Zur Rekonstruktion der Pyramide verwendet man den EXPAND Kern, der den Speicherbedarf minimal werden läßt. Um eine höhere Ebene dem Betrachter in Originalgröße und ohne Artefakte zu zeigen, verwendet man einen aus positiven Gewichten bestehenden Kern (z.B.: $c = 0.4$ in Gleichung (6)).

4 Quantisieren

Die Entropie eines Bildes ist abhängig von der Anzahl der verschiedenen Grauwertstufen. Ein Bild mit 16 verschieden Grauwertstufen läßt sich mit maximal 4 bit/pixel, ein Bild mit 256 verschiedenen Grauwertstufen mit 8 bit/pixel repräsentieren. Verringert man die Anzahl der verschiedenen Grauwertstufen pro Ebene, so verringert sich die Entropie. Den Vorgang zur Verringerung der Grauwertstufen nennt man *Quantisieren*. Allerdings führt man durch Quantisieren *Quantisierungsfehler* ein.

Ein *Quantisierer* ist eine Einheit, die zu jeder Eingabevariable X einen Wert Y liefert, und zwar wenn $a_{i-1} < X \leq a_i$ dann ist $Y = \gamma_i$. Die a_i werden *Quantisierungsschwellwerte* genannt, die γ_i sind die *Rekonstruktionswerte*. Bei dieser Arbeit werden Quantisierer verwendet, deren a_i gleichen Abstand zueinander haben ($a_{i+1} - a_i = $ const.) und deren γ_i genau der Mitte zwischen den Schwellwerten entsprechen ($\gamma_i = (a_{i-1} + a_i)/2$). Berger [1] zeigte, daß solche uniformen Quantisierer optimal sind, um bei gegebener oberer Schranke für die Entropie den Quantisierungsfehler minimal zu halten.

Bei der Laplace Pyramide haben die meisten Pixel den Wert 0, da die Funktion EXPAND eine sehr gute Näherung G'_i von G_i liefert. Die Differenz der beiden Bilder G'_i und G_i ist daher an den meisten Stellen null. Deshalb ist es wichtig, die a_i des Quantisierers so zu wählen, daß ein γ_i genau null ist.

Tabelle 2 zeigt, daß sich die Entropie der einzelnen Ebenen durch Quantisieren verringern läßt. Bei der Entropie der quantisierten Ebenen ist auch die Anzahl der verschiedenen Rekonstruktionswerte γ_i angegeben. Der Wertebereich des Originalbildes umfaßt dabei $M = 256$ verschiedene Grauwertstufen. Die Anzahl der

$3 \times 3/2$ Laplace Pyramide	original	quantisiert	
mit $c = 0.6$	H	H	Stufen
Ebene 0	3.243	0.234	15
Ebene 1	3.233	0.422	19
Ebene 2	3.243	0.614	23
Ebene 3	3.443	0.895	27

Tabelle 2: „girl" - Bild: Reduktion der Entropie H durch Quantisieren.

γ_i ist deshalb ungerade, um ein $\gamma_k = 0$ zu erhalten.

Die Anzahl der Rekonstruktionswerte in einer bestimmten Pyramidenebene bezieht sich auf die besondere Fähigkeit des menschlichen Sehens, Fehler innerhalb der Frequenzbänder der jeweiligen Ebene zu erkennen. Der Mensch kann Störungen in hohen Frequenzen kaum erkennen, wohl aber in mittleren und tiefen Frequenzen [4,12].

5 Progressive Übertragung

Die Laplace Pyramide ist für die *progressive Übertragung* von Bildern geeignet. Das folgende Schema (7) zeigt links, wie die Pyramide $G_n, \ldots, G_0$ durch Übertragen der Laplace-Ebenen $L_n, \ldots, L_0$ erzeugt werden kann und rechts, wie höhere Ebenen durch wiederholtes Anwenden von EXPAND auf Originalgröße vergrößert werden können.

$$
\begin{array}{llll}
\text{Schritt 0} & G_n = L_n & G_0^n = \text{EXPAND}^n(G_n) \\
\text{Schritt 1} & G_{n-1} = \text{EXPAND}(G_n) + L_{n-1} & G_0^{n-1} = \text{EXPAND}^{n-1}(G_{n-1}) \\
\vdots & \vdots \quad \vdots \quad \vdots & \vdots \quad \vdots \\
\text{Schritt } n & G_0 = \text{EXPAND}(G_1) + L_0 &
\end{array}
\tag{7}
$$

Bei progressiver Übertragung wird zuerst nur die kleine Datenmenge der obersten Pyramidenebene G_n übertragen und beim Empfänger expandiert. Dadurch erhält man rasch eine grobe Darstellung G_0^n des Bildes. Danach wird die nächste, tieferliegende Ebene L_{n-1} übertragen und zum vorhergehenden Bild EXPAND(G_n) addiert. Dieser Vorgang kann bis zur untersten Ebene wiederholt oder auch nach einer bestimmten Ebene abgebrochen werden. Der Betrachter am Empfangsgerät hat den Eindruck eines erst verschwommenen, unscharfen Bildes, welches mit Fortdauer der Übertragung immer schärfer wird.

6 Ergebnisse

Die quantisierten Ebenen der Laplace Pyramide wurden mit verschiedenen Verfahren komprimiert. Verwendet wurden Huffman Code [6], adaptiver Huffman Code [5] und ein adaptiver Lempel-Ziv-Welch (LZW) Algorithmus [17]. Der Hauptunterschied zwischen Huffman Code und LZW liegt darin, daß Huffman einzelne Wörter (Bytes, Pixel) in Codes variabler Länge umwandelt, LZW mehrere (variable Anzahl) Wörter in Codes fixer Länge umwandelt.

Tabelle 3 zeigt die Ergebnisse der Komprimierung des Bildes „girl". Angegeben sind die Anzahl der Bytes, die notwendig sind, um bis zu einer bestimmten Ebene progressiv zu übertragen, sowie die umgerechneten Kosten in bits/pixel bezogen auf die Größe des Originalbildes (512×512). Die Ebenen der $3 \times 3/2$ Laplace Pyramide wurden dabei von unten nach oben mit 15, 19, 23, 27, 31, 47, 63, 95 und 127 Stufen quantisiert. Die Ebenen der $5 \times 5/4$ Laplace Pyramide wurden mit 15, 23, 31, 63 und 127 Stufen quantisiert.

Wie aus Tabelle 3 zu ersehen ist, liefert LZW die besseren Ergebnisse. Dies ist damit zu erklären, daß Huffman Code mindestens ein Bit als Code pro Pixel benötigt und deshalb keinen Code mit einer Entropie $H < 1$ erzeugen kann. Die untersten Ebenen der Laplace Pyramide verfügen aber über solche

Übertragene Ebenen	3 × 3/2						5 × 5/4		Übertragene Ebenen
	Huffman		adapt. Huff.		LZW		LZW		
	Bytes	bits/pel	Bytes	bits/pel	Bytes	bits/pel	Bytes	bits/pel	
9,8,7,6,5,4,3,2,1,0	76086	2.3220	75525	2.3048	32403	0.9889	29915	0.9129	4,3,2,1,0
9,8,7,6,5,4,3,2,1	41756	1.2743	41204	1.2574	24200	0.7385	—		
9,8,7,6,5,4,3,2	23686	0.7228	23144	0.7063	17763	0.5421	17114	0.5223	3,2,1,0
9,8,7,6,5,4,3	14027	0.4281	13491	0.4117	12404	0.3785	—		
9,8,7,6,5,4	8656	0.2642	8128	0.2480	8592	0.2622	7892	0.2408	2,1,0
9,8,7,6,5	5653	0.1725	5129	0.1565	5915	0.1805	—		
9,8,7,6	3634	0.1109	3114	0.0950	3763	0.1148	3535	0.1079	1,0
9,8,7	2429	0.0741	1918	0.0585	2295	0.0700	—		
9,8	1536	0.0469	1038	0.0317	1201	0.0367	1295	0.0395	0
9	512	0.0156	512	0.0156	512	0.0156	—		

Tabelle 3: „girl" - Bild: Vergleich von Komprimierungsverfahren bei Übertragung von quantisierten Ebenen.

niedrigen Entropiewerte (siehe Tabelle 2). Der Code von LZW hingegen kommt diesen niedrigen Werten sehr nahe.

Burt [4] errechnet in seiner Arbeit mit der 5 × 5/4 Pyramide die Komprimierungsraten nur aus der Entropie. Um genaue Vergleichszahlen zu erhalten, führten wir diesselben Komprimierungstests auch mit der von Burt beschriebenen 5 × 5/4 Pyramide aus. Die Ergebnisse (Tabelle 3) zeigten, daß vor allem mit LZW Codierung bei der Komprimierungsrate kein nennenswerter Unterschied auftritt. Dies obwohl die 3 × 3/2 Pyramide doppelt soviele Ebenen besitzt wie die 5 × 5/4 Pyramide. Auch in der optischen Qualität der rekonstruierten Bilder konnten kaum Unterschiede festgestellt werden. Die 3 × 3/2 Pyramide hat aber gegenüber der 5 × 5/4 Pyramide den Vorteil, daß bei progressiver Übertragung der Betrachter am Empfangsgerät mehr Verfeinerungsstufen zu sehen bekommt.

7 Schluß

Es wurde eine Methode zur progressiven Übertragung und komprimierten Speicherung von digitalen Bildern präsentiert. Das Prinzip wurde von Burt und Adelson [4] für die 5×5/4 Pyramide eingeführt. Wir verwenden als Datenstruktur die 3 × 3/2 Pyramide.

Nach Definition der Laplace Pyramide wurde ein Faltungskern gefunden, der die Entropie und damit den Speicherbedarf (= Übertragungszeit) minimal hält. Um die sich ergebenden Artefakte bei mehrmaligen EXPAND einer höheren Ebene mit dem optimalen Kern zu verhindern, werden für mehrmaliges EXPAND Faltungskerne ohne negatives Gewicht verwendet. Die Komprimierungsraten, die sich bei den Versuchen ergaben, liegen zwischen 1 : 10 und 1 : 20.

Die Laplace Pyramide kann aber nicht nur zur progressiven Übertragung verwendet werden, sondern ist auch im Gebiet des dynamischen Sehens von Bedeutung [2].

Literatur

[1] T. Berger. Optimum Quantizers and Permutation Codes. *IEEE Transactions Information Theory*, IT-18(No. 6):pp. 759–765, November 1972.

[2] P. J. Burt. Attention Mechanisms for Vision in a Dynamic World. In *IEEE Proceedings of the 9th International Conference on Pattern Recognition*, pages 977–987, International Association for Pattern Recognition, November 1988.

[3] P. J. Burt. Fast Filter Transforms for Image Processing. *Computer Vision, Graphics, and Image Processing*, Vol. 16(No. 1):pp. 20–51, May 1981.

[4] P. J. Burt and E. H. Adelson. The Laplacian Pyramid as a Compact Image Code. *IEEE Transactions on Communications*, COM-31(No. 4):pp. 532–540, April 1983.

[5] R. G. Gallager. Variations on a Theme by Huffman. *IEEE Transactions on Information Theory*, IT-24(No. 6):pp. 668–674, November 1978.

[6] D. A. Huffman. A Method for the Construction of Minimum-Redundancy Codes. *Proceedings of the I.R.E.*, Vol. 40(No. 9):pp. 1098–1101, September 1952.

[7] W. G. Kropatsch. Ein Konzept mit zwei sich ergänzenden Pyramiden. In W. G. Kropatsch and P. Mandl, editors, *Mustererkennung'86*, pages 158–171, R. Oldenburg, 1986.

[8] W. G. Kropatsch. A pyramid that grows by powers of two. *Pattern Recognition Letters*, Vol. 3:pp. 315–322, 1985.

[9] W. G. Kropatsch and H. Herbst. Effiziente Störungsreduktion in der Gauss-Pyramide. In F. Pichler, editor, *Mustererkennung'87*, pages 22–47, Oldenbourg, 1987.

[10] H. F. Mayer and W. G. Kropatsch. Kompakte Bildkodierung mit der $3 \times 3/2$ Pyramide. 13. ÖAGM-Arbeitstreffen, Wien, 26.–27. Mai 1989.

[11] P. Meer, E. S. Baugher, and A. Rosenfeld. Frequency Domain Analysis and Synthesis of Image Pyramid Generating Kernels. *IEEE Transactions on Pattern Analysis and Machine Intelligence*, Vol. PAMI-9(No. 4):pp. 512–522, July 1987.

[12] A. N. Netravali and J. O. Limb. Picture Coding: A Review. *Proceedings of the IEEE*, Vol. 68:336–406, March 1980.

[13] W. K. Pratt. *Digital Image Processing*. John Wiley & Sons, New York, 1978.

[14] H. Samet. The Quadtree and Related Hierarchical Data Structures. *Computing Surveys*, Vol. 16(No. 2):pp. 187–260, June 1984.

[15] K. R. Sloan, Jr. and S. L. Tanimoto. Progressive Refinement of Raster Images. *IEEE Transactions on Computers*, C-28(No. 11):pp. 871–874, November 1979.

[16] G. S. van der Wal and J. O. Sinninger. Real time pyramid transform architecture. In *Proceedings of SPIE - The International Society for Optical Engineering*, pages 300–305, Cambridge, Massachusetts, September 1985.

[17] T. A. Welch. A Technique for High-Performance Data Compression. *IEEE Computer*, Vol. 17(No. 6):pp. 8–19, June 1984.

Bildanalytische Qualitätskontrolle in der Mikrofertigung
Ein vollautomatisches, hochflexibles und schnelles Strukturmeßsystem

Bernhard Bürg, Helmut Guth, Andreas Hellmann
Kernforschungszentrum Karlsruhe GmbH, Institut für Datenverarbeitung in der Technik
Postfach 3640, D-7500 Karlsruhe 1

1   Einführung

Im Rahmen des Arbeitsschwerpunktes **Mikrotechnik** werden im Kernforschungszentrum Karlsruhe (KfK) Prozeßtechniken zur Herstellung von mechanischen Mikrostrukturen entwickelt, die durch ihr außerordentlich hohes Aspektverhältnis, das Verhältnis von Strukturwandhöhe zu Strukturwandbreite, stabile Strukturen im mikroskopischen Bereich ermöglichen. Mit dem sog. LIGA-Verfahren (**Li** thographie, **G** alvanik, **A** bformung) können Strukturen im Mikrometerbereich mit einem Aspektverhältnis bis zu 100 und Abweichungen in der Strukturwanddicke im Submikrometerbereich hergestellt werden. Dabei sind beliebige Geometrien realisierbar [Becker85]. Beschleunigungssensoren, optische Multiplexer, hochpräzise Düsen, Mikrostecker oder Filter (Abb. 1) sind eine kleine Auswahl möglicher Kandidaten für die Fertigung mit dem LIGA-Verfahren.

Der Herstellungsprozeß beginnt mit dem Entwurf der Struktur in einem CAD-System. Die CAD-Daten dienen zur Steuerung eines Elektronenstrahlschreibers, mit dem eine primäre Maske hergestellt wird. Über die aus der primären Maske gefertigten Röntgenmaske wird mit Sychrotronstrahlung ein Resist bestrahlt, dessen nichtresistente Teile weggeätzt werden. Durch verschiedene Galvanik- und Abformungsschritte entsteht ein "Werkzeug", mit dem durch weitere Abformung die Endprodukte in Serie hergestellt werden.

Sowohl alle Zwischenprodukte als auch die Endprodukte müssen einer zerstörungsfreien Oberflächenprüfung unterzogen werden. Dazu eignet sich in hervorragender Weise die digitale Bildanalyse.

2   Anforderungen

Die Qualitätskontrolle soll zu Aussagen gelangen, mit denen neben einer Ja-Nein-Entscheidung über das Endprodukt der Herstellungsprozeß selbst überwacht werden kann. Dazu ist ein Meß- und Kontrollsystem notwendig, das detaillierte Informationen über einen Strukturfehler liefert. Es soll **beliebige geometrische Formen** vermessen, um eine aussagekräftige Beschreibung der tatsächlich vorliegenden Strukturform geben zu können.

Langfristiges Ziel der Fertigung von Mikrostrukturen mit dem LIGA-Verfahren ist die Serienherstellung in einer Produktionsstraße. Der Kontrollprozeß muß deshalb ab der Justierung des Substrats unter dem Mikroskop über die Vermessung der Strukturen bis zur Auswertung der ermittelten Formparameter **vollautomatisch** ablaufen, und darf dabei eine bestimmte Taktzeit des Prozesses nicht überschreiten.

Das unterschiedliche Reflexionsverhalten der durch die Vielfalt der Prozeßschritte bedingten Materialien darf die Messergebnisse nicht verfälschen.

Weiterhin soll die Möglichkeit bestehen, sich ohne langwieriges Einlernen unter Berücksichtigung des vollautomatischen Prozeßablaufs auf wenige **kritische**, zu vermessende **Bereiche** zu beschränken.

3   Konzept

Die Randbedingung des vollautomatisch ablaufenden Kontrollprozesses verbietet ein Einlernen von Referenzstrukturen und kritischen Bereichen. Da die CAD-Daten ohnehin vorhanden sind, soll das CAD-System die Daten zur Steuerung des Kontrollsystems liefern. Dazu wird zunächst eine hierarchische Beschreibung der Strukturform definiert, in der Art, daß eine Struktur durch einen geschlossenen Kantenzug beschrieben wird. Eine Kante kann eine gerade Linie oder ein Kreisbogen sein. Jede Kante wird durch zwei Ecken bestimmter Ausprägung begrenzt. Sowohl Kanten als auch Ecken erhalten Attribute, durch die ihre eindeutige Beschreibung möglich ist [Bürg87/1].

Im CAD-System werden auf einer zusätzlichen Informationsebene (Layer) Hinweise auf kritische Bereiche und Dimensionen eingetragen, die örtlich mit den Geometriedaten in der Original-Layer korrespondieren. Dem Entwurfsingenieur bietet sich damit schon beim Design der Struktur die Möglichkeit, Kontrollwünsche, individuell auf jede Struktur abgestimmt, zu formulieren, die Strukturen in beliebige Bildabschnitte zu zerlegen und für jeden Bildabschnitt durch die Wahl seiner Größe Auflösung und Genauigkeit frei zu bestimmen.

In einer Datenaufbereitungsphase werden aus den korrespondierenden Informationen beider "Layer" des CAD-Systems Bildverarbeitungsbefehle formuliert und in eine Datei (Soll-Datei) geschrieben, die die Grundlage für die Steuerung des kompletten Qualitätskontrollprozesses bildet. Die so aufgebaute Daten- und Informationsbasis zwischen CAD- und BV-System garantiert die einfache Erweiterbarkeit des Befehlssatzes.

Das Kontrollsystem fährt die in der Soll-Datei beschriebenen Bildabschnitte auf dem Substrat an, zieht ein Bild mit einer Videokamera ein und arbeitet sukzessiv die BV-Befehle ab, indem die Strukturmerkmale in diesem Bild verifiziert werden. Die detaillierten Ergebnisdaten werden in einer zweiten Datei (Ist-Datei) gespeichert. Die Auswertung erfolgt durch den Vergleich von Soll- und Ist-Datei (Abb. 2).

Durch die Unabhängigkeit der Auswertung der Bildabschnitte untereinander lassen sich Strukturen und Strukturteile an beliebiger Position und ohne direkte Verbindung in Form von z.B. Kantenzügen zum Substratkoordinatensystem ausmessen. In einem frei wählbaren Rahmen können durch die Art der Detektion und Mustererkennung auch ungenau positionierte Strukturen vermessen werden.

Die notwendige Auflösung und Genauigkeit erfordern die Verarbeitung von Grauwertbildern der Größe 512 * 512 -Pixel mit 256 Grauwertstufen pro Pixel.

4   Systemaufbau

4.1   Funktion

Für den Aufbau des Systems wurde eine Multi-Prozessor-Architektur gewählt, bei der sich die Systemleistung auf einfache Weise durch Zuschaltung weiterer Prozessoren steigern läßt.

Die Aufgaben werden funktional in zwei Gebiete aufgeteilt, in die Verwaltung, die vom Masterrechner übernommen wird, und in die Auswertung von Strukturbildern, die den einzelnen Slaves zugeteilt wird.

Zu Beginn einer Substratkontrolle ermittelt der Masterrechner automatisch die Position von drei Markierungskreuzen, die auf dem Substrat an vorbestimmtem Ort aufgebracht sind und das Substratkoordinatensystem definieren. Durch die Berechnung einer drei-dimensionalen Transformationsmatrix wird der Bezug zwischen Substrat- und Tischkoordinaten hergestellt. Weiterhin übernimmt der Masterrechner die Steuerung des softwarebasierten Autofokus, die Positionierung des Mikroskoptisches, den automatischen Objektivwechsel, den Bildeinzug, das Lesen des Soll-Daten-Abschnitts bezüglich des aufgenommenen Bildes, das Senden von Auswertungs-Aufträgen an die Slaves, das Entgegennehmen und Abspeichern der Ergebnis-Daten und das Führen einer Gesamt-Statistik über die wichtigsten detektierten Fehler.

Die Aufgabe eines jeden Slaves besteht in der Bildvorverarbeitung, der positionsunabhängigen Strukturverifikation und der Vermessung der Struktur eines kompletten Bildes. In der Bildvorverarbeitung werden aus dem Original-Grauwertbild ein Kantengrauwertbild und ein Gradientenrichtungsbild hoher Auflösung berechnet. Das hieraus ermittelte Linienbild mit 1-Pixel breiten Linien ist Grundlage für die Strukturverifikation, die mittels der Abarbeitung der "BV-Befehle" aus der Soll-Datei durchgeführt wird. Nach erfolgreicher Detektion der Strukturecken und -kanten erfolgt die Vermessung der Struktur, die ebenfalls durch "BV-Befehle" aus der Soll-Datei gesteuert wird [Bürg89].

4.2   Hardware

Als Rechner-System wurde ein Industrie-Standard VME-Bus-System verwendet. Masterprozessor ist ein Motorola MC68020/68881 25 MHz, 4 MByte Hauptspeicher und 128 MByte-Platte. Als Slave-Rechner-Verband wird ein Transputer-Netz (Inmos T800 20 MHz mit je 4 MByte Hauptspeicher) eingesetzt, das sich für eine einfache Erweiterbarkeit hervorragend eignet (Abb. 3). Der Bildeinzug erfolgt mit einer normalen Video-Karte. Eine Erweiterung des Systems um spezielle Bildverarbeitungs-Hardware, die die Bildvorverarbeitung durchführen kann, ist vorgesehen. Als Kamera wird eine CCD-Kamera mit Sony-Sensor mit einer Auflösung von 756 * 581 Bildpunkte eingesetzt [Bürg87/2].

Das Mikroskop erlaubt Vergrößerungen von 50 bis 1000. Es ist mit 5 Objektiven ausgerüstet, hell- und dunkelfeldgeeignet und arbeitet im Durch- und Auflicht. Der Mikroskoptisch verfügt über einen Fahrbereich von 150 mm * 150 mm und kann mit einer reproduzierbaren Genauigkeit von 3 µm positioniert werden. X-Y-Z-Position des Tisches und der Objektivrevolver lassen sich vom Rechner aus steuern.

4.3   Software

Das Betriebssystem für den Masterrechner ist OS-9 V2.2. Die Programme auf den Slaves laufen stand-alone. Das System wurde in der Programmiersprache C implementiert.

5   Erfahrungen mit dem System und Ergebnisse

Ziel war der Aufbau eines Prototyps, der durchschnittlich ein Bild einer Mikrostruktur auf einer Substratmaske vollständig in ca. 6 Sekunden ausmißt. Diese Entwicklung ist abgeschlossen. Mit insgesamt 7 Transputern wird die geforderte Performance erreicht.

Für die Entwicklung des Systems sind Teststrukturen entworfen worden, die alle Schwierigkeitsgrade beinhalten, die für die Bildvorverarbeitung und Mustererkennung von Bedeutung sind [Deiß87]. Es handelt sich dabei sowohl um Strukturen, deren Kanten aus aufeinanderfolgenden Kreisbögen gleicher oder unterschiedlicher Drehrichtung und unterschiedlichen Radien bestehen (Abb. 4), als auch um Strukturen, die mit dem LIGA-Verfahren schon hergestellt wurden wie Spinndüsen, Filter in Wabenform (Abb. 1), Balken unterschiedlicher Abmessungen, Spiralen, konzentrische Ringe, Vielecke, Kreuze und Kreisscheiben. Zudem ist die Auswertung der Testmaske im CAD-System so konzipiert worden, daß im BV-System mit unterschiedlichen Auflösungen und damit Objektiven gearbeitet werden muß.

Vielfache Testläufe haben ergeben, daß das System alle Teststrukturen vollautomatisch in der genannten Zeit ausmessen kann. Die Reproduzierbarkeit der Ergebnisse liegt bei konstanter Tischposition bei +- 1 Pixel, das bedeutet bei einer 50-fachen Vergrößerung (kleinstes verwendetes Objektiv) +- 1.1 µm, bei einer 1000-fachen Vergrößerung (größtes verwendetes Objektiv) +- 50 nm (rein rechnerisch).

Die Ungenauigkeit des Mikroskoptisches beträgt 3 µm. In Kombination mit einem hochpräzisen Tisch, der eine reproduzierbare Positioniergenauigkeit von etwa 50 nm aufweist, könnte das System Maskenverzüge mit hoher Genauigkeit über das gesamte Substrat vermessen.

6 Zusammenfassung und Ausblick

Das beschriebene System erfüllt alle gestellten Anforderungen: In einem vollautomatischen Prozeß vermißt es alle Strukturen eines vollständigen Substrats unabhängig von ihren geometrischen Formen und ihrer Lage auf dem Substrat. Optimale Strukturen und kritische Bereiche werden schon beim CAD-Entwurf definiert. Langwieriges Suchen und Einlernen von Referenzstrukturen entfällt. Das System liefert eine detaillierte Beschreibung von Strukturfehlern, die eine abgestufte Beurteilung der Qualität des Produkts zuläßt. Trotz Verwendung eines kostengünstigen Tisches wird eine hohe Präzision der Meßergebnisse erreicht.

Die weiteren Entwicklungsarbeiten konzentrieren sich zunächst auf die Ausdehnung des Kontrollprozesses auf andere Prozeßmaterialien. Langfristig wird eine durchschnittliche Verarbeitungszeit von 1 Sekunde pro Bild angestrebt. Dieses Ziel ist unter dem Aspekt der Systemerweiterung durch Transputer und Spezial-Hardware für die Bildvorverarbeitung und einer weiteren Optimierung einzelner Algorithmen erreichbar.

7 Literaturhinweise

[Becker85] E.W. Becker, W. Ehrfeld, P. Hagmann, A. Maner, D. Münchmeyer: *Herstellung von Mikrostrukturen mit großem Aspektverhältnis ...,*. KfK-Bericht 3995, Kernforschungszentrum Karlsruhe 1985.

[Bürg87/1] B. Bürg, B. Deiß, H. Guth, U. Klein, D. Schmidt: *Konzept für die Informations- und Datenbasis zwischen CAD und BV-System zur Kontrolle kritischer Bereiche von Mikrostrukturen.* Unveröffentlichter Bericht, Kernforschungszentrum Karlsruhe (1987).

[Bürg87/2] B. Bürg, A. Hellmann, H. Guth: *Konzept für den Aufbau eines automatischen, bildanalytischen Qualitätskontrollsystems für die Herstellung von Mikrostrukturen.* Unveröffentlichter Bericht, Kernforschungszentrum Karlsruhe (1987).

[Bürg89] B. Bürg, H. Guth, A. Hellmann: *Bildanalytische Strukturvermessung in der Mikrofertigung.* Unveröffentlichter Bericht, Kernforschungszentrum Karlsruhe (1989).

[Deiß87] B. Bürg, B. Deiß, U. Klein, D.Schmidt: *Entwurf einer Testmaske für Untersuchungen zur automatischen Kontrolle von Mikrostrukturen mittels digitaler Bildverarbeitung.* Unveröffentlichter Bericht, Kernforschungszentrum Karlsruhe (1987).

Abb. 1: Wabenförmige LIGA-Struktur, die beispielsweise als Filter eingesetzt werden kann (Stegbreite 8 μm). Zum Test der Bildverarbeitung wurden definierte Fehler eingearbeitet.

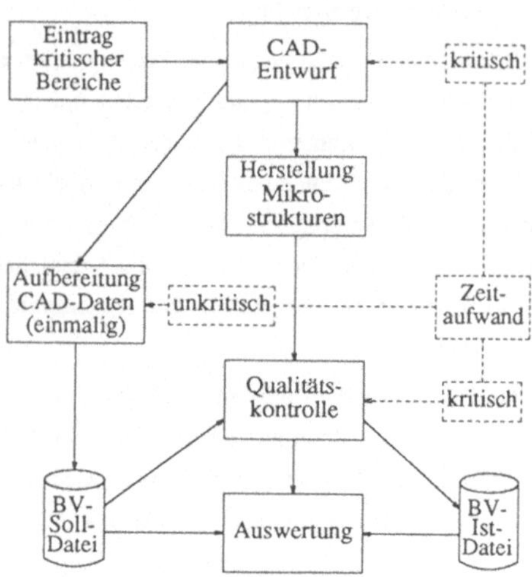

Abb. 2: Konzept des Qualitätskontroll- und Meßsystems. Die gestrichelten Einträge verdeutlichen die zeitlichen Restriktionen.

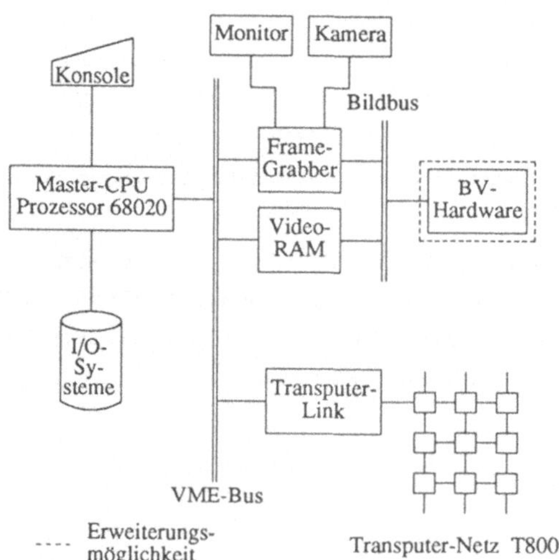

Abb. 3: Rechnerarchitektur des Qualitäts-kontroll- und Meßsystems.

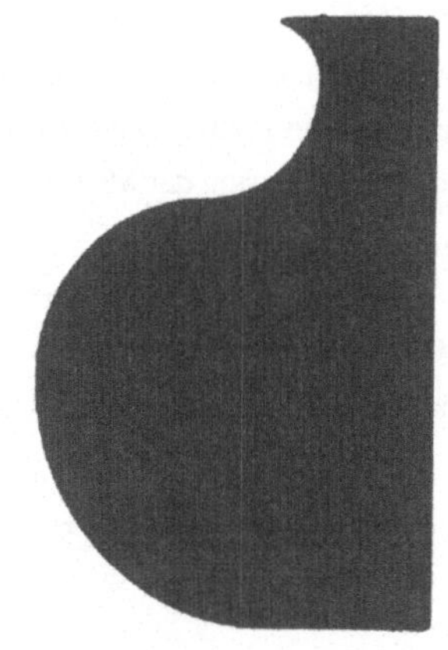

Abb.4: Teststruktur zur Detektion und Vermessung von Strukturkanten als Folge von Kreisbögen unterschiedlicher Radien und Drehrichtung.

AUTOMATISCHE AUSWERTUNG VON MIKROSKOPBILDSEQUENZEN BEIM KRISTALLWACHSTUM

V. Märgner
Institut für Nachrichtentechnik, TU-Braunschweig
Schleinitzstr. 23, 3300 Braunschweig

W. Beckmann, M. Rauls
Institut für Physikalische Chemie, TU-Braunschweig
Hans-Sommer-Str. 10, 3300 Braunschweig

1. Einleitung

Am Institut für Physikalische Chemie der TU werden die Mechanismen des Wachtums und der Auflösung von organischen Molekülkristallen am Beispiel der Fettsäure Stearinsäure untersucht /1/. Bestimmungsgröße bei diesen Untersuchungen ist die Wachstums- bzw. Auflösungsgeschwindigkeit der Kristallränder. Bei der Auflösung bilden sich zusätzlich auf der Kristalloberfläche Ätzgruben, deren Randausbreitungsgeschwindigkeit erfaßt werden soll. Dies erfordert die lichtmikroskopische Beobachtung der Kristalle über typischerweise 5-60 Minuten. Da insbesondere die Dynamik der untersuchten Prozesse als Funktion der Zeit bestimmt und mehrere Kanten und Ätzgruben mit je 4 Rändern gleichzeitig beobachtet werden sollen, ist eine direkte manuelle Vermessung am Mikroskop oder der auf einem Videorecorder aufgezeichneten Bildsequenz nicht möglich bzw. zu zeitaufwendig und fehleranfällig. Eine umfangreichere Untersuchung ist nur möglich, wenn es gelingt, mit Hilfe von Bildverarbeitungsverfahren die Messung zu automatisieren. Deshalb wurde am Institut für Nachrichtentechnik in Zusammenarbeit mit dem Institut für Physikalische Chemie ein Verfahren entwickelt, welches eine automatische Auswertung ermöglichen soll.

2. Versuchsaufbau

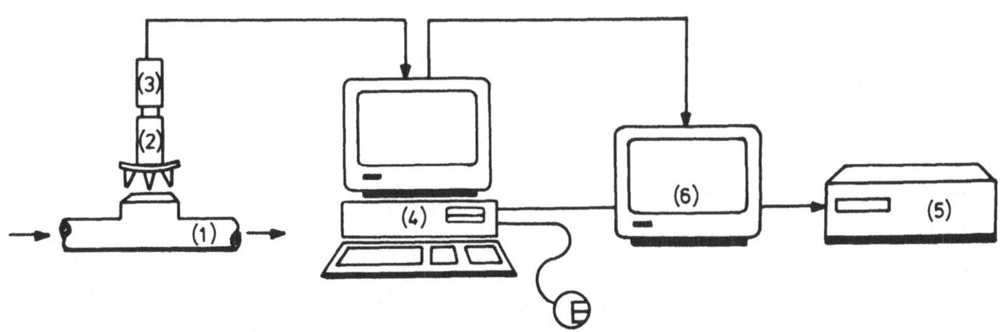

Bild 1: Versuchsaufbau zur Auswertung von Mikroskopbildsequenzen

Der Versuchsaufbau zur Auswertung der Mikroskopbildsequenzen (<u>Bild 1</u>) besteht aus der lösungsmittelumströmten Kristallprobe (1) und dem Lichtmikroskop (2) mit aufgesetzter Videokamera (3). Das Videosignal wird mit Hilfe einer "frame grabber" Karte in einem PC-AT (4) digitalisiert. Das Videosignal wird außerdem zu Kontrollzwecken kontinuierlich auf einem Videorecorder (5) aufgezeichnet. Mit Hilfe des Kontrollmonitors (6) wird das digitalisierte Bild beobachtet. Über Tastatur und Maus kann der Benutzer am PC den Ablauf der Auswertung steuern.

3. <u>Bildanalyse</u>

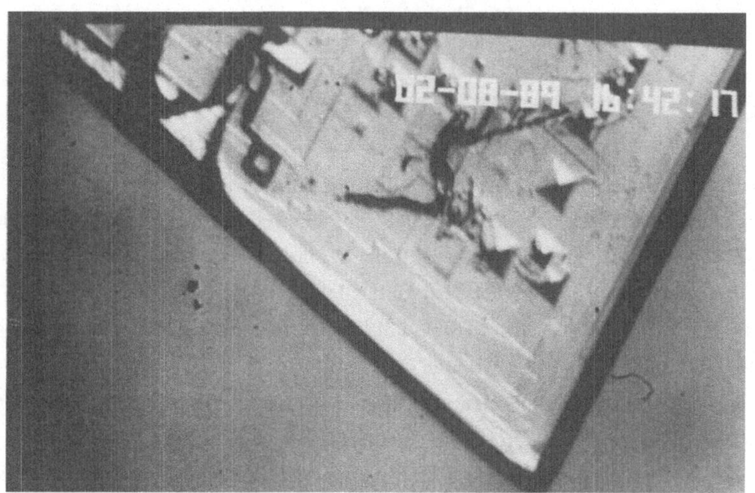

<u>Bild 2</u>: Mikroskopbild eines Kristalls

<u>Bild 2</u> zeigt ein typisches Mikroskopbild eines Kristalls, wie es am Anfang einer Meßreihe zur Verfügung steht. Während einer Meßreihe sollen z.B. die linke Kristallkante und die Kanten einiger Ätzgruben beobachtet und deren Positionsveränderungen gemessen werden. Eine Analyse der Mikroskopbilder ergibt die folgenden für die Bildauswertung wesentlichen Eigenschaften:

o   Aufgrund der Struktur der Kristalle kann davon ausgegangen werden, daß die Kanten gerade sind und nicht in jeder beliebigen Richtung verlaufen. Die untersuchten Kristalle weisen zwei kristallographische Vorzugsrichtungen auf zu denen alle Kanten und Oberflächenstrukturen parallel verlaufen.

o   Die Beleuchtung des Kristalls (Durchlicht) kann nicht beliebig variiert bzw. für die Bildauswertung optimiert werden. Daraus folgt eine nicht korrigierbare stark unterschiedliche Helligkeit der Ätzgrubenränder.

Diese Eigenschaften und die Tatsache, daß ein teilautomatisches Vorgehen am besten den gestellten Anforderungen genügt, haben zu dem im folgenden beschriebenen Verfahren geführt.

4. Verfahren

Das Verfahren zur Auswertung der Mikroskopbildsequenzen unterteilt sich in eine Start- bzw. Ein-
lernphase und in die Auswertphase. In der Startphase werden am ersten Bild der Bildsequenz die für die
Auswertung benötigten Parameter eingegeben, die Bearbeitungsfenster werden festgelegt, und es wird der
zeitliche Abstand der auszuwertenden Bilder bestimmt.

Folgende Angaben werden gemacht:

- falls möglich: Winkellage beider Kristallkanten
- Fenster, in denen Kanten beobachtet werden sollen
- Kennzeichnung, ob in einem Fenster eine Kante oder eine Ätzgrube vorliegt
- zeitlicher Abstand der Meßbilder (kann von Bild zu Bild variieren)

Danach werden in den eingegebenen Fenstern die Kanten detektiert und dem Originalbild überlagert. Ist
das Ergebnis zufriedenstellend, so wird die Auswertephase zur Bearbeitung der Sequenz gestartet,
andernfalls wird erneut eine Startphase begonnen.

Die Auswertephase führt in jedem definierten Auswertefenster die Arbeitsschritte Bildglättung, Kantendetektion und Binarisierung, Liniendetektion und Datenausgabe durch.

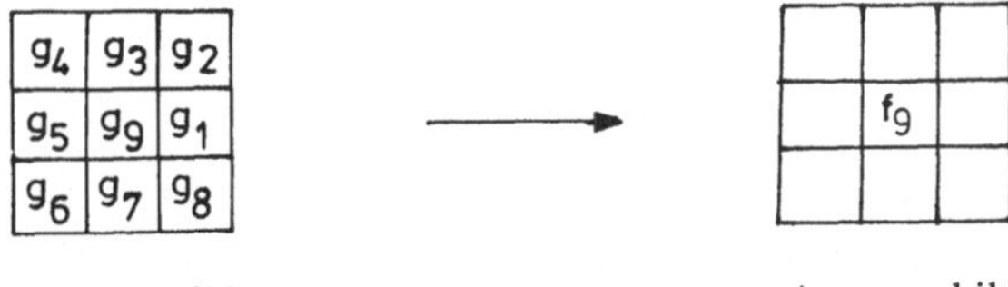

Eingangsbild   Ausgangsbild

$$f_9 = \begin{cases} 1 & \text{für} \quad g_9 < \dfrac{1}{9} \displaystyle\sum_{i=1}^{9} g_i \\[2ex] 0 & \text{sonst} \end{cases}$$

g_i - Grauwert im geglätteten Eingangsbild

f_i - Wert (0 oder 1) im binären Ausgangsbild

Bild 3: Beschreibung der Binärwandlung

Die Vorverarbeitungsschritte Glättung und Kantendetektion werden für den Fall, daß die Richtungen der
Kristallkanten nicht bekannt sind, mit einem universellen Tiefpaßfilter der Fenstergröße 3x3 und einer
nachfolgenden lokaladaptiven Binärwandlung (siehe Bild 3) durchgeführt. Von größerem Interesse ist der
Fall, daß die Vorzugsrichtungen der Kristalle bekannt sind. Z.B. können diese Vorzugsrichtungen am
Beginn der Verarbeitung einer Sequenz interaktiv dem System mitgeteilt werden. In diesem Fall können
sowohl die Glättung als auch die Kantendetektion mit Hilfe von gerichteten Operatoren durchgeführt

werden, die nach der Eingabe der Kantenrichtungen berechnet werden (siehe als Beispiel <u>Bild 4</u>). Da je nach Beleuchtung bzw. nach Lage der Ätzgrube die Ränder sowohl hell als auch dunkel sein können, werden an jedem Rand zwei Kanten detektiert. Anschließend wird mit einer von der Dynamik abhängigen Schwelle binarisiert und die jeweils außenliegenden Kanten werden dann zur Berechnung herangezogen.

1 0 0 0 0		1 0 0 0 0
0 1 0 0 0		0 1 0 0 0
0 0 1 1 0		-1 0 1 1 0
0 0 0 0 1		0 -1 0 0 1
0 0 0 0 0		0 0 -1 -1 0
0 0 0 1 0		0 0 1 0 -1
0 0 1 0 0		0 1 0 -1 0
0 0 1 0 0		0 1 0 -1 0
0 1 0 0 0		1 0 -1 0 0
0 1 0 0 0		1 0 -1 0 0
5x5-Glättungsoperatoren		Kantenoperatoren

<u>Bild 4</u>: Beispiele für richtungsselektive Glättungs- und Kantenoperatoren

Die Liniendetektion wird dann nach einer Hough-Transformation /2/ des jeweiligen Fensterausschnittes im Houghraum durch einfache Maximumsuche durchgeführt, wobei auch hier bei Kenntnis der Richtungen der Kristallkanten nur noch nach dem Maximum bei der jeweiligen Richtung gesucht werden muß. Der Umstand, daß hier nach Geraden von bekannter Richtung in einem kleinen Ausschnitt des Bildes gesucht wird, macht den Einsatz der Hough-Transformation auch bei dem verwendeten einfachen PC-AT Rechner praktikabel und zeiteffektiv.

Als Ergebnis wird schließlich für jedes Fenster aus dem Zeitpunkt der Messung und den Orten der detektierten Geraden nach einer Kalibrierung über eine interne Zeitbasis die jeweilige Wachstums- bzw. Auflösungsgeschwindigkeit berechnet und in einer Datei abgespeichert.

5. Ergebnisse

Die <u>Bilder 5a und 5b</u> veranschaulichen am Beispiel des Kristalles von Bild 2 das Ergebnis einer Kantendetektion und der daraus berechneten Linien. Umfangreiche Tests mit dem hier beschriebenen Verfahren zur Auswertung von Mikroskopbildsequenzen haben die Eignung des realisierten Systems unter Beweis gestellt. Das Verfahren arbeitet mit großer Sicherheit und ausreichender Genauigkeit. Die Verarbeitungsgeschwindigkeit von ca. 1 Sekunde pro Kante (PC-AT, 8 MHz Takt) ist ausreichend um umfangreiche, bisher nicht durchführbare, quantitative Auswertungen durchführen zu können /3/. Die Benutzung des Systems ist einfach und beinhaltet zudem Kontroll- und Korrekturmöglichkeiten, um z.B. die Lage unsicher detektierter Kanten interaktiv zu korrigieren oder zu bestätigen. Dadurch läßt sich die Sicherheit der Ergebnisse deutlich steigern.

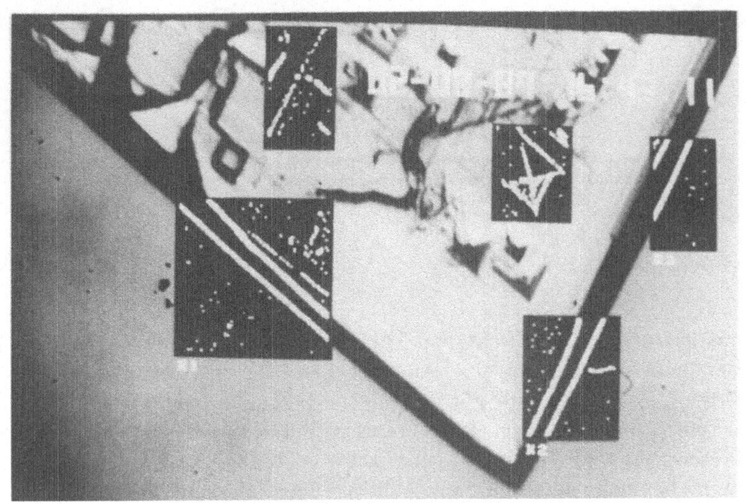

<u>Bild 5a</u>: Mikroskopbild nach Glättung, Kantendetektion und Binarisierung in einigen Fenstern

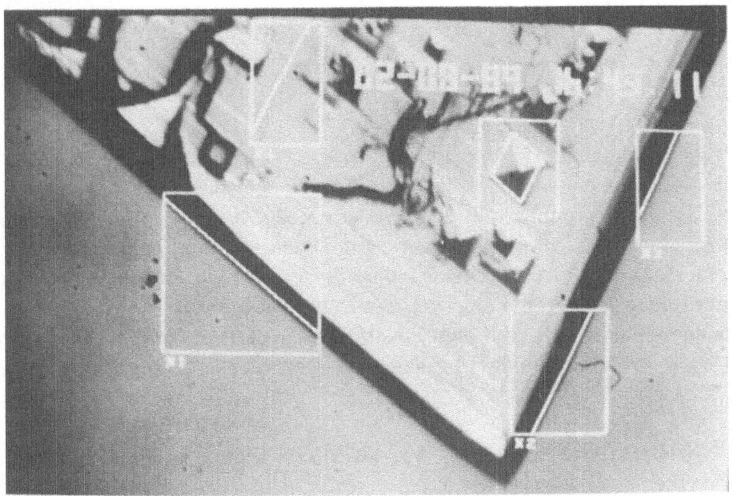

<u>Bild 5b</u>: Mikroskopbild mit eingeblendetem Ergebnis

<u>Literatur</u>

/1/   Beckmann, W.: J. Crystal Growth 86 (1987) 84

/2/   Duda, R.O. and Hart, P.E.: Use of the hough transformation to detect lines and curves in pictures. Comm. of ACM, Vol. 15, No. 1, Jan. 72, pp. 11-15

/3/   Rauls, M. and Beckmann, W.: The development of surface features and kinetics of the dissolution of stearic acid. To be published in J. of Crystal Growth special issue: proceedings of the 9th. Int. Conf. Crystal Growth, Sendai, Japan, Aug. 1989

On-Line-Mustererkennung von Prozeßzuständen

Harald Peters
Bergische Universität Wuppertal
Fachbereich Elektrotechnik
Lehrstuhl f. Automatisierungstechnik (Prof.Heidepriem)
Fuhlrottstr.10 , 5600 Wuppertal 1

1. Einleitung

In der Automatisierungstechnik tritt das Problem auf, daß ein aktueller Prozeßverlauf beurteilt werden muß. Sei es, um eine Aussage über die zu erwartende Güte des Endproduktes zu machen, um eventuelle Störungen vorhersagen zu können oder als Orientierungshilfe für einen menschlichen Prozeßfahrer. Bei kontinuierlichen technischen Prozessen, deren Systemparameter eventuell nur schwer zugänglich sind (z.B. in der Verfahrenstechnik) und die sich außerdem nur schwer über mathematische Modelle beschreiben lassen, besteht eine Möglichkeit im Vergleich mit in der Vergangenheit aufgetretenen und damit bereits bekannten Systemzuständen. Hierbei sollen Methoden der Mustererkennung eingesetzt werden, die schritthaltend eine Wissensbasis aufbauen und diese parallel zum Prozeßgeschehen fortlaufend anpassen.

2. Problematik

Die hier in Frage kommenden technischen Prozesse können in der Regel durch analoge Meßwertsignale beschrieben werden. Zur ausreichend genauen Beschreibung des Prozeßzustandes müssen je nach Komplexität des Prozesses 20 bis 100 Signale herangezogen werden. In der Literatur wird die Erkennung von Mustern aus eindimensionalen Zeitreihen bereits eingehend behandelt ; eine Erweiterung dieser Verfahren auf die Untersuchung von Mustern, die aus einer größeren Zahl von Zeitreihen aufgebaut sind, ist aber bisher nicht erfolgt. Im Zusammenhang mit der hier beschriebenen Anwendung und ausgehend von der Annahme, daß sich Prozeßzustände überhaupt wiederholen , treten dabei folgende Probleme auf :

* Wann sind Prozeßzustände ähnlich ? Wie ist vorzugehen , wenn sich zwei Zustände nur in einem Meßwertkanal unterscheiden ?
* Welche prinzipiellen Vorgehensweisen kommen für die Erkennung vieldimensionaler Zeitreihenmuster in Frage ? Können Methoden aus der Verarbeitung eindimensionaler Verläufe übertragen werden ?
* Wie können in einem kontinuierlichen Prozeßverlauf ohne prozeßbedingte Triggerzeitpunkte vergleichbare Muster gewonnen werden ?
* Welche Merkmale kommen angesichts der Tatsache, daß keine Aussagen über Eigenschaften der Meßwertverläufe getroffen werden können, in Frage ?
* Da die zu erwartende Anzahl von Klassen (= unterschiedliche Prozeßzustände) sehr groß ist (Größenordnung 1000 - 5000) und die Aufnahme einer genügend großen Stichprobe zeitlich unmöglich ist , stellt sich die Frage nach einem geeigneten Klassifikator , der zum einen sehr schnell arbeiten und zum anderen einen schritthaltenden Klassenaufbau ermöglichen muß.
* Sollen die Klassenparameter nach dem anfänglichen Aufbau weiter adaptiert oder konstant gehalten werden ? Hier ist die Änderung der Prozeßparameter durch Alterung von

Meßwertgebern sowie durch Verschiebung sonstiger Randbedingungen genauso zu berücksichtigen wie die Notwendigkeit der Klassenanpassung durch neu hinzukommende Merkmalvektoren zur Verbesserung der Klassifikationssicherheit.

All diese Punkte sind unter Berücksichtigung der Realzeitfähigkeit des Gesamtsystems zu betrachten. Hier muß durch geeignete Software- und Hardwarekonzepte die Verarbeitungsgeschwindigkeit möglichst maximiert werden, wobei allerdings der Kostenumfang von heute üblichen Rechnern für den Automatisierungseinsatz nicht überschritten werden darf.

3. Lösungsansätze

Das Grundprinzip zur Bildung eines Prozeßzustandsmusters besteht in einem verschiebbaren Zeitfenster , dessen Länge durch die Dynamik des untersuchten Prozesses festgelegt wird. *Bild 1* zeigt symbolisch den Vergleich zwischen dem aktuellen Zustandsbild und den in einer Datenbasis abgelegten Klassenprototypen von Abbildern vergangener Prozeßzustände. Anhand eines ausgewählten Prozeßsignals wird über ein Triggerverfahren der Endpunkt eines Zeitfensters festgelegt. Dieses Zeitfenster bildet die Grundlage für die anschließende Merkmalbildung. Das Triggerverfahren mß hierbei gewährleisten, daß ein ähnlicher Signal-

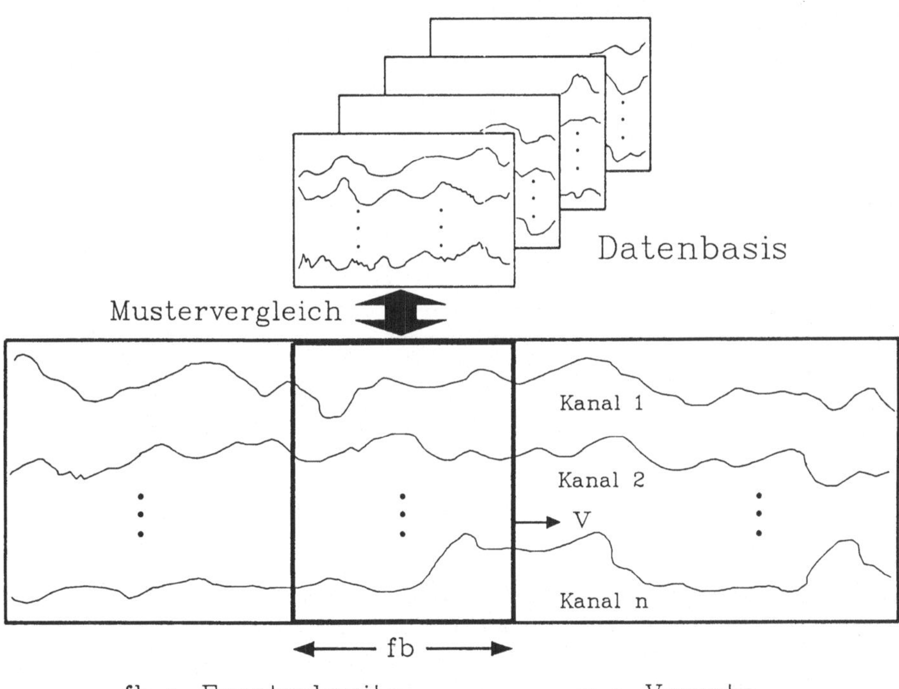

Bild 1

verlauf ohne größere Abweichungen zu denselben Triggerpunkten führt. Derzeit werden unterschiedliche Verfahren auf ihre Eignung untersucht.
Im Anschluß an die erfolgreiche Triggerung erfolgt die Musterbildung. Hier stehen augenblicklich drei unterschiedliche Methoden zur Verfügung, deren Prinzipien in *Bild 2* dargestellt sind.

Ein-Kanal-Zeitfenster :

Zunächst wird für jedes Analogsignal des Prozesses einzeln die Merkmalberechnung durchgeführt. Hierbei werden sowohl statistische Kennwerte als auch aus orthogonalen Transformationen gewonnene Merkmale untersucht. Die bisher ermittelte optimale Anzahl liegt je nach Merkmalart bei 6 bis 12 Werten je Kanal, ist allerdings abhängig von der Fensterbreite. Die aus diesen Merkmalen und für jedes Signal gebildeten Vektoren werden einzeln klassifiziert (=> N Klassifikationsebenen wobei N = Anzahl der Kanäle) . In einem zweiten Schritt wird dann aus den vorliegenden Klassifikationsergebnissen mit Hilfe einer Häufigkeitsuntersuchung auf den Zeitpunkt in der Vergangenheit geschlossen, der dem aktuellen Zustandsmuster am ähnlichsten ist. Dabei kann eine Entscheidungsschwelle angegeben werden, die zwischen Zuweisung und Rückweisung differenziert. Als Klassifikator wird ein "Nächster-Prototyp"-Klassifikator eingesetzt, der einen guten Kompromiß zwischen Rechenzeitbedarf, Güte der Klassifikationsresultate und Eignung für den hier anzuwendenden schritthaltenden Klassenaufbau darstellt.

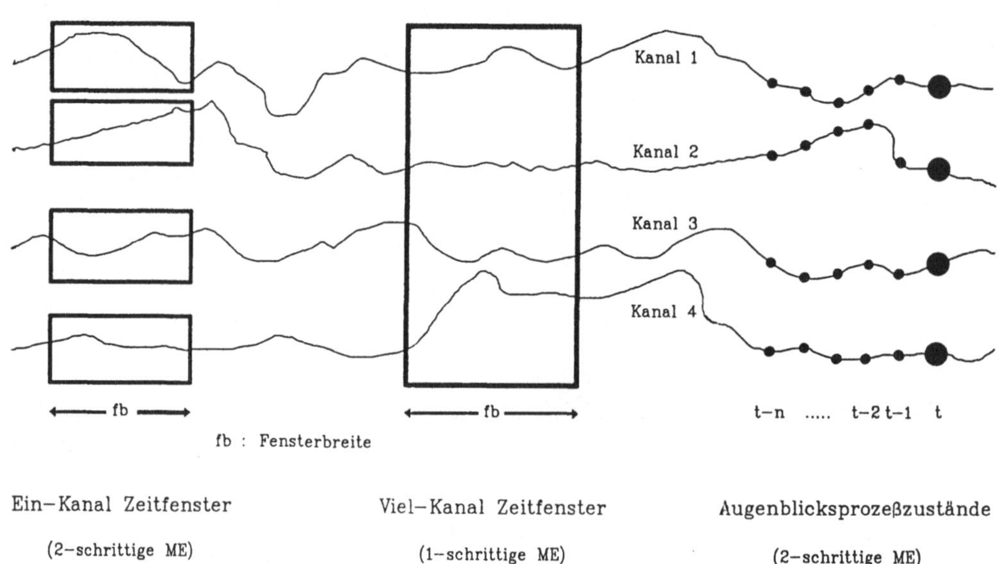

Bild 2

Die Vorteile dieser Methode liegen in der einfachen Merkmalberechnung, der Unempfindlichkeit gegenüber einem zeitlichen Versatz der Signale untereinander und der relativ kleinen Klassenanzahl je Klassifikationsebene . Schwierig ist hingegen die Auslegung des zweiten Klassifikationsschrittes zur Gesamtbeurteilung des Prozeßzustandes.

Viel-Kanal-Zeitfenster :

Hierbei werden die Merkmale direkt aus der Gesamtheit aller in dem Zeitfenster enthaltenen Daten gebildet. Es besteht eine prinzipielle Ähnlichkeit zur Bildverarbeitung : die einzelnen Kanäle des Prozeßzustandsbildes können als Zeilen eines Videobildes betrachtet werden und jeder Meßwert als Codierung der Graustufe eines Bildpunktes. Ein wesentlicher Unterschied besteht allerdings in der Anzahl der Quantisierungsstufen , die hier

deutlich größer ist. Es bieten sich zur Merkmalberechnung also Verfahren aus der optischen Mustererkennung an , die nicht auf spezielle Eigenschaften optischer Informationen zugeschnitten sind. Insbesondere sollen Transformationen wie die Diskrete-Cosinus- oder die Hadamard-Transformation auf ihre Eignung untersucht werden. Aber auch die bereits oben erwähnten statistischen Kennwerte werden berücksichtigt. Die Anzahl der Merkmale je Klassifikationsebene (hier nur eine) ist deutlich größer anzusetzen als bei der Verwendung des Ein-Kanal-Zeitfensters, da die Information hier in einem Schritt klassifiziert werden soll. Dementsprechend muß auch der Klassifikator ausgewählt werden. Aufgrund der zu erwartenden grossen Klassenanzahl und der geforderten Realzeitfähigkeit des Systems erscheinen auch hier besonders zeitunkritische Verfahren untersuchenswert.
Der Vorteil dieser zweiten untersuchten Methode liegt in der ganzheitlichen Betrachtung des Prozeßzustandes und den damit stärker berücksichtigten Korrelationen der Signale untereinander. Ferner ist bei der Merkmalberechnung ein deutlicher Zeitgewinn gegenüber dem ersten Verfahren zu verzeichnen. Die Trefferrate hingegen zeigt keine signifikanten Unterschiede gegenüber der Verwendung des 'Ein-Kanal-Zeitfensters'. Ein Nachteil besteht in der grösseren Empfindlichkeit gegenüber unzureichend genauer Triggerung.

Augenblicksprozeßzustand :

Der dritte Ansatz zur Muster- und Merkmalbildung geht von der Vorstellung aus, daß ein Zustandsbild auch generiert werden kann durch eine Folge von Augenblicksprozeßzuständen. Diese werden aus den Kurzzeitmittelwerten der einzelnen Prozeßsignale gebildet. Im ersten Klassifikationsschritt werden die so erzeugten Merkmalvektoren einzeln klassifiziert. Anschliessend wird die Folge der Klassifikationsergebnisse der N letzten Augenblicksvektoren herangezogen, um den aktuellen Prozeßzustand , einschließlich seiner Vorgeschichte , einem Zeitpunkt der Vergangenheit zuzuordnen oder als unbekannt zurückzuweisen. Der Ansatz des verschiebbaren Zeitfensters wird hier also nur indirekt verwendet. Probleme liegen bei dieser Methode in der Anzahl der aufgenommenen Muster und im Vergleichsverfahren für den zweiten Klassifikationsschritt. Hier können zum einen metrische Konzepte verfolgt werden, zum anderen ist der Einsatz von syntaktisch orientierten Verfahren denkbar.

4. Simulation realer Prozeßverläufe

Zur Entwicklung der angegebenen Methoden ist es unbedingt notwendig geeignetes Datenmaterial zu verwenden. Werden hierzu reale Prozeßverläufe eingesetzt , so stellt sich das Problem, die Ergebnisse des Klassifikationssystems zu überprüfen. Es müssen hierbei Zusatzinformationen vorhanden sein, zu welchem Zeitpunkt welcher Prozeßzustand vorlag, d.h. es muß sich um eine klassifizierte Stichprobe handeln.
Eine andere Möglichkeit besteht in der Simulation solcher Prozeßverläufe. Hierbei muß es möglich sein, gleiche bzw. ähnliche Zustände zu bestimmten Zeitpunkten erzeugen und diese mit in der Realität auftretenden Veränderungsformen beaufschlagen zu können. Im Rahmen des hier vorgestellten Projektes wurde ein Programmsystem entwickelt, mit dem Prozeßverläufe erzeugt werden können, die aus einer Kombination von realen Daten (jeweils eines Prozeßzustandes) einer technischen Anlage, Zufallsverläufen und über generierende Verfahren erzeugten Signalen, die bestimmten dynamischen Eigenschaften genügen, bestehen. Ferner können diese Daten kanalweise mit Rauschen, zeitlicher und amplitudenmäßiger Dehnung/Stauchung verändert werden. Auf diese Art und Weise können beliebige Situationen als Datenstruktur simuliert werden, die dann für die notwendigen Tests des Mustererkennungsverfahrens herangezogen werden können.

5. Software- und Hardwarestruktur des Entwicklungssystems

Auf Basis eines VME-Bus Rechners (Prozessor 68020) unter dem Realzeitbetriebssystem OS-9 und in der Programmiersprache MODULA-2 ist ein umfangreiches Softwarepaket entstanden, das die oben erläuterten Konzepte mit den unterschiedlichsten Arten von Triggermethoden, Musterbildungskonzepten, Vorverarbeitungsarten, Merkmalarten und Klassifikatoren zusammenfassend realisiert. Durch die Aufteilung in fünf verschiedene TASKs ist weiterhin die Beschleunigung der Verarbeitung durch Parallelisierung der einzelnen Schritte (im Sinne eines *Pipelining*) auf unterschiedliche Prozessoren möglich. Hier wird am Konzept eines Systems aus drei CPU-Karten gearbeitet , mit dem dann die Blöcke "Meßwertaufnahme und allgemeine Steuerung", "Muster- und Merkmalbildung" sowie "Klassifikation" in einer Pipeline-Struktur parallel ablaufen. In den beiden letzten Teilgebieten kann je nach Bedarf an den zusätzlichen Einsatz von Signalprozessoren (z.B. für eine 2-dimensionale FFT) oder die Verwendung eines TRANSPUTER-Netzwerkes gedacht werden, um die Geschwindigkeit weiter zu steigern. Besondere Aufmerksamkeit ist außerdem dem Dateikonzept zu widmen, da der Festplattenzugriff trotz der großen Hauptspeicherkapazität bei der enormen Menge zu verwaltender Daten einen wesentlichen Anteil an der Verarbeitungszeit hat.

6. Zusammenfassung

Es wurde die Entwicklung eines Testsystems vorgestellt, das in der Lage ist , aus einem technischen Prozeß gewonnene Zustandsbilder in Realzeit auf Ähnlichkeit zu untersuchen. Basierend auf einer Fenstertechnik wurden drei Arten der Musterbildung erläutert. Die eingesetzten Methoden zur Merkmalberechnung und Klassifikation gehen im wesentlichen auf Verfahren zurück, die sich in anderen Bereichen der Mustererkennung bereits bewährt haben. Weiterhin wurde eine Software- und Hardwarekonfiguration diskutiert , die einen umfangreichen Test der entwickelten Methoden gewährleistet und dabei optimal auf die gestellten Anforderungen zugeschnitten ist.

FERNERKENNUNG IN DER ORTUNG [*]

D.Ruser

TU Gdansk, Inst. f. Telekommunikation

deleg.: Universität WPU Rostock

1. Einführung

Die im Rahmen der Ortungs-Fernerkennung von der Mustererkennung zu lösende Aufgabe besteht darin, aus den empfangenen Echosignalen Wahrscheinlichkeitsaussagen über in Störungen vorhandene Objekte in Form von Entscheidungen abzuleiten. Dazu werden bekannte vordem statistische Eigenschaften der Nutzsignale und der Störungen genutzt. In der Literatur zur Objekterkennung besteht die Tendenz zur Prozeßautomatisierung, [1], [2]. Wenn es auch auf diesem Gebiet in der Telekommunikation große Errungenschaften gibt, so betrifft das jedoch in geringem Maße die Ortungssysteme. Beweis dafür ist die Tatsache, daß sich am Ausgang der meisten Ortungssysteme ein Monitor befindet, dessen Beobachtung einem Operateur es ermöglicht, eine Entscheidung über das Vorhandensein eines Objektes und über die Bestimmung seiner Lage im Raum zu treffen.

Diese durch den Operateur getroffenen Entscheidungen sind nicht frei von Fehlern. Dabei beruhen die Fehler auf der Nichterkennung vorliegender Objekte oder auf der Beobachtung von in Wirklichkeit nicht vorhandenen Objekten. Ursache der fehlerhaften Entscheidungen sind im Ortungssystem auftretende Störungen. Diese Störungen werden durch Systemelemente und durch den Übertragungskanal hereingetragen.

Eine Qualitätsverbesserung der zu treffenden Entscheidungen kann man erreichen, indem man den Störpegel begrenzt, also filtert. Das Ziel der Filterung in mit einem Monitor ausgestatteten Ortungssystem ist es, das Schirmbild von Störungen zu befreien und gleichzeitig die Nutzsignale hervorzuheben, die von den sich im durchsuchten Raum befindlichen Objekten ausgehen. Im Vortrag werden die Filterungsprinzipien besprochen und ein Simulierungssystem des tatsächlichen Betriebes der Ortungseinrichtung vorgestellt, das mit Entscheidungsschaltungen ausgerüstet ist, die der Verbesserung der Bildqualität dienen.

---

[*]   Die Arbeit entstand im Rahmen des polnischen Zentralplanes 02.16.4.

2. KRITERIEN ZUR MUSTERERKENNUNG

Der Erfolg von Filterungsmethoden zur Mustererkennung ist vor allem davon abhängig, wie weit sich Nutz- und Störsignale unterscheiden. Wirkungsvolle Filterungskriterien ergeben sich somit beim Auffinden unterschiedlicher Eigenschaften von Nutzsignalen und Störungen. Solche in klassischen Filterungsmethoden ausgenutzten (distinktiven) Eigenschaften sind meistens die Amplitude A(t) (Amplitudendiskriminatorschaltungen) und das Spektrum (Filterung im Frequenzbereich) sowie seltener die Signaldauer (Zeitdiskriminierung).

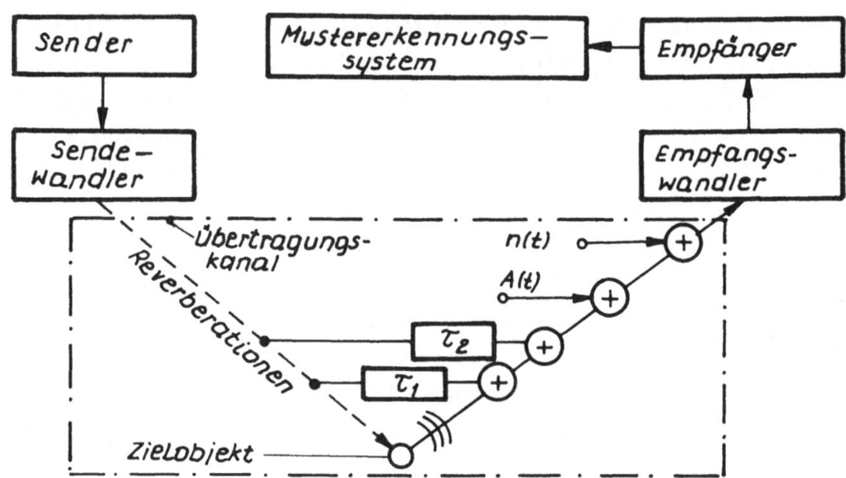

Bild 1: Grundstruktur von Ortungssystemen mit Mustererkennung

In Ortungssystemen mit Mustererkennung, Bild 1,[3], kann man noch andere Eigenschaften finden, die zur Unterscheidung der Nutzsignale von den Störungen n(t) dienen können. Eine dieser Eigenschaften ist die Periodizität bzw. Quasiperiodizität der Nutzechosignale vom Objekt. Diese Periodizität ergibt sich aus der periodischen Erzeugung der Sondierungssignale im Sender und deren Reflektion an ruhenden oder relativ langsam sich bewegenden Zielobjekten. Falls wegen Bewegung der Signalquelle keine Periodizität vorliegt, kann man sie durch Kompensierung der infolge der Quellenbewegung hervorgerufenen Verzögerungen erreichen, wenn man die Bewegungsparameter kennt. Die für Nutzsignale typische Periodizität tritt nicht im Falle von Rauschen n(t) auf. Im Falle von Reverberationen ist sie schwach oder existiert nicht, da Reverberationen die Summierung der Echosignale von zahlreichen, zufällig im Ortungsraum angeordneten Elementen darstellen, deren Ergebnis zufällig wird, weil es stark von den Signalphasenverschiebungen abhängt.

Eine andere die Nutzsignale charakterisierende Eigenschaft ist die Verlängerung des Echosignals im Verhältnis zum Sondierungssignal. Die Verlängerung ist ein Effekt der endlichen geometrischen Abmaße der Ziele, die oft vergleichbar sind mit der räumlichen Länge der Sondierungsimpulse. Im Rahmen des verlängerten Echosignals können Amplitudenfluktuationen auftreten, welche die Bestimmung der tatsächlichen Länge des Echosignales erschweren. Im betrachteten Zeitintervall kommen größere Signalamplituden mit bedeutend größerer Wahrscheinlichkeit vor. Man kann somit von einer charakteristischen Ansammlung großer Signale um den Ort des Auftretens eines wirklichen Zieles sprechen.

3. Simulationsuntersuchungen für die Fernerkennung

Im erarbeiteten und mit Simulationsmethoden untersuchten Empfänger des Ortungssystems wird eine Filterung entsprechend den distinktiven Eigenschaften der Echonutzsignale angewendet. Die klassischen Detektionskriterien nutzend wird so im Signalweg ein Schmalbandfilter mit einer Bandbreite gleich der effektiven Spektralbreite des Sondierungssignales verwendet. Dieses Filter beseitigt die außerhalb des Nutzsignalbandes liegenden Anteile des Störspektrums. Außerdem ist ein Amplitudendiskriminator eingesetzt, der Signale mit geringerer Amplitude als die vorgesehenen Amplituden des Nutzsignales eliminiert.

Außer diesen Aktivitäten ist eine Mehrfunktionsentscheidungsschaltung entwickelt, deren Tätigkeit sich auf die spezifischen Eigenschaften der Nutzsignale stützt. Dieser Schaltung eliminiert nichtperiodische Signale, Signale von zu kurzer Zeitdauer und Signale von zu großem Amplitudenspiel, über die Zeit gestreut, [4]. Dies Entscheidungssystem ist parallel in den Signalweg geschaltet, und an seinem Ausgang entsteht ein Binärsignal, wobei der Zustand "1" die Entscheidung über die Anwesenheit eines Nutzsignales bedeutet und der Zustand "0" das Nichtvorhandensein eines solchen Signales. Das in Reihe zum Signalweg eingefügte Tor wird durch den Zustand "1" geöffnet und durch den Zustand "0" geschlossen. So gelangen durchs Tor nur die Teile des Empfangssignales, die das Entscheidungssystem als Echosignal eines Objektes qualifizierte. Diese Signale werden auf dem Farbbildschirm angezeigt, dessen Farben verbunden sind mit dem Momentanwert der Echosignalamplituden. Die endgültige Entscheidung über das Vorhandensein eines Objektes trifft der Ortungstechniker als Operateur aufgrund der Bildbeobachtung am Monitor.

4. Untersuchungsergebnisse

Die Forschungen zur Fernerkennung wurden in zwei Etappen durchgeführt. In der ersten Etappe wurde die Entscheidungsschaltung ohne Bildschirmbeobachtung untersucht. Dazu wurden die Empfangswahrscheinlichkeit P(D) und die Falschalarmwahrscheinlichkeit P(FA) auf automatische Weise mit Hilfe eines Rechners bestimmt. Der Rechner erzeugte zufällige Echosignale und bestimmte ihre Anzahl und Lage. Die Echosignale wurden mit Störungen summiert und auf die Filterschaltungen, den Amplitudendetektor sowie die Entscheidungsschaltungen gegeben. Die Wahrscheinlichkeiten P(D) und P(FA) wurden an den jeweiligen Ausgängen des Mehrentscheidungsschwellensystems gekennzeichnet. Parameter war jedesmal das Signalstörverhältnis. Die zahlenmäßigen Werte von P(D) und P(FA) wurden berechnet als Quotienten der tatsächlich entdeckten zur Anzahl der existierenden Objekte sowie der Anzahl der falschen zur Anzahl der möglichen Objekte, die im Beobachtungsraum auftreten. Ein Beispiel dieser auf automatische Weise bestimmten Ergebnisse zeigt Bild 2.

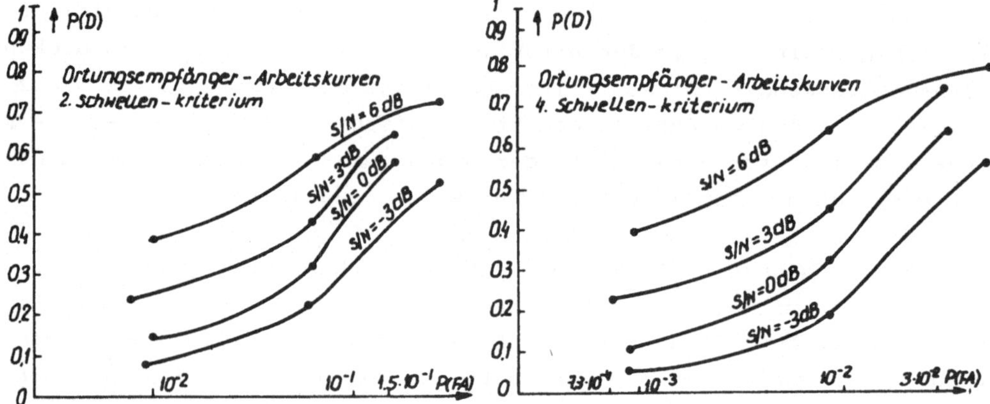

Bild 2: Empfängerbetriebskurven nach der 2.u.4. Entscheidungsschwelle

In der zweiten Etappe werden die Empfangswahrscheinlichkeit und die Falschalarmwahrscheinlichkeit aufgrund der Bilddeobachtung bestimmt. Auch in diesem Falle erzeugt der Rechner die Echosignale und ordnet sie zufällig verteilt im Beobachtungsraum an. Auf dem Bildschirm sind ebenfalls Störsignale sichtbar, die Scheinobjekte darstellen können und auch die Bilder vom Nutzobjekt deformieren können. Aufgabe des Beobachters ist es, die Anzahl und die Lage der Ziele zu bestimmen. In der Abschätzung des Beobachters treten Fehler auf, deren Anzahl mit der Verschlechterung des Signalstörverhältnisses wächst.

Das Berechnungsprinzip von P(D) und P(FA) ist das gleiche wie im Falle der automatischen Untersuchung der Empfangsschaltung. Die Simulationsschaltung ermöglicht die Ausnutzung weiterer Entscheidungsglieder und die Untersuchung ihrer Einflußnahme auf die Bildqualität und damit auf die Werte der Wahrscheinlichkeiten P(D) und P(FA). Die Wirkung einzelner Entscheidungsschaltungen auf die am Monitor zu analysierende Darstellung illustriert Bild 3. Diese Arbeiten stellen einen Teil der Vorbereitungen auf die Automatisierung der Fernerkennung in der Unterwasserortung dar.

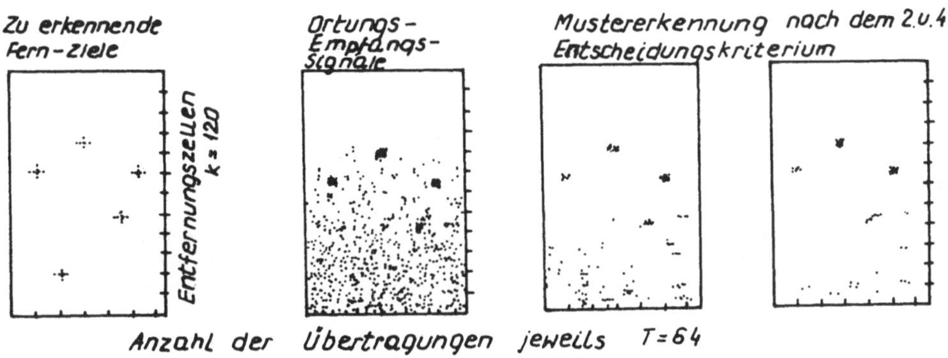

Bild 3: Tatsächliche Simulierungssituation und das Ergebnis der Fern-Erkennung durch den Beobachter am Ortungs-Monitor

Literatur:

[1] M.Heuser, C.E.Liedtke Ein hierarchisches Relaxationsverfahren zur Lageerkennung dreidimensionaler Objekte. Proceedings 9.DAGM-Symposium , S.140, 1987.

[2] M.Köhler, S.Kinzel Automatische Klassifikation multispektraler Bilddaten aus der Fernerkundung. Proceedings 10.DAGM-Symposium, S.172, 1988.

[3] R.Salamon, H.Lasota S.Kubica Überblick moderner Probleme der Hydroakustik (poln.). Telekommunikations-Konferenz Polens KST'88. Proceedings A, S.124, 1988.

[5] D.Ruser Bildqualitätsverbesserung in der Ultraschall-ortung. Proceedings.15.DAGA-Symposium 1989, Fortschritte der Akustik.

Skalenmessungen in der Eis-Fernerkundung

Markus Schmidt, Thomas Viehoff
Alfred-Wegener-Institut für Polar- und Meeresforschung
Am Handelshafen, D-2850 Bremerhaven

Zusammenfassung

Eine zentrale Größe des globalen Klima ist die Eisbedeckung der polaren Ozeane. Diese muß in unterschiedlichen Skalen beobachtet werden, um flächendeckende Aufnahmen sowie statistisch sichere Aussagen über kleinskalige Prozesse zu erhalten. Die quantitative Größe der Modifikation von Wärmeverlust und Verdunstung wird vor allem von der Eis-Konzentration, d.h. der prozentualen Eisbedeckung, sowie der morphologischen Beschaffenheit der Schollen bzw. der eisfreien Rinnen, bestimmt. Morphologische Größenmessungen liefern Aussagen über die vorkommenden Skalen. Sie können auf spezifische Bildregionen eingegrenzt werden und liefern für Eis- (hohe Albedo) und eisfreie Flächen (niedriger Albedo) getrennte Werte. Durch eine geeignete Wahl der Größenmasken sowie deren Zerlegung in eine Reihe von Zweipunktoperationen können die Größenverteilungen sehr effizient berechnet werden.

1.   Eis-Fernerkundung

Die genaue Kenntnis der Eisbedeckung bzw. des prozentualen Anteils der eisfreien Flächen sowie deren räumliche und zeitliche Variabilität ist für das Verständnis des gesamten Klimasystems unerläßlich. Bei durchschnittlichen Eiskonzentrationen von mehr als 80 Prozent im Inneren der Packeisgebiete ist gerade die großräumige exakte Bestimmung der verbleibenden eisfreien Flächen eine bislang noch nicht zufriedenstellend gelöste Aufgabe. Die Verwendung passiver Radiometer im Mikrowellenbereich ist aufgrund der Unabhängigkeit von atmosphärischen Störeinflüssen (Wolken, Streuung, etc.) sowie aufgrund der klaren Signaturunterschiede zwischen Wasser und Eis in diesem Spektralbereich ein praktikables Verfahren (Cavalieri et.al., 1984). Die geringe räumliche Auflösung dieser Verfahren (≥ 25 km) macht es jedoch notwendig, statistische Aussagen über die typischen Bedeckungssituationen im sub-pixel Bereich zu treffen. Dazu werden Fernerkundungsmethoden im sichtbaren und infraroten Spektralbereich verwendet, die eine wesentlich bessere geometrische Auflösung (0.1-1 km) besitzen. Mit ihrer Hilfe ist es möglich, die Größenverteilungen der Schollen und offenen Wasserflächen in Abhängigkeit von den lokalen und regionalen Gegebenheiten zu untersuchen.

2.   Radiometerdaten

Es werden die Daten des Advanced Very High Resolution Radiometer (AVHRR) verwendet. Instrumente dieser Baureihe werden auf den operationellen, polar umlaufenden Satelliten der TIROS-N/NOAA Serie eingesetzt [Schwalb 1978]. Für die Eisfernerkundung werden die beiden Kanäle verwendet, die im sichtbaren Spektralbereich bzw. an der Grenze zum Infrarot bei 0.58-0.68 µm und 0.72-1.1 µm liegen und im wesentlichen die zurückgestreute solare Strahlung messen. Das Rauschniveau der Kanäle ist geringer als die digitale Auflösung der Daten, so daß die radiometrische Auflösung nur durch die 10-Bit Struktur der Daten begrenzt wird. Daraus resultiert eine dem Reflexionsverhältnis von 0.1% entsprechende Auflösung. Die Ergebnisse werden i.a. in Prozent Albedo angegeben, wobei dies jedoch lediglich als relative Einheit zu verstehen ist [Rao 1988]. Das Radiometer tastet die Erdoberfläche mit Hilfe eines rotierenden Spiegels ab. Insgesamt werden pro Scanlinie 2048 Pixel registriert, wobei die Auflösung ca 1.1×1.1 km beträgt.

Ein Problem bei der Analyse sind die geringen Unterschiede der spektralen Charakteristiken von Eis und Wolken. Eine zufriedenstellende Klassifikation von Wolken und Eis ist nur unter Verwendung auch der Infrarot-Spektralkanäle möglich. Bis auf wenige Ausnahmen (dünne Cirruswolken) sind Wolken im sichtbaren wie im infraroten Spektralbereich nicht transparent. Eisbeobachtungen können demnach nur nach einer umfassenden Wolkenerkennung durchgeführt werden. In einem zweiten Schritt werden dann auch diejenigen Ozeangebiete mitberücksichtigt, die unter einer dünnen, semitransparenten Wolkendecke liegen. Die Trennung zwischen eisbedeckten und eisfreien Flächen wird durch die großen Unterschiede der Albedowerte von typischerweise 6% für Wasser und >25% für Eis möglich. Kleinräumige Strukturen im Subpixelbereich (<1.1 km) führen jedoch zu einem fließendem Übergang zwischen beiden Oberflächentypen in den Häufigkeitsverteilungen.

3. Regionenorientierte morphologische Skalenmessung

Zur Bestimmung der Größenskalen von Eisschollen und eisfreien Meeresflächen sind nach dem oben gesagten die folgenden drei Bedingungen zu erfüllen:

a) Die Größenmessung darf nur auf den von Wolken freien Gebieten erfolgen, d.h. sie ist regional einzuschränken, muß mithin im Ortsraum vorgenommen werden.

b) Wegen der sanften Übergange in den Oberflächenmerkmalen durch die beschränkte Auflösung sollen die Skalen ohne vorherige Eis/Wasser-Klassifizierung bestimmt werden.

c) Die Größenmessung soll getrennt für Eis- und Wasserflächen möglich sein.

Alle drei Bedingungen werden durch morphologische Größenmessung realisiert. Sie basiert auf den Operationen Erosion und Dilation d.i. Minimums- und Maximumsfaltung, aus deren Zusammensetzung sich das Opening-Filter ergibt [Serra 1982]. Wenn I die Grauwertbildfunktion und B eine Binärmaske darstellt, sind sie durch die folgenden Definitionen gegeben:

$$(B \text{ ero } I)(x) := \min\{I(x-u)| \, u \in B\} \qquad \text{und} \qquad (B \text{ dila } I)(x) := \max\{I(x-u)| \, u \in B\} \qquad (1)$$

$$B \text{ opening } I := B \text{ dila } (B^\wedge \text{ ero } I) \qquad \text{mit der Spiegelung am Ursprung} \quad B^\wedge := \{-u| \, u \in B\} \qquad (2)$$

Das Opening hat die Eigenschaften:

$$(B \text{ opening } I) \leq I \qquad \text{und} \qquad I \leq I' => (B \text{ opening } I) \leq (B \text{ opening } I') \qquad (3)$$

Es ist die Einhüllende aller unter dem Grauwertgebirge I liegenden Verschiebungen von B. Wenn B ein Kreis mit Radius r ist (K_r), dann werden alle hellen Strukturen mit kleinerem Radius durch das Opening beseitigt. Die duale Operation zum Opening, das Closing {vertausche "ero" und "dila"} in (2)) ist analog für dunkle Strukturen spezifisch. Die Differenz des Openings mit K_{r-1} zu dem mit K_r liefern gerade die Strukturen mit größerem Radius als r-1 und kleinerem als Radius r. Mit

$$D_r := (K_{r-1} \text{ opening } I) - (K_r \text{ opening } I) \qquad D_{-r} := (K_r \text{ closing } I) - (K_{r-1} \text{ closing } I) \qquad (4)$$

können Eisschollen- (helle Strukturen) bzw. Wasserflächengrößen (dunkle Strukturen) bestimmt werden. (Es bezeichnet K_0 opening $I = I = K_0$ closing I.) Wesentlich hierbei ist die sogenannte Sieb-Eigenschaft

$$K_r \text{ opening } (K_s \text{ opening } I) = K_{\max(r,s)} \text{ opening } I \qquad (5)$$

Sie garantiert, daß $(K_r$ opening I) monoton mit r fällt oder D_r nichtnegativ ist.
Opening und Closing sind rauschempfindliche Filteroperationen insbesondere bei Punkt-Rauschen. Hier können sie wegen des minimalen Rauschens ohne rauschmindernde Vorverarbeitung auf die empfangenen Daten angewandt werden, allerdings müssen fehlerhafte Daten, wie Scanlinienausfälle vorher korrigiert werden.

Mit (4) ist an jedem Pixel die Größe durch die Veränderung gemessen, die der Grauwert im Verlauf der anwachsenden Filtergröße durchläuft. Damit kann aber keine komplette Verteilung geschätzt werden. Die einfachste Möglichkeit, eine verläßliche Größenaussage über die Strukturen im Bild zu gewinnen, ist das globale Aufsummieren der Differenzen D_r. Das Interesse konzentriert sich hier jedoch auf spezifische Regionen im Bild, nämlich solche, die wolkenfrei oder nur mit dünner Wolkenschicht bedeckt sind. Deshalb wird hier die regionenoriertierte Größenverteilung d_r^L als Summe über die Region L definiert:

$$d_r^L := \text{Sum} \{ D_r(i,j)| \, i,j \in L\} \qquad (6)$$

Durch Normierung über die Größen r ergeben sich die regionenorientierten Größenverteilungen
$p^+(R)^L = (p_1^L, \ldots p_R^L)$ und $p^-(R)^L = (p_{-1}^L, \ldots p_{-R}^L)$ mit

$$p_r^L := d_r^L / \text{Sum} \{ d_s^L \, | \, s = 1 \text{ bis } R \text{ bzw. } s = -1 \text{ bis } -R \text{ falls } r<0\} \qquad (7)$$

Von p^+ und p^- können auf die übliche Weise Erwartungswert und Varianz der (positiven oder negativen) Größe angegeben werden.

4. Maskenzerlegung

Um Eigenschaft (5) auf dem digitalen Gitter zu erfüllen, werden keine rein kreisförmigen Masken wie in Dengler et al [1988] verwendet. Serra gebraucht für das hexagonale Gitter Sechsecke H_r [Serra 1982]. Hier werden Achtecke A_r verwendet, die gegenüber Quadraten eine höhere Isotropie aufweisen.

Die Masken A_r werden durch die folgende Vorschrift definiert:

$$A_{2i} = A_1 \text{ dila } A_{2i-1} \quad \text{und} \quad A_{2i+1} = A_1^t \text{ dila } A_{2i} \quad \text{mit } A_1 = \begin{smallmatrix} & 1 & 1 & \\ 1 & 1 & 1 \\ & 1 & 1 & \end{smallmatrix} \quad A_1^t = \begin{smallmatrix} 1 & 1 & \\ 1 & 1 & 1 \\ 1 & 1 & \end{smallmatrix} \qquad (8)$$

Definition (8) hat eine wichtige praktische Bedeutung. Aufgrund ihrer Assoziativität kann die Dilation (resp. Erosion) mit der Maske A_r durch eine Serie von Dilationen abwechselnd mit A_1 bzw. A_1^t ausgeführt werden [Schmidt 1988]. Die mit der Maskenfläche quadratisch in r anwachsende Anzahl von Minimums- und Maximumsoperationen zur Berechnung eines Opening mit A_r kann so auf eine in r lineare Anzahl reduziert werden. Wie man sofort sieht werden A_1 und A_1^t ihrerseits durch 3 Zweipunktmasken erzeugt, so daß für eine so zerlegte Faltung mit A_n gerade n1=3n Operationen benötigt werden. Tatsächlich ist sogar noch eine wesentlich effizientere Zerlegung möglich, wie das folgende Beispiel der Zerlegung von A_6 in 10 Zweipunktmasken (anstatt 18) zeigt. Dabei sind jeweils die Gruppen von Zweipunktmasken einer Richtung in Serien von Zweipunktmasken mit abfallendem Durchmesser zusammengefasst.

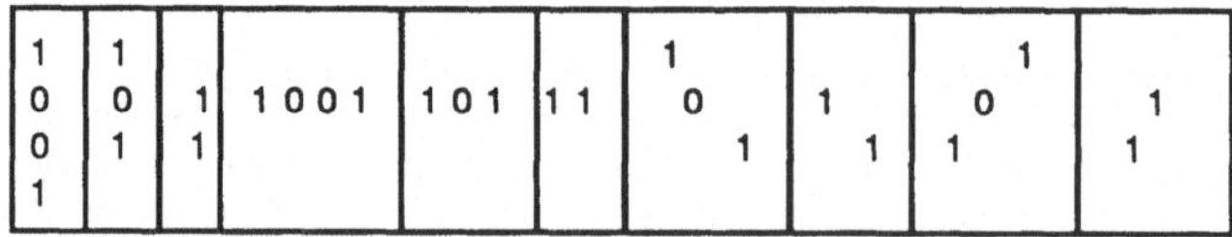
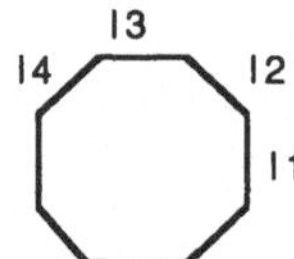

Die Berechnung dieser Zerlegungen ist einfach: Die Seitenlängen l_i (siehe oben) des Achtecks A_{r-1} ($l_1=l_3=r$, $l_2=$ Int((r+1)/2), $l_4=$ Int(r/2)) liefern die Zerlegung in vier Linienelemente, die aus je l_i benachbarten Einsen in die jeweilige Richtung bestehen. Die Zerlegung eines Linienelements der Länge 1 besteht aus einer Zweipunktmaske der Länge Int((l+1)/2) und den Zerlegungen des Linienelements der Länge m=Int (l/2). Int(x) ist hier die kleinste ganze Zahl größergleich x. Diese einfache rekursive Vorschrift erspart die aufwendige Untersuchung aller möglichen Zerlegungen der Maske [Zhuang et al 1986].

In der nachstehenden Tabelle sind die Größen der Masken, die Komplexitäten der einfachen Zerlegung n1 und der effizienteren Zerlegung n2 angegeben, Bemerkenswert ist, daß n2(r) nur logarithmisch mit r wächst. Mit G(r) schließlich ist angegeben, wieviele Operationen auszuführen sind, um sämtliche Openings (oder Closings) bis zur Größe r, also eine skalenabdeckende Größenverteilung zu berechnen.

<u>Tabelle zu Maskengrößen und Zerlegungsaufwand</u>

r	1	2	3	4	5	6	8	10	12	14	16	18	20	25	30	35	40	50
F	7	21	41	69	103	145	249	381	541	729	945	1189	1461	2263	3241	4393	5721	8901
n1	3	6	9	12	15	18	24	30	36	42	48	54	60	75	90	105	120	150
n2	3	6	7	10	10	10	14	14	14	14	18	18	18	18	18	22	22	22
G	6	15	25	38	51	64	95	129	163	197	236	278	320	425	530	652	777	1027

r: Radius der Achteck-Maske / F: Flaeche (#Pixel)
n1/n2: Anzahl der Zweipunktmasken bei erzeugender/effizientester Zerlegung
G: Anzahl der Operationen bei Berechnung einer Größenverteilung bis Radius r

Es gelten die einfach nachzurechnenden Beziehungen (lb: Logarithmus zur Basis 2)

$$F(r)= \text{Int}(0.5+ 3r + 3.5r{*}2) \qquad\qquad n1(r)=3r \qquad\qquad (9)$$
$$G(r)= n1(r) + n2(1)+n2(2)+ \dots +n2(r) \qquad n2(r) = 2\,\text{Int}(\,lb(1+r)) +\text{Int}(lb(1+r/2)) +\text{Int}(\,lb(0.5 + r/2))$$

Größenverteilung können auf diese Weise außerordentlich effizient berechnet werden. Die Wahl der Skalenbereiche, also von R, ist natürlich problemspezifisch. Wenn a priori sicher ist, daß kleine Skalenbereiche für die Größenmessung nicht relevant sind, kann durch Reduktion der Auflösung zusätzlich beträchtliche Arbeit gespart werden.

5. Ergebnisse

Die Anwendbarkeit der Verfahren auf die Interpretation von Meereis-Verteilungen wurde anhand einer NOAA-7 AVHRR Szene (Kanal 1) vom 29.6.1984 überprüft (Abb. 2). Die Szene zeigt einen Ausschnitt des Ostgrönlandstromes zwischen 78°N und 82°N sowie 5°E und 15°W. Die Szene kann in 4 große Regionen unterteilt werden: dichte Wolkenbedeckung über eisfreiem Ozean (1), Eiskante und wolkenfreie Eisbedeckung im Ostgrönlandstrom (2), dichteres Packeisfeld unter semitransparenten Wolken (3), und ein Mischgebiet aus Wolken, einzelnen Eis- und Wasserflächen sowie der grönländischen Küste (4). Die Verteilungen zeigen deutliche Unterschiede vor allem zwischen den Regionen 2 und 3 einerseits und 1 und 4 andererseits (Abb. 1).   Wolken- und Mischgebiete zeigen eine sehr flache Verteilung und keine Präferenzen für bestimmte Größen was im wesentlichen durch die grossen typischen Skalen von Wolken und Festeisflächen bedingt ist. Die Meereisflächen zeigen hingegen scharfe Maxima in den Verteilungen bei Radien von +2 und -2 Pixel, entsprechend ca. 4-5 km Durchmesser (sehr kleine Eisschollen und noch mehr kleine Zwischenräume). Im positiven Bereich ist der Knick bei +10 bemerkenswert. Er ist auf die drei- bis vier großen Eisschollen in der Region 2 zurückzuführen und markiert somit die obere Grenze der einzelnen Schollengrößen.   Die Region 3 mit der dünnen Bewölkung läuft im wesentlichen parallel zur Meereiskante und zum Haupteisstrom   und bildet offenbar diese ab, wenn auch in abgeschwächtem Kontrast. Die frappierende Übereinstimmung der Verteilungen von Region 2 und 3 im negativen Bereich läßt sich so interpretieren, daß Wolken vorallem die in der Helligkeit ähnlichen Eisstrukturen maskieren, während die dunklen, eisfreien Flächen auch durch die Wolkenschicht erkennbar bleiben. Dies wird auch durch die Erwartungswerte dokumentiert, die für die positiven Verteilungen bei 6.69 (Meereis) und 8.19 (dünne Bewölkung) gegen 8.82 (Wolken) liegen, während sie sich für die negative Verteilungen mit 5.59 (Meereis) und 5.37 (dünne Bewölkung) deutlich von den Wolken (8.62) unterscheiden.

Die Verteilung zeigt deutlich: Die eisfreien Rinnen sind deutlich kleinskaliger als die Eisschollen. Dies ist auch in Abbildung 3 zu sehen. Sie zeigt einen Ausschnitt des Meereises aus der linken Bildmitte in verschiedenen Verarbeitungsstufen. Von links oben nach rechts unten zeilenweise Closing mit A_{16}, A_{12}, A_8, und A_4, Original und Opening mit A_4, A_8, A_{12}, und A_{16}. Das Closing mit A_4, das die dunklen Rinnen beseitigt, ist schon fast so hell wie bei den größeren Masken, beim Opening, das die Eisschollen wegnimmt, zeigen sich auch noch in den Schritten von A_4 nach A_8 und A_8 nach A_{12} deutliche Unterschiede.

Diese erste Anwendung von morphologischer Größenmessung auf Meereisverteilungen zeigt die gute Interpretierbarkeit dieser Methode. Für die praktische Auswertung ist vorallem die Vermessung der eisfreien Rinnen auch unter semitransparenten Wolken von großer Bedeutung.

6. Abbildungen

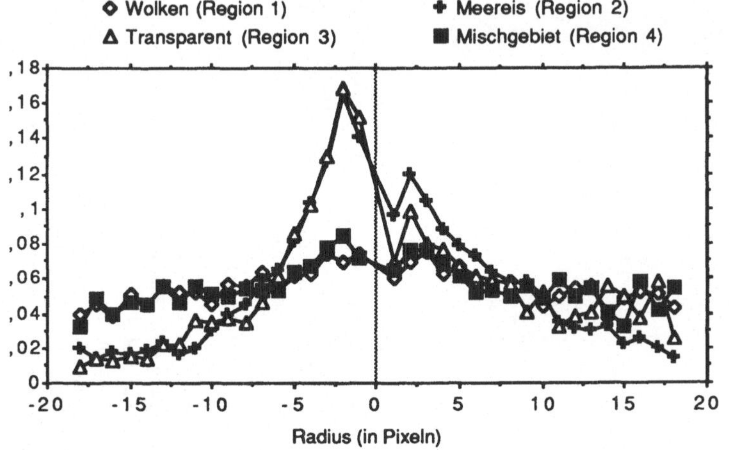

Abb. 1: Größenverteilung in den vier Regionen aus Abbildung 2: Häufigkeiten für die Größen, links die Verteilung $p^-(R=18)$, rechts $p^+(R=18)$

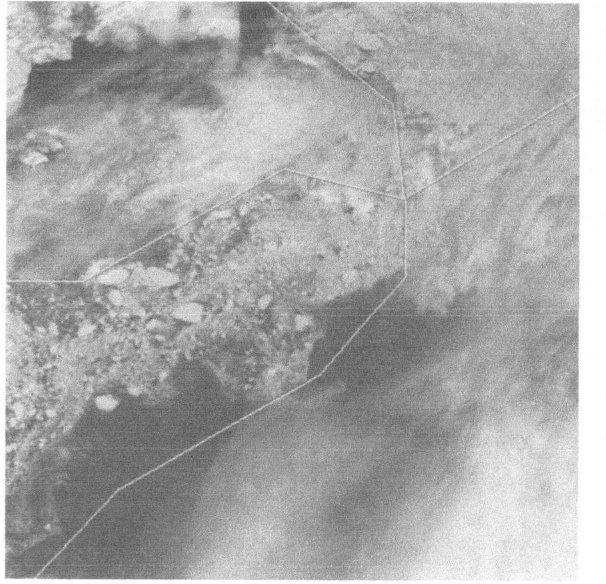 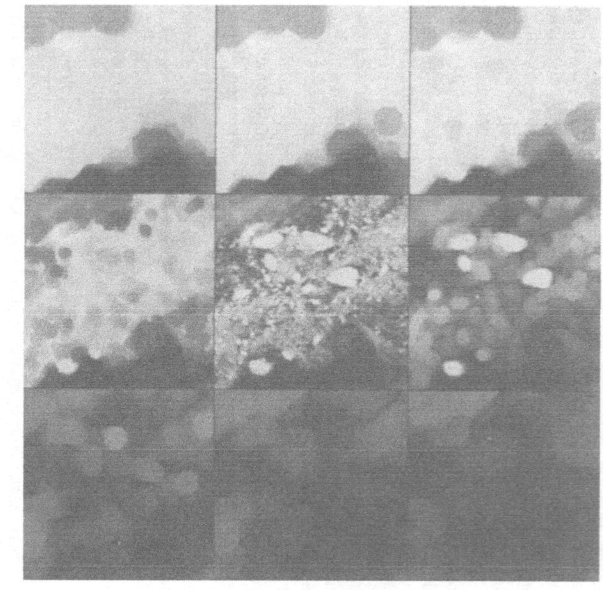

Abb. 2: Eis-, Wasser- und Wolkenverteilung im nördlichen Ostgrönlandstrom aus den Kanal 1-Daten (Albedowerte) des NOAA-7 AVHRR. Szene vom 29.06.1984 und Markierung der vier Regionen, für die eine Größenverteilung bestimmt wurde.

Abb. 3: Closing mit A16, A12, A8, und A4, Original und Opening mit A4, A8, A12, und A16 für einen Bildausschnitt (Mitte links von Abb. 2) des Meereisgebietes.

7. Literatur

Cavalieri D.J., Gloersen P.,Campbell W.J. (1984): Determination of sea-ice parameters with NIMBUS-7. J. Geophys. Res., Vol 89, 5355-5369
Dengler J., Bertsch H., Desaga J., Schmidt M. (1988). New Trends of Image Analysis. Methods of Inf. in Medic., Vol 27, 53-57
Rao C.R.N. (1988): Pre-Launch calibration of channels 1 and 2 of the AVHRR. NOAA Tech. Rep. NESS 36 US Dep of Commerce, Washington D.C
Schmidt M. (1988): Grundlegende Operationen der Bildverarbeitung. APL-Journal, Jg. 7, Nr. 1, 14-31
Schwalb A. (1978): The TIROS-N/NOAA A-G satellite series. NOAA Technical Memorandum NESS 95. U.S. Dep. of Commerce, Washington D.C
Serra J. (1982): Image Analysis and Mathematical Morphology. London, Academic Press
Zhuang X., Haralick R.M. (1986): Morphological Structuring Element Decomposition.   Comp. Vision Graph. Imag. Process. 35, 370-382

Ein lernendes System zur Zellbildanalyse

R. Dörrer, J. Fischer, W. Greiner, W. Schlipf, P. Schwarzmann

Institut für Physikalische Elektronik,Universität Stuttgart
Pfaffenwaldring 47, D–7000 Stuttgart 80

Zielsetzung

Im Rahmen von Maßnahmen zur Krebsfrüherkennung fallen große Mengen von Zellausstrichpräparaten zur mikroskopischen Begutachtung an. In der ersten Stufe der Auswertung werden aus der Gesamtmenge der Präparate, in der die negativen, das heißt *gesunden*, bei weitem überwiegen, die eindeutig negativen Fälle ausgesondert. Der Rest, bestehend aus positiven und möglicherweise positiven Präparaten wird einer genaueren Inspektion zugeführt. Dieses visuelle Vorsortieren (*Prescreening*) ist sehr ermüdend und mit der Gefahr von fehlerhaften Entscheidungen behaftet, da positive Präparate häufig nur sehr wenige eindeutig positive Zellen enthalten. Bild 1 zeigt ein typisches Mikroskopgesichtsfeld eines Abstrichpräparates der Zervix (Gebärmutterhals).

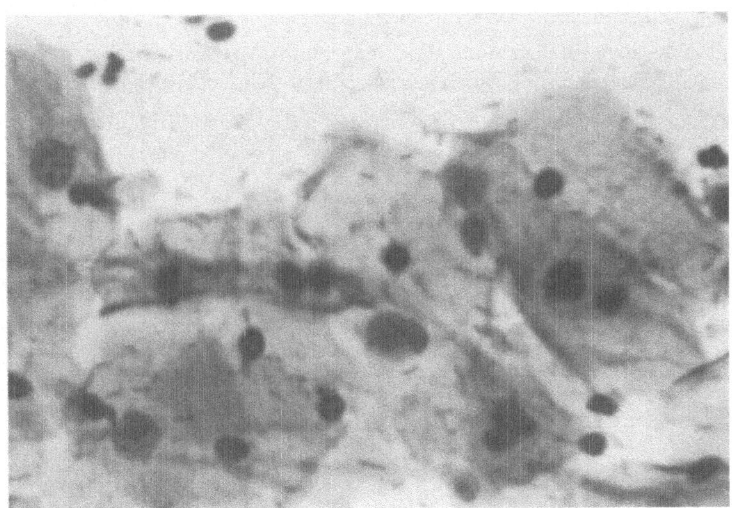

Bild 1: Mikroskopisches Bild eines Zervix-Abstrichpräparates

Der vorliegende Beitrag beschreibt ein System, das dieses *Prescreening* automatisch leistet. Bedingt durch die Aufgabenstellung — großer Präparatedurchsatz — ist für die Realisierung ein Vollautomat mit Präparatewechsler, Autofokus am Mikroskop und Überwachung aller Baugruppen und Funktionsparameter aufgebaut worden. Die vom Automaten als verdächtig eingestuften Präparate müssen vom Zytologen visuell begutachtet werden. Das bedeutet, daß das Zellbild dem vertrauten Zellbild in der konventionellen Zytologie ähnlich sein muß. Dadurch scheiden fast alle maschinengerechten, quantitativen Zellfärbemethoden aus und es wird der herkömmlichen Färbung nach PAPANICOLAOU [SCHW83] der Vorzug gegeben.

Auswertestrategie

Der Versuch, die Vorgehensweisen visueller Auswertungen auf maschinellen Systemen nachzubilden stößt sehr schnell auf Schwierigkeiten. Die Gründe hierfür sind insbesondere:

- viele menschliche Auswerteschritte laufen unbewußt ab und sind daher nicht formulierbar.
- selbst präzise Begriffe des menschlichen Erkennens, wie zum Beispiel *Granularität des Zellkerns* sind sehr schwer in Algorithmen zu fassen
- die Vielfalt der Erscheinungsformen von Objekten in einem Zellabstrich ist sehr groß. Ihre formale Beschreibung, selbst wenn sie eindeutig möglich wäre ist daher sehr umfangreich.
- der menschliche Betrachter schließt sehr viel semantisches Vorwissen ein, das der Maschine verschlossen bleibt.

Da die Modellierung des menschlichen Erkennens nahezu unmöglich ist, das umfangreiche Problemwissen und die Erfahrung des Zytologen jedoch unbedingt Eingang in die maschinelle Auswertung finden müssen, bietet sich hier die bewährte Methode der *statistischen Klassifikation* an. Sie erlaubt es, große Mengen von Meßgrößen, gewonnen an sehr vielen Bildern, mit dem Expertenwissen des Menschen über diese Bilder zu verknüpfen. Weil die Vielfalt der Erscheinungsformen in allen Ebenen der Auswertung sehr groß ist, werden deshalb in vielen Verarbeitungsstufen Klassifikationsmethoden eingesetzt. Dadurch bleibt das System flexibel und kann durch andere Stichproben an die verschiedensten zytologischen Aufgaben angepaßt werden

Der Vorgang der Präparatauswertung wird in vier aufeinanderfolgende Schritte aufgeteilt:

- Bildaufnahme mit Präparatewechsler, Etikettenleser und Autofokuseinrichtung
- Objektsuche in der digitalisierten Szene und Vorsortierung der Objekte (Segmentation).
- Auswertung von Zellbildern
- Bewertung der Zellpopulation des Präparats

Abbildung 2 zeigt diese Prozeßschritte als Blockdiagramm. Mit Ausnahme der ersten Stufe (Bildaufnahme) werden zur Entscheidung durchweg adaptierte Klassifikatoren eingesetzt.

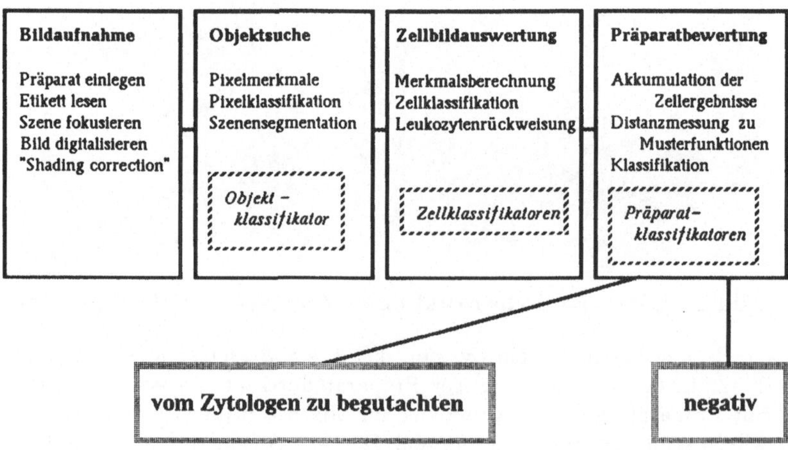

Bild 2: Ablauf der Präparatauswertung

Betrachtet man den Auswerteprozeß unter den Gesichtspunkten *ikonischer* und *symbolischer* Verarbeitung, so muß man die erste Stufe der rein ikonischen Verarbeitung zurechnen. In der zweiten Stufe und mehr noch in der dritten findet der Übergang zur symbolischen Auswertung statt: den ikonischen Bildern entnommenen Merkmalen werden Symbole (Zellklassen) zugeordnet und diese werden mit Attributen versehen (Grad der Klassenzugehörigkeit). In der vierten Stufe, der Präparatklassifikation) findet eine ausschließlich symbolische Auswertung statt, die nur auf den symbolischen Ergebnissen der vorhergenden Stufe aufbaut.

Der Auswerteprozeß (Arbeitsphase) läßt sich als *Top-down* Strategie beschreiben: dem ikonischen Bild entnommene Maße werden durch Klassifikatoren, die hier als Repräsentation eines Modells dienen, abgefragt. Dem Lernprozeß (Trainingsphase), das heißt, der Adaption der Klassifikatoren liegt eine *Bottom-up* Strategie zu Grunde: aus ikonischen Daten wird über einen Satz von Beispielen ein Modell, die Klassifikatorkoeffizienten, bestimmt, das anhand von Fehlerraten auf seine Akzeptanz geprüft wird. In den folgenden Abschnitten werden die Zell- und die Präparatklassifikation detaillierter dargestellt. Die Objektsuche wurde bereits auf der DAGM 1987 vorgestellt [DAGM87]

Zellauswertung

Jedes einzelne, vom Automaten aufgefundene Objekt wird verschiedenen Verfahrensschritten unterworfen, die im folgenden erläutert werden.

Aus zytologischer Sicht konzentriert sich die Information über den Malignitätsgrad einer Zelle nicht allein im Zellkern. Die Frage ob hauptsächlich densitometrische oder morphologische Eigenschaften des Materials die Entscheidung herbeiführen ist ebenfalls nicht ad hoc beantwortbar. Daher wurde zunächst ein großer *Merkmalspool* erstellt. Durch verschiedene Vorverarbeitungsschritte erhöht sich die Zahl der Merkmale und Verfahrensparameter noch weiter. Danach wurden schrittweise die signifikantesten Merkmale ermittelt, und die Einflußgrößen optimiert. Folgende Merkmalsgruppen und Parameter wurden untersucht:

- Zellbild mit automatisch segmentiertem Zellkern
    - o MINKOWSKI-Maße im absoluten und normierten Grauwertbereich
    - o TEXTUR-Merkmale nach Haralick [HARA73] im reduzierten, normierten Grauwertbereich
- nicht segmentiertes, grauwertnormiertes Zellbild
    - o Einfluß der Bildgröße
    - o Einfluß der Graustufenzahl und der Normierungsparameter
    - o MINKOWSKI-Maße (im Folgenden: MM)
    - o Histogramme und HISTOGRAMM-Merkmale (HM genannt)[PRAT78]
    - o TEXTUR-Merkmale (TM) [HARA73]

Wegen der gegen Schwankungen im Grauwertbereich sehr empfindlichen Zellkernsegmentation werden mit den so bestimmten Merkmalen keine brauchbaren Ergebnisse erzielt. Im weiteren werden nur noch die Merkmale am nicht segmentierten, grauwertnormierten Zellbild untersucht. An einer Testdatenbank aus positiven (Q) und negativen (D) Zellen wurde der Einfluß von Bildgröße und Graustufenzahl auf das Klassifikationsergebnis, getrennt für MM, HM und TM ermittelt. Sowohl für MM als auch für HM erweist sich eine Graustufenzahl von 64 als optimal. TM liefert bei 16 Graustufen die besten Resultate. Eine Bildgröße von 48 × 48 Pixeln (Pixelabstand: $0.5\mu m$) ergibt die niedrigsten Fehlerraten. Der Rechenaufwand ist dabei noch vertretbar.

Nach Festlegung der Parameter *Bildgröße* und *Graustufen* wurden mehrere Zellklassifikatoren adaptiert. Die Merkmalsselektion wurde dabei systematisch vorgenommen. Durch sukzessives Einbeziehen von Merkmalen ließen sich nahezu optimale Merkmalssätze für alle Klassifikatoren finden, die bereits bei geringer Merkmalszahl geringe Fehlerraten liefern (Bild 3). Zur Absicherung der Ergebnisse wurde dem 3600 D/Q-Zellen umfassenden Kollektiv jeweils zufällig 1%, 2%, ... zur Klassifikatoradaption entnommen und mit dem verbliebenen Rest getestet. Bild 4 zeigt die gute Übereinstimmung von Lern- und Testsatz bereits bei kleinen Lernstichproben.

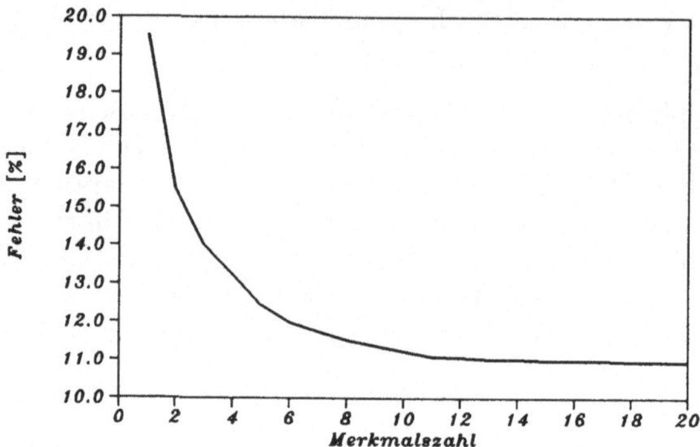

Bild 3: Klassifikationsfehler als Funktion der Merkmalszahl bei sukzessiver Einbeziehung neuer Merkmale (D/Q-Klassifikator)

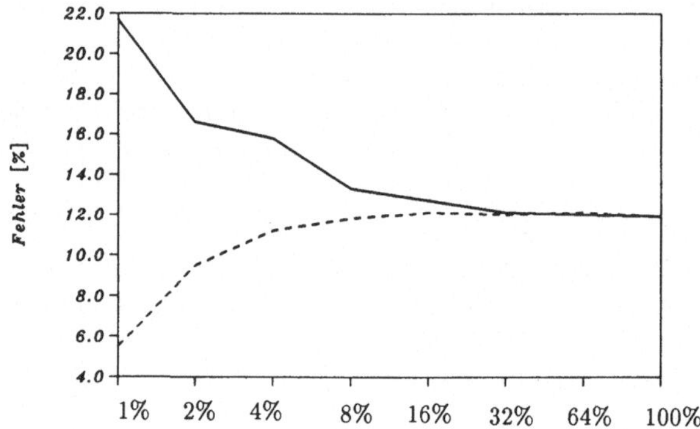

Bild 4: Fehlerrate in Lern (---) und Testdatenbank (—)

Die folgenden Verfahrensschritte gelangen als Kompromiss aus Leistung und Aufwand im realisierten System zur Anwendung.

- das eingelesene 64 × 64 große, 256 Graustufen umfassende Bild wird in ein 48 × 48 großes, um den Zellkern zentriertes, auf 64 Graustufen mittels Histogramm normiertes Bild reduziert.
- obere und untere Normierungsschwelle werden als einzige absolute Grauwertmaße in den Merkmalsatz übernommen (2 Merkmale)
- in diesem normierten Bild werden Histogramme aufgenommen und daraus Histogrammerkmale nach [PRAT78] berechnet (100 Merkmale).
- ebenfalls im normierten Bild werden die MINKOWSKI-Maße *Fläche, Umfang* und daraus *Umfang²/Fläche* bei mehreren Grauwertschwellen gemessen. Dies wird wiederholt nachdem das Bild lokalen Minimum- und Maximumoperationen (Erosion/Dilatation) unterworfen wurde (72 Merkmale)

196

• aus dem 174 Merkmale umfassenden Zellmerkmalssatz werden jeweils 7 bis 20 Merkmale zur Klassifikation in verschiedenen 2-Klassenfällen ausgewählt und deren erste Entscheidungsvektorkomponenten zur Präparatanalyse gesammelt.

Präparatanalyse

In jedem Präparat werden etwa 1000 Zellen ausgewertet und wie beschrieben klassifiziert. Leukozyten werden dabei nicht mitberücksichtigt. Die Einzelentscheidungen der Zellklassifikatoren führen zu den Präparatmerkmalen. Der Präparatmerkmalssatz setzt sich zusammen, aus den Dichtefunktionen der Zellklassifikatorentscheidungen und aus funktionalen Abstandsmaßen zu den *typischen Klassendichtefunktionen*, die als Mittelwertfunktionen aus den Präparatgruppen N (= PAP I + PAP II), D (= PAP IIId), A (= PAP IVa) und P (= PAP IVb + PAP V) gebildet werden. Bild 5 zeigt diese Funktionen gemessen an einer 1500 Präparate beinhaltenden Datenbank.

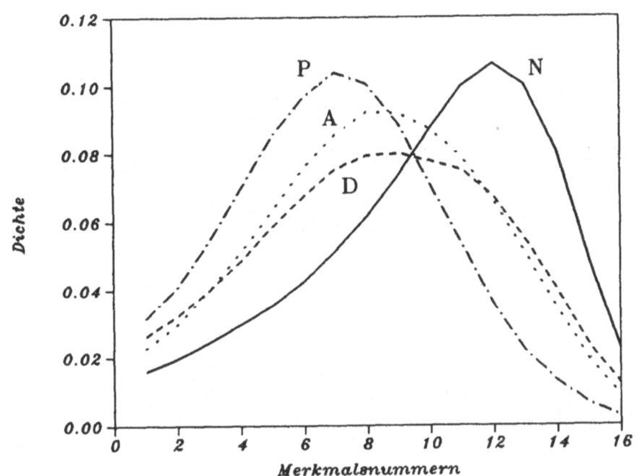

Bild 5: typische Klassendichtefunktionen erzeugt aus den Entscheidungen des D/Q-Zellklassifikators

Zur Präparatklassifikation wird ein mehrstufiges Entscheidungsschema eingesetzt. In der ersten Stufe werden die Präparate der Gruppe P von allen anderen abgetrennt. Vom verbliebenen Rest werden die der Gruppe A abgesondert und der letzte Schritt trennt dann die Gruppen D und N.

Realisierung in einem Gerätekonzept

Forderungen an ein Prototypsystem, das die beschriebenen Ziele erreichen soll sind neben der Einhaltung der aus medizinischer Sicht geforderten Fehlerraten:

• ausreichender Präparatedurchsatz (ca. 10 Minuten/Präparat)
• kurze Entwicklungszeit für den Prototyp
• flexible und übersichtliche Programmierung der Systemkomonenten
• Adaptierbarkeit an andere zytologische Aufgabestellungen

Daraus resultiert ein Systemkonzept, das weitgehend auf marktgängige Komponenten zurückgreift. Außerdem laufen möglichst viele Aufgaben auf Universalrechnern (68020 Prozessoren unter Unix) ab, die in einer leicht portierbaren Hochsprache (C) programmiert werden. Abbildung 6 zeigt die Gerätekonzeption. Den einzelnen Einheiten werden folgende Aufgaben zugewiesen:

- Ferngesteuertes Mikroskop mit Bildaufnahme:
  - o Autofokuseinstellung
  - o Leerfelder und Felder mit Zellhaufen erkennen und übergehen
  - o Präparatwechsler
  - o Beleuchtungskontrolle
  - o Filterwechsler
  - o Präparat positionieren und verschieben
  - o Präparatetikett lesen
- Schneller Bildverarbeitungsprozessor mit Bildspeicher
  - o Digitalisierung des TV-Signals in 512×512 Bildpunkte mit 8 bit Auflösung
  - o "Online-Shadingcorrection" zum Ausgleich von Ungleichmäßigkeiten der Gesichtsfeldausleuchtung des Mikroskops
  - o Schnelle Segmentierung des Gesichtsfeldes in Gebiete des Hintergrundes und unerwünschter Bildbestandteile (z.B. Zellen ohne diagnostischen Wert, Schmutz etc.) und Gebiete, die ein Zellbild enthalten
  - o Entnahme von Unterbildern (64×64 Bildpunkte), die diagnostisch relevante Zellen enthalten und ihre Zwischenspeicherung
- "Slave"-Rechner (gegenwärtig sind maximal 4 zuschaltbar)
  - o Berechnung der Zellmerkmale
  - o Klassifikation der Zellen
  - o Rückweisung unerwünschter Objekte
- Zentraler Steuerrechner
  - o Steuerung und Überwachung aller Mikroskopfunktionen
  - o Steuerung des Bildverarbeitungsprozessors
  - o Verteilung der Zellbilder an die "Slave"-Prozessoren und Akkumulation der Zellklassifikationsergebnisse
  - o Präparatklassifikation
  - o Speicherung der Daten auf Magnetband oder Festplatte
  - o Dokumentation der Präparatbewertung

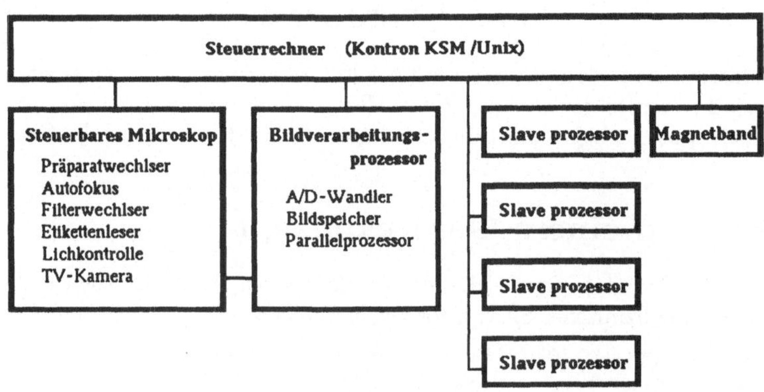

Bild 6: Gerätekonzeption

Die Wechselkassette des Geräts faßt 100 Präparate, die bei einer mittleren Bearbeitungszeit von 10 Minuten/Präparat in weniger als 17 Stunden abgearbeitet sind. Damit sind bei entsprechender Beschickung mehr als 700 Präparate pro Arbeitswoche verarbeitbar.

Experimentelle Ergebnisse

Die Experimente zur Zellklassifikation wurden an einer vom zytologischen Experten gekennzeichneten Stichprobe von ungefähr 38000 Objekten durchgeführt.* Die folgende Tabelle gibt einen Überblick über die Zusammensetzung dieser Datenbank.

Zellgruppe	Symbol	Anzahl
negative Zellen	D	15978
D-Zellen aus negativen Präparaten	N	6234
dysplastische Zellen	I	833
positive Zellen	Q	1805
Leukozyten	L	3536
Artefakte	Z	14594

Mit den oben genannten Objektklassen (D/N/I/Q/Z/L) wurden lineare Polynomklassifikatoren adaptiert. Die notwendige Größe der Lernstichproben und die benötigte Merkmalszahl wurden jeweils ermittelt (Bild 3 und 4 zeigt dies für den D/Q-Klassifikator). Adaptiert wurden die Klassifikatoren dann an gleichverteilten Stichproben. Die erzielten Resultate zeigt die nachstehende Tabelle:

Klassen	Lernsatzgröße	Merkmale	Fehlerrate [%]
D/Q	3600	11	12,7
N/Q	3600	8	10,2
I/Q	1700	11	19,4
D/Z	6000	20	15,9
Q/L	1700	9	11,4

Der Q/L-Klassifikator dient zur Rückweisung von Leukozyten, deren Anteil dadurch von 30% – 50% auf 5% – 10% reduziert wird. Die übrigen Klassifikatoren dienen der Erzeugung von Präparatmerkmalen. Die folgende Tabelle zeigt die in einem Vorexperiment an 600 Präparaten gewonnenen Erkennungsleistungen:

Präparatklassen	Symbol	Erkennungsleistung
negativ (PAP I + PAP II)	N	62%
dysplastisch (PAP IIId)	D	85%
Carcinoma in situ (PAP IVa)	A	96%
invasives Carcinom (PAP IVb + PAP V)	P	100%

Die Erkennungsleistung für die positiven Klassen (PAP IIId – PAP V) ist hinreichend. Durch verbesserte Artefakterkennung wird auch eine bessere Klassifikation der negativen Präparate erreicht werden.

---

* Die Sammlung der Präparate, sowie deren Präparation, zytologische Begutachtung und Dokumentation wurde von Prof. Dr. E. Sprenger, Abteilung Zytopathologie im Klinikum der Universität Kiel durchgeführt

Ausblick

Mit dem beschriebenen Gerätekonzept steht erstmals eine Möglichkeit zur Verfügung in übersehbaren Zeiten Versuche mit praxisgerechten Präparatemengen durchzuführen und die Einführung von maschinellen Prescreeningtechniken in die zytologische Routine zu beurteilen. Die realisierte Konzeption erlaubt auch, ohne wesentliche Änderungen, die Anpassung des Automaten an andere Zellmaterialien. Dazu müssen lediglich neue Lerndatenbanken erstellt werden und die dann signifikanten Merkmale und Parameter ermittelt werden. Eine erneute Programmierung des Systems ist nicht notwendig.

Gegenwärtig werden Experimente zur Verbesserung der Klassifikationsleistung an einem Präparatekollektiv von etwa 3000 Fällen durchgeführt. Ziel dieser Arbeiten ist in erster Linie die Erhöhung der Erkennungsleistung für negative Präparate (die Senkung der "Falsch-Positiv-Rate") um die Zahl der vom Zytologen zu begutachtenden Fälle weiter zu vermindern.

Literatur

[SCHW83] Schwarz, G., Ein Konzept für die routinemäßige Herstellung von automatengerechten gynäkologischen Präparaten
Acta Microscopica, Suppl. 6, 29-36 (1983)

[DAGM87] R. Dörrer, Automatische Auffindung von Zellkernen in mikroskopischen Zellbildszenen, Proc. 9.DAGM Symposium Mustererkennung, Informatik Fachberichte 149, 197 (1987)

[PRAT78] Pratt, W.K., Digital image processing,
Wiley-Interscience, 410-416 (1978)

[HARA73] Haralik, R.M., Textural features for image classification
IEEE transactions on Systems, Man and Cybernetics,
vol. SMC-3, 610-621 (1973)

Der vorliegende Beitrag wurde vom BMFT und der Firma Kontron Bildanalyse im Rahmen des Projekts FAZYTAN gefördert.

Adernextraktion durch iteratives Gradientenmatching in stark verrauschten medizinischen Bildern

K. Waidhas, R. Kutka
Siemens AG, Zentralabteilung Forschung und Technik
Otto-Hahn-Ring 6, D-8000 München 83

Abstract - Wir stellen einen Ansatz zur Extraktion schmaler Adern in verrauschten medizinischen Einzelbildern vor. Es werden im wesentlichen Eigenschaften wie die nahezu konstante Aderdicke, die Parallelität und gleiche Steilheit der Ränder als Vorwissen benutzt. Obwohl an Verzweigungsstellen noch Erkennungsfehler auftreten, reicht die Auflösung außerhalb dieser Bereiche bis zu einer Aderbreite von 2-3 Bildpunkten.
Der Algorithmus ist klinisch noch nicht erprobt und nicht als Produkt erhältlich.

1. Einleitung

In diesem Beitrag stellen wir einen speziellen Ansatz zum Finden und Verfolgen feinster Adern in medizinischen Röntgenbildern vor. Unser Anliegen ist es, ohne Benutzerführung dünne Adern von wenigen (etwa 2) Pixeln Durchmesser noch zu erfassen, selbst wenn sich ihr Gradient nicht über den des Hintergrundrauschens erhebt. Auf dieses Ziel konzentrieren wir uns zunächst, sodaß andere Eigenschaften, wie die Topologie des Aderbaumes (Verzweigungen, Überkreuzungen) unberücksichtigt bleiben. Als weiterer Grundsatz soll vorausgesetzt sein, daß die Antwort des Adernfinders keine künstlichen Strukturen enthält, d. h. der Operator soll im wesentlichen aus einem Kantenbild diejenigen Bildpunkte auswählen, die zu Adern gehören, aber keine Unterbrechungen, die beispielsweise aufgrund von Überdeckungen im Originalbild bestehen, künstlich überbrücken.

Der Algorithmus wurde auf Einzelbilder aus Subtraktionangiographie-Serien angewandt, mit gutem Ergebnis, was die Auflösung feiner Strukturen betrifft. Jedoch bestehen noch positive Antwortsfehler in Form von Strukturen, die keine Adern sind. Der Ansatz ließe sich in Zukunft mit bestehenden Algorithmen z. B. zur 3-d-Adernrekonstruktion verbinden oder als Vorverarbeitung zur quantitativen Auswertung von Bilddaten (Adernvermessung) benutzen.

Die aus der Literatur bekannten Arbeiten lassen sich nach Eigenschaften wie Auflösungsvermögen, Rauschrobustheit, Benutzervorselektion oder Rechenzeit beurteilen. Je nach Anwendung sind diese Eigenschaften verschieden ausgeprägt, und es wird unterschiedliches Vorwissen benutzt.

Ein sehr umfassendes wissensbasiertes System mit bis zu 11 Regeln zur Adernerkennung wird von Smets et al. (1988) vorgestellt. Es benutzt allgemeine physikalische Eigenschaften von Blutgefäßen, wie z. B. die Kenntnis des Luminanzverlaufs in Querrichtung. Es liefert eine polygonale Approximation der Mittellinien und Ränder samt Intensitäten, sowie die Topologie des Aderbaumes. Von ähnlichen Voraussetzungen gehen Collorec und Coatrieux (1988) aus. Sie modellieren Blutgefäße abschnittsweise durch Zylinder mit elliptischem Querschnitt. Annahmen über die längliche Form von Adern und ihre räumliche Lage werden getroffen. Der prinzipielle Aufbau eines Expertensystems zur Adernfindung auf Subtraktionsbildern werden von Catros et al. (1988) untersucht. Sie unterscheiden zwischen medizinischem (physikalischem) und menschlichem (physiologischem) Wissen. Das Expertensystem ANGY - von Stansfield (1986) vorgestellt - trennt in der ersten Stufe die Strukturinformation von Hintergrund und selektiert in der zweiten die Adern.

Viele Algorithmen setzen eine Vorselektion durch den Benutzer voraus, wie z. B. die Methode von Reiber (1986), womit nach Vorgabe des Startpunktes der gesamte Coronarienbaum rekonstruiert wird. Da die Suche nur von gefundenen Adern aus weiterführt, funktioniert dies jedoch nicht, falls Blutgefäße durch Überdeckung oder starkes Rauschen nicht zusammenhängend abgebildet sind. Die Verfolgungsalgorithmen von Stevenson (1987) und Parker (1987), ebenso von Rake (1987), insbesondere auf Herzkranzgefäße angewandt, setzen ebenfalls Benutzerführung voraus: Nachdem die Verzweigungspunkte von Hand markiert worden sind, werden von dort aus die Adermittellinien erstellt.

Eine weitere Spezialisierung von Adererkennungssystemen ist die 3-dimensionale Rekonstruktion aus mehreren Ansichten. Diese Zusatzinformation kann zur Kontrolle der auf Einzelbildern erstellten Ergebnisse benutzt werden: Suetens et al. (1982) benutzen im wesentlichen die Aderskelette und Kruger et al. (1987) das Luminanzprofil, um darauf das Matching zwischen den Projektionsbildern durchzuführen. Suetens et al. (1983) beweisen, daß es nicht möglich ist, ohne Vorwissen aus zwei orthogonalen Aufnahmen irgend einen Teil der Gehirnblutgefäße zu rekonstruieren.

Ohne die Forderung nach einem Verzweigungszusammenhang kommt der Algorithmus von Nguyen und Sklansky (1986) aus, im wesentlichen ein schneller Kantenfinder und -verfolger.

Unser Ansatz stützt sich auf die Regeln, daß Adern nahezu gleich dick, ihre Ränder nahezu parallel und gleich steil sind. Aufbauend auf ein Kantenbild werden entsprechend dieser Regeln Kantenpunkte als Kandidaten für Aderränder ausgewählt. Davon ausgehend werden iterativ entlang der Kanten immer neue Feinauswahlen getroffen unter Lockerung der Anforderungen (Schwellen). Es werden jedoch keine Punkte hinzugefügt, die nicht in der Antwort des Adernfinders enthalten sind.

2. Benutztes Vorwissen

Auf folgende Anforderungen soll der Algorithmus aufgebaut werden:

1. Auf den Röntgennegativfilmen sind Adern heller als der Hintergrund und im wesentlichen symmetrisch.

2. Etwa in der Mitte des Intensitätsprofils des Aderquerschnitts finden wir ein Maximum.

3. Aderränder sind lokal nahezu parallel.

4. Die Dicke der Adern ändert sich nur langsam.

5. Aderränder sind etwa gleich steil

Wir verzichten auf die Modellierung des Luminanzverlaufs quer zur Ader, da dies bei schmalen Strukturen (2 bis 3 Bildpunkte) hinfällig ist. Aufgrund der angestrebten hohen Auflösung verfolgen wir auch den Aderbaum nicht von einem bestimmten Startpunkt aus, da feinste Blutgefäße durch Röntgenrauschen stellenweise unterbrochen abgebildet sind und in diesem Fall nicht mehr erkannt werden könnten.

3. Algorithmus

3.1. Elementarer Kantendetektor

Wir benutzen einen Kantenfinder, der sämtliche Aderränder wiedergibt. Als geeignet erweist sich der Canny-Operator in der einfachsten Version, ohne Schwellwert (Canny 1983, Waidhas 1989). Ähnliche, wenn auch nicht so glatte Kanten liefert das Verfahren von Haralick (1984). Die Wahl des primären Kantenfinders ist für unser Verfahren nicht sehr maßgebend, wenn nur die Gradienten und die Entscheidung, ob Kante oder nicht an jedem Punkt zur Verfügung stehen.

3.2. Gradientenmatching

In jedem Kantenpixel sind Gradient und Gradientenlänge bekannt. Es sollen nun Parallelität und Kantenintensität getestet werden. Die Regeln 3 und 5 entsprechen der Forderung, daß Gradienten v_1 und v_2 an gegenüberliegenden Aderrändern etwa entgegengesetzt gleich sind. Wir testen die Ähnlichkeit der Längen über eine Schwelle. Eine Winkelschwelle begrenzt den Cosinus des Zwischenwinkels, sodaß dieser nur geringfügig von 180° abweicht:

$$\frac{\big|\, \|v_1\| - \|v_2\| \,\big|}{\max\{\|v_1\|\ \|v_2\|\}} < \text{Längenschwelle} \qquad (1.1)$$

$$\frac{<v_1,v_2>}{\|v_1\|\ \|v_2\|} < \text{Winkelschwelle} \qquad (1.2)$$

Beide Forderungen sind skaleninvariant.
Der Algorithmus läuft zunächst von einem Kantenpixel aus in Gradientenrichtung (also in die Ader hinein) solange, bis er auf ein Luminanzmaximum trifft oder eine vorgegebene Suchweite überschreitet. Im ersten Fall haben wir die vermutlich halbe Aderdicke bestimmt, im zweiten Fall ist das Kantenpixel nicht relevant.
Bei erfolgreicher Suche testen wir nun alle Bildpunkte, die in Gradientenrichtung des Ausgangspixels liegen und von diesem in etwa die geschätzte Aderdicke als Abstand haben, ob sie die Bedingungen (1.1) und (1.2) erfüllen. Treffen wir auf ein Pixel, das diese Bedingung erfüllt, so haben wir vermutlich eine Gegenkante gefunden. Wir markieren das Ausgangspixel und sein Gegenüber und haben die Aderdicke bestimmt. Derartige Paare von Kantenpixeln sollen *einander zugeordnet* heißen. Es kann vorkommen, daß einem Kantenpixel mehrere Bildpunkte zugeordnet werden. Speziell bei Aderschleifen werden die an der Schleifeninnenseite liegenden Kantenpunkte i. a. mehrmals von auf der Außenseite liegenden getroffen und diesen zugeordnet.

Schematische Darstellung des Zuordnungsprozesses:

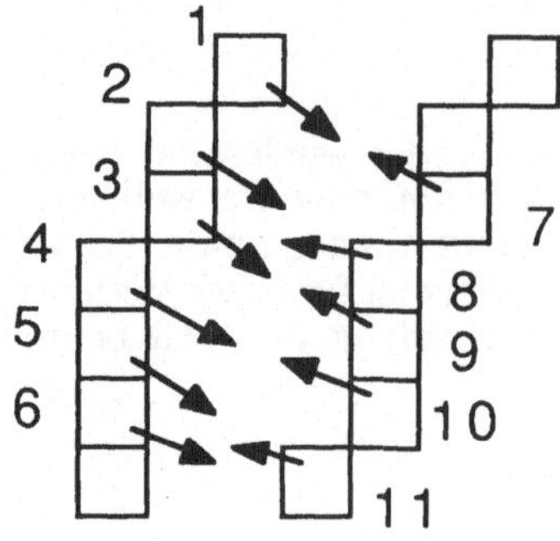

Zuordnung hier z. B.:

1 -> 7	7 -> 1
2 -> 8	8 -> 3
3 -> 9	9 -> 3
4 -> 10	10 -> 4
5 -> 11	11 -> 6
6 -> 11	

3.3. Weglassen kleiner Zusammenhangskomponenten

Durch Rauschen kann es vorkommen, daß Kantenpunkte, die nicht an Adern liegen, einander zugeordnet werden, allerdings sind diese Störungen meist nur wenige Pixel groß. Wir berechnen nun von der Menge aller zugeordneten Bildpunkte die Zusammenhangskomponenten und entfernen daraus solche, deren Größe unter einer Schwelle liegt.

3.4. Berechnung neuer Ausgangspunkte für die weitere Zuordnung

Die beschriebene Auswahl von Aderrandpunkten läßt sich iterativ fortsetzen, indem man die Schwellen für die Zuordnung zweier Kantenpixel immer weiter erniedrigt, also immer weniger Anforderungen an die jeweils gefundenen Pixel stellt, aber in jedem Iterationsschritt nur Kantenpixel betrachtet, die auf der gleichen Zusammenhangskomponente wie bereits zugeordnete Punkte liegen und von diesen einen vorgegebenen Höchstabstand haben.

Zwischen zwei Schritten k-1 und k des Zuordnungsprozesses wird folgendes Wachstum durchgeführt:

Sei E_{k-1} die bereits zugeordnete Kantenmenge, also alle Bildpunkte, die zum (k-1)-ten Schritt als Aderkandidaten bestimmt wurden und E die Menge aller vorhandenen Kanten, so bestimmen wir daraus E_k:

Setze zur Initialisierung $E_k := E_{k-1}$

Für eine vorgegebene Anzahl von Wachstumsschritten W berechne:

$$E_k := \text{Region Grow}(E_k) \cap E$$

E_k sind jetzt alle Kantenpixel aus E, die von der Menge E_{k-1} längs gefundener Kantenzüge einen Abstand von maximal W Pixel haben (siehe auch Waidhas, 1989).

3.5. Dickentest

Mit fortschreitender Iterationszahl k werden die Anforderungen an eine erfolgreiche Zuordnung immer geringer, sodaß auch immer mehr Rauschen als Ader vorgeschlagen werden könnte. Diesen Effekt vermeiden wir durch Überprüfung der Aderdicke (Forderung 4). Die Dicken der bereits zugeordneten Kantenpunkte lassen sich aufsummieren und wir erhalten eine lokale Durchschnittsdicke, die wir mit derjenigen der neuen Kandidaten vergleichen. Bei geringer Abweichung erfolgt die Zuordnung.

3.6. Verwendung von verschiedenen Filtergrößen im Verfahren von Canny

Der Algorithmus aus 3.1. bis 3.5. läßt sich selbst wieder iterativ anwenden. Ein Filter, das stärker glättet findet i. a. weniger Kanten bzw. verursacht lückenhafte Adern, ist aber auch weniger rauschanfällig als ein schwächeres. Daher benutzten wir das Ergebnis des stark glättenden Filters als Ausgangsmenge für die Suche entlang den Kanten des schwach glättenden. Dieses Erweiterung liefert auf unserem Testmaterial (Abb. 1) bei gleich guter Erkennung von Adern weniger Rauschen außerhalb der Aderbereiche.

4. Experimentelle Ergebnisse

Es wurden Bilder mit einer Auflösung von 512 x 512 Bildpunkten und einer Tiefe von 10 bit aus verschiedenen medizinischen Röntgenszenen zur digitalen Subtraktions- angiographie als Testmaterial herangezogen. Der Algorithmus wurde aber nicht auf Differenz- sondern auf Originalbilder angewandt, was die Extraktion erschwert, aber den Anwendungsbereich des Verfahrens auch auf Einzelbilder ausdehnt.
Es wurden folgende Parameter benutzt:

Breite des Gaußfilters $\sigma = 1.3$ für 1. Iteration, $\sigma = 0.85$ für 2. Iteration, Wachstumsschritte jeweils W = 10.
Die Schwellen für den normierten Cosinus der Winkeldifferenz (Formel 1.2) wurden in 6 Iterationen wie folgt gelockert: 0.1, 0.15, 0.35, 0.5, 0.6 und 0.8. Der letzte Schritt geschieht mit Dickenprüfung auf Abweichung kleiner gleich 1 Pixel.
Die Rechenzeit von 1 Stunde pro Bild wird größtenteils von der Prozedur des Region Grow verbraucht, da die Algorithmen ohne Hardwareunterstützung simuliert wurden.

Das Testbild (Abb. 1) zeigt eine Aufnahme des Bauchraums mit einer von unten ins Bild ragenden Kanüle, durch die Kontrastmittel in die Blutbahn geleitet wird. Es sind einige sehr feine, für den menschlichen Betrachter jedoch gut sichtbare Adern zu erkennen. Rechts der Mitte in Abb. 1 ist beispeisweise eine vertikal verlaufende Ader mit Durchmesser 2-3 Pixeln zu sehen (in Abb. 5 vergrößert), links der Mitte eine stark geschlängelte, ebenso schmale. Diese werden vom Canny-Operaror erfaßt (Abb. 2). In Abb. 3 sind die Gradientenlängen als Helligkeitswerte dargestellt. Man erkennt, daß die Verlängerung feiner Adern sehr schwach ausgeprägte Kanten (Gradientenlängen) besitzt, welche stellenweise nicht höher als die des Rauschens in der Umgebung sind. Die Antwort des Adernfinders (Abb. 4 und Abb. 6) zeigt, daß solche schmalen Strukturen über weite Strecken verfolgt worden sind. Das Ergebnis enthält jedoch zusätzlich noch viele positive Antwortsfehler. Es werden zufällig aderförmige Strukturen, wie z. B. Wirbelfortsätze mit ausgegeben.

5. Ausblick

Der gegenwärtige Stand des Verfahrens kommt mit sehr wenig Vorwissen aus und findet bereits einen großen Teil des Aderbaumes, selbst feine Strukturen.
In Zukunft sollten neben dem Gradientenfeld auch die absoluten Helligkeitswerte mit einbezogen werden. Experimente mit Rückenlinien zeigen, daß diese in vielen Fällen den bisherigen Ansatz ergänzen. Damit ließe sich sowohl noch vorhandenes Rauschen in der Antwort des Adernfinders verringern, als auch Verzweigungsstellen ausfindig machen.

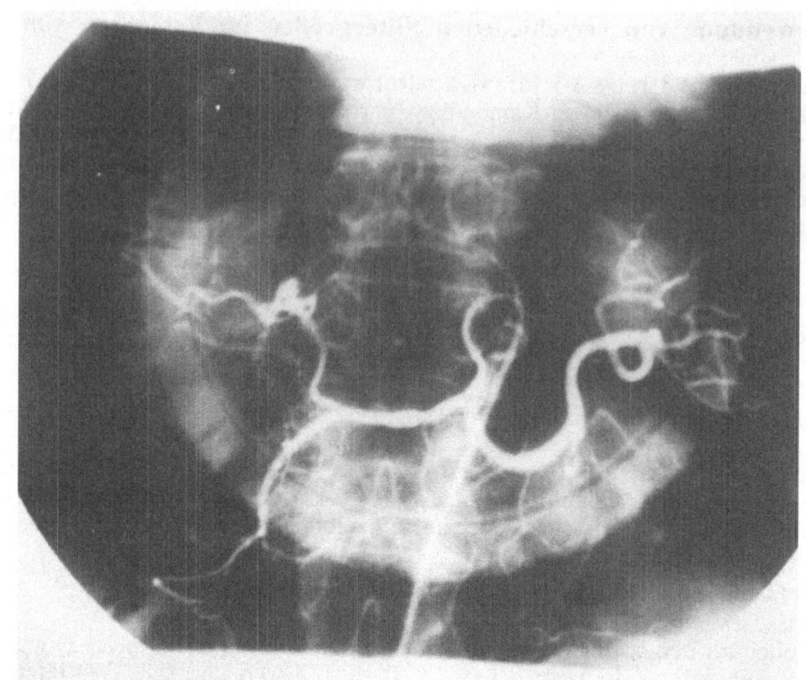

Abb. 1: Originalbild einer DSA-Serie, 512 x 512 Bildpunkte

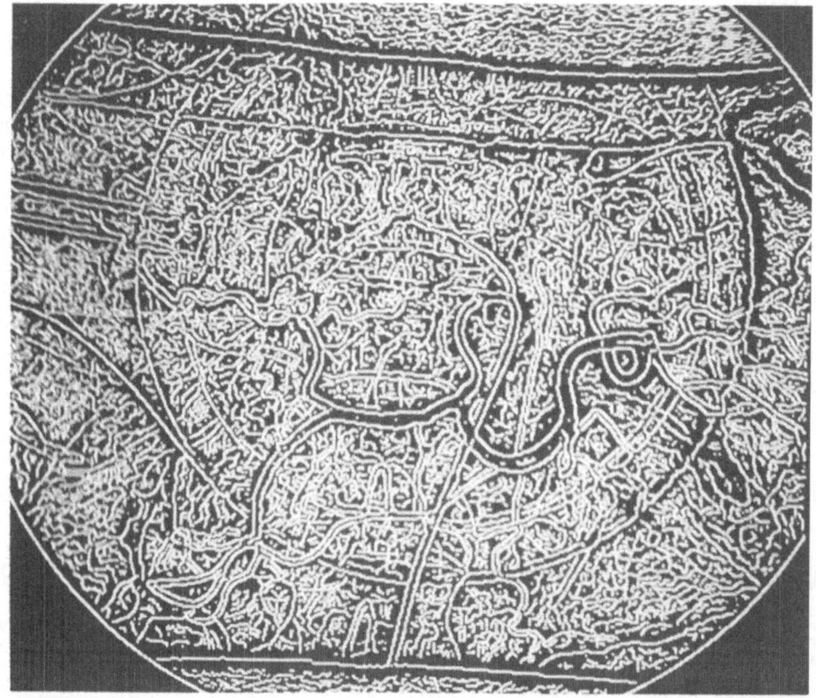

Abb. 2.:   Kantendetektion nach dem Verfahren von Canny

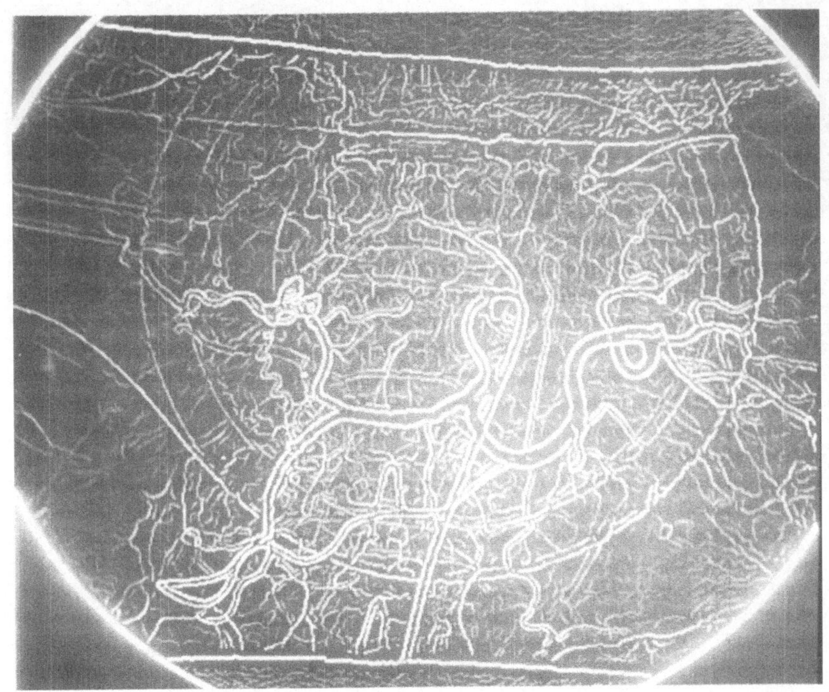

Abb. 3.: Gradientenlängen der gefundenen Kantenpunkte als Helligkeitswerte dargestell

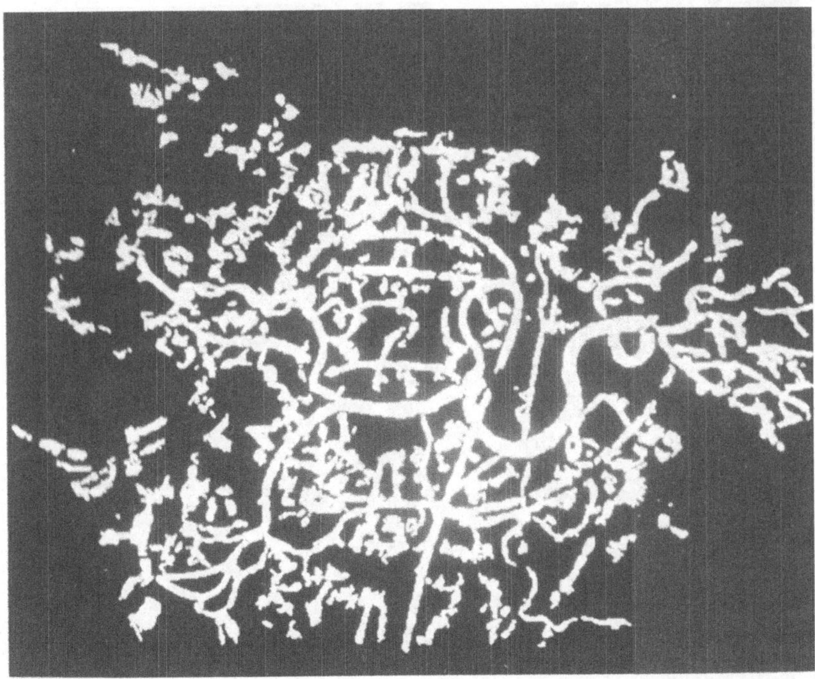

Abb. 4: Ergebnis des Adernfinders: Erkennung feiner Adern im Pixelbereich, jedoch auch Markierung von zufällig aderförmigen Strukturen

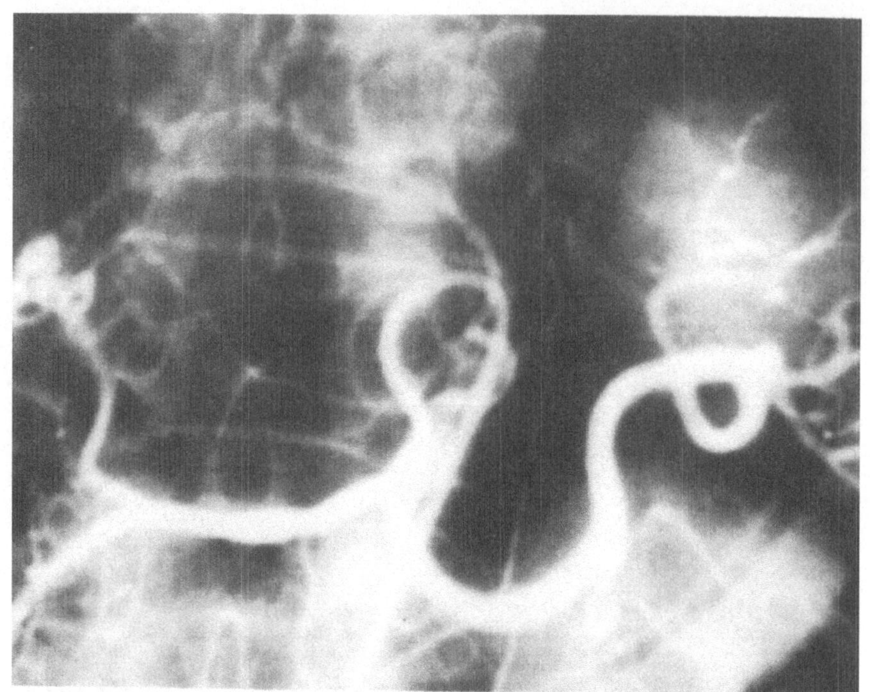

Abb. 5: Vergrößerter Ausschnitt aus Abb. 1   (Original)

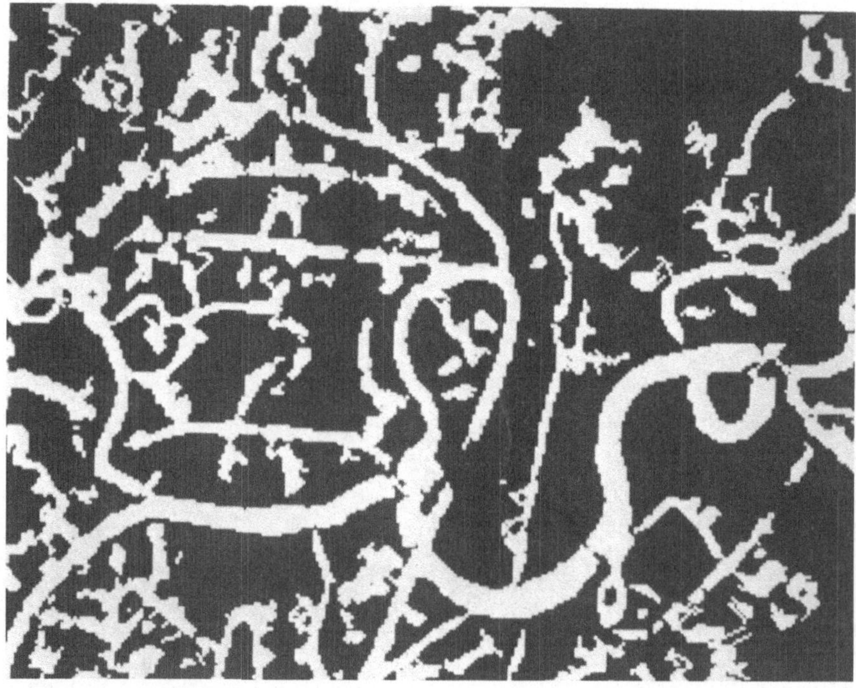

Abb. 6: Vergrößerter Ausschnitt aus Abb. 4   (Antwort des Kantenfinders)

6. Literatur

Akita K., Kuga H., "A computer method for understanding ocular fundus images", *Pattern Recognition* **15**, 431-443 (1982)

Canny, "Finding Edges and Lines in Images", *Technical Report AI TR 720 Massachusetts Institute of Technology* (1983)

Catros J. Y., Mischler D., "An artificial intelligence approach for medical picture analysis", *Pattern Recognition Letters* 8, 123-130 (1988)

Collorec R., Coatrieux J. L., "Vectorial tracking and directed contour finder for vascular network in digital subtraction angiography", *Pattern Recognition Letters* **8**, 353-358 (1988)

Haralick R. M., "Digital Step Edges from Zero Crossing of Second Directional Derivates", *IEEE Trans. on Pattern Analysis and Machine Intelligence, PAMI* **6** (1984)

Hoffman K. R., Doi K., Chan H. P., Fencil L. Fujita H. and Muraki A., "Automated tracking of vascular tree in DSA images using a double-square-box region-of-search algorithm", *SPIE 626, Med XIV, PACS IV*, 323-333 (1986)

Kruger R. A. , Reinecke D. R. ,Smith S. W. Ning R., "Recognition of blood vessels from x-ray subtraction projections: Limited Angel Geometrie, *Med. Phys.* **14**, 940-949 (1987)

Nguyen T. V., Sklansky J, "A fast skeleton-finder for coronary arteries", *Proc 8th ICPR, Paris*, 481-483 (1986)

Parker D. L., Pope D. L., Van Bree R. and Marshall H., "Three-dimensional reconstruction of moving arterial beds from digital subtraction angiography", *Comp. Biomed. Res* **20**, 166-185 (1987)

Rake S. T., Smith L. D. R., "The interpretation of x-ray angiograms using a blackboard control architecture", *Proceedings of the Int. Symposium CAR,* 681-686 (1987)

Reiber J. H. C., Serruys P. W. and Slager C. J.,"Structural analysis of the coronary and retinal arterial tree", In: *Quantitative Coronary and Left Ventricular Cineangiography. Martinus Nijhoff, Dordrecht,* 185-213 (1986)

Smets C., Verbeeck G., Suetens P., Oosterlinck A., "A knowledge-based system for the delineation of blood vessels on subtraction angiograms", *Pattern Recognition Letters* **8**, 113-121 (1988)

Stansfield S. A., "ANGY: A rule based expert system for automatic segmentation of coronary vessels from digital subtracted angiograms", *IEEE Pattern Anal. Mach. Intell.* **8** (2), 188-189 (1986)

Stevenson D. J., Smith L. D. R. and Robinson G., "Working towards the automatic detection of blood vessels in X-ray angiograms", *Pattern Recognition Lettes* **6**, 107-112 (1987)

Suetens P., Haegemans A., Oosterlinck A., Gybels J., "An attempt to reconstruct the cerebral blood vessels from a lateral and a frontal angiogram", *Pattern Recognition* **16**, 517-524 (1983)

Suetens P., Oosterlinck A., Haegmans A., Gybels J., "Three-dimensional reconstruction of the blood vessels of the brain", *Proceedings of the ISMIII, Int. Symp. on Medical Imaging and Image Interpretation Berlin,* 429-435 (1982)

Waidhas K, "Funktionalanalytische Untersuchung des Gaußfilters bei Kantendetektions-verfahren", *Diplomarbeit am Mathematischen Institut der LMU, München* (1989)

Segmentierung biologischer Objekte aus CT– und MR– Schnittserien ohne Vorwissen

F. Saurbier, D. Scheppelmann, H. P. Meinzer

Deutsches Krebsforschungszentrum Heidelberg
Abt.: Medizinische und Biologische Informatik
Leiter: Prof. Dr. C. O. Köhler

Zusammenfassung

Durch die Entwicklung von Raytracingverfahren [2,3], welche ohne Oberflächeninformation arbeiten, ist es möglich geworden, biologische Objekte aus CT- und MR–Schnittserien transparent erscheinen zu lassen. In den rekonstruierten Bildern konnte aufgrund der transparenten Darstellung die Größe und Lage von Organen 'erahnt' werden. Eine klare Trennung bzw. Abbildung der Organe kann mit dem Raytracingverfahren alleine nicht eziet werden, da die Information der Grauwerte von solchen Schnittserien nicht ausreichend ist. Allerdings ist in vielen Fällen gerade die genauere Betrachtung der einzelnen Organe für eine eindeutige Diagnosefindung von Bedeutung. Bei tumorösen Veränderungen interessiert zusätzlich auch noch das Organinnere.

Daher suchen wir in unserer Arbeitsgruppe nach geeigneten Segmentations- bzw. Klassifikationsverfahren, um bei der dreidimensionalen Visualisierung von Schnittserienbildern eine klar erkennbare Organdifferenzierung durchführen zu können. Zu diesem Zweck benutzen wir die selbstlernende topologische Merkmalskarte als geeignetes Hilfsmittel.

1 Die topologische Merkmalskarte

Die topologische Merkmalskarte [4] ist ein Algorithmus zur Simulation eines Neuronalen Netzwerkes, der von Teuvo Kohonen zur Spracherkennung eingeführt wurde. Dieser Algorithmus wurde für die Anwendung auf medizinische Bilddaten modifiziert und optimiert [1]. Die Karte eignet sich zur Klassifikation innerhalb hochdimensionaler Merkmalsräume, indem sie diese auf einen zweidimensionalen Bereich abbildet. Um eine Klassifikation durchführen zu können bedarf es keinerlei Vorinformation über die Art und Anzahl der zu klassifizierenden Objekte. Die Karte lernt selbstständig aus dem zur Verfügung gestellten Merkmalssatz, der aus dem original Grauwertbild und daraus abgeleiteten Maßen besteht.

Dazu wird zu jedem Vektor aus dem Merkmalsraum der ähnlichste Kartenvektor gesucht und eine gaußgewichtete Angleichung der Nachbarpunkte innerhalb einer definierten, rechteckigen Umgebung vorgenommen. Als Ähn lichkeitsmaß wird die euklidische Distanz verwendet. Durch diese Vorgehensweise wird eine Gruppierung der sich ähnlicher Merkmalsvektoren erreicht. Dieser Vorgang entspricht einer Diskriminanzfunktion, die sanfte Klassenübergänge zuläßt, was dem Erscheinungsbild von Übergängen zwischen einzelnen Organen in medizinischen Schnittbildern sehr nahe kommt, da hier meist keine scharfen Texturkanten erkennbar sind. Eine Interpretation solcher Bilder steht und fällt mit dem anatomischen Vorwissen des Betrachters.

2 Der Merkmalsraum für die Lernphase

Die Klassifizierung der verschiedenen biologischen Objekte in einem Schnittbild ist abhängig von den zur Lernphase verwendeten Eingangsdaten, dem Lernfaktor und der Anzahl der Lernschritte. Dabei soll in diesem Abschnitt hauptsächlich auf die verwendeten Eingangsdaten eingegangen werden.

Zusätzlich zum original Grauwertbild können hier lokale Texturmaße, grauwertmorphologische Operatoren citeserra und Filterfunktionen Verwendung finden. Dabei hat sich gezeigt, daß die Verwendung von diesen Parametern besonders problematisch ist, da diese nur mit relativ kleinen Masken aus dem Originalbild ermittelt werden können, um nicht feinere Srukturen eines Organs völlig zu eliminieren.

Damit ergibt sich vor allem bei den Texturparametern das Problem des zu kleinen Stichprobenumfanges. Es ist einfach nicht sinnvoll aus 'mager' besetzten Histogrammen oder Coocurrenzmatrizen irgend einen Texturparameter berechnen zu wollen. Daher wurde in diesem Zusammenhang versucht auf die Dreidimensionalität der Objekte zurückzugreifen, indem die benachbarten Schnittbilder einer Serie zur Berechnung mit herangezogen werden. Dieses Verfahren ist aber nur solange sinnvoll, wie die einzelnen Schnittbilder dicht beieinander liegen, was in der Praxis aber nicht immer gewährleistet werden kann.

Die Verwendung von grauwertmorphologischen Operatoren erscheint in diesem Zusammenhang sinnvoll, da man dadurch zusätzlich Information über die Größe und Form von einzelnen Organen erhalten kann. Der einzige kritische Punkt in diesem Zusammenhang ist der, daß ein Organ nicht in jedem Schnitt der Bildserie eine konstante Größe bzw. Form aufweist. Diese Vergrößerungen und Verkleinerungen müßten bei der Klassifikation mit berücksichtigt werden, da sonst gerade die sich verjüngenden Randbereiche nicht mehr zu dem entsprechenden Organ gezählt würden. Genau dieser Problematik kann man eigentlich nur durch die Verwendung von Vorwissen zuleibe rücken. Will man diese Parameter möglichst allgemeingültig verwenden, so bleibt auch hier nur die vorsichtige Wahl kleiner Masken, wodurch natürlich die Wirksamkeit einer solchen Waffe erheblich in Mitleidenschaft gezogen wird.

Der Einsatz von Hoch–, Tief– und Bandpaßfiltern ist bei dem Einsatz als Merkmalsebene für die topologische Karte nicht ganz so kritisch, da sie keine so starken Auswirkungen auf die Bildinformationen haben. Dafür zeichnen sie sich durch eine recht hohe Korrelation mit dem Originalbild aus, was bedeutet, daß sie nicht sonderlich viel an neuer Information bei der Segmentierung beisteuern können.

3 Lernverhalten der Topologischen Karte

Neben den verwendeten Merkmalen für die Bildung eines Merkmalsraumes spielt für das Lernverhalten der topologischen Karte die Lernumgebung, Lernschritte und der Lernfaktor eine wesentliche Rolle. Der Einfluß verschiedener Lernfaktoren wurde bereits von Bertsch [1] dargestellt. Dabei zeigte sich, daß ein Lernfaktor von 1 zu Beginn einer Lernphase den Merkmalsraum gut auf die gesamte Karte abbildet. Dies ist für die Berechnung einer neuen Karte sinnvoll. Je kleiner der Startlernfaktor gewählt wird, desto weniger Kartengebiete zeigten sich von den Eingangdaten beeinflußt. Ein kleinerer Startlernfaktor kann also für den sanften Angleich einer bereits berechneten Karte an einen etwas veränderten Merkmalsraum verwendet werden.

Als Lernphase wird hier immer der Abschnitt des Lernprozesses bezeichnet, in dem die Größe der Karte und der Lernumgebung unverändert bleiben. Eine Lernphase wiederum besteht aus mehreren Lernschritten. Dabei wird innerhalb einer Lernphase der Lernfaktor, abhängig von der Häufigkeit der Merkmalsvektoren zu den ermittelten Kartenvektoren, von Lernschritt zu Lernschritt individuell für jede Kartenposition verringert. Dadurch wird der Einfluß der besonders oft auftretenden Merkmalsvektoren auf die Karte klein gehalten, so daß keine Verdrängung von selteneren Merkmalsvektoren stattfinden kann. Alle Lernphasen, die zu einer berechneten Karte geführt haben, werden mit dem Begriff Lernprozeß zusammengefaßt.

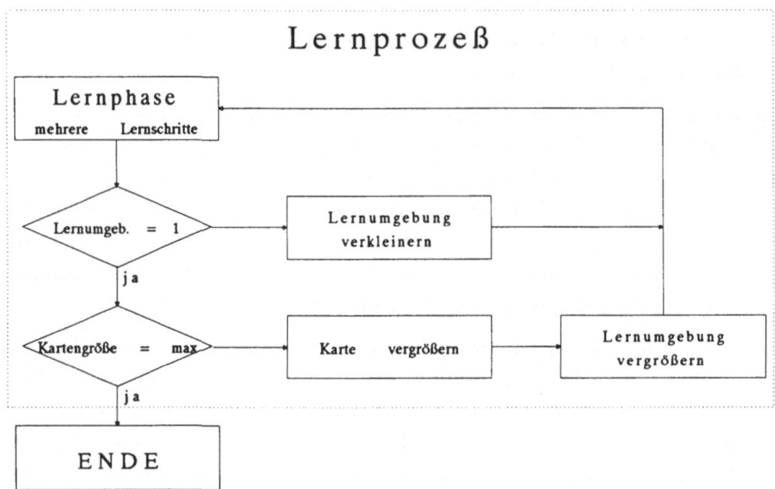

Abbildung 1: Schematischer Ablauf des Lernprozesses der topologischen Merkmalskarte

Nach einer Lernphase erfolgt entweder eine Vergrößerung der Karte, eine Verkleinerung der Lernumgebung oder der gesamte Lernprozeß wird beendet. Dies geschieht nach dem in Abbildung 1 dargestellten Schema.

3.1 Abbruchkriterium einer Lernphase

Der Lernprozeß der topologischen Karte ist in mehrere von einander abhängige Lernphasen eingeteilt. Er beginnt mit einer kleinen 3 × 3 Karte. Diese Startkarte wird nach der Hauptkomponententransformation initialisiert. Dadurch wird schon vor dem eigentlichen Lernprozess eine gute Repräsentation des Merkmalsraumes auf dieser, neun Vektoren umfassenden Karte erreicht. Um einen Eindruck von der Nützlichkeit dieser Initialisierung zu erhalten, ist eine Zählung der Merkmalsvektoren, die von Lernschritt zu Lernschritt wieder die gleiche Kartenposition einnehmen, denkbar.

Die Zählung der Merkmalsvektoren ist ein stabiles Kriterium. Um von der Größe eines Merkmalsraumes unabhängig zu sein, wurde daraus die Wiederzuweisungsrate W abgeleitet.

$$W = \frac{Anzahl\ wiederzugewiesener\ Vektoren}{Anzahl\ aller\ Merkmalsvektoren}$$

Eine Verifikation der Initialisierung der Karte mittels der Wiederzuweisungsrate ergab nach dem ersten Lernschritt einen Wert von über 75%. Bereits nach zehn Lernschritten betrug die Wiederzuweisungrate über 98%. Daraus resultierte für die Wiederzuweisungsrate als Abbruchkriterium folgender Wertebereich.

$$0.97 \leq W < 1$$

Damit war ein geeignetes Maß für den Abbruch einer Lernphase gefunden. Wenn die Wiederzuweisungsrate der Merkmalsvektoren einen angegebenen Wert erreicht oder überschritten hat, kann eine Lernphase des Lernprozesses als abgeschlossen betrachtet werden.

3.2 Vergrößerung der Karte nach einer Lernphase

Ist die endgültige Kartengröße eines Lernprozesses noch nicht erreicht, so muß die Karte aus der zuvor abgeschlossenen Lernphase vergrößert werden. Dazu werden durch Vergrößerung neu entstehende Kartenpositionen durch Interpolation mit Werten belegt. Um dabei den Einfluß der Interpolation auf der Karte so gering wie möglich zu halten, wird zwischen zwei Kartenpositionen jeweils eine neue Position berechnet. Damit ergibt sich für die Dimension der neuen Karte folgender Formalismus.

$$DIM(Karte) = 2 \cdot DIM(Karte) - 1$$

Dadurch wird die innere Ordnung der Karte nach einer Vergrößerung beibehalten.

3.3 Angleichen der Lernumgebung

Nach einer Vergrößerung der Karte einer abgeschlossenen Lernphase sind zwischen den Ursprungspositionen immer interpolierte Werte vorhanden. Um diese neuen Kartenpositionen möglichst schnell dem Merkmalsraum anzugleichen, wird die Lernumgebung am Anfang der Lernphase vergrößert.

Da bei einer größeren Lernumgebung der Einfluß einer Kartenposition auf ihre Nachbarpunkte zunimmt, sollte gleichzeitig eine Verminderung der angestrebten Wiederzuweisungrate erfolgen. Denn die Veränderung der Umgebung einer Kartenposition führt zu häufigeren Positionswechseln der Merkmalsvektoren. Auf diese Weise kann sichergestellt werden, daß eine Lernphase immer zu einem determinierten Ende findet. Damit sind alle Kriterien, die für einen automatischen Lernprozeß benötigt werden, gefunden.

4 Ergebnisse der automatischen Segmentierung

Trotz der Nachteile, die bei dem Versuch einen möglichst allgemeingültigen Parametersatz für die Lernphase der topologischen Karte zu verwenden, entstehen, ist es möglich, durch die Verwendung einer Mischung aus den zuvor genannten Bereichen, eine Segmentierung mit der Karte vorzunehmen. Ist der Lernprozeß der Karte abgeschlossen, so sind in verschiedenen Kartenregionen unterschiedliche Organe bzw. Organteile repräsentiert. Die Karte kann dann zur Klassifizierung der übrigen Schnittbilder einer Schnittbildserie verwendet werden. Allerdings handelt es sich dabei nur um eine Grobklassifizierung, daß heißt die gewonnenen Bilder sind noch verrauscht, was aber durch eine Markierung der zusammenhängenden Regionen unterbunden werden kann. Diese Markierung dient dann in der gesamten Schnittserie zur Identifizierung eines Organs.

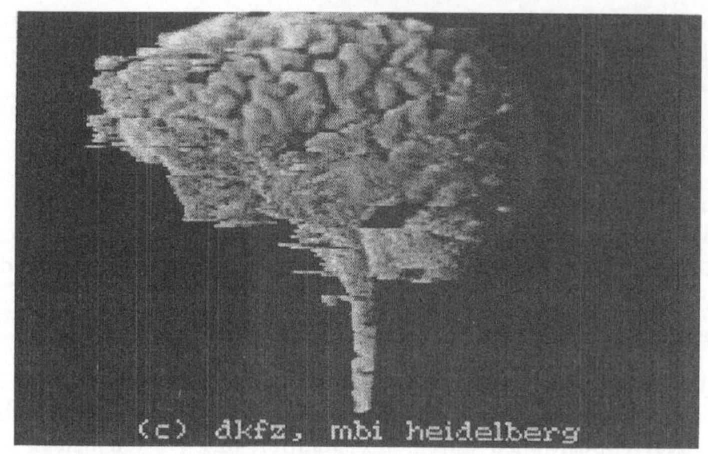

Abbildung 2: 3D–Rekonstruiertes Gehirn

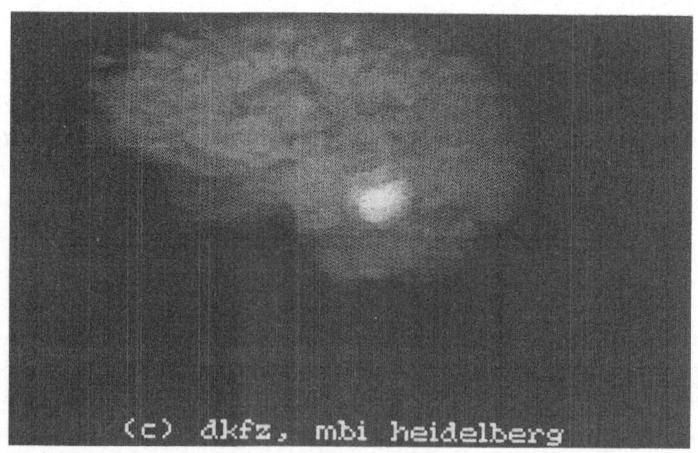

Abbildung 3: Transparentes Gehirn mit Tumor

5 Weiterführende Anwendung

Wie die Abbildung 2 zeigt, bedingt einer Segmentierung, die vorerst nur auf die zweidimensionalen Einzelschnitte eine Serie zurückgreift, daß es noch zu kleinen Fehlern in den Randbereichen führen kann. Daher wird eine Umstellung des Verfahrens auf die Dreidimensionalität der Objekte nötig werden. Zusätzlich sollte für eine bessere Ausnutzung der einzelnen Parameter die Möglichkeit, Vorwissen bei der Segmentierung zu verwenden, geschaffen werden. Dazu wäre eine formale Beschreibung der Anatomie des Menschen nötig, um damit eine Verbindung von der pixelorientierten zu einer objektorientierten Verarbeitung medizinischer Bilddaten zu schaffen.

Die Abbildung 3 zeigt sehr gut, wie durch oberflächenfrei arbeitende Raytracingalgorithmen ineinanderliegende Bildobjekte dargestellt werden können, wenn eine eingeutige Segmentierung erfolgt ist.

Literatur

[1] Bertsch H.: Die selbstlernende topologische Merkmalskarte zur Bildsegmentierung und Klassifikation. Technical Report 23/1988, Deutsches Krebsforschungszentrum Heidelberg Abteilung MBI.

[2] Heckbert P. S.: Ray Tracing JELL-O Brand Gelatin. Computer Graphics, Volume 21, Number 4 (1987) 73–74.

[3] Heyers V.: Raytracing in Grauwertvoxelräumen zur Visualisierung medizinischer Schichtdaten. Technical Report 20/1988, Deutsches Krebsforschungszentrum Heidelberg Abteilung MBI.

[4] Kohonen T.: Clustering, Taxonomy and topological maps of patterns. Proceedings of the 6th Int. Conf. on Pattern Recognition 1982, 114– 128. Computer Society Press, Silver Spring.

[5] Serra J.: Introduction to Mathematical Morphology. In: Computer Vision, Graphics and Image Processing 35 (1986) 114–128

On Scale-Space Edge Detection in Computed Tomograms

Svend Back, Heiko Neumann, H.Siegfried Stiehl[1]

Universität Hamburg, FB Informatik
Bodenstedtstr.16, D-2000 Hamburg 50

Summary. We describe research toward a general multi-scale edge detection scheme. Since local discontinuity profiles at arbitrary loci in two-dimensional discrete images can be characterized as possessing prior unknown scales along the gradient direction, convolution of the discrete intensity function with kernels of fixed spatial operator support evidently results in only locally suboptimal and thus non-reliable operator responses. Hence we propose a multi-scale scheme following Marr's and Korn's approach. We present the theoretical framework for the 1-D case as well as results from an initial experimental implementation for the two-dimensional discrete case of X-ray computed tomograms.

Introduction

Computational segmentation of primitives such as edge pixels as well as subsequent primitive grouping, to e.g. smooth contours, has to be considered the starting point of a processing cascade which maps raw CT or MR image data into geometric, iconic, or symbolic representations of physical properties of the human body ([1]). Such representations are the methodological prerequisite to specific classes of computer-assisted tools for supporting clinical routine tasks, e.g. morphometry, display of organ geometry representations, quantitative stereotaxy and radiation therapy planning. Accurate segmentation of discrete noisy computed tomograms must confront some basic problems associated with the domain: a) the specificity of the imaging modality (e.g. X-ray versus MR computed tomography), b) the anisotropic resolution of spatial image sequences from clinical routine examinations, and c) the imperfect image structure itself (e.g. discrete noisy discontinuities and different junction types both of prior unknown spatial scales). Moreover one must be aware of the fact that "real world" computed tomogram image data in daily clinical routine - which is in sharp contrast to quasi-perfect CT data used for experimental laboratory processing say, for 3-D graphics display - is in the main still of anisotropic resolution and degraded by noise and artifacts. As a consequence, and by also taking into account deficiencies of segmentation schemes in operational systems (see [2] for a brief discussion of medico-technical problems), research has to be directed toward general approaches to segmentation which a) allow for rapid, parallel, and precise primitive extraction and grouping with minor interaction burden and b) exhibit utmost invariance against the boundary conditions as explicated above. The claim for generality is motivated by the need for fast, reliable, and automated, or at least semi-automated, segmentation of *arbitrary* objects in computed tomograms, e.g. anatomical landmarks, therapeutical target volumes, and organs at risk with regard to interventional therapeutic procedures. Moreover we are concerned with the detection of 2-D structure as a precursor to morphology-sensitive reconstruction of isotropic 3-D voxel arrays from anisotropic spatial image sequences as opposed to a "blind", e.g. simply linear, interpolation between 2-D computed tomograms such as to generate discrete isotropic 3-D X-ray or MR data arrays suitable for voxel-based graphics architectures.

Whereas we focus on multi-scale edge detection and grouping, Pizer and coworkers investigated advanced multi-scale region segmentation for computed tomograms ([3]).

---

[1]Parts of this research have been carried out while the authors have been with Fachbereich Informatik at the Technische Universität Berlin. S.Back is now with Sun Microsystems GmbH, D-8011 Grasbrunn 1.

Image Structure in Computed Tomograms

A spatial sequence of computed tomograms (or, respectively, an either isotropic or anisotropic voxel cube) is a three-dimensional discrete noisy estimate of continuous physical properties of the human body, e.g. smooth morphology mapped onto a discrete spatial grid with associated "intensities" related to tissue absorption coefficients (as in the case of X-ray computed tomography). A 2-D noisy Houndsfield unit ("intensity") function, $f(x,y)$, defining a single anisotropic computed tomogram is again only a numerical single-valued estimate of a continuous 3-D distribution of different tissues within a slice of finite thickness. Discontinuities of the intensity function indicate both the locus and the spatial extent of an intersection of a 3-D object surface patch with the slice. The type (e.g. a step or an arbitrarily scaled ramp edge) and the local characteristics (e.g. slope and local contrast) of a discontinuity directly depend on both the slice thickness and the local geometric properties of the intersecting surface patch. Apart from single discontinuities, a variety of junction types may appear in the 2-D intensity function such as a T-junction whose constituents again may have different slopes and contrasts. As a conclusion, we define basic image structure in 2-D computed tomograms (aside from textured or homogeneous regions) as differently scaled discrete discontinuities, arbitrarily shaped smooth contours, and a set of different junction types at prior unknown discrete spatial loci. Clearly, the 2-D image structure is intrinsically related to spatial properties of single 3-D organ surface patches (or even locally touching arrangements thereof) intersecting a computed tomography slice (see [2] for further details).

Korn's Quantitative Multi-Scale Edge Detection Scheme

In [4] and [5] Korn published a multi-scale edge-detection scheme[2] based upon the first order partial derivatives G_x and G_y of a normalized Gaussian distribution $G(x,y,\sigma_i)$ (σ_i: space constant) which are the components of the gradient vector $\nabla G = (\frac{\partial G}{\partial x}, \frac{\partial G}{\partial y})$. The sampled normalized partial derivatives $\hat{G}_x$ and $\hat{G}_y$ serve as linear shift-invariant discrete convolution kernels to numerically estimate $|\nabla G(x,y,\sigma_i) * f(x,y)|$. Parametrization of the Gaussian by appropriately varying the space constant σ_i yields a family of differently scaled operators $|\nabla G(x,y,\sigma_1)|, ..., |\nabla G(x,y,\sigma_n)|$ such that any operator has a different local support but similar overall shape of the constituent impulse response functions $\hat{G}_x$ and $\hat{G}_y$.

From $|\nabla G(x,y,\sigma_i) * f(x,y)|$ ($i = 1,..,n$) we yield a three-dimensional scale space composed of filtered intensity functions. The gradient magnitude at edge pixels along the σ_i-axis in scale space asymptotically converges to a constant maximum value. Tracking of local maxima of the gradient magnitude through scale space along the σ_i-axis per pixel at (x,y) allows for determination of a particular scale σ_Φ for which the operator response is optimal with respect to the underlying scaled discontinuity (see also next paragraph). Korn proposed two different termination criteria for scale-space search, e.g. either relative scale-to-scale increase of operator response less than 4% or determination of an absolute maximum (see [4] and [5] for further details).

Scale-Space Integration Based Upon the Error Integral Curve as 1-D Edge Model

Implicit or explicit edge models play an important role in computational theories of edge detection, e.g. Canny's step edge model [7] or Nalwa-Binford's *tanh* model [8]. The normalized error integral curve (with space constant $^1\sigma$; $H(x)$: Heaviside function)

$$\Phi(x,{}^1\sigma) = H(x) * G(x,{}^1\sigma) = \frac{1}{\sqrt{2\pi}\,{}^1\sigma} \int_{-\infty}^{x} \exp[-\frac{z^2}{2\,{}^1\sigma^2}]dz \tag{1}$$

is an appropriate explicit 1-D discontinuity model for the following three reasons

---

[2]The late David Marr at first qualitatively reported on such an approach (see [6], Fig.1 and pp.487). He coined the graph of the operator response as a function of mask width (which is directly related to σ_i) as 'signature'.

- the complete range of differently scaled discontinuities (with arbitrary local contrast), e.g. in computed tomograms, can be derived from the generalization of (1)[3] under the assumption that the true discontinuity is smooth (as opposed to the corresponding noisy edge in the image data) and is caused by locally smooth anatomical morphology,

- the optimum scale σ_Φ (and thus local support) of the 1-D operator $\hat{G}_x(x,{}^2\sigma)$ can be precisely determined with respect to the unknown local width, viz Marr's edge 'fuzziness' ([6], p.488), of the underlying discontinuity, and

- the intrinsic local parameters of a prior unknown 1-D discontinuity, which are the space constant ${}^1\sigma$ (determining width and steepness, see (1)) and the contrast c, can be estimated via scale-space processing.

Using Korn's normalized partial derivative (${}^2\sigma$: space constant of the impulse response function in (2)) for the 1-D case with $k({}^2\sigma) = \sqrt{2\pi}\,{}^2\sigma$

$$\hat{G}_x(x,\,{}^2\sigma) = k({}^2\sigma) \cdot G_x(x,{}^2\sigma) \tag{2}$$

we derive the operator response (Marr's 'signature', see Fig.1)

$$M(x,{}^1\sigma,{}^2\sigma) = \Phi(x,{}^1\sigma) * \hat{G}_x(x,{}^2\sigma) \tag{3}$$

Using standard calculus [9] we get

$$M(x,{}^1\sigma,{}^2\sigma) = k({}^2\sigma) \cdot G(x,\sqrt{{}^1\sigma^2 + {}^2\sigma^2}) = \frac{{}^2\sigma}{\sqrt{{}^1\sigma^2 + {}^2\sigma^2}} \exp[-\frac{x^2}{2\,({}^1\sigma^2 + {}^2\sigma^2)}] \tag{4}$$

and, at $x = 0$ (the assumed edge locus),

$$M(0,{}^1\sigma,{}^2\sigma) = \frac{{}^2\sigma}{\sqrt{{}^1\sigma^2 + {}^2\sigma^2}} \tag{5}$$

From the generalization of (1) follows the final discontinuity model (with $c = b - a$, $a < b$ and a, b: local plateaus)

$$\Phi_g(x,{}^1\sigma) = a + c\,\Phi(x,{}^1\sigma) \tag{6}$$

as well as the operator response (see (5)) at $x = 0$

$$M_g(0,{}^1\sigma,{}^2\sigma) = \frac{c\,{}^2\sigma}{\sqrt{{}^1\sigma^2 + {}^2\sigma^2}} = cM(0,{}^1\sigma,{}^2\sigma) \tag{7}$$

with the limits $\lim_{{}^2\sigma\to\infty} M_g(0,{}^1\sigma,{}^2\sigma) = c$ and $\lim_{{}^2\sigma\to 0} M_g(0,{}^1\sigma,{}^2\sigma) = 0$. For a single known scale ${}^2\sigma$ of the impulse response function of (2), we cannot determine the two unknowns ${}^1\sigma$ and c directly from $M_g(0,{}^1\sigma,{}^2\sigma)$. However, for two different scales[4] ${}^2\sigma_i$ and ${}^2\sigma_{i+1}$ we can compute

$$t = \frac{M_g(x_{max},{}^1\sigma,{}^2\sigma_i)}{M_g(x_{max},{}^1\sigma,{}^2\sigma_{i+1})} = \frac{{}^2\sigma_i\sqrt{{}^1\sigma^2 + {}^2\sigma_{i+1}^2}}{{}^2\sigma_{i+1}\sqrt{{}^1\sigma^2 + {}^2\sigma_i^2}} \tag{8}$$

and consequently it follows from (8)

$$\frac{1}{\sigma} = {}^2\sigma_i\,{}^2\sigma_{i+1}\sqrt{\frac{1 - t^2}{t^2\,{}^2\sigma_{i+1}^2 - {}^2\sigma_i^2}} \tag{9}$$

---

[3] Apart from the class of computed tomograms, we also investigate local concatenation of differently parametrized error integral curves to model arbitrary 1-D discontinuities for the general case.

[4] ${}^2\sigma_{i+1} > {}^2\sigma_i$ must hold. The local coordinate system with the origin at $x = 0$ in (7) has been translated to a locus of local maximum, $x = x_{max}$, in $M_g(\cdot)$ without loss of generality.

and from (7) (for $x = x_{max}$)

$$c = \frac{M_g(x_{max}, {}^1\sigma, {}^2\sigma_i)\sqrt{{}^1\sigma^2 + {}^2\sigma_i^2}}{{}^2\sigma_i} \tag{10}$$

However, this solution for the general continuous case of discontinuities extending over the open interval $(-\infty, \infty)$ is not valid for the real case: Image discontinuities of a spatially limited (but prior unknown) width may be bordered by further image structure (such as neighbouring discontinuities) which in turn distort the operator response if the spatial support of the operator (governed by the space constant ${}^2\sigma$) is larger than the local width (governed by the space constant ${}^1\sigma$). Consequently, only a scale space search approach allows for the selection of an optimum operator scale such that both $M_g(x_{max}, {}^1\sigma, {}^2\sigma)$ is optimal and the operator support fits the local width of the discontinuity, e.g. ${}^1\sigma = {}^2\sigma = \sigma_\Phi$ in our case.

From (7) we get the operator response for the optimum scale $\sigma_\Phi = {}^1\sigma = {}^2\sigma$

$$M_g(x_{max}, {}^1\sigma, {}^2\sigma) = \frac{c\,{}^2\sigma}{\sqrt{{}^1\sigma^2 + {}^2\sigma^2}} = \frac{c}{\sqrt{2}} \tag{11}$$

Differentiating $M_g(x_{max}, {}^1\sigma, {}^2\sigma)$ (see (7)) with respect to ${}^2\sigma$ yields

$$\frac{d\,M_g(\cdot)}{d\,{}^2\sigma} = \frac{c\,{}^1\sigma^2}{(\sqrt{{}^1\sigma^2 + {}^2\sigma^2})^3} \tag{12}$$

or, respectively, for $\sigma_\Phi = {}^1\sigma = {}^2\sigma$

$$M_g'(\cdot) = \frac{c}{2\sqrt{2}\sigma_\Phi} \tag{13}$$

Substitution of c in (13) by $c = \sqrt{2}M_g(x_{max}, {}^1\sigma, {}^2\sigma)$ (from (11)) yields

$$M_g'(\cdot) = \frac{M_g(\cdot)}{2\sigma_\Phi} \tag{14}$$

Since $M(\cdot)$ in (4) and $M_g(\cdot)$ in (7) are monotically increasing functions of ${}^2\sigma$ (the latter of which asymptotically converges to c), $M_g'(\cdot)$ is a monotically decreasing function such that $M_g''(\cdot) < 0$. Consequently, for two arbitrary different scales ${}^2\sigma_i <^2 \sigma_{i+1}$ it follows that $M_g(x_{max}, {}^1\sigma, {}^2\sigma_{i+1}) > M_g(x_{max}, {}^1\sigma, {}^2\sigma_i)$ and $M_g'(x_{max}, {}^1\sigma, {}^2\sigma_{i+1}) < M_g'(x_{max}, {}^1\sigma, {}^2\sigma_i)$ must hold. Assuming both discrete sampling of the scale space and peacewise linear interpolation we get for a particular discrete scale ${}^2\sigma_{i+1}$

$$\frac{\Delta M_g(\cdot)}{\Delta\,{}^2\sigma} = \frac{M_g(x_{max}, {}^1\sigma, {}^2\sigma_{i+1}) - M_g(x_{max}, {}^1\sigma, {}^2\sigma_i)}{{}^2\sigma_{i+1} - {}^2\sigma_i} \approx M_g'(\cdot) \tag{15}$$

Assuming $\sigma_\Phi = {}^1\sigma = {}^2\sigma_i$ for the optimum scale, we derive (from (13) and (14))

$$M_g'(\cdot) = \frac{M_g(\cdot)}{2\sigma_\Phi} = \frac{c}{2\sqrt{2}\sigma_\Phi} \tag{16}$$

Because of monotonicity we can conclude that for all non-optimum scales (Fig.2)

$$\frac{\Delta M_g(\cdot)}{\Delta\,{}^2\sigma} > M_g'(\cdot) \tag{17}$$

must hold. Inserting (15) and (16) into (17) leads (after some rewriting) to a quantitatively precise criterion for terminating scale space search at the optimum scale σ_Φ

$$\frac{{}^2\sigma_{i+1} - {}^2\sigma_i}{2\,{}^2\sigma_{i+1}} > \frac{\sqrt{2}\,M_g(x_{max}, {}^1\sigma, {}^2\sigma_{i+1}) - M_g(x_{max}, {}^1\sigma, {}^2\sigma_i)}{c} \tag{18}$$

e.g. ${}^2\sigma_{i+1} >^1\sigma$ must then hold and therefore $\sigma_\Phi = {}^2\sigma_i = {}^1\sigma$.

For the case of the optimum scale equation (10) yields

$$c = \sqrt{2}\, M_g(x_{max}, {}^1\sigma = {}^2\sigma_i, {}^2\sigma_i) \tag{19}$$

and, finally, substitution of the unknown c in (18) by (19) leads to

$$\frac{{}^2\sigma_{i+1} - {}^2\sigma_i}{2\,{}^2\sigma_{i+1}} > \frac{M_g(x_{max}, {}^1\sigma = {}^2\sigma_i, {}^2\sigma_{i+1}) - M_g(x_{max}, {}^1\sigma = {}^2\sigma_i, {}^2\sigma_i)}{M_g(x_{max}, {}^1\sigma = {}^2\sigma_i, {}^2\sigma_i)} \tag{20}$$

Consequently, the unknown 1-D discontinuity at $x = x_{max}$ has the parameters c in (19) and scale ${}^1\sigma = {}^2\sigma_i = \sigma_\Phi$. In the case of a sampled (discrete) scale space, however, the optimum scale σ_Φ may lie in the interval ${}^2\sigma_i$ and ${}^2\sigma_{i+1}$ and thus c and ${}^1\sigma = {}^2\sigma_i = \sigma_\Phi$ from above are numerical estimates only which can be further improved using (9) and (10).

Results and Research Prospects

Experiments have been carried out for both the 1-D and the 2-D case. For instance, in the case of 1-D discrete noiseless generalized error integral curves, e.g. ${}^1\sigma = 3.4$ and $c = 12$, scale-space processing terminated at scale ${}^2\sigma_{i+1} = 3.5$ with $M_g(x_{max}, {}^1\sigma, {}^2\sigma_{i+1}) = -7.9376$ and $M_g(x_{max}, {}^1\sigma, {}^2\sigma_i) = -8.6052$ such that ${}^2\sigma_i = 3.0$ had been selected as σ_Φ (with a discrete scale space sampling at ${}^2\sigma_1 = 0.5, {}^2\sigma_2 = 1.0, {}^2\sigma_3 = 1.5$, etc. and using the termination criterion in (20); relative scale-to-scale increase is 8.41% (!) in this case). The improvement of the estimated parameters of the error integral curve using (9) and (10) results in ${}^2\sigma_i = 3.292$ and $c = 11.897$.

However, the 2-D generalization of the theory for the 1-D continuous case is as usual a non-trivial problem, e.g. in the presence of arbitrarily curved smooth organ contours in noisy computed tomograms. The current experimental implementation has been restricted both to scaled support-limited operators based upon the norm $\|\cdot\|$ of $\nabla G = (\hat{G}_x, \hat{G}_y)$ using Korn's discrete normalized convolution kernels and to a coarse sampling of the scale space (${}^2\sigma_1 = 0.75, {}^2\sigma_2 = 1.0, {}^2\sigma_3 = 1.5, {}^2\sigma_4 = 2.0$, etc., see [5] for details). The major drawbacks of these simplifying assumptions are a) $\|\cdot\|$ is a non-linear operator, b) the operator response is only a local estimate of the magnitude along gradient direction thus lacking orientation specificity and localization preciseness, and c) the loci of local maxima at $(x, y, {}^2\sigma_i)$ in the scale space are not constant thus a search along the ${}^2\sigma_i$ axis must extend from $(x, y, {}^2\sigma_i)$ to a 1-D neighborhood centered at $(x, y, {}^2\sigma_{i+1})$ with an orientation estimated by $\varphi = \tan^{-1}(\hat{G}_y/\hat{G}_x)$ (as proposed in [5]). Moreover, we assumed modality-specific isotropic Gaussian noise the parameters of which have been estimated from homogeneous regions in the Houndsfield unit function. Prior to scale-to-scale search non-maximum operator responses with high statistical evidence of being caused by noise have been suppressed on the basis of a scale-dependent statistical significance level. Despite of these assumptions, the experimental implementation of the theoretical framework resulted in the detection of even subtle local contrast related to e.g. white/gray matter transitions in computed tomograms e.g. from patients to whom no contrast agent has been injected (as can be seen from the arbitrarily chosen noisy X-ray CT of the head of a corpse in Fig.3).

In addition, we investigated the problem of detecting loci of organ surface patches intersecting a computed tomogram. Detecting edges in the case of anisotropic tomograms with prevailing partial-volume effects on the basis of either gradient magnitude maxima or zero-crossings alone is simply a sub-optimal procedure with respect to the localization capability. The precise loci evidently correspond with the extrema of the second directional derivative $\frac{\partial^2 f}{\partial n^2}$ along the gradient direction and consequently partial volume ribbons along organ contours can be estimated such as to improve the localization of true organ contours induced by organ surface intersections of the computed tomogram slice. The tangent vectors at the true organ contour loci then may contribute to primitive grouping. As a consequence, our future mid-term research will concentrate on precise noise estimation and regularization, directionally sensitive scaled operators and curvature extrema localization, massively parallel grouping and junction detection, as well as image structure based 3-D organ reconstruction.

Conclusion

We presented an in-depth analysis of the complex problem of segmenting 2-D discrete noisy computed tomograms of anisotropic resolution. A theoretical framework has been proposed for scale-space edge detection based upon an explicit 1-D discontinuity model incorporating a physical smoothness constraint related to anatomical morphology. Initial experiments with 2-D X-ray computed tomograms resulted in a detection performance being clearly superior to conventional approaches using single operators with fixed local support. Open questions related to future research have been also addressed.

Literature

[1] Stiehl, H.S.; Jackél, D: On a Framework for Processing and Visualizing Spatial Images. in: Lemke, H.U.; Rhodes, M.L.; Jaffe, C.C.; Felix, R. (Eds.): Computer Assisted Radiology (CAR '87). Berlin: Springer (1987), 665-670.

[2] Back, S.; Neumann, H.; Stiehl, H.S.: On Segmenting Computed Tomograms. in: Lemke, H.U.; Rhodes, M.L.; Jaffe, C.C.; Felix, R. (Eds.): Computer Assisted Radiology (CAR '89). Berlin: Springer, 691-696.

[3] Pizer, S.M.; Gauch, J.M.; Lifshitz, L.M.: Interactive 2D and 3D Object Definition in Medical Images. Proc. SPIE Vol. 914 "Medical Images II" (1988).

[4] Korn, A.: Das visuelle System als Merkmalsfilter. in: Syrbe, M.; Thoma, M. (Eds.): Fachberichte Messen, Steuern, Regeln 13. Berlin: Springer (1985), 112-165.

[5] Korn, A.: Towards a Symbolic Representation of Intensity Changes in Images. IEEE Trans. on Pattern Analysis and Machine Intelligence 10 (5) (1988) 610-625.

[6] Marr, D.: Early Processing of Visual Information. Phil. Trans. of the Royal Society of London 276 (Series B) (1976) 483-519.

[7] Canny, J.F.: A Computational Approach to Edge Detection. IEEE Trans. on Pattern Analysis and Machine Intelligence 8 (6) (1986) 679-698.

[8] Nalwa, V.S.; Binford, T.O.: On Detecting Edges. IEEE Trans. on Pattern Analysis and Machine Intelligence 8 (6) (1986) 699-714.

[9] Bracewell, R.N.: The Fourier Transform and Its Applications. McGraw Hill (1978).

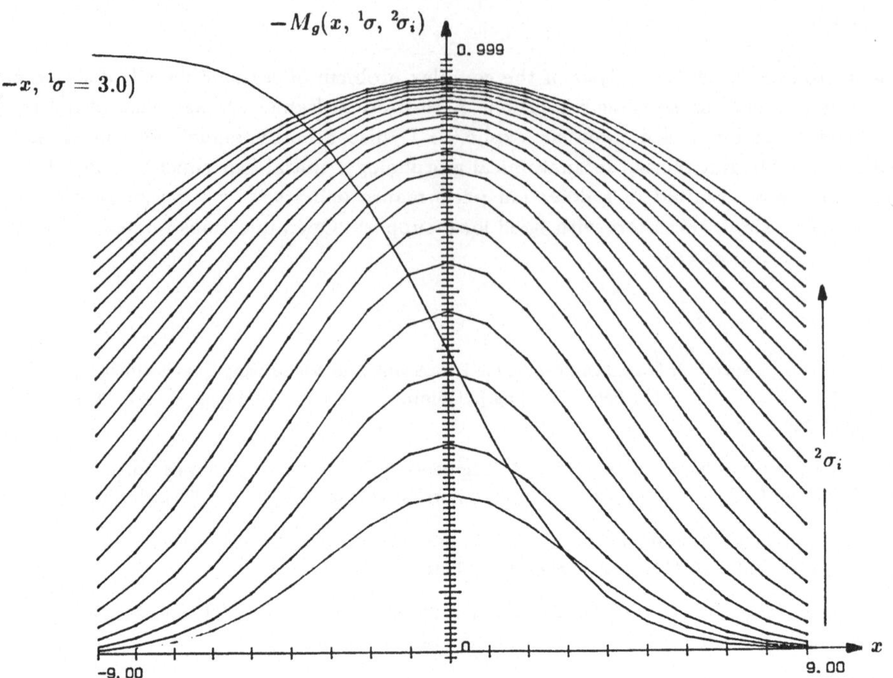

Fig.1: Operator response $M_g(\cdot)$ for normalized error integral curve $\Phi(-x,\ {}^1\sigma = 3.0)$
$({}^2\sigma_1 = 0.5, {}^2\sigma_2 = 1.0, ..., {}^2\sigma_{20} = 10.0)$

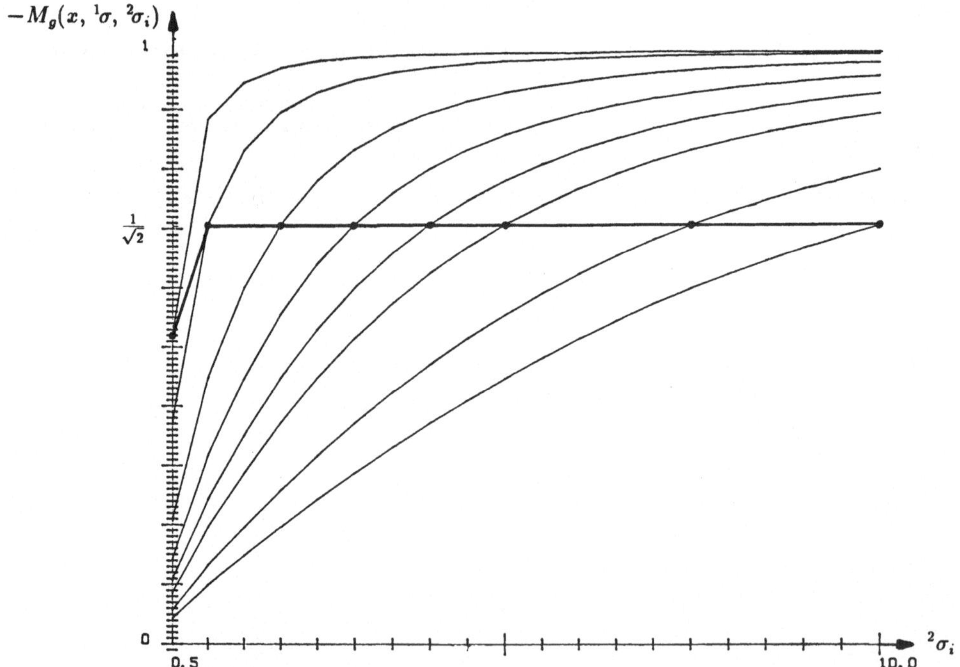

Fig.2: Operator responses $M_g(\cdot)$ for different normalized error integral curves $\Phi(\cdot)$
$({}^1\sigma = 0.5, ..., 10.0)$ in scale space $({}^2\sigma_1 = 0.5, ..., {}^2\sigma_{20} = 10.0)$; Note: Bold type
curve indicates optimum scale ${}^1\sigma = {}^2\sigma_i$

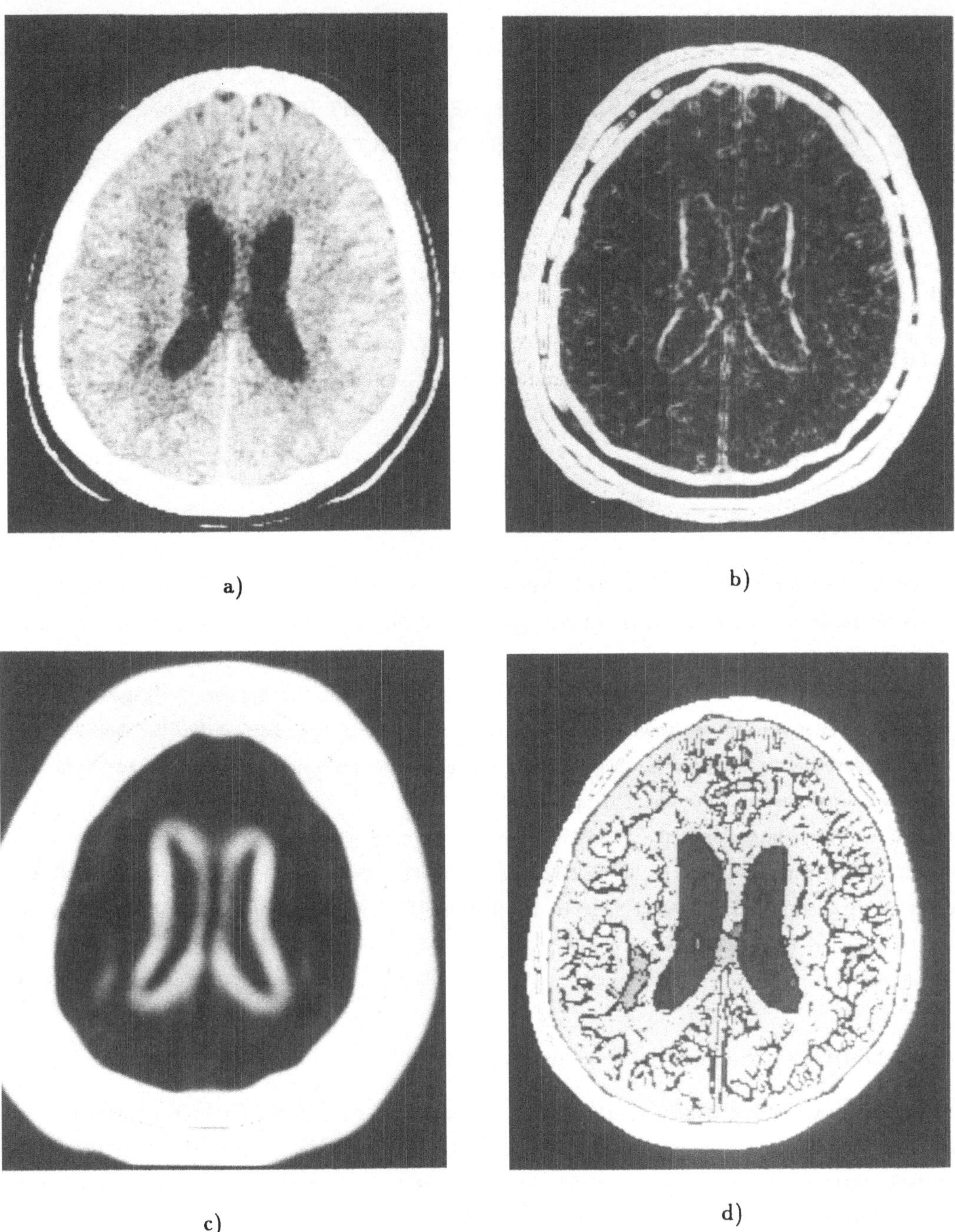

Fig.3: a) X-ray cranial computed tomogram of corpse (256^2 matrix, 8mm slice thickness)
     b) Operator response in finest scale $^2\sigma = 0.75$ (prior to non-maximum suppression)
     c) Operator response in coarse scale $^2\sigma = 4.0$ (prior to non-maximum suppression)
     d) Local evidence for contrast discontinuities (accumulated through scale-space) prior to grouping

ERKENNUNG VON BILDMUSTERN MIT HILFE VON INVARIANTEN MOMENTEN

Norbert Harendt[1], Werner Döler[1], Andreas Jäger[2]

[1] *Institut für Medizinische Physik und Biophysik,*

[2] *Abt. Kieferorthopädie des Zentrums ZMK,*

Georg-August-Universität Göttingen

1. Einleitung

In der Röntgendiagnostik tritt häufig die Fragestellung auf, anhand einer im Laufe eines Beobachtungszeitraumes entstandenen Röntgenbildserie zeitabhängige Veränderungen bestimmter anatomischer Strukturen zu erkennen. Aufgrund der hohen optischen Dichte der zu vergleichenden Röntgenbilder führt eine direkte Überlagerung der Filme auf einem Leuchtkasten oft zu unbefriedigenden Ergebnissen. Die gegenseitige Zuordnung der Bilder anhand von stabilen anatomischen Strukturen stellt ein zusätzliches Problem dar. Ein Anwendungsbeispiel ist die Überlagerung von Fernröntgenseitenbildern in der Kieferorthopädie zum Studium von Wachstumsvorgängen und Therapieverläufen.

Zur Problemlösung lassen sich nach Digitalisierung der zu vergleichenden Bilder Methoden der digitalen Bildverarbeitung anwenden. Prinzipiell können die Bilder anhand von Referenzstrukturen (z.B. der Schädelbasis) durch Translation und Rotation im Bildspeicher zur Deckung gebracht werden, um so die Strukturveränderungen in Relation zu den Referenzstrukturen quantitativ zu erfassen.

2. Template Matching mit Invarianten Momenten

Liegt ein Muster in zwei Bildern verschoben und rotiert vor, so lassen sich die bestehenden Standardverfahren zur Lokalisation des Musters [1] nicht anwenden. Um den Lokalisationsprozeß des template matching invariant gegenüber Rotation des zu lokalisierenden Musters zu gestalten, empfiehlt es sich, kreisförmige templates zu verwenden und aus diesen einen rotationsinvarianten Merkmalsvektor zu extrahieren.

Als Merkmalsvektor einer Grauwerteverteilung $g(i,j)$ bieten sich die in [2] angegebenen Invarianten φ_i an :

$$\varphi_1 = \eta_{20} + \eta_{02} \qquad\qquad \varphi_3 = (\eta_{30} - 3\eta_{12})^2 + (3\eta_{21} - \eta_{03})^2$$

$$\varphi_2 = (\eta_{20} - \eta_{02})^2 + 4\eta_{11}^2 \qquad\qquad \varphi_4 = (\eta_{30} - \eta_{12})^2 + (\eta_{21} + \eta_{03})^2$$

$$\varphi_6 = (\eta_{20} - \eta_{02})\left[(\eta_{30} + \eta_{12})^2 - (\eta_{21} + \eta_{03})^2\right] + 4\eta_{11}(\eta_{30} + \eta_{12})(\eta_{21} + \eta_{03})$$

$$\varphi_5 = (\eta_{30} - 3\eta_{12})(\eta_{30} + \eta_{12})\left[(\eta_{30} + \eta_{12})^2 - 3(\eta_{21} + \eta_{03})^2\right]$$
$$+ (3\eta_{21} - \eta_{03})(\eta_{21} + \eta_{03})\left[3(\eta_{30} + \eta_{12})^2 - (\eta_{21} + \eta_{03})^2\right]$$
$$\varphi_7 = (3\eta_{21} - \eta_{03})(\eta_{30} + \eta_{12})\left[(\eta_{30} + \eta_{12})^2 - 3(\eta_{21} + \eta_{03})^2\right]$$
$$- (\eta_{30} - 3\eta_{12})(\eta_{21} + \eta_{03})\left[3(\eta_{30} + \eta_{12})^2 - (\eta_{21} + \eta_{03})^2\right]$$

$$(1)$$

die sowohl gegenüber Rotation als auch (bis auf φ_7) gegenüber Spiegelung invariant sind. Dabei sind die η_{pq} aus den Zentralmomenten μ_{pq} abgeleitet :

$$\mu_{pq} = \sum_i \sum_j (i - i_0)^p (j - j_0)^q g(i,j) \quad , \quad i_0 = \frac{\sum_i \sum_j i\, g(i,j)}{\sum_i \sum_j g(i,j)} \quad , \quad j_0 = \frac{\sum_i \sum_j j\, g(i,j)}{\sum_i \sum_j g(i,j)} \tag{2}$$

Für η_{pq} setzt man entweder $\eta_{pq} = \mu_{pq}$ oder

$$\eta_{pq} = \frac{\mu_{pq}}{\mu_{00}^\gamma} \quad , \quad \gamma = \frac{p+q+2}{2} \tag{3}$$

Bei Verwendung der durch (3) definierten η_{pq} werden die Invarianten φ_i zusätzlich invariant gegenüber Skalenänderung. Da nur 6 der 7 Invarianten voneinander unabhängig sind, und in Röntgenbildern auftretende gespiegelte Muster meist anderen Strukturen angehören, besteht der hier verwendete Merkmalsvektor aus :
$$\vec{\varphi} = [\varphi_1, \varphi_2, \varphi_3, \varphi_4, \varphi_6, \varphi_7]^t.$$

3. Normierung

Als Maße für die Ähnlichkeit zweier Merkmalsvektoren bieten sich die Summe der absoluten Differenzen (SAD) und der Korrelationskoeffizient (KK) an :

$$SAD = \sum_i |\varphi_i(i_0, j_0) - \varphi_i(i_1, j_1)| \tag{4}$$

$$KK = \frac{\sum_i \varphi_i(i_0, j_0) \cdot \varphi_i(i_1, j_1)}{\sum_i \varphi_i^2(i_0, j_0) \cdot \sum_i \varphi_i^2(i_1, j_1)} \tag{5}$$

Die Verwendung dieser Ähnlichkeitsmaße erfordert, daß die Werte der Invarianten von gleicher Größenordnung sind. Zur Normierung sind bereits Verfahren vorgeschlagen worden [3], [4], die jedoch die Verteilung der Werte der Invarianten über ein Bild und die großen Unterschiede der Größenordnungen, die die Wertebereiche der einzelnen Invarianten überstreichen, nicht berücksichtigen. Deshalb wird folgendes Verfahren zur Normierung vorgeschlagen :
Unter der Annahme, daß die Zentralmomente 2. und 3. Ordnung von gleicher Größenordnung sind, lassen sich die Größenordnungen der Invarianten als Potenzen der Zentralmomente abschätzen, so daß man durch die Wurzeltransformation

$$\phi_i = \sqrt[n_i]{\varphi_i} \quad , \tag{6}$$

$$n_1 = 1 \quad , \quad n_2 = 2 \quad , \quad n_3 = 2 \quad , \quad n_4 = 2 \quad , \quad n_5 = 4 \quad , \quad n_6 = 3 \quad , \quad n_7 = 4 \quad ,$$

Terme von gleicher Größenordnung erhält. Zur Überprüfung dieser Aussage wurde eine Abschätzung der Verteilungsdichte der Werte der Invarianten vorgenommen. Abb. 1 zeigt exemplarisch die Verteilungsdichte im gesamten Wertebereich von 2 Invarianten ohne und mit Normierung.

Weiterhin wirkt sich diese Transformation der Verteilungsfunktion günstig auf die Lokalisation von Regionen unter nicht idealen Bedingungen aus. Möchte man den noch bestehenden kleinen Unterschied der Größenordnungen vollends eliminieren, so läßt sich das durch eine Merkmaltransformation

$$S_i = \frac{1}{\sigma_i}(\phi_i - \overline{\phi}_i) \tag{7}$$

erreichen; dabei ist $\overline{\phi}_i$ der Mittelwert und σ_i die Standardabweichung der Verteilung der i-ten Invarianten ϕ_i.

4. Ergebnisse

Zur Abschätzung der Verteilungsdichte des Wertebereiches der Invarianten wurden die Invarianten für ein Bild innerhalb eines Kreises mit $r = 16$ Pixel berechnet und die Histogramme ermittelt (siehe Abb.1).

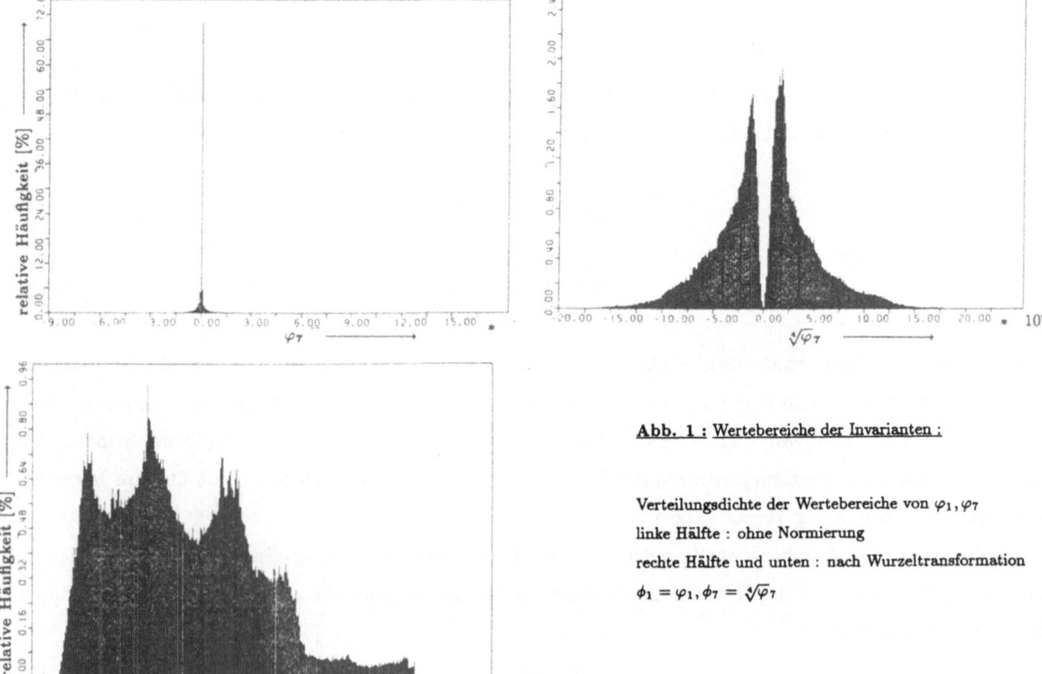

Abb. 1 : <u>Wertebereiche der Invarianten :</u>

Verteilungsdichte der Wertebereiche von φ_1, φ_7
linke Hälfte : ohne Normierung
rechte Hälfte und unten : nach Wurzeltransformation
$\phi_1 = \varphi_1, \phi_7 = \sqrt[3]{\varphi_7}$

Zur Untersuchung der Rauschanfälligkeit des Algorithmus wurden folgende Simulationen vorgenommen :
Eine kreisförmige Region wurde vorgewählt, die Standardabweichung σ_0 ihrer Grauwerte bestimmt und in
100 Versuchen mit gleichförmigem Rauschen der Standardabweichung σ_R verrauscht. Es wurde versucht,
dieses Template im Originalbild unter Verwendung der Ähnlichkeitsmaße (4),(5) zu lokalisieren, wobei der
Merkmalsvektor nach dem in [3] und durch die in Gleichung (6) und (7) angegebenen Verfahren normiert
worden war.

Die Anzahl der Fehllokalisationen betrug bei den hier vorgestellten Verfahren bei $\sigma_R = \sigma_0/4$ durchschnittlich
3%, bei $\sigma_R = \sigma_0/2$ durchschnittlich 25%, bei dem in [3] vorgestellten Verfahren 44% bzw. 67%.

Zur Demonstration der Leistungsfähigkeit des Algorithmus wurde das Testbild um 10° gedreht, die Achsen wurden mit dem Faktor 1.3 gestreckt, und das gesamte Bild wurde mit gleichförmigem Rauschen der
Standardabweichung $\sigma_R = \sigma/4$ belegt. Zur Überlagerung der Bilder wurden 6 Punkte vorgewählt und im
Originalbild automatisch lokalisiert (siehe Abb. 2). Die Menge der möglichen Punkte wurde dabei nicht
auf ein Teilgebiet des Bildes eingeschränkt. Das Differenzbild zeigt, daß mit dem verwendeten Verfahren
eine optimale Bildüberlagerung erzielt werden kann. Die Kantenanhebung in den Randbereichen kommt
teilweise durch die Rundung auf ganzzahlige Pixelwerte bei der Rotation der Bildmatrizen zustande.

Schlußfolgerungen

In der Theorie stellen Momente unter rauschfreien Bedingungen sehr gute Merkmale dar. Hier wurde
gezeigt, daß bei geeigneter Normierung Regionen trotz leichten Rauschens wiedergefunden werden können.
Bei der praktischen Anwendung muß jedoch große Sorgfalt auf die Vorverarbeitung gelegt werden, insbesondere hinsichtlich der Rauschbefreiung und Angleichung der Grauwertskalen der zu vergleichenden Bilder.

5. Literatur

[1] D. I. BARNEA UND H. F. SILVERMANN, *A class of algorithms for fast digital image registration,* IEEE
Trans. Comput., 1972, C-21, *179 - 186*

[2] M.-K. HU, *Visual pattern recognition by moment invariants,* IRE Trans. Inform. Theory, 1962, *179 -
187*

[3] A. GOSHTASBY, *Template matching in rotated images,* IEEE Trans. Pattern Anal. Machine Intell., 1985,
PAMI-4, *338 - 344*

[4] R. Y. WONG UND E. L. HALL, *Scene matching with invariant moments,* Comput. Graphics Image
Processing, 1978, vol. 8, *16 - 24*

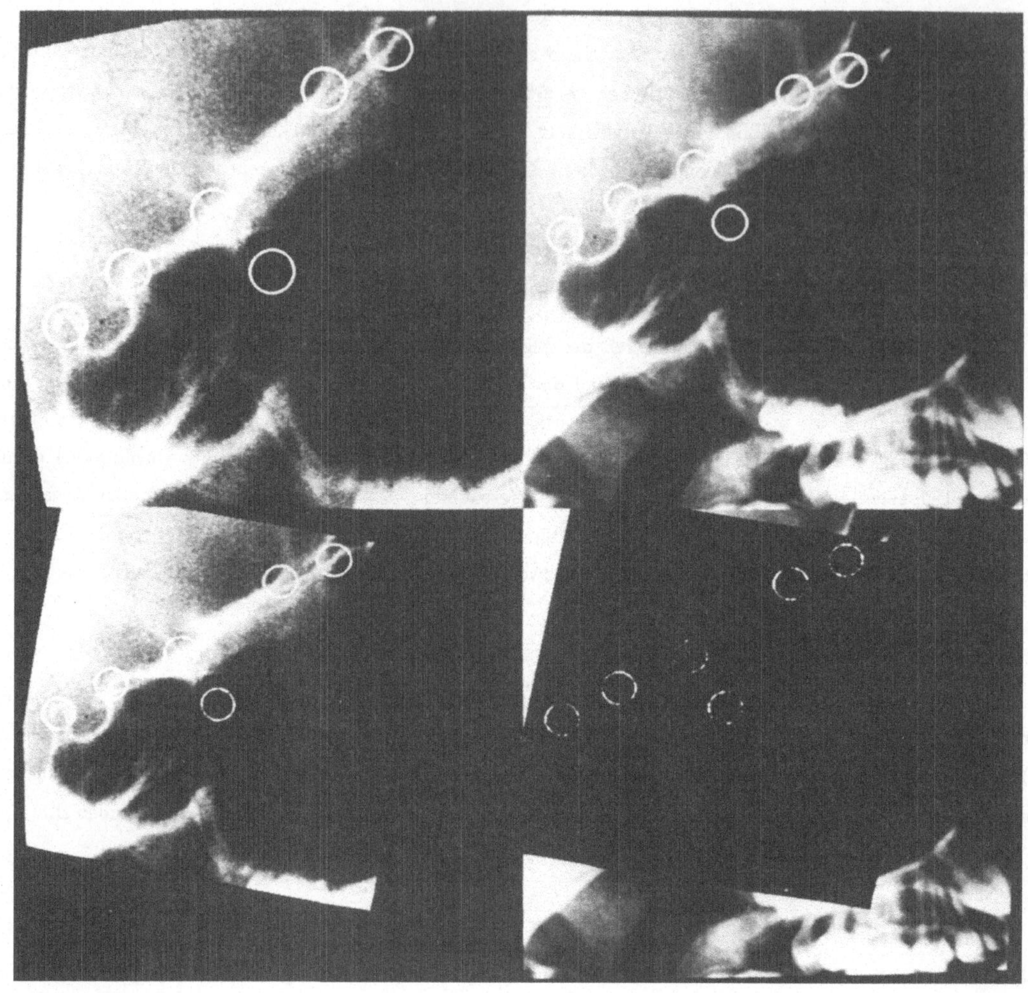

Die Kreise im Bild links oben markieren die ausgewählten Regionen, die im Bild rechts oben die durch den Algorithmus automatisch gefundenen Regionen.

Das Bild links oben stellt einen verdrehten, gedehnten und verrauschten Ausschnitt des Bildes rechts oben dar. Das Bild links unten zeigt das Ergebnis der Überlagerungsroutine, die anhand der ermittelten 6 Punktepaare die optimale affine Abbildung ermittelt. Im Bild rechts unten ist schließlich der Absolutbetrag der Differenz der überlagerten Bilder zu sehen.

MULTISCALE, GEOMETRIC IMAGE DESCRIPTIONS
FOR INTERACTIVE OBJECT DEFINITION

Stephen M. Pizer, John M. Gauch[*], James M. Coggins,
Timothy J. Cullip, Robin E. Fredericksen, Victoria L. Interrante

Medical Image Display Research Group
University of North Carolina
Chapel Hill, NC 27599-3175, USA

ABSTRACT

A means is described of analyzing two- and three-dimensional images into a directed acyclic graph of visually sensible, coherent regions and of using this DAG as the basis for interactive object definition. The image analysis is in terms of the geometry of the intensity surface via a multiscale approach with a focus on symmetry properties about ridges. The image analysis method, a system for interactive object definition, and results of their use on two-dimensional images are reported.

1. INTRODUCTION

The definition of objects in images can lead to measurements of the objects, display of the objects, or actions based on their properties. Much work in computer vision has taken as its objective the automatic definition of these objects from image data. It seems to us that in an interactive environment a more promising approach is to have the computer analysis provide a framework in which a human who understands the semantics of the scene can quickly define the necessary objects using the image data.

We suggest that without any scene semantics it is possible by automatic computation to describe an image as a related collection of visually sensible, coherent image regions (e.g., see figure 1b) that can provide the basis for quick object definition by the interacting human. More precisely, a method is set forth for automatically computing a directed acyclic graph (DAG) of such image regions with arcs specifying region containment. We demonstrate an interactive system that allows the user, with his or her knowledge of scene semantics, to use this description in the definition of regions containing semantically relevant objects, by interactive selection of DAG regions and possibly some modification of the DAG and *in extremis* the regions defined in its nodes.

Our objective is to demonstrate the power of this approach and the possibility of automatically computing visually sensible, coherent regions without the use of semantics, in particular by a geometric focus on the image as an intensity surface. On the way to this demonstration a number of choices within the analysis approach had to be made. These included the particular axis of symmetry and the means of varying scale. In most cases the

---

[*] Presently at College of Computer Science, Northeastern University

229

alternative chosen was the one which most simplified the initial evaluation. In the following we report encouraging results concerning the power of the overall approach, along with indications of the need to explore some alternatives better matched to the human visual system's function. The paper closes with a discussion of the alternatives needing exploration.

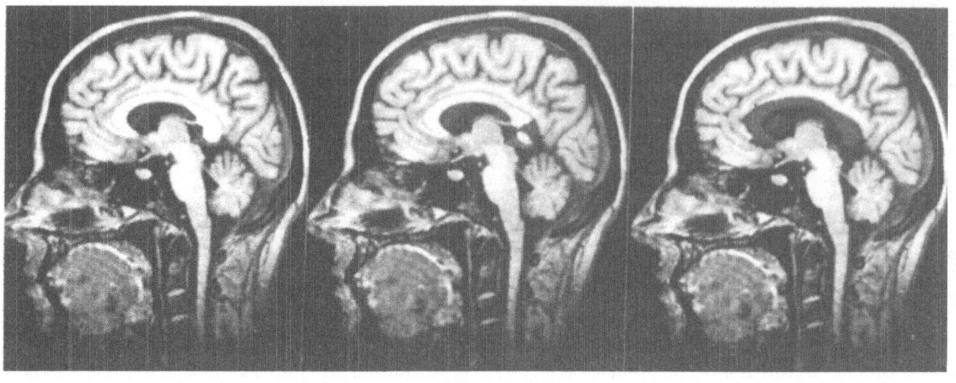

a   b   c

Figure 1. Visually sensible, coherent image regions: a) the MRI head image used as an example throughout this paper; b) a primitive region;   c) a meaningful region.

Our area of application of this work is 3D medical image display. In this area actually defining the part of space making up a meaningful object, or equivalently the surface which encloses that part of space, provides the surface locations and orientations that are the basis for shaded display techniques, whether by surface or volume rendering [Fuchs, 1988; Levoy, 1988].   But once a region has been found that contains the object and parts of its background at contrasting intensities, other more local analysis is available for determining these surface locations and orientations. Therefore, it is also of use simply to define such regions of object containment (*object surrounds* - see figure 1c). In this paper we focus on methods that will lead to such object surrounds while also describing how the same methods can produce the object surfaces themselves.

2. IMAGE DESCRIPTION

We take our goal to be the description of a scalar image in terms of light objects on darker backgrounds. All of the following can be directly transformed for dark objects on lighter backgrounds. The combination of these two analyses is also of interest but is beyond the scope of this paper.

We aim to find image regions lighter than their background that match reasonably well what humans would choose as sensible image regions if they had no knowledge of the semantics of the image scene. We suggest that if the image is viewed as an intensity surface (see figure 2), where height corresponds to image intensity, then shape properties of this surface will determine the region definitions. The need for these regions to be coherent and sensible implies that these shape properties be global properties of the intensities

230

within and around them and not just a collection of local properties such as intensity
gradient. That is, the properties will be region-based rather than edge-based. In particular,
humans seem to use ridges of intensities as organizing features. Yet, in determining these
ridges they are somehow insensitive to rotations of the co-ordinate system and to scaling or
monotonic transformations of intensity. (Ideally they should also be insensitive to the
context sensitive transformations of intensity effected by the human visual system.) The
analysis that we propose must have these characteristics.

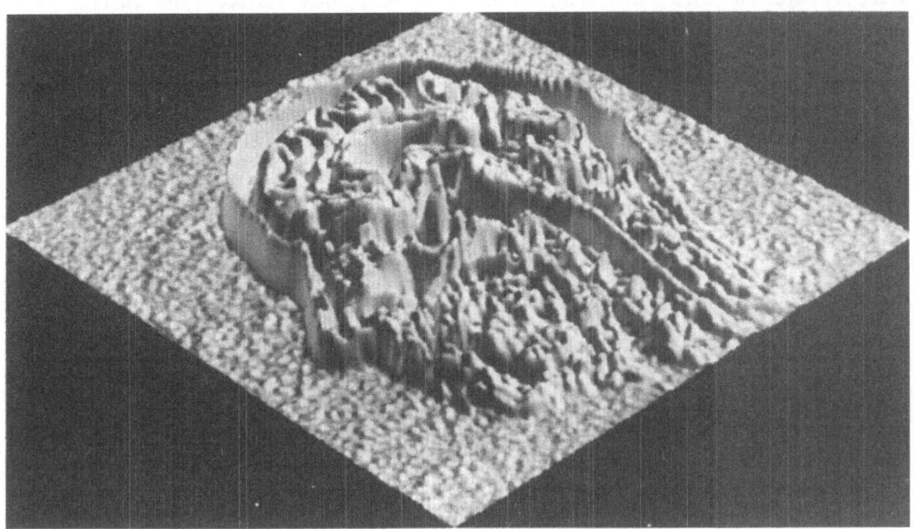

Figure 2. The MRI image of the brain of figure 1, smoothed and viewed as an intensity
surface

We suggest that another basic visual property is symmetry -- regions are two sided, so that
for each point on one flank of a ridge there is another point on the other flank to which it
has some sort of symmetry relation. As a result we define an intensity axis of symmetry
(IAS). This axis, made of a forest of branching sheets, fits under the ridges midway between
the two symmetric flanks of the ridge (we use the medial axis, though other axes of
symmetry are also possible). In order to avoid sensitivity to any monotonic transformation
of intensity and also to avoid the incommensurability of intensity and space, the IAS cannot
be defined by a symmetry operation in more than the spatial dimensions. Instead, the image
(intensity surface) must be considered as a one-parameter family of slices in the intensity
dimension.   That is, the IAS is the one-parameter family of axes of symmetry of the
intersections of the intensity surface with a series of slicing surfaces (see figure 3).
Temporarily we have been slicing at isointensity levels, so that each axis of symmetry in
the family is the medial axis of a level curve of intensity, even though this slicing focuses
too greatly on intensity levels and too little on local image structure. The present IAS has
the advantage that there is a 1-1 relation between each branch and a ridge top as defined by
the locus of maxima of positive curvature of intensity level curves (these loci are called
vertex curves) [Gauch, 1988b].

As illustrated in figure 3, associated with each point on the IAS are two points in the image where the maximal disk centered at the IAS point is tangent to a level curve at its intensity. The basic sheets that are the twigs in the IAS forest thus have associated with them a set of image points that are taken as the primitive regions of the image.

The computation of the IAS uses the fact that the medial axis of a boundary is the locus of maximum distance from that boundary. The IAS is computed by collapsing the image's intensity surface toward the locus of maximum distance from that surface. The surface is allowed to move in the spatial direction but held fixed in intensity as it is collapsed toward the IAS.

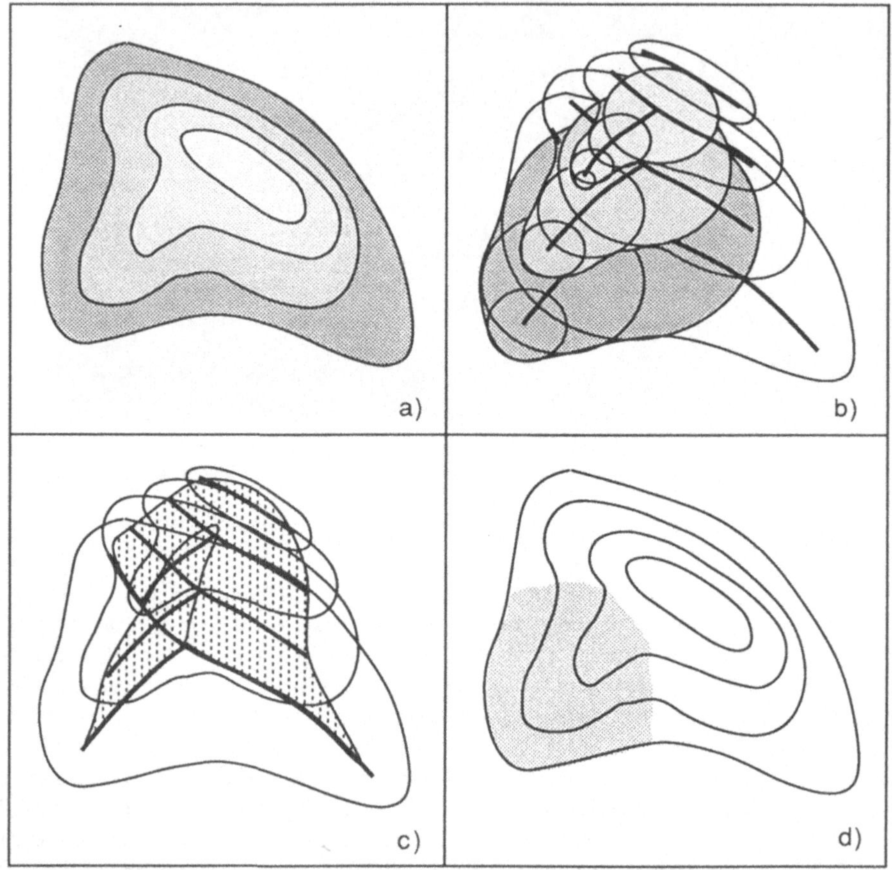

Figure 3. The IAS as a one parameter family of slice axes of symmetry. a) Level curves of a simple image; b) Level curves on intensity surface, their medial axes, and selected maximal disks; c) Level curves on intensity surface, and IAS; d) Image region associated with an IAS branch

In order to maintain surface smoothness while migrating the surface toward the IAS, we extend the concept of active contours [Kass, 1987] to handle surfaces (producing a method of *deformable surfaces*). By relaxation a minimization is obtained of the sum over surface points of a combination of the magnitudes of first and second derivatives of the deforming surface and the negative of the distance the point has moved away from the isointensity contour on which it started. The derivative constraints maintain the integrity of the surface, while the distance function drives the surface toward the IAS.

Once the surface has been collapsed to the IAS, we have a mapping from the original surface to the IAS. We next determine for each point in the image its involute, the point on the other side of the IAS that mapped to the same point on the IAS. These involute pairs are then used in a marching fashion to determine the individual branches of the IAS, and thence the region of the image associated with each branch [Bookstein, 1979]. To compute a branch, the algorithm starts with an arbitrary point on that branch, finds its involute, and then adds neighboring points if their involutes are proper neighbors of the involute of the original point. This operation continues until the current involute is not a neighbor of the previous involute (this occurs at IAS branch points). The result of this is a segmentation of the image that gives one region per branch of the IAS.

For a 256 x 256 image on a 10 MIPS, 5 MFLOPS computer (SUN 4), computing the distance function requires 5 minutes, iterative collapse of the surface requires 20 minutes, finding involutes requires 2 minutes, and identifying branches and regions requires 10 minutes.

Since surround regions for meaningful objects are frequently formed from many of the primitive regions associated with the basic IAS branches, these regions need to be grouped into larger visually sensible, coherent regions. Our strategy is to compute the region containment relations induced by the connectivity of the IAS branches and the annihilation of one branch into another as scale is increased. Since a single IAS branch can attach to more than one other branch (IAS sheets can form a loop around a pit in the intensity surface [Gauch, 1988a]), a single branch can have more than one parent, so a DAG is computed, where the parent-child relation is region containment.

Scale increase is performed by Gaussian blurring, since this form of scale increase most guarantees against the creation of structure [Yuille, 1983]. For simplicity we now use isotropic and stationary Gaussian blurring, but ultimately we expect to vary scale with a blurring kernel that is sensitive to local image structure in both orientation and size. To follow the IAS branches in scale space requires computing the IAS simultaneously at all scales. To accomplish this, we have begun to develop a method of deformable surfaces that applies to surfaces in physical space, intensity, and scale. However, the DAGs that we have computed to date are based on the more computationally efficient approximation of the vertex curves at IAS tops by watershed boundaries and of the IAS branch containment relations by those induced by watershed combination as critical points merge under Gaussian blurring.

All of the above applies not only in two spatial dimensions but also in 3D. To date we have implemented the calculations fully only in 2D. Sample results are given in figure 4.

To this point in the discussion object edges have not been relevant. The bright regions defined by IAS branches are delimited by a locus of a sort of curvature minima of the intensity surface, i.e., by the adjacent bright regions. As such they go past the steepest part of the ridge sides that might be taken to define the object edge -- for regions corresponding to objects they include part of the object surround. The IAS provides the basis for also defining an edge strength that reflects symmetry properties and for computing that edge as a continuous closed contour (in 2D) or surface (in 3D). Instead of the normal edge strength that is based on intensity steepness as the spatial arguments vary, we define the symmetry-dependent edge strength to be the steepness in intensity as the radius of the maximal disk touching the image point varies most strongly, i.e. the component of ∇I in the ∇r direction: $\nabla I \cdot \nabla r / |\nabla r|$. This is a sort of generalization to the IAS of Blum's object angle of the medial axis [Blum, 1978]. The two sides of such a symmetry-dependent edge can be computed together, and the edge is easily followed by following along the associated IAS sheet.

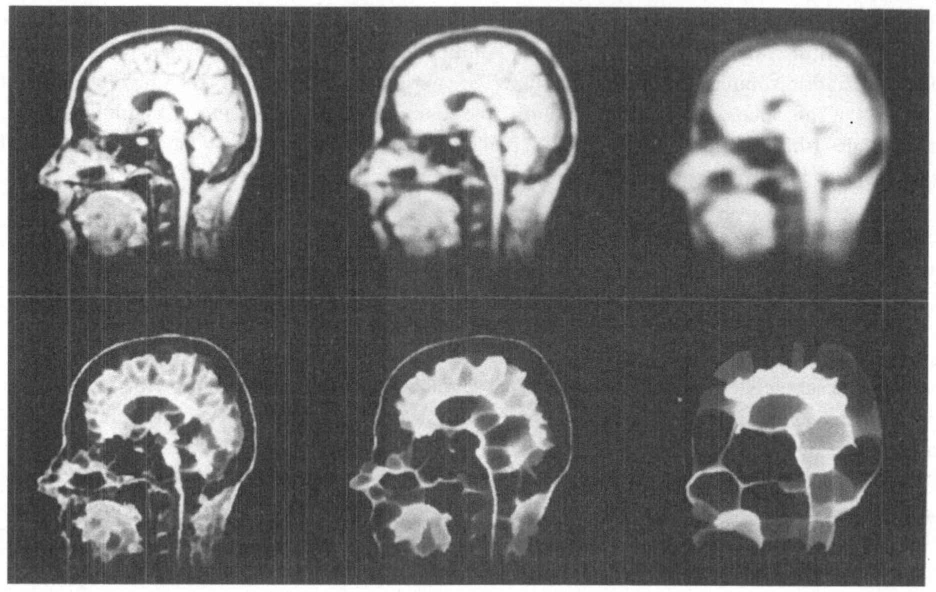

Figure 4. Multiscale sequence of 2D images and their IAS's, based on a 2D MRI head image

3. INTERACTIVE OBJECT-REGION DEFINITION

Automatic image description methods such as the multiscale IAS describe the structure of the image as a DAG of nested regions. Our approach is designed to let the human user label object surrounds of interest by creating regions based on traversals of the region hierarchy. Toward this end, we have developed a tool, the Image Hierarchy Editor (IHE), that we will describe in this section.

To define an object using IHE, the user typically begins by selecting a color for the region to be defined and then selecting a pixel in the desired region. IHE then identifies from the DAG the primitive region containing that pixel and redisplays the pixels in the region using the selected color. Usually, the primitive region is a small component of the object of interest. The user may click on a button to enlarge the labeled region by traversing up the region DAG and labeling the region corresponding to the parent (containing) node. This process may continue until the region labeled matches the desired object. At this point, the user may label this node of the hierarchy as being a "brain stem" or "cerebellum" or any other meaningful object. This region may be retrieved by name at any point in the future.

Alternatively, one can label portions of the object as separate regions and combine them by logical or set operations. While IHE provides all of the logical functions, we have found set union (logical A or B) and set difference (logical A or NOT B) to be most helpful. When a region of interest is created that does not already exist in the hierarchy, the act of labeling the region causes it to be "posted" to the region hierarchy. At present, a posted region is added to the hierarchy as a child of the root node.

The representation of the region DAG in IHE permits traversals from a pixel to its containing region and from a node in the hierarchy to all pixels contained in the node's corresponding region. Manipulation of the image using the DAG is supported by a set of region masks, that allow manual specification of region pixels, color labeling of regions defined in the DAG, and the performance of logical operations between regions.

We have developed a display protocol that permits assignment of a color label at each location while still portraying, at lower contrast, the original gray scale structure of the labeled region. Each region mask is associated with a color. In case a pixel is marked in more than one region mask, a priority ordering is imposed on the region masks; the pixel is displayed in the color corresponding to the highest-priority region mask. An enable byte permits the user to turn off region masks if desired in order to see the lower-priority colors.

We found in testing IHE that the user needed some additional information to determine whether to enlarge the current region by moving to the parent node in the hierarchy or to enlarge it by growing a new region and using logical OR operations to combine the separate regions into one object. We provide this information in a second display in which the region corresponding to the parent node is labeled in a different color. Thus, the user can see the effect of performing the "parent" operation before requesting it. This capability has proven useful since the alternative involves backtracking when the "parent" operation deviates from the desired object - a more frustrating and complex sequence of operations. This presentation is illustrated in figure 5. In the right part of the screen the image, with the selected region colored (shaded, in this grey-scale photograph), appears. In the left part of the screen the selected region and its parent are shaded with two different colors (that cannot be distinguished in grey scale).

The amount of work required to construct a region depends on the quality of the image hierarchy produced by the image description tools. We hope to minimize that work by creating more sophisticated image descriptions. In the meantime the work required to define objects of interest serves as a benchmark for how good the image descriptions are. An example of regions selectable by these operations are given in figure 6. The tongue region selected in figure 6a required 4 parent operations. The brain cortex region selected in figure 6b required the union of 18 nodes, each requiring 2-4 parent operations. The brain stem region selected in figure 6c required 5 parent operations followed by 4 difference operations.

The current implementation of IHE works only with the primitive regions defined by the image description method. Interactive pixel editing is possible within the framework of IHE

but is not implemented.   Pixel editing would complicate the system since it could modify leaf
node   structures.

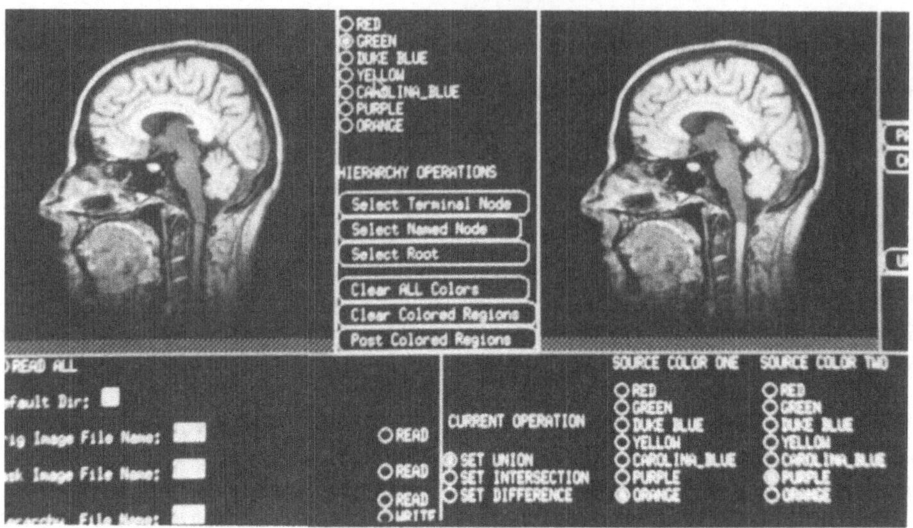

Figure 5. IHE interactive display: region selected by parent relations (darkened, top portion
of brain stem in image on right) from 2D MRI head image and same including the parent of
the selected region (bottom portion of brain stem in image on left)

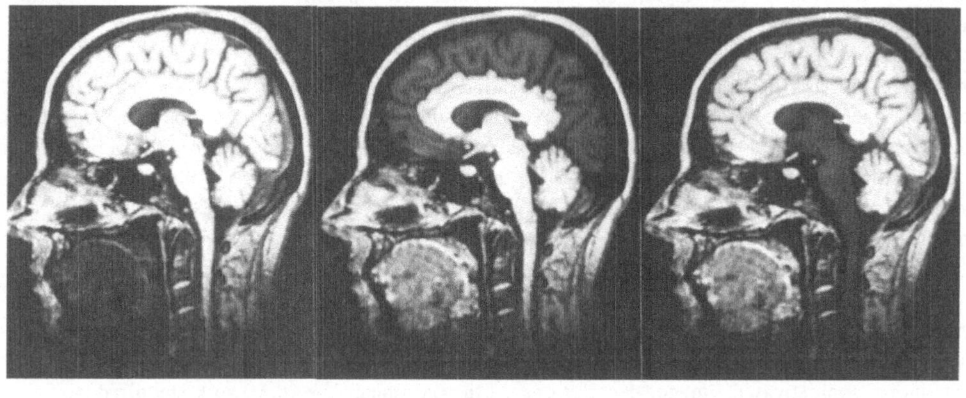

a   b   c

Figure 6. Regions selected by IHE on 2D MRI head image: a) by parent relations alone; b)
using unions of regions selected by parent relations; c) using differences among a region
selected by parent relations and primitive region

4. EVALUATION

Our results to date suggest that the primitive regions, in both 2D and 3D, based on IAS
branches are sensible to humans and can provide the basis for the interactive region

definitions needed for display and analysis. The containing super-regions computed to date, while frequently satisfactory, fail too frequently to accord with human choices.

Despite some weaknesses in the image description DAG, our early experience with the interactive approach to defining meaningful object regions is very positive. The image hierarchy editor and its user interface provide a quick way to define object regions in medical images with only a reasonably good image description. This experience is limited to a few tens of images.

Our experience with symmetry-dependent object edges is even more limited. An example of the edge strengths computed using this measure is shown in figure 7. While this result is encouraging, it has weaknesses related to inaccuracies in the IAS on which it is based.

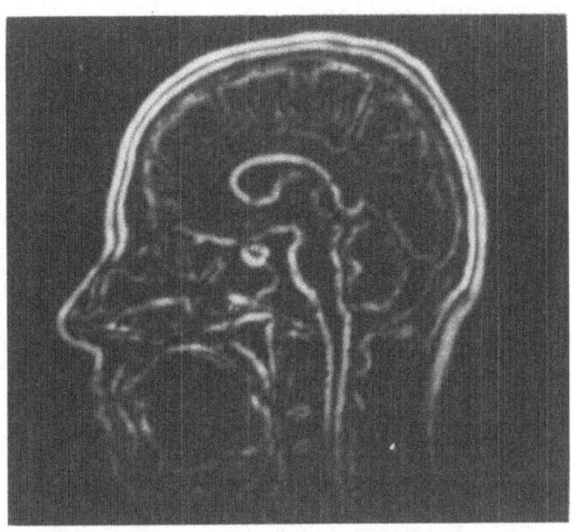

Figure 7. Symmetry-dependent edge strength for 2D MRI head image. The edge strengths shown correspond only to intensity ridges

5. SUMMARY AND FUTURE DIRECTIONS

We have discussed an approach in which human users can easily define objects and object surrounds using an interactive tool with an image description based on ridge symmetries and connectivities in the intensity surface of the image. Our initial results suggest that the idea of interactive object definition is a powerful one and that it is possible, using these geometric analyses, automatically to compute primitive image regions on which such object definitions can be based. Even the DAG based on the watershed approximation to ridge connectivities and annihilations under isotropic, stationary Gaussian blurring seems frequently to produce containment relationships that are useful in combination with the interactive object definition tools. Thus the early results seem to confirm the basic thrust of our approach. However, the methods that we have developed have a number of weaknesses, and these will require some modification in some of the structures and computing methods within our approach.

The Image Hierarchy Editor needs to be improved so that new regions defined by the user can be sensibly placed in the DAG of regions.

Other alternatives to the structure of the IAS must be considered. First, we expect that the calculation of the region parent relations using the actual IAS connectivity and annihilations, rather than the approximation based on watersheds, will give better axes. Second, some redefinition of the IAS should be found to produce a connected DAG of regions, since in a greyscale image the human viewer perceives luminance-based regions with a DAG structure. In contrast, the presently defined IAS forms a forest of disconnected components, sometimes broken at ridge ends. In fact, small scale bumps in the image may be lost in our deformable surfaces approach to computing the IAS precisely because the connectivity enforced by the computing method is not a property of the true IAS. Part of the problem will be solved with a combination of the analysis of light-on-dark with dark-on-light patches, but the solution seems also to be related to the idea that the slicing in the intensity dimension on which our method is based should be determined by the local intensity context, perhaps by the local symmetry information. We expect to investigate transformations of intensity according to local symmetries.

A related issue is the inadequacy of a form of scale change that is independent of local image structure. Some research [e.g., Grossberg, 1985; Koenderink, 1984, 1989] suggests that human vision involves a diffusion of intensity attenuated by edges or local objects. It is known that nonstationary, nonisotropic diffusions retain the guarantees against structure formation available with stationary diffusion [Yuille, 1983; Lifshitz, 1987]. Investigations of ways of generating sensible IAS annihilations by scale-increasing intensity diffusions that are affected by local edge strengths are underway. Means of calculating edge strengths that are related to ridge symmetry will therefore be especially important, providing impetus for the needed development of a multiscale theory of edges based on ridge symmetry. Among the tasks awaiting us are detailed comparison of our symmetry-dependent edge strengths to one-sided edge strengths, the development of a means of computation of a symmetry-dependent closed edge (perhaps by finding the locus of maxima of our symmetry-dependent edge strength along integral curves of IAS radius gradient, as in [Canny, 1983]), and the inclusion of the scale under attention as a parameter of the edge.

Both the need to compute nonstationary intensity slicing and nonstationary scale increase, as well as the need to compute IAS's simultaneously at all scales and intensities leads to a major concern for computing time. Even though the image analysis step is completely automatic and only the interactive object region definition based on this analysis need be done in real time, the analysis computations today in three spatial dimensions plus intensity require tens of hours on a 10 MIPS computer. Adding the dimension of scale and making the two above-mentioned operations nonstationary suggest that only a connectionist scheme will ultimately allow the required computations to be done in a reasonable time. Nevertheless, the attractiveness of the underlying theory leads us to develop the approach and ultimately to find forms of the computations that can be done in a few minutes, even for three-dimensional images.

Finally, the results of our methods based on geometric analysis and annihilations under scale increase must be compared to other multiscale methods with attractive possibilities, including the pyramid-based methods [Burt, 1981; Meer, 1988; Rosenfeld, 1984] or methods based on the sequence of primitive template match strengths across scale, including those based on successive Gaussian templates [Stiehl, 1989; Blom, 1988], those based on Gaussian and orientation templates [Coggins, 1986], and those based on Difference of Gaussian templates [Crowley, 1984].

ACKNOWLEDGEMENTS

We gratefully acknowledge the collaboration of Drs. Fred Bookstein, Robert Gardner, Jan Koenderink, and Bart ter Haar Romeny on the theory of multiscale IAS representations. We thank Dr. Bookstein for his comments on this paper and the UNC Med3D 3D display team, including its leader, Dr. Henry Fuchs, and the designer of our volume rendering methods, Dr. Marc Levoy, for their support and stimulation. We thank Carolyn Din, for help in manuscript preparation and Bo Strain for photographic assistance. The research reported herein was carried out with the partial support of NIH grant number P01 CA47982 and the NASA Center of Excellence in Space Data and Information Science.

REFERENCES

Blom, J, ter Haar Romeny, BM (coaches), Bookelmann, MS, "Edge Behaviour in Scale Space." Internal Report, Dept. of Radiology, State Univ. of Utrecht, 1988.
Blum, H, and Nagel, RN, "Shape Description Using Weighted Symmetric Axis Features." *Pattern Recognition* 10:167-180, 1978.
Bookstein, FL, "The Line-Skeleton." *Comp. Graphics and Im. Proc.* 11(2):123-137, 1979.
Burt, P, Hong, TH, Rosenfeld, A, "Segmentation and Estimation of Image Region Properties Through Cooperative Hierarchical Computation." *IEEE Trans SMC* 11:802-809, 1981.
Canny, JF, "Finding Edges and Lines in Images." Tech Report #720, Artificial Intell. Lab., MIT, Cambridge, MA, 1983.
Coggins, J, Fay, F, Fogarty, K, "Development and Application of a Three-dimensional Artificial Visual System." *Comp. Meth and Prog. in Biomed.* : 69-77, 1986.
Crowley, J & Parker, A, "A Representation for Shape Based on Peaks & Ridges in the Difference of Low-Pass Transform." *IEEE Trans. PAMI* 2(2):156-169, 1984.
Fuchs, H, Pizer, SM, Creasy, JL, Renner, JB, Rosenman, JG, "Interactive, Richly Cued Shaded Display of Multiple 3-D Objects in Medical Images." *Medical Imaging II*, SPIE **914**, Part B: 842-849, 1988.
Gauch, J, Oliver, W, Pizer, S, "Multiresolution Shape Descriptions and Their Applications in Medical Imaging." *Information Processing in Medical Imaging* (IPMI X, June 1987): 131-150, Plenum, New York,1988a.
Gauch, J and Pizer S, "Image Description via the Multiresolution Intensity Axis of Symmetry." *Proc. 2nd Int. Conf. on Comp. Vis.*, IEEE Catalog #88CH2664-1:269-274, 1988b.
Grossberg, S and Mingolla, E, "Neural Dynamics of Perceptual Grouping: Textures, Boundaries, and Emergent Segmentations." *Perception and Psychophysics* **38**(2):141-171, 1985.
Kass, M, Witkin, A, and Terzopoulos, D, "Snakes: Active Contour Models." *Proc 1st Int. Conf. on Comp. Vis.*, IEEE Catalog #87CH2465-3:259-268, 1987.
Koenderink, JJ., *Solid Shape*. MIT Press, Cambridge, 1989.
Koenderink, JJ "The Structure of Images." *Biological Cybernetics* **50**:363-370, 1984.
Levoy, M, "Display of Surfaces from Volume Data." *IEEE Computer Graphics and Applications* 8(3):29-37, 1988.
Lifshitz, LM, *Image Segmentation Using Global Knowledge and A Priori Information*. Ph.D. Dissertation, Tech. Report 87-012, UNC Chapel Hill, 1987.
Meer, P, Jiang, S-N, Baugher, ES, Rosenfeld, A, "Robustness of Image Pyramids under Structural Perturbations." *Comp. Vis., Graphics, and Image Proc.* 44(3):307-331, 1988.
Rosenfeld, A, *Multiresolution Image Processing and Analysis*. Springer-Verlag, Berlin, 1984.
Stiehl, HS, Neumann, H, Back, S, "On Scale-Space Edge Detection in Computed Tomograms." In this volume, 1989.
Yuille, AL and Poggio, T, "Scaling Theorems for Zero-Crossings." A.I. Memo 722, MIT, 1983.

Ein neues Verfahren zur Kontursegmentierung als Grundlage für einen maßstabs- und bewegungsinvarianten Strukturvergleich bei offenen, gekrümmten Kurven

Stephan Frydrychowicz

Lehrstuhl für Informatik 5 (Mustererkennung)
Universität Erlangen-Nürnberg
Martensstraße 3, 8520 Erlangen

Zusammenfassung

Intention für den hier dargelegten, neuen Ansatz zur Segmentierung von gekrümmten Kurven war, eine Methode zu finden, mit der es möglich ist, Polygonzüge miteinander zu vergleichen. Ähnlichkeiten sollten auch dann festgestellt werden, wenn die Polygonzüge sich zwar hinsichtlich der Stützstellen erheblich unterscheiden, wenn aber die Polygonzüge als Approximation derselben stetig gekrümmten Kontur entstanden sind. Mit der hier dargelegten Segmentierungsmethode können diese Vorgaben erfüllt werden. Es ist darüber hinaus sogar möglich die Polygonzüge unabhängig von ihrem Maßstab zu vergleichen.

Eine Kontur wird in Abschnitte zerlegt, die invariant sind gegenüber Verschiebungen, Drehungen und Größenänderung der Kontur. Diese Abschnitte können aus dem die Kontur approximierenden Polygonzug näherungsweise berechnet werden. Damit können für zwei Polygonzüge, die dieselbe Kontur approximieren, Segmente mit einer Vielzahl von Merkmalen bestimmt werden. An einem Beispiel wird gezeigt, daß mit diesen Segmenten offene, sich unter Umständen nur überlappende Kurven verglichen werden können.

Einführung

In der Bildanalyse werden sowohl bei dem Erkennungproblem als auch bei dem Korrespondenzproblem, aus den linienhaften Strukturen die wesentlichen Merkmale gewonnen. Während sich zwei gerade Liniensegmente leicht miteinander vergleichen lassen, so ist der Vergleich von gekrümmten Konturen um ein vielfaches schwieriger. Nun sind in der Regel die auftretenden Kontursegmente keine Geradenstücke sondern gekrümmte Kurven. Ferner handelt es sich häufig nicht um geschlossene sondern um offene Konturlinien. Vorteilhaft bei geschlossenen Konturen ist, daß sie sich durch globale Merkmale, wie eingeschlossene Fläche, Länge der Umrißlinie, Formfaktor, etc. beschreiben lassen. Auch läßt sich das Leistungsspektrum der durch die Bogenlänge parametrisierten Umrißlinie als Merkmalsvektor für die Klassifikation verwenden ([bur79],Kap.3). Für offene Konturen, die sich wohlmöglich nur teilweise überlappen, lassen sich keine vergleichbaren Merkmale angeben.

Für die Aufgabe, ein offenes Konturstück mit einer geschlossenen Kontur zur vergleichen und zu entscheiden, ob es sich um ein Stück der geschlossenen Kontur handelt, ist in [dav79] eine Methode angegeben worden. Für diesen Strukturvergleich werden die Konturen durch Polygonzüge approximiert. Verglichen werden die Stützstellen zusammen mit ihren geometrischen Relationen. Wesentlich für das Gelingen dieses Vergleichs ist, daß die Stützstellen für die Polygonzüge bei der Approximation an vergleichbaren Stellen der Konturen liegen. Bei Konturen mit nur wenigen Knickpunkten läßt sich diese Voraussetzung nicht mehr sicherstellen.

Ein anderer Ansatz wird in [che82] vorgestellt. Bei dieser Methode wird die Kontur in konvexe, konkave oder gerade Teile zerlegt. Für diese Teile werden Merkmale berechnet, die dann zusammen mit den geometrischen Relationen der Teile zueinander für den Strukturvergleich verwendet werden.

Eine graphentheoretische Clustering-Methode zur Zerlegung geschlossener Konturen in konvexe bzw. "annähernd konvexe" Bereiche wird in der Arbeit [sha79] dargelegt.

Im folgenden soll eine neue Methode dargelegt werden, eine unter Umständen auch offene Kontur in Teile zu zerlegen. Für diese Teile werden Merkmalssätze berechnet, die dann zusammen mit ihren geometrischen und topologischen Relationen für einen Strukturvergleich zur Objekterkennung aber auch zum Auffinden von Korrespondenzen in Bildpaaren oder -folgen verwendet werden können.

Lokale Formelemente

Um auch bei Kurven mit keinen oder nur wenigen markanten Punkten, wie z.B. Knickpunkten, dennoch lokale Merkmale extrahieren zu können, kann man versuchen, statt Punkten kleine Kurvenstücke zu verwenden. Für die Beschreibung eines Kurvenstücks wird die Region gewählt, die durch das Kurvenstück zusammen mit der die Endpunkte verbindenden Sehne eingeschlossen wird. Damit lassen sich dem Kurvenstück u.a. folgende Merkmale zuordnen:

b   die Bogenlänge des Kurvenstücks

$\vec{s}$ die Sehne zwischen den beiden Kurvenendpunkten

f   die Fläche der Region zwischen Kurve und Sehne

$\vec{c}$ die Koordinaten des Flächenschwerpunktes

Ein solches Kurvenstück wird im weiteren als *lokales Formelement* bezeichnet (siehe Abb.1).

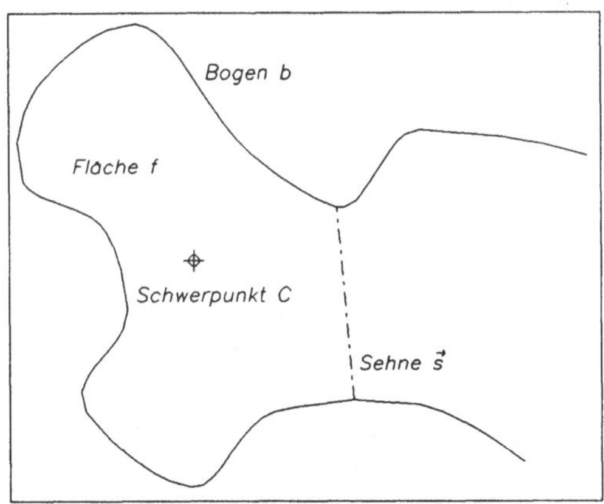

Abbildung 1: Ein lokales Formelement

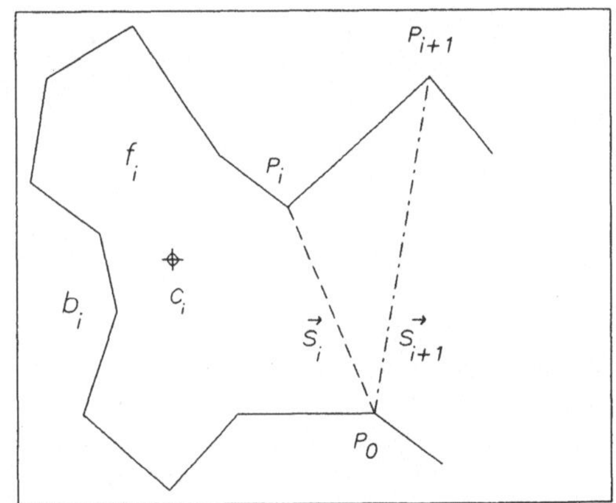

Abbildung 2: Iterative Berechnung der Formelemente

Für die Beschreibung der ganzen Kontur werden nur die Kurvenstücke verwendet, die bezüglich eines Formkriteriums lokal optimal sind. Das Kriterium sollte translations-, rotations- und maßstabsinvariant sein. Die Merkmale f, b und die Länge der Sehne $\| \vec{s} \|$ sind translations- und rotationsinvariant. Bei dem Kriterium ist folglich nur noch die Maßstabsinvarianz sicherzustellen. Das

241

Kriterium

$$\mu = \frac{f}{\kappa_1 \cdot b^2 + \kappa_2 \cdot b \cdot \|\vec{s}\| + \kappa_3 \cdot \|\vec{s}\|^2} \tag{1}$$

bzw.

$$\mu = \frac{b}{\|\vec{s}\|} \tag{2}$$

hat die gewünschten Eigenschaften. Es soll im weiteren *lokaler Formfaktor* genannt werden. Die Gewichtungsparameter κ_1, κ_2 und κ_3 im Kriterium (Gl.1) sind geeignet zu wählen. Für $\kappa_1 = \kappa_3 = 1$ und $\kappa_2 = 2$ erhält man den globalen Formfaktor ([nie83],S.106) des durch Sehne und Bogen eingeschlossenen Flächenstücks.

Berechnung

Für alle Kombinationen von Konturpunkten werden nun die Merkmale $b, \vec{s}, f$ und $\vec{c}$ bestimmt, wobei $\vec{c}$ der Ortsvektor des Schwerpunktes C sei. Die Merkmale der Formelemente mit demselben Anfangspunkt P_0 lassen sich iterativ berechnen (siehe Abb.2).

Im folgenden sei mit $\vec{p_i}$ der Ortsvektor von P_i, mit $\vec{c_i}$ der Ortsvektor des Schwerpunktes C_i des i-ten Formelementes und mit $\vec{e_z}$ der Einheitsvektor, der aus der Bildebene herauszeigt, bezeichnet. Ferner seien

$$\vec{q_i} = \vec{p_{i+1}} - \vec{p_i}$$

$$f_\Delta = \frac{(\vec{q_i} \times \vec{s_i})\vec{e_z}}{2}$$

$$\vec{c_\Delta} = \frac{\vec{q_i} + 2 \star \vec{s_i}}{3} + \vec{p_0}$$

Mit diesen Hilfgrößen lassen sich die Merkmale der Formelemente und mithin der lokale Formfaktor (Gl.1) bzw. (Gl.2) berechnen. Für die Merkmale gilt:

$$\vec{s_{i+1}} = \vec{s_i} + \vec{q_i}$$

$$b_{i+1} = b_i + \|\vec{q_i}\|$$

$$f_{i+1} = f_i + f_\Delta$$

$$\vec{c_{i+1}} = \frac{f_i \cdot \vec{c_i} + f_\Delta \cdot \vec{c_\Delta}}{f_{i+1}}$$

Berechnung und Interpolation bei Polygonzügen

Um nicht für jede Kombination zweier Konturpunkte ein Formelement berechnen zu müssen und um damit erheblich Rechenaufwand einzusparen, kann die Kontur durch ein Polygonzug approximiert werden ([gon87], S.394). Für alle Kombinationen von Stützstellen des Polygonzugs werden die Sehne, die Bogenlänge, die Fläche, der Schwerpunkt und der lokale Formfaktor (Gl.1) bzw. (Gl.2) berechnet. Als nächstes wird für jedes Formelement der Anfangs- und der Endpunkt der Sehne auf den anschließenden Strecken variiert. Falls ein Extremwert des lokalen Formfaktors existiert, wird dieser berechnet. Aus der Menge der so berechneten relativen Extrema und der Menge der Formelemente zwischen den Polygonstützstellen werden die Formelemente ausgewählt, die im Vergleich zu ihren Nachbarn extremal sind. Die ausgewählten Formelemente stimmen näherungsweise mit den extremalen Formelementen der ursprünglichen Kontur überein. Mögliche Extrema werden durch Differentation des lokalen Formfaktors analytisch berechnet. Im folgenden werden die zugrunde liegenden Formeln entwickelt.

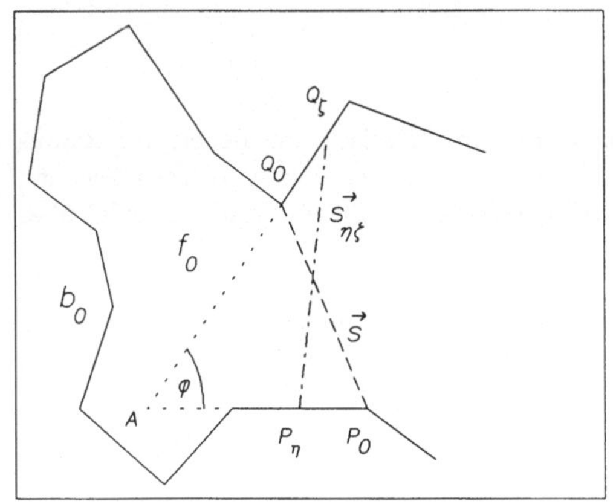
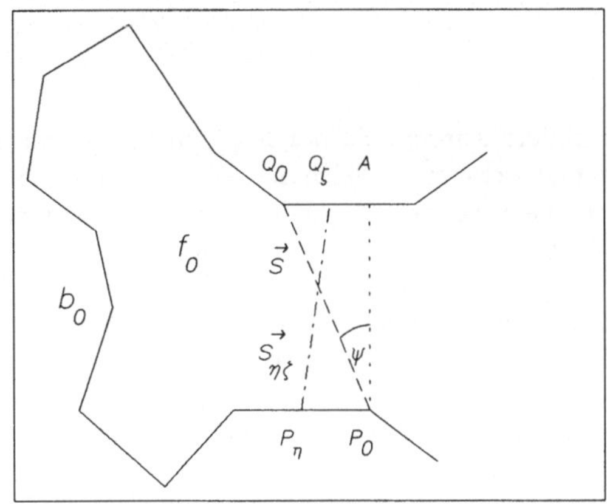

Abbildung 3: Interpolation lokaler Formelemente

Abbildung 4: Interpolation bei parallelen Anschlußstrecken

Mit den Bezeichnungen wie in Abbildung 3, sowie mit den Längen $u_0 = \bar{AP_0}$, $v_0 = \bar{AQ_0}$, $u_\eta = \bar{AP_\eta}$ und $v_\zeta = \bar{AQ_\zeta}$ gelten die Gleichungen

$$f_{\eta\zeta} \;=\; f_0 - u_0 v_0 \frac{\sin\phi}{2} + u_\eta v_\zeta \frac{\sin\phi}{2}$$

$$b_{\eta\zeta} \;=\; b_0 - u_0 - v_0 + u_\eta + v_\zeta$$

$$\|s_{\eta\zeta}\|^2 \;=\; (u_\eta + v_\zeta)^2 \sin^2 \frac{\phi}{2} + (u_\eta - v_\zeta)^2 \cos^2 \frac{\phi}{2}$$

Durch Einsetzen von $f_{\eta\zeta}$, $b_{\eta\zeta}$ und $\|s_{\eta\zeta}\|$ in die Gleichung 1 bzw. 2 erhält man den lokalen Formfaktor in Abhängigkeit von u_η und v_ζ.

Der Fall, daß die beiden Anschlußstrecken parallel sind, muß gesondert betrachtet werden. Mit den Bezeichnungen wie in Abbildung 4, sowie mit den Längen $u_\eta = \bar{P_0 P_\eta}$ und $v_\zeta = \bar{Q_0 Q_\zeta}$ gelten die Gleichungen

$$f_{\eta\zeta} \;=\; f_0 + (v_\zeta - u_\eta)\frac{\|\vec{s_0}\| \sin\psi}{2}$$

$$b_{\eta\zeta} \;=\; b_0 + v_\zeta - u_\eta$$

$$\|\vec{s_{\eta\zeta}}\|^2 \;=\; (u_\eta + v_\zeta)^2 + \|\vec{s_0}\|^2 - 2(u_\eta + v_\zeta)\|\vec{s_0}\| \cos\psi$$

Durch Einsetzen von $f_{\eta\zeta}$, $b_{\eta\zeta}$ und $\|s_{\eta\zeta}\|$ in die Gleichung 1 bzw. 2 erhält man den lokalen Formfaktor in Abhängigkeit von u_η und v_ζ.

Durch Differentation können lokale Extremwerte des lokalen Formfaktors $\mu(u_\eta, v_\zeta)$ berechnet werden. Die Optimierungsgleichung für den Formfaktor bei einem festgehaltenen Sehnenendpunkt in einer Polygonstützstelle erhält man aus den Gleichungen durch Wahl von $u_\eta = 0$ bzw. $v_\zeta = 0$.

Ergebnisse

Testkonturen

Erprobt wurde diese neue Methode an synthetischen Konturen, die zu einem frei wählbaren Parametersatz $n, a_0, \phi_0, a_1, \phi_1, ..., a_n, \phi_n$ durch

$$r(\phi) \;=\; \sum_{\nu=0}^{n} a_\nu \sin(\nu\phi - \phi_\nu)$$

$$x(\phi) = r(\phi)\cos\phi$$
$$y(\phi) = r(\phi)\sin\phi$$

definiert wurden. Es handelt sich bei diesen Testkonturen um "Kreise", bei denen der Radius gemäß der ersten Gleichung abhängig vom Winkel ϕ variiert wird. Da es sich bei dieser Gleichung um die nach dem n-ten Glied abgebrochene Fourier-Reihe handelt, kann jedes "Radiusprofile" über $[0, 2\pi]$ beliebig genau approximiert werden.

Als Definitionsbereich wurde $[\phi_a, \phi_e] \subset [-2\pi, 2\pi]$ gewählt, um auch sich u.U. überlappende Kurvenstücke definieren zu können. Der Definitionsbereich wurde in äquidistante Intervalle der Länge $\Delta\phi = 2\pi/N$ aufgeteilt. Die Stützstellen lassen sich mit $\xi \in [0, 1)$ um $\hat{\phi} = \xi\Delta\phi$ verschieben. Die Testkurven $\vec{x} = (x(\phi), y(\phi))^T$ können mit dem Maßstabsfaktor λ, dem Rotationswinkel ψ und dem Translationsvektor $\vec{t} = (t_x, t_y)^T$ gemäß

$$\vec{x'} = \lambda D_\psi \vec{x} + \vec{t} \tag{3}$$

transformiert werden. Damit kann einerseits eine Testkurve in das "Bildfenster", $0 <= x, y <= 511$, abgebildet werden. Anderseits kann sie durch diese Transformation innerhalb des "Bildfensters" verschoben, rotiert oder in der Größe verändert werden. Die Abtastung im diskreten Bildraster wird durch Ab/Aufrunden der reellen Koordinaten simuliert.

Größen- und bewegungsinvarianter Konturvergleich

Daß mit den lokalen Formelementen ein maßstabs-, rotations- und translationsinvarianter Kurvenvergleich möglich ist, wird exemplarisch an der Kontur mit dem Parametersatz

$$a_0 = 2, a_1 = 1, a_4 = a_6 = 0.5, \phi_1 = \phi_4 = -90°, \phi_6 = 30°, a_\nu = \phi_\nu = 0 \; sonst$$

gezeigt. Aus dieser Kontur wurden zwei verschiedene Abschnitte ausgewählt und verschieden transformiert. Als Definitionsbereiche wurden $[-120°, 210°]$ bzw. $[-90°, 180°]$ gewählt.

Die Kurven wurden mit den Parametern, wie in Tabelle 1 angegeben, transformiert und approximiert. Anschließend wurden die Koordinaten der Polygonzüge auf ganze Zahlen gerundet.

Kurve	N	ξ	λ	ψ	$\vec{t} = (t_x, t_y)^T$	
7	40	0	70	0°	256	256
8	40	0.5	52.5	60°	256	256

Tabelle 1. Transformations- und Approximationsparameter für zwei Testkurven

	Maßtab	Rotation	Translation	
real	0.75	60.0°	326.3	-6.3
geschätzt	0.77	60.5°	329.6	-14.7

Tabelle 2. Reale und geschätzte Lageparameter

Die erste Kurve besteht aus 37, die zweite aus 30 Strecken. Die relativen Lageparamter zischen den Kurven sind in Tabelle 2 angegeben. Für jede der beiden Kurven wurden die Formelemente berechnet. Hierbei lag das Kriterium (Gl.1) mit $\kappa_1 = \kappa_2 = 0$ and $\kappa_3 = 1$, zugrunde. Die Merkmale der einzelnen Formelemente sind in den Tabellen 3 und 4 angegeben. In Abbildung 5 sind die Polygonzüge zusammen mit den Formelementen (Sehnen) skizziert. Die Formelemente wurden verglichen und bestmöglich einander zugeordnet (Tab.5). Mit dieser Assoziation wurden die Lageparameter geschätzt (Tab.2). Die zweite Kurve wurde mit den geschätzten Lageparametern zurücktransformiert. Das Ergebnis ist in Abbildung 6 dargestellt.

Das Beispiel wurde auf einer $CADMUS$ Workstation berechnet. Die Formelemente wurden in ca. 1.2 CPU-Sekunden berechnet. Für das Suchen der besten Assoziation und für die Lagebestimmung wurden zusammen ca. 0.5 CPU-Sekunden benötigt.

Index	Formfaktor	Fläche	Bogen	Sehnenlänge	Schwerpunkt	
0	3.89978	122180	910	177	337.5	275.3
1	2.56379	112935	835	210	346.2	277.9
2	1.35724	59709	483	210	401.2	249.4
3	0.90864	28277	316	176	433.3	295.3
4	1.12372	25713	376	151	268.4	374.6
5	1.76421	17755	272	100	279.8	398.6
6	0.39874	7468	167	137	389.2	152.5
7	0.15722	5248	222	183	196.7	311.0
8	0.54678	2671	98	70	346.3	363.1
9	0.34873	1826	86	72	313.0	120.1
10	0.63665	1195	65	43	198.1	300.9
11	0.10993	1040	99	97	452.7	208.7
12	0.44238	816	56	43	214.8	326.6
13	0.48633	520	44	33	188.6	273.5

Tabelle 3: Formelemente der Kurve Nr.1

Index	Formfaktor	Fläche	Bogen	Sehnenlänge	Schwerpunkt	
0	3.86838	66923	671	132	273.4	318.3
1	2.64225	61427	613	152	275.3	325.3
2	1.40941	36081	378	160	313.1	344.1
3	1.17400	15802	293	116	186.8	307.0
4	0.88910	12811	210	120	297.5	393.0
5	1.77318	11570	224	81	176.8	321.7
6	0.38472	4180	127	104	373.2	310.1
7	0.54368	3314	112	78	216.9	362.7
8	0.33419	1902	89	75	370.3	242.8
9	0.68081	706	50	32	204.0	237.3
10	0.09518	656	84	83	363.9	361.3
11	0.40267	253	31	25	196.6	257.2

Tabelle 4: Formelemente der Kurve Nr.2

Kurve	Index der Formelemente											
1	0	1	2	4	3	5	6	8	9	10	11	12
2	0	1	2	3	4	5	6	7	8	9	10	11

Tabelle 5. Die beste Assoziation

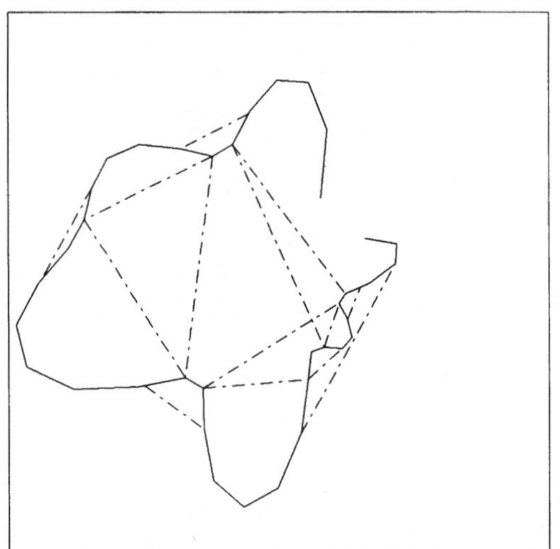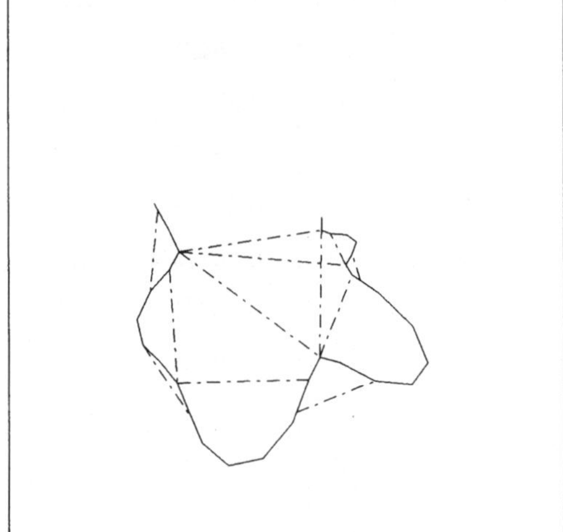

Abbildung 5. Polygonzüge und Formelemente für zwei Testkurven

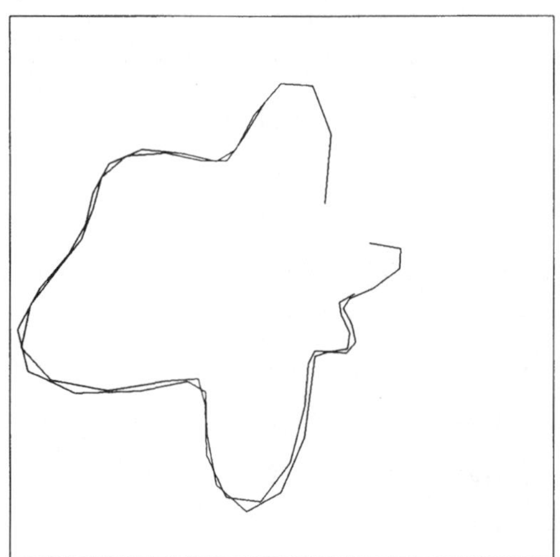

Abbildung 6. Ergebnis des Kurvenvergleichs

Literatur

[bur79]   Hans Burkhardt, Transformationen zur lageinvarianten Merkmalsgewinnung, Fortschrittbericht (Reihe 10, Nr.7), VDI-Verlag, 1979

[che82]   J.K. Cheng; T.S. Huang; Recognition of Curvilinear Objects by Matching Relational Structures, Proc. Conf. on Pattern Recognition and Image Processing, 1982, pp. 343-348

[dav79]   Larry S. Davis; Shape Matching Using Relaxation Techniques, IEEE Trans. on Pattern Analysis and Machine Intelligence, vol. PAMI-1, no. 1, jan. 1979, pp. 60-72

[gon87]   Rafael C. Gonzalez; Paul Wintz; Digital Image Processing, Addison-Wesley Publishing Company, Reading, Massachusetts, 1987

[low86]   David G. Lowe; Perceptual Organization and Visual Recognition, Kluwer Academic Publishers, Boston, 1986

[nie83]   Heinrich Niemann; Klassifikation von Mustern, Springer-Verlag, Berlin, 1983

[sha79]   Linda G. Shapiro; Robert M. Haralick; Decomposition of Two-Dimensional Shapes by Graph-Theoretic Clustering, IEEE Trans. on Pattern Analysis and Machine Intelligence, vol. PAMI-1, no. 1, jan. 1979, pp. 10-20

Matched Median Filter zur Bildsegmentierung

Hans-Ullrich Döhler
Institut für Nachrichtentechnik, Technische Universität Braunschweig
Schleinitzstr. 23, 3300 Braunschweig

Zusammenfassung: Das Ziel der vorliegenden Arbeit ist die Bestimmung der Fensterform und -größe zweidimensionaler Medianfilter für deren Anwendung in der Bildsegmentierung. Ausgangspunkt ist der Begriff der *Ursignale* von Medianfiltern (engl. *root signals*). Ursignale eines gegebenen Medianfilters sind solche Signale, die gegenüber der Anwendung dieses Medianfilters invariant sind. Zunächst wird gezeigt, wie sich für ein gegebenes Medianfilter ein spezielles binäres Ursignal bestimmen läßt, das das *kleinste überlebensfähige Objekt* des Medianfilters (engl. *smallest surviving object, SSO)* genannt wird. Danach wird ein Verfahren beschrieben, das für ein beliebiges gegebenes Bild, einen interessierenden Auschnitt davon oder für spezielle Objekte eines Bildes, das flächengrößte Fenster eines Medianfilters generiert, das dieses Bild bzw. diese Objekte in der Menge seiner Ursignale enthält. Dieses Medianfilter wird das *Optimalmedianfilter* oder engl. *matched median filter* bezüglich des gegebenen Bildes bzw. der gegebenen Objekte genannt. Schließlich wird anhand einiger Beispiele gezeigt, wie sich das Verfahren zur Bestimmung des Optimalmedianfilters nutzbringend zum Zweck der Bildsegmentierung einsetzen läßt.

1. Einleitung

Die wiederholte Anwendung eines Medianfilters auf ein zunächst beliebiges zweidimensionales Signal führt im allgemeinen auf ein stabiles Ausgangssignal, das sich bei weiterer Anwendung des gleichen Filters nicht mehr ändert. Solche Signale werden *Ursignale* (engl. *root signals* oder *fixed points*) des zugehörigen Medianfilters genannt. Im Bereich der Bildanalyse bzw. -segmentierung können diese Ursignale zur Modellierung von Bildsignalen eingesetzt werden. Die Strukturen von Ursignalen für eindimensionale sowie separierte Medianfilter wurden bereits untersucht [1,2,3,5,7]. Zur Beschreibung der Ursignale echter zweidimensionaler Medianfilter gibt es dagegen bis heute relativ wenig bekannte Literatur. In [1] werden nur einige wenige Beispiele für einfache Filterformen und relativ kleine Fenstergrößen bis zu 5 × 5 Bildpunkten angegeben.

Rangordnungsoperationen auf orts- und amplitudendiskrete mehrstufige Signale und damit insbesondere auch das Medianfilter liefern das gleiche Ergebnis, wie die entsprechenden Operationen auf die einzelnen Schichten der *Höhenschichtenzerlegung* des Signals und die anschließende Überlagerung der gefilterten Schichten zum Ausgangssignal. Solche Filter werden *Höhenschichtenfilter* genannt (engl. *stack filter*) [4,6]. Die folgende Darstellung kann sich daher auf die Untersuchung binärer Signale beschränken, obwohl die Anwendung der folgenden Ergebnisse insbesondere für die Modellierung und Segmentierung von Grautonbildern von Interesse ist.

2. Digitale Region, digitales Geradensegment und Konvexität

Zur nachfolgenden Beschreibung der Ursignalstrukturen zweidimensionaler Medianfilter sind einige Begriffsdefinitionen aus dem Bereich der digitalen Geometrie notwendig, die hier in komprimierter Form gegeben werden sollen [8,9,10].

Die Zerlegung der kontinuierlichen Ebene in orthogonale disjunkte Zellen mit der Fläche 1 heißt digitales Mosaik. Die Menge der Zentralpunkte P dieser Zellen mit ganzzahligen Koordinaten (x, y) heißt digitales Raster. Bei quadratischer Berandung der einzelnen Zellen handelt es sich um ein quadratisches Raster, bei hexagonaler Berandung der Zellen um ein hexagonales Raster. Zwei Zellen heißen direkt benachbart, wenn sie eine gemeinsame Seite besitzen und indirekt benachbart, wenn sie nur eine gemeinsame Ecke besitzen.

Ein endlicher im allgemeinen orthogonaler Auschnitt aus dem digitalen Raster heißt digitales Bild B, wobei jedem Punkt des Rasters bzw. jeder Zelle (engl. pixel) eine Farbe zugeordnet wird. Gibt es nur zwei verschiedene Farben, schwarz (1 Objekt) und weiß (0 Hintergrund), so handelt es sich um ein Binärbild. Ein Untermenge von benachbarten Zellen oder Rasterpunkten gleicher Farbe in einem Binärbild heißt digitale Region R. Die Anzahl der schwarzen Rasterpunkte und damit die Anzahl von schwarzen Zellen in R wird Fläche $A(R)$ genannt. Die Menge von Objektzellen der Region R, die in ihrer direkten Nachbarschaft mindestens einen Nachbarn haben, der zum Hintergrund gehört, heißt Kontur $K(R)$. Der Weg durch eine Folge von Rasterpunkten direkt oder indirekt benachbarter Konturzellen kann auf bekannte Weise durch die Konturkette (z.B. Freemann-Konturcode) beschrieben werden. Der Umlaufsinn durch eine Folge von Konturzellen heißt positiv, wenn dabei die Objektzellen stets links und Hintergrundzellen stets rechts vom Umlaufweg liegen. Eine digitale Region heißt digital konvex, wenn für zwei beliebige Konturpunkte P_1 und P_2 gilt, daß die Fläche zwischen dem kontinuierlichen Geradensegment g_{12} und dem Rand der Konturzellen keinen Rasterpunkt enthält der zum Hintergrund gehört (Bild 1 a)). Diese Eigenschaft digital konvexer Regionen wird in [8,10] als Sehneneigenschaft oder engl. *chord property* bezeichnet. Eine positiv durchlaufene Folge von Konturzellen einer digitalen Region vom P_1 nach P_2 heißt digitales Geradensegment G_{12}, wenn für sie die chord property gilt und alle ihre Rasterpunkte links oder höchstens auf der kontinierlichen Geraden g_{12} liegen (Bild 1 b),c)). Eine digitale Region R heißt digitales Polygon, wenn sich ihre Kontur aus digitalen Geradensegmenten zusammensetzt.

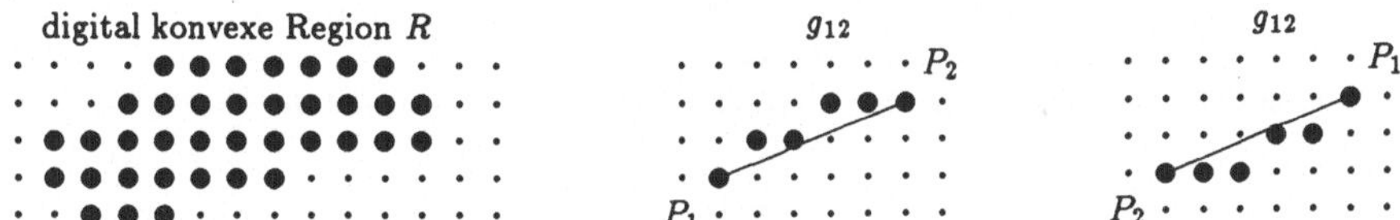

Bild 1: a) Digitale Region, b) und c) Beispiel für die Diskretisierung des Geradensegmentes g_{12} von P_1 nach P_2.

2.1 Bestimmung der Konturpunkte digitaler Geradensegmente

Für die weiteren Betrachtungen wird eine Diskretisierungsregel zur Abbildung eines kontinuierlichen Geradensegmentes auf das digitale Raster benötigt. Es gibt zahlreiche verschiedene Algorithmen zur Lösung dieses Problems z.B. [9]. Diese liefern aber im allgemeinen keinen umkehrbar eindeutigen Zusammenhang zwischen dem kontinuierlichen Geradensegment und seiner digitalen Abbildung. Das hier verwendete Verfahren weist einen solchen eindeutigen Zusammenhang auf und basiert auf dem Begriff der bereits eingeführten *chord property* digitaler Geradensegmente (Bild 1. b) und c)).

Es sei g_{12} eine kontinuierliches Geradensegment von $P_1(x_1, y_1)$ nach $P_2(x_2, y_2)$. Die diskreten Punkte $P_i(x_i, y_i)$ des positiv durchlaufenen digitalen Geradensegmentes G_{12} von P_1 nach P_2 werden bestimmt durch:

$$x_i = \begin{cases} x_1 \cdots x_2 & \mid \Delta y \mid \le \mid \Delta x \mid \\ x_1 + \lfloor (y_i - y_1) \cdot \Delta x/\Delta y \rfloor & \mid \Delta y \mid > \mid \Delta x \mid \quad \text{und} \quad \Delta y > 0 \\ x_1 + \lceil (y_i - y_1) \cdot \Delta x/\Delta y \rceil & \mid \Delta y \mid > \mid \Delta x \mid \quad \text{und} \quad \Delta y < 0 \end{cases}$$

$$y_i = \begin{cases} y_1 \cdots y_2 & \mid \Delta y \mid > \mid \Delta x \mid \\ y_1 + \lceil (x_i - x_1) \cdot \Delta y/\Delta x \rceil & \mid \Delta y \mid \le \mid \Delta x \mid \quad \text{und} \quad \Delta x > 0 \\ y_1 + \lfloor (x_i - x_1) \cdot \Delta y/\Delta x \rfloor & \mid \Delta y \mid \le \mid \Delta x \mid \quad \text{und} \quad \Delta x < 0 \end{cases}$$

mit

$$\Delta x = x_2 - x_1 \quad \text{und} \quad \Delta y = y_2 - y_1.$$

und

$$v = \lfloor u \rfloor \text{ nächster ganzzahliger Wert mit } v \le u$$
$$v = \lceil u \rceil \text{ nächster ganzzahliger Wert mit } v \ge u.$$

Alle Punkte eines mit dieser Regel erzeugten digitalen Geradensegmentes liegen unter der Voraussetzung eines positiven Umlaufsinnes links oder höchstens auf dem erzeugenden kontinuierlichen Geradensegment.

3. Vom Medianfilter zum kleinsten überlebensfähigen Objekt

Es sei R eine digitale Region und es sei W ein Medianfenster mit der Fläche

$$A(W) = 2n + 1 \quad \text{mit } n \in \{1, 2, \cdots\}. \tag{1}$$

Die Elemente von W seien weiterhin zentralsymmetrisch um den Fenstermittelpunkt angeordnet. Damit R Ursignal des Medianfilters mit dem Fenster W ist muß für das auf den Punkt $P(x, y)$ zentrierte Medianfenster die Bedingung

$$\begin{aligned} A(W \cap R)_P &> n \quad \text{für } P = 1 \\ A(W \cap \overline{R})_P &> n \quad \text{für } P = 0 \end{aligned} \tag{2}$$

gelten.

Laute die Frage nun nach der Gesamtheit binärer Ursignale eines zweidimensionalen Medianfilters, so könnte ein mögliches Verfahren zunächst sämtliche stabile Muster innerhalb des Medianfensters bestimmen und diese Muster dann so miteinander kombinieren, daß für zwei benachbarte Muster die Medianbedingung ebenfalls gilt. Allerdings kann sofort aus der Vielzahl z möglicher Muster, die die Bedingung (2) erfüllen, mit

$$z = \binom{2n}{n}$$

bereits für Fenster mit kleiner Fläche geschlossen werden, daß die Bestimmung aller stabilen Kombinationen dieser Muster, etwa mit Methoden der vollständigen Suche, kaum in endlicher Zeit lösbar ist. Daher beschränkt sich die weitere Betrachtung auf ein spezielles, für die Anwendung aber um so interessanteres Ursignal, das das *kleinste überlebensfähige Objekt* oder engl. *smallest surviving object (SSO)* genannt wird. Dieses SSO ist ein lochfreies, digital konvexes Polygon. Dazu werden zunächst zwei Sätze über die Eigenschaft einfacher digitaler Regionen formuliert.

Satz 1: Jede unendlich ausgedehnte digitale Halbebene, die von einer digitalen Geraden mit beliebiger Steigung begrenzt wird, ist Ursignal jedes beliebigen zentralsymmetrischen Medianfensters.

Beweis: Aufgrund der Symmetrie der beiden Halbebenen kann die Betrachtung auf Konturpunkte mit $P = 1$ beschränkt werden. Die die Halbebene begrenzende digitale Gerade G werde auf Basis der o.a. Diskretisierungsvorschrift aus einer kontinuierlichen durch das Fensterzentrum verlaufenden Geraden g erzeugt. Die Größe $k \geq 0$ sei die Zahl von Punkten in G die außerhalb des Zentralpunktes exakt auf der kontinuierlichen Geraden g liegen. Dann gilt für die Zahl von schwarzen Punkten innerhalb des Fensters

$$A(W \cap R)_P = \frac{1}{2}(2n + 1 - k - 1) + k + 1 = n + \frac{1}{2}k + 1 > n$$

und (2) ist erfüllt.

Satz 2: Es seien R_1 und R_2 zwei digitale Halbebenen, die durch zwei digitale Geraden G_1 und G_2 begrenzt werden und durch Diskretisierung aus g_1 und g_2 hervorgehen. Es gelte weiterhin, daß sich g_1 und g_2 im Punkt P schneiden und mit dem auf den Punkt P zentrierten Fenster $2k_1 \geq 2$ bzw. $2k_2 \geq 2$ Rasterpunkte außerhalb des Zentrums gemein haben. Für die Schnittmenge aus R_1 und R_2 gilt ebenfalls die Medianbedingung, wenn die Geraden g_1 und g_2 zwei in der Steigung direkt aufeinanderfolgende Geraden sind, die durch das Zentrum und einen weiteren zunächst beliebigen Rasterpunkt des Medianfensters verlaufen (Bild 2).

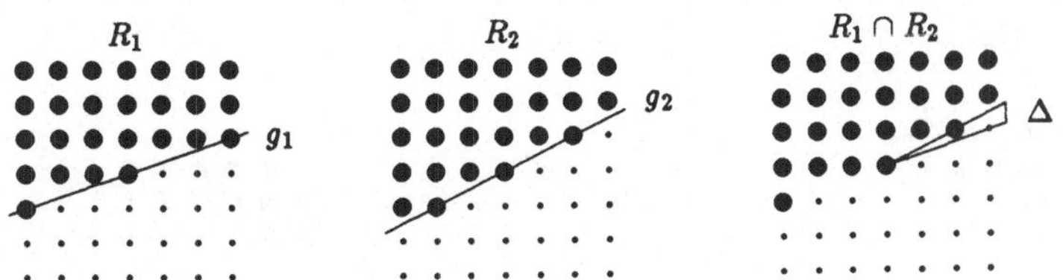

Bild 2: Medianbedingung am Schnittpunkt zweier digitaler Halbebenen R_1 und R_2.

Beweis: Analog zum Beweis des Satzes 1 folgt für

$$A(W \cap R_1)_P = n + 1 + k_1$$

und

$$A(W \cap R_1 \cap R_2)_P = A(W \cap R_1)_P - \epsilon - k_1,$$

wobei $\epsilon \geq 0$ die Zahl von Rasterpunkten angibt, die echt innerhalb der kontinuierlichen Fläche Δ liegen, die von g_1, g_2 und dem Rand von W begrenzt wird (Bild 2). Die Medianbedingung lautet damit für den Punkt P

$$A(W \cap R_1 \cap R_2)_P = n + 1 + k_1 - \epsilon - k_1 > n.$$

Damit folgt

$$\epsilon < 1 \text{ also } \epsilon = 0,$$

d.h innerhalb der Fläche Δ, die von g_1, g_2 und dem Rand von W begrenzt wird, darf kein Rasterpunkt liegen. Damit müssen g_1 und g_2 zwei ihrer Steigung direkt aufeinanderfolgende Geraden innerhalb des Medianfensters sein.

Aus den Sätzen 1 und 2 folgt die Regel zur Beschreibung des kleinsten überlebensfähigen Objektes eines Medianfilters mit gegebenem zentralsymmetrischen Fenster:

Satz 3: Die Kontur des SSO ergibt sich aus der Verkettung aller möglicher Geradensegmente ausgehend vom Zentrum zu einem weiteren Punkt des Medianfensters. Die Verkettung ist dabei derart vorzunehmen, daß sämtliche Geradensegmente bezüglich ihrer Steigung sortiert auftreten. Gibt es Segmente gleicher Steigung, so ist nur das jeweils längste zu verwenden (Bild 3).

3.1 Algorithmus zur Erzeugung des SSO

Wir sind nun in der Lage, den Algorithmus zur Erzeugung des kleinsten überlebensfähigen Objektes SSO für ein gegebenes Medianfilter mit punktsymmetrischer Fensterform bezüglich des Fensterzentrums anzugeben (Bild 3):

Algorithmus 1: 1. Berechne Steigung und Länge aller kontinuierlichen Geradensegmente innerhalb des Fensters, die vom Fensterzentrum ausgehen.

2. Sortiere diese Geradensegmente bezüglich ihrer Steigung. Gibt es Segmente gleicher Steigung, so benutze nur die jeweils längsten Exemplare und lösche alle anderen.

3. Verkette diese sortierte Liste von Geradensegmenten zu einem Polygon. Aufgrund der vorausgesetzten Zentralsymmetrie des Fensters ist dieses Polygon notwendigerweise geschlossen.

4. Berechne die Konturpunkte des zugehörigen digitalen Polygons unter Berücksichtigung der o. a. Regel. Die so erzeugte Kontur ist die Kontur des gesuchten SSO.

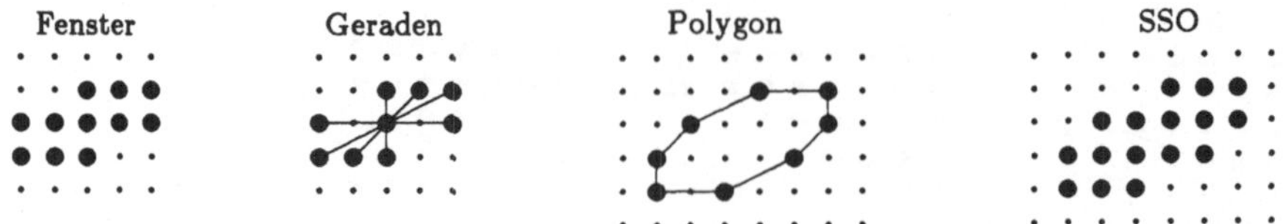

Bild 3: Beispiel zum Algorithmus zur Generierung des SSO.

In [11] werden für zahlreiche verschiedene Fensterformen und Größen die zugehörigen SSO angegeben. Für die Beschreibung der Gesamtheit aller Ursignale eines gegebenen Medianfilters läßt sich der folgende sehr nützliche Satz formulieren:

> **Satz 4:** In jedem binären Ursignal ist an keiner Stelle der Kontur zwischen weißen und schwarzen Regionen die Krümmung der Kontur stärker als die entsprechende Krümmung des SSO.

Jedes Binärbild läßt sich durch die Codierung der Konturen der im Bild enthaltenen Objekte vollständig beschreiben. Die Objekte des Bildes werden dabei jeweils durch einen Aufpunkt und eine Folge von Kettenelementen (Freemann-Code) repräsentiert. Die Implementation der zweidimensionalen Medianfilterung läßt sich, unter Ausnutzung des Satzes 4, nun derart modifizieren, daß die Kontur eines jeden Objektes des Bildes jeweils entlang seiner Konturkette gefiltert wird. Solange es Konturpunkte gibt, die die in Satz 4 formulierte Bedingung verletzen, wird an diesen Punkten die Kontur entweder lokal erodiert oder dilatiert, je nachdem ob die Richtungsänderung der Kontur an diesem Punkt positiv oder negativ ist. Der zeitaufwendige Prozeß der zweidimensionalen Medianfilterung, wird dadurch auf eine quasi eindimensionale Implementation abgebildet, die wir *konturgestützte Medianfilterung* nennen. Die Laufzeit bis zum Erreichen des Ursignals wird hierbei gegenüber der echten zweidimensionalen Implementierung drastisch reduziert.

Damit stellt das Medianfilter ein einfach zu implementierendes Werkzeug zur Glättung von Konturen dar. Aber auch die Segmentierung von Konturen, die dabei nicht länger in sich geschlossen sein müssen, läßt sich mit über die konturgestützte Medianfilterung realisieren. Bezeichnet man die Bereiche einer Kontur als glatt, in denen eine definierte Maximalkrümmung nicht überschritten wird, so segmentiert ein Medianfilter dessen SSO genau diese Maximalkrümmung einhält, eine beliebige gegebene Kontur in stückweise glatte Bereiche. Allgemein gilt, daß mit der konturgestützten Medianfilterung unter Ausnutzung der Kenntnisse der SSO dieser Filter verschiedene Aufgaben, wie sie z.B. bei Konturanalysen auftreten, gelöst werden können (siehe Kap. 5).

4. Vom Binärbild zum Optimalmedianfilter

Bildsignale sind häufig aus zwei Anteilen zusammengesetzt, von denen der eine stückweise glatt ist. Die Trennlinien zwischen den einzelnen glatten Stücken können dabei auf Objektkonturen, Textur- und Schattengrenzen und dgl. zurückgehen. Die Konturen der stückweise glatten Anteile lassen sich in vielen Fällen hinreichend genau durch Ursignale von Medianfiltern annähern. Häufig kann sogar die Kontur einzelner Objekte des Bildes als Ursignal und damit als Überlagerung gegeneinander verschobener Exemplare ein und desselben SSO beschrieben werden. Besteht die Aufgabe der Bildsegmentierung in der Trennung der so beschreibaren Objekte vom Rest des Bildes, so liefert die wiederholte Filterung mit dem auf die Konturen der Objekte angepaßten Medianfenster die optimalen Ergebnisse. Solche angepaßten Medianfilter werden von uns *matched median filter* oder *Optimalmedianfilter* genannt.

Die folgende Definition erklärt noch einmal diesen Zusammenhang:

> **Def. 1:** Für ein gegebenes Bild, einen Bildauschnitt oder eine Menge von Objekten daraus ist das Optimalmedianfilter das flächengrößte Medianfilter, das das gegebene Bild in der Menge seiner Ursignale enthält, d.h die Objekte des Bildes sind die Überlagerung gegeneinander verschobener SSO des Medianfilters.

Der folgende Algorithmus bestimmt für ein beliebiges Binärbild, oder einen Auschnitt daraus, das flächengrößte Medianfilter, das dieses Bild in der Menge seiner Ursignale enthält. Im Falle der Untersuchung eines mehrstufigen Bildes (Graubild) muß der Algorithmus gegebenenfalls für einige oder alle Höhenschichten durchlaufen werden (Bild 4):

Algorithmus 2:

1. Initialisiere alle Punkte eines quadratischen Medianfensters W_0 der Kantenlänge l mit Einsen (schwarz), so daß gilt:

$$A(W_0) = l^2 = 2n_0 + 1$$

Die ungerade Kantenlänge l ist mindestens so groß zu wählen, daß das erwartete zu generierende Medianfenster hineinpaßt. Setze weiterhin $i = 0$.

2. Suche zeilenweise im Binärbild B schwarze (1) oder weiße (0) Konturpunkte P. Für jeden gefundenen Punkt gehe zu 3. Sonst gehe zu 6.

3. Bilde die auf den Punkt P zentrierte Region R_P durch

$$R_P = \begin{cases} (B \cap W_0)_P & P = 1 \\ & \text{für} \\ (\overline{B} \cap W_0)_P & P = 0 \end{cases}$$

4. Bilde die am Zentrum punktweise gespiegelte Region $\tilde{R}_P$. Gilt $A(R_P \cap \tilde{R}_P) = 1$ inkrementiere i und gehe zu 5. Sonst gehe zu 2.

5. Erzeuge das aktuelle W_P und daraus das W_i durch

$$W_P = (R_P \cup \tilde{R}_P) \quad \text{und} \quad W_i = W_{i-1} \cap W_P$$

und gehe zu 2.

6. Ende.

Beweis: Ist die Fläche der Schnittmenge aus $R_P \cap \tilde{R}_P$ größer als 1, d.h. überlappen sich R_P und $\tilde{R}_P$ auch außerhalb des Zentralpunktes, so kann über W_P keine Aussage getroffen werden. Im anderen Fall überlappen sich R_P und $\tilde{R}_P$ nur im Punkt P. Daher ist für das Medianfenster aus der Vereinigungsmenge von $R_P \cup \tilde{R}_P$ die Medianbedingung (2) gerade erfüllt, d.h.

$$A(W_P \cap R_P) = \frac{1}{2}A(W_P) + 1 = n_P + 1 \quad > \quad n_P \quad \text{für } P = 1$$

$$A(W_P \cap \overline{R}_P) = \frac{1}{2}A(W_P) + 1 = n_P + 1 \quad > \quad n_P \quad \text{für } P = 0$$

Jeder Punkt außerhalb von $R_P \cup \tilde{R}_P$ würde zur Verletzung der Medianbedingung führen und kann daher nicht zum Fenster des zu bestimmenden Medianfilters gehören. Die Forderung 2., daß P Konturpunkt sein muß, könnte eigentlich entfallen, da an Nichtkonturpunkten 4. nicht erfüllbar ist. Die Einführung dieses Punktes führt aber zu einer deutlichen Beschleunigung des Algorithmus.

5. Experimente

Die Anwendung der wiederholten Medianfilterung bis zum stabilen Ursignal als Werkzeug für die Bildsegmentierung soll in den folgenden Beispielen verdeutlicht werden. Bild 4. a) zeigt verschieden große kreisförmige Objekte mit überlagertem Rauschen. Die Segmentieraufgabe bestehe hier beispielhaft darin, Kreise deren Durchmesser eine gewisse Größe überschreiten vom Rest des Bildes zu trennen. Wie die Ergebnisse in Bild 4. b) zeigen, läßt sich diese Mindestgröße auf Basis der Kenntnis der SSO verschiedener Medianfilter gut einstellen.

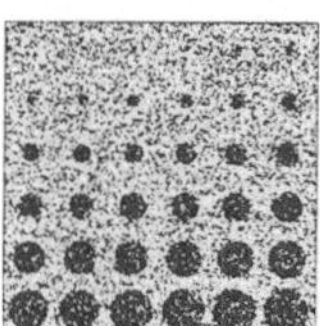
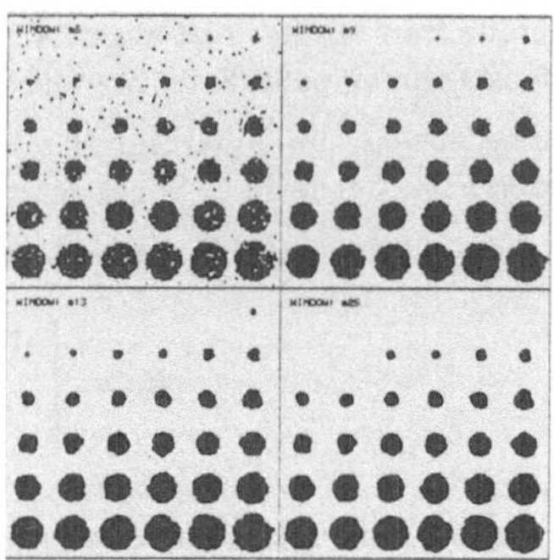

Bild 4: a) Kreisförmige Objekte mit überlagertem Rauschen, b) Filterergebnisse.

Die nächsten zwei Beispiele zeigen die Anwendung der konturgestützten Medianfilterung zur Segmentierung von Objektkonturen. Bild 5. a) ist das Grautonbild einer Rechnertastatur. Die Aufgabe ist hier die Extraktion möglichst langer Kontursegmente, die die Berandung der Tasten beschreiben. Zur Lösung dieser Aufgabe wurde das Grautonbild mit einer geeigneten Schwelle binarisiert und in ein konturcodiertes Bild konvertiert. Dieses Konturbild wurde nun zunächst mit einem Medianfilter (5 × 5) geglättet (Bild 5. b)). Danach wurde durch Anwendung eines Filters mit größerer Fensterfläche (7 × 7) die Kontur in stückweise glatte Bereiche zerlegt. Schließlich wurden diese glatten Bereiche einer Geradenapproximation unterworfen. Die damit in eine Vektordarstellung umgesetzten Konturen der Tasten wurden auf einem Plotter ausgegeben und sind in Bild 5. c) dargestellt.

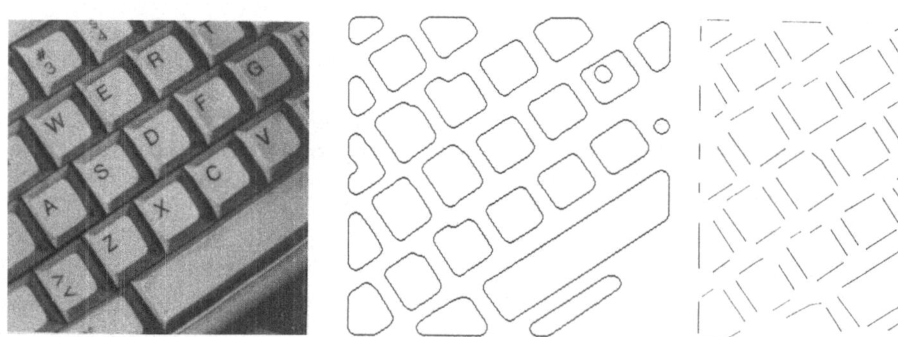

Bild 5: a) Grautonbild, b) geglättete Objektkonturen, c) Extrahierte Geradensegmente.

Das Grautonbild einer Szene mit verschiedenen einfach geformten Bausteinen zeigt Bild 6 a). Durch ein Gradientenfilter (Sobeloperator) wurden die Konturen des Grautonbildes extrahiert und durch einen Konturverfolgungsalgorithmus in ein Konturbild umgesetzt (Bild 6 b)). Mittels konturgestützter Medianfilterung wurde danach zunächst eine gewisse Grundglättung der Konturen vorgenommen und anschließend die so geglätteten Konturen in stücksweise glatte Bereiche segmentiert. Die glatten Bereiche der Konturen wurden abschließend durch Geraden- und Kreissegmente approximiert. Das Ergebnis wurde auf einem Plotter ausgegeben und ist in Bild 6 c) dargestellt.

Bild 6: a) Grautonbild, b) Objektkonturen nach Konturverfolgung, c) Extrahierte Geraden- und Kreissegmente.

Diese Arbeit wurde von der DFG im Rahmen des Schwerpunktprogramms „Methoden und Prozessorarchitekturen für hochkomplexe Signalverarbeitung in intelligenten Systemen "gefördert.

Literatur

[1] S.G. Tyan, "Median Filtering: Deterministic Properties", in: T.S. Huang ed., *Two-Dimensional Digital Signal Processing II*, Springer, New York, Berlin, 1981, Seite 205-209.

[2] Th.A. Nodes, N.C. Gallagher, "Median Filter: Some Modifications and Their Properties", *IEEE Trans. ASSP*, Vol. ASSP-30, 1982, Seite 739-746.

[3] Th.A. Nodes, N.C. Gallagher, "Two Dimensional Root Structures and Convergence Properties of the Separable Median Filter", *Trans. ASSP* Vol. ASSP-31, 1983, Seite 1350-1365.

[4] J. Fitch, et al., "Median Filtering by Threshold Decomposition", *IEEE Trans. ASSP*, Vol. ASSP-32, 1984, Seite 1183-1188.

[5] J. Astola et al., "On Root Structures of Median and Median-Type Filters", *IEEE Trans. ASSP*, Vol. ASSP-35, 1987, Seite 1199-1201.

[6] P. Maragos, R.W. Schafer, "Morphological Filters - Part II: Their Relations to Median, Order-Statistic, and Stack Filters", *IEEE Trans. ASSP*, Vol. ASSP-35, 1987, Seite 1170-1184.

[7] A. Nieminen et. al., "A New Class of Detail-Preserving Filters for Image Processing", *IEEE Trans. PAMI*, Vol. PAMI-9, 1987, Seite 74-90.

[8] H.-U. Döhler, P. Zamperoni, "Compact Contour Codes for Complex Binary Patterns", *Signal Processing*, Vol. 8, North-Holland, 1985, Seite 23-29.

[9] R. Brons, "Linguistic Methods for Description of a Straight Line on a Grid", *Computer Graphics and Image Processing*, Vol. CGIP-3, 1974, Seite 48-62.

[10] C.E. Kim, A. Rosenfeld, "On the Convexity of Digital Regions", *Proc. 5th ICPR*, Miami Beach, 1980, Seite 1010-1015.

[11] H.-U. Döhler, "Generation of Root Signals of Two Dimensional Median Filters", *Signal Processing*, North-Holland, vorauss. Nov 1989.

A Competitive/Cooperative (Artificial Neural) Network Approach to the Extraction of N-th Order Edge Junctions

Heiko Neumann, H.Siegfried Stiehl

Universität Hamburg, Fachbereich Informatik
Bodenstedtstr.16, D-2000 Hamburg 50

Abstract. We describe research toward accurate and reliable general-purpose segmentation of two-dimensional static monocular images of 3-D scenes. The computational architecture composed of competitive/cooperative processes is an extension of the Grossberg/Mingolla scheme which is a partial outcome of Grossberg's theory for preattentive visual computation. Our experimental software testbed has been designed to incorporate an initial linear filtering cascade consisting of a) firstly applying isotropic center-surround operators and b) secondly applying appropriately scaled directionally sensitive anisotropic operators. The use of different types of orientationally sensitive operators allows for both extraction of different classes of primitives and simple emergent groupings based on spatial proximity. In this article we focus on the extraction of junctions which may occur in any order n (i.e. n is the number of terminating edges) and which' constituents may have prior unknown oblique orientation.

1. Introduction

The central goal of early visual processing of images is the detection and localization of image structure. Image primitives, e.g edges and lines, either can be made explicit in a representation such as the primal sketch proposed by Marr ([13]) or - in the sense of Grossberg ([5]) - are implicitly contained in a holistic form-in-depth representation. In any such representation, these primitives along with their significant groupings should act - explicitly or implicitly - as precursors to the recognition of the three-dimensional structure of a viewed scene.

The first step for the extraction of line-like image domain primitives is the process of edge - or more precisely *contrast* - detection. Since no general purpose detection scheme has been defined so far, an overwhelming number of contributions has been published. With respect to reliable, i.e. robust, detection of edge candidate sets from discrete raw image data with prior unknown structural content, the discussion concentrates on the optimality of isotropic vs. directionally sensitive operators (e.g. [1], [10], [14], [18]). Alternatively, based on mathematical investigations as well as on results of neurophysiological studies, one of the authors has developed a framework for *hierarchical* linear feed-forward processing ([16]). The initial filtering hierarchy - which has been shown to be less noise sensitive as compared to a zero-crossing detection scheme - allows a) for the detection of a candidate set of primitives such as edges and lines, b) for the estimate of vector fields of isophotes and c) for simple proximity grouping in a discrete noisy static monocular image. Inherent positional as well as directional uncertainty of operator responses makes further processing of the outputs from the initial filtering steps necessary. In order to deal also with artifacts and noise in the output from the initial feed-forward processing scheme, subsequent feed-back network processing is required such as to both suppress insignificant operator responses and produce emergent segmentation results even in the case of distorted, thus erroneous, initial data. The design of the feed-back network described in Ch.3 is a modification and extension of the model proposed by Grossberg and Mingolla ([6], [7]). In this article we describe results of ongoing research toward a cooperation scheme for extraction of edge junctions (for initial research on the extraction of single lines only, see [16]). The need for such a computational junction detection can be straightforwardly motivated: Junctions play a crucial role in three-space inference but, on the other hand, differential operators for e.g. T-junctions may not be developed on the basis of differential geometry models of the underlying (necessarily idealized) intensity surface ([15]). Our approach differs significantly from others which have been defined e.g. on a) either the basis of special linear filters ([4]) or analysis of the local differential geometry of the intensity surface for corner detection only ([3]), b) local analysis of Gabor filter based energy representation for junction key-point detection and subsequent feature synthesis based on vector calculus ([8]), and c) comparison of spatially distributed operator responses in a local neighborhood ([9]).

The hierarchical filtering cascade is briefly reviewed in Ch.2. Ch.3 describes the components of the iterative feed-back scheme and first results of processing will be documented in Ch.4.

2. Linear Filtering Cascade

The linear filtering hierarchy for processing the raw intensity image consists of

(1) applying isotropic band-pass filters, derived from a scaled rotationally symmetric 2D Gaussian, i.e. scaled Laplacian-of-Gaussian (LoG), and

(2) filtering the LoG-output with anisotropic orientationally sensitive operators, derived from appropriately scaled and rotated 1st and 2nd order partial derivatives of a Gaussian.

Interestingly, it has been shown by Linsker ([12]) that such a cascade could emerge from self-organization of receptive field profiles by applying feed-forward multiple layer networks utilizing Hebbian weight modification.

The LoG-operators decompose the full spectrum of the 2D image function into bands of different frequencies. These band-pass operators are based on rotationally symmetric Gaussian regularization filters ([18]) and have been logarithmically scaled by the set of $\sigma \in \{1.41, 3.18, 6.01, 12.37\}$ (such as to minimize the trade-off between space and frequency window location). The impulse response functions of the discrete filter kernels have been truncated at a radius of $\frac{3}{2}\omega_{2-D}$ ($\omega_{2-D} = 2\sqrt{2}\sigma$).

Based on neurophysiological findings of cortical filters, as summarized in e.g. [2] and [11], the characteristic spatial distribution of LoG-operator responses itself is directly forwarded to a subsequent processing step rather than zero-crossings alone. In [16] zero-crossing detection has been analyzed by particularly considering the necessarily involved finite differencing operations in frequency domain. It has been shown that, with respect to the direction of contrast, the optimally suited operator for the detection of an edge-shaped intensity variation is a first derivative of a Gaussian with a space constant equal to the space constant of the corresponding LoG operator. Orthogonal to the direction of contrast, the parametrization can be varied such as to narrow the frequency components that contribute to the detection of an edge primitive. For the subsequent simulations, operators with axis ratios of 1:1 and 1:2, respectively, have been used. Examples of both LoG and first derivatives of a Gaussian are shown in Fig.1. The discrete operators were derived from appropriately sampled partial derivatives of a Gaussian (with the operator's major axis oriented along the vertical image coordinate axis). The operators were rotated by discrete angular increments of $\Delta\varepsilon = \pi/8$, resampled by using a bilinear interpolation and subsequently rescaled. Details of mathematical considerations as well as a discussion of the detection of line primitives can be found in [16].

For the computation of operator responses (viz potentials) of directionally sensitive operators, two different operations have been defined. The first was designed such that the operator only respond at loci of zero-level crossings in LoG operator output, this means at loci where a significant sign change (loci in the $\nabla^2 G * I$-array flanked by regions of maximum operator responses of opposite sign) occur. In order to implement this mechanism, excitatory and inhibitory parts of the receptive fields of edge and line detectors are splitted into separate parts. Both parts must contribute positive net input to excite an artificial cell in the sense of generating a potential, $J_{xy\varepsilon}$, at locus (x, y) for orientation ε (see [16]). An artificial cell is here meant to be an e.g. computational operator receiving weighted inputs from cells located in a neighborhood defining an artificial receptive field such as e.g. in Fig.1. Secondly, in order to let the cells also respond to smoothly varying contrasts (e.g. for extraction of vector fields of isophotes), the excitatory and inhibitory parts of the receptive fields are combined in such a way that a potential is generated whenever the sum of both parts contribute a positive net input (see [16]).

The J-potentials generated by the filter hierarchy define the net input for the feed-back network described in the following chapter.

3. Competitive/Cooperative Intra-Scale Processes

Goals of further processing are, to name a few, a) to thin out the response distribution in the neighborhood of a true primitive position, b) to close gaps within the distribution of spatially aligned operator responses and c) to extract junctions.

The design of the feed-back network is a modification and extension of the model proposed by Grossberg and Mingolla (see e.g. [6] and [7]). The basic processing stages can be summarized as follows (see Fig.2):

Stage 1: Positional competition between potentials (in a defined local neighborhood) of direction-ally sensitive operators with anisotropic receptive field profiles,

Stage 2: Directional competition between potentials of different orientations utilizing

    (a) mutual inhibition of potentials of nearly orthogonal directions at one spatial location, and

    (b) directional competition of potentials within a local neighborhood

Stage 3: Directional cooperation of spatially aligned potentials and subsequent generation of feed-back potentials

The first competitive processing stage realizes the mutual inhibition of potentials in an on-center off-surround interaction of directional sensitive cells. The excitatory component currently used is restricted to the position of a cell itself while the inhibitory neighborhood is weighted inversely proportional to the distance of a cell position with respect to the local reference cell. The size of the neighborhood is equal to the size of the diameter of the filter kernels from the initial pro-cessing cascade. Thus, the weighting functions can be easily modeled by using $\lambda^{+}_{xyij} = 2 \cdot \delta(x,y)$ (excitation; δ is the dirac impulse) and $\lambda^{-}_{xyij} = 1/(1 + d((x,y),(i,j)))$ (inhibition), respectively, with $d((x,y),(i,j)) = ((x-i)^2 + (y-j)^2)^{1/2}$. The coordinate pairs (x,y) and (i,j) in these and the following equations indicate local reference cell locus and neighborhood loci, respectively. The competition producing a new potential W has been defined via the differential equation

$$\dot{W}_{xyc} = \mathrm{net}^{+} - W_{xyc}(1 + \mathrm{net}^{-})$$

with $\mathrm{net}^{+} = I + f(J_{xyc}) \cdot \lambda^{+}_{xyij} + V_{xyc}$ and $\mathrm{net}^{-} = \sum_{ij} f(J_{ijc}) \cdot \lambda^{-}_{xyij}$ (after [6] and [7]). Since W_{xyc} converges to a stable equilibrium state, the determination of the competition potentials W_{xyc} can be derived from the equilibrium equation, $\dot{W}_{xyc} = 0$. The components of the excitatory net input are a) the tonical activity I (which is set to zero in our simulations (see [16])), b) the input potentials transformed by a signal function, $f(J_{xyc})$, and c) the feed-back activity (initially set to zero), V_{xyc}. The inhibitory net input is defined by the sum of transformed J-potentials weighted by the function λ^{-}_{xyij}.

The second competitive stages consist of a) an inhibition of potentials of nearly orthogonal direc-tions and b) a competition of potentials for all directions (in the same neighborhood as defined for the first competitive stage). The inhibition process, which generates potential X, was designed in [6] in order to reduce the directional uncertainty of operator responses at locations of image structure such as corners and terminations. For this purpose, the inhibition of potentials for orthogonal directions (including potentials of neighboring orientations weighted by a distance dependent function γ) defines an on/off-cell dipole (using the operations $X_{xy\epsilon} = \max[W_{xy\epsilon} - W_{xy\bar\epsilon}, 0]$ and $X_{xy\bar\epsilon} = \max[W_{xy\bar\epsilon} - W_{xy\epsilon}, 0]$ as well as $\bar X_{xy\epsilon} = X_{xy\bar\epsilon}$ and $\bar X_{xy\bar\epsilon} = X_{xy\epsilon}$ ($\bar\epsilon = \epsilon + \frac{\pi}{2}$)). The definition of the dipole field is necessary to avoid unintentional creation of cooperation potentials (see below). The directional competition acts like a normalization process for the potentials within a local neighborhood. The size of this neighborhood is defined to be equal to the one for positional competition. The equilibrium formula used is

$$Y_{xy\epsilon} = \frac{X_{xy\epsilon}}{1 + \sum_\theta \sum_{ij} X_{ij\theta} \cdot \lambda^-_{xyij}}$$

The previous three processing stages define the mechanisms of mutual inhibition of a spatial arrangement of potentials which are originally generated by the hierarchical filtering steps. The purpose of the following processing steps is to increase potentials at loci (x, y) where the associated orientation of a local edge segment matches the orientation of a virtual straight line extending from (x, y) to a spatially distant edge segment with appropriate orientation at (i, j). The cooperation should not be activated in cases where gaps in a field of potentials are caused by another arrangement of activations in a nearly orthogonal direction (see [7]). In order to deal with this problem, the dipoles of Y-potentials enter into the cooperation step. Thus, a Y-potential in direction ϵ, $Y_{xy\epsilon}$, occurs together with $-\bar Y_{xy\epsilon}$ and this coupled pair enters into the cooperation step.

The computation of a cooperation potential Z is based on a) the spatial weightings for the locations in the neighborhood of a point (x, y), b) a compatibility function to evaluate the support of a particular potential generated at a particular location for a given orientation, and c) the magnitude of a potential at a considered location. The compatibility between a potential at location (x, y) associated with an orientation ϵ and a potential at location (i, j) in the spatial neighborhood of (x, y), but with orientation θ, is evaluated by $|\cos(\phi - \theta)|^T$ (which measures the deviation between the virtual line ($\phi = \tan^{-1}(\frac{y-j}{x-i})$) and the associated orientation of the potential at location (i, j); T is a tuning parameter). For this cooperation process, the spatial weighting function in a local circular neighborhood has been defined as a full rosette composed of orientationally sensitive fans of equi-angular spacing (see Fig.3). The definition of the spatial weighting takes into account the requirement for crisp separation between single cooperation fans such as to minimize the overlap between two neighbors. For convenience, the generation of the fans is based on a polar coordinate system. The first part, w^ϵ_{i1}, for generating the spatial weightings is based on a shifted cosine with frequency f which determines the number of discrete orientations (fans). In order to get well-separated fans, the two following equations for odd and even directions define the first part of the weighting:

$$w^\epsilon_{i1} = \begin{cases} (\frac{1}{2}[1 + \cos(\frac{n}{2}\psi)])^m & \text{if } i \bmod 2 = 0 \\ (\frac{1}{2}[1 - \cos(\frac{n}{2}\psi)])^m & \text{if } i \bmod 2 = 1 \end{cases}$$

with $\psi \in [\frac{i-1}{n}360, \frac{i+1}{n}360]$, $i = \frac{n}{360}\epsilon$, n: number of discrete orientations, and m: control parameter for tuning of the width of each fan.

The second part of the function, w_{i2}, represents the weighting of a spatial location within a single fan with respect to its distance from the reference location. This part of the weighting function is simply

$$w_{i2} = \exp[-(\frac{4d_{xyij}}{3\omega_{2-D}} - 1)]$$

with $\omega_{2-D} = 2\sqrt{2}\sigma$ for the parametrization of the original filter kernels. The total weighting of the influence of a particular cell at location (i, j) on (x, y) is $w^\epsilon_{xyij} = w^\epsilon_{i1} \cdot w_{i2}$. Based upon this formulation, the spatial weights of a discrete fan are defined by

$$F^{\epsilon\theta}_{xyij} = w^{\epsilon}_{xyij} \cdot |\cos(\phi - \theta)|^{T}.$$

The local compatibility of an operator response at location (i,j) for direction θ with the response for direction ϵ at (x,y) is evaluated on the basis of the product of $i)$ the spatial weighting associated with the virtual line between (x,y) and (i,j) and $ii)$ the difference of local directions, weighted by the strength of operator response represented in the dipole $(Y_{ij\theta}, \bar{Y}_{ij\theta})$. The cooperative support of an operator potential at location (x,y) for direction ϵ from other operator responses within the local neighborhood defined for a particular fan is computed by

$$S_{xy\epsilon} = \sum_{ij} \sum_{\theta} (Y_{ij\theta} - \bar{Y}_{ij\theta}) \cdot F^{\epsilon\theta}_{xyij}$$

In order to activate a cooperation potential Z (as a function of S), which is subsequently fed back into the network loop, at least two distinct fans of the full rosette must generate positive excitation at a spatial location (x,y). Therefore, the rosette cooperation mechanism can adaptively act as both an nth-order junction detector (for e.g. T-, Y- or L-junction, see Fig.4) and, in case of noise-induced artifacts and gaps, a mechanism for enhancement of potentials related to straight linear contrast variations and gap filling.

Finally, the Z-potentials serve as input to a feed-back competition step. The competition process to produce V-potentials (feedback activities) is activated separately for each direction ϵ (see [16] for details).

Moreover, the analysis of conditions for convergence of the feed-back loop, a proposal for a simple winner-take-all inter-scale inhibition scheme as well as a mathematical analysis of the relationship between estimated isophote fields and the corresponding surface geometry in the scene can be found in a previous paper ([16]).

4. Processing Results and Prospects

An assessment of the network capabilities has been undertaken on the basis of an experimental strategy which can be easily realized within a carefully defined testbed ([17]). For instance, Figure 5 shows the capability of the cooperation mechanism to close gaps in the simple case of a set of synthetically generated potentials arranged to form an L-junction. Figure 6 shows resulting potentials from the iterative feed-back network processing of a T-junction configuration which may be input potentials to a multi-level architecture either for preattentive emergent visual computation in the sense of Grossberg or for token labelling à la Marr's primal sketch.

In this article the potentiality of iterative feed-back processing for the extraction of a class of image domain primitives has been demonstrated. Further research has been initiated toward a) a general-purpose processing scheme for preattentive segmentation, e.g. the detection of curved edge and line segments, interaction of cells of different specificity, scale-space processing, perceptual organization, etc., and b) the establishment of a computerized perception laboratory.

Literature

[1] Canny, J.F.: A Computational Approach to Edge Detection. IEEE Trans. on Pattern Analysis and Machine Intelligence 8 (6) (1986), 679-698.

[2] Frisby, J.P.: Seeing. Oxford: Oxford University Press (1980).

[3] Dreschler, L.S.; Nagel, H.-H.: On the Selection of Critical Points and Local Curvature Extrema of Region Boundaries for Inference Matching. Proc. 6th IJCPR, Munich, FRG, Oct.19-22 (1982), 542-544.

[4] Gennert, M.A.: Detecting Half-Edges and Vertices in Images. Proc. Conf. on CVPR, Miami Beach, USA, June 22-26 (1986), 552-557.

[5] Grossberg, S.: Cortical Dynamics of Three-Dimensional Form, Color, and Brightness Perception: I. Monocular Theory. Perception & Psychophysics 41 (1987), 87-116.

[6] Grossberg, S.; Mingolla, E.: Neural Dynamics of Form Perception: Boundary Completion, Illusiory Figures, and Neon Color Spreading. Psychological Review 92 (2) (1985), 173-211.

[7] Grossberg, S.; Mingolla, E.: Neural Dynamics of Perceptual Grouping: Textures, Boundaries, and Emergent Segmentations. Perception & Psychophysics 38 (1985), 141-171.

[8] Heitger, F.; Gerig, G.; Rosenthaler, L.; Kübler, O.: Extraction of Boundary Key-Points and Completion of Simple Figures. Proc. 6th Scandinavian Conf. on Image Analysis (1989), preprint.

[9] Hsieh, C.-H.: A Connectionist Algorithm for Image Segmentation. University of North Carolina at Chapel Hill, Dept. of Computer Science: TR89-008 (1989).

[10] Korn, A.: Towards a Symbolic Representation of Intensity Changes in Images. IEEE Trans. on Pattern Analysis and Machine Intelligence 10 (5) (1988), 610-625.

[11] Lindsay, P.H.; Norman, D.A.: Human Information Processing - An Introduction to Psychology (2nd Ed.). New York: Academic Press (1977).

[12] Linsker, R.: Self-Organization in a Perceptual Network. IEEE Computer 21 (1988), 105-117.

[13] Marr, D.: Vision. San Francisco: W.H.Freeman & Co (1982).

[14] Marr, D.; Hildreth, E.C.: Theory of Edge Detection. Proc. of the Royal Society of London 207 (B) (1980), 187-217.

[15] Nagel, H.H.: Principles of (Low Level) Computer Vision. Proc. Conf. on Fundamentals in Computer Understanding: Speech, Vision, Natural Language (1985).

[16] Neumann, H.: Extraction of Image Domain Primitives with a Network of Competitive/Cooperative Processes. in: Brauer, W. (Ed.): Informatik Fachberichte 181 "Künstliche Intelligence" (Proc. GWAI-88) (1988), 265-274.

[17] Neumann, H.; Stiehl, H.S.: Toward a Testbed for Evaluation of Early Visual Processes. Mathematical Research 40: "Computer Analysis of Images and Patterns", Proc. CAIP'87 (1987), 202-208.

[18] Torre, V.; Poggio, T.: On Edge Detection. MIT, AI Laboratory: AI Memo 768 (1984).

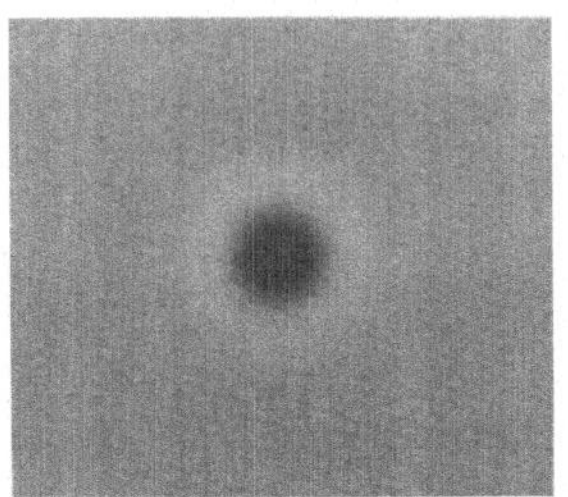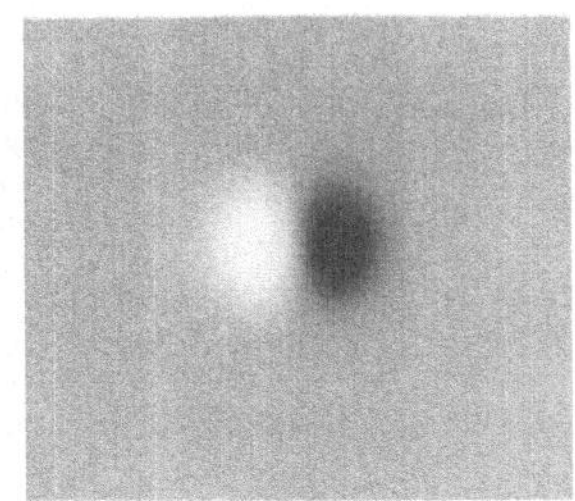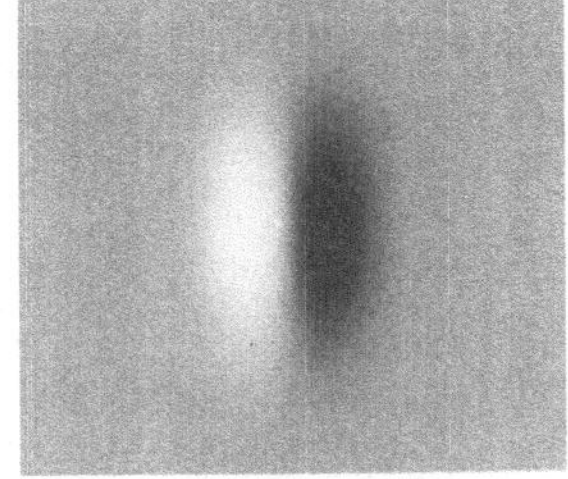

Fig.1: Examples of receptive fields for LoG (first processing step) and first derivative operators (based on isotropic and anisotropic Gaussians, respectively)

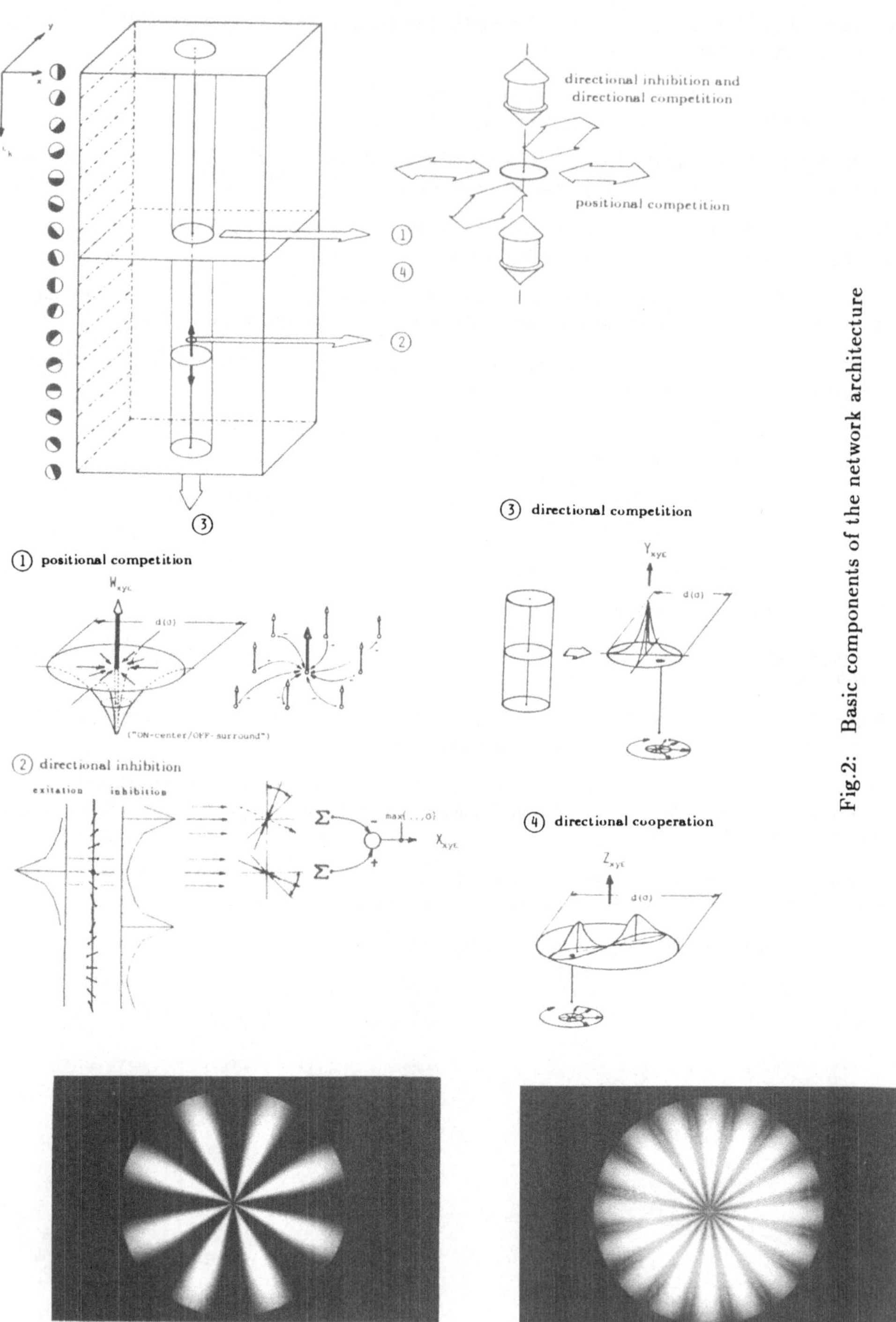

Fig.2: Basic components of the network architecture

Fig.3: Spatial weightings defined for cooperation (left: for odd numbered orientations only; right: for the full rosette)

Fig.4: Cooperation fans activated for different junction types

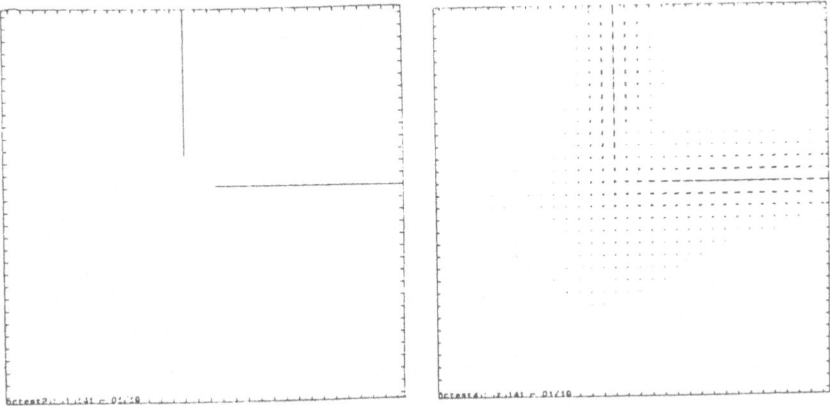

Fig.5: Synthesized J-potentials for an L-junction with a gap at central junction key point and its vicinity; Z-potentials after three iterations

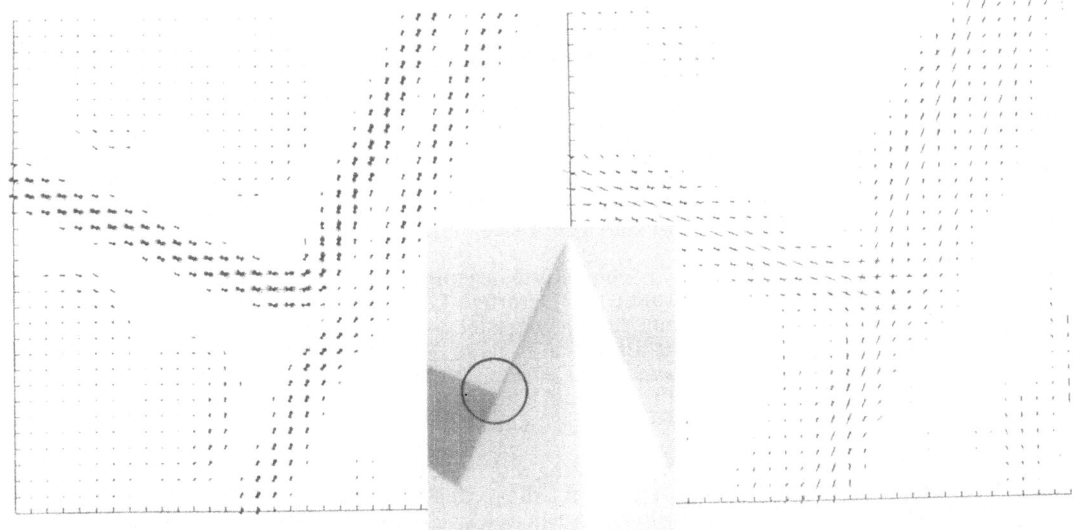

Fig.6: Computational results of J- and V-potentials for a detail of a discrete (CCD) camera image of a an illuminated physical 3D object in a laboratory scene

STEUERUNG VON ERKENNUNGSPROZESSEN DURCH BAUMSUCHVERFAHREN

Astrid Banzhaf
Ulrich Boes
Manfred Kramer

AEG Electrocom GmbH
Bücklestr.1-5, D-7750 Konstanz

Zusammenfassung

Es wird ein allgemeiner Ansatz zur maschinellen Bearbeitung von Bürodokumenten mit frei gestaltetem Layout beschrieben. Der Lösungsansatz beruht auf einer ergebnisabhängigen Steuerung der für Teilaufgaben vorgesehenen Spezialisten innerhalb des gesamten Erkennungsprozesses. Hierbei werden mögliche Lösungen als Blätter eines zu entwickelnden Suchbaumes aufgefaßt und ein im Sinne einer geeignet vorzugebenden Kostenfunktion optimaler Lösungsweg schnittweise ausgewählt. Die praktische Durchführbarkeit des Verfahrens wird anhand einer Fallstudie zur Segmentierung verklebter Schriftzeichen aufgezeigt und durch mehrere Anwendungsbeispiele erhärtet.

1. Einleitung

Die Informations- und Computertechnik dringt immer weiter vor in alle Bereiche der Arbeits- und Privatwelt. Computer sollen den Menschen unterstützen und ihn von Routinearbeiten entlasten. Von allen Arbeitsvorgängen ist allerdings die Dateneingabe am wenigsten automatisiert. Nach wie vor werden Daten über Tastatur, eventuell mittels Maus oder Digitalisiertablett eingeben. Zunehmend werden Texterkennungsprogramme auf dem Markt angeboten. Bei näherer Betrachtung stellt sich heraus, daß viele angebotene Systeme nicht voll den Benutzererwartungen entsprechen. Beispielsweise kann nur Text gelesen werden, der maschinell auf eine saubere Vorlage gedruckt wurde.
Hier wird ein systematischer Ansatz vorgestellt, der es - aufbauend auf bisherigen Erfahrungen - ermöglicht, Dokumente allgemeiner Art, die Graphiken und Bilder enthalten können, zu analysieren und deren Inhalt zu erkennen. Insbesondere werden Störungen, wie Verschmutzungen, verklebte und zerfallende Schrift und anderes mehr berücksichtigt.

2. Ein Konzept zum automatischen Lesen von Dokumenten

Die Dokumente, die im Büroalltag verwendet werden, sind jedem aus eigener Erfahrung bekannt. Briefe, Notizen, Berichte zeigen ein vielfältiges Layout und enthalten neben Texten auch Bilder und Graphiken. Das Layout, das heißt die geometrische Gestaltung eines Dokumentes kann standardisiert und damit vorher bekannt sein, es kann aber auch völlig frei aufgebaut sein. Soll eine derart auf Papier vorliegende Information automatisch oder zumindest halbautomatisch in ein Rechensystem eingegeben werden, sind komplexe Erkennungsprozesse anzuwenden.

Dazu wurden in der Vergangenheit verschiedene Konzepte entwickelt [1, 2] und Methoden der künstlichen Intelligenz wurden angewandt [3]. Derartige Überlegungen führen zu regelbasierten Erkennungssystemen, wie zum Beispiel in [3] vorgestellt. Sie sind Gegenstand der Forschung. Zielrichtung hier ist eine Umsetzung in verkaufsfähige Realisierungen. Deswegen wurde aus diesem allgemeinen Konzept ein Systemaufbau abgeleitet, der praktischen Bedürfnissen entgegen kommt. Dieses System besteht aus einer Ablaufsteuerung, einer Datenbasis und den einzelnen Erkennungsalgorithmen oder Spezialisten [4].

Die Spezialisten werden von der Ablaufsteuerung aufgerufen, und greifen auf die Datenbasis zu, um Daten und Zwischenergebnisse zu lesen, zu verändern und neue einzutragen. Zu den Spezialisten gehören Module wie Bildvorverarbeitung, Extraktion von photographischen Bereichen, Zusammenhangsanalyse, Unterscheidung von Inhaltskategorien, Segmentierverfahren, Zeichenklassifikation und Kontextanalyse. Alle Daten und Zwischenergebnisse sind in der Datenbasis abgelegt. Sie ist entsprechend einem hierarchisch organisierten Datenbank organisiert.

Der hierarchische Aufbau entspricht der geometrischen Gestaltung - dem Layout - des Dokumentes und ist so auch in der standardisierten Architektur von Bürodokumenten enthalten [5]. Die verwendete Datenbasis ist objektorientiert und gemäß der Aufgabenstellung flexibel strukturiert, Objekte können gelöscht oder neu erzeugt werden. Die Hierarchie kann jederzeit umgebaut werden, wie es während des Erkennungsvorgangs notwendig werden kann.

Wesentliches Element ist die Ablaufsteuerung; wie bereits angedeutet, wurden hierfür verschiedene Modelle entwickelt. Der am meisten verbreitete Ansatz ist die sogenannte Blackboard-Architektur nach dem Expertensystem HEARSAY [6]. In der Praxis herrscht jedoch nach wie vor der sequentielle Ablauf der Einzelalgorithmen vor. Dennoch hat sich die Erkenntnis durchgesetzt, daß der sequentielle Ablauf nicht ausreicht, um alle Fälle der Dokumentanalyse zu bearbeiten, zumal dann, wenn Störungen vorhanden sind, wie etwa Kopierränder, handschriftliche Korrekturen, Stempelaufdrucke. Für einen praktisch wichtigen Fall - die Erkennung verklebter Schriftzeichen - wurde diese starre Verarbeitungsfolge aufgebrochen, um durch mehrmaligen Aufruf einzelner Spezialisten die Erkennungsgüte zu erhöhen. Das wird in den nächsten Kapiteln gezeigt.

Der geschilderte Ansatz basiert darauf, daß der Erkennungsvorgang durch einen Baum als spezielle Art eines Graphen modelliert wird. Die Verarbeitung wird durch einen Algorithmus gesteuert, der einen optimalen Weg durch diesen Baum finden soll. Dieses Modell konnte für den einen Anwendungsfall erfolgreich angewendet werden, so daß eine Verallgemeinerung angestrebt und auch in diesem Beitrag vorgestellt wird.

3. Anforderungen einer erweiterten Dokumentanalyse

Eine erweiterte Dokumentanalyse muß Dokumente mit den Bestandteilen Photos, Graphiken, Text, Logos, Tabellen, etc. verarbeiten können. Dazu ist es notwendig, die Bestandteile Photo, Text und Graphik getrennt voneinander zu bearbeiten und abzuspeichern, so daß sie später wieder auf Bildschirm oder Drucker bei Wunsch zusammengefügt werden können.
Hierzu werden im ersten Schritt eventuell vorhandener Schmutz auf dem Dokument entfernt und eine Auftrennung von Photos und den restlichen Bestandteilen des Dokuments vorgenommen. Der Bildbestand eines Dokuments wird sodann als Pixelbild kodiert und komprimiert auf einem Datenträger gespeichert. Das restliche Dokument wird in Text und Graphik aufgespalten. Die Graphikteile werden als Graphikprimitive, also Vektoren, gespeichert. Der verbleibende Textteil wird der Einzelzeichenerkennung zugeführt, so daß der Text anschließend als ASCII-Text ebenfalls abgespeichert werden kann.
Diese zu verarbeitenden Dokumente dürfen mit allen möglichen Schrifterzeugungsgeräten, wie Schreibmaschinen, Tintenstrahl-, Laser- und Matrixdrucker erzeugt werden. Auch Handschriften und Satzmaschinen sollten zugelassen sein.
Bei der Bearbeitung solcher Dokumente tauchen Gestaltungsmerkmale wie Spaltendruck, Unterstreichungen, Fettdruck, etc. sowie Störfälle, wie Schmutz, Stempel, Unterschriften und Textkorrekturen, auf. Diese erweisen sich bei der Dokumentanalyse als Problemfälle. Und diese Problem- und Störfälle stellen an die Segmentierung gesteigerte Anforderungen. Mit folgenden Segmentierungsproblemen muß man also zurechtkommen können:
- Graphiken erkennen und vom Text trennen;
- verklebte Unterstreichungen (durchgezogen oder gestrichelt) vom Text trennen und entfernen;
- miteinander verklebte Zeilen separieren;
- verklebte Zeichen je nach Teilungsart mit verschiedenen Algorithmen voneinander trennen;
- zerfallene Schriftzeichen miteinander verschmelzen und dann mit den oben erwähnten
    Algorithmen wieder trennen;

Wie durch die Sammlung umfangreichen Stichprobenmaterials festgestellt werden konnte, wird sich die Problematik der verklebten Zeichen in Zukunft weiter häufen durch die zunehmende Verbreitung von Proportionalschriften, zu eng gedruckter fester Teilung, Handbeschriftung und Buchdruck (variable Teilung).

Eine Studie zum Thema Verklebungen beschäftigte sich mit allen Segmentierproblemen, die durch Verklebungen, bzw. Zerfallen entstehen. Insbesondere wurden Verklebungen von Zeichen, Zeilen, von Zeichen mit ihren Unterstreichungen und von Zeichen mit Graphikelementen auf Dokumenten untersucht. Die Ergebnisse sollen hier kurz festgehalten werden:

Oftmals werden Unterstreichungen, die mit Text verschmolzen sind, als Graphik erkannt. Dies muß durch eine neue Definition der Graphikerkennung verhindert werden. So sollten z.B. nur "gleichmäßige" (ohne "Ausbuchtungen" durch Buchstaben) Linien als Graphiken gewertet werden.

Bereiche, die linienhafte Objekte enthalten, müssen mit Hilfe von geometrischen Kenndaten identifiziert und dann als Histogramm der Überlappungen der Scanzeilen aufgezeichnet werden. Überschreitet das so gewonnene Höhen-Breiten-Verhältnis der Histogrammspitze einen vorgegebenen Schwellwert, so liegt ein waagerechter Strich im Pixelbild vor, der dann "ausradiert" werden kann. Dieses Verfahren ist unempfindlich gegenüber Graphikobjekten, die nicht auf einer Waagerechten liegen. Bei unterbrochenen Unterstreichungen wird nach der Klassifikation auf erkannte Unterstreichungen geprüft und deren Lage im Pixelbild festgehalten. Bei nicht gut erkannten Zeichen in der Nachbarschaft unterstrichener Zeichen wird die vermutliche Unterstreichung, deren Lage jetzt bekannt ist, entfernt und das Zeichen dann erneut klassifiziert.

Verklebte Schriftzeichen werden an ihrer Überbreite erkannt. Um zu bestimmen, was Überbreite ist, werden mit Hilfe der oben erwähnten Zeichenanalyse maximale und minimale Zeichenbreite aus dem Kontext entnommen. Sehr schmale verklebte Zeichen können jedoch nur dadurch erkannt werden, daß alle schlecht erkannten Zeichen, die breiter als zweimal minimale Breite sind, ebenfalls als verklebt betrachtet werden.
Sodann werden die jeweils geeigneten Schnitte durchgeführt, d.h. bei fester Teilung werden die Schnitte dort angesetzt, wo sie durch das Teilungsverhältnis bereits vorgegeben sind. Bei proportionaler Teilung werden die Schnittstellen durch verschiedene aufwendigere Verfahren (beschrieben im nachfolgenden Kapitel) ermittelt. Es ist zunächst beabsichtigt, nur einfache, vertikale Schnitte durchzuführen. Bei ausgeprägter Zeichenschräglage, z.B. bei stark geneigter Handschrift oder Kursivschrift, kann es jedoch nötig werden, Schnitte entsprechend der Schräglage zu führen.

4. Lösungsansatz für ein Teilproblem

4.1 Beschreibung des Baumsuchverfahrens

Die Steuerung von Erkennungsprozessen durch Baumsuchverfahren wird im folgenden anhand der Segmentierung verklebter Schriftzeichen illustriert [7], [8]. Die Auflösung der Verklebungen erfolgt dadurch, daß alle möglichen Segmentierungsalternativen in einen Suchbaum eingefügt und jeweils nur die aktuell beste Lösung weiterverfolgt wird.

Schnittpositionen und Bewertungsmaß
Im einfachsten Fall werden miteinander verklebte Schriftzeichen durch senkrechte, gerade Schnitte in ihre mutmaßlichen Bestandteile getrennt. Es sei $S = (S_1,...,S_k)$ eine derartige Folge von Zeichen mit $n(S)$ als der Position des zuletzt geführten Schnittes. Durch Ausschneiden eines weiteren möglichen Zeichens S erhält man eine erweiterte Folge (S,S). Die Erkennungsqualität des entstandenen Zeichens S als auch dessen statistischer, typographischer und sprachlicher Kontext fließen zunächst in ein lokales Bewertungsmaß $g(S,S)$ ein. Um jedoch auch die gesamte Zeichenfolge nach Hinzufügen dieses Zeichens bewerten zu können, wird zusätzlich eine globale Bewertungsfunktion rekursiv durch $f(S,S) = f(S)*g(S,S)$ definiert.

Uniform-Cost-Verfahren
Das im folgenden beschriebene Baumsuchverfahren verwendet nun einen Suchbaum, an den bei jedem weiteren Schnitt die möglichen neu erzeugten Einzelzeichen S als Blätter angefügt werden. Die zugehörige gesamte Zeichenfolge erhält man dann aus dem eindeutig bestimmten Pfad zum Wurzelknoten des Baumes, der das ursprüngliche verklebte Schriftzeichen darstellt. Der Algorithmus läßt sich nunmehr mit den vorstehend eingeführten Bezeichnungen formal wie folgt beschreiben:

Erzeuge den Wurzelknoten des Suchbaums und initialisiere die Liste OPEN. Setze $S = ()$ und $f() = 1$. Aktuell bester Knoten K ist die Wurzel.

WHILE n(S) < = Zeichenende DO
  (i)   Füge alle durch Schnitte entstandenen neuen Zeichen S mit n(S,S) > n(S) als Söhne von K
        im Baum an.
  (ii) Entferne den Blattknoten K aus der Liste OPEN und füge die in (i) erzeugten Nachfolger von
        K dort ein.
  (iii) Bestimme für jeden Blattknoten in OPEN die globale Kostenfunktion f(S,S) = f(S)*g(S,S).
  (iv) Suche den (für alle Baumtiefen) optimalen Blattknoten K* in der Liste OPEN mit maximaler
        Bewertung f(K*) und setze K = K*.
END

Eine wichtige Eigenschaft des Verfahrens ist, daß nach endlich vielen Schritten stets ein globales
Maximum der Bewertungsfunktion über dem Suchraum gefunden wird.

Die Menge der Lösungen wächst exponentiell mit der Anzahl der Bildspalten. Da für jede
Zeichenklassifikation ein erheblicher Rechenaufwand erforderlich ist, sind zusätzliche heuristische
Verfahren erforderlich, um die Menge der bei der Expansion eines Knotens entstehenden
Nachfolgeknoten so klein als möglich zu halten.

Bei Schriften mit fester Teilung, wo Verklebungen durch zu eng gedrucktem Text entstehen, ist eine
ziemlich genaue Abschätzung der Zeichenbreite möglich. Der Suchraum ist dann durch die
minimale und maximale Zeichenbreite wesentlich eingeschränkt.
Bei Schriften mit proportionaler Teilung kann man auch die Projektion der Schwärzungsverteilung
auf die Horizontale ausnutzen, um potentielle Trennstellen zu lokalisieren. Allerdings genügt es
nicht, die Minimalwerte der Schwärzung allein dafür zu verwenden, vielmehr bieten sich die
zweiten Differenzen als zusätzliches Kriterium zur Schnittfindung an.
Eine weitere Möglichkeit zur Einschränkung des Suchraums ergibt sich aus der Verschmelzung
aller benachbarten Blattknoten mit demselben Bedeutungsinhalt des zugehörigen Zeichens.

4.2 Ergebnisse

Mit Hilfe der Verklebungsanalyse konnte bei unseren Testdokumenten eine Senkung der Sub-
stitutionsrate um 12 % erreicht werden.
Alle Verklebungen auf den getesteten Dokumenten wurden als solche sicher erkannt. Von den
verklebten Strings konnten 54% richtig aufgelöst, 30% nur zum Teil richtig (d.h. ein oder mehrere
Zeichen waren auch nach der Auftrennung noch substituiert) und 16% aller verklebten Strings
waren völlig falsch aufgelöst.

Es sind durchschnittlich 30 Schnitte und Klassifikationen notwendig, um zwei Zeichen bei
Schnittschritten von einem Pixel und einer Differenz von 5 Pixeln zwischen minimaler und maximaler Zeichenbreite zu trennen.

5. Ein systematischer Ansatz zur Steuerung des Gesamtlesesystems

Wie vorausgegangenen Abschnitt darstellt, konnte ein Teilproblem der Dokumentenanalyse erfolgreich durch ein Baumsuchverfahren gelöst werden. Dabei handelt es sich um einen systematischen Ansatz zur Steuerung der Einzelalgorithmen, der Spezialisten. Die Frage stellt sich, ob
dieser Ansatz erweitert werden kann, um den gesamten Erkennungsvorgang zu steuern. Dieser
Grundgedanke soll hier umrissen werden. Ziel ist, von einem systematischen und theoretisch
fundierten Ansatz auszugehen, der schließlich zu einer effizienten und eleganten Implementierung
führt. Wir gehen von der Voraussetzung aus, daß zunächst nur der Dokumenteninhalt und die
Layoutstruktur erkannt wird.

Eine weitere Voraussetzung sei, daß bestimmte - hier nicht näher spezifizierte - Bildverarbeitungsoperationen und eine Zusammenhangsanalyse abgelaufen sind, bevor die eigentliche
Erkennung aktiviert wird. Die Eingangsdatenstruktur besteht also aus sogenannten Zusam-
menhangsobjekten, die im folgenden kurz Primärobjekten genannt werden.

Zunächst muß definiert werden, was die Knoten (Blätter) und die Kanten (Äste) des Baumes
bedeuten. Der Knoten repräsentiert ein Objekt, das mit einer Hypothese - dem Erkennungs-

ergebnis - und einer Güte - dem Schätzwert - belegt wird. Im allgemeinen Fall müssen die folgenden Hypothesen eines Objektes ermittelt werden:

- Kategorie (Text, Graphik, Schmutz)
- Wert (Erkennungsergebnis)
- Relation zu anderen Objekten
- Eigenschaften (Einzelobjekt, zusammengesetztes Objekt, Teilobjekt)

Diejenige Hypothese mit dem höchsten Schätzwert wird ausgewählt.

Den Kanten wurde bei der Verklebungsanalyse die Operation "Schneiden" zugewiesen. Hier müssen beliebige Segmentieroperationen zugelassen werden um von einem Knoten zum nächsten zu gelangen. Die Kanten stellen somit Ereignisse dar.
Bei der Aufzählung der Hypothesen, die zu einem Objekt gehören, wurde bereits angedeutet, daß einem Objekt nicht immer ein Inhaltswert zugeordnet werden kann. Dazu wurden zusammengesetzte Objekte - zum Beispiel verklebte Schriftzeichen - und Teilobjekte - Punkte einer Matrixschrift - eingeführt. Beide Fälle erfordern nachfolgende Aktionen, nämlich das Zerlegen von Objekten, bzw. das Zusammenfassen zu größeren Einheiten. Eine Übertragung in die Datenbasis erfolgt nur dann, wenn dem Objekt als Einzelobjekt ein sinnvoller Wert zugewiesen wurde. Einzelne Erkennungsmodule laufen nacheinander ab, solange es sinnvoll ist. So kann der Einzelzeichenklassifikator aktiviert werden, wenn festgestellt wurde, daß ein Primärobjekt genau ein Schriftzeichen repräsentiert. Wird während der Vorverarbeitung entschieden, daß es sich um ein zusammengesetztes Objekt handelt, wird nicht mehr klassifiziert, da diesem Objekt kein sinnvoller Wert zugewiesen werden kann. Stattdessen wird - wie früher beschrieben - das zusammengesetzte Objekt zerschnitten, um in der nächsten Ebene des Abarbeitungsbaums sinnvolle Objekte zu erzeugen. Wird ein "Teilobjekt" erkannt, werden Nachbarn des gleichen Typs gesucht, so daß ein Einzelobjekt zusammengebaut werden kann.

Um die Layoutstruktur zu bestimmen, müssen größere Einheiten wie Wort, Zeile, Abschnitt gebildet werden. Dazu dient die Nachbarschaftsbeziehung des zu analysierenden Objektes. Sobald die Beziehung zu den Nachbarn feststeht und etwa eine größere Lücke erkannt wurde (Wortabstand oder Zeilenende) kann die entsprechende Relation in die Datenbasis übertragen werden. Ein neuer Knoten im Abarbeitungsbaum wird nicht erzeugt. Während der Suche nach dem optimalen Weg durch den Baum kommt es vor, daß ein fälschlich eingeschlagener Weg wieder zurückgegangen werden muß. Folglich müssen entsprechende Korrekturen auch in der Datenbasis nachgezogen werden, nämlich Objekte müssen gelöscht oder neu in die Hierarchie eingehängt werden. Das ist aber bereits in der Grundkonstruktion der Datenbasis vorgesehen.

Diese Überlegungen weisen einen Weg auf, komplexe Probleme der Dokumentanalyse durch einen systematischen Ansatz zu lösen. Im Verlauf einer weitergehenden Untersuchung und Implementierung müssen verbleibende offene Fragen geklärt werden.

Bibliographie

[1]   K.Y. Wong, R.G. Casey, F.M. Wahl: "Document Analysis System", IBM Journal of Research and Development 26 (1982), S. 647

[2]   J. Schürmann: "Reading Machines", Proc. Eth ICPR 1982, München, S. 1031-1044

[3]   Th. Bayer: "Dokumentinterpretation und Analysestrategie in einem Frame-System" Proceedings 10. DAGM Symposium, Zürich vom 27.-29.09.88, S. 284 -S. 290

[4]   J. Angele, U. Boes, N. Schultes: "Moderne Datenerfassung: Automatisches Lesen von Dokumenten", Informatik- Fachberichte 156, 17. Jahrestagung der GI; Oktober 1987

[5]   ISO-International Standard 8613: "Information Processing - Text and Office Systems - Office Document Architecture (ODA) and Interchange Format", März 1988

[6]   L.D. Ermann, F. Hayes-Roht, V.R. Lesser, D.R. Reddy: "The HEARSAY-II speech understanding system: Integrating knowledge to resolve uncertainty, Computing Surveys 12(2):213-253

[7]   Th. Bayer: "Segmentierung von verklebten Einzelzeichen mit einer heuristischen Suchstrategie", Mustererkennung 1987, Paulus E. (ed) Springer Verlag Berlin 1987

[8]   J. Pearl: "Heuristics: Intelligent Search Strategies For Computer Problems Solving", Addison-Wesley Publishing Company, 1984

Image Segmentation for the Recognition of Characters on Different Materials

M. Dehesa, C.-E. Liedtke

Institut für Theoretische Nachrichtentechnik und Informationsverarbeitung
Universität Hannover
Appelstr. 9 A, D–3000 Hannover 1
Fed. Rep. Germany

Abstract

Image segmentation is a fundamental problem in scene analysis. In optical character recognition (OCR), uniform illumination and good contrast between characters and background are desirable to achieve a useful segmentation. These conditions cannot be guaranteed for industrial and outdoor scenes. In this paper we present a segmentation method for grey–scale images which has been successfully applied on rubber wheels, glass bottles and traffic signs as preprocessing before the actual character recognition is done. The method is based on a model of human vision. A measure for the segmentation error of characters is proposed. Using this measure, a knowledge based segmentation procedure is implemented. Segmentation and classification results are presented and discussed.

Index Terms

Segmentation; local adaptation, character recognition.

1. Introduction

Image segmentation is a fundamental problem in scene analysis. A meaningful segmentation simplifies enormously the task of interpreting image contents. Many approaches to image segmentation have been proposed in the past ([1], [2], [3], [4],[5]). Usually in OCR the characters to be read are suitably illuminated before scanning them with some optical sensor. It's thus easy by thresholding to get proper blobs which can be used for recognition purposes. In industrial applications the same can be done if there is enough contrast between the characters and the background. In general, it is difficult to achieve a homogeneous illumination if the object has a complicated geometry. Furthermore, illumination can be controlled to some extent in industrial setups but hardly in outdoor scenes.

2. The segmentation system

The system used for image segmentation is shown Fig. 1. The solid lines connecting the blocks indicate flow of control or parameter information. The dashed lines indicate flow of image data.

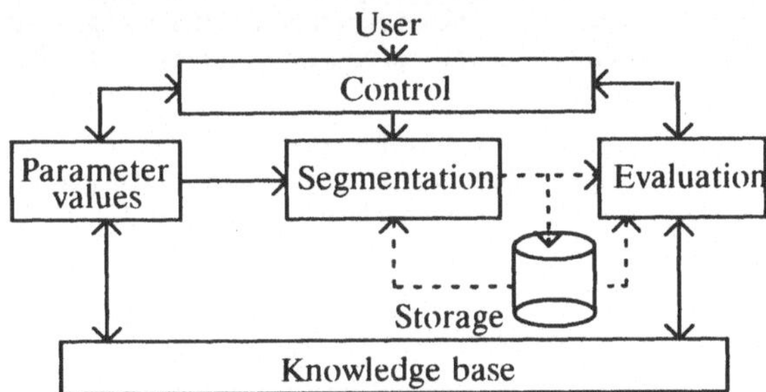

Fig. 1. The segmentation system.

Initially the user provides information to the system about the characters to be read (e.g. their width, their relative contrast to the background, etc.). Using a *knowledge base*, the *control unit* determines the appropriate initial *parameter values* for the segmentation process. The values are provided to the *segmentation unit*. This unit processes an image which is taken from the storage. The segmented image is stored and also provided to the *evaluation unit*. This unit judges the quality of segmentation using the gradient of the original image and the *knowledge base* (see sections 4 and 5). Depending on the evaluation result, the *control unit* repeats the whole process or only parts of it, until an optimal quality is achieved.

The image processing for the proposed system was implemented on a digital computer using the programming language C. The knowledge base was implemented in the language for the programming of expert systems OPS5.

3. The segmentation model

The proposed segmentation model which is used in the system presented in section 2 is shown in Fig. 2.

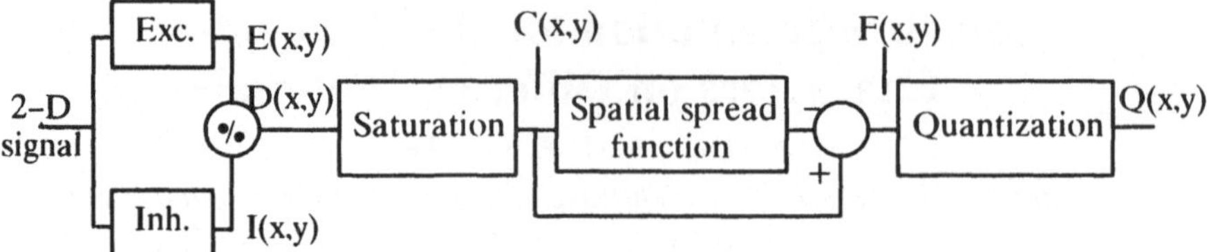

Fig. 2. The proposed segmentation model

The first stage of the model consists of a filter with parallel excitation and inhibition paths. Such a filter has been used by different authors to model part of the signal processing perfomed by the human eye. This excitation–inhibition filter (*E–I filter*) has a band–pass characteristic which can be modeled as

$$D(x,y) = \frac{U_e(x,y)}{U_i(x,y)} \quad . \tag{1}$$

where $U_e(x,y)$ and $U_i(x,y)$ are spatial spread functions which are given in polar coordinates by

$$U_e(r) = e^{-r^2/2\sigma_e^2} \quad . \tag{2}$$

$$U_i(r) = e^{-r^2/2\sigma_i^2} \quad . \tag{3}$$

Based on experimental data, Girod [7] proposed a *saturation stage* after the *E–I filter* to model a decrease in sensitivity when the signal $D(x,y)$ overrides certain limiting values. For this work, the signal $D(x,y)$ was clipped according to the following rule:

$$C(x,y) = \begin{cases} c_{min} & \text{if} & D(X,Y) < c_{min} \quad . \\ D(x,y) & \text{if} & c_{min} <= D(X,Y) <= c_{max} \quad . \\ c_{max} & \text{if} & D(X,Y) > c_{max} \quad . \end{cases} \tag{4}$$

The constants c_{min} and c_{max} are determined using a priori knowledge given by the user (section 5).

In the model of Lukas and Budrikis [6] an additional signal corresponding to the background is assumed to be available for predicting image quality. For pattern recognition applications this is a very limiting condition. Therefore, in the present work an estimation of the background is made via a filter with a *spatial spread function*. In this filter, a mean operator of constant amplitude and with circular support was used. Let the radius of the support be denoted by r_{mean}.

Finally, the amplitude values of the difference signal between the input and the output of the *spatial spread function* filter are quantized. The *quantizer* used is scalar and in general non–uniform. The curve of the quantizer is shown in Fig. 3. The location of the decision levels z_1 and z_2 is variable. This helps to control the sensitivity to low amplitude disturbances.

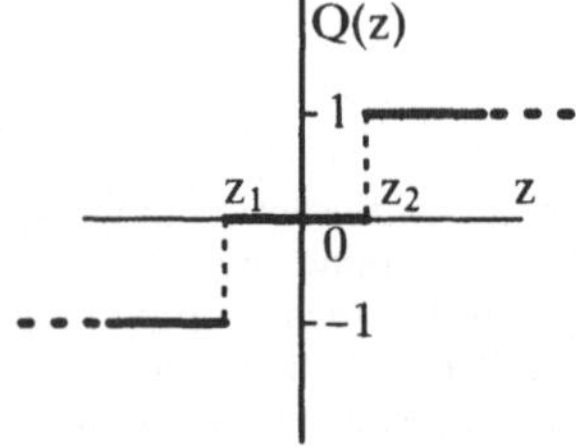

Fig. 3. Curve for the quantization of the amplitude of D(x,y) (see text).

Resuming, the model has 7 free parameters. namely: the standard deviations σ_e and σ_i for the *E–I filter*, the limiting constants c_{min} and c_{max} for the *saturation*, the radius r_{mean} for the *spatial spread function*, and the decision levels z_1 and z_2 for the *quantizer*.

4. Segmentation error

For an automated adaptation to changing viewing conditions it is desirable to quantify the quality of segmentation. A measure of the discrepancy between the borders found by segmentation and the actual ones will now be introduced. This measure is called *segmentation error*. We begin our discussion by considering the one–dimensional case.

Let the smooth curve shown in Fig. 4 represent the edge of an object. Suppose that after segmentation the border of the object is at point B. Assuming that the actual border should be at the point with maximal gradient magnitude (point A), the error of missing the correct border position may be quantified by

$$e(B) = |\,Grm(B) - Grm(A)\,|, \tag{5}$$

where $Grm(x)$ denotes the gradient magnitude at the point x. The location of point A is usually unknown, so we have to make an estimation of it. This is achieved by examining the values of the gradient magnitude in a neighborhood of radius δ around the point B. Actually, using the information of the gradient phase the search must be done only in the direction of the gradient at point B (indicated by a row in Fig. 4). The maximal value found is used in (5) instead of $Grm(A)$. In general, then,

$$e(x) = |\,Grm(x) - \max_{w \in N(x)}\{Grm(w)\}\,|, \tag{6}$$

where

$$N(x) = \{w\,|\ w \text{ lays in the direction of the gradient at point } x$$
$$\text{and } |B - w|) < \delta\}. \tag{7}$$

In the two–dimensional case the gradient phase information is used to get the proper direction in which the maximum needed for equation (6) should be searched. The total error for a given contour is computed by adding the error values of its individual points and dividing the sum through the number of contour points. The points for which no greater value is found are called here *true contour points*.

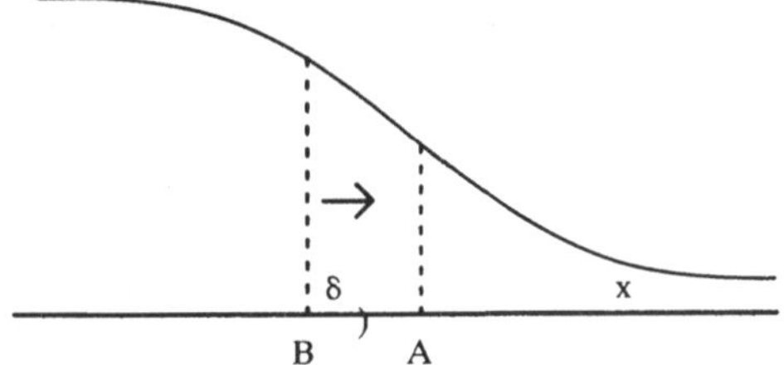

Fig. 4. Smooth edge. The ideal border should be at point A and the border found is at point B. The maximum gradient value is searched in the direction indicated by the row (see text).

5. Knowledge based segmentation

The user must provide the following a priori infomation to the system: 1) the size, and 2) the thickness of the characters, 3) whether the characters are darker or brighter than the background, 4) the quality of the character contours (good, middle, bad), and 4) whether there is no, some, much or severe texture in the image. In the *knowledge base* some heuristics are stored in form of rules for:
1. selecting appropriate parameters for the segmentation, and
2. judging the quality of the result.

Some examples of rules for the first task are:
if (the characters have a thickness T) then (use $r_{mean} = 2{*}T$)
if (the quality of the contours is good and there is less noise in the image) then
(use $\sigma_e = 1$ and $\sigma_i = 5$)
if (the characters are darker than the background)   then (use $c_{min} = 0.2$ and $c_{max} = 0.9$)
if (the characters are brighter than the background) then (use $c_{min} = 0.9$ and $c_{max} = 0.2$)

The first time, the segmentation is done with $z_1 = 1$ and $z_2 = -1$. For judging the quality of the result the *segmentation error* is computed. Afterwards a set of rules is applied for deciding if the whole segmentation or parts of it should be repeated. Examples of the rules applied in this case are:
if (the percentage of *true contour points* is greater than 0.9) then stop
if (the difference between old and new *segmentation error* values is less than 0.005) then stop
if ($z_1 > 40$ or $z_2 < -40$) then stop.

If no such rule applies, then z_1 is incremented and z_2 is decremented both by 1, and the *quantization* is repeated. Other rules detect the case when in the resulting image no meaningful blobs are found (based for example on the size of the characters). In this case the *control unit* changes the initial values of the parameters and the whole segmentation is repeated.

6. Results

The system proposed was used to segment different kinds of images. The images shown have different sizes and 8 bits/pel. Three output quantization levels were used (see Fig. 3). In the segmented images the output value 1 of the quantizer is displayed in white, the value 0 in grey, and the value –1 in black. In two of the application cases some classification results are presented. The detailed discussion about the classification methods used lies beyond the scope of this paper.

In Fig. 5(a) the image of a rubber wheel with some characters on its surface is shown. Note the variations of the brightness in the areas corresponding to characters. Based on the gradient magnitude of the original image, the areas where possibly characters are present are preselected. Only in this areas the segmentation is perfomed (Fig. 5(b)). The blobs which were automatically selected as possible characters were extracted, rotated, and displayed again on the original image for comparison purposes. The result is shown in Fig. 5(c).

(a)

(b)

(c)

Fig. 5. (a) Original. (b) Segmentation. (c) The extracted and rotated characters are displayed on the original image.

Another case of interest is that of characters on glass bottles (Fig. 6(a)). Here the optical properties of the material are responsible for undesirable transparency and reflection effects. The characters are not very clear to see and the surface of the bottle presents different grey tones. In the segmented image (Fig. 6(b)), the characters can be easily differentiated from the background.

Sometimes the characters may be embedded in a complex environment where it is difficult to decide automatically what parts of the image could correspond to characters. Such a case is shown in Fig. 7(a). The segmented image is shown in Fig. 7(b). In comparison to the original image, the segmented one is much simpler. Nevertheless, it is still difficult to select the blobs which correspond to characters. Using a contour analysis method similar to that of Scherl [8], the contents of the segmented image are classified into text and graphic components. The text component is shown in Fig. 7(c). The characters have small size and appear more distorted than they actually are because of the dithering algorithm used to obtain printable images.

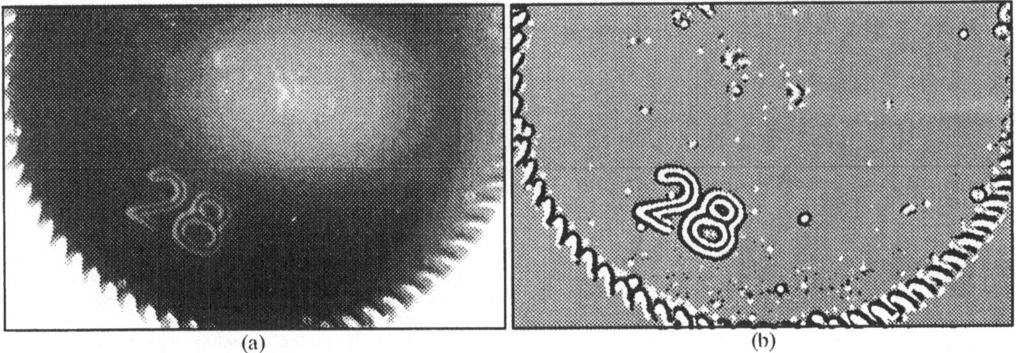

Fig. 6. (a) Original (332x512 pels). (b) Segmentation.

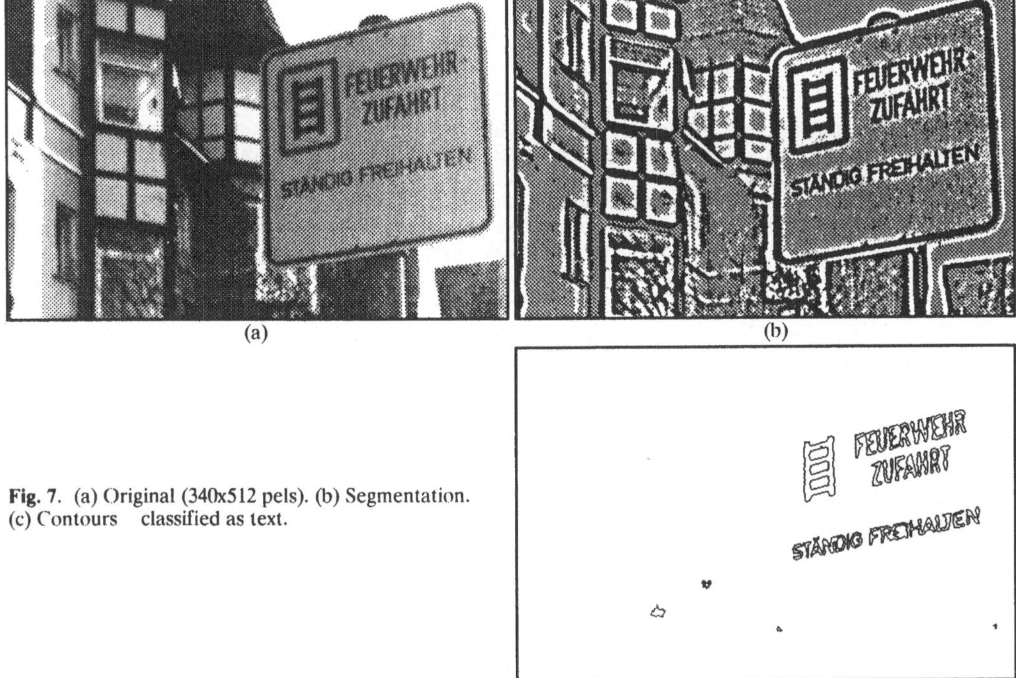

Fig. 7. (a) Original (340x512 pels). (b) Segmentation.
(c) Contours   classified as text.

References

1. Rosenfeld, A. & Davis, L.S., "Image Segmentation and Image Models", *Proc. IEEE*, vol. 67, pp. 764–772, May 1979.
2. Pérez, A. & Gonzalez, R.C., "An Iterative Thresholding Algorithm for Image Segmentation", *IEEE Trans. Pattern Anal. and Machine Intell., vol. PAMI-9*, pp.742–751, Nov. 1987.
3. Coleman, G.B. & Andrews, H.C., "Image Segmentation by Clustering", *Proc. IEEE*, vol. 67, pp. 773–785, May 1979.
4. Michel, X., Leonardi, R. & Gersho, A., "Unsupervised Segmentation of Texture Images", in *Proc. SPIE Conf. on Visual Communications and Image Processing*, vol. 1001, pp. 633–640, 1988.
5. Beaulieu, J.-M. & Goldberg, M. "Hierarchy in Picture Segmentation: A Stepwise Optimization Approach", *IEEE Trans. Pattern Anal. and Machine Intell.*, vol. PAMI-11, pp. 150–163, Feb. 1988.
6. Lukas, F.X.J. & Budrikis, Z.L. "Picture Quality Prediction Based on a Visual Model", *IEEE Trans.on Communications*, vol. COM-30, pp. 1679–1692, July 1982.
7. Girod, B. *Ein Modell der menschlichen visuellen Wahrnehmung zur Irrelevanzreduktion*. Düsseldorf: VDI-Verlag, 1988. (Reihe 10: Informatik/Kommunikationstechnik, Nr. 84)
8. Scherl, W., *Bildanalyse allgemeiner Dokumente*. Berlin, Heidelberg, New York: Springer-Verlag, 1987.

Rotationswinkelbestimmung in abgetasteten Dokumentbildern

ANDREAS DENGEL & EBERHARD SCHWEIZER

Deutsches Forschungszentrum für Künstliche Intelligenz (DFKI)
Erwin-Schrödinger-Straße, Postfach 20 80,
6750 Kaiserslautern

Zusammenfassung: Das Papier behandelt ein wenig angesprochenes, jedoch essentielles Problem im Bereich der Low-Level Bildverarbeitung von Dokumenten: Die Bestimmung des optimalen Rotationswinkels, um ein abgetastetes Dokumentbild rechnerintern in eine Lage zu bringen, sodaß die enthaltenen Textzeilen waagrecht verlaufen. Das Papier stellt zwei aus der Literarur bekannte Ansätze vor und stellt sie einem von uns entwickelten neuen Verfahren gegenüber, das hinsichtlich der Geschwindigkeit der Winkelbestimmung wesentliche Verbesserungen bringt. Dabei werden alle drei Verfahren kritisch beeuchtet und an Beispielen ein Vergleich bezgl. ihrer Laufzeit und Genauigkeit aufgezeigt.

1 Einleitung

Zur automatischen Verarbeitung von Papierdokumenten ist es oft erforderlich, das durch einen Scanner abgetastete Bild rechnerintern zu justieren. Viele Segmentierungsverfahren sind zwar in der Lage Zusammenhangsgebiete schwarzer Pixel zu finden, berücksichtigen dabei jedoch die Drehlage des abgetasteten Dokuments nicht. Wenn es dann aber darum geht - im Falle von Text - die Zusammenhangsgebiete von Zeichen zu Worten zusammenzufassen, sind sie abhängig von waagrecht abgetasteten Textzeilen.

Daher geht ein Großteil von Segmentierungsverfahren von der Idealvorstellung bereits justierter Vorlagen aus und das Problem wird nicht weiter betrachtet. Die Notwendigkeit dieses Verarbeitungsschrittes scheint auf den ersten Blick nicht sonderlich zwingend, doch einem Menschen gelingt es, ohne besonders vorsichtig zu sein, nur selten ein Dokument mit einer Schieflage von weniger als 3° unter einen Scanner zu legen /1/. Auch die Praxis hat gezeigt, daß es besonders bei gefalteten Dokumenten selbst bei Scannern mit Anschlagleisten nicht möglich ist, diese sauber in den Scanner zu legen. Dazu kommt die "natürliche Drehung" eines Dokuments, die z.B. beim Einspannen des Papiers in eine Schreibmaschine entstehen kann. Daher sollte ein Justierungsverfahren auf einige wünschenswerte Eigenschaften hin überprüft werden:

- gutes Laufzeitverhalten;
- Toleranz gegenüber Verschmutzungen, d.h. Knicke, Flecken (z.B. durch Lochungen oder Büroklammern) u.ä. haben keinen Einfluß auf das Ergebnis der Justierung;
- Graphik und Bilder verfälschen das Ergebnis der Justierung nicht;
- Toleranz gegenüber unterschiedlichen Schriftarten und -größen, unterstrichene, fettgedruckte, kursive Zeichen, Zeichen aus anderen Alphabeten (z.B. kyrillische oder chinesische Schriftzeichen).

In diesem Papier werden zwei bekannte, verschieden vorgehende Justierungsverfahren beschrieben, zum Teil weiterentwickelt und ihre spezifische Eigenschaften erläutert. Die Verfahren werden einem neu entwickelten Ansatz gegenübergestellt. Beispiele sollen dabei die jeweiligen Vor- und Nachteile aufzeigen

2 "Simulated Skew Scan"-Methode

Ein Verfahren, das /5/ zur Justierung vorschlägt verwendet zur Bestimmung des Rotationswinkels die gesamte Bildmatrix des abgetasteten Dokuments. Postl nennt sein Verfahren *"Simultated skew scan"*-Methode. Damit soll ausgedrückt werden, daß das Verfahren den Scanner in einer Weise simuliert, in der die Bildmatrix mit verschiedenen Winkeln durchlaufen wird (vgl. Abb. 1).

Für jeden dieser Winkel wird dabei eine Zahl (*"premium"*) berechnet. Diese Zahl ist umso höher, je genauer der Winkel des simulierten Abtastvorgangs mit dem tatsächlichen Rotationswinkel übereinstimmt. Zur Berechnung des Premiums $P(\alpha)$ für einen vorgegebenen Winkel α werden zunächst $L+1$ verschiedene Zeilen aus der Bildmatrix herausgegriffen. Diese werden dann um den Winkel α gedreht. Den erhaltenen Linien durch die Bildmatrix (vgl. Abb. 2) werden nun *"kumulierte Grauwerte"* $T(y_i,\alpha)$ zugeordnet (y_i ist die Bildzeile, aus der die i-te ausgewählte Linie durch Rotation hervorgegegangen ist). Dazu greift man einzelne Punkte der Linie heraus und addiert ihre Grauwerte. Das *"premium"* berechnet sich daraus als *Summe der quadrierten Differenzen "kumulierter Grauwerte" benachbarter Linien*, formal durch

$$P(\alpha) = \sum_{i=-\infty}^{+\infty} (T(y_i,\alpha) - T(y_{i-1},\alpha))^2 \tag{1}$$

ausgedrückt. Wobei $T(y_i,\alpha)=0$ gesetzt wird, falls die um α gedrehte Bildzeile y_i außerhalb der Bildmatrix liegt. Dabei sind sowohl der Abstand der Bildzeilen $\Delta y = y_{i+1}-y_i$, als auch der Abstand der Punkte entlang einer Zeile $\Delta x = x_{i+1}-x_i$ äquidistant. Die Entwicklung des gesamten Formalismus ist in /7/ beschrieben.

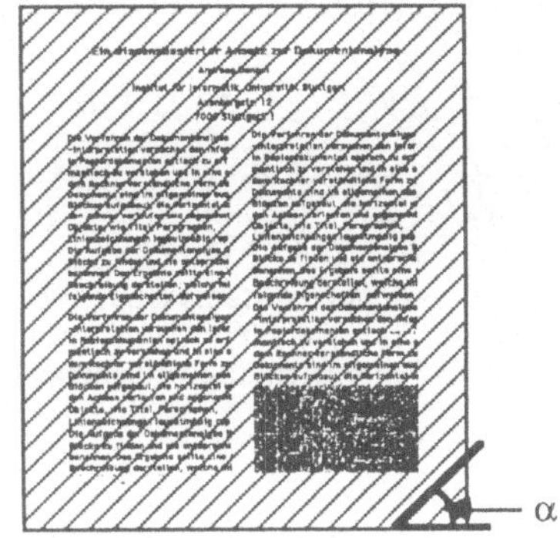

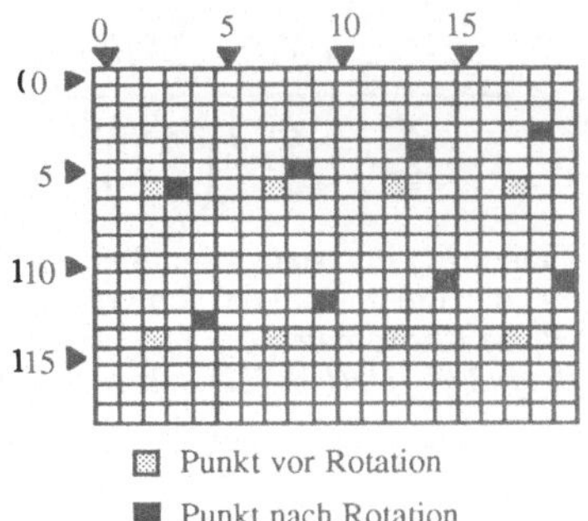

Abb. 1: Simulation des Scanners mit Winkel α.   **Abb. 2:** Bsp. einer gedrehten Bildzeile ($\alpha = +10°$)

Eigenschaften der "*Simulated skew scan*"-Methode:

• Die Digitalisierung ist nicht auf binäre Grauwerte beschränkt;

• die Laufzeit ist unabhängig vom Dokumentaufbau und hängt lediglich von den Größe der Bildmatrix ab;

• Verschmutzungen beeinträchtigen die Ergebnisse nur, wenn sie in zu starkem Ausmaß vorkommen;

• das Verfahren ist invariant gegenüber verschiedenen Schriftarten; Graphik beeinträchtigt die Justierung nur, wenn sie eine andere Richtung als der Text hat;

• Schwierigkeiten bereitet die Wahl von geeigneten Parametern: Sie bestimmt jedoch den Grad der Effizienz und die Genauigkeit des Verfahrens;

• können Grenzen bestimmt werden, zwischen denen der Wert des Rotationswinkels liegt, so kann die Suche des Rotationswinkels auf dieses Intervall beschränkt werden (bewirkt Effizienzsteigerung);

• die Parameter Δx und Δy können in Abhängigkeit von der Auflösung gewählt werden, so daß die Laufzeit des Verfahrens unabhängig von der Genauigkeit der Rasterauflösung wird.

3 Methode der kleinsten Quadrate

Ein anderes Verfahren /6/ arbeitet nur mit dem linken Rand der Bildmatrix. Der Rotationswinkel α eines gescannten Dokuments, das in Form einer $n \times m$ Bitmatrix M vorliegt wird wie folgt berechnet:

1.Schritt: Berechnung von Vektors V mit $dim(V)=n$ und $V[i]=j \Rightarrow (\forall 0 <j_l< j: M[i,j_l]=0)\wedge (M[i,j]=1)$.

D.h. V enthält als i-ten Eintrag den Abstand des ersten Schwarzpixels der Zeile i vom Rand. Falls Zeile i kein Schwarzpixel enthält, so wird $V[i]=0$ gesetzt.

2.Schritt: Bestimmung von Geradensegmenten $S : y = ax + b$

Hierzu werden die Punkte $P (i / V[i])$ als Stützpunkte herangezogen. Zur Berechnung des die Punkte $P_j, P_{j+1}, ..., P_k$ approximierenden Geradensegments $S_{j,k}$ wird das *Kriterium der kleinsten Quadrate /2/* verwendet. D.h. die Geradenparameter a und b werden so bestimmt, daß gilt:

$$\forall a,b \in \Re: \quad \sum_{i=j}^{k} (a_{j,k}*i + b_{j,k} - V[i])^2 \leq \sum_{i=j}^{k} (a*i + b - V[i])^2 \tag{2}$$

Die Anzahl der zur Berechnung von $S_{j,k}$ verwendeten Stützstellen $j-k+1$ ist nicht fest, sondern sie hängt von einem Schwellwert Δ ab. Es wird gefordert, daß die Summe $F_{j,k}$ der Fehlerquadrate kleiner als Δ ist:

$$F_{j,k} := \sum_{i=j}^{k} (a_{j,k}*i + b_{j,k} - V[i])^2 \leq \Delta \tag{3}$$

Zur Bestimmung der Länge eines Geradensegments $S_{j,k}$ wird die Halbierungsmethode verwendet, d.h. beginnend mit $j = 0$ und $k = n - 1$. Falls $F_{j,k} > \Delta$, so werden für zwei neue Segmente $S_{j, j+(k-j+1)/2}$ und $S_{j+1+(k-j+1)/2, k}$ jeweils a, b und der Fehler F berechnet. Die Geradenparameter a und b können dabei mit folgendem allgemeinen Ansatz gewonnen werden:

geg.: $n+1$ Punkte $P_0, ... , P_n$ mit $P_i = (x_i / f_i)$ &

Menge $\{\phi_j / 0 \leq j \leq m\}$ von $m+1$ linear unabhängigen Funktionen $(m < n)$

ges.:   Linearkombination $\sum_{j=0}^{m-1} c_j\phi_j(x) =: \Phi(x)$, die den Fehler $F := \sum_{i=0}^{n-1} (\phi(x_i) - f_i)^2$ minimiert.

Durch Verwendung sogenannter Normalgleichungen (vgl. /7/) erhält man daraus

$$a_{j,k} = \frac{\sum_{i=j}^{k} 1 \cdot \sum_{i=j}^{k} i * V[i] - \sum_{i=j}^{k} V[i] \cdot \sum_{i=j}^{k} i}{\sum_{i=j}^{k} 1 \cdot \sum_{i=j}^{k} i^2 - \left(\sum_{i=j}^{k} i\right)^2} \quad , \quad b_{j,k} = \frac{\sum_{i=j}^{k} i^2 \cdot \sum_{i=j}^{k} V[i] - \sum_{i=j}^{k} i \cdot \sum_{i=j}^{k} i * V[i]}{\sum_{i=j}^{k} 1 \cdot \sum_{i=j}^{k} i^2 - \left(\sum_{i=j}^{k} i\right)^2} \qquad (4)$$

Da der Fehler $F_{j,k}$ invariant gegnüber einer Verschiebung des Koordinatensystems ist, läßt sich unter der Annahme, daß die Anzahl der Punkte j-k+1 ungerade ist, folgende Transformationvornehmen. Anstatt dem Punkt P_i die Koordinaten *(i / V[i])* zuzuweisen, wird er um *(j+k)/2* in *x*-Richtung verschoben und somit zu *(i+ (k+j)/2 / V[i +(j+k)/2])* mit *i= -(k-j)/2, ... , (k-j)/2*.

Damit ergibt sich: $\quad a_{j,k} = \dfrac{\sum\limits_{i= -(k-j)/2}^{(k-j)/2} i * V[i+(k+j)/2]}{\sum\limits_{i= -(k-j)/2}^{(k-j)/2} i^2} \quad , \quad b_{j,k} = \dfrac{\sum\limits_{i= -(k-j)/2}^{(k-j)/2} i * V[i+(k+j)/2]}{k\text{-}j\text{+}1}$ $\qquad (5)$

In /6/ werden die Formeln $\quad a = \dfrac{\sum i * V[i]}{\sum i^2}\quad$ und $b = \dfrac{\sum V[i]}{\textit{Anzahl der Elemente von V}}\quad$ angegeben, ohne dabei

die Summationsgrenzen zu benennen. *b* ist jedoch offensichtlich falsch, da der *y*-Achsenabschnitt der Approximationsgeraden des Segments $S_{j,k}$ sicher nicht von der Gesamtzahl der Punkte abhängt.

3. Schritt: Berechnung des notwendigen Rotationswinkels:

Dazu werden die Steigungen $a_{j,k}$ der Segmente $S_{j,k}$ in Winkel $\alpha_{j,k}=arctan(-1/a_{j,k})$ (in Grad) umgerechnet. In /6/ wird dabei $arctan(-1/a_{j,k})$ anstatt $arctan(a_{j,k})$ benutzt, um zu vermeiden, daß die Punkte *(i, 0)* eine Verfälschung des Resultats bewirken, d.h. die Winkel werden bzgl. der Horizontalen statt der Vertikalen benutzt (vgl. Abb. 3). Jedem Punkt P_i ($j \leq i \leq k$) des Segments $S_{j,k}$ wird somit durch die Steigung $a_{j,k}$ ein Winkel $\alpha_{j,k}$ zugeordnet. Die Häufigkeiten der auf ganzzahlige Werte gerundeten Winkel zwischen -$45°$ und +$45°$ werden im Vektor *häuf* festgehalten. Der maximale Wert bestimmt den notwendigen **Rota**tionswinkel für das abgetastete Dokumentbild.

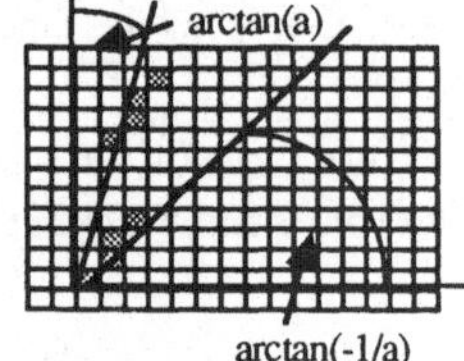

Abb. 3: Winkel bzgl. der Horizontalen bzw. Vertikalen.

Unter Beachtung nachfolgender Punkte kann man bei Verwendung **des** in /6/ genannten Verfahrens eine Genauigkeit bis zu $\pm 0.5°$ erreichen:

- Die Quantisierung liefert eine Binärmatrix
- der linke Rand des Dokuments darf nicht stark verschmutzt sein, da sonst der Vektor *V* verfälscht wird;
- Graphik im Dokument kann das Ergebnis nur beeinflußen, wenn Schwarzpixel der Graphik in die Berechnung von *V* einfließen;
- die Wahl des Schwellwerts Δ hängt von der Auflösung des Rasterbildes und der Größe der verwendeten Schrift ab;
- Das Dokument sollte etliche Absätze enthalten (ansonsten kann sich die Berechnung des Winklels nur auf verhältnismäßig wenig Punkte stützen, da durch *arctan(-1/a)* der Horizontalwinkel bestimmt wird.

Die Laufzeit des Verfahrenskann nachhaltig verbessert werden, wenn die Punkte P_i *(i , 0)* zur Berechnung der Segmente *S* ausgeschlossen werden, d.h. die Geradenparameter werden durch die Formeln (4) und nicht durch (5) bestimmt. Das derartig modifizierte Verfahren besitzt dieselben Eigenschaften wie das ursprüngliche Verfahren, jedoch kann der Schwellwert nicht ohne Änderung übernommen werden.

4 Linke Blockrandsuche

Das Verfahren der *linken Blockrandsuche*, das in /4/ zur Justierung vorgeschlagen wird, ist ein Verfahren, das den Rotationswinkel unter Betrachtung des linken Randes berechnet. Dabei wird im ersten Schritt ebenfalls der Vektor *V* bestimmt , der im *i*-ten Eintrag die Position des ersten Schwarzpixels der *i*-ten Bildzeile enthält. Jedoch werden hier - im Gegensatz zu /6/ - schon von vornherein diejenigen Punkte eliminiert, die für die Bestimmung des Rotationswinkels nicht in Frage kommen und darüberhinaus das Ergebnis der Justierung in Effizienz und Qualität beeinträchtigen können.

Im nächsten Schritt werden aus dem Vektor *V* Geradensegmente gebildet, die jedoch so konstruiert sind, daß sie aus einer Menge sinnvoll zusammenhängender Punkte bestehen. Dies geschieht dadurch, daß versucht wird, Geradensegmente aus einzelnen Dokumentteilen zu bilden, die für sich genommen (im justierten Zustand) linksbündige Ränder haben. Die Geradensegmente werden bei dem Verfahren der *linken Blockrandsuche* sozusagen "bottom-up" ermittelt, während sie in /6/ "top-down" festgelegt werden.

Im einzelnen besteht das Verfahren aus folgenden Schritten:

1. Schritt: Berechnung des Vektors V (s. Abschnitt 3)

2. Schritt: Bestimmung von Geradensegmenten $S : y = ax + b$

a) Bestimmung der Punkte, die zur Berechnung eines Geradensegments S_k herangezogen werden

Der Punkt P_{i+1} *(i+1 / V[i+1])* (wobei $V[i+1] \neq 0$, d.h. Zeilen ohne Schwarzpixel finden keine Berücksichtigung) wird dem Geradensegment S_k zugerechnet, falls er eine der folgenden Bedingungen erfüllt

- P_{i+1} ist benachbart zu seinem Vorgänger P_i, d.h. Abstand $d(P_i,P_{i+1}) = 1$
- P_{i+1} ist in der "Nähe" von P_i, d.h. $d(P_i,P_{i+1}) \leq 1.5 * \vartheta$

ϑ ist die Zeilenhöhe in Pixeln. Zu Beginn der Bestimmung eines Geradensegments ist ϑ die in Pixel angegebene Höhe einer Zeile bei 6-Pixel-Schrift. Der absolute Anfangswert von ϑ hängt somit nur von der Rasterauflösung ab. Werden im Zuge der Berechnung der Punktmenge, die zu einem Geradensegment gehört, mehr als ϑ direkt benachbarte Punkte gefunden, so wird ϑ entsprechend erhöht. Damit liegt P_{i+1} bei Erfüllung des zweiten Kriteriums höchstens eine Dokumentzeile weiter als P_i. Sobald P_{i+1} keine der beiden Aussagen erfüllt, sind alle Punkte des Segments S_k bestimmt. P_{i+1} wird dem nächsten Segment S_{k+1} zugeteilt.

b) Bestimmung des Parameters a

Der Geradenparameter a wird wie bei dem in Abschnitt 3 beschriebenen Verfahren mit Hilfe der Methode der kleinsten Quadrate berechnet:

$$a = \frac{\sum\limits_{P_i \in S} 1 * \sum\limits_{P_i \in S} i*V[i] - \sum\limits_{P_i \in S} V[i] * \sum\limits_{P_i \in S} i}{\sum\limits_{P_i \in S} 1 * \sum\limits_{P_i \in S} i^2 - \left(\sum\limits_{P_i \in S} i \right)^2} \tag{6}$$

3. Schritt: Berchnung des Rotationswinkels

Wie in /6/ wird nun aus den Steigungen a_k ein Winkel α_k berechnet. Jedoch stützt sich das Verfahren der *linken Blockrandsuche* nicht auf die Zeilenstruktur, sondern auf den linken Rand von Blöcken. Somit gilt

$$\alpha_k = arctan\ a_k\ (\alpha_k\ in\ Grad)$$

Das Ergebnis der Winkelhäufigkeiten wird ebenfalls in einen Vektor *häuf* eingetragen und der dominante Winkel als notwendiger Rotationswinkel angenommen.

Das Verfahren der *linken Blockrandsuche* weist folgende Eigenschaften auf:

- Das Verfahren setzt linksbündig geschriebene Textblöcke voraus. Es schlägt daher bei ausschließlich zentriert geschriebenen Dokumenten fehl (kommt jedoch in der Praxis selten vor);
- das Dokument muß in Form einer Binärmatrix vorliegen;
- bezüglich Verschmutzungen und Graphiken im Dokument gelten dieselben Aussagen wie in Abschn. 3;
- dagegen berechnet das Verfahren nur so viele Geradensegmente, wie auch tatsächlich vorhanden sind (in /6/ muß u.U. halbieren, d.h. aus $S_{i,k}$ werden $S_{i,j}$ und $S_{j,k}$);
- das Verfahren ist unabhängig von Schrifttyp und -größe, insbesondere wegen der dynamischen Anpassung der Zeilenhöhe;
- die Geradensegmente sind i.d.R. kürzer als die, die in /6/ bestimmt werden, was sich in seltenen Fällen nachteilig auf die Genauigkeit des Ergebnisses auswirken kann;
- die Bestimmung des y-Achsenabschnitts b eines Geradensegments entfällt, da die Berechnung der Punkte eines Geradensegments unabhängig ist vom Fehler

$$F = \sum\limits_{P_i \in S} (a*i + b - V[i])^2.$$

Das in diesem Abschnitt vorgestellte Justierungsverfahren kann unter Beibehaltung seiner wesentlichen Eigenschaften durch Vorschalten eines *Filters* verändert werden /7/ Der *Filter* eliminiert alle Punkte aus dem Vektor V, die keinen Nachbarpunkt haben. Ziel der Filterung ist es, die Bildung sehr kleiner, das Ergebnis negativ beeinflußender Geradensegmente zu verhindern.

5 Vergleich der Verfahren

Die in diesem Papier erläuterten Verfahren sollen in diesem Abschnitt unter den Aspekten Effizienz (CPU-Zeit), Ergebnis und Ergebnissicherheit einander gegenübergestellt werden. Das Ergebnis resultiert dabei aus dem maximalen Vertrauensmaß für einen bestimmten Winkel. Ergebnissicherheit soll die Deutlichkeit ausdrücken, mit der sich das Vertrauensmaß des Ergebniswinkels von dem anderer Winkel abhebt.

Der Vergleich wird mit Hilfe von drei Beispieldokumenten vorgenommen. Es handelt sich dabei um einen Geschäftsbrief, einen Zeitungsausschnitt (relativ verschmutzt) und einen Artikel (vgl. Abb. 4 bis 6). Alle drei Dokumente wurden mit jeweils 2 unterschiedlichen Winkeln betrachtet. Die grauen, waagrechten Linien, die von der rechten Seite des Dokumentbilds ausgehen, stellen die Elemente des Vektors V dar.

Tabelle 1 zeigt, daß das Verfahren der *linken Blockrandsuche* hinsichtlich der Effizienz am besten abschneidet. Die *"Simultated skew scan"*-Methode ist mit Abstand das langsamste Verfahren, jedoch können hier durch Vergrößerung der Parameter Δx und Δy noch Verbesserungen erreicht werden. Die Tabelle zeigt weiter das Verhalten der drei Verfahren bezüglich der Genauigkeit der berechneten Ergebnisse auf. Dabei wird ersichtlich, daß die *"Simultated skew scan"*-Methode die zuverlässigsten Resultate liefert. In Bezug auf die Ergebnissicherheit schneiden die beiden anderen Verfahren jedoch besser ab.

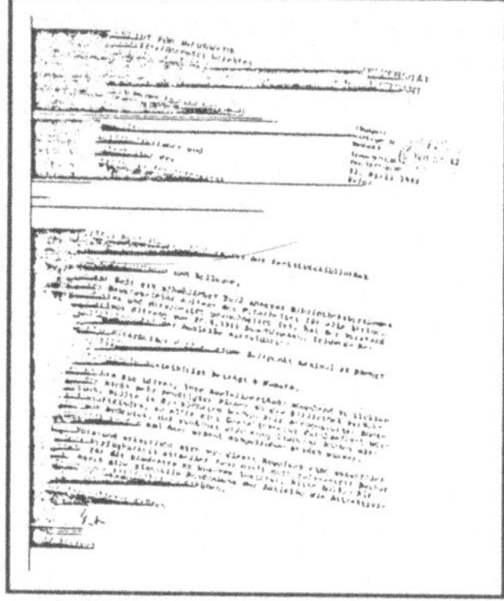

Abb. 4: Dokument 1 (-10°).

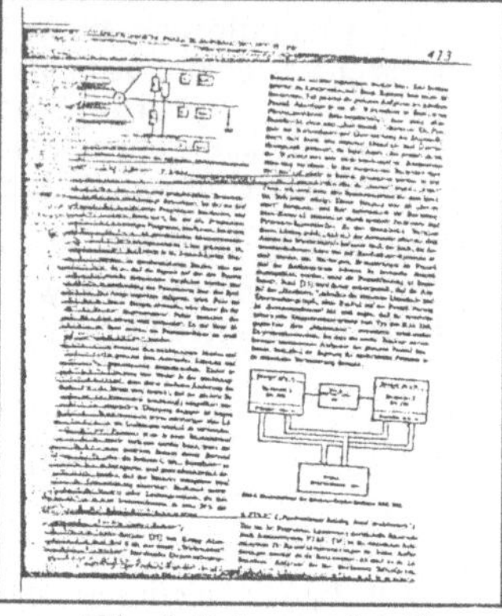

Abb. 5: Dokument 2 (-19°).

Abb. 6: Dokument 3 (-4°).

Dokument	Größe	exakter Winkel	Kleinste Quadrate		Kleinste Quadrate (mod.)		Linke Blockrandsuche		Simulated Skew Scan	
			Winkel	Zeit	Winkel	Zeit	Winkel	Zeit	Winkel	Zeit
1	3507x2480	zw. -10 u. -9	-10	6199	-10	4866	-9	2816	-10	13282
1	3507x2480	0	0	5899	0	4383	0	2500	0	13332
2	1727x1423	19	19	1749	20	1299	20	533	19	8066
2	1755x1445	-19	-19	2149	-19	1633	-15	683	-19	8149
3	2333x1650	9	9	3033	9	2299	9	799	9	13332
3	2333x1650	ze. -4 u. -3	-4	3233	-4	2416	-4	900	-4	13349

Tabelle 1: Dokumentbildgröße (in Pixel), exakter Winkel im Dokumentbild und ermittelter Winkel der drei Verfahren (in Grad[°]), sowie benötiget CPU-Zeit (in msec).

Alle drei Verfahren sind in das Dokumentanalyse-System ANASTASIL /3/ eingebettet und speziell für die Analyse von Bürodokumenten hervorragend einsetzbar.

Literatur

/1/ H. S. Baird, *The Scew Angle of Printed Documents*, Advanced Printing of teh SPSE´s 40th Annual Conf. and Symposium on Hybrid Imaging Systems, Rochester NY, Mai 1987, S 22

/2/ *Taschenbuch der Mathematik*, 19. Auflage, Verlag Harri Deutsch, Frankfurt, S. 339

/3/ A. Dengel und G. Barth, *ANASTASIL: Hybrid Knowledge-based System for Document Image Analysis*, erscheint in Proceedings der IJCAI´89, Detroit, MI, Aug. 1989

/4/ A. Dengel, *Automatische Visuelle Klassifikation von Dokumenten*, Ph.D. Dissertation, Fakultät Informatik, Universität Stuttgart, Februar 1989

/5/ W. Postl, *Detection of Linear Oblique Structures and Scew Scan in Digitized Documents*, Proceedings of the 8th Intl. Conference on Pattern Recogntion, Paris 1986, S. 240

/6/ J.-P. Trincklin, *Conception d´un Sytème d´Analyse de Documents: Etude et Realisation d´un Module d´Extraction de la Structure Physique de Documents à Support visuel*, Ph.D. Dissertation, Universität de France-comte, Besançon, 1984

/7/ E. Schweizer, *Erfassung, Justierung und Segmentierung von Dokumentstrukturen*, Diplomarbeit im Fachbereich Informatik, Universität Stuttgart, 1989

Visualisation and Three Dimensional Presentation in Orthopaedics and Traumatology

K.-H. Englmeier, A. Wieber, C. Hamburger*, T. Mittlmeier*

GSF – Institut für Medizinische Informatik und Systemforschung, Ingolstädter Landstr. 1, 8042 Neuherberg

*Klinikum Grosshadern der Ludwig–Maximilians–Universität München, Marchioninistr. 15, 8000 München 70

Introduction: Since the discovery of the X–ray, sagittal and frontal tomography has been introduced to medical practices, for more exact definitions of opacities for survey radiographs. These tomographies occur by only highlighting a particular area of the body and blurring the background through smear techniques. A computer supported evaluation, such as a tumour volume assessment, is however very inadequatly predictable by this method. Only since the conception of X–ray–computer tomography at the beginning of the 1970's has it been made possible, to further process x–rays digitally. The rotational tomography equipment used today, consist of a fanbeam whose radiation angle is so concipated that it encompasses the whole body section. The weekened rays are taken from a respectively large detector array (more than 500 monitoring probes) simultaneously /1/. A higher radiation speed, with equivalent mechanical stability, is supplied by tomography equipment of the fourth generation, by which a 360 degree fixed assembled set of detectors (over 1000 monitoring probes) radiographs the rays from the X–ray tube rotating around the body. With dedicated hardware and special image reconstruction algorithms (Radon Transformation or Back Projektion) the sectional image of the radiated layer from the low intensity samples is calculated within approximately 1–3 secs. The local resolution of these radiographs is typically 256•256 pixels or even 512•512 pixels. The image halftone resolution is 12 bit /15/.
From the consecutive tomograms, the radiologist must mentally reconstruct a three dimensional scene of the skeleton structure, together with the concerned pathologic modifications. This information concerning the three dimensional shape of organs and skeleton parts, is achieved by the physician evaluating the individual serial films. The more complex the structure of the radiographed organs are, (e.g. after traumatic influences or modifications caused by carcinomas), the more difficult it is for the examiner to mentally reconstruct the three dimensional scene of the effected part of the body /3/.

Data Acquisition and Hardware Configuration: For examination of a defined volume of the human body at first a scout view is acquisitioned by the help of the fixed x–ray–tube and the moved ct–couch. This topogram enables the determination of the starting and ending cross–section, the slice distance of the images (typical 2, 4, 8 or 12 mm), as well as the gantry angle of the ct–scanner /1/. After acquisition of the raw data files, the calculation of the exact density units (Houndsfield Units) is executed by specialized software (back projection) and dedicated hardware. The calculated images can be displayed on monitor by the window technique. Finally the 12–bit pixel data are compressed and stored on disk in combination with the patient and examination data. The pseudo–3–dimensional presentation of human organs as well as the computer assisted design of custom made prosthesis are based on a segmentation process of regions of interest and the surface calculation of the evaluated contours. For these reasons the images are transferred to the image analysis system via magnetic tape or floppy disks. The image analysis system consists of a vectorizing computer (CONVEX C–210) with 64 MB working storage, a tape unit and an image display system. Using the UNIX operating system there are FORTRAN, ADA and C as programming languages available. Via Ethernet and Network File System the vectorizing computer is connected with an IRIS 4D/70 GT Silicon Graphics workstation, which is mainly used for 3–dimensional presentations and 3–dimensional surface interpolations with Cardinal Splines.

Computer-Supported Segmentation: X–ray computer–tomographies, which are acquisitioned in the above mentioned manner, can be used as a basis of therapy control and therapy planning of surgical operations. For on the one hand x–ray computer images include the morphometric information of human organs and skeleton parts. On the other the density measurements of the soft tissue enable the control of the applied therapy /3/. Both applications – the morphometry as well as the density measurements – are based on the process of segmentation and contour determination. We use the possibility of transfer the image data to the above mentioned processing system and an automated segmentation process, which is divided into the following steps:
- image transfer and decompression
- image filtering, segmentation and contour definition of bony regions
- surface calculation via triangulation
- optional: therapy planning
- pseudo–3–dimensional presentation

After transfer of the images to the image analysis system, the images are decompressed into their original format of a 12 bit gray value resolution. The region of interest of the body cross–section is detected automatically by simple thresholding /5/, /9/. The advantage of the definition of the body cross–section is the reduction of the data volume and the subsequent presentation of the body surface for clearly arranged presentations of complex anatomical structures (fig. 2). In the following step of the procedure the raw areas of the bony regions are determined by the above mentioned thresholding algorithm. But these regions are overlayed with disturbances such as contour indentations as well as artificial regions of calcifications in blood vessels. But by the following method of the mathematical morphology a partial elimination of these disturbances can be executed /13/: Suppose $b(x,y)$ as a binary image function with $x,y=1,\ldots,256$ and $b(x,y)=1$ if $b(x,y)$ is evaluated by the above mentioned segmentation procedure else $b(x,y)=0$. In addition to that the structuring element $B(x,y)$ is defined in a $3{\cdot}3$–environment. The result of the erosion $O \ominus B(x,y)$ of the segmented object O by $B(x,y)$ is the set of all the points (x,y) such that $b(x,y)$ is included in O. The result of the dilation $O \oplus B(x,y)$ – the complementary function of the erosion – is the set of all the points (x,y) such that $B(x,y)$ hits O. The series connection of both morphological transformation erosion and following dilation is designated by the opening function; the inversion is designated by the closing function /13/. Closing enables the connection of objects contacting in one point,whereas, the opening function separates the object.

The result of the opening function is the region of the interesting object, defined by the Houndsfield Units of compacta and spongiosa. For the process of surface calculation uses only the contour information of the objects, the binary images have to be differentiated. The most commonly used method of differenciation in image processing applications is the gradient.A useful approximation, sometimes called the Robert's Gradient, uses the cross–difference.For only the location of edges is of interest a binary image is calculated where the edges and the background are displayed in any two specified gray levels. Following that a simple contour detection algorithm stores the contour points counterclockwise on disk.

Within the scope of the visual control of automatic contour definition process, the detected contours are overlayed with the original image and displayed on the color monitor. For in case of fractures of bones or in the critical regions of joints the minimized discrimination of gray values yields to insufficient separation of the medical objects /12/. Therefore an interactive correction of the contours is necessary. After definition of the contour points bye the user, the correct contours are interpolated by Cardinal splines and the contour points are stored in the above mentioned manner.

Surface–Calculation and Pseudo–3–Dimensional Presentation: After the application of the gradient method and the automatic contour detector, the contour informations of the object of interest are stored in form of two dimensional point sequencies. The given distance of the cross–sectional images yields to the z–coordinate. These informations enable the calculation of the surface models of the detected   structures.

For this the selected triangulation method connects contours of one level to the next level. Therefore the applied algorithm is divided into two hierarchical levels: On the upper level it has to be decided which loop of level i has to be connected with loop of level i+1: In the simple case there is only one loop per level which can be easily connected by triangles. But if there is more than one loop per level available or if there is a contour bifurcation, the user or a knowledge based system must assign the loops of level i to loops of level i+1.

On the lower level the points of the assigned loops are connected. By the help of the construction of closed polygons the contour of both levels are connected in such a manner, that patches are being created which solve only the defined condition of optimization. Then the optimization is executed according to the minimization of the surface. The result of this calculation process is a data structure which enables the presentation of the surface as a wireframe and the colored presentation of opaque and transparent objects under consideration of the following lightening model:

The development of a visualisation model refers to the presentations as realistic as possible of objects in imitation of the physical laws of optics and photometrics. Therefore the three–dimensional scene is divided into the following elements:

– the objects being presented and the background
– the color of light sources and the position of lights

The procedure of lighting model is divided into the following steps: At first the color values of the vertices of the actual patches are calculated. Within the scope of the polygon filling a so called shading model is applied. Therefore we distinguish between the lighting model and the shading method.

The presented method applies a commonly used lighting model combined with the Phong lighting method /10/ and the Gouraud shading method /7/. For on the one hand there are time efficient algorithms of the above mentioned models available, and on the other they produce a realistic impression of complex anatomical structures. The use of the above mentioned model of an illuminated scene yields to the common light equation: The color of a point at position p is C and is the emitted light plus the sum of the ambient, diffuse and specular light reflected by the point towards the eye of the observer /4/, /6/, /9/:

$$C = C_1 + C_2 + C_3 + C_4$$

$C_1 :=$ color of the emitted light $C_2 :=$ color of ambient reflected light
$C_3 :=$ color of the diffuse reflected light $C_4 :=$ color of the specular reflected light

The results of this light equation are the intensities for the points along the edges of the polygons. The polygons themselves are the patches of the surface wireframe. Based on this calculations the polygon filling algorithm as described by Gouraud is applied /7/. This algorithm guarantees a continuous color shading inside of the polygons and to the adjacent patches. The color values of all points positioned inside of the polygon are determined by interpolating linearely between the pairs of vertex–points that lie along the actual scan line

Examples:

The Three Dimensional Reconstruction of the Geometry of the Tibial Medullary Space – Possibilities of the Anatomical Intramedullary Osteosynthesis: The intramedullary osteosynthesis has gained in significance because of the concept of nailing and dynamisation. The stability of the intramedullary osteosynthesis is determined primarily by the material characteristics of the axial contact length. In spite of boring out the medullary space, only a limited compatibility between the geometry of the nail and the bone is achieved with currently available nails, because the relatively robust tube does not follow the phisiological form of the curvature of the medullary space.

On the example of the tibia, it is attempted to reconstruct the medullary cavity on the basis of segmented CT–images, and therefore to determine the pre–requisites for an anatomically formed medullary nail. In addition to the cortical, the spongy structure is calculated and subsequently the triangulation of the bone surface together with the medullary space of the tibia is determined. The design of an anatomically formed medullary nail then occurs in the following way:

From the center point of the medullary cavity at each level and the minimum diameter of the medullary space, the input data for the approximation process with Cardinal Splines are determined. Under consideration of the typical nail crossection, the computation of the coordinates of the adapted nail occurs in respect of curvature and thickness of the medullary space. A visual quality check of the compatibility of the designed nail occurs with three dimensional representations (Fig. 1).

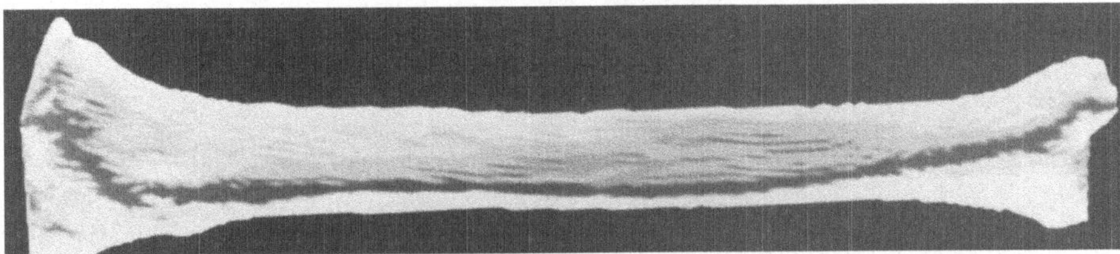

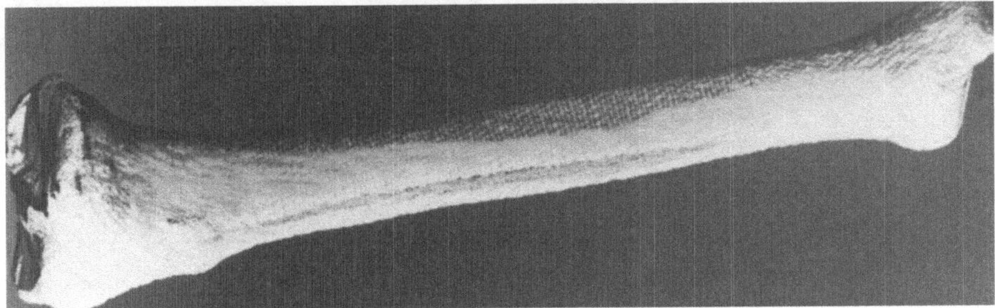

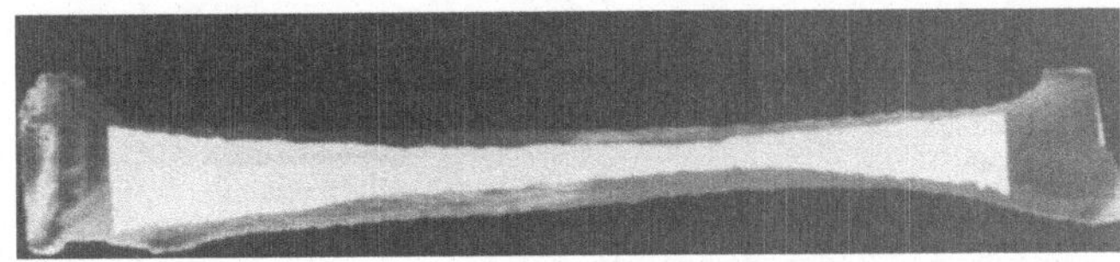

Fig. 1:   3–dimensional presentation of the tibia (solid model), the medullary space (solid model) and the tibia as a wireframe and a presentation of the transparent tibia and medullary space with the implanted nail

The Three Dimensional Reconstruction of Tumours and Fractures of the Spinal Column: The computer tomography enables the exact determination of bony and soft tissue modifications in the region of the spinal column. Often the degree of bony or tissue changes in the region of the spinal column can only be exactly represented by the computer tomogramphy in conjunction with the conventional radiographs. Especially with regard to larger traumatic and tumorous processes, the exaxt pre–operative planning of the therapeutic and surgical procedures is essential. The three dimensional reconstruction permits the volumic representation of the spreading of tumours and traumatic changes to the spinal column and enables improved planning for the operative therapy /14/.

In addition to the automatic contour definition of the bone structure in the region of the spinal column as described above, the tumour contour is determined interactively. With the help of lighting and shadow models the spinal column, the vertebral canal and the tumour can be presented pseudo–three–dimensionally (fig. 2). For a clearer representation of the tumour within the vertebrae, the possibilities of the transparent object presentation are employed.

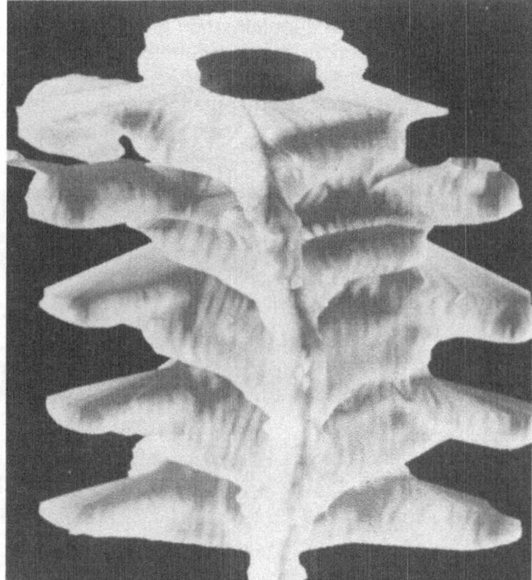

Fig. 2:   Pseudo–3–dimensional presentation of the vertebra, the myelon and a neurinoma (white) as wireframe and solid model

Conclusion: The 3–dimensional presentation of human bones and joint structures is a well known experimantal method. As mental recognition of 3–dimensional structures is very difficult in transversal CT–images, 3–dimensional presentation is an important and very useful part of processing CT–scans; for on the one hand it gives the user a visual impression of the bone's shape in the 3–dimensional scene (especially bone's deformations caused by trauma influences and pathologic modifications) and on the other it enables the individual design of the prostheses used in orthopedic surgery.

References:

1. **Alexander J., Kalender W.A., Linke G.**: Computertomographie, Bewertungsmerkmale – Gerätetechnik – Anwendungen, Siemens AG, Berlin – München, (1985)
2. **Bunke H.**: Modellgesteuerte Bildanalyse, Teubner Verlag, Stuttgart, (1985)
3. **Englmeier K.-H., Pöppl S. J.**: A New Imaging Method and Its Application in Gynecological Treatment Planning, Proc. of the Fifth Conference on Medical Informatics, Washington, Oct. 26–30,(1986), 745–751
4. **Foley J. D., Van Dam A.**: Fundamentals of Interactive Computer Graphics, Addison – Wesley Publishing Co., Amsterdam – Sydney – Singapore – Tokyo – Madrid – Bogota – Santiago – San Juan, (1982)
5. **Gonzalez R. C., Wintz P.**: Digital Image Processing, Addison – Wesley Publishing Company, Amsterdam – Sydney – Singapore – Tokyo – Madrid – Bogota – Santiago – San Juan, (1987)
6. **Gordon D., Reynolds R. A.**: Image Space Shading of 3-dimensional Objects, Computer Vision, Graphics and Image Processing 29, (1985), 361–376
7. **Gouraud H.**: Continuous Shading of Curved Surfaces, IEEE Transactions on Computers, vol. C–20, No. 6, (1971), 623–629
8. **Granholm J. W., Robertson D.D., Walker P.S. Nelson P.C.**: Computer Design of Custom Femoral Stem prostheses, IEEE Computer Graphics and Applications 2, (1987), 26 – 35
9. **Newman W. A., Sproull, R. F.**: Principles of Interactive Computer Graphics, McGraw–Hill, New York (1979)
10. **Phong B. T.**: Illumination for Computer Generated Pictures, Communications of the ACM,6, Vol. 18, (1975)
11. **Rhodes R.L., Azzawi Y.-M., Chu E.S., Pang A.T.**: Computer Communications and Graphics for Clinical Radiology, Proc. of the Int. Symposium C.A.R '85, Berlin 1985, Springer–Verlag, Berlin – Heidelberg – New York – Tokyo
12. **Rothuizen P., Van Erning L., Huiskes R.**: The Accuracy of Criteria for Automatic 3–D Graphics Reconstruction of Bone from Computer Tomography, Proc. of the 5. Meeting of the European Society of Biomechanics, Berlin, 9.1986, Martinus Nijhoff Publishers, Dordrecht – Boston– Lancaster, (1987), 109–114
13. **Serra J.**: Image Analysis and Mathematical Morphology, Vol 1, Academic Press, London– San Diego – New York – Berkely – Boston – Sydney – Tokyo – Toronto, (1988)
14. **Wojcik W. G., Edeiken-Monroe B. W., Harris J. H.**: Three–dimensional Computed Tomography in Acute Cervical Spine Trauma. A Preliminary Report, Skeletal Radiology,16 (1987), 261–269
15. **Zonneveld F. W., Albrecht C.**: Computed Tomography: A Review of the Past and Present and the Perspective of the Future, Medicamundi 26,(1981),81

Verfahren zur graphisch-interaktiven Objektselektion in gespeicherten Bildern

Georg Rainer Hofmann [1], *Axel Hildebrand* [2]
[1] *Fraunhofer-Arbeitsgruppe für Graphische Datenverarbeitung*
Abt. Simulation und Animation
[2] *Technische Hochschule Darmstadt*
Fachgebiet Graphisch-interaktive Systeme (GRIS)
Wilhelminenstr. 7, 6100 Darmstadt, FR Germany
Telex: 4 197 367 agd d   Telefax: +49 6151 1000 99

1. Einleitung

Unter dem Begriff des "Imaging" werden Methoden und Applikationen der digitalen Verarbeitung von Bildern subsummiert. Das orts-, amplituden-, und zeitdiskrete Bild erscheint dabei gleichsam als eine Schnittstelle zwischen den beiden klassischen Bereichen der Bildgenerierung (Bildsynthese, Visualisierung von Objekten) und des Bildverstehens (Bildanalyse, Objekterkennung).

Während die digitale Synthese von Bildern aus selektierten Komponenten nur relativ einfache (meist affine Transformationen in der Ebene) Operationen verlangt (siehe /Hofm88/, /Hofm89/), erscheint der umgekehrte Prozeß der Selektion von Objekten aus diskreten Bildern als Gegenstand intensiver Forschung.

Zwei Wege der Objektselektion aus gespeicherten Bildern scheinen vorgegeben: Entweder man sucht ein vollautomatisches Verfahren zu entwickeln oder man implementiert halbautomatische Verfahren.
Da vollautomatische Verfahren nicht ohne eine umfangreiche Beschreibung der zu selektierenden Objekte und meist nicht ohne Methoden der Künstlichen Intelligenz auskommen, sind halbautomatische Verfahren hier im Vorteil, da sie den Eingriff des Benutzers überall da nutzen können

- wo das Objekt als solches erstmals erkannt und bestimmt werden muß, und

- wo - in Zweifelsfällen - das vom Benutzer "Gemeinte" nicht algorithmisch nachvollzogen werden kann.

2. Einfache Verfahren zur Objektselektion und deren Nachteile

Bei der Bearbeitung digitalisierter Photographien wird in verschiedenen Anwendungsbereichen ein Werkzeug benötigt, um ein Rasterbild in sinnvolle Teilrasterbilder zu zerlegen. Hierbei heißt "sinnvoll", daß z.B. ein Teilrasterbild genau ein vom Bediener bzw. Benutzer ausgesuchtes Objekt zeigt. Extrahierte Objekte werden affin transformiert und in neuen Bildern eingefügt. Zudem sind die gewonnenen photographischen Komponenten als Textur verwendbar.
Bei der Implementierung eines Werkzeuges zur Teilbildselektion kann man sich auf die Simulation elementarer Wahrnehmungsprozesse (z.B. nach /Marr79/) beschränken und diese interaktiv überwachen. Die verwendeten einfachen graphisch-interaktiven Verfahren zur Objektselektion in digitalisierten Photographien basierten meistens auf einer polygonalen Ausschnittsdefinition. Eine polygonale Objektkonturspezifizierung weist allerdings einen wesentlichen Nachteil auf. Dieser zeigt sich beim Versuch Objekte mit nicht-kanonischer Kontur, so z.B. eine Wolke oder einen Baum, mithilfe von Liniensegmenten zu begrenzen: Dies führt trotz langer Bearbeitungszeiten (und entsprechender Mühe des Benutzers) zu "unnatürlichen" Bildausschnitten.

3. Durch lineare Ortsfilterung unterstütztes graphisch-interaktives Verfahren

Das hier implementierte Verfahren wird anhand eines Beispiels vorgestellt und beschrieben.
Das Bild wurde durch Abtastung mittels einer CCD-Kamera gewonnen, es zeigt ca. 120 x 150 Bildelemente (Pixel).

3.1. Objektdefinition durch den Benutzer

Der Benutzer identifiziert das Objekt im Bild und grenzt es grob durch einen Polygonzug ein.

Der Benutzer ist an einer Bildkomponente interessiert, welche den Leuchtturm zeigt.

3.2. Lineare Ortsfilterung mit dem Sobeloperator

Wird bei den verwendeten Farb-Bildern zur nachfolgenden linearen Ortsfilterung ein skalares (Luminanz-)Bild zugrundegelegt, so ist es nicht möglich, ein Objekt von seinem Hintergrund zu trennen, falls sich beide nur in der Chrominanz unterscheiden. Das Verfahren berücksichtigt in einem ersten Schritt nur die Luminanz, falls dies nicht ausreicht, wird auf den evaluierten Abstand im empfindungsgemäß gleichabständigen Lab-Farbmodell /DIN 5033/, /DIN 6174/ zurückgegriffen.

Die Reaktion des visuellen Systems des Menschen auf Helligkeits und Farbdiskontinuitäten ist am ehesten vergleichbar mit der lokalen Faltung mithilfe eines Laplaceoperators, welcher aufgrund seiner Rauschempfindlichkeit meistens in Verbindung mit einer Glättungsoperation verwendet wird /Marr79/.
Die Eigenschaft des visuellen Systems, im Gesichtsfeld Helligkeitsunterschiede zu detektieren, läßt sich durch Gradientenoperatoren simulieren /Korn82/, /Gilc79/. Ein Gradientenoperator besitzt gegenüber dem Laplaceoperator wesentliche Vorteile: Auf der einen Seite ist die Rauschempfindlichkeit geringer und zudem läßt sich die Richtung der Lichtveränderung feststellen. Bei der Verwendung von Masken zur Realisierung der Gradientenoperation erhält man Näherungswerte des Gradientenbetrags, bzw. des Winkels. Vergleiche hinsichtlich des entstehenden Fehlers (siehe /Wall83/ und /Davi84/) ergaben, daß der Sobel-Operator eine gute Approximation bezüglich der gesuchten Größen darstellt.

- 1	0	1
- 2	0	2
- 1	0	1

1	2	1
0	0	0
- 1	- 2	- 1

Filtermatrizen des verwendeten Sobeloperators zur Approximation der partiellen Richtungsableitungen nach x und nach y.

Der Betrag und die Richtung des approximierten Sobeloperators werden berechnet zu:

$$\text{Betrag:} \quad \left| \frac{df}{dxy} \right| = \left[\left(\frac{df}{dx}\right)^2 + \left(\frac{df}{dy}\right)^2 \right]^{\frac{1}{2}} \qquad \text{Richtung:} \quad \phi\left(\frac{df}{dxy}\right) = \arctan\left[\frac{\frac{df}{dx}}{\frac{df}{dy}} \right]$$

Das Ergebnis der Sobel-Operation (ermittelte Konturen) wird als Binärbild dargestellt. Der Benutzer hat die Möglichkeit, Verzweigungsstellen zu erkennen, welche bei der anschließenden Konturverkettung zu Fehlern führen könnten.

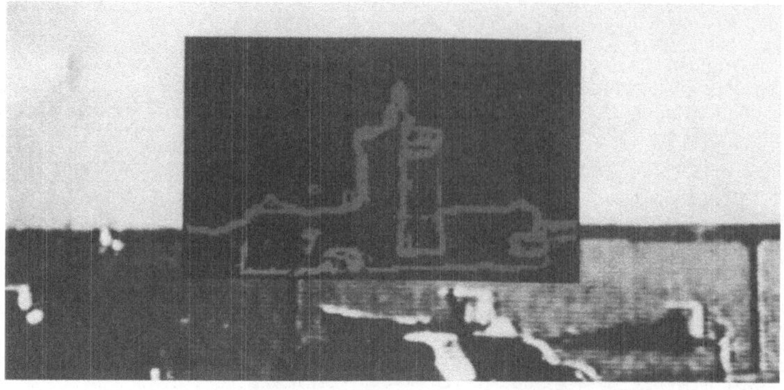

Der Benutzer markiert einen Konturaufsatzpunkt auf der als solche erkannten äußeren Kontur des Objektes. Da die Konturbestimmung allein aufgrund von Farbdiskontinuitäten erfolgt, existiert keine sonstige Information über die Lage der äußeren Kontur.

3.3. Konturverkettung

Bei der Konturverkettung werden *automatische* und *graphisch-interaktive* Methoden kombiniert.

Beim *automatischen* Verkettungsprozess werden Betrag und Richtung des Gradientenoperators verwendet. Ausgehend von dem interaktiv gewonnenen Konturaufsatzpunkt wird in Abhängigkeit von der Richtungsinformation ein Nachbarpunkt aus der 8-er Nachbarschaft des aktuellen Punktes ermittelt. Hierzu wird die Richtungsinformation entsprechend der Lage in einem von 8 festgelegten Bereichen bewertet.

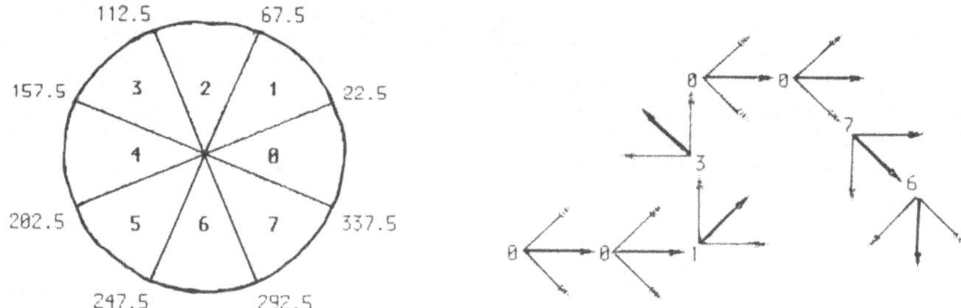

Automatische Konturverkettung

Es werden jeweils drei Nachbarpunkte untersucht. Derjenige Punkt mit dem größten Gradientenbetrag wird mit dem aktuellen Punkt verbunden und die weitere Verkettung erfolgt analog unter Berücksichtigung der Richtungsinformation des so gewonnenen Nachfolgepunktes. Die Breite der durch den Sobel-Operator gewonnenen Kontur beträgt im allgemeinen 1-2 Pixel /Wahl79/. Hier bei dieser Realisierung ist eine Konturausdünnung überflüssig, denn die gewonnene Kontur ist maximal ein Pixel breit.

Interaktives Eingreifen ist nötig, wenn das automatische Verfahren verkehrt läuft, weil eine starke Kontur in das Objektinnere führt.
Der Benutzer hat nun die Möglichkeit, die Kontur bis zu einer Verzweigungsstelle zurückzusetzen. Danach kann die automatische Konturverkettung vom neu gewählten Aufsatzpunkt aus fortgesetzt werden.

3.4. Weiterverarbeitung der gewonnenen photographischen Komponente

Nachdem die Kontur des Objektes vollständig bestimmt wurde, kann der Bildausschnitt extrahiert und mit seinem Transparenzanteil gespeichert werden.

Insgesamt waren fünf interaktive Eingriffe des Benutzers an den markierten Stellen notwendig. Die gefundene Umgrenzung des Objekts besteht aus ca. 85 einzelnen Geradenstücken. Es besteht die Möglichkeit, das gewonnene Objekt affin zu transformieren und an einer anderen Bildposition (bzw. auch in einem anderen Bild) einzufügen.

Durch eine lokale Filterung entlang der Einfügekante wird der neue Hintergrund in die Randpixel des eingefügten Bildteils einbezogen, um das Aliasing zu reduzieren.

4. Bewertung des Verfahrens

Der Aufwand der Objektselektion verringert sich erheblich gegenüber dem einfachen polygonalen Ausschneiden. Im dokumentierten Beispielbild war die Bearbeitungszeit ca. 2 Minuten; ein rein interaktives Verfahren, welches ohne Methoden der Konturerkennung die ca. 85 Polygonzüge "von Hand" setzte, würde ca. eine halbe Stunde Bearbeitungszeit in Anspruch nehmen.

Der verwendete Sobel-Operator ist relativ schnell berechenbar, darum ist er für interaktive Verfahren geeignet. Nachteile der Sobel-Operation (lediglich lokaler Operator; Konturbreite nicht ideal dünn; nur eine Approximation des Gradienten) werden durch das hier angewandte Verfahren der Konturverkettung bzw. der Benutzer-Interaktion kompensiert.

5. Literatur

/Davi84/   E.R. Davis, *Circularity - a new principle underlaying the design of accurate edge orientation*
IVC 2, No. 3, 134-142; 1984

/DIN5033/   DIN 5033: Farbmessung
Deutsches Institut für Normung; Berlin; März 1977

/DIN6174/   DIN 6174: Farbmetrische Bestimmung von Farbabständen bei Körperfraben nach der CIE-Lab Formel
Deutsches Institut für Normung; Berlin; Januar 1979

/EnHo89/   J.L. Encarnacao, G.R. Hofmann, *Computergenerierte, naturgetreue Bilder*
in: FhG-Berichte 1/89; Fraunhofer-Gesellschaft; München; 1989

/Gilc79/   A.L. Gilchrist, *Die Wahrnehmung schwarzer und weißer Flächen*
in: Wahrnehmung und visuelles System (Spektrum der Wissenschaft); Heidelberg

/Hild89/   A. Hildebrand, *Objektselektion in gespeicherten Bildern*
Diplomarbeit; Technische Hochschule Darmstadt, FG GRIS; Darmstadt; 1989

/Hofm88/   G.R. Hofmann, *The Calculus of the Non-exact Perspective Projection*
in: D. A. Duce, P. Jancene (Eds.): EUROGRAPHICS '88; Amsterdam, New York, Oxford, Tokyo; 1988

/Hofm89/   G.R. Hofmann, *Non-planar Polygons and Photographic Components for Naturalism in Computer Graphics*
to appear in: EUROGRAPHICS '89; Hamburg; 1989

/HoKr89/   G.R. Hofmann, D. Krömker, *FTCRP: File format for the device-independent transfer of colored raster pictures*
in: SPIE/SPSE Symposium on Electronic Imaging; Los Angeles; 1989

/Korn82/   A. Korn, *Bildverarbeitung durch das Visuelle System*
Berlin, Heidelberg, New York; 1982

/Marr79/   D. Marr, *Theory of Edge Detection*
Proc. Royal Soc.; London; 1979

/Wahl79/   F. Wahl, J. Kugler, *Kantendetektion mit lokalen Operatoren*
in: Angewandte Szenenanalyse, Informatik-FB 20; Berlin, Heidelberg, New York; 1979

/Wall83/   A.M. Wallace, *Grey-scale image processing for industrial applications*
IVC 1, No. 4, 178-188; 1983

Cognitive Texture Parameters – the Link to Artificial Intelligence

D. Scheppelmann[1], F. Saurbier[1], H.P. Meinzer[1],
J. Klemstein[2]

[1] German Cancer Research Center
Department of Medical and Biological Informatics
(Head: Prof. Dr. C.O. Köhler)
Im Neuenheimer Feld 280, D–6900 Heidelberg, F.R.G.

[2] Neuropathologisches Histologielabor
Nervenklinik Spandau
Griesinger Straße 27–33, 1000 Berlin

The experience with todays texture shows, that it helps only little for the interpretation of CT images in the medical field, since first it is not easy to choose the correct texture parameter from the zoo of exsisting ones and second the texture parameters in general do not match human visual impressions. The following concept shows, how texture parameter can be grouped in families, which gives a better insigth into texture analysis. Each class of texture parameter is represented by a complete set of texture parameters avoiding redundancies. Those families and their texture parameters are adopted to the human texture impressions and are therefore called 'cognitive texture parameters'.

1. Einleitung

Das im folgenden vorgestellte Konzept der kognitiven Texturparameter entstand aus der Notwendigkeit heraus Vorwissen, und damit symbolische Wissensverarbeitung (i.a. als künstliche Intelligenz bezeichnet), mit der Bildverarbeitung zu koppeln.

Hierbei ist eine zentrale Schwierigkeit, die sich besonders im medizinischen Umfeld und in der Texturanalyse zeigt, das Wissen eines Experten zu formulieren und darüber hinaus in numerische Operationen auf Bildern zu übersetzen – d.h. den Engpaß zwischen 'know how' und 'say how' zu überwinden. Dies hat im wesentlichen folgende Gründe:

o   der Zoo der exsistierenden Texturparameter [1] ist so groß, unübersichtlich und hochgradig redundant, daß es zu schwierig ist die relevanten Texturparameter zu finden.

o   die Sprache der Experten (hier Mediziner) kann nicht unmittelbar in Texturparameter übersetzt werden, denn die Texturanalyse orientiert sich zur Zeit nur an der Frage 'was kann ich rechnen' und nicht 'was kann ich sehen'.

o die exsistierenden Texturparameter sind nur für eine Klassifikation per Diskriminanzanalyse aber nicht für KI–Systeme geeignet, da sie kaum gruppiert oder gar hierarchisch organisiert sind. Eine Klassifikation ist aber nicht immer ausschließlich mit Methoden der Diskriminanzanalyse zu bewerkstelligen [2].

Die kognitiven Texturparameter sind hingegen den menschlichen Sehempfindungen angepasst. Sie sind nicht für eine spezielle Anwendung gestrickt und sie enthalten auch keine a priori Information über die Auflösung im Bild, wie das bei anderen Texturparametern z.B. den Coocurrencemaßen der Fall ist. Die Einteilung der Texturen in Familien impliziert eine Hierarchie, die sich ganz natürlich aus der Komplexität der entsprechenden Texturparameter ergibt.

2. Familien der kognitiven Texturparameter

Bild 1 gibt eine graphische Übersicht und zeigt die Schachtelung der Texturparameter. Die Nummern geben aufsteigend den Grad der Komplexität an, wodurch sich eine natürliche Hierarchie der Texturparameter ergibt. Diese Hierarchie kann von der symbolischen Wissensverarbeitung (KI) genutzt werden, indem man z.B. eine Suche nach isotropen Texturen aufgibt, weil es bereits unmöglich war amorphe Texturen zu detektieren.

I) Homogene Texturparameter sind die primitivsten, die noch keinerlei Information über die Struktur in einer ROI geben. Sie spiegeln lediglich Effekte wie Helligkeit und Kontrast wieder, die Lage der Pixel zueinander ist irrelevant. Solche Texturen werden vollständig durch das lokale Histogramm beschrieben! Andere Texturparameter wie z.B. die Entropie sind darin implizit enthalten. Die Parameterisierung der lokalen Histogramme kann auf kognitiv vernünftige Weise durch die statistischen Momente [3] erfolgen.

II) Versuchen wir ein wenig Struktur in die Szene zu bringen ohne jedoch irgendwie eine erkennbare Ordnung zu erzeugen, landen wir bei den amorphen Texturen. Stellt man sich die Grauwerte eines solchen Bildes als physikalische Massen vor, so würde diese geringfügige Fluktuation von Grauwerten gegenüber dem homogenen Fall eine Änderung der Massenträgheitsmomente [4] verursachen. Kenneth Laws [3] experimentierte mit dem allgemeineren Ansatz der 'spatial moments'. Im Ergebnis erwiesen sich jedoch nur die physikalischen Momente als relevant! Damit ist es also möglich, amorphe Texturen von homogenen zu trennen um sie der weiteren Texturanalyse zu unterziehen.

III) Der nächste Schritt ist dann die Klasse der isotropen Texturen, die erste Gesetzmäßigkeiten in ihrer Struktur aufweisen, die z.B. als Funktion des Radius geschrieben werden können, oder die sich in Symmetrien äußern. Die Texturen haben aber noch keine ausgezeichnete Richtung und die entsprechenden Texturparameter sind somit rotationsinvariant.

IV) Bei den anisotropen Texturen hingegen finden wir Vorzugsrichtungen, welche z.B. durch die
Eigenwerte des Massenträgheitstensors [4] bestimmt werden können. Auch hier wieder die
Interpretation der Grauwerte als Massen. Sollen auch anisotrope Texturen lageinvariant
bestimmt werden, so muß dies explizit eingebaut werden, etwa durch mehrfache Messung
unter verschiedenen Winkeln. Über die Vorzugsrichtung hinaus können diese Texturen auch
noch eine Orientierung haben, was zu einem unsymetrischen Verhalten längs der Hauptrich-
tung führt.

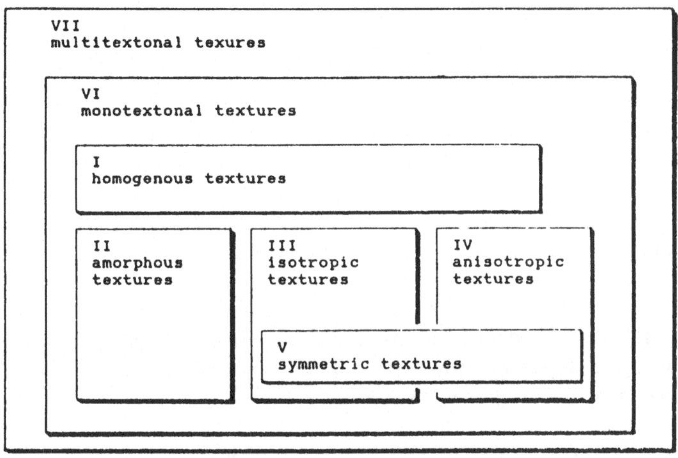

Bild 1   Textur Familien

V) Eine weitere Steigerung der Komplexität ist die Einführung von Symmetrien. Das können
primitive sein wie Axen– oder Rotationssymmetrie oder komplexere, die die Regelmäßigkeit
der Anordnung der Textone wiederspiegeln. Solche Symmetrien sind im dreidimensionalen
durch die Bravaisgitter [5] beschrieben, was sich natürlich auch auf zweidimensionale Gitter
reduzieren läßt.

VI) Bis zu diesem Punkt haben wir uns nicht um das einzelne Texton gekümmert. Wollen wir
aber Texturen trennen, die in allen Stufen übereinstimmen, so müssen wir in der Lage sein
die einzelnen Textone zu identifizieren. Wir nennen dann eine Textur monotextonal, wenn
sie sich durch wiederholte Anordnung nur eines Textons konstruieren läßt.

VII) Dem gegenüber stehen die multitextonalen Texturen, die sich nur aus verschiedenartigen Textonen konstruieren lassen. Reicht die Auflösung tatsächlich aus einzelne Textone auszumachen, sind wir vor das Problem gestellt diese zu finden und zu untersuchen. Dies kann einmal durch die mathemathische Morphologie [6] geschehen, aber auch durch eine erneute Suche nach Texturen mit allerdings erheblich verkleinerten Masken, wobei wir wieder bei Stufe I anfangen. Dies bezeichnet man dann i.a. als Suche nach Mikrotextur.

3. Anwendung, Ziel

Kognitive Texturparameter sollen helfen sich selbst oder dem Experten die richtigen Fragen zu stellen, die selben Fragen aber auch an das Bild stellen zu können.

Als Anwendung werden Computer Tommogramme (und MR Bilder) untersucht, insbesondere im Schädelbereich und dort die Trennung von Ödem/Hämatom/Tumor.

Das Konzept der kognitiven Texturparameter ist keine Texton–Theorie sondern betrachtet die 'region of interest' global. Die Identifikation von Textonen in Computer Tomogrammen kann man wohl ausschließen, wenn selbst schon eine Texturanalyse, die über die Grauwertstatistik zweiter Ordnung [3] hinausgeht, i.a. angezweifelt wird.

Ferner ist anzunehmen, daß nur Texturparameter der Klassen I, II und III mit schwachen Symmetrieeigenschaften anzutreffen sind, während Texturen mit strenger Symmetrie (IV mit V) auf systematische Gerätefehler zurückzuführen wären.

Mit den kognitiven Texturparametern soll, bei gegebenem Signal/Rausch–Verhältnis, geklärt werden, ob Computertomogramme Strukturinformation (Größenordnung mm) liefern können, und ob sich erwartete Strukturen mittels Texturanalyse finden lassen. Das Ziel ist die Verbesserung der Diagnose von Hirntumoren.

Literatur

[1] Van Gool, L.; Dewaele, P.; Oosterlinck, A.: Texture Analysis Anno 1983. In: Computer Vision, Graphics and Image Processing 29 (1985) 336–357.

[2] Gernert, D.: Advanced definitions of similarity and their use in classification and related fields. TU München. In Gaul, W.; Schader, M. (eds.): Classification as a tool of research. North–Holland: Elsevier Science Publisher B.V. 1986.

[3] Laws, K.: Textured Image Segmentation. Technical Report, Jan 1980, USCIPI Report 940. Los Angeles, CA 90007: Image Processing Institute, University of Southern California.

[4] Gerthsen, Ch.; Kneser, H.; Vogel, H.: Physik. Berlin: Springer 1977.

[5] Kittel, Charles: Einführung in die Festkörperphysik (Introduction to solid state physics). 6.Aufl. München u.a.: Oldenburg 1983. ISBN 3–486–32766–6.

[6] Serra, J.: Introduction to Mathematical Morphology. In: Computer Vision, Graphics and Image Processing 35 (1986) 114–128.

<h1 style="text-align:center">Zur Schätzung von Geschwindigkeitsvektorfeldern in Bildfolgen mit einer richtungsabhängigen Glattheitsforderung</h1>

C. Schnörr

Fraunhofer - Institut für Informations- und Datenverarbeitung (IITB)
Fraunhoferstrasse 1, 7500 Karlsruhe 1

Kurzfassung

In dem vorliegenden Beitrag wird gezeigt, daß für das von *Nagel 87* formulierte Minimierungsproblem zur Bestimmung von Geschwindigkeitsvektorfeldern in Bildfolgen eine eindeutige, stetig von den Eingangsdaten abhängige Lösung existiert. Aus den Eigenschaften des Ansatzes folgt, daß die Matrix des linearen Gleichungssystems, welches sich aus der Diskretisierung der Aufgabe mit finiten Elementen ergibt, symmetrisch und positiv definit ist. Darüber hinaus ist sie nur sehr dünn besetzt. Eine Näherungslösung kann deshalb problemlos mit einem beliebigen Abstiegsverfahren berechnet werden. Prinzipiell können diese Aufgaben effizient von parallel arbeitenden Prozessoren ausgeführt werden. Der Einfluß der Richtungsabhängigkeit der Glattheitsforderung (*Nagel 87*) wird an einer Bildfolge einer realen Szene demonstriert.

1. Übersicht

Bei der Interpretation von Bildfolgen spielen Geschwindigkeitsvektorfelder, die die zeitlichen Grauwertänderungen aufgrund einer Relativbewegung zwischen Kamera und Umwelt beschreiben, eine große Rolle. Durch die Auswertung dieser Vektorfelder läßt sich prinzipiell ein großer Teil der Information, die zu einer rechnerinternen Erstellung einer Beschreibung der Umwelt benötigt wird, gewinnen (vgl. Arbeiten, die sich mit der Auswertung von Geschwindigkeitsvektorfeldern beschäftigen; siehe z.B. *Maybank 87*, *Nagel 88*, *Aggarwal/Nandhakumar 88* und die jeweils dort zitierte Literatur).

Horn/Schunck 81 leiteten eine Beziehung zwischen der zeitlichen Änderung des Grauwertverlaufes g(x,y,t) und der gesuchten Verschiebungsgeschwindigkeit $\mathbf{v}$ = (v1,v2)^T derselben an der Stelle (x,y)T der Bildebene her:

$$g_x v1 + g_y v2 + g_t = (\nabla g)^T \mathbf{v} + g_t = 0$$

$$(1.1)$$

Da mit (1.1) lokal nur die Komponente in der Richtung des Grauwertgradienten bestimmt werden kann und die partiellen Ableitungen der Bildfunktion g(x,y,t) durch die Digitisierung der Bildfolge sowie durch das Sensorrauschen unsicher sind, schlugen *Horn/Schunck 81* zur Rekonstruktion des gesuchten Geschwindigkeitsvektorfeldes die Minimierung eines quadratischen Funktionals vor:

$$\int_\Omega \left\{ ((\nabla g)^T \mathbf{v} + g_t)^2 + \lambda^2(|\nabla v1|^2 + |\nabla v2|^2) \right\} dx \;\to\; min. \;,\; \lambda^2 > 0$$

$$(1.2)$$

Nagel 83 verallgemeinerte diesen Ansatz, führte eine richtungsabhängige Glattheitsforderung ein und arbeitete einen iterativen Ansatz zur numerischen Lösung der zugehörigen, nunmehr nichtlinearen Euler-Lagrangeschen Differentialgleichungen aus. Das intensive Studium dieses Ansatzes an Bildern realer Szenen führte zu verschiedenen Modifikationen der Iterationsvorschrift (*Nagel/Enkelmann 83,84,86; Enkelmann 85*). *Enkelmann 85,86* bettete das Verfahren in einen Mehrgitteransatz ein, um die Konvergenz des Iterationsverfahrens zu beschleunigen. Für bestimmte Wertebereiche der Parameter wird das Verfahren jedoch numerisch instabil (*Schnörr 87*).
Eine sorgfältige Diskussion der Gleichung (1.1) führte *Nagel 87* zu einer Vereinfachung seines Ansatzes:

$$\int_\Omega \left\{ ((\nabla g)^T \mathbf{v} + g_t)^2 + \lambda^2((\nabla v1)^T W(\nabla v1) + (\nabla v2)^T W(\nabla v2)) \right\} dx$$

$$(1.3)$$

$$= \int_\Omega \left\{ ((\nabla g)^T \mathbf{v} + g_t)^2 \right.$$

$$\left. + \lambda^2 \left[\frac{((\nabla v1)^T(\nabla g'))^2}{|\nabla g|^2 + 2\gamma} + \frac{((\nabla v2)^T(\nabla g'))^2}{|\nabla g|^2 + 2\gamma} + \gamma \frac{|\nabla v1|^2 + |\nabla v2|^2}{|\nabla g|^2 + 2\gamma} \right] \right\} dx \;\to\; min$$

mit

$$W = \frac{1}{|\nabla g|^2 + 2\gamma} \left[(\nabla g')(\nabla g')^T + \gamma I \right] \quad , \quad \nabla g' = (\nabla g)' = (g_y, -g_x)^T$$

Gegenüber dem Ansatz (1.2) wird hier bei kleinem γ und nichtverschwindendem Grauwertgradienten lokal die Variation des Vektorfeldes nur noch in der Richtung senkrecht zum Grauwertgradienten, also entlang der Konturlinie konstanten Grauwertverlaufs gemessen. Bei verschwindendem Gradienten hingegen wirkt der richtungsunabhängige Glattheitsterm wie in (1.2).

Ausgehend von einer Diskussion des Ansatzes von *Horn/Schunck 81* (1.2) in Abschnitt 2 wird in Abschnitt 3 für den Ansatz von *Nagel 87* (1.3) gezeigt, daß ein eindeutiges Vektorfeld $\mathbf{u} = (u1, u2)^T$ existiert, welches dem Funktional (1.3) sein Minimum erteilt, und daß diese Lösung stetig von den Eingangsdaten abhängt. Die Diskretisierung der Aufgabe mit der Methode der finiten Elemente wird kurz in Abschnitt 4 skizziert. Auf die Bildvorverarbeitung und die Schätzung der Eingangsdaten wird in Abschnitt 5 eingegangen. Der Einfluß der richtungsabhängigen Glattheitsforderung wird anhand einiger berechneter Vektorfelder deutlich (Abschnitt 6). In Abschnitt 7 werden die Ergebnisse noch einmal zusammengefaßt.

2. Der Ansatz von Horn/Schunck 81

In diesem Abschnitt wird der Ansatz von *Horn/Schunck 81* zur Rekonstruktion eines Geschwindigkeitsvektorfeldes abstrakt als Minimierung eines quadratischen Funktionals formuliert.

Für das Funktional $J:V \to \mathbb{R}$ (1.2) kann man schreiben:

$$J(\mathbf{v}) = \frac{1}{2} a(\mathbf{v}, \mathbf{v}) - f(\mathbf{v}) + c \tag{2.1}$$

mit der stetigen Bilinearform $a(\cdot,\cdot): V \times V \to \mathbb{R}$

$$a(\mathbf{u},\mathbf{v}) = 2 \int_\Omega \left\{ g_x^2 u1 v1 + g_x g_y (u1 v2 + u2 v1) + g_y^2 u2 v2 + \lambda^2 (u1_x v1_x + u1_y v1_y + u2_x v2_x + u2_y v2_y) \right\} d\mathbf{x} \tag{2.2}$$

der stetigen Linearform $f(\cdot): V \to \mathbb{R}$

$$f(\mathbf{v}) = -2 \int_\Omega \left\{ g_t g_x v1 + g_t g_y v2 \right\} dx \tag{2.3}$$

und der Konstanten

$$c = \int_\Omega g_t^2 \, d\mathbf{x} \tag{2.4}$$

welche für das folgende unwesentlich ist und nicht mehr weiter beachtet wird.
Der Raum V wird festgelegt durch

$$V := \left\{ \mathbf{u} = \binom{u1}{u2} \in H^1(\Omega) \times H^1(\Omega) \right\}$$

mit dem Skalarprodukt

$$(\mathbf{u},\mathbf{v})_V := (u1, v1)_{H^1} + (u2, v2)_{H^1}$$

und der Norm

$$\|\mathbf{u}\|_V := [(\mathbf{u},\mathbf{u})_V]^{1/2} = \left[\int_\Omega \left\{ |\mathbf{u}|^2 + |\nabla u1|^2 + |\nabla u2|^2 \right\} dx \right]^{1/2}$$

$H^1(\Omega)$ bezeichnet wie üblich den Sobolevraum

$$H^1(\Omega) := \{ u \in L^2(\Omega) \mid D^\alpha u \in L^2(\Omega) \text{ für } |\alpha| \le 1 \}$$

Es gelten nun die folgenden Aussagen (Lax-Milgram-Lemma (*Aubin 79*), *Hackbusch 86*):

Die Bilinearform a($\cdot,\cdot$) sei symmetrisch und V-elliptisch, d.h.

$$a(\mathbf{v},\mathbf{v}) \geq C_E \|\mathbf{v}\|_V^2 \quad \textit{für alle } \mathbf{v} \in V \textit{ mit } C_E > 0$$

(2.5)

Dann nimmt J($\mathbf{v}$) sein eindeutiges Minimum für die Lösung folgender Aufgabe an:

$$\textit{Suche } \mathbf{u} \in V \textit{ mit } a(\mathbf{u},\mathbf{v}) = f(\mathbf{v}) \quad \textit{für alle } \mathbf{v} \in V$$

(2.6)

Denn sei für ein beliebiges $\mathbf{w} \in V$ $\mathbf{v} := \mathbf{w}\text{-}\mathbf{u}$. Dann folgt bei Beachtung von (2.6)

$$J(\mathbf{w}) = J(\mathbf{u}+\mathbf{v}) = \frac{1}{2} a(\mathbf{u}+\mathbf{v},\mathbf{u}+\mathbf{v}) - f(\mathbf{u}+\mathbf{v})$$

$$= \frac{1}{2}\{a(\mathbf{u},\mathbf{u})+a(\mathbf{u},\mathbf{v})+a(\mathbf{v},\mathbf{u})+a(\mathbf{v},\mathbf{v})\} - f(\mathbf{u})-f(\mathbf{v})$$

$$= J(\mathbf{u}) + a(\mathbf{u},\mathbf{v})-f(\mathbf{v}) + \frac{1}{2}a(\mathbf{v},\mathbf{v}) \;\; = \;\; J(\mathbf{u}) + \frac{1}{2}a(\mathbf{v},\mathbf{v})$$

$$\geq J(\mathbf{u}) + \frac{C_E}{2}\|\mathbf{v}\|_V^2 \qquad\qquad = \;\; J(\mathbf{u}) + \frac{C_E}{2}\|\mathbf{w}-\mathbf{u}\|_V^2$$

also J($\mathbf{w}$) > J($\mathbf{u}$) für alle $\mathbf{w} \neq \mathbf{u}$.

Bezeichne V' den Dualraum, $<\cdot,\cdot>_{V' \times V}$ die durch $<\mathbf{v}',\mathbf{v}> = \mathbf{v}'(\mathbf{v})$ definierte Dualform auf V×V' und A:V→V' den der stetigen Bilinearform a($\cdot,\cdot$) durch

$$a(\mathbf{u},\mathbf{v}) = (A\mathbf{u})(\mathbf{v}) = <A\mathbf{u},\mathbf{v}>_{V' \times V} \quad \textit{für alle } \mathbf{u},\mathbf{v} \in V$$

eineindeutig zugeordneten linearen, beschränkten Operator. Unter der oben genannten Voraussetzung existiert A^{-1} mit $\|A^{-1}\| \leq 1/C_E$. (2.6) läßt sich in der Form

$$<A\mathbf{u}-f,\mathbf{v}>_{V' \times V} = 0 \quad \textit{für alle } \mathbf{v} \in V \qquad d.h.$$

$$A\mathbf{u} = f \quad in \; V'$$

schreiben und hat genau die Lösung

$$\mathbf{u} = A^{-1}f$$

welche stetig von f abhängt:

$$\|\mathbf{u}\|_V \leq \frac{1}{C_E}\|f\|_{V'}$$

Unter schwachen Bedingungen bzgl. des Grauwertgradienten sind die o.g. Voraussetzungen erfüllt (*Schnörr 89*).

3. Der Ansatz von Nagel 87

Dieses Ergebnis läßt sich nun leicht auf den Ansatz von *Nagel 87* übertragen, denn mit (2.3), (2.4) läßt sich (1.3) in der Form (2.1) schreiben, wobei

$$a(\mathbf{u},\mathbf{v}) = 2 \int_\Omega \Big\{ g_x^2 u1v1 + g_x g_y(u1v2+u2v1) + g_y^2 u2v2$$

$$+ \frac{\lambda^2}{g_x^2+g_y^2+2\gamma} \Big[(g_y^2+\gamma)(u1_x v1_x + u2_x v2_x) + (g_x^2+\gamma)(u1_y v1_y + u2_y v2_y)$$

$$- g_x g_y (u1_x v1_y + u1_y v1_x + u2_x v2_y + u2_y v2_x) \Big] \Big\} \, dx$$

Schreibt man für $a(\cdot,\cdot)$ gemäß (2.2) $a(\cdot,\cdot)_{\text{Horn/Schunck}}$ und ersetzt dort λ^2 durch

$$(\lambda')^2 := \frac{\lambda^2 \gamma}{|\nabla g|^2 + 2\gamma} \, ,$$

so kann man schreiben

$$a(\mathbf{v},\mathbf{v}) = a(\mathbf{v},\mathbf{v})_{\text{Horn/Schunck}} + \frac{2(\lambda')^2}{\gamma} \int_\Omega \Big\{ ((\nabla g')^T (\nabla v1))^2 + ((\nabla g')^T (\nabla v2))^2 \Big\} \, dx$$

Für $\gamma > 0$ folgt damit (2.5). Damit gelten die Aussagen aus Abschnitt 2.

4. Diskretisierung der Aufgabe

Zur Diskretisierung der Aufgabe (2.6) wird das Ritz-Galerkin-Verfahren angewendet, welches bei spezieller Wahl der Basisfunktionen auf die Methode der Finiten Elemente führt (*Ciarlet 78, Hackbusch 86*).

Zunächst wird der Raum V durch einen endlich-dimensionalen Raum V_N ersetzt:

$$V_N := span\{\mathbf{\Phi}_1, \dots, \mathbf{\Phi}_N\} \subset V$$

wobei $\{\mathbf{\Phi}_i\}$ eine Basis von V_N darstellt. Als finites Element wurde der bilineare Ansatz im Quadrat (zwischen jeweils vier Pixeln) für jede Komponentenfunktion gewählt. Die gemäß (4.2) durchzuführenden Integrationen werden wegen der rationalen Koeffizientenfunktionen numerisch mit einem Gaußverfahren (*Schwarz 80*) durchgeführt. Für jeden Koeffizientenvektor $v = (v_1, \dots, v_N)^T$ (nicht-fett gedruckte, nichtindizierte Buchstaben bezeichnen in diesem Abschnitt Vektoren aus dem $\mathbf{R}^N$) wird die Abbildung definiert:

$$P : \mathbf{R}^N \to V_N, \quad Pv := \sum_{i=1}^{N} v_i \mathbf{\Phi}_i$$

Die v_i bezeichnen dabei die z.B. zeilenweise zu dem Vektor v zusammengefassten Werte der beiden Komponentenfunktionen von $v \in V_N$ an den Pixelpositionen.

(2.6) lautet nun:

$$\text{Suche } \mathbf{u} \in V_N \text{ mit } a(\mathbf{u},\mathbf{v}) = f(\mathbf{v}) \quad \text{für alle } \mathbf{v} \in V_N$$

$$(4.1)$$

Es ergeben sich die Darstellungen

$$a(\mathbf{u},\mathbf{v}) = a(Pu,Pv) = \sum_{i,j} u_j v_i a(\mathbf{\Phi}_j, \mathbf{\Phi}_i) = u^T L v \quad \text{mit } L_{ij} = a(\mathbf{\Phi}_j, \mathbf{\Phi}_i) \, , \; i,j = 1,\dots,N$$

$$f(\mathbf{v}) = f(Pv) = \sum_i v_i f(\mathbf{\Phi}_i) = b^T v \quad \text{mit } b_i = f(\mathbf{\Phi}_i) \, , \; i = 1,\dots,N$$

$$J = \frac{1}{2} v^T L v - b^T v$$

$$(4.2)$$

Mit $a(\cdot,\cdot)$ ist auch L symmetrisch und aus der V-Elliptizität folgt, daß L positiv-definit ist. Denn für jedes $v \neq 0$ ist auch $Pv \neq 0$ und

$$v^T L v = a(Pv,Pv) \geq C_E \|Pv\|_V^2 > 0$$

Aus dem bilinearen Ansatz im Quadrat ergibt sich, daß die Variablen lokal über ein 3x3 Pixel großes Fenster miteinander verkoppelt sind. In jeder Zeile der Matrix L stehen somit (von Randvariablen entsprechenden Zeilen abgesehen) $2 \times 9 = 18$ von Null verschiedene Einträge. L ist somit sehr dünn besetzt. Der Lösungsvektor u - und damit die Ritz-Galerkin-Lösung $\mathbf{u} = Pu$ - ergibt sich damit aus dem linearen Gleichungssystem

$$Lu = b \qquad (4.3)$$

(4.3) entspricht somit (4.1).

5. Bildvorverarbeitung, Schätzung der Eingangsdaten

Da die beiden Halbbilder eines Vollbildes verschiedenen Zeitpunkten zugeordnet sind, ergeben sich in Bildbereichen von Objektgrenzen stark verzerrte Grauwertverläufe (Abb. 5.1). Deshalb wird jeweils nur das erste Halbbild benutzt. Die Auflösung in Zeilenrichtung wird nach einer Tiefpaßfilterung auf die Hälfte des Vollbildes reduziert, so daß bei einem (512x512)-Vollbild das Format anschließend (256x256)-Pixel beträgt (Abb. 5.2). Die durch das Abschneiden der beiden Ränder auftretenden Effekte werden vernachlässigt.
Als Tiefpaß wurde eine "equiripple"-Approximation des benötigten Halbbandfilters verwendet. Bei festen Eingangsdaten (u.a. Länge der Impulsantwort) ist das Frequenzverhalten des resultierenden Filters in einem gewissen Sinn optimal (vgl. *Oppenheim/Schafer 75*, Abschnitt 5.6.2).
Die Daten des Filters wurden von *Meer et al. 87* übernommen, die ein von der eindimensionalen Signalverarbeitung her bekanntes Programm benutzten (*McClellan et al. 73*; Länge der Impulsantwort: $k = 13$; Bandfrequenzen: $\omega_{pass} = 1/2(\pi{-}\Delta\omega)$, $\omega_{stop} = 1/2(\pi + \Delta\omega)$ mit $\Delta\omega = 0.2\pi$; Toleranzen $\delta_{pass} = \delta_{stop} = 5\%$).
Die Impulsantwort des idealen Filters hat bekanntlich die angenehme Eigenschaft, daß die Nulldurchgänge sich an jeder zweiten Rasterposition befinden. Die aufgrund der Approximation von Null verschiedenen Werte der Impulsantwort an diesen Positionen werden zu Null gesetzt. Die übrigen Werte werden anschließend normiert. Die Frequenzeigenschaften des Filters ändern sich dadurch so gut wie nicht (*Meer et al. 87*).

Zur Schätzung der Ableitungen der Bildfunktion wurde ein Gaußfilter mit $\sigma = 1.0$ gewählt. Die diskrete Approximation erfolgte gemäß *Hashimoto/Sklansky 87*, da bei annähernd gleichem Verhalten im Frequenzbereich die Größe der Impulsantwort kleiner ist als die des Filters, das man durch Abtasten der kontinuierlichen Impulsantwort erhält.

6. Einige Ergebnisse

Die Abbildungen 6.3 und 6.4 zeigen die für das Bild 6.1 nach dem Ansatz von *Horn/Schunck 81* bzw. *Nagel 87* berechneten Vektorfelder. Die in Abb. 6.5 und 6.6 gezeigten Vektorfelder wurden entsprechend für das Bild 6.2 berechnet. Die Parameterwerte betrugen $\lambda^2 = 100$, $\gamma = 0.01$. Das Iterationsverfahren zur Lösung von (4.3) wurde jeweils bei $\| u^k - u^{k\text{-}1} \|_\infty < 10^{\text{-}3}$ abgebrochen.
Wie man sieht, wird das Ausbreiten der Information über Grauwertkanten hinweg durch die richtungsabhängige Glattheitsforderung eingeschränkt. Dadurch werden die Bewegungsgrenzen als lokale Variationen des Vektorfeldes in Abb. 6.6 deutlicher sichtbar. Dazu vergleiche man auch die Detailansichten Abb. 6.7, 6.8. Dies müßte eine a-posteriori Segmentation des Vektorfeldes erleichtern.

7. Zusammenfassung

Es wurde gezeigt, daß das von *Nagel 87* formulierte Minimierungsproblem zur Bestimmung von Geschwindigkeitsvektorfeldern korrekt gestellt (well posed) ist. Die Eigenschaften des Operators A (vgl. Abschnitt 2) bilden eine gute Grundlage für weitere Untersuchungen. In dieser Arbeit münden sie in ein wohldefiniertes Approximationsverfahren (finite Elemente) zur Berechnung einer Näherungslösung. Die Eigenschaften des kontinuierlichen Ansatzes übertragen sich dabei auf den diskreten Fall (positiv definite Matrix L) und garantieren damit globale Konvergenz. Eine drastische Konvergenzbeschleunigung sowie eine Verbesserung der Konditionierung des Problems läßt sich mit bekannten Mehrgittermethoden (*Brandt 80*, *Hackbusch 85*) erreichen. Die Tatsache, daß die Matrix des zu lösenden Gleichungssystems nur sehr dünn besetzt ist, ermöglicht die schnelle Berechnung duch geeignete Prozessorstrukturen.
Es muß nun überprüft werden, welche Aufgaben schon durch die Auswertung dieser Vektorfelder gelöst werden können.

Danksagung

Herrn Prof. H.-H. Nagel danke ich für hilfreiche Diskussionen und die kritische Durchsicht dieses Beitrages.

Diese Arbeit wurde im Rahmen des Projektes PROMETHEUS gefördert.

Literatur

Aggarwal/Nandhakumar 88 : On the Computation of Motion from Sequences of Images - A Review, Proceedings of the IEEE, Vol.76, No.8, August 1988, 917-935

Aubin 79 : Applied Functional Analysis, Wiley + Sons, New York, 1979

Brandt 80 : Multi-Level Adaptive Finite-Element Methods, in: Special Topics of Applied Mathematics, J.Frehse, D.Pallaschke, U.Trottenberg (eds.), North-Holland Publishing Company, 1980

Ciarlet 78 : The finite element method for elliptic problems, P.G.Ciarlet, North - Holland Publ. Comp., Amsterdam, 1978

Enkelmann 85 : Mehrgitterverfahren zur Ermittlung von Verschiebungsvektorfeldern in Bildfolgen, Dissertation (Juli 1985), Fachbereich Informatik der Universität Hamburg

Enkelmann 86 : Investigations of multigrid algorithms for the estimation of optical flow fields in image sequences, Proc.Workshop on Motion : Repr. and Analysis, May 7 - 9 (1986), Charleston, South Carolina, 81 - 87

Hackbusch 85 : Multi-Grid Methods and Applications, Springer-Verlag, Berlin Heidelberg New York Tokyo, 1985

Hackbusch 86 : Theorie und Numerik elliptischer Differentialgleichungen, W.Hackbusch, B.G.Teubner Stuttgart 1986

Hashimoto/Sklansky 87 : Multiple-Order Derivatives for Detecting Local Image Characteristics, Computer Vision, Graphics, and Image Processing 39, 28 - 55 (1987)

Horn/Schunck 81 : Determining optical flow, B.K.P.Horn, B.G.Schunck, Artif. Intell. 17, 1981, 185 - 203

Maybank 87 : A Theoretical Study of Optical Flow, PhD. thesis, Univ. of London, Nov. 1987

McClellan et al. 73 : A computer program for designing optimum FIR linear phase digital filters, J.H.McClellan, T.W.Parks, L.R.Rabiner, IEEE Trans. Audio Electroacoust., Vol. AU-21, 1973, 506-525

Meer et al. 87 : Frequency domain analysis and synthesis of image pyramid generating kernels, P.Meer, E.S.Baugher, A.Rosenfeld, IEEE Trans. on Patt. Analysis and Mach. Int., Vol. PAMI-9, No.4, July 1987, 512 - 522

Nagel 83 : Constraints for the estimation of displacement vector fields from image sequences, H.-H.Nagel, Proc.Eighth Int.Joint Conf.Artificial Intell., Karlsruhe, 1983, 945 - 951

Nagel 87 : On the estimation of optical flow : relations between different approaches and some new results, H.-H.Nagel, Artificial Intelligence 33, 1987, 299 - 324

Nagel 88 : Image Sequences - Ten (octal) Years - From Phenomenology towards a Theoretical Foundation, Int.J.of Patt.Rec. and Artif.Intelligence Vol.2, No.3 (1988), 459 - 483

Nagel/Enkelmann 83 : Iterative estimation of displacement vector fields from TV - frame sequences, Proc. Second European Signal Processing Conf. (EUSIPCO-83), H.W.Schüssler, Ed., Erlangen, Sept.12-16, 1983, 299 - 302

Nagel/Enkelmann 84 : Berechnung von Verschiebungsvektorfeldern in Bildbereichen mit linienhaften oder partiell homogenen Grauwertverteilungen, DAGM-Symposium Mustererkennung 1984, W.Kropatsch, Hrsg., Graz, Oct.2-4, 1984, 154 - 160

Nagel/Enkelmann 86 : An investigation of smoothness constraints for the estimation of displacement vector fields from image sequences, H.-H.Nagel, W.Enkelmann, IEEE Transactions on Pattern Analysis and Machine Intelligence, Vol. PAMI-8, No. 5, September 1986, 565 - 593

Oppenheim/Schafer 75 : Digital signal processing, A.V.Oppenheim, R.W.Schafer, Prentice-Hall, Inc., Englewood Cliffs, New Jersey, 1975

Schnörr 87 : Untersuchung von modifizierten Ansätzen zur Ermittlung von örtlichen und zeitlichen Ableitungen bei der Schätzung des optischen Flusses in Bildfolgen, C.Schnörr, Diplomarbeit, November 1987, Fakultät für Elektrotechnik der Universität Karlsruhe

Schnörr 89 : Zum Ansatz von Horn/Schunck 81 zur Schätzung von Geschwindigkeitsvektorfeldern in Bildfolgen, Hausbericht Nr. 10165, Fraunhofer-Institut für Informations- und Datenverarbeitung (IITB), Karlsruhe

Schwarz 80 : Methode der finiten Elemente, B.G. Teubner, Stuttgart, 1980

Abb. 5.1 Ausschnitt des Vollbildes

Abb. 5.2 1.Halbbild, Zeilen tiefpassgefiltert und vergrößert dargestellt

Abb. 6.1

Abb. 6.2

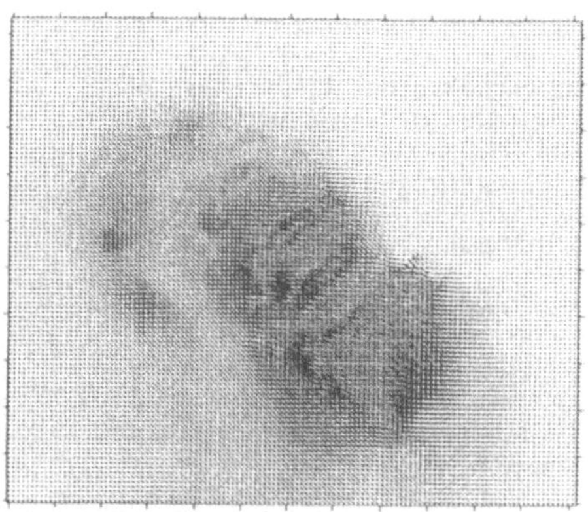

Abb. 6.3 : nach *Horn/Schunck 81* für Abb. 6.1 berechnetes Vektorfeld

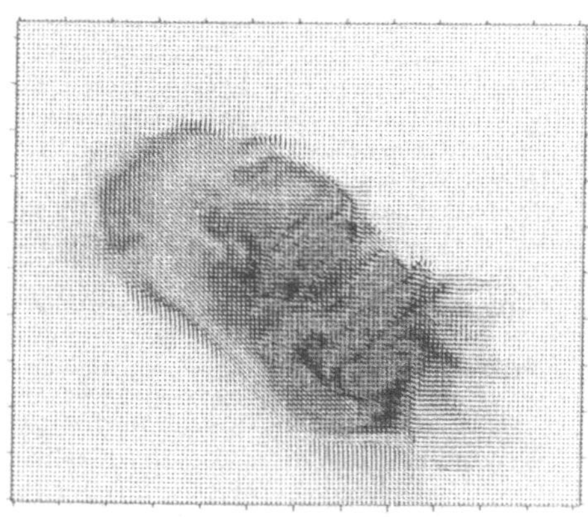

Abb. 6.4 : nach *Nagel 87* für Abb. 6.1 berechnetes Vektorfeld

Abb. 6.5 : nach *Horn/Schunck 81* für
Abb. 6.2 berechnetes Vektorfeld

Abb. 6.6 : nach *Nagel 87* für Abb. 6.2
berechnetes Vektorfeld

Abb. 6.7 : Detail aus Abb. 6.5   **Abb. 6.8** : Detail aus Abb. 6.6
(Rechteckiger Ausschnitt um die Stelle, an der sich der dunkle Golf links unten und das helle Taxi in der
Mitte am nächsten kommen.)

A comparison of two non-linear motion estimation methods

Z. Houkes and M.J. Korsten
University of Twente
POB 217, 7500 AE Enschede/NL

Summary

This paper examines two distinct methods for the estimation of motion from successive images. The first one is the modified Newton–Raphson method, used by Burkhardt, Moll and Diehl [1], [2], [3]. The second one is the method used by Houkes [4], [5]. It is shown, that the results of both methods differ only slightly.

Keywords: motion, motion estimation, motion prediction, image sequence analysis, computer vision, image processing, non–linear estimation.

General Method

Both methods estimate motion parameters from image pairs. The motion model can be expressed by:

$$\mathbf{V}' = \mathbf{h}(\mathbf{V}; \boldsymbol{\alpha}) \tag{1}$$

Here $\mathbf{V}'$ is the position of a pixel in the first image, given its position $\mathbf{V}$ in the second image. $\mathbf{V}$ is the independent variable constrained by a Region of Interest (ROI), where the motion model is supposed to apply. The "Constant Brightness Assumption" states, that the intensity I_i ($i=1, 2$) of corresponding pixels is the same in both images, therefore:

$$I_2(\mathbf{V}) = I_1(\mathbf{V}') \tag{2}$$

In order to estimate the motion parameters a cost criterion is defined, to be minimized with respect to the parameter vector $\boldsymbol{\alpha}$:

$$J(\boldsymbol{\alpha}) = \tfrac{1}{2}E[\{I_1(\mathbf{h}(\mathbf{V}; \boldsymbol{\alpha})) - I_2(\mathbf{V})\}^2] \tag{3}$$

where $E[]$ stands for the expectation operator. In practice the expectation is computed in the spatial domain from the two images, a process discussed more deeply in [3] and [4]. The minimum of $J(\boldsymbol{\alpha})$ is computed iteratively using successive linearization. The difference between the two methods lies in the way of linearizing (3).

The modified Newton–Raphson method

The Newton–Raphson method requires the cost criterion $J(\alpha)$ to be expanded in a Taylor series about an estimate $\hat{\alpha}_k$ according to:

$$
\begin{aligned}
J(\alpha) &= J(\hat{\alpha}_k + d\alpha) \\
&= J(\hat{\alpha}_k) + g(\hat{\alpha}_k)^T d\alpha + \tfrac{1}{2} d\alpha^T H(\hat{\alpha}_k) d\alpha + \dots
\end{aligned}
\tag{4}
$$

where $g(\hat{\alpha}_k)^T$ is a row vector and $H(\hat{\alpha}_k)$ the Hessian matrix reflecting the first and second order derivative of the cost criterion (3) with respect to $d\alpha$. Minimizing (4) with respect to $d\alpha$ requires the first order derivative of $J(.)$ with respect to $d\alpha$ to be zero. Ignoring the higher order terms in (4) we obtain an estimate $d\hat{\alpha}$ for $d\alpha$:

$$
d\hat{\alpha} = -H^{-1}(\hat{\alpha}_k) g(\hat{\alpha}_k)
\tag{5}
$$

where $H^{-1}(\hat{\alpha}_k)$ exists if the elements of the parameter vector $\hat{\alpha}_k$ are independent. The update of the estimate $\hat{\alpha}_k$ is:

$$
\hat{\alpha}_{k+1} = \hat{\alpha}_k + d\hat{\alpha}
\tag{6}
$$

There are two drawbacks to the Newton–Raphson method. In the first place the method is numerically unfriendly, especially because of the appearance of the Hessian matrix. In the second place its convergence range is restricted to the range between the real parameter magnitude and the first maximum of the gradient $g(\hat{\alpha}_k)$ [3], [4]. Burkhardt and Moll [1] already considered the possibility of replacing the Hessian matrix $H(\hat{\alpha}_k)$ by its value $H(\alpha^*)$ in the optimum α^*. If it is assumed, that $H(\alpha^*)$ in the optimum of J is maximal, the convergence range is extended to the first zero of the gradient. The Hessian matrix in the optimum can be shown from (3) and (4) to be:

$$
H(\alpha^*) = E[(\frac{\partial I_1(h(V;\alpha^*))}{\partial \alpha})^T \frac{\partial I_1(h(V;\alpha^*))}{\partial \alpha}]
\tag{7}
$$

In order to compute $H(\alpha^*)$ in (7) without knowing the optimal parameter estimate α^*, Burkhardt and Diehl perform a coordinate transformation on the position vector V according to the definition:

$$
h(V;\hat{\alpha}_{k+1}) = h(h(V;\hat{\alpha}_k);\tilde{\alpha}_{k+1})
\tag{8}
$$

Expression (8) defines the innovation $\tilde{\alpha}_{k+1}$ in the coordinates $\hat{V}_k = h(V;\hat{\alpha}_k)$ and only indirectly in the coordinates V. The cost criterion $J(.)$ from (4) is now a function of $\hat{\alpha}_k$ and $\tilde{\alpha}_{k+1}$. Assuming that the optimum has been reached at the k–th iteration step, and performing the linearization with respect to $\tilde{\alpha}_{k+1}$ about zero, one obtains:

$$J(\alpha) = J(\hat{\alpha}_k, \tilde{\alpha}_{k+1}) = (^1/_2)E[\{I_1(h(h(V;\hat{\alpha}_k);\tilde{\alpha}_{k+1}))-I_2(V)\}^2]$$

$$= J(\hat{\alpha}_k,0) + \tilde{g}(\hat{\alpha}_k)^T \tilde{\alpha}_{k+1} + \frac{1}{2}\tilde{\alpha}_{k+1}^T \tilde{H} \tilde{\alpha}_{k+1} + \dots \tag{9}$$

with:

$$\tilde{g}(\hat{\alpha}_k)^T = E[\{I_1(h(h(V;\hat{\alpha}_k);\tilde{\alpha}_{k+1}))-I_2(V)\}\frac{\partial I_1(h(h(V;\hat{\alpha}_k);\tilde{\alpha}_{k+1}))}{\partial \tilde{\alpha}_{k+1}}]\Big|_{\tilde{\alpha}_{k+1}=0} \tag{10}$$

$$\tilde{H} = E[\frac{\partial I_1(h(h(V;\hat{\alpha}_k);\tilde{\alpha}_{k+1}))}{\partial \tilde{\alpha}_{k+1}}(\frac{\partial I_1(h(h(V;\hat{\alpha}_k);\tilde{\alpha}_{k+1}))}{\partial \tilde{\alpha}_{k+1}})^T]\Big|_{\tilde{\alpha}_{k+1}=0} \tag{11}$$

The expression between brackets [] in (11) depends on $\hat{\alpha}_k$ via the coordinate $\hat{V}_k=h(V;\hat{\alpha}_k)$. The expectation E[] is (see [2]) independent of an arbitrary coordinate system, therefore the Hessian matrix in the optimum can be expressed by the derivatives of I_1 with respect to the coordinates in the first image.

This method requires the existence of a coordinate transformation according to (8). An affine transformation meets this condition. The model in this paper is described by a translation vector $d=(d_1,d_2)^T$ and a rotation angle φ:

$$\alpha = (\varphi,d_1,d_2)^T \tag{12}$$

$$h(V;\alpha) = \begin{bmatrix} \cos\varphi & -\sin\varphi \\ \sin\varphi & \cos\varphi \end{bmatrix}\begin{bmatrix} V_1 \\ V_2 \end{bmatrix} - \begin{bmatrix} d_1 \\ d_2 \end{bmatrix} \tag{13}$$

The required update $\hat{\alpha}_{k+1}$ is obtained by combining the equations (8), (12) and (13), and replacing $\tilde{\alpha}_{k+1}$ by the value obtained when (9) is minimized with respect to $\tilde{\alpha}_{k+1}$:

$$\hat{\alpha}_{k+1} = A_{k+1}\hat{\alpha}_k - \tilde{H}^{-1}\tilde{g}(\hat{\alpha}_k) \tag{14}$$

with A_k the matrix:

$$A_k = \begin{bmatrix} 1 & 0 & 0 \\ 0 & \cos\tilde{\varphi}_k & -\sin\tilde{\varphi}_k \\ 0 & \sin\tilde{\varphi}_k & \cos\tilde{\varphi}_k \end{bmatrix} \tag{15}$$

As an example the derivative of the intensity function I_1 to $\tilde{\varphi}_{k+1}$, appearing in the Hessian matrix (11), is computed to be:

$$\frac{\partial I_1(h(h(V;\hat{\alpha}_k);\tilde{\alpha}_{k+1}))}{\partial \tilde{\varphi}_{k+1}}\Big|_{\tilde{\alpha}_{k+1}=0} = \frac{\partial I_1(\hat{V}_k)}{\partial V_k}\begin{bmatrix} -\hat{V}_{k,2} \\ \hat{V}_{k,1} \end{bmatrix}\Big|_{\hat{V}_k=h(V;\hat{\alpha}_k)} \tag{16}$$

with expressions for the derivatives with respect to $d_{k+1,1}$ and $d_{k+1,2}$ analogous to (16). Therefore the dependence on $\hat{\alpha}_k$ only exists via the coordinate $\hat{V}_k$ and disappears when taking the expectation.

The above derived expressions are obtained again, when the direct linearization method, as will be described below, is applied to the reformulated problem using the innovation $\tilde{\alpha}_{k+1}$ (see [6]).

Direct linearization of the grey value function

The alternative for the above given solution, proposed by Houkes [4], [5] is to linearize the predicted intensity function I_1 instead of the cost criterion. We obtain from (1) and (2):

$$I_1(h(V;\alpha)) = I_1(h(V;\hat{\alpha}_k+d\alpha)) = I_1(h(V;\hat{\alpha}_k)) + \frac{\partial I_1(h(V;\hat{\alpha}_k))}{\partial \alpha_k}d\alpha + \ldots \ldots \tag{17}$$

Substituting (17) in (3) we have:

$$J(\alpha) = J(\hat{\alpha}_k,d\alpha) = \tfrac{1}{2}E[\{I_1(h(V;\hat{\alpha}_k))-I_2(V)\}^2] + g(\hat{\alpha}_k)d\alpha + \tfrac{1}{2}d\alpha^T H_1(\hat{\alpha}_k)d\alpha +.. \tag{18}$$

with $g(\hat{\alpha}_k)^T$ the row vector defined as:

$$g(\hat{\alpha}_k)^T = E[\{I_1(h(V;\hat{\alpha}_k))-I_2(V)\}\frac{\partial I_1[h(V;\hat{\alpha}_k)]}{\partial \alpha_k}] \tag{19}$$

and $H_1(\hat{\alpha}_k)$ as:

$$H_1(\hat{\alpha}_k) = E[(\frac{\partial I_1(h(V;\hat{\alpha}_k))}{\partial \alpha_k})^T \frac{\partial I_1(h(V;\hat{\alpha}_k))}{\partial \alpha_k}] \tag{20}$$

Taking the gradient with respect to $d\alpha$ and neglecting the higher order terms in (18) we obtain:

$$d\hat{\alpha} = -H_1^{-1}(\hat{\alpha}_k)g(\hat{\alpha}_k) \tag{21}$$

The expression (21) differs from (5), because the matrix $H_1(\hat{\alpha}_k)$ is used instead of the Hessian matrix in the optimum $H(\alpha^*)$. The difference between $H_1(\hat{\alpha}_k)$ in (20) and $H(\alpha^*)$ in (7) is, that in $H_1(.)$ the estimate $\hat{\alpha}_k$ has been used instead of the optimal value α^*. Therefore $H_1(\hat{\alpha}_k)$ is known and no coordinate transformation is necessary. In the next section the difference between the modified Newton–Raphson method and the direct linearization used by Houkes is investigated.

Comparing the modified Newton–Raphson method and the direct linearization.

To compare the two methods it is observed, that in both cases an update $\hat{\alpha}_{k+1}$ is computed from an estimate $\hat{\alpha}_k$. The performance of a coordinate transformation in the modified Newton–Raphson method is not important in itself, because the resulting $\tilde{\alpha}_{k+1}$ is transformed back to $\hat{\alpha}_{k+1}$ in (14). The point is that the coordinate transformation causes the Hessian to be independent of the true motion vector. Therefore the Hessian can be computed in advance, instead of computing it every iteration. To investigate the difference with the direct expansion, we eliminate $d\alpha$ from (18) in favor of $\tilde{\alpha}_{k+1}$. Comparing (9) to (11) with (18) to (20) it can be seen that only two terms have to be considered:

$$D_1 = \left.\frac{\partial I_1(\mathbf{h}(\hat{\mathbf{V}}_k;\tilde{\alpha}_{k+1}))}{\partial \tilde{\alpha}_{k+1}}\right|_{\substack{\tilde{\alpha}_{k+1} \\ \tilde{\alpha}_{k+1}=0 \\ \hat{\mathbf{V}}_k = \mathbf{h}(\mathbf{V};\hat{\alpha}_k)}} \qquad (\text{from } (9) \text{ to } (11))$$
$$\text{Burkhardt, Moll, Diehl}$$

$$\text{vs:}\quad D_2 = \frac{\partial I_1(\mathbf{h}(\mathbf{V};\hat{\alpha}_k))}{\partial \hat{\alpha}_k} d\hat{\alpha} \qquad (\text{from } (18) \text{ to } (20)) \qquad (22)$$
$$\text{Houkes}$$

First $d\hat{\alpha}$ is expressed as a function of $\tilde{\alpha}_{k+1}$ and $\hat{\alpha}_k$. We have from (6):

$$d\hat{\alpha} = \hat{\alpha}_{k+1} - \hat{\alpha}_k$$

$$= (A_{k+1}-I)\hat{\alpha}_k + \tilde{\alpha}_{k+1} \qquad (23)$$

where the last expression has been obtained from (14) and I is the identity matrix. With (23) D_2 can be computed in terms of $\tilde{\alpha}_{k+1}$, according to:

$$D_2 = \frac{\partial I_1(\hat{\mathbf{V}}_k)}{\partial \hat{\mathbf{V}}_k} \left[\begin{matrix} -(\hat{V}_{k,2}+\hat{d}_{k,2}) & -1 & 0 \\ (\hat{V}_{k,1}+\hat{d}_{k,1}) & 0 & -1 \end{matrix}\right] \left[\begin{matrix} \tilde{\varphi}_{k+1} \\ (\cos\tilde{\varphi}_{k+1}-1)\hat{d}_{k,1} - \sin\tilde{\varphi}_{k+1}\hat{d}_{k,2} + d_{k+1,1} \\ \sin\tilde{\varphi}_{k+1}\hat{d}_{k,1} + (\cos\tilde{\varphi}_{k+1}-1)\hat{d}_{k,2} + d_{k+1,2} \end{matrix}\right]_{\hat{\mathbf{V}}_k = \mathbf{h}(\mathbf{V};\hat{\alpha}_k)}$$

$$= \frac{\partial I_1(\hat{\mathbf{V}}_k)}{\partial \hat{\mathbf{V}}_k} \left[\begin{matrix} -\tilde{\varphi}_{k+1}(\hat{V}_{k,2}+\hat{d}_{k,2}) - (\cos\tilde{\varphi}_{k+1}-1)\hat{d}_{k,1} + \sin\tilde{\varphi}_{k+1}\hat{d}_{k,2} - d_{k+1,1} \\ \tilde{\varphi}_{k+1}(\hat{V}_{k,1}+\hat{d}_{k,1}) - \sin\tilde{\varphi}_{k+1}\hat{d}_{k,1} - (\cos\tilde{\varphi}_{k+1}-1)\hat{d}_{k,2} - d_{k+1,2} \end{matrix}\right]_{\hat{\mathbf{V}}_k = \mathbf{h}(\mathbf{V};\hat{\alpha}_k)} \qquad (24)$$

where (14) has been used for $\tilde{\alpha}_{k+1}$, while A_{k+1} has been obtained from (15). D_1 can be computed straightforwardly to be:

$$D_1 = \frac{\partial I_1(\hat{\mathbf{V}}_k)}{\partial \hat{\mathbf{V}}_k} \left[\begin{matrix} -\hat{V}_{k,2}\tilde{\varphi}_{k+1} - d_{k+1,1} \\ \hat{V}_{k,1}\tilde{\varphi}_{k+1} - d_{k+1,2} \end{matrix}\right]_{\hat{\mathbf{V}}_k = \mathbf{h}(\mathbf{V};\hat{\alpha}_k)} \qquad (25)$$

From (24) and (25) it is observed, that the direct linearization does not compute the Hessian matrix precisely in the optimum. Therefore, in general the modified Newton–Raphson method and the direct linearization of Houkes do not yield the same results. However in two special cases (24) and (25) are equal. In the first place if $\hat{\alpha}_k$ is zero. If the initial estimate is zero, the first iteration yields in both cases the same result. In the second place if $\tilde{\varphi}_{k+1}$ is so small, that:

$$\sin\tilde{\varphi}_{k+1} \approx \tilde{\varphi}_{k+1} \qquad \text{and} \qquad \cos\tilde{\varphi}_{k+1} \approx 1 \tag{26}$$

This result could have been expected as Houkes [4] already showed, that the two methods lead to exactly the same result if only a translation and no rotation is involved. In fact the result, presented here, is an extension of this result. In order for the two methods to coincide $\tilde{\varphi}_{k+1}$ does not have to be zero, but only should be so small, that (26) holds. Therefore the similarity between the two methods is an order of magnitude better, than could have been expected. Summarizing it must be concluded that the behaviour of both estimation methods is similar. Especially when the starting estimate is zero the results of the first iteration in both cases coincide. In many cases the first iteration already gives a strong shift in the correct direction. Important differences only appear if the innovation angle $\tilde{\varphi}_{k+1}$ is very large, but this may not be expected because of the restricted area of convergence.

Some results from simulations

Both algorithms have been implemented at the University of Twente by W. Oelen [6]. Two results are reproduced here. In both cases image pairs have been produced. The first image consisted of a sine wave in two directions in a region of interest (ROI), according to:

$$I(V) = 128 + 63\sin(V_1/5) + 63\cos(V_2/5) \tag{27}$$

The region outside the ROI consisted of a background, that was varied during the experiments. The second image was obtained by rotating and translating the ROI, according to the image motion model described by (12) and (13), and with parameters: $d_1=d_2=5$ pixels and $\varphi=10°$. In both cases image data from the complete ROI were used for the estimations. No measures against model errors were taken.

It can be concluded, that the differences between both methods are very minor. They might occur from the different ways of computing the expectation values in the Hessian matrices (20) and (7) respectively. However this has not been investigated thoroughly. The results from both methods coincide sufficiently in order not to contradict the theoretical agreement between the two methods.

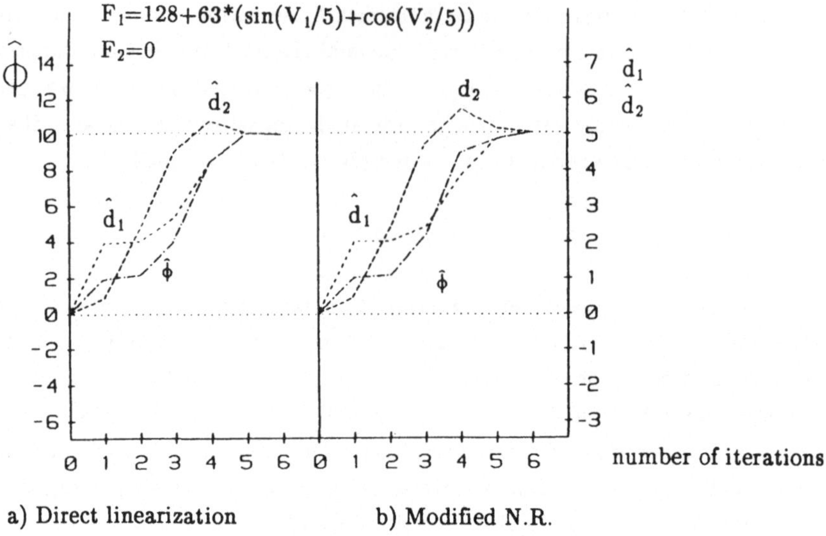

Fig. 1   Motion estimation, uniform background with zero greylevel

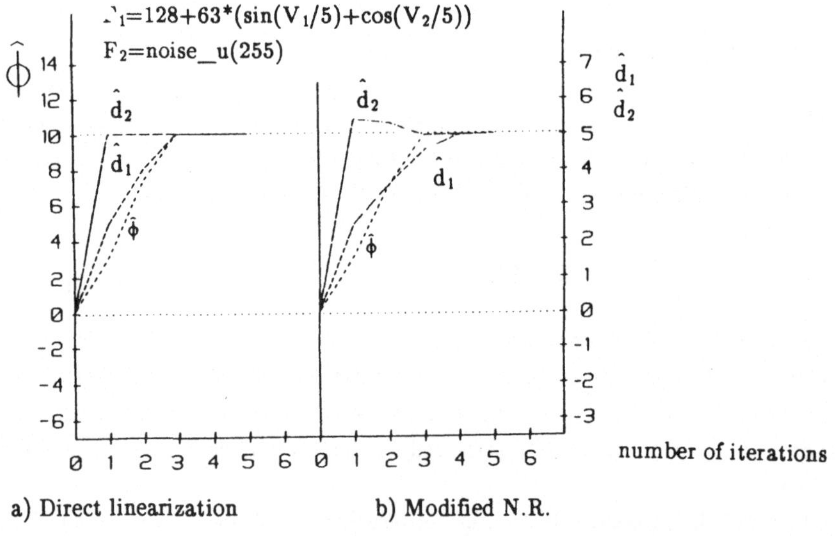

Fig. 2   Motion estimation, background consisting of white uniform noise

<u>Summary and conclusion</u>

In this paper two motion estimation methods have been compared. The difference between the two methods lies in the way of linearizing the non linear estimation problem. In theory both methods must coincide if the innovation angle $\tilde{\varphi}_{k+1}$ is small enough in order to equal its sine. Results from simulations seem to affirm this result.

<u>References</u>

1: Burkhardt H., Moll H. (1979), A modified Newton–Raphson search for the model–adaptive identification of delays, IFAC–Conference on identification and system parameter identification, p1279–1286
2: Burkhardt H., Diehl N., (1986), Simultaneous estimation of rotation and translation in image sequences, Proc. of the European Signal Processing Conference, EUSIPCO–86, Den Haag, p821–824
3: Diehl N. (1988), Methoden zur Allgemeinen Bewegungsschätzung in Bildfolgen, Ph. D. dissertation, Fortschrittberichte VDI, Reihe 10, Nr. 92, VDI Verlag, Düsseldorf, FRG
4: Houkes Z. (1985), Distance measurement by stereo vision, Proc. of the 5th Int. Conf. on Robot Vision and Sensory Control (RoViSeC 5), 29–31 oct 1985, Amsterdam, Holland (IFS Publications Ltd., Bedford, UK, 1985), p393–401
5: Houkes Z. (1983), Motion parameter estimation in TV–pictures, In: NATO ASI series, Vol F2: Image processing and dynamic scene analysis, Ed. by Th.S. Huang, Berlin Springer Verlag, 1983, p249–263
6: Houkes Z., Oelen W, (1988), Onderzoek naar de convergentie eigenschappen van twee algoritmen voor het schatten van beweging in videobeelden, Internal report no. 88V159, University of Twente.

Zuordnung von Bewegungsverben zu Trajektorien in Bildfolgen von Straßenverkehrsszenen

N. Heinze, W. Krüger, H.-H. Nagel

Institut für Algorithmen und Kognitive Systeme
Fakultät für Informatik der Universität Karlsruhe (TH)
per Adresse IITB, Fraunhoferstr. 1, 7500 Karlsruhe 1

1. Zusammenfassung

Die maschinelle Interpretation von Bildfolgen erfordert häufig eine Beschreibung der erfaßten Vorgänge auf höheren Abstraktionsstufen. Dazu bieten sich natürlich-sprachliche Begriffe an, die auch ein menschlicher Beobachter zur Beschreibung benutzen würde, bei Vorgängen also insbesondere Verben. Untersuchungen in diese Richtung für Verkehrsszenen wurden beispielsweise bereits von *Neumann & Novak 86* publiziert, allerdings noch für synthetische Bildfolgendaten. Für Bildfolgen von einem Fußballspiel gibt es erste Resultate bei *Herzog & Rist 88* sowie *Herzog et al. 88*. Eine breitere Einführung in die Problematik findet sich z. B. in *Nagel 88*.

Hier soll ein Ansatz zur Ermittlung von Bewegungsverben aus Bildfolgen vorgestellt werden, die Verkehrsgeschehen beschreiben. Fahrzeugtrajektorien, die nach *Sung 88* aus Bildfolgen einer Szene am Durlacher-Tor-Platz in Karlsruhe ermittelt worden sind, werden Verben zugeordnet, wobei alle Verben zu berücksichtigen sind, die im Rahmen von Verkehrsgeschehen einfache Vorgänge im Sekundenbereich beschreiben. Der Erkennungsalgorithmus baut auf einem Verbdefinitionsformalismus von *Cahn v. Seelen 88* mit entsprechenden Verbdefinitionen auf, die in dieser Untersuchung präzisiert und weiterentwickelt wurden.

2. Einführung

Ziel dieser Untersuchungen ist die experimentelle Erprobung von Ansätzen, die Zuordnungen zwischen einer rechnerinternen Repräsentation (einer Auswahl) deutscher Verben und Trajektorien herstellen, die - weitestgehend - ohne interaktive Eingriffe aus Bildfolgen von Verkehrsszenen extrahiert worden sind. Der Betrachtungsrahmen wurde dabei zunächst auf deutsche Bewegungsverben beschränkt, die sich auf Verkehrsszenen mit PKWs beziehen. Weiterhin sollen sich die Verben nur auf ein Singular-Subjekt beziehen, daneben noch zusätzlich auf die Fahrbahn, ein anderes Fahrzeug oder einen Ort. Zur möglichst vollständigen Erfassung aller relevanten Verben, die diesen Kriterien genügen, wurden ca. 9000 Verben aus einem Wörterbuch der deutschen Sprache (*Duden 67*) gesammelt. Für die so gefundenen Verben wurden Kriterien aufgestellt, die garantieren, daß die Verben einer einheitlichen Komplexitätsebene angehören und für die Beschreibung von Verkehrsgeschehen geeignet sind. Diese Kriterien werden von 119 Verben erfüllt, von denen 67 nach Aussonderung der Synonyme für die weiteren Untersuchungen verblieben. Für diese Verben wurde von *Cahn v. Seelen 88* ein Definitionsformalismus entwickelt, der in *Heinze 89* weiterentwickelt wurde, um ihn in einem realen System zur Erkennung von solchen Verben implementieren zu können. Dabei kamen noch weitere Verben hinzu, aber andere entfielen dafür, so daß gegenwärtig 90 verschiedene Verbdefinitionen zur Verfügung stehen.

Das von Cahn v. Seelen entwickelte Definitionsschema besteht aus der Angabe von drei Bedingungen, denen die aus Trajektoriendaten von Fahrzeugen ermittelten Attribute genügen müssen, nämlich einer Vorbedingung (VB), einer Monotoniebedingung (MB) und einer Nachbedingung (NB). Die Vorbedingung legt die Anfangswerte eines Geschehens fest. Die Monotoniebedingung soll für einen gleichmäßigen Übergang sorgen, dessen Abschluß die Nachbedingung erfüllt. Die VB und NB schränken die Werte von einzelnen Attributen auf diskrete Wertebereiche ein. Die Monotoniebedingung gibt die erlaubten Übergangsrichtungen und den Verlauf der Übergänge für die Attributwerte an.

3. Attribute zur Charakterisierung von Trajektorien

Um aus den Trajektoriendaten die relevanten Informationen herausfiltern zu können, werden aus den Trajektoriendaten eine Reihe von Attributen berechnet. Je nach Bezugstyp des Verbs werden unterschiedliche Gruppen von Attributen zur Definition benutzt. Aus Platzgründen wird hier nur ein Beispiel für die Attribute von Agens- und Ortsverben angegeben:

- Für Verben, die sich nur auf das **Agens** selber beziehen, wie z.B. *bremsen* oder *sich drehen*, werden drei Attribute definiert, die auch für die anderen drei Bezugsklassen benutzt werden.

A-v-Betrag	Betrag der Geschwindigkeit des Agens. Diskrete Wertebereichsstufen: *null, klein, mittel, normal, schnell, sehr schnell*
A-Richtung	Richtung des Geschwindigkeitsvektors des Agens. Wertebereich: *0° - 360°*.
A-Fahrmodus	Ausrichtung des Geschwindigkeitsvektors des Agens in bezug auf die Orientierung des Fahrzeuges. Diskrete Wertebereichsstufen: *vorwärts, rückwärts, verdreht*.

- Für Verben, die neben dem Agens einen **Ort** einbeziehen, sind neben den oben beschriebenen Agensattributen drei weitere Attribute definiert.

Ort-Kurs	Bewegungsrichtung des Agens in bezug auf den Ort. Diskrete Wertebereichsstufen: *zulaufend, passierend, ablaufend*.
Ort-Entfernung	Abstand zwischen Agens und Ort. Diskrete Wertebereichsstufen: *null, klein, normal*.

Für die zugehörigen Monotoniebedingungen der Attribute werden Werte wie „kleiner (werdend)", „gleich (bleibend)" und „größer (werdend)" eingeführt. Das dadurch erhaltene Definitionsschema kann nun zur Definition von Verben eingesetzt werden, zum Beispiel für **verlassen**:

	Vor- bedingung	Monotonie- bedingung	Nach- bedingung
Agensbezug			
A-v-Betrag:	(0)	( )	(> 0)
A-v-Richtung:	( )	( )	( )
A-Fahrmodus:	( )	(gleich)	( )
Ortsbezug			
Ort-Kurs:	( )	( )	( )
Ort-Entfernung	(0)	(größer)	(> 0)

4. Geschehenstypen und Geschehenserkennung

Bewegungsverben können in vier verschiedene Klassen aufgeteilt werden. Diese ergeben sich aus der Betonung des Verlaufsaspektes, des Zustandsaspektes, des Beginnzeitpunktes einer Handlung (also des inchoativen Aspektes) oder des Abschlußzeitpunktes einer Handlung (also des resultativen Aspektes).

Die ursprüngliche Annahme, daß inchoative und resultative Verben Spezialfälle des Verlaufstyps darstellen und sich durch Fortlassung der Anfangs- bzw. Endzeitpunkte aus diesen ableiten lassen, erwies sich im Rahmen dieses Definitionsformalismus als nicht richtig.

Für die Darstellung der vier Geschehensmodelle werden Zustandsübergangsdiagramme benutzt. Sie sind besonders geeignet, die schrittweise simultane Erkennung der Verben, aber auch das Neuaufsetzen des Erkennungsalgorithmus nach einem Erkennungsfehlversuch darzustellen, siehe beispielweise Abbildung 1).

Das Prüfen der Bedingungen zu einem diskreten Zeitpunkt wird mit den Pfeilen und den ihnen zugeordneten Bedingungen dargestellt. Die Zustände stehen für die erreichten Stufen der Verberkennung in den Intervallen zwischen zwei diskreten Zeitpunkten. Die Kanten sind insofern vollständig, als für jeden Fall eine zu durchlaufende Kante mit den entsprechenden Bedingungsangaben existiert. Die Auswertung beginnt im „wartend"-

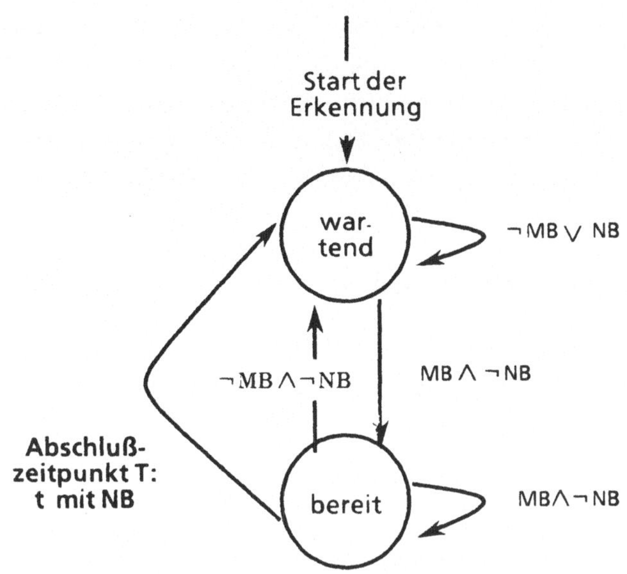

Abbildung 1: resultatives Geschehen
Der Abschlußzeitpunkt T wird durch den Zeitpunkt definiert, in dem die entsprechend markierte Kante von „bereit" nach „wartend" durchlaufen wird.

Knoten, wenn die Bedingungen definierte Werte besitzen und die Erkennung gestartet wird. Ein Endzustand existiert nicht, da immer zu durchlaufende Kanten zur Verfügung stehen und auch nach erfolgreicher Erkennung eines Geschehens weitere nachfolgen können. Das Durchlaufen von besonders gekennzeichneten Kanten führt zur Ausgabe von Gültigkeitszeitpunkten der damit erkannten Verben.

Der Erkennungsgraph kann als Dämonprozeß gedeutet werden, der im „wartend"-Zustand auf die Startbedingung wartet. Der Dämonprozeß ist dabei so gestaltet, daß ineinandergreifende Geschehen erkannt werden können, weil schon vorliegende Vorbedingungen noch während des alten Erkennungsprozesses berücksichtigt werden. Daher braucht nur ein einziger Dämonprozeß für jede Geschehenskonstellation, bestehend aus dem zu erkennenden Verb, dem Agens und wenn nötig dem Bezugs-objekt, installiert zu werden, um alle Geschehen dieser Konstellation erkennen zu können. Der Dämonprozeß kann nur von außen beseitigt werden. Die Beseitigung ist notwendig, wenn die Attribute, die der Berechnung der Bedingungen zugrunde liegen, keine definierten Werte mehr besitzen. Dies tritt insbesondere dann auf, wenn ein Objekt den Beobachtungsraum verlassen hat.

Mit Hilfe dieses Ansatzes wurde ein System zur Extraktion von Verkehrsverben aus Bildfolgen entwickelt. Dieses System wertet die von einem Bildfolgenauswertungssystem (*Sung 88*) berechneten Fahrzeugtrajektorien an einem Karlsruher Verkehrsknoten aus und bestimmt die auf die jeweilige Szene anwendbaren deutschen Verkehrsverben.

Bei der Entwicklung des Definitionsformalismus wurde angestrebt, Geschehensprimitive zu erhalten, deren Gültigkeit direkt an die Definition gebunden ist. Die berechneten Geschehensprimitive weisen eine einheitliche Komplexitätsstufe auf und sind in ihren Definitionen unabhängig voneinander. Damit haben Anpassungen der Definition einzelner Verben keine Auswirkungen auf andere Geschehensprimitive.

5. Beispiel eines Analyseergebnisses für ein einzelnes Fahrzeug

Es folgen als Beispiel die Ausgaben für ein Fahrzeug (Nr. 66 der Trajektoriendaten), die das Ergebnis der Analyse der Fahrt eines einzelnen Fahrzeuges und eines Ortes ist. Damit enthält sie nur Agens- und Ortsgeschehen.

Bildnr	Geschehensname	Kommentar
44-45	abbremsen	
44-45	auffahren (Straße)	Das Fahrzeug befand sich vorher außerhalb des Wegenetzes. Dies ist auf die noch unvollkommene Übereinstimmung der Trajektoriendaten aus der Projektion mit dem aus einem Katasterplan gewonnenen Wegemodell zurückzuführen.
44-45	bremsen	
46-47	abbremsen	
46-47	bremsen	
44-47	geradeausfahren	
47-47	langsam fahren	
48-49	abbiegen	Fahrzeug biegt am Punkt A in einen neuen Straßenzug ein
48-49	abbremsen	
48-49	bremsen	
47-49	kurven	
47-49	langsam fahren	
44-50	sich nähern (Ort 1)	Der Ort ist durch ein Rechteck markiert worden.
48-51	passieren (Ort 1)	
48-52	sein, nahe bei Ort 1	
50-52	entfernen (Ort 1)	
45-52	befahren	Auf einer Straße fahren
51-52	beschleunigen	
45-52	durchfahren	In Längsrichtung auf einer Straße fahren
44-52	fahren	
49-52	geradeausfahren	

6. Eigenschaften des Ansatzes:

- Die Definitionen bestehen aus Kombinationen von Vor-, Nach- und Monotoniebedingungen. Diese Bedingungen beziehen sich auf diskretisierte Attributwerte, die aus Bildfolgen berechnet werden.

- Die durch die Verben definierten Geschehen werden entsprechend vier verschiedener Geschehensmodelle (Verlaufsgeschehen, Zustandsgeschehen, inchoative Geschehen, resultative Geschehen) behandelt.

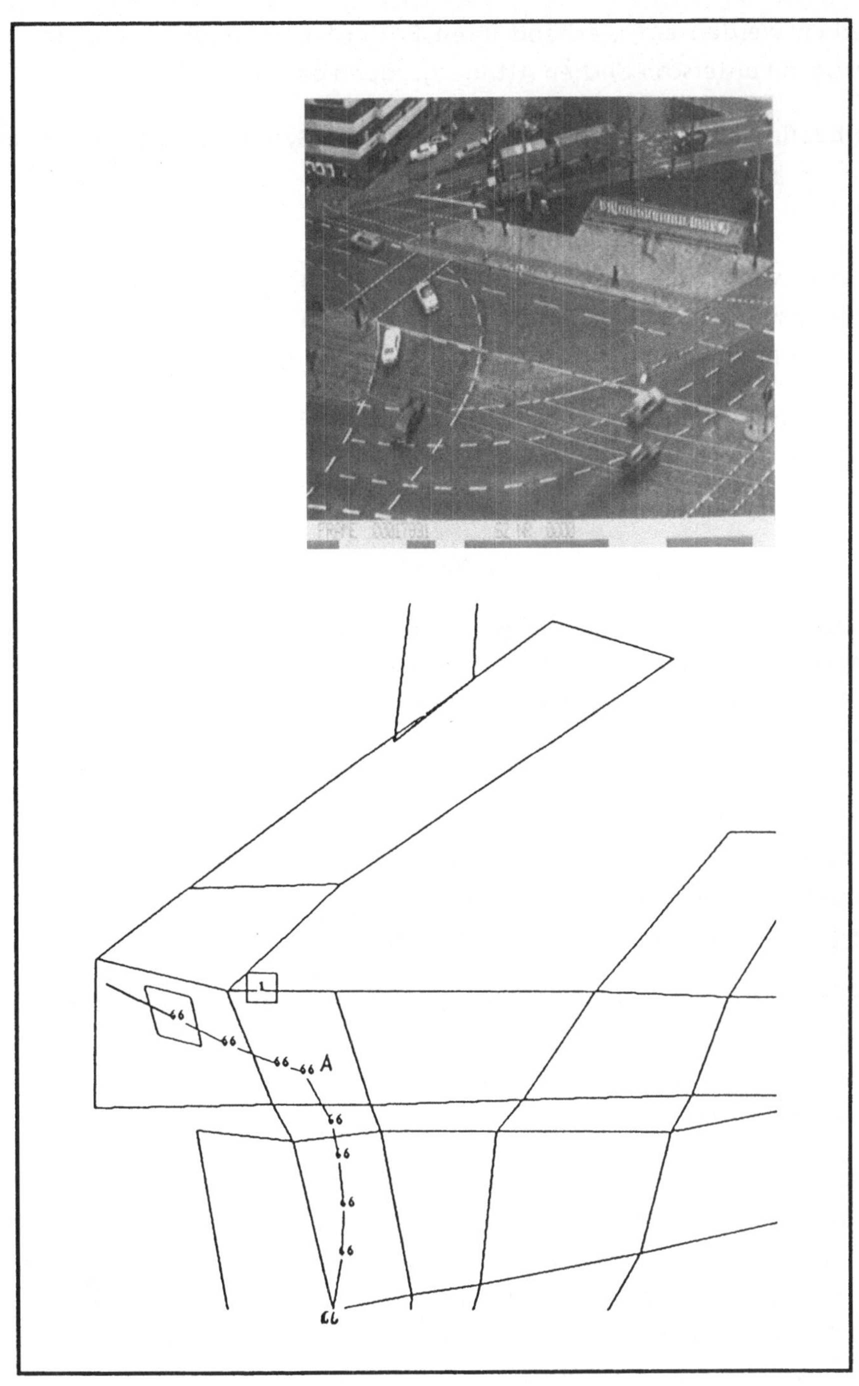

Abbildung 2: Kreuzung im Zeitraum 44- 52 mit analysiertem
Fahrzeug 66 und betrachtetem Ort 1.

- Die Verben werden entsprechend ihren Bezugstypen: Agens-, Fahrbahn-, Objekt-, Ortsbezug mit unterschiedlichen Attributgruppen definiert.

- Die Verbdefinitionen werden im implementierten System dazu benutzt, um parallel alle auf die (von einem Karlsruher Verkehrsknoten stammenden) Trajektoriendaten anwendbaren Verben zu bestimmen.

- Die Verben werden unabhängig voneinander definiert und stellen eine einheitliche Komplexitätsebene dar. Die Geschehen können im realisierten System flexibel um- und neu-definiert werden.

- Es wird eine hohe Zahl (90) von Verkehrsverben unterschieden, die sämtliche erfaßten Verben aus diesem Bereich darstellen.

7. Ausblick

Bei dem hier beschriebenen System wurde das Wegemodell aus einem Katasterplan erstellt. Wesentlich günstiger wäre es hingegen, das Wegemodell direkt aus den Bilddaten zu gewinnen. Damit könnte die Geschehenserkennung dann auch einfacher an neue Beobachtungbereiche angepaßt werden, ohne daß das Wegemodell jeweils per Hand neu erstellt werden muß. Ebenso wäre es interessant, statt einer stationären eine mobile Kamera direkt in einem Fahrzeug zu benutzen. Zur Erstellung des jeweilig aktuellen Wegemodells gehört es dann, die Bewegungsdaten für die Kamera in die Berechnungen mit einzubeziehen sowie Verkehrsschilder mit Geschwindigkeits- begrenzungen zur Bestimmung der Normalgeschwindigkeit der Wegabschnitte zu lesen.

Das System dient zur Bestimmung von Geschehensprimitiven, deren Benutzung durch die Benennung mit angemessenen Verben erleichtert wird. Dabei sollten auch Geschehensprimitive definiert und betrachtet werden, die sich nicht durch ein einzelnes Verb bezeichnen lassen.

Das hier beschriebene Verfahren kann zur Strukturierung von Trajektoriendaten benutzt werden. Mit Hilfe der bestimmten Verben (als hoher Abstraktionsstufe einer Bewegungs- beschreibung) können Rückschlüsse auf die aus den Bildfolgen extrahierbaren Daten gezogen werden, so daß eine modellgesteuerte Auswertung ermöglicht wird. Daraus ergäben sich Anwendungsmöglichkeiten im Bereich der fahrerunterstützenden Systeme und der autonomen mobilen Systeme.

Literaturverzeichnis

Cahn v. Seelen U.
Ein Formalismus zur Beschreibung von Bewegungsverben mit Hilfe von Trajektorien
Diplomarbeit, Universität Karlsruhe, Fakultät für Informatik, August 1988

Heinze, N.
Realisierung eines Ansatzes zur Berechnung von Bewegungsverben aus Trajektorien
Diplomarbeit, Universität Karlsruhe, Fakultät für Informatik, April 1989

Herzog G., Rist T.
Simultane Interpretation und natürlichsprachliche Beschreibung zeitveränderlicher Szenen: Das System Soccer
Diplomarbeit, Universität des Saarlandes, Saarbrücken
Fachbereich Informatik 1.7.1988

Herzog G., Sung C.-K., André E., Enkelmann W., Nagel H.-H., Rist T., Wahlster W., Zimmermann G.
Incremental Natural Language Description of Dynamic Imagery
Interner Bericht
Fachbereich Informatik Universität Saarbrücken, Fraunhofer-Institut für Informations- und Datenverarbeitung Karlsruhe, Deutsches Forschungszentrum für KI GmbH Saarbrücken, Fakultät für Informatik der Universität Karlsruhe, Dezember 1988

Nagel, H.-H.
From Image Sequences towards Conceptual Descriptions
Image and Vision Computing 6 (1988) 59-74

Neumann B., Novak H.-J.
NAOS: Ein System zur natürlichsprachlichen Beschreibung zeitveränderlicher Szenen.
Informatik Forschung und Entwicklung 1 (1986) 83-91

Sung C.-K.
Extraktion von typischen und komplexen Vorgängen aus einer langen Bildfoge einer Verkehrsszene. In: H. Bunke, O. Kübler und P. Stucki (Herausg.), Mustererkennung 1988, Informatik-Fachberichte 180, Berlin: Springer-Verlag 1988, pp. 90-96

Modellgestützte 3D Bewegungs- und Formanalyse unter Verwendung eines Parallelrechners

Hans Busch, Jörg Uthoff

Institut für Theoretische Nachrichtentechnik und Informationsverarbeitung
Universität Hannover

Einleitung

Kappei und Liedtke /1 + 2/ haben ein Verfahren zur modellgestützten Bewegungs- und Formbestimmung dreidimensionaler Objekte vorgestellt bei dem aus einer Bildfolge eines bewegten Objektes durch Vergleichen der Abbildung des 3D Modells mit den Bildern der Sequenz die Modellanalyse durchgeführt wird.

Die Oberfläche des von einem Dreiecksnetz aufgespannten Modells ist mit dreiecksförmigen Bildausschnitten gefüllt. Zur Abbildung des 3D Modells in eine gedachte Bildebene, im folgenden Rekonstruktion genannt, wird die perspektivische Projektion verwendet.

Die Algorithmen zur Bestimmung der Bewegungsparameter sowie zur Gewinnung der Form basieren auf dem Vergleich zwischen Bildern des Objektes und der Rekonstruktion. Als Gütekriterium wird die Grauwertdifferenz zwischen den Bildpunkten des Sequenzbildes und der Rekonstruktion verwendet.

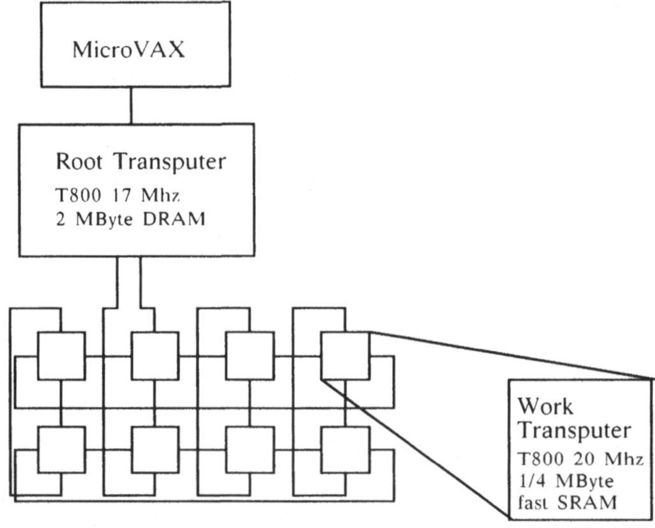

Abb. 1, Konfiguration des Parallelrechners

Um den beim Vergleich anfallenden Rechenaufwand in Grenzen zu halten, werden nur Bildpunkte an Orten mit ausreichendem Kontrast ausgewählt. Diese Punkte werden Oberflächenelemente genannt. Aus einem Bild das im vorliegenden Fall 352 x 288 Bildpunkte umfaßt werden typischerweise 5000 Oberflächenelemente extrahiert. Jedem wird sein Ort im Raum, der korrespondierende Grauwert und ein Luminanzgradientenpaar zugeordnet. Mit diesen Informationen kann sowohl die Bewegung als auch die Form des Modells mittels eines Gradienten-Optimierungsverfahrens an das Objekt angepasst werden.

Diese Arbeit wird vom Forschungsinstitut der DEUTSCHEN BUNDESPOST beim FTZ gefördert

Die Algorithmen sollen unter anderem zur Echtzeitanalyse von Bildtelefonsequenzen eingesetzt werden, was eine hohe Rechenleistung erfordert. Möglichkeiten zur Parallelisierung der Algorithmen wurden mit Hilfe eines Transputer-Parallelrechners untersucht, dessen Vorzüge in der im Vergleich zu einer dedizierten Hardware freien Programmierbarkeit liegen. Außerdem ist er für die hier auftretenden komplexen Operationen bei relativ geringen Datenmengen ausgelegt. Die Konfiguration des Parallelrechners zeigt Abb. 1.

Das Verfahren zur Bestimmung der Modellparameter

Die Änderung der Form und der Lage des 3D Modells kann für jeden Punkt der Modelloberfläche wie folgt beschrieben werden:

$$P' = [R] * (P + V * s) + T \qquad (1)$$

mit
- P = alter Ortsvektor
- P' = neuer Ortsvektor
- V = Sichtlinie des Oberflächenelements
- s = Formparameter
- $[R]$ = Rotationsmatrix
- T = Translationsvektor

Der Formparameter s bestimmt die Lage des Oberflächenelementes auf der sogenannten Sichtlinie. Letztere ist die Verbindungslinie zwischen Oberflächenelement und Beobachter zum Zeitpunkt der Bestimmung des Oberflächenelementes. Da der Formparameter s für jeden Stützpunkt des Dreiecksnetzes getrennt ermittelt wird, handelt es sich um eine lokale Formadaption.

Die Rotationsmatrix [R] und der Translationsvektor T bestimmen die Bewegung des Oberflächenelementes im Raum. Die Bewegungsparameter werden im Gegensatz zu den Formparametern global für alle Oberflächenelemente eines Objektes bestimmt.

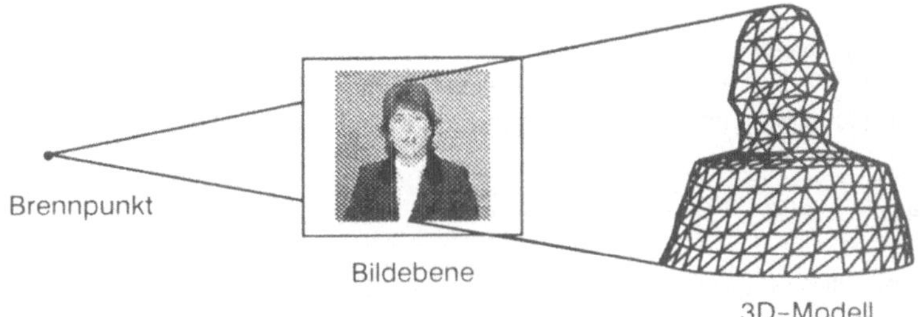

Abb. 2, Projektion des 3D-Modells auf die Bildebene

Kappei /3/ hat gezeigt wie die Bewegungsparameter R und T beziehungsweise der Formparameter s aus den Bilddaten mit Hilfe der Oberflächenelemente geschätzt werden können. Als Gütemaß wird dazu die Methode der kleinsten Fehlerquadrate verwendet, die nur dann geschlossen gelöst werden kann, wenn linear von den Parametern abhängige Fehlerfunktionen vorliegen. Im folgenden ist am Beispiel der Bewegungsschätzung der Lösungsweg skizziert:

Die linearisierten 3D Bewegungsgleichungen für ein Oberflächenelement lauten:

$$
\begin{aligned}
P'_x &= P_x - P_y * R_z - P_z * R_y + T_x \\
P'_y &= P_x * R_z - P_y - P_z * R_x + T_y \\
P'_z &= P_x * R_y - P_y * R_x - P_z + T_z
\end{aligned} \qquad (2)
$$

wobei R_x, R_y und R_z die Rotationswinkel bezeichnen, aus der die Rotationsmatrix [R] gewonnen werden kann.

Die perspektivische Projektion eines Raumpunktes lautet:

$$B_x = P_x * F / P_z$$
$$B_y = P_y * F / P_z \tag{3}$$

mit B = Position in der Bildebene
        F = Brennweite der Kamera

Die linearisierte Luminanzfunktion am Projektionsort ergibt sich zu:

$$I' = I + G_x * (B_x' - B_x) + G_y * (B_y' - B_y) \tag{4}$$

mit I = Luminanz am Punkt B
        I' = Approximierte Luminanz am Punkte B'

Durch Einsetzen von Gleichungen 2 und 3 in 4 sowie anschließende Linearisierung ergibt sich folgende Fehlerfunktion eines Oberflächenelementes in Abhängigkeit der 6 Bewegungsparameter:

$$
\begin{aligned}
I' - I = {} & F * G_x / P_z * T_x \\
& + F * G_y / P_z * T_y \\
& + F * (P_x * G_x + P_y * G_y) / P_z^2 * T_z \\
& + F * (P_x * G_x * P_y + P_y^2 * G_y - P_z^2 * G_y) / P_z^2 * R_x \\
& + F * (P_y * G_y * P_x + P_x^2 * G_x - P_z^2 * G_x) / P_z^2 * R_y \\
& + F * (G_x * P_y - G_y P_x) / P_z^2 * R_z
\end{aligned}
\tag{5}
$$

Mit der Methode der kleinsten Fehlerquadrate lassen sich die N Gleichungen für N Oberflächenelemente zusammenfassen und lösen. In symbolischer Schreibweise lautet die einzelne Fehlerfunktion von Gleichung 5:

$$F_i = a_{1i} * P_1 + a_{2i} * P_2 + \dots \tag{6}$$

mit F_i = Fehlermaß einer Gleichung
        P_j = Optimierungsparameter, hier die gesuchte Translations- und Rotationsparameter
        a_{ji} = Koeffizienten der Gleichungen des überbestimmten Gleichungssystems

Das Ziel der Optimierung ist die Minimierung der Fehlerquadrate:

$$\sum_{i=1}^{N} (F_i - (a_{1i} * P_1 + a_{2i} * P_2 + \dots))^2 -> \text{Min.} \tag{7}$$

Durch Differenzieren nach den Parametern und Null setzen der Ableitungen erhält man ein lineares Gleichungssystem:

$$
\begin{aligned}
& \sum_{i=1}^{N} (F_i * a_{1i} - (a_{1i}^2 * P_1 + a_{1i} * a_{2i} * P_2 + \dots)) = 0 \\
& \sum_{i=1}^{N} (F_i * a_{2i} - (a_{2i} * a_{1i} * P_1 + a_{2i}^2 * P_2 + \dots)) = 0 \\
& \dots
\end{aligned}
\tag{8}
$$

Die Gesamtsummen können in Einzelsummen aufgespalten werden, sodaß man für jeden Koeffizienten des Gleichungssystems einen Summenterm erhält:

$$
\begin{aligned}
& P_1 * \sum_{i=1}^{N} (a_{1i}^2) + P_2 * \sum_{i=1}^{N} (a_{1i} * a_{2i}) + \dots = \sum_{i=1}^{N} a_{1i} * F_i \\
& P_1 * \sum_{i=1}^{N} (a_{2i} * a_{1i}) + P_2 * \sum_{i=1}^{N} (a_{2i}^2) + \dots = \sum_{i=1}^{N} a_{2i} * F_i \\
& \dots
\end{aligned}
\tag{9}
$$

Wird die oben beschriebene Methode der Minimierung kleinsten Fehlerquadrate auf die globale 3D-Bewegungsschätzung angewendet, führt die Optimierung zu einem Gleichungssystem mit 6 Gleichungen und 6 Unbekannten. Da die Gleichungen der Oberflächenelemente auf den Transputern verteilt berechnet werden, können auf jedem Transputer auch gleich die Teilsummen des Gleichungssystems (9) der lokal zu berechnenden Oberflächenelemente bestimmt werden, sodaß diese zum Schluß nur noch zusammengefaßt werden müssen.

Die Parallelisierung der Modellparameterschätzung

Das hier vorgestellte Schätzverfahren für die Bewegungs- und Formbestimmung läßt sich in die folgenden Teilschritte zerlegen:

1. Berechnung der Gradientenbilder mit den Komponenten G_x und G_y.

2. Aufstellen der linearen Gleichung jedes Oberflächenelementes entsprechend Gleichung 5.

3. Zusammenfassen der Gleichungen nach der Methode der kleinsten Fehlerquadrate entsprechend Gleichung 9.

4. Lösen des linearen Gleichungssystems.

5. Update der Modellparameter.

Die ersten beiden Schritte lassen sich prinzipiell vollständig parallelisieren, da bei beiden eine große Anzahl gleicher Operationen auf voneinander unabhängige Daten ausgeführt werden muß. Dagegen lassen sich die Schritte 3 bis 5 nur bedingt parallelisieren, da die dabei auszuführenden Operationen recht elementar sind und der Hauptaufwand im reinen Datenzugriff auf die gemeinsame Datenbasis liegt.

Trotz der für einen Algorithmus mit einfachen Operationen und großen Datenmengen ungünstigen Struktur des Parallelrechners wurde die **Berechnung der Gradientenbilder** auf dem Parallelrechner implementiert, um den Parallelisierungsgewinn in Abhängigkeit von der Anzahl der Parallelrechner zu studieren und mit dem Parallelisierungsgewinn bei der Parameterbestimmung zu vergleichen. Für eine Hardware, die nur diese Filteroperationen ausführen soll, bieten sich 2D-Filterprozessoren /4/ an, die einfache Bildoperationen wie die Berechnung des Sobelgradienten in Echtzeit berechnen.

Die Gradientenbildberechnung wurde durch Zerschneiden des Bildes in Streifen parallelisiert, indem jeder Prozessor einen Bildstreifen berechnet und die Streifen zum Schluß wieder zu einem kompletten Gradientenbild zusammengesetzt werden.

Beim **Aufstellen der Gleichungen der Oberflächenelemente** muß für jedes Oberflächenelement bei der Bewegungsbestimmung die Gleichung 4 aufgestellt werden. Dies kann in drei Teilschritte zerlegt werden:

- Projektion der Raumposition des Oberflächenelementes in die Bildebene.

- Auslesen des Grauwertes und der Gradienten an dem Ort, auf den das Oberflächenelement projiziert wurde.

- Formulieren der Gleichung in Abhängigkeit der zuvor gewonnen Bild- und Oberflächenelement-Information.

Die für die Projektion der Oberflächenelemente notwendigen Daten sind dezentral auf den Work-Transputer untergebracht, sodaß die Projektion vollständig parallelisiert werden kann.

Für eine schnelle Ausführung der Optimierungsverfahren ist neben der Verfügbarkeit der Daten des Modells ein schneller Zugriff auf die Bilddaten beim Auslesen der Luminanz- und der Gradientenwerte des Szenenbildes wünschenswert. Letztere können aufgrund der mit zu kleinem Speicher ausgerüsteten Work-Transputer nicht auf jedem Transputer in voller Größe gehalten werden, sondern nur innerhalb eines Ausschnitt von ca 30% des Gesamtbildes. Dadurch kann es vorkommen, daß der benötigte Bildausschnitt auf dem Work-Transputer nicht vorliegt und vom Root-Transputer angefordert werden muß, wodurch der entsprechende Rechenschritt bis zum Eintreffen der benötigten Daten unterbrochen werden muß. Durch Anordnen mehrerer gleicher Rechenprozesse auf einem Transputer ist dafür gesorgt, daß die einzelnen Prozessoren trotzdem stets ausgelastet sind.

Beim **Zusammenfassen der Gleichungen** muß zwischen globalen Optimierungen, wie der Bewegungsparameterbestimmung und lokalen Optimierungen, wie der Formbestimmung, unterschieden werden. Bei
der lokalen Optimierung müssen viele kleine überbestimmte Gleichungssysteme gelöst werden, sodaß eine Parallelisierung unproblematisch ist. Anders bei der globalen Optimierung, wo die Methode der kleinsten Fehlerquadrate selbst parallelisiert werden muß wie oben erläutert wurde.

Ergebnisse

Bild 3 gibt einen Überblick über die Rechenzeiten zur Berechnung von Gradientenbildern. Mit steigender
Anzahl Transputer werden die komunikationsbedingten Verluste größer, sodaß die Rechenleistung nicht
proportional zur Anzahl der eingesetzten Transputer steigt.

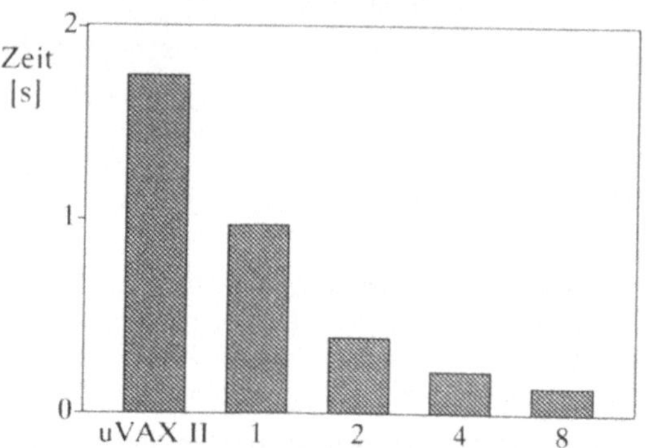

Abb. 3, CPU–Zeit für die Berechnung eines Gradientenbildpaares in Abhängigkeit von der Anzahl Transputer (MicroVAX II zum Vergleich).

Bei der Bewegungsparameterbestimmung setzt sich der sequentiell auszuführende Teil im wesentlichen
aus der Aufstellung der Gleichungen und der Lösung des linearen Gleichungssystems zusammen. Der Par-
allelisierungsoverhead entsteht durch die Verteilung der für die Bewegungsoptimierung relevanten Daten
des Szenenmodells. Vergleicht man die Rechenzeiten zwischen MicroVAX II und einem Transputer, so ist
eine Leistungssteigerung um den Faktor 3.8 zugunsten des Transputers zu verzeichnen. Die Verbesserung
ist hier erheblich deutlicher als bei der Grandientenbildberechnung, da im Unterschied zu jener viel Gleit-
kommaarithmetik benötig wird. (siehe Abb. 4).

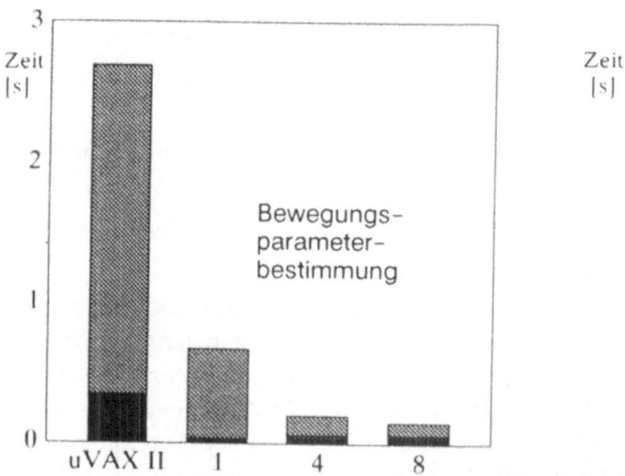
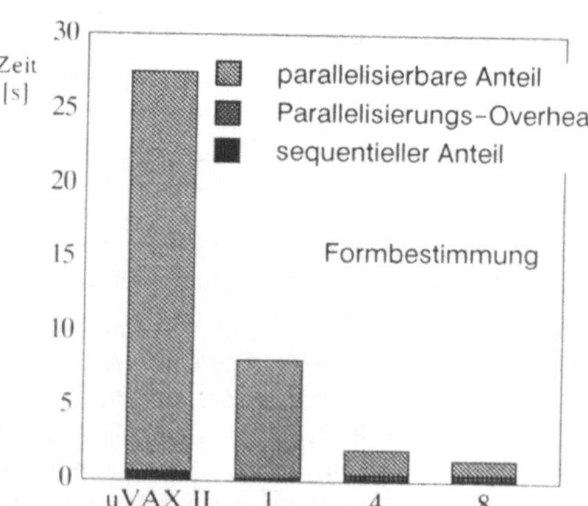

Abb. 4, Rechenzeiten in Abhängigkeit der Anzahl eingesetzter Transputer
(MicroVAX II zum Vergleich)

Beim Einsatz von 8 Transputern zeigt sich bereits ein Sättigungseffekt, der darauf zurückzuführen ist, daß der Root-Transputer die angeforderten Bildausschnitte nicht mehr schnell genug liefern kann. Bei der Betrachtung der Beschleunigung, die für die gesamte Bewegungsparameterbestimmung erzielt wird, fällt auf, daß der sequentielle Anteil als additive Konstante in der Gesamtrechenzeit mit wachsender Anzahl Transputer eine immer größere Bedeutung bekommt und bei 8 Transputern einen Anteil von 44% erreicht! Die Beschleunigung der Formbestimmung übertrifft die bei der Bewegungsparameterbestimmung erzielten Ergebnisse, da der geringere sequentielle Anteil sowie die größere Komplexität der Rechenoperationen das Verhältnis zwischen Rechenbedarf und Datentransfer positiv beeinflussen.

Abbildung 5 zeigt die benötigten Rechenzeiten für einen vollständigen Optimierungszyklus, wie er beim Anpassen des Modells an ein Bild einer Sequenz durchlaufen wird.

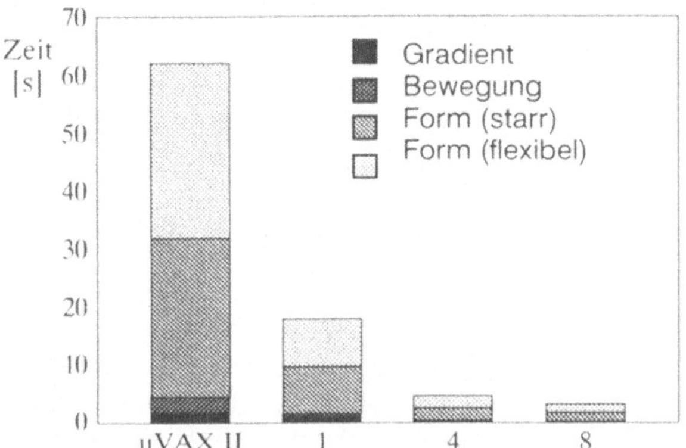

Abb 5, Rechenzeiten für einen vollständigen Optimierungszyklus in Abhängigkeit der Anzahl eingesetzter Transputer (MicroVAX II zum Vergleich)

Das Transputernetzwerk wurde anhand der Videotestsequenzen "Claire" und "Miss America" getestet, die von CCITT für die Simulation von Bildtelefonsystemen zur Verfügung gestellt werden. Für diese beiden Sequenzen führte der Einsatz von 8 Transputern im Vergleich zu einem Transputer bei der Bewegungskompensation der Person der Testsequenzen zu einer Beschleunigung um den Faktor 6 von 0.7 s auf 0.15 s pro Bild. Die Gesamtzeit für die Form- und Bewegungskompensation eines Bildes reduziert sich von 18.2 s auf 3.5 s. Der Einsatz einer größeren Anzahl von Prozessoren ist nur dann zweckmäßig, wenn für die Verteilung der benötigten Luminanz- und Gradientenbilder statt der mit 1.7 MByte/s zu langsamen seriellen Transputer-Kanäle ein spezieller Videobus eingesetzt wird und die Work-Transputer mit zusätzlichem Speicher versehen werden um diese Bilder lokal zu halten.

Literatur

/1/ F. Kappei, C.-E. Liedtke, 1987. Modelling of a natural 3D scene consisting of moving objects from a sequence of monocular TV images. Proc. SPIE Volume 860, pp 126.

/2/ F. Kappei, C.-E. Liedtke, 1987. Ein Verfahren zur Modellierung von 3D-Objekten aus Fernsehbildfolgen. Proc. of the 9. DAGM-Symposium, Oktober 1987

/3/ F. Kappei, 1988. Modellierung und Rekonstruktion bewegter dreidimensionaler Objekte in einer Fernsehbildfolge.
PhD. Thesis, University of Hannover, 1988.

/4/ INMOS IMS A110 Image and Signal Processing Sub-system
Advanced Information June 1988.

Auf dem Wege zu modellgestütztem Erkennen von bewegten nicht-starren Körpern in Realweltbildfolgen

K. Rohr

Institut für Algorithmen und Kognitive Systeme
Fakultät für Informatik der Universität Karlsruhe (TH)
Postfach 6980, D-7500 Karlsruhe 1

Kurzfassung

Die meisten bisherigen Ansätze in der Bildfolgenauswertung machen die Annahme, daß es sich bei den bewegten Objekten in der Szene um starre Körper (z. B. Autos) handelt. Dieser Beitrag beschäftigt sich mit der Erkennung von gehenden Menschen (Fußgängern), d. h. mit nicht-starren Körpern, in Realwelt-bildfolgen. Beschrieben wird ein Ansatz, bei dem Wissen über die menschliche Körperform und über die Bewegungen des Körpers während des Gehens vorausgesetzt wird. Verwendet werden Daten aus Bewegungsanalysen nach *[Murray et al. 64]* und *[Murray 67]*.

1. Einleitung

Der Mensch ist sowohl in der Lage, die Form und örtliche Anordnung von physikalischen Objekten wahrzunehmen, als auch zu erkennen, ob sich ein Objekt bewegt und wie es sich bewegt. Nach *[Marr & Vaina 80]* müssen dazu zwei Dinge vorhanden sein. Zum einen ein symbolisches System, um Form- und Bewegungsinformationen im Gehirn repräsentieren zu können, und zum anderen eine Reihe von Prozessen, um solche Informationen aus dem "Gesehenen" (aus Bildern) abzuleiten. Es läßt sich z. B. nur dann zwischen Geh- und Laufbewegungen unterscheiden, wenn Wissen über diese beiden Bewegungsformen und die Unterschiede zwischen ihnen vorhanden ist.

Ein System zur Interpretation von Bildfolgen, das die Aufgabe hat, menschliche Bewegungsabläufe in Bildern zu erkennen und zu charakterisieren, sollte daher explizite Modelle der Form und Bewegung von Objekten intern zur Verfügung haben (gelernte oder eingegebene Modelle), um mit Hilfe dieses expliziten Wissens Bildfolgen auszuwerten. Innerhalb der Diskurswelt, d. h. innerhalb des Ausschnitts der Welt, für den das System konzipiert ist, erfordert eine solche Vorgehensweise nach *[Nagel 88]* eine erschöpfende Modellierung der Formen und Bewegungsarten von bewegten Objekten sowie eine Beschreibung der stationären Szenenanteile. Ein derartiger modellgestützter Ansatz sollte bei Analyse einer Bildfolge die wahrgenommene Szene nicht nur durch quantitative geometrische Beschreibungen charakterisieren, sondern darüberhinaus die Szene durch Begriffe beschreiben, die der natürlichen Sprache entsprechen *[Nagel 88, Nagel 89]*.

Wählt man als Diskurswelt Ausschnitte von Fußballspielen, so stellt sich, unter den oben skizzierten Annahmen, beispielsweise die Aufgabe, Fernsehbildfolgen in der Art einer Radioreportage automatisch zu beschreiben. Das Ziel hierbei ist die Entwicklung eines Systems zur natürlich-sprachlichen Beschreibung von Fußballspielen, wobei ein Zusammenwirken von bildverstehenden und sprachverstehenden Systemen nötig ist *[André et al. 88, Herzog et al. 89, Schirra et al. 88, Sung 88]*. Eine Schwierigkeit für das Bildauswertesystem bei diesem speziellen Diskursbereich ergibt sich daraus, daß Fußballspieler keine starren Körper sind. Zusätzlich treten häufig gegenseitige Verdeckungen der Körper auf. Die Modellierung aller vorkommenden Bewegungsabläufe ist aufgrund der großen Anzahl an Bewegungen, die Fußballspieler ausführen (z. B. gehen, rennen, drehen), ein äußerst weit gestecktes Ziel. Deshalb wird man sich (zunächst) auf die Modellierung einzelner Bewegungsformen beschränken müssen.

Ein Schritt in diese Richtung soll die vorliegende Arbeit sein, bei der es um die Erkennung von bewegten nicht-starren Körpern in zeitlichen Bildfolgen geht. Gemeint sind damit Körper, die aus mehreren gelenkig miteinander verbundenen starren Körperteilen bestehen. Ein solcher Körper stellt z. B. der menschliche Körper dar, wenn man die Anzahl seiner möglichen Bewegungsarten einschränkt (z. B. keine Verdrehungen des Oberkörpers). Unter der Vielfalt der menschlichen Fortbewegungsarten (z. B. rennen, hüpfen, gehen, springen) wird im folgenden das aufrechte Gehen des Menschen betrachtet. Diese spezielle Fortbewegungsart ist bei den meisten Menschen sehr ähnlich. Das Wissen über die ungefähre Gestalt des

Menschen und über die Gehbewegung soll dazu verwendet werden, um gehende Menschen (Fußgänger) in Bildfolgen zu erkennen. Dieser modellgestützte Ansatz setzt voraus, daß sowohl ein Körpermodell als auch ein Bewegungsmodell des Menschen im Rechner vorhanden ist.

Da wissenschaftliche Analysen von menschlichen Bewegungen schon seit langer Zeit durchgeführt werden, liegt es nahe, die Ergebnisse von solchen Untersuchungen für die Modellierung zu verwenden. Deshalb werden im nächsten Abschnitt Bewegungsstudien an Menschen beschrieben. Im 3. Abschnitt wird dann eine Modellierung von Fußgängern angegeben, die Daten von Bewegungsanalysen zugrundelegt. Der letzte Abschnitt zeigt auf, wie mit Hilfe dieses Modells Fußgängerbewegungen in monokularen Realwelt-bildfolgen erkannt werden sollen.

2. Bewegungsstudien an Menschen

Die ersten Beobachtungen und Untersuchungen der Bewegung des Menschen wurden schon im 4. Jhd. v. Chr von Aristoteles und seinen Mitarbeitern gemacht. Mitte des 19. Jhd. führten *[Weber & Weber 1836]* erste allgemeinere Experimente bezüglich der Gangart des Menschen durch und weckten damit das Interesse an Bewegungsstudien. Erst Ende des 19. Jhd. ermöglichten photographische Methoden genauere Untersuchungen *[Muybridge 1887]*. Um die Grundmuster des aufrechten normalen Gehens bei Menschen zu ermitteln, werteten *[Murray et al. 64]* Aufnahmen von 60 Personen in 5 verschiedenen Altersgruppen zwischen 20 und 65 Jahren aus. Für die Gesamtbewegung und für die Bewegung der einzelnen Körperteile erhalten die Autoren sehr charakteristische Kurven, die einen Hinweis darauf geben, daß die Grundelemente des normalen Gehens bei den meisten Menschen gleich sind. Das Anliegen von *[Murray et al. 64]* ist es, mit Hilfe dieser Daten einen Standard zu schaffen, um einen Vergleich mit Messungen von abnormalen Gangarten zu ermöglichen, also zur besseren Diagnose bei Bewegungsstörungen.

Ein Merkmal des Gehens ist z. B. die zyklische Bewegung sowohl der Körperteile als auch des gesamten Körpers. Jedes Bein dient abwechselnd als Standbein oder als Schwingbein (Spielbein), wobei mindestens ein Fuß Bodenkontakt hat. Der einzelne Gehzyklus kann in vier Phasen eingeteilt werden. Definiert man den Beginn des Zyklus zu dem Zeitpunkt, an dem das rechte Bein nach dem Schwingen gerade wieder den Boden berührt, so ergibt sich zunächst eine Phase, in der beide Beine Bodenkontakt haben. Diese erste Phase nimmt ca. 10% der Gesamtzykluszeit ein. Darauf folgt die Schwungphase des linken Beines (ca. 40% der Gesamtzykluszeit). In der dritten Phase berühren wieder beide Beine gleichzeitig den Boden (ca. 10%), und die vierte Phase ist die Schwungphase des rechten Beines (ca. 40%). Dieser Zyklus wiederholt sich *[Murray 67]*. Bemerkenswert ist die Symmetrie innerhalb eines Zyklus, die wohl aufgrund des symmetrischen Körperbaus zustandekommt. Die Bewegungen des Armes und des Beines der einen Körperhälfte sind die gleichen wie die Bewegungen der anderen Körperhälfte. Lediglich der Zeitpunkt der Ausführung ist um die Hälfte der Zykluszeit verschoben. Wenn eine Person schneller als üblich geht, nimmt der zeitliche prozentuale Anteil der gleichzeitigen Bodenberührung (erste und dritte Phase) ab. Beim Laufen verschwinden diese beiden Phasen völlig und werden durch kurze Phasen ersetzt, in denen beide Beine vom Boden frei sind und somit der Körper in der Luft schwebt.

3. Modellierung von Fußgängern

3.1 Modellierung der menschlichen Körperform

[Marr & Nishihara 78] formulieren drei Kriterien, die eine statische 3D-Representation erfüllen muß, um mit Hilfe dieser Repräsentation Objekte in Bildern erkennen zu können. Unter Zugrundelegung dieser Kriterien kommen sie zu dem Ergebnis, daß für jedes Objekt ein eigenes Objektkoordinatensystem zu benutzen ist, das entsprechend den natürlichen Achsen des Objekts ausgerichtet ist und nicht vom Beobachterstandpunkt abhängt. Außerdem soll das Objekt aus Volumenprimitiven zusammengesetzt sein. Die Beschreibung des gesamten Objekts soll modular und hierarchisch aufgebaut sein. Als Beispiel geben *[Marr & Nishihara 78]* ein 3D-Modell der menschlichen Gestalt an, wobei Zylinder als Volumenelemente dienen.

In der vorliegenden Arbeit wird der menschliche Körper ebenfalls durch ein Volumenmodell dargestellt. Die verwendeten 14 Volumenprimitive (Kopf, Oberkörper und jeweils drei Primitive für Arme und Beine) sind gerade Zylinder mit elliptischem Querschnitt, die durch drei Parameter charakterisiert werden: ein Parameter für die Länge des Zylinders, und zwei Parameter für die Größe der beiden Halbachsen. Für die Lage der einzelnen Körperteile zueinander wird der hierarchische Aufbau nach *[Marr & Nishihara 78]* verwendet (Abb. 1). Die Teile sind gelenkig miteinander verbunden. Jedes Körperteil und jedes Gelenk besitzt ein eigenes Koordinatensystem, das entsprechend den natürlichen Achsen der jeweiligen Körperteile ausgerichtet ist. Die Koordinatensysteme sind kartesisch und rechtshändig. Das Koordinatensystem des Gesamtobjektes hat seinen Ursprung im Zentrum des Oberkörpers. Der Übergang ins jeweilige Bezugssystem wird durch homogene Koordinaten des Raumes, d. h. durch 4x4 Matrizen, beschrieben. Dadurch können sowohl Bewegungen im mathematischen Sinn (bijektive Abbildungen eines Raumes auf sich, bei der alle Längen erhalten bleiben) als auch Projektionen dargestellt werden.

3.2 Modellierung der Fußgängerbewegung

[Hogg 83] teilt den Gehzyklus explizit in vier Phasen ein. Diese diskrete Approximation des kontinuierlichen Gehzyklus verwendet er, um mit Hilfe von Parametereinschränkungen den Bewegungszustand eines Fußgängers aus Bildern bestimmen zu können. Das verwendete Modell des menschlichen Körpers besteht aus elliptischen Zylindern. Sein Programm ist in der Lage, einen einzelnen Fußgänger zu erkennen und zu verfolgen. Die Leistungsfähigkeit seines Programms zeigt er durch Überlagerung des projizierten Modells mit den realen Bildern. In *[Hogg 84]* wird die explizite Einteilung des Gehzyklus aufgegeben. Die Positionsänderungen der Gelenke an Schulter, Ellbogen, Hüfte und Knie werden jetzt durch kontinuierliche Funktionen beschrieben. Die benötigten Funktionswerte hat *[Hogg 84]* durch Analyse der Bildfolge aus *[Hogg 83]*, in der sich ein einzelner Fußgänger bewegt, interaktiv ermittelt. Diese Funktionswerte interpoliert er durch periodische kubische B-Splines. Der Vorteil der kontinuierlichen Funktionen ist, daß *[Hogg 84]* einen Teil der Parametereinschränkungen zur Bestimmung des Bewegungszustandes des abgebildeten Fußgängers nicht mehr benötigt.

Das im folgenden verwendete Bewegungsmodell benutzt die experimentell ermittelten Daten aus *[Murray et al. 64, Murray 67]*, da diese Daten repäsentativ für eine große Anzahl von Personen sind. Winkeländerungen des Hand- und Fußgelenkes werden nicht berücksichtigt. Für die Gelenke an Schulter, Ellbogen, Hüfte und Knie wird vorausgesetzt, daß sie sich nur um eine einzige Achse drehen, wobei Winkel in mathematisch positiver Richtung angegeben werden. Für jedes dieser Gelenke werden an 10 Stützstellen innerhalb des Zyklus die zugehörigen Funktionswerte der Winkel den empirischen Daten entnommen. Die Daten nach *[Murray et al. 64, Murray 67]* liegen als Meßkurven vor, wobei die Meßwerte markiert und linear interpoliert wurden. Die abgelesenen Funktionswerte werden nun durch periodische kubische Splines interpoliert. Bei dieser Interpolationsart wird gefordert, daß die gesuchte Funktion und ihre erste und zweite Ableitung überall stetig ist (d. h. auch an den beiden Intervallrändern). Die Zykluszeit (Periode) wird auf 1 normiert. Aufgrund der symmetrischen Gehbewegung benötigt man die Bewegungskurven nur für eine Körperhälfte (Die Bewegungskurve des Kniegelenkes ist in Abb. 2 dargestellt). In gleicher Weise wird die vertikale Verschiebung des Gesamtkörpers modelliert. Abb. 3 zeigt den Bewegungsablauf des simulierten Gehens für einen halben Zyklus. Der Fußgänger ist hier um die Körperlängsachse um 90^0 gegenüber seiner Ausgangsposition (Abb. 1) gedreht. Die Bewegung beginnt damit, daß der rechte Fuß nach seiner Schwungphase gerade wieder den Boden berührt. Der rechte Arm ist nach hinten ausgerichtet. Gezeigt werden die Positionsstellungen zu den Zeitpunkten 0.0, 0.1, 0.2, 0.3, 0.4 und 0.5 der normierten Zykluszeit. Die abgebildeten Konturen sind die sichtbaren Berandungslinien der elliptischen Zylinder des 3D-Körpermodells, die sich bei Zentralperspektive ergeben. Das Wissen über das 3D-Modell und den relativen Abstand der Körperteile zum Beobachter wird ausgenutzt, um (Teile von) Konturen von sich gegeneinander verdeckenden Zylindern zu unterdrücken.

4. Ausblick

Um die Bewegung von Fußgängern automatisch zu erkennen, sollen die Kanten der auszuwertenden Grauwertbilder verwendet werden. Bei genügend großem Kontrast zwischen Objekt und Hintergrund geben diese Grauwertkanten deutliche Hinweise auf die Gestalt der sich bewegenden Objekte. Die einzelnen Kantenpunkte sollen zu Kantenzügen verkettet und durch Geradenstücke approximiert werden. Das generische Körpermodell des "Zylinder-Männchens" wird mit den absoluten Größen einer durchschnittlich großen Person ausgeprägt. Vorausgesetzt werden soll, daß sich der Fußgänger mit konstanter Geschwindigkeit auf einer Ebene in der 3D-Welt bewegt. Eine wesentliche Annahme, die man machen kann, besteht darin, daß der Fußgänger aufgrund der Schwerkraft und seiner Bewegungsart senkrecht auf dieser Ebene steht, wobei die Füße auf dem Boden sind und nicht der Kopf (Bei der Werkstückerkennung ist eine Annahme dieser Art bzgl. der Orientierung von Objekten im allgemeinen nicht gerechtfertigt).

Abb. 4 zeigt eine einzelne Aufnahme aus der Video-Bildfolge eines Fußgängers. In Abb. 5 wurden die Kontursegmente, die sich aus der Projektion des Modells in die Bildebene ergeben, interaktiv überlagert. Trotz des relativ groben Körpermodells ergibt sich eine einigermaßen gute Übereinstimmung. Der Kopf des Fußgängers wird jedoch sehr schlecht approximiert. Das Modell gibt den Bewegungszustand zum Zeitpunkt 0.44 der Gesamtzykluszeit wieder.

Mit Hilfe der ermittelten Grauwertkanten und der Modellkonturen soll versucht werden, die Position und Orientierung des Fußgängers in der 3D-Welt zumindest näherungsweise zu bestimmen. Das Bewegungsmodell soll dann dazu verwendet werden, um eine Prädiktion für die Position des Fußgängers im nächsten Bild zu machen. Durch diese "Analyse durch Synthese" Methode könnte es möglich sein, die Position des Fußgängers auch dann zu schätzen, wenn in einzelnen Bildern der Kontrast zwischen Fußgänger und Hintergrund gering ist, oder der Fußgänger teilweise verdeckt ist. Da die verwendeten Daten für das Bewegungsmodell repräsentativ für eine größere Anzahl von Personen sind, ist zu hoffen, daß die Erkennung der Fußgängerbewegung mit Hilfe dieser Daten für eine größere Anzahl verschiedener Fußgänger möglich ist.

Dank

Diese Arbeit wird von der DFG im Rahmen des Sonderforschungsbereiches 314 "Künstliche Intelligenz - Wissensbasierte Systeme" gefördert. H.-H. Nagel danke ich für hilfreiche Anregungen und eine kritische Durchsicht des Beitragsentwurfs.

Literatur

[André et al. 88] : *On the Simultaneous Interpretation of Real World Image Sequences and Their Natural Language Description: The System SOCCER*, E. André, G. Herzog, T. Rist, Proc. ECAI-88, 8th European Conference on Artificial Intelligence, München/FRG, 1-5 August 1988, 449-454

[Herzog et al. 89] : *Incremental Natural Language Description of Dynamic Imagery*, G. Herzog, C.-K. Sung, E. André, W. Enkelmann, H.-H. Nagel, T. Rist, W. Wahlster, G. Zimmermann, zur Veröffentlichung eingereicht, 1989

[Hogg 83] : *Model based vision: a program to see a walking person*, David Hogg, Image and Vision Computing, Vol. 1, No. 1 (1983), 5-20

[Hogg 84] : *Interpreting Images of a Known Moving Object*, David Hogg, PhD dissertation, University of Sussex, 1984

[Marr & Nishihara 78] : *Representation and recognition of the spatial organization of three-dimensional shapes*, David Marr, H.K. Nishihara, Proc. R. Soc. Lond. B 200 (1978), 269-294

[Marr & Vaina 80] : *Representation and recognition of the movements of shapes*, David Marr, Lucia Vaina, AI Memo 597, Artificial Intelligence Laboratory, MIT, Cambridge, MA, USA (October 1980)

[Murray 67] : *Gait as a total pattern of movement*, M.P. Murray, American Journal of Physical Medicine, Vol. 46, No. 1 (1967), 290-332

[Murray et al. 64] : *Walking Patterns of Normal Men*, M.P. Murray, A.B. Drought, R.C. Kory, The Journal of Bone and Joint Surgery, Vol. 46-A, No. 2 (March 1964), 335-360

[Muybridge 1887] : *Muybridge's complete Human and Animal Locomotion. All 781 Plates from the 1887 'Animal Locomotion'*, E. Muybridge, Volume 1, Dover Publications, Inc., New York 1979

[Nagel 88] : *From image sequences towards conceptual descriptions*, H.-H. Nagel, Image and Vision Computing, Vol. 6, No. 2 (May 1988), 59-74

[Nagel 89] : *Zur Erkennung von Situationen durch Auswertung von Bildfolgen*, H.-H. Nagel, Fraunhofer-Gesellschaft Bericht 1/1989, 25-33

[Schirra et al. 88] : *From Image Sequences to Natural Language: A first step toward Automatic Perception and Description of Motion*, J.R.J. Schirra, G. Bosch, C.-K. Sung, G. Zimmermann, Applied Artificial Intelligence 1 (1987) 287-305

[Sung 88] : *Extraktion von typischen und komplexen Vorgängen aus einer langen Bildfolge einer Verkehrsszene*, C.-K. Sung, DAGM Symposium Mustererkennung 1988, Zürich, 27.-29.9.1988, Informatik-Fachberichte 180, H. Bunke, O. Kübler und P. Stucki (Hrsg.), Springer-Verlag, Berlin-Heidelberg-New York-London-Paris-Tokyo, 1988, 90-96

[Weber & Weber 1836] : *Mechanik der menschlichen Gehwerkzeuge*, W. Weber, E. Weber, Dietrichsche Buchhandlung, Göttingen 1836

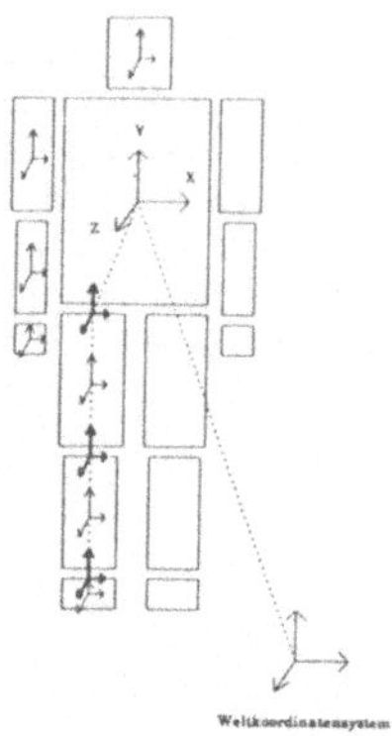

Abb. 1: Geometrisches Modell als
approximative Beschreibung
des menschlichen Körpers

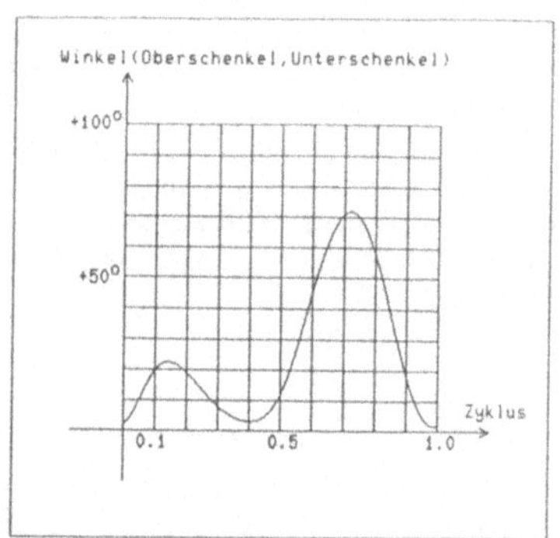

Abb. 2: Bewegungskurve des
Kniegelenkes

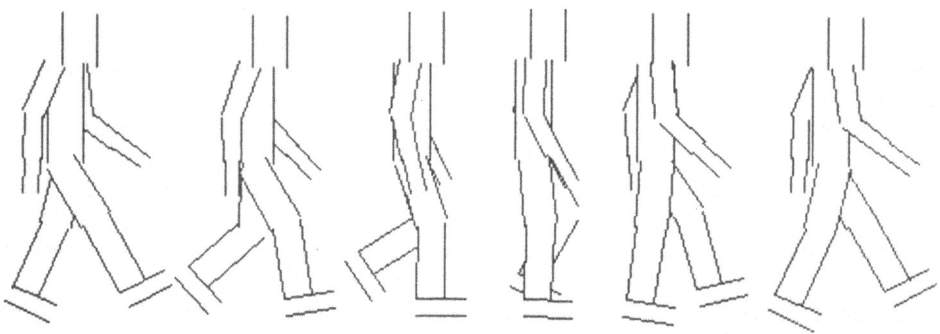

Abb. 3: Bewegungsablauf des simulierten Gehens für die Hälfte der Zykluszeit

Abb. 4: Aufnahme eines Fußgängers

Abb. 5: wie Abb. 4, jedoch interaktiv über-
lagerte Kontursegmente des
projizierten Modells

Registrierung und Wiedererkennen eines Straßenzuges durch komplexe Bildmerkmale

Georg Zimmermann

Fraunhofer-Institut für Informations- und Datenverarbeitung (IITB)

Fraunhoferstr. 1, D-7500 Karlsruhe 1

Zusammenfassung

Die Bildfolge einer PKW-Fahrt durch die Karlsruher Innenstadt wird automatisch durch sehr allgemeine, zusammengesetzte Bildmerkmale beschrieben. Die Bilder einer zweiten Fahrt, die etwa 10 Minuten später stattfand, wird in derselben Weise behandelt und auf Übereinstimmungen mit der ersten Fahrt überprüft. Es zeigt sich, daß in den meisten Fällen an den Stellen eine Maximalzahl der Korrespondenzen gefunden wird, wo auch der menschliche Betrachter eine Bildzuordnung feststellt. Damit ist demonstriert, daß diese Art von Bildmerkmalen automatisch eine sehr spezifische Beschreibung von Bildinhalten liefert.

Einleitung

Eines der Ziele der automatischen Bildauswertung besteht darin, die von einem Bildsensor aufgenommene Umgebung zu erfassen und komprimiert als rechnerinterne Repräsentation abzuspeichern /1/. Mithilfe dieser Repräsentation sollen dann Aufnahmen zugeordnet werden, die zu einem anderen Zeitpunkt bzw. aus einem ähnlichen Blickwinkel gemacht wurden, um so z.B. eine Hilfe bei der Objekterkennung oder der automatischen Navigation zu bieten.

Ein Weg zu diesem Ziel ist der Aufbau von Modellen aus konstruktiv-geometrischen Bildmerkmalen wie Kanten, Ecken etc., die zu Abbildern größerer Einheiten wie Quader, Zylinder etc. zusammengefasst werden. Wünschenswert bei diesem Weg ist, daß die aus dem Bild extrahierten Strukturen vom Menschen interpretierbar sind. Beim heutigen Stand der Forschung ist der automatische Aufbau eines internen Modells dieser Art jedoch nur in den Fällen möglich, wo ausreichend Vorwissen über die aufgenommene Szene vorliegt /2/.

Mit der vorliegenden Arbeit wird dagegen versucht, bildbezogene rechnerinterne Repräsentationen von Szenen zu erstellen, die nicht im Vorhinein bekannt sind. Dazu wird der Modellierungsvorgang in mehrere Schritte aufgeteilt, wobei der erste mithilfe von sehr allgemeinen Bildmerkmalen eine komprimierte Bildbeschreibung einer Szene erstellt. Die Beschreibung soll ausreichend spezifisch sein, um sie von Beschreibungen anderer Szenen zu unterscheiden. Dabei wird auf die anschauliche Interpretation durch den Menschen bewußt verzichtet. Zukünftig soll dann darauf aufbauend z.B. 3D-Information bzw. konstruktiv interpretierbare Merkmale eingebracht werden.

Methode

Die Aufgabe besteht darin, aus Bildern verschiedener Szenen Merkmale zu extrahieren, die typisch für die in dem jeweiligen Bild dargestellten Strukturen sind. Dies können naturgemäß keine einfachen Merkmale wie Kanten, Ecken oder Flecken sein, da solche in jedem Bild in vielfältiger Weise vorkommen können und somit beliebig verwechselbar sind. Es wird deshalb so vorgegangen, daß nicht einzelne, sondern Gruppen von Merkmalen untersucht werden.

Als Grundmerkmale werden die durch den Monotonie-Operator aus Bildern extrahierten Flecken zugrundegelegt. Sie beschreiben lokale Maxima und Minima im Grauwertgebirge, die in einer durch die zuvor durchgeführte Bandpassfilterung festgelegten Suchumgebung in Bildfolgen meist eindeutig und somit individuell zugeordnet werden können. Sie wurden zur Schätzung von Verschiebungsvektorfeldern vielfältig eingesetzt /3..6/.

Falls die dargestellten Objekte in der Bildfolge weiter auseinander liegen, muß man größere Suchbereiche zulassen und dazu weitere identifizierende Eigenschaften heranziehen. Eine Möglichkeit besteht darin, daß man die zweidimensionale Anordnung von mehreren Flecken betrachtet und mit den Anordnungen im nächsten Bild vergleicht. Diese Methode wurde bei der Detektion und Verfolgung von Fahrzeugen in einer Verkehrsszene mit einer stationären Kamera (Durlacher Tor in Karlsruhe) erfolgreich angewandt /7,8/.

Für das Wiederfinden von visuell erkennbaren Objekten in einer statischen Umgebung bei einer bewegten Kamera liegt der Nachteil dieser Verfahrensweise darin, daß gleichartige Fleckanordnungen in einer Vielzahl von Fällen vorkommen können. Dies liegt einmal daran, daß die Flecken völlig vom Grauwert und vom Kontrast abstrahieren und zum anderen daran, daß in den obigen Methoden die Flecken alle aus demselben Ortsfrequenzbereich stammen. Die Verwendung des Grauwertes hat sich als ungünstig erwiesen, da er sehr instabil sein kann. Die naheliegende Erweiterung der Methode besteht somit darin, daß man Flecken aus verschiedenen Ortsfrequenzbereichen extrahiert und zu größeren Einheiten zusammenfasst. Das hier vorgestellte Verfahren besteht also aus folgenden Schritten:

A) Extraktion von Flecken aus einem Bild, das mit großer, mittlerer und kleiner örtlicher Wellenlänge gefiltert wurde (große, mittlere und kleine Flecken).

B) Jedem großen Flecken werden die innerhalb einer vorher festgelegten Umgebung liegenden mittleren Flecken zugeordnet.

C) Jedem mittleren Fleck werden analog zu B) kleine Flecken zugeordnet.

Die Flecken sind dabei jeweils durch ihren Schwerpunkt repräsentiert. Das Verfahren kombiniert also die bekannten Methoden der Auflösungspyramide und der Bandpassfilterung von Bildern sowie der Baumstruktur von Daten mit den Eigenschaften des Monotonie-Operators.

Der Anschaulichkeit halber werden im nachfolgenden die so gefundenen Fleckstrukturen als Sternkonstellationen oder kurz Konstellationen bezeichnet, da hier wie am Sternenhimmel eine Anordnung von Punkten in einer gewissen Nachbarschaft als eine 'Gestalt' gesehen und mit einer individuellen Bezeichnung versehen wird. In beiden Fällen ist dies eine rein visuelle Betrachtungsweise, die mit physikalischen oder Objektzusammenhängen zunächst nichts zu tun hat. Es wird sich jedoch zeigen, daß die hier vorgestellten Konstellationen sich ebenfalls zu Orientierungszwecken eignen.

Das weitere Vorgehen besteht darin, daß die Konstellationen, die aus einem Bild gewonnen wurden, mit denen des zu untersuchenden Bildes verglichen werden. Der Vergleich erfolgt analog zu bekannten Korrelationsverfahren so, daß für jede Konstellation, ausgehend von einem großen Fleck im Ausgangsbild die großen Flecken im Vergleichsbild daraufhin untersucht werden, wieviele mittlere Flecken relativ dazu annähernd an derselben Stelle liegen (Suchbereich +/- 3 Pixel). Falls solche gefunden werden, wird mit den kleinen Flecken weitergesucht. Die Wahl des Suchbereichs für die großen Flecken wird dadurch bestimmt, daß in dem hier untersuchten Anwendungsfall in einer neuen Aufnahmesituation große Verschiebungen in der Horizontalen und kleine in der Vertikalen zu erwarten sind. Deswegen wurde waagrecht die ganze Bildbreite und senkrecht: +/- 100 Pixel genommen.

Das Maß der Übereinstimmung ist die Anzahl der Flecken, die in der Ausgangs- und Vergleichskonstellation übereinstimmen. Eine Korrespondenz wird dann angenommen,

wenn die Übereinstimmung maximal ist und(!) auch bei umgekehrter Suchrichtung
dieselben Konstellationen gefunden wurden. Alle anderen Zuordnungen werden verworfen.

Bildmaterial und Auswertung

Die Bildserien wurden von einer auf einem Pkw montierten CCD- Kamera aufgenommen
und auf einer U-Matic-Kassette zwischengespeichert. Die Fahrt führte im Zentrum
Karlsruhes durch die Karl-Friedrich-Straße in Richtung zum Marktplatz. Zum Zeitpunkt
der Aufnahme herrschte diesiges Winterwetter. Der Straßenverkehr war nicht sehr dicht,
so daß mit normaler Geschwindigkeit gefahren werden konnte.

Die erste Bildfolge (ca. 34 Sekunden) wurde als Referenzserie genommen und jedes vierte
Bild digitalisiert (VTE-MBV, 512*512 Pixel). Daraus wurden die Konstellationen berechnet
und auf Platte abgelegt. Von der zweiten Fahrt, die ca. 10 Minuten später durch dieselbe
Straße in derselben Richtung führte, wurde aus Aufwandsgründen nur jedes 75. Bild
ausgewertet, so daß in der zweiten Bildserie ein Abstand von drei Sekunden zwischen den
Aufnahmen liegt. Ebenfalls aus Aufwandsgründen wurden nur die Konstellationen
ausgewertet, deren großer Fleck von einem lokalen Minimum des Grauwertgebirges
stammt.

Ergebnisse

Pro Bild wurden in der gegenwärtigen Programmrealisierung zwischen 100 und 150
Konstellationen gefunden. Der Speicherbedarf für die Konstellationen eines Bildes beträgt
etwa ein Zehntel dessen eines Grauwertbildes.

Die Konstellationen jedes ausgewerteten Bildes der zweiten Bildserie wurden mit den-
jenigen aller 210 Bilder der ersten Serie verglichen und die Anzahl der korrespondierenden
Konstellationen über der Bildnummer aufgetragen. Es zeigte sich, daß bei 9 der 10 Bilder
der zweiten Serie ein ausgeprägtes Maximum in der Größenordnung von 15 bis 40 Übereinstimmungen entsteht. In Abb. 1 ist links ein Bild dargestellt, das von der zweiten Fahrt
stammt. Rechts daneben ist das aufgrund der darunter aufgetragenen Kurve ausgewählte
Bild der früheren Fahrt. Die Kurve gibt die Anzahl der Konstellationen der früheren Fahrt
an (Abszisse = Bildnummer), die mit den Konstellationen des Bildes der zweiten Fahrt
übereinstimmen. Als Kriterium für die Auswahl wird das Maximum der Kurve genommen.

Die Anzahl der vom Algorithmus gefundenen Übereinstimmungen sagt natürlich noch
nichts darüber aus, ob diese richtig in dem Sinne sind, wie sie ein menschlicher Betrachter
definieren würde. Zur Überprüfung wurden deshalb die Bilder der ersten Fahrt, die eine
maximale Zahl von übereinstimmenden Konstellationen aufweisen, verglichen mit den
zugehörigen Bildern der zweiten Fahrt. In allen 9 Fällen, in denen ein ausgeprägtes
Maximum gefunden wurde, ist die Übereinstimmung ausgezeichnet.

Abb. 2 zeigt zwei weitere Beispiele von der Mitte und dem Ende der Karl-Friedrich-Straße.
Links sind die Bilder der zweiten Fahrt, rechts die automatisch als zugehörig gefundenen
der ersten Fahrt dargestellt. Die eingetragenen Vektoren in Nadeldarstellung zeigen die
Stellen, an denen korrespondierende Konstellationen gefunden wurden. Im linken Bild gilt
die Spitze, im rechten der Kopf der Nadel. Die Länge der Vektoren gibt die Verschiebung
der korrespondierenden Bildstrukturen relativ zum Bildrahmen an.

Man sieht also, daß das sehr einfache Maß 'Anzahl der Übereinstimmungen' schon ausreicht, um aus 210 Bildern dasjenige herauszusuchen, das auch ein menschlicher Betrachter als sehr ähnlich ansieht. Weiterhin konnten bei einer visuellen Inspektion der Korrespondenzen der zugeordneten Bilder nur vier fehlerhafte gefunden werden.

Diskussion

Die vorliegende Untersuchung zeigt, daß die Zusammenfassung von Monotonieflecken aus
verschieden bandpassgefilterten Bildern zum komplexen Merkmal 'Konstellation' ein sehr
selektives Mittel darstellt, Bildteile und ganze Bilder individuell zu beschreiben. Es ist

offensichtlich möglich, verschiedene Ansichten eines Straßenzuges komprimiert abzuspeichern und ähnliche Ansichten wiederzufinden. Weitere Untersuchungen müssen zeigen, inwieweit die einzelnen Konstellationen bereits als Objektbeschreibung im Sinne eines menschlichen Betrachters taugen bzw. welche Informationen dafür noch hinzugenommen werden müssen (z.B. Kanten, 3D). Die Anwendung des Verfahrens auf Stereobildauswertung sollte leicht möglich sein, da diese einfacher ist als die vorliegende Aufgabe (Ausnutzung der Epipolarengeometrie und damit eine Verkleinerung des Suchbereichs von fast einem halben Bild auf eine Gerade).

Die Grenzen des Verfahrens werden durch die vorliegenden Experimente ebenfalls angedeutet. Sie liegen zum einen in großen Aspektwinkeländerungen (in einem Bildpaar wird in der ersten Fahrt ein Radfahrer überholt) und zusätzlich Bildkippungen (das Bildpaar bei dem kein ausgeprägtes Maximum gefunden wurde). Hier sind ebenfalls Untersuchungen notwendig, ob sich durch ein anderes Maß für die Übereinstimmung der Konstellationen Verbesserungen erzielen lassen.

Schlußbemerkung

Die vorliegende Arbeit wurde mit Mitteln des BMVg durchgeführt. Herrn Wolff sei an dieser Stelle besonders für die stetige Förderung gedankt, die eine kontinuierliche Arbeit ermöglicht hat. Ebenso möchte ich meinem früheren Kollegen Ralf Kories für die kollegiale Zusammenarbeit und die vielen breitgefächerten Diskussionen danken, die die Entwicklung des Monotonie-Operators zu einem faszinierenden Erlebnis gemacht haben.

Literatur

/1/ Zheng, Y.Z., Asada, M., Tsuji, S.: Color-Based Panoramic Representation of Outdoor Environment for a Mobile Robot. Proc. 9th Int. Conf. on Pattern Recognition, Rome, Italy, Nov. 1988, pp. 801- 803.

/2/ Tsuji, S.: Continuous Image Interpretation by a Moving Viewer. Proc. 9th Int. Conf. on Pattern Recognition, Rome, Italy, Nov. 1988, pp. 414- 519.

/3/ Zimmermann, G., Kories, R.: Die Eignung spezifischer Bildstrukturen für die Bewegungsbestimmung in Bildfolgen. In: Muster- erkennung 1983, 5. DAGM-Symposium Karlsruhe. VDE-Fachberichte 35, pp. 60-65; VDE-Verlag Berlin, 1983.

/4/ Kories, R., Zimmermann, G.: Motion Detection in Image Sequences: An Evaluation of Feature Detectors. Proc. 7th Int. Conf. on Pattern Recognition, Montreal, Canada, July 1984, pp. 778-780.

/5/ Zimmermann, G., Kories, R.: Eine Familie von Bildmerkmalen für die Bewegungsbestimmung in Bildfolgen. In: Mustererkennung 1984, Informatik-Fachberichte 87, pp. 147-153, Springer-Verlag Heidelberg, New York, Tokyo, 1984.

/6/ Kories, R., Zimmermann, G.: A Versatile Method for the Estimation of Displacement Vector Fields from Image Sequences. Proc. Workshop on Motion: Representation and Analysis., May 7-9, 1986, Kiawah Island Resort, Charleston, SC, pp. 101-106.

/7/ Schirra, J.R.J., Bosch, G., Sung, C.K., Zimmermann G.: From Image Sequences to Natural Language: A First Step Toward Automa- tic Perception and Description of Motions. Applied Artificial Intelligence 1 (1987), pp. 287-305.

/8/ Sung, C.K.: Extraktion von typischen und komplexen Vorgängen aus einer langen Bildfolge einer Verkehrsszene. In: Mustererkennung 1988, Informatik-Fachberichte 180, pp. 90-96. Springer- Verlag Berlin, Heidelberg, New York, London, Paris, Tokyo, 1988.

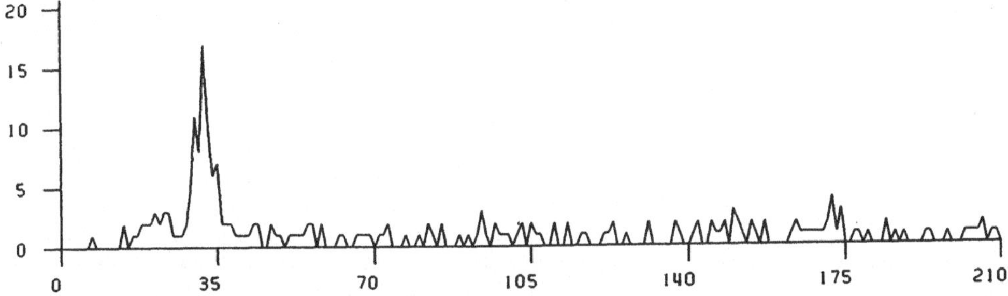

Abb. 1: Automatische Zuordnung zweier Bildserien durch komplexe Bildmerkmale.
Links ist ein Bild vom Anfang einer Fahrt durch die Karl-Friedrich-Straße in Karlsruhe
dargestellt; rechts das 33. Bild einer zweiten Fahrt, die 10 Min. später stattfand. Dieses Bild
ist eines von 210 und hat am meisten übereinstimmende Merkmale mit dem linken Bild
Der Graph zeigt diese Anzahl als Funktion der Bildnummer der zweiten Fahrt.

Abb. 2: Zwei weitere Beispiele von der Mitte und dem Ende der Bildserie.

Generierung von Entscheidungsbäumen aus CAD-Modellen für Erkennungsaufgaben

Thomas Glauser, Horst Bunke
Institut für Informatik und angewandte Mathematik
Länggassstrasse 51
3012 Bern

Zusammenfassung

Diese Arbeit beschreibt Aufbau und vollautomatische Erzeugung von Entscheidungsbäumen aus CAD-Modellen für die stufenweise Erkennung von Objekten durch Zusammensetzung von Flächenklassen. Die erzeugten Bäume für verschiedene Modelle lassen sich einfach kombinieren. Sie gestatten zudem bei ungestörten Bildern ein Erkennungsverfahren mit linearem Zeitaufwand.

1 Einleitung

In der Fabrik der Zukunft werden Bauteile mit Hilfe eines CAD-Systems konstruiert, automatisch gefertigt und von Robotern mit Sichtsystemen montiert, sortiert oder anderweitig behandelt. Um die Bauteile erkennen zu können, müssen die Roboter-Sichtsysteme über Merkmale und Verfahren zur Erkennung dieser Teile verfügen. Besonders effizient ist es, wenn diese Merkmale und Verfahren vollautomatisch aus den schon vorhandenen CAD-Modellen extrahiert werden können. Es wurden daher in letzter Zeit verschiedene Verfahren für diese Extraktion beschrieben.

Ikeuchi [6] erzeugt Ansichten des Modells aus allen Richtungen, wie ursprünglich in [4] vorgeschlagen wurde, und verwendet zur Unterscheidung dieser Ansichten viele verschiedene Merkmale. Ein Entscheidungsbaum dient dazu, die geeigneten Merkmale zur Unterscheidung auszuwählen und die Erkennung zu steuern. Dieser Entscheidungsbaum wird jedoch im wesentlichen für jedes Modell von Hand aufgebaut.

Hansen und Henderson [5] verwenden ebenfalls Ansichten und Entscheidungsbäume. Bei ihnen dient die erste Entscheidung im Baum zur Formulierung einer Hypothese bezüglich der Lage des Bildobjektes, welche dann mit verschiedenen Kriterien gemäss den Angaben im entsprechenden Unterbaum erhärtet wird. Die Auswahl von Merkmalen für die Hypothesenbildung ist jedoch recht schwierig.

In der vorliegenden Arbeit beschreiben wir ein einfaches Verfahren zur vollautomatischen Extraktion und Darstellung von Modelldaten aus einem CAD-System. Wir verwenden dabei alle Ansichten eines Modells und erzeugen einen Entscheidungsbaum, der die Konstruktion jeder Ansicht aus Flächenklassen und Vorschriften über deren flächenweise Zusammensetzung repräsentiert. Die hier beschriebenen Entscheidungsbäume zeichnen sich dadurch aus, dass gleichartige Konstruktionen von Ansichten nur einmal enthalten sind. Zudem lassen sich Bäume verschiedener Modelle problemlos zu einem einzigen Baum zusammensetzen, der bei der Erkennung Verwendung findet. Das hier vorgestellte Verfahren erlaubt eine rasche Objektidentifikation und ist tolerant gegenüber Verdeckungen und anderen Störungen.

Die vorliegende Arbeit beschränkt sich auf die Behandlung von Modellen, deren Oberflächen aus ebenen Polygonen ohne Löcher bestehen. Erweiterungen auf allgemeine Modelle sind geplant.

2 Struktur der Entscheidungsbäume und Flächendarstellung

Die Idee für den Aufbau der Entscheidungsbäume ist folgende: Bei der Erkennung eines Objekts (d.h. beim Suchen eines Matchings) soll flächenweise vorgegangen werden. Zuerst soll eine einzelne Fläche eines bestimmten Typs, d.h. einer bestimmten Klasse, gefunden werden. Anschliessend sollen schrittweise weitere, benachbarte Flächen im Bild gesucht werden, wobei bestimmte Bedingungen über die Position der jeweiligen neuen Fläche relativ zu den schon gefundenen Flächen erfüllt sein müssen.

Der Entscheidungsbaum beschreibt also die Zusammensetzung von Flächenklassen zu Ansichten. Jeder Pfad von der Wurzel des Baumes bis zu einem Blatt beschreibt eine bestimmte Ansicht. Jede Kante von einem Baumknoten zu einem Nachfolger beschreibt die Hinzunahme einer weiteren Fläche. Da nicht klar ist, welche Fläche im Bild zuerst erkannt wird (bzw. nicht gestört oder verdeckt ist), ist jede mögliche Zusammensetzungsreihenfolge von Flächen im Entscheidungsbaum enthalten. Abb. 1 stellt die Entscheidungsbäume für einen Würfel und für ein Prisma dar. Ausgefüllte Knoten repräsentieren Flächenzusammenstellungen, die einer vollständigen möglichen Ansicht entsprechen. (Im vorliegenden Beispiel sind dies alle Knoten mit Ausnahme der beiden Wurzelknoten.) Die Anschriften (Attribute) der Kanten sind folgendermassen zu verstehen: die erste Zahl beschreibt die gewünschte Flächenklasse der neuen Fläche (hier 0 = Parallelogramm, 1 = Dreieck); jedes der folgenden Tripel ist eine Bedingung für die Adjazenz der neuen Fläche in der Form: Nummer der alten Fläche in der Erkennungsreihenfolge, Kantennummer der alten Fläche, Kantennummer der neuen Fläche. Die Numerierung der Kanten einer Fläche erfolgt im Gegenuhrzeigersinn. Neben den Bäumen sind je eine Ansicht gezeichnet und mit Nummern markiert, die dem dick ausgezogenen Pfad im Baum entsprechen. Dabei sind die Flächen in der Reihenfolge der Erkennung mit Zahlen in Kreisen numeriert. Kanten im Baum werden nun als äquivalent bezeichnet, wenn sie denselben Vater haben, eine Fläche derselben Klasse beschreiben und alle Adjazenzbedingungen gleich sind bis auf eine Verschiebung der Kantennumerierung, die mit der entsprechenden Flächenklasse verträglich ist (bei einem Parallelogramm beispielsweise sind alle Kanten gleichwertig und damit kann jede Kante mit der Nummer 0 markiert werden). Das in Kapitel 3 beschriebene Verfahren liefert Entscheidungsbäume, bei denen keine zwei Baumkanten äquivalent sind; die Bäume sind also in Bezug auf unser Äquivalenzkriterium minimal. (Es ist jedoch ohne weiteres möglich, dass gleiche Flächenkonfigurationen mehrmals im Baum auftreten, allerdings mit verschiedener Erkennungsreihenfolge der Einzelflächen.)

Die Stärke dieses Ansatzes wird ersichtlich, wenn wir den Baum in Abb. 2 betrachten, der aus den beiden Bäumen von Abb. 1 zusammengesetzt wurde. Er beschreibt alle Ansichten von Würfel und Prisma. Jeder Knoten ist markiert mit den Modellen, für die eine entsprechende Teilansicht existiert (c für Würfel (cube) und p für Prisma). Neben jedem Knoten ist ein Beispiel einer entsprechenden Teilansicht gezeichnet. Da die Entscheidungsbäume verschiedener Modelle beliebig zusammengesetzt werden können, wird ersichtlich, dass durch einen geeigneten Interpreter der zusammengesetzte Baum gemäss den gefundenen Bildflächen durchlaufen und der Klassifizierungsentscheid für das betrachtete Objekt getroffen werden kann, sobald die Flächenzusammensetzung keine Ansicht eines anderen Objektes mehr darstellen kann. Es ist somit möglich, Teilansichten zu erkennen, unabhängig davon, welches Objekt die ganze Ansicht darstellt.

Zur Darstellung der Flächen in einem Entscheidungsbaum wie in Abb. 2 verwenden wir die Momentinvarianten nach [1], welche sich dadurch auszeichnen, dass sie nicht nur gegenüber Bewegungen der betrachteten Fläche in der Ebene sondern auch gegenüber Bewegungen derselben im Raum bei nachfolgender Projektion invariant sind. Wie eine Untersuchung an unserem Institut gezeigt hat [7], sind diese Invarianten wenig anfällig gegenüber Störungen an den Flächenkanten und trotzdem gut geeignet zur Unterscheidung verschiedener Flächenklassen.

Als zweites Flächenmerkmal neben den Invarianten werden spezielle Schnittverhältnisse verwendet. Wir berechnen dabei das Verhältnis der Länge einer Kante zum Abstand des Schnittpunkts der

Kantenverlängerung mit der übernächsten bzw vorletzten Kante. In Abb. 3 ist das linke Verhältnis
für die Kante a gleich dem Verhältnis AB zu BC; das rechte Verhältnis entspricht CD zu BC. Diese
beiden genannten Verhältnisse sind bei beliebigen Bewegungen jeder betrachteten Fläche im Raum
ebenfalls invariant. Wir verwenden nun für jede Kante die beiden genannten Schnittverhältnisse und
erhalten somit eine Menge weiterer Flächenmerkmale zur Klassifikation. Die beschriebenen Schnittverhältnisse dienen zur Numerierung von Flächenkanten. (Eine eindeutige Numerierung ist zwar
nicht immer möglich, da bei gewissen Flächenklassen, z.B. bei den Parallelogrammen, alle Kanten
gleichwertig sind. Bei Flächen solcher Klassen ist jedoch gerade wegen der Gleichwertigkeit der Kanten keine Unterscheidung notwendig.) Des weiteren können diese Verhältnisse verwendet werden, um
bei Verdeckungen eine Fläche zu rekonstruieren, sofern von ihr drei aufeinanderfolgende Ecken und
eine zusätzliche Kantenrichtung bekannt sind. Es ist vorgesehen, dieses Wissen für die Korrektur
der unvermeidlichen Störungen der Bilder bei der Erkennung zu verwenden.

An sich wären auch andere Längenverhältnisse denkbar, z.B. der Abstand eines Eckpunktes zum
Schwerpunkt einer Fläche im Verhältnis zum Abstand des Eckpunktes vom Schnittpunkt des Strahls,
definiert durch Schwerpunkt und Eckpunkt, mit einer Kantenverlängerung. Das Problem besteht
aber hier darin, dass bei Verdeckungen der Schwerpunkt der sichtbaren Teilfläche anders liegt, also
die Verhältnisse störungsanfälliger sind als diejenigen, die wir verwenden.

Die beschriebenen Schnittverhältnisse sind vorerst nur bei Polygonflächen definiert. Eine Erweiterung auf beliebige Flächen ist jedoch denkbar bei Verwendung von Kurventangenten. Zudem ist
klar, dass die Verhältnisse für Dreiecke trivial und bedeutungslos sind, was jedoch nicht stört, da alle
Dreieck zueinander affin sind.

3 Generierung der Entscheidungsbäume aus CAD-Modellen

Die Modelle für die zu erkennenden Objekte werden mit dem CAD-System MEDUSA auf einem
PRIME-Computer generiert. Innerhalb von MEDUSA sind die Modelldaten mittels Begrenzungsli-
nien und -flächen dargestellt. Für die graphische Darstellung können jedoch auch beliebig Drahtmo-
delle, Oberflächen- oder Volumenmodelle verwendet werden. MEDUSA verfügt über eine Schnittstelle, mit der die Geometrie- und Topologiedaten eines Modells auf ein ASCII-File ausgegeben
werden können.

3.1 Generierung von Entscheidungsbäumen für einzelne Modelle

Wir generieren eine Anzahl gleichmässig verteilter Sichtpunkte mit wählbarer Auflösung auf einer
dem Modell umschriebenen Würfeloberfläche. Für jeden Sichtpunkt führen wir dann die folgenden
8 Schritte durch:

1. Bestimme die prinzipiell sichtbaren Flächen. Es sind diejenigen, deren Normalenvektor mit
dem Vektor zum Betrachtungspunkt einen Winkel von weniger als 90 Grad bildet. Die anderen
Flächen sind prinzipiell unsichtbar, da sie vom Betrachter weggewandt sind.

2. Zu jeder prinzipiell sichtbaren Fläche bestimmen wir eine Menge OCCL, die alle Flächen
enthält, die potentiell die Fläche verdecken können. Dies sind alle Flächen, die eine Ecke haben, die
näher beim Betrachter liegt als die potentiell verdeckte Fläche.

3. Zu jeder prinzipiell sichtbaren Fläche bestimmen wir den sichtbaren Teil indem wir sie gegen
jedes Element der zugehörigen Menge OCCL clippen. Dies erreichen wir mit einem hidden-line-
Algorithmus. Der Algorithmus berücksichtigt, dass das Resultat dieses Clippings aus mehreren
Flächen bestehen kann, weil es möglich ist, dass eine Fläche vor der Fläche liegt, die wir clippen
wollen, und diese in zwei unzusammenhängende Teilflächen zerschneidet.

4. Die in Schritt 3 erhaltenen Flächen werden unter Weglassung der z-Koordinate in eine neue,
temporäre Datenstruktur eingetragen. In dieser Datenstruktur haben wir jetzt die Daten, die unsere
gewählte Ansicht beschreiben.

5. Für jede der neu erhaltenen Flächen berechnen wir die Schnittverhältnisse jeder Kante mit der übernächsten und der vorletzten Kante. (Dreiecke werden gesondert behandelt, weil das Verfahren dort sinnlos wäre.)

6. Wir klassifizieren die neuen Flächen mit Hilfe von Momentinvarianten und Schnittverhältnissen aus Schritt 5. Bei Selbstverdeckung des Modells wird das durch Verdeckung entstandene Flächenstück so klassifiziert, wie es sichtbar ist und nicht gemäss der eigentlichen Klasse der vollständigen Fläche.

7. Nun müssen wir noch die Topologie der Flächen berechnen, die sich ja von derjenigen des Modells unterscheidet, denn bei Selbstverdeckung sind in der Ansicht Flächen adjazent, die es im Modell nicht sind. Das Resultat dieses Schrittes besteht aus einer Beschreibung zu jeder Fläche, mit welchen anderen Flächen diese adjazent ist, wobei die Nummer der beiden beteiligten Kanten ebenfalls festgehalten wird.

8. Im letzten Schritt wird die betrachtete Ansicht in den Entscheidungsbaum eingefügt. Wir erzeugen dazu durch schrittweises Hinzunehmen von Flächen alle zusammenhängenden Teilansichten, erzeugen eine Kante, die die Hinzunahme der jeweils neuen Fläche beschreibt und tragen diese Kante im Baum ein, wenn noch keine äquivalente Kante im Baum enthalten ist.

Haben wir schliesslich alle Ansichten verarbeitet, schreiben wir die Beschreibungen der Flächenklassen und den Erkennungsbaum auf eine Datei.

3.2 Zusammenfügen verschiedener Entscheidungsbäume

Durch die Erzeugung von Entscheidungsbäumen für verschiedene Modelle erhalten wir mehrere Dateien, die den Aufbau der Ansichten für diese Modelle enthalten. Für die Erkennung sollen diese Entscheidungsbäume nun zu einem einzigen Baum zusammengesetzt werden, wobei äquivalente Kanten im zusammengesetzten Baum nur einmal vorkommen sollen.

Zu diesem Zweck werden der erste Modellbaum eingelesen und dann nacheinander die gewünschten Modellbäume eingefügt. Das Einfügen geschieht in zwei Schritten: Zuerst werden äquivalente Flächenklassen identifiziert und die Verschiebungen der Kantennumerierungen festgestellt. Anschliessend werden der zweite Modellbaum traversiert und die Kanten im Resultatbaum eingefügt, sofern unter Berücksichtigung von Flächenklassenübereinstimmung und Verschiebung der Kantennumerierungen keine äquivalente Kante vorhanden ist.

Wenn alle gewünschten Modellbäume eingefügt sind, sortieren wir jeden Unterbaum so, dass bei einer Traversierung Unterbäume mit wenig Knoten zuerst besucht werden. Dies bewirkt, dass beim Erkennungsverfahren der Rechenaufwand möglichst klein gehalten wird.

Der zusammengesetzte Baum wird auf eine Datei geschrieben und beschreibt auf minimale Art den Aufbau verschiedener Ansichten verschiedener Modelle.

4 Resultate und Ausblick

Die Erzeugung und Zusammensetzung von Entscheidungsbäumen für ebenflächige Modelle ohne Löcher ist vollständig implementiert in COMMON LISP auf einem IBM PC RT. Die Algorithmen wurden mit einer Reihe von Objekten getestet. Die Laufzeit für die Generierung von Entscheidungsbäumen aus 24 bis 96 Ansichten beträgt zwischen einer und neun Minuten. Bei der Zusammensetzung verschiedener Entscheidungsbäume liegt die Rechenzeit im Sekundenbereich.

Es hat sich gezeigt, dass die Knotenzahl im Entscheidungsbaum mit der Komplexität der Modelle sehr rasch wächst; so ergeben sich für den Baum für das L-förmige Objekt in Abb. 4 wegen der Selbstverdeckung bereits 9 Flächenklassen und 477 Baumknoten. Dies ist der Preis für die Möglichkeit, Flächen in beliebiger Reihenfolge zu erkennen. Der Baum hat jedoch nur eine Tiefe von 5 Stufen, was eine schnelle Objektidentifikation ermöglicht. Die theoretische Komplexität des Durchlaufens eines beliebigen Entscheidungsbaumes ist linear in der Anzahl der Flächen im Bild, sofern wir voraussetzen, dass Fehlentscheidungen nach spätestens einer festen Anzahl von Schritten sicher als solche erkannt werden. (Man beachte, dass Fehlentscheidungen nur bei stark gestörten Flächen überhaupt auftreten können.) Unter dieser Annahme ist der hier vorgestellte Ansatz anderen Methoden, die auf speziellen Graphvergleichsverfahren beruhen, überlegen, da diese polynomiale (Verfahren in [2] und [3]) oder sogar exponentielle (Verfahren in [1] und [5]) Zeit benötigen.

Momentan wird der Interpreter entwickelt, der die Brauchbarkeit der hier vorgestellten Entscheidungsbäume für die Erkennung von ganz oder teilweise sichtbaren Objekten aufzeigen soll. Als Eingabe für den Interpreter dienen einerseits der zusammengesetzte Baum und andererseits eine oder mehrere vorverarbeitete Bilddateien. Die Vorverarbeitung geschieht mit Hilfe von Programmen des Bildverarbeitungs- und Erkennungssystems PHI-1, das in [2] und [3] beschrieben wurde. Diese Programme extrahieren Linien aus Grauwertbildern und finden Flächen und Adjazenzen derselben mit Hilfe von Kantenverfolgung. Es ist geplant, das Verfahren zu erweitern für Flächen mit Löchern und ebenfalls für gekrümmte Flächen.

Danksagung

Der verwendete Rechner sowie der COMMON-LISP-Interpreter/Compiler wurden uns freundlicherweise von der Firma IBM-Schweiz zur Verfügung gestellt.

Literaturverzeichnis

[1] B. Bamieh, R.J.P. de Figueiredo,   A General Moment Invariants Attributed Graph Method for Three Dimensional Object Recognition from a Single Image,   IEEE Journal of Robotics and Automation, Vol. RA-2, No. 1, März 1986, pp. 31–41

[2] E. Gmuer,   Ein Roboter-Sichtsystem basierend auf CAD-Modellen,   Doktorarbeit, Universität Bern, 1988

[3] E. Gmuer, H. Bunke,   PHI-1: Ein CAD-basiertes Roboter Sichtsystem,   in H. Bunke, O. Kübler, P. Stucki (ed.), Mustererkennung 88, Informatik Fachberichte 180, Springer Verlag, September 1988, pp 240–247

[4] C. Goad,   Special Purpose Automatic Programming for 3D Model Based Vision,   Proc. Image Understanding Workshop 1983, pp 94–104

[5] C. Hansen, T. Henderson,   CAGD-Based Computer Vision,   Proc. IEEE Workshop on Computer Vision, Miami Beach, 1987, pp 100–105

[6] K. Ikeuchi,   Generating an Interpretation Tree from a CAD Model for 3D-Object Recognition in Bin-Picking Tasks,   International Journal of Computer Vision, 1987, pp 145–165

[7] D. Möri,   2-D Objekterkennung mit Moment-Invarianten,   Projektbericht 121, Institut für Informatik und angewandte Mathematik der Universität Bern, 1988

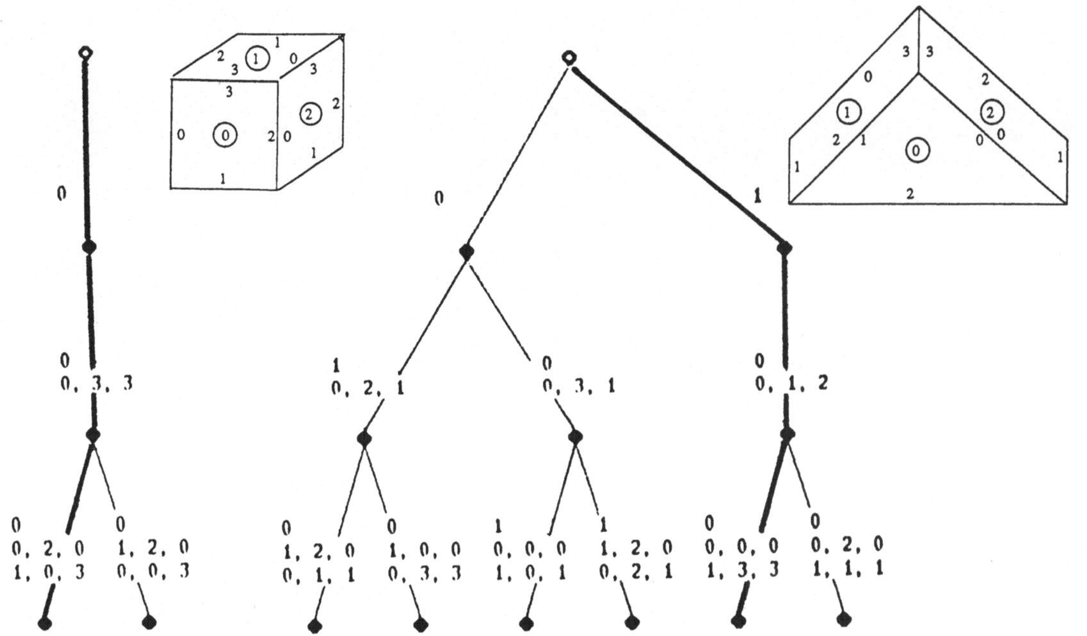

Abbildung 1: Entscheidungsbäume zu einem Würfel bzw zu einem Prisma

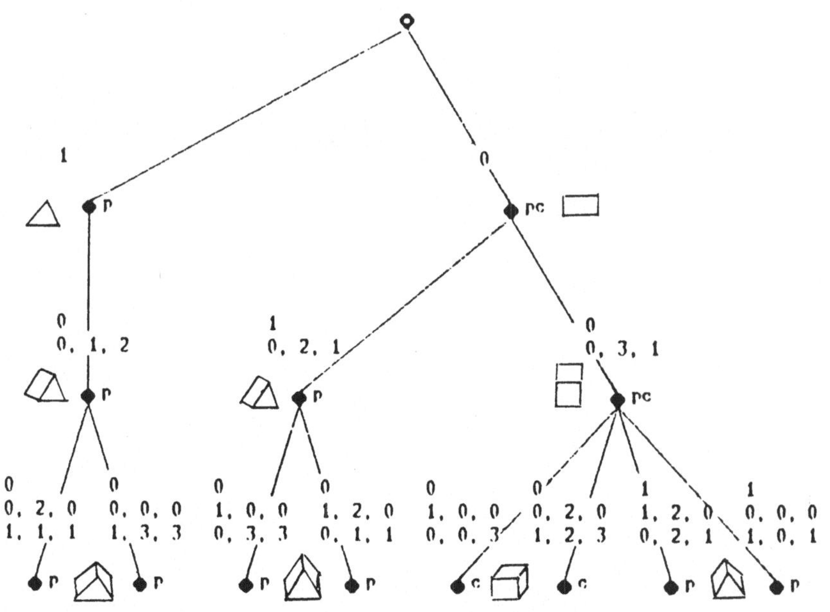

Abbildung 2: Kombinierter Entscheidungsbaum zu Würfel und Prisma

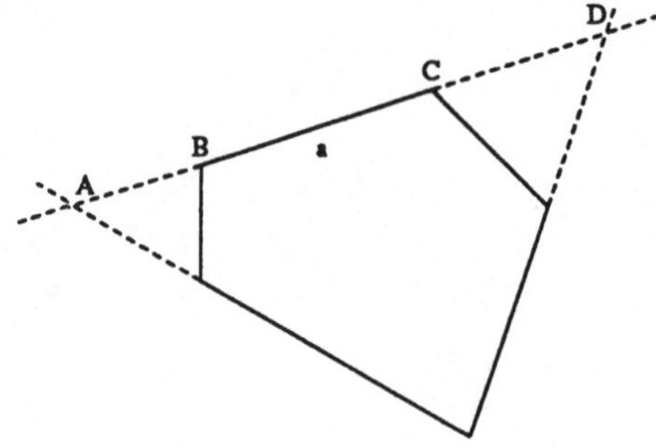

Abbildung 3: Beispiel zu den Schnittverhältnissen

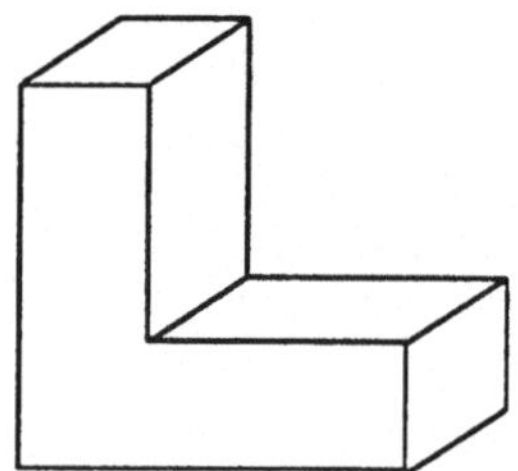

Abbildung 4: Ein Testkörper

<u>Automatischer Wissenserwerb für ein Bildanalysesystem
auf der Basis des Hierarchischen Strukturcodes</u>

Bärbel Mertsching, Georg Hartmann
Universität-Gesamthochschule-Paderborn, Fachbereich Elektrotechnik
Pohlweg 47-49, D-4790 Paderborn

<u>Zusammenfassung:</u>
Die Hierarchische Strukturcodierung ist eine kontextunabhängige
Transformation für Grauwertbilder. Auf verschiedenen Auflösungsebenen
werden Strukturelemente (Kanten-, Linien-, Flächenelemente) detek-
tiert und unter Kontinuitätsprüfung verallgemeinert. Hierbei entsteht
eine hierarchische Datenstruktur. Ein wissensbasiertes System zur
Analyse komplexer Bildszenen, welches auf dem Hierarchischen
Strukturcode (HSC) aufbaut und eine semantische Netzwerkstruktur zur
Repräsentation des Wissens verwendet, konnte bereits vorgestellt
werden. In diesem Beitrag wird über erste Ergebnisse berichtet, wie
der Erwerb von Wissen, der bisher interaktiv erfolgte, durch das
Lernen von Strukturbeschreibungen automatisiert werden kann. Hierbei
werden zuerst elementare Beschreibungen aus einer Serie von Trai-
ningsbildern bestimmt und generalisiert, die dann durch topologische
Untersuchungen zu Substrukturen und Objekten gruppiert werden.

<u>Einführung:</u>
Bei der Hierarchischen Strukturcodierung werden beliebig verlaufende
Konturen und beliebig geformte Flächen in Grauwertbildern auf ein
Multiresolutionssystem, die sogenannte HSC-Datenbasis, abgebildet.
Die derart abgelegte Strukturinformation beschreibt zweidimensionale
Ansichten von dreidimensionalen Objekten und stellt die Eingangsin-
formation für ein wissensbasiertes Bildanalysesystem auf der
Grundlage des Hierarchischen Strukturcodes dar (vgl. [HAR87, DRÜ88]).
Ein Vorteil dieses Verfahrens besteht darin, daß die Generierung des
HSC problemunabhängig ist. Aus einer HSC-Datenbasis können lage- und
größeninvariante Merkmale extrahiert werden, auf denen eine modulare
Modellierung von Objekten und Szenen aufgebaut werden kann. Die Merk-
male werden zu symbolischen Primitiven, den sogenannten Attributier-
ten Strukturtypen, zusammengefaßt. Sie beschreiben Objekte auf einer
niedrigen Abstraktionsebene. Die Konjunktion von Attributierten
Strukturtypen ergibt Substrukturen von Objekten, die Konjunktion von
Substrukturen die Objekte selbst. Szenen, Objekte, Substrukturen und
Attributierte Strukturtypen bilden Knoten in einem semantischen Netzwerk, die durch Konzepte beschrieben und durch Kanten verbunden sind
[MER88a, MER88b]. Das semantische HSC-Netzwerk bildet eine Art Skelett, welches die allgemeine Struktur einer deklarativen Wissensbasis
darstellt. In jedem Knoten sind Verweise auf lokales, Aufgaben-spezi-
fisches prozedurales Wissen in Form von HSC-Operationen enthalten.
Die HSC-Operationen selbst werden in einer Methodenbasis bereitgestellt. Das deklarative und prozedurale Wissen wird durch einen
anwendungsunabhängigen Kontrollalgorithmus [BUS89] aktiviert. Der
hier verwendete Wissensrepräsentationsformalismus ist an [SAG85] und
[NIE87] angelehnt, er ist aber direkt auf die Modellbildung von
hierarchisch-strukturcodierten Objekten zugeschnitten. Die modulare
Modellierung von Szenen auf verschiedenen Abstraktionsebenen erlaubt
nicht nur die Erkennung von singulären kompletten Objekten, sondern
auch die Interpretation von Szenen mit teilverdeckten oder sich
überlappenden Objekten. Ein Blockschaltbild des HSC-basierten Bild-
analysesystems zeigt Fig. 1. Das System ist in PASCAL unter VAX/VMS
realisiert. Eine weitere Implementation wird zur Zeit in PROLOG (Wissensbasis, Kontrolle)/ C (Methoden) unter ULTRIX erstellt.

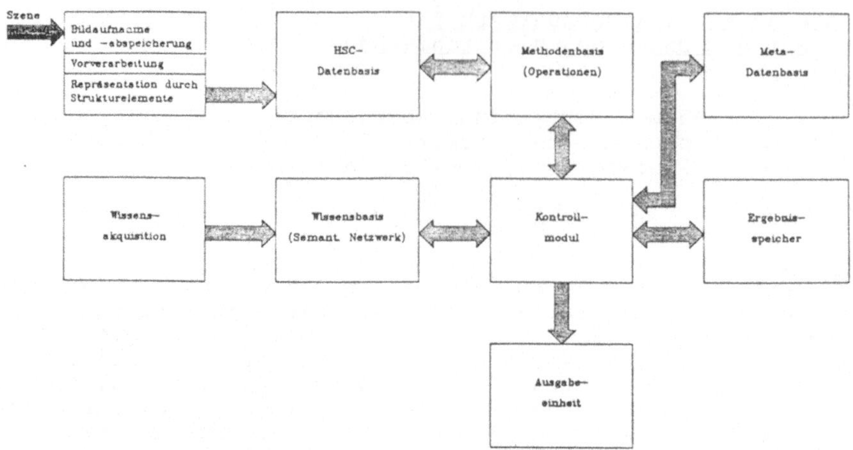

Fig 1: Blockschaltbild des wissensbasierten Erkennungssystems

Wissenserwerb

Der Wissenserwerb für das Bildanalysesystem erfolgte bisher interaktiv. Bei der Entwicklung einer Wissensbasis ist es möglich, weitgehend ohne Kenntnis des von der Maschine genutzten Repräsentationsformalismus auszukommen; das zu repräsentierende Wissen muß aber nach wie vor ausgewählt und strukturiert, Beobachtungen müssen generalisiert werden. Im folgenden wird ein allgemeiner Algorithmus für überwachtes Lernen von einem Objekt oder einer Klasse von Objekten vorgestellt. Es sollen Strukturbeschreibungen aus dem Hierarchischen Strukturcode extrahiert und die Bestandteil-Hierarchie des semantischen HSC-Netzwerks automatisch generiert werden. Spezielle Relationen, wie z. B. die Orientierung von Strukturen zueinander, müssen vorläufig noch interaktiv eingegeben werden.

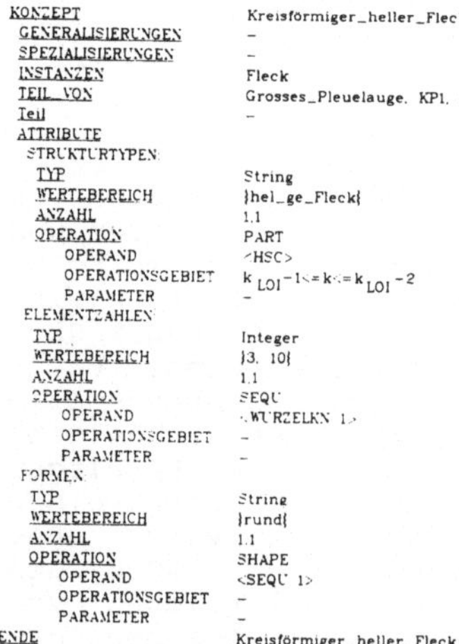

Fig. 2: Textuelle Beschreibung eines Attributierten Strukturtyps

Bei der Analyse der interaktiv erzeugten Konzepte des Netzwerks fällt auf, daß alle Knoten, die Attributierte Strukturtypen beschreiben, den gleichen Aufbau aufweisen. Die textuelle Beschreibung eines typischen Konzepts zeigt Fig. 2. Ein solches Konzept besitzt drei Attribute: Das erste Attribut 'Strukturtypen' beschreibt einen Wurzelknoten im HSC, als prozedurales Wissen ist die Operation PART angegeben, die den im Attribut definierten Wurzelknoten aus der HSC-Datenbasis extrahieren soll. Im zweiten Attribut 'Elementzahlen' ist eine Codeelementsequenz beschrieben, die aus dem im ersten Attribut gefundenen Wurzelknoten entwickelt werden soll. Das dritte Attribut 'Formen' beschreibt den Formverlauf der Sequenz, der durch die Operation SHAPE analysiert wird. Die Slots der Attributdefinitionen sind fast alle festgelegt. Lediglich die Wertebereiche und das Operationsgebiet des ersten Attributs variieren zwischen den einzelnen Attributierten Strukturtypen. Das Operationsgebiet der Operation PART gibt an, in welchen Ebenen höherer oder niedriger

Auflösung als die der Gesamtstruktur des modellierten Objekts der Wurzelknoten des Attributierten Strukturtyps zu finden sein muß.

Hieraus ergibt sich nun die Idee für den automatischen Wissenserwerb: Die Operationen, die
bisher ausschließlich zur Instanziierung der
Konzepte eingesetzt wurden, werden nun zur
Identifikation von Strukturen eines zu lernenden Objekts verwendet. Da die Strukturen in
einem bestimmten festen Größenverhältnis zur
Gesamtstruktur eines Objekts stehen, kann das
Suchgebiet bezogen auf die Ebene, in der das
Gesamtobjekt codiert ist (Level Of Interest),
angegeben werden. Um auszuschließen, daß virtuelle Strukturen gelernt werden, wird eine
Serie von Trainingsbildern (hier: fünf) betrachtet. Virtuelle Strukturen können entstehen, wenn ein Objekt nahe am Bildrand liegt
und der Raum zwischen Objekt und Rand codiert
wird.

Das Lernen von Strukturbeschreibungen eines
Objekts für ein semantisches HSC-Netzwerk geschieht in den folgenden Schritten:

- Aufnahme und Hierarchische Strukturcodierung
  einer Serie von Trainingsbildern,
- Bestimmung der Bezugsebene (Level Of
  Interest) und des Suchgebiets (Volume Of
  Interest) für jedes Bild,
- Generierung von hypothetischen Attributier-
  ten Strukturtypen,
- Generalisierung der Attributierten
  Strukturtypen durch
  - Elimination von ungeeigneten Hypothesen
  und
  - Konjunktion von Wertebereichen und Opera-
  tionsgebieten,

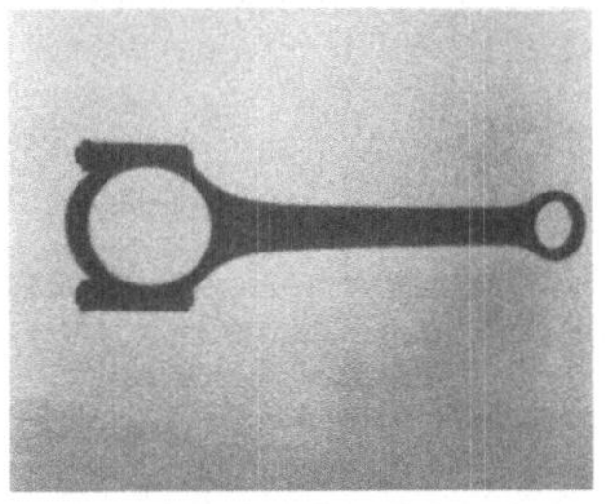

Fig. 3: Trainingsbil-
der eines zu
modellieren-
den Pleuels

- Gruppierung von verallgemeinerten Attributierten Strukturtypen zu
  Substrukturen durch Nachbarschaftsuntersuchungen,
- Konjunktion von Substrukturen zur Objektbeschreibung,
- Erzeugung von TEIL-Relationskanten und inversen TEIL_VON-Kanten
  zwischen Konzepten,
- Fusion von gleichartigen Konzepten.

Die einzelnen Schritte zur automatischen Ermittlung von Attributierten Strukturtypen und die anschließende Generalisierung der Beobachtungen wird nun an dem Beispiel einer Pleuelstange gezeigt. Fig. 3
zeigt drei Grauwertbilder aus einer Serie von Bildern des zu lernenden Objekts.

<u>Lernen von Attributierten Strukturtypen</u>
Zuerst wird für jedes Bild der Trainingsserie die Detektorebene k_{LOI}
als Bezugsebene (Level of Interest) ermittelt, in der die ganze
Pleuelstange in der höchsten Auflösung codiert ist. Hierzu wird der
am höchsten verknüpfte Wurzelknoten der Gesamtstruktur gesucht. Im
Beispielbild in Fig. 4 befindet sich dieser in der Ebene |4;4>. Da
der dunkle Pleuel auf dem hellen Hintergund sowohl linien- als auch
flächenhafte Komponenten aufweist, ist der Wurzelknoten vom
Strukturtyp 'Dunkler Vertex'. Nachdem der Level of Interest zu
$k_{LOI}=4$ bestimmt worden ist, werden alle Elemente der Codebäume, die

343

das komplette zu lernende Objekt beschreiben, aus der Bild- in eine Meta-Datenbasis kopiert und markiert. So kann vermieden werden, daß diese Codeelemente zum Lernen von Attributierten Strukturtypen benutzt werden. In Fig. 4 sind diese Strukturen (dunkle Flecken und Vertices der Ebenen $|4;n>$, $|5;n>$ und $|6;n>$) mit (1) gekennzeichnet. Lediglich die linienhaften Komponenten der Gesamtstruktur werden zur Bestimmung von Attributierten Strukturtypen herangezogen. Soll ein Objekt bis in das kleinste Detail modelliert werden, so ist es notwendig, alle Ebenen, die bei der Hierarchischen Strukturcodierung erzeugt werden, in die Modellbildung einzubeziehen. Reicht die Genauigkeit der Modellierung aus, wie sie in [MER88a] und [MER88b] beschrieben wurde, so ist es sinnvoll, das Suchgebiet (Volume Of Interest) zu beschränken, um die Anzahl der Zugriffe auf die HSC-Datenbasis zu reduzieren. Im folgenden werden die Detektorebenen $|k;n=0>$ und die sich daraus ergebenden Verknüpfungsebenen $|k;n\neq0>$ betrachtet, für die gilt: $k_{LOI}-1\leq k\leq k_{LOI}+2$.

Im nächsten Schritt werden zur Gewinnung der Definition des jeweils ersten Attributs der Attributierten Strukturtypen die Wurzelknoten aller noch nicht markierten Strukturen im Suchgebiet durch die Operation PART ermittelt. Zur Vereinfachung der Beschreibung wird die Menge der zu suchenden Strukturen auf dunkle offene und geschlossene Linien und helle geschlossene Flecken beschränkt. (Fig. 5 zeigt zur Verdeutlichung des Ablaufs einen Ausschnitt von Fig. 4, der nur noch die Codierungsebenen des Kanten-, des dunklen Linien- und des hellen Fleckencodes enthält.) Bei der Wurzelknotensuche gelten die folgenden Regeln:

R1:   Wenn ein Wurzelknoten gefunden wird, so wird er und der zugehörige Codebaum in der Meta-Datenbasis als bereits abgearbeitet markiert.

R2:   Wenn eine Struktur in mehr als einer Detektorebene codiert ist, so wird der am höchsten verknüpfte Wurzelknoten weiter betrachtet, die anderen werden als virtuelle Wurzelknoten verworfen.

R3:   Auch die Codebäume virtueller Wurzelknoten werden in der Meta-Datenbasis markiert.

Für das erste Trainingsbild werden sieben Wurzelknoten ermittelt. Sie sind in Fig. 5 mit (1)...(7) gekennzeichnet, während virtuelle Wurzelknoten die Nummer des Hauptwurzelknotens mit angehängtem 'a' tragen. Lediglich fünf der Wurzelknoten stammen von reproduzierbaren Strukturen: (1) und (7) beschreiben die hellen Flecken im großen und kleinen Pleuelauge, (2) gehört zum Schaft, (3) und (6) sind die Wurzelknoten der Linienbogen im großen und kleinen Pleuelauge. (4) und (5) gehören nicht zu dem zu lernenden Pleuel; sie entstehen, da bei dieser Aufnahme zwischen Objekt und Bildrand die Bedingungen zur Erzeugung von Code vorhanden sind. Sie treten nicht in jedem Trainingsbild auf und sind deshalb für eine Modellbildung nicht geeignet. In diesem Stadium des Lernens kann diese Entscheidung allerdings noch nicht getroffen werden. Durch die Operation SEQU werden aus den Linienwurzelknoten Liniensequenzen und aus dem Fleckenwurzelknoten die den Fleck berandenden Kantensequenzen auf der jeweiligen Detektorebene entwickelt. Es wird die Anzahl der Codeelemente der Sequenzen bestimmt und im jeweils zweiten Attribut der Attributierten Strukturtypen eingetragen. Im weiteren wird der Formverlauf der entwickelten Sequenzen mit der Operation SHAPE auf die Zugehörigkeit zu einem elementaren Formensatz analysiert und das Ergebnis im dritten Attribut ebenfalls eingetragen. Hiermit sind die sieben hypothetischen Attributierten Strukturtypen für das Testbild bestimmt. Im Anschluß werden auf die gleiche Weise die Attributierten Strukturtypen der weiteren Trainingsbilder generiert. Nachdem nun Serien von Attributierten Strukturtypen vorliegen, müssen aus ihnen diejenigen

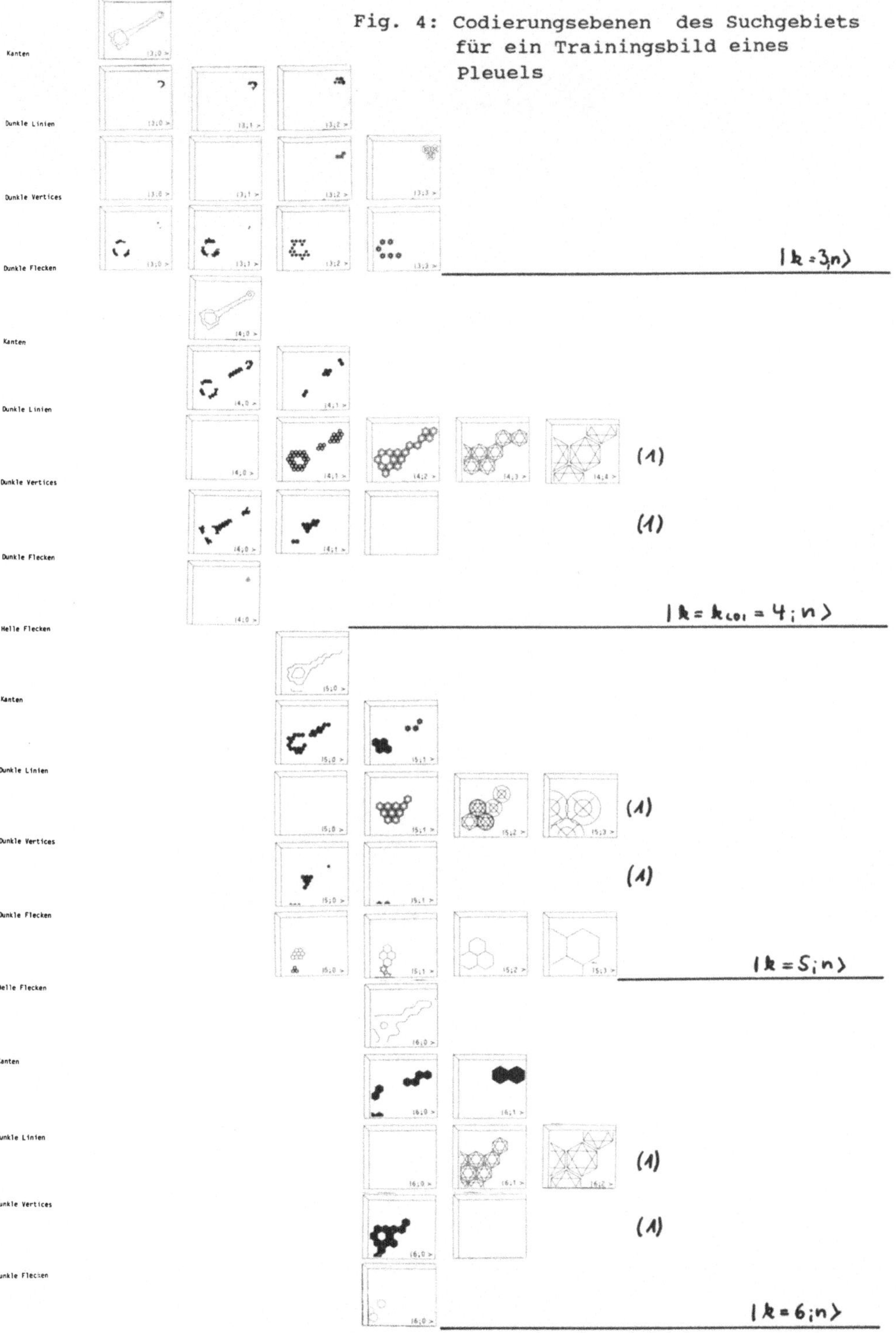

Fig. 4: Codierungsebenen des Suchgebiets für ein Trainingsbild eines Pleuels

Fig. 5: Ausschnitt von Fig. 4 (Erläuterung im Text)

generalisiert werden, die den aufgenommenen Pleuel typischerweise charakterisieren und, die restlichen werden eliminiert. Wenn angenommen werden kann, daß es sich bei den Trainingsbildern um einfache Grauwertbilder des Objekts ohne starke Schatten und Glanzlichter handelt, gilt die Regel:

R4: Wenn ein Attributierter Strukturtyp in jeder Trainingsserie mindestens einmal auftritt (Wertebereiche des ersten und dritten Attributs stimmen überein), so ist er ein generalisierter Attributierter Strukturtyp des zu lernenden Objekts.

Die Attributierten Strukturtypen, die aus den Wurzelknoten (4) und (5) in Fig. 5 erzeugt werden, können nun gelöscht werden. Beim Vergleich der Attributierten Strukturtypen der verschiedenen Serien wird die Reihenfolge des Auffindens berücksichtigt. So wird der helle Flecken im großen Pleuelauge (Wurzelknoten (1) in Fig. 5 mit anderen hellen Flecken von großen Pleuelaugen anderer Bilder gematcht. In dieser Phase werden die hellen Flecken im großen und kleinen Pleuelauge noch als getrennte Knoten betrachtet. Nun werden die Wertebereiche und die Operationsgebiete von jedem Attributierten Strukturtyp verallgemeinert. Es wird für jeden Attributierten Strukturtyp aus jedem Trainingsbild die Detektorebene des Auftretens der Struktur bestimmt. Die jeweils maximale und minimale Detektorebene wird auf den Level Of Interest bezogen im Operationsgebiet des ersten Attributs jedes verallgemeinerten Attributierten Strukturtyps eingetragen. Entsprechend wird für die Wertebereiche verfahren. Nach der Generalisierung liegen nun fünf verallgemeinerte Attributierte Strukturtypen vor: Zwei beschreiben kreisförmige helle Flecken (Pleuelaugen), einer ein gerades Linienstück (Schaft) und wiederum zwei beschreiben kreisförmige Linienbögen. Sie bilden die ersten fünf Konzepte des semantischen HSC-Netzwerks zur Beschreibung von Pleuelstangen.

<u>Lernen von Substrukturen</u>
In einem zweiten Schritt werden die Attributierten Strukturtypen zu Substrukturen gruppiert. Es wird ein Trainingsbild betrachtet. Als Kriterium, welche Attributierten Strukturtypen zu einer Substruktur gehören, wird das Nachbarschaftsverhältnis von Strukturen untersucht. Es gelten die folgenden Regeln:

R5: Wenn zwei Attributierte Strukturtypen A1 und A2 direkte benachbarte Strukturen beschreiben, dann gehören sie zu einer Substruktur.
R6: Wenn der Attributierte Strukturtyp A1 direkt benachbart zu dem Attributierten Strukturtyp A2 und der Attributierte Strukturtyp A2 direkt benachbart zu dem Attributierten Strukturtyp A3 ist, dann gehören A1, A2 und A3 zu einer Substruktur.

Die Vorgehensweise wird an den Codierungsebenen des Trainingsbildes in Fig. 5 demonstriert. Es wird mit der Operation NEIGHBOUR die Struktur auf der Detektorebene analysiert, die zu dem Wurzelknoten des ersten Attributierten Strukturtyps gehört. In dem Beispiel ist sie die helle Fläche in der Ebene |5;0>, die aus dem Wurzelknoten (1) entwickelt wird. In der Nachbarschaft werden Codeelemente vom Typ dunkler Liniencode gefunden, die zu dem Wurzelknoten (3) gehören. Weitere direkt benachbarte Strukturen können in der Detektorebene |5;0> weder zu der hellen Fläche noch zu dem dunklen Linienbogen gefunden werden. Da unterschiedlich große, benachbarte Strukturen nicht in derselben Detektorebene codiert sein müssen, werden auch die Strukturen, die zu den zugehörigen virtuellen Wurzelknoten gehören, auf Nachbarn untersucht. Für das Beispiel sind das die Strukturen, die aus den mit (1a) und (3a) bezeichneten Wurzelknoten entwickelt werden können. Auch auf diesem Weg können keine weiteren Nachbarn gefunden werden, also bilden die Attributierten Strukturtypen mit den Wurzelknoten (1) und (3) eine Substruktur (das große Pleuelauge besteht aus einem hellen Fleck mit einem berandenden Linienbogen).

Die restlichen Attributierten Strukturtypen werden entsprechend analysiert. Es ergibt sich, daß aus den Wurzelknoten (6) und (7) der Fig. 5 eine Substruktur (kleines Pleuelauge) gebildet werden kann. Zu dem geraden Linienstück aus Wurzelknoten (2) werden keine direkt benachbarten Strukturen gefunden: Es bildet eine Substruktur (Schaft), die nur aus einem Attributierten Strukturtyp besteht. Die Konjunktion der drei Substrukturen ergibt die Objektbeschreibung. Die Attributierten Strukturtypen, die Substrukturen und das Objekt 'Pleuel' werden durch Konzepte beschrieben. Zwischen den Knoten auf den Abstraktionsebenen 'Attributierte Strukturtypen', 'Substrukturen' und 'Objekte' werden TEIL-Relationskanten und ihre inversen TEIL_VON-Kanten erzeugt.

Aus den bisher generierten Knoten ergibt sich die TEIL/TEIL_VON-Hierarchie eines semantischen HSC-Netzwerks mit drei Beschreibungsebenen, die das zu lernende Objekt beschreibt. In dieser Form kann das Teilnetzwerk bereits durch den Kontrollalgorithmus ausgewertet werden. Das Netzwerk enthält allerdings noch redundante Information: Jeweils zwei Attributierte Strukturtypen beschreiben kreisförmige helle Flächen bzw. dunkle Linienbögen. In einem Fusionsschritt werden die Attributierten Strukturtypen miteinander verschmolzen, ihre Wertebereiche und Operationsgebiete vereinigt. Zur weiteren Unterscheidung von Substrukturen (z. B. den Pleuelaugen untereinander) können noch Attributierte Strukturtypen auf der Basis von Teilkanten des Objekts gelernt werden. Nachdem nun ein Objekt gelernt wurde, können durch eine zweite Serie von Trainingsbildern Spezialisierungen dieses Objekts ebenfalls automatisch generiert werden.

Literatur

[BUS89]   Busemann, Martin: Implementierung eines Kontrollalgorithmus zur Auswertung eines semantischen HSC-Netzwerks. Diplomarbeit (unveröffentlicht), Paderborn 1989

[DRÜ88]   Drüe, Siegbert: Wissensbasiertes Erkennungssystem für hierarchisch-strukturcodierte linienhafte Objekte. Dissertation, Paderborn 1988

[HAR87]   Hartmann, Georg: Recognition of Hierarchically Encoded Images by Technical and Biological Systems. In: Biological Cybernetics 56, 1987, 593-604

[MER88a]   Mertsching, Bärbel; Hartmann, Georg: Wissensbasierte Erkennung im HSC. Bericht zum DFG-Projekt Ha 1314/4-2 (unveröffentlicht), 1988

[MER88b]   Mertsching, Bärbel, Hartmann, Georg: Modulare Modellierung von hierarchisch-strukturcodierten Objekten und Szenen durch ein semantisches Netzwerk. In: Bunke, H. (Hg.): Mustererkennung. Informatik-Fachberichte 180. Berlin u. a. (Springer-Verlag) 1988, 158-164

[NIE87]   Niemann, Heinrich; Bunke, Horst: Künstliche Intelligenz in Bild- und Sprachanlyse. Stuttgart (Teubner-Verlag) 1987

[SAG85]   Sagerer, Gerhard: Darstellung und Nutzen von Expertenwissen für ein Bildanalysesystem. Informatik-Fachberichte 104. Berlin u. a. (Springer-Verlag) 1985

Zur Automatischen Entwicklung von Objektmodellen durch Imitation

W. Burger *

Johannes Kepler Universität
Institut für Systemwissenschaften
A-4045 Linz, Austria
email K320440@AEARN.Bitnet

Abstract: Die Verfügbarkeit von Modellen ist eine wesentliche Voraussetzung zur effizienten Erkennung von Objekten. Wir präsentieren einen neuen Ansatz zur Konstruktion von Objektmodellen durch Interaktion zwischen einem menschlichen Instruktor und einem Lernsystem anhand konkreter Bildbeispiele. Die Auswahl charakteristischer Objekteigenschaften und die Bildung der der Modellstruktur wird dem System überlassen. Die Erprobung des deklarativen Modells erfolgt durch Umsetzung in ein lauffähiges Erkennungsprogramm und wiederholte Anwendung desselben.

1. Problemstellung

Die Erkennung von Objekten ist ein zentrales Thema der visuellen Informationsverarbeitung. Traditionell erfolgt die Programmierung eines Vision-Systems durch einen (menschlichen) Experten, der versucht, aus den subjektiven Eigenheiten des Bildmaterials eine brauchbare Erkennungsstrategie und in der Folge ein entsprechendes Programm zu erzeugen. Dies erfordert eine erhebliche Menge an Spezialkenntnissen und Erfahrung, sowie einen hohen Experimentieraufwand für die Auswahl geeigneter Verfahren und deren Parameter. Das dabei gesammelte Wissen wird in der Regel nicht dokumentiert bzw. ist aus dem fertigen Programm nicht mehr ableitbar und ist dadurch für nachfolgende Anwendungen nicht unmittelbar verfügbar. Die damit verbundenen Entwicklungskosten verteuern den Einsatz von Comouter Vision bzw. machen viele potentielle Anwendungen unwirtschaftlich.

Die Erkennung eines Objekts bedingt die Zuordnung der visuellen Daten zu einem bestehenden Referenz-Modell, das in geeigneter Weise die Eigenschaften der Objektklasse beschreibt. Die Modellbildung nimmt daher in der Konstruktion von Vision Maschinen eine Schlüsselrolle ein, und es erscheint unumgänglich, diesen Vorgang zu automatisieren oder zumindest effizient zu unterstützen.

---

* Diese Arbeit wird im Rahmen einer Forschungskooperation durch die Siemens AG. München (Zentralbereich Technik ZTI) in dankenswerter Weise unterstützt.

2. Automatische Modellbildung

Bisherige Arbeiten zur Automatisierung der Modellbildung gingen vorwiegend in zwei Richtungen. Zum ersten, versucht man, ausgehend von vorhandenen 3-D Modellen der Objekte, automatisch Prozeduren zur Erkennung dieser Objekte in 2-D Abbildungen zu erzeugen [5, 8, 6, 7]. Während jedoch bei vielen Objekte (z.B. bei Früchten) die exakte Beschreibung ihrer dreidimensionalen Form nicht möglich ist, scheitert sie bei anderen Objekten (z.B. Autos) an der großen Zahl der möglichen Ausprägungen.

Demgegenüber geht man in der Mustererkennung direkt von konkreten Bildbeispielen aus. Die Auswahl relevanter Merkmale und die Partitionierung des mehrdimensionalen Merkmalsraums in Objektklassen erfolgt häufig in einer *Trainingsphase*, interaktiv oder auch automatisch. Statistische Erkennungsmethoden sind allerdings, u.a. wegen der fehlenden Strukturinformation, für viele Anwendungen ungeeignet.

Winston [12] versuchte als einer der Ersten den Einsatz von Lernen zur Bildung von Objektkonzepten, allerdings anhand abstrakter Beschreibungen. Das Lernen von konkreten Bildern bietet demgegenüber den Vorteil, daß nicht nur ein künstlicher Ausschnitt der Realität in Form von abstrakten Beschreibungen als Grundlage für Schlußfolgerungen dienen kann, sondern ein vollständiges Abbild der Realität zur Verfügung steht. Außerdem sind Bilder i.A. hoch redundant, sodaß ein lernendnes System die Möglichkeit hat, durch Experimentieren selbst die für das System subjektiv "interessantesten" Merkmale zu suchen.

3. Das Visual Modeling Tool (VMT)

Unser Ziel ist die semi-automatische Ableitung von Objektmodellen für die strukturorientierte Objekterkennung durch eine lernfähige Systemkomponente, die wir als *Visual Modelling Tool* bezwichnen (Abb.1). Wir gehen davon aus, daß eine Sammlung *Visueller Basis-Routinen* [11] zur Verfügung steht, die wir in ihrer Gesamtheit als *Primitive Vision Engine* (PVE) bezeichnen. Die Bildung der Objektmodelle, einer geeigneten Erkennungsstrategie und des lauffähigen Erkennungsprogramms erfolgt während der *Modellierungsphase*. Endprodukt ist ein fertiges Laufzeitsystem RVE (für *Runtime Vision Engine*) für die jeweilige Anwendung. In der *Anwendungsphase* erfolgt dann der routinemäßige Einsatz des Laufzeitsystems, ohne weitere Veränderung der internen Modelle und Strategien.

Die *Modellierungsphase* besteht aus der Analyse von Bildbeispielen der wesentlichen Objekte unter Führung durch einen menschlichen Instruktor. Diesem obliegt es, die relevanten Objekte auszuwählen und zu benennen, sie eventuell zu markieren, subjektiv auffällige Strukturen anzudeuten und offensichtliche Fehler in der Modellbildung interaktiv zu korrigieren. In der Terminologie des maschinellen Lernens (Michalsky [10]) könnte man diesen Vorgang mit *Learning by Example* bzw. *Learning by Exploration* bezeichnen. Ein wesentliches Element dabei ist die Möglichkeit, einen Großteil des Dialogs zwischen dem Instruktor und dem System auf der visuellen Ebene selbst ablaufen zu lassen und damit dem schwierigen Problem der Verbalisierung visueller Konzepte auszuweichen.

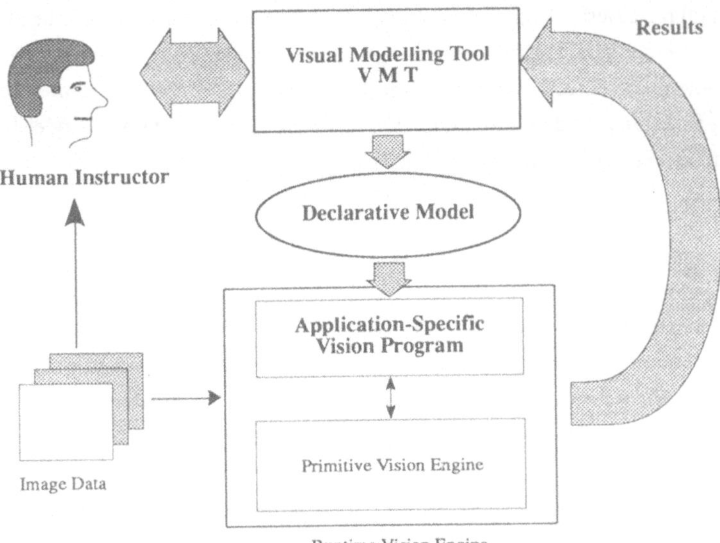

Abb.3. Modellierungsphase des Vision-Systems. Zentrale Komponente ist das *Visual Modelling Tool* (VMT), das anhand von Bildbeispielen ein deklaratives Modell der Objekte erzeugt. Wesentlich dabei ist die Führung durch den menschlichen Instruktor während des gesamten Modellierungsvorgangs. Aus dem deklarativen Modell wird ein anwendungsspezifisches Vision-Programm "kompiliert", dessen Anwendung zur Beurteilung und evtl. Modifikation des deklarativen Modells verwendet wird. Der Instruktor nimmt in der Regel keinen direkten Eingriff auf das erzeugte Modell.

Während der Modellierungsphase werden Objektmodelle schrittweise entwickelt. Die Anforderungen an die erzeugte Darstellung sind, daß sie *(1) spezifisch* genug ist, um verschiedene Objekte mit ausreichender Zuverlässigkeit richtig zu klassifizieren und daß sie *(2) allgemein* genug ist, um eine möglichst große Klasse äquivalenter Objekte zu beschreiben. Die Erfüllung dieser gegenläufigen Forderungen ist das Hauptproblem der Modellierung schlechthin.

Das deklarative Modell einer Anwendungsumgebung (*world*) besteht aus einer Menge von Objektmodellen (*object*), die wiederum durch verschiedene Ansichten (*aspect*) beschrieben sind (Abb.2). Die Spezifikation der *aspects* erfolgt mithilfe von realisierbaren Bildmerkmalen (*aspect-description*). Im Folgenden bedeuten $\{x\}$ Mengen von Objekten vom Typ x und $(a_1, a_2, ..., a_n)$ N-Tupel mit Elementen vom Typ a_i.

world	=	{*object*}
object	=	(*object-name, aspect-collection*)
aspect-collection	=	{*aspect*}
aspect	=	(*aspect-description, prototypes*).

Zusätzlich zur abstrakten, strukturellen Beschreibung stehen *aspects* in Beziehung zu einer Menge von Prototypen (*prototypes*). Dies sind konkrete Ausprägungen von Ansichten des betreffenden Objekts, die zu einem früheren Zeitpunkt in der Modellierungsphase klassifiziert wurden. Sie dienen dazu, bei Veränderungen im Modell sicherzustellen, daß diese Veränderungen konsistent mit früheren Klassifizierungen sind. Sie stellen also eine Art *ikonisches Gedächtnis* dar, bzw. die semantische Beziehung zur Realität.

Aspects sind strukturelle Beschreibungen konkreter Ansichten von Objekten, bestehend aus einer Menge von Primitiven und einer darauf definierten Menge von Prädikaten. Interpretiert man die Primitiven und Prädikate als Knoten bzw. Kanten, so ergibt sich ein Graph ähnlich einem *Semantischen Netzwerk*. Die Darstellung ist *operational* in dem Sinn, daß alle verwendeten Prädikate direkt oder indirekt durch das Laufzeitsystem (*Runtime Vision Engine*) verifizierbar sind.

4. Modellierungsvorgang

Der eigentliche Modellierungsvorgang geschieht durch die interaktive Analyse von konkreten Bildbeispielen durch das VMT zusammen mit einem menschlichen Instruktor . Dabei soll das System versuchen, primär durch Beobachtung des Instruktors ein Modell zu erzeugen, mithilfe dessen es letztendlich das Verhalten des Instruktors *imitieren* kann. Dazu bedarf es mehrer Voraussetzungen:

(1) Das Lernsystem muß in der Lage sein, die visuellen Vorgänge des Instruktors bei der Objekterkennung zumindest teilweise nachvollziehen zu können. Im wesentlichen geschieht dies dadurch, daß der Instruktor seine Interpretation (z.B. die Umrisse von Objekten) graphisch auf das Bild zeichnet und dabei vom System *beobachtet* wird.

(2) Das System muß in der Lage sein, die graphischen Ausführungen des Instruktors mit dem ursprünglichen Bild in Zusammenhang zu bringen, auch wenn die Graphik nicht exakt ausgeführt ist.

(3) Das System muß von sich aus nach weiteren Merkmalen suchen können, die zur Charakterisierung des Objekts in Frage kommen, ohne daß diese Merkmale vom Instruktor explizit markiert wurden.

(4) Das System muß Korrekturen von außen akzeptieren. Dabei sollte in der Regel ausreichen, daß der Instruktor falsch kategorisierte Objekte markiert und das System daraufhin selbst geeignete Korrekturen am Modell vornimmt. Nur in Ausnahmefällen sollte durch den Instruktor ein direkter Eingriff in das Modell erfolgen.

Während der Modellierung werden folgenden Hauptschritte iterativ durchlaufen:
(1) Vorgaben und Korrekturen durch den Instruktor (Instruktionsphase),
(2) Herstellen des Bezugs zum Realbild und bestehenden Modell (Explorationsphase),
(3) Änderung des deklarativen Modells,
(4) Umsetzung des deklarativen Modells in ein lauffähiges Programm,
(5) Ausführung des lauffähigen Programms,
(6) Bewertung der Ergebnisse.

Deklarative Modell besteht ausschließlich aus Primitiven bzw. aus Gruppierungen von Primitiven, die durch das Laufzeitsystem tatsächlich realisiert werden können. Dadurch wird sichergestellt, daß diese die erzeugte Darstellung in jedem Fall "operational" ist. Bei der Überführung des Deklarativen Modells in das ausführbare Programm muß darüberhinaus die zugrundeliegende Erkennungsstrategie berücksichtigt werden. Für einen typischen Hypothesize-and-Test Ansatz müssen getrennte Entscheidungsbäume für den Indexing- und den Verifikationsschritt erzeugt werden, die zum einen von bestimmten Features und zum anderen von bestimmten Objekthypothesen ausgehen.

5. Zusammenfassung

Ziel dieser Entwicklung ist es, die Konstruktion von Modellen für die Objekterkennung effizienter zu gestalten, und zwar für eine möglichst breite Klasse von Objekten. Unser Ansatz stützt sich insbesondere auf die Anwendung von Lerntechniken sowie graphisch/visuelle Dialogmechanismen zum Wissenstransfer zwischen Mensch und Maschine. Dadurch sollen erstens die Details der internen Modellierung dem Benutzer weitgehend verborgen bleiben und außerdem Konzepte vermittelbar werden, die sonst nur umständlich oder überhaupt nicht zu verbalisieren sind. Grundlage der Modellbildung sind konkrete Bildbeispiele der relevanten Objekte, die durch das *Visual Modelling Tool* mithilfe eines menschlichen Instruktors analysiert werden. Als Beginn einer Implementierung arbeiten wir derzeit an einem *Vision Kernel System* [3, 4], das die wesentliche Dialogmechanismen, Visuellen Basis-Routinen sowie Werkzeuge zur Manipulation von Bildern auf Graphik-Workstations zur Verfügung stellt. Dieses System ist zur problemlosen Einbindung in wissensbasierte Umgebungen in LISP implementiert.

Literatur

[1] Bolles R.C., Cain R.A., "Recognizing and Locating Partially Visible Objects: The Local-Feature-Focus Method," *The International Journal of Robotics Research* 1 (3), pp.57-82 (1982).

[2] Brooks R.A., "Symbolic Reasoning Among 3-D Models and 2-D Images," *Artificial Intelligence* 17, pp.285-348 (1981).

[3] Burger W., Hellwagner H., Müller-W. T., "VKS - A Vision Kernel System," Interner Bericht, Johannes Kepler Universität, Systemtheorie und Informationstechnik, A-4040 Linz (1988).

[4] Burger W., "Towards Interactive Development of Object Models for Machine Vision," erscheint in *Wissensbasierte Mustererkennung*, R.Albrecht und A.Pinz (Hrsg.), OCG Schriftenreihe, R. Oldenbourg, Wien (1989).

[5] Goad C., "Special Purpose Programming for 3D Model-Based Vision," Proc. DARPA Image Understanding Workshop, Arlington, Virginia, pp.94-104 (June 1983).

[6] Hansen C., Henderson T., "Towards the Automatic Generation of Recognition Strategies," Proc. International Conference on Computer Vision ICCV´88, Tampa FL, IEEE Computer Society, pp.275-279 (Dec. 1988).

[7] Horaud R., Skordas T., "Model-Based Strategy Planning for Recognizing Partially Occluded Parts," IEEE *Computer* 20 (8), pp.58-64 (Aug. 1987).

[8] Ikeuchi K., Kanade T., "Automatic Generation of Object Recognition Programs," *Proceedings of the IEEE* 76 (8), pp.1016-1035 (Aug.1988).

[9] Lowe D.G., "Three-Dimensional Object Recognition from Single Two-Dimensional Images, " *Artificial Intelligence*, 31, pp. 355-395 (1987).

[10] Michalski R.S., Carbonell J.G., Mitchell T.M., *Machine Learning - An Artificial Intelligence Approach*, Vol.II, Morgan Kaufmann Publishers, Los Altos CA (1986).

[11] Ullman S., "Visual Routines," in *Readings in Computer Vision: Issues, Problems, Principles, and Paradigms*, Fischler & Firschein (Eds.), Morgan Kaufmann Publishers, pp. 298-328 (1987).

[12] Winston P.H., "Learning Structural Descriptions from Examples," in *The Psychology of Computer Vision*, P.H.Winston (Ed.), pp. 157-209, McGraw-Hill, New York (1975).

Wissensbasiertes Verstehen von Strassenkarten

Markus Ilg und Olaf Kübler
Institut für Kommunikationstechnik
Fachgruppe Bildwissenschaft
ETH–Zentrum, CH–8092 Zürich

Zusammenfassung

Am Beispiel des Verstehens von Strassenkarten wird gezeigt, dass sich dieses Spezialgebiet der Dokumentenanalyse geradezu ideal zum Kennenlernen der Methoden aus dem Bereich der Künstlichen Intelligenz und somit als Vorstufe zum viel komplexeren Problem des Verstehens von allgemeinen Szenen eignet. Aus den Bilddaten werden Primitive extrahiert, die einerseits möglichst wissensfrei, d.h. möglichst unabhängig vom Inhalt des Dokuments gewonnen werden, andererseits aber doch die Aufgabe des Aufbaus komplexerer Strukturen und die Rückgewinnung des semantischen Gehalts dieser Strukturen möglichst gut unterstützen. Der Aufbau komplexerer Strukturen erfolgt über mehrere Zwischenstufen; dabei wird zunehmend die Verwendung von Bereichswissen zugelassen. Operatoren und Prädikate auf den interessierenden Entitäten können mit einem objekt-orientierten Ansatz sehr elegant formuliert werden. Deren zeitliche und örtliche Anwendung wird mit Hilfe von Bereichswissen gesteuert, welches vorgängig in eine 'expert system shell' gebracht wurde.

1   Einleitung

Die Dokumentenanalyse setzt sich die Interpretation ganzer Dokumente (Texte, Graphiken und Bilder oder eine beliebige Mischung davon) mit Hilfe des Computers zum Ziel. Ein Grund, sich damit zu beschäftigen, ist sicher pragmatischer Art. Das Bedürfnis, die Information auf bestehenden Dokumenten digital zu erfassen, ist jedenfalls ausgewiesen. Es gibt aber noch einen zweiten Grund, der für unsere Arbeiten im Zentrum steht. Das Verstehen graphischer Dokumente ist ein ausgezeichnetes Übungsfeld, in dem fast alle Probleme der allgemeinen Computer Vision in nicht-trivialer Form angesprochen werden. Dabei ist es möglich, sich auf die höheren Ebenen der Architektur eines allgemeinen Computer-Vision-Systems (z.B. [3]) zu konzentrieren.

Im folgenden werden allgemeine Grundsätze und Methoden, wie sie in der Dokumentenanalyse zur Anwendung gelangen, anhand eines Ausschnitts aus einer 'Schicht' einer Strassenkarte dargestellt. Diese 'Situation' genannte Schicht ist eine Linienzeichnung und enthält Symbole in Form von strukturierten Linien, die verschiedene Strassentypen als Interpretation haben, sowie Symbole für Agglomerationsgrenzen in der Form von Polygonzügen. So gibt es zum Beispiel Doppellinien, strichlierte Linien, parallel-strichlierte Linien, 3- oder 4-fach parallele Linien etc., deren Verlauf naturgemäss sehr unregelmässig sein kann.

Die abgetastete Vorlage wird mittels Schwellwertoperation wieder in ein Binärbild übergeführt, wobei der Vordergrund das digitale Abbild der ursprünglichen Linienzeichnung ist. Anschliessend wird die Pixel-Darstellung in eine erste symbolische Darstellung übergeführt. Dazu wird, wie in [1, 2] beschrieben, ein Verfahren in zwei Etappen angewendet. Zuerst wird im verdünnten und gelabelten Rasterbild der Liniengraph des Vordergrunds extrahiert. Dann wird das Einflusszonenskelett (EZS, dualer Graph) der gelabelten Liniensegmente extrahiert. Eine so erhaltene duale Linie hat Verweise auf die am nächsten liegenden Liniensegmente links und rechts von ihr (explizite Nachbarschaftsinformation).

Wegen gewissen Unzulänglichkeiten des EZS wurden andere Verfahren zur Mittelachsengewinnung untersucht. Eine detaillierte und vergleichende Bewertung aller Verfahren steht noch aus, doch lässt sich bereits jetzt sagen, dass sich eine auf dem allgemeinen Voronoi-Diagramm basierende Lösung [5] als vielversprechend abzuzeichnen beginnt.

Soweit bediente man sich bekannter Verfahren der digitalen Bildverarbeitung (Low Level). Die nachfolgenden Schritte zum eigentlichen Verstehen einer Strassenkarte sind unten aufgeführt und werden in den folgenden Kapiteln kurz beschrieben:

- Die bereinigten Primitive werden mit wenig Bereichswissen (i.w. den Zeichenregeln) zu Segmenten eines bestimmten Linientyps verknüpft (zusammengesetzte oder aggregierte Primitive), im folgenden auch Strassenstücke genannt (Intermediate Level).

- Auf der obersten Stufe (High Level) versucht man, unter Zuhilfenahme von immer mehr und spezifischerem Bereichswissen, Segmente von unteren Stufen zu einer semantisch noch bedeutungsvolleren Einheit, wie der eines Strassenbogens (von Strassenknoten zu Strassenknoten), zu verbinden, und somit schrittweise das gesamte Strassennetz aufzubauen.

Am Ende soll eine vollständige und kompakte symbolische Beschreibung der Kartenentitäten sowie deren Beziehungen vorliegen, d.h. das gesamte Strassennetz soll bekannt sein bezüglich Geometrie und Topologie, sowie der Priorität sich überlagernder Entitäten.

2 Erkennen zusammengesetzter Symbole (Strassenstücke)

Für das Ergebnis der Primitiven-Extraktion spielte der semantische Gehalt des Dokuments noch keine Rolle. Ausgehend von diesen Primitiven versucht man nun, diese in mehreren Stufen auf ein immer höheres semantisches Niveau zu bringen. Hier ist das Vorgehen sehr stark von der Art der Dokumente, u.a. auch von ihrem Inhalt, abhängig.

Leider stellt man fest, dass sich die extrahierten Daten trotz der hohen Qualität des Originals dessen Topologie nicht hinreichend genau wiedergeben. Es wird daher eine Nachbehandlung auf den Linienstücken durchgeführt. Als Grundlage dazu dient ein einfaches 'Linienzeichnungsmodell'. Der Name soll andeuten, dass es sich hier um Linien ohne jede Strukturierung und minimalem semantischen Gehalt, nämlich um die Elemente einer Linienzeichnung, handelt. Mit Hilfe eines dreiteiligen Satzes von Regeln (zum Säubern, Auftrennen von Berührungen und Verbinden von unterbrochenen Liniensegmenten) wird das fehlerhafte Bild in eine zum Modell konforme Form übergeführt.

Nachdem die Möglichkeiten des Linienzeichnungsmodells ausgeschöpft sind, muss spezifischeres Wissen herangezogen werden. Auf dem 'Intermediate Level' gruppiert man nun die bereinigten Primitive mit Hilfe der aus der Karte abgeleiteten Zeichenregeln zu der nächst höheren Entität, nämlich zu einem Segment eines bestimmten strukturierten Linientyps, das einem Strassenstück entspricht. Dabei repräsentiert eine solchermassen gewonnene Entität i. a. noch keine sinnvolle Einheit aus der richtigen Welt, insbesondere nicht ein Strassenstück, das von Verzweigung zu Verzweigung geht.

Am Beispiel der strichlierten Linie sollen die in den relevanten Zeichenregeln enthaltenen Bedingungen aufgezeigt werden: die Striche haben eine vorgegebene Länge und Dicke, müssen aber nicht gerade verlaufen; zwei aufeinanderfolgende Striche müssen einen bestimmten Abstand haben; des weiteren besteht eine strichlierte Linie aus mindestens zwei Strichen. In der Folge wird ein mehrstufiges Vorgehen angewendet, indem, die elementaren Linien analysierend, zuerst Striche der richtigen Länge und Dicke gesucht, davon mit Hilfe der dualen Linien (Abstand!) Strich-Doubletten geformt, welche schliesslich zu maximal langen Listen verknüpft werden.

Das Gros der in unserer Vorlage vorkommenden Linientypen basiert auf parallelen Linienstücken. Für die Erkennung eignet sich dazu besonders das eingeschlossene Flächenstück, da dessen Mittelachsentransformation etwas über den Abstand der beiden Randlinien aussagt. Eine (duale) Linie, die als Mittellinie eines parallelen Liniensegments gedeutet werden soll, muss folgenden Bedingungen genügen:
- Sie muss eine minimale Länge haben
- Die Varianz der Distanzwerte der inneren Linienpunkte darf einen bestimmten Betrag nicht überschreiten. Mittlere Werte deuten darauf hin, dass es u.U. mehrere parallele Sektionen innerhalb des Linienstücks hat und daher ein split-and-merge Vorgehen beim Bestimmen der Werte angezeigt ist.
- Der Mittelwert der Distanzwerte der inneren Linienpunkte muss in einem vordefinierten Bereich liegen (entsprechend dem Abstand der drei verschiedenen Klassen von parallelen Linien: schmal, mittel und breit).
- Der parallele Anteil einer Linie muss gross sein (z.B. mehr als 90 %). Nur an den Enden nehmen die Distanzwerte zu, weil sie dort von den Verzweigungszonen beeinflusst werden.

Die Achsen von Autobahnen, dargestellt durch 4 parallele Linien, werden als solche erkannt, wenn folgende Voraussetzungen erfüllt sind: Erstens gibt es ein paralleles Linienpaar (beschrieben durch seine duale Linie), das beidseitig als Nachbarn ebenfalls je ein paralleles Linienpaar so hat, dass zwei benachbarte duale Linien Verweise auf die gleiche dazwischenliegende Linie haben. Zweitens müssen sich alle parallelen Linienpaare in Laufrichtung genügend überlappen. Bei Autostrassen (3 parallele Linien) gelten ähnliche Bedingungen, jedoch muss nur auf einer Seite ein benachbartes Linienpaar vorhanden sein. Die Mittelachse fällt dann auch gerade mit einer Linie des Vordergrunds zusammen.

Ebenfalls aus parallelen, aber kurzen Segmenten aufgebaut ist die parallel-strichlierte Linie, d.h. eine durchgehende Linie von einer parallelen strichlierten Linie begleitet. Die dualen Linien, welche solche Strukturelemente beschreiben, müssen eine charakteristische Länge, und die benachbarten Linien müssen auf mindestens einer Seite (die Seite mit den Unterbrüchen kann wechseln) einen kurzen isolierten Strich haben.

Es ist bekannt, dass 'bottom-up'-Methoden besonders bei störungsfreien, vollständigen Daten angezeigt sind. Man nimmt dann oft an, dass Interpretationen nicht revidiert werden müssen. In unserem Fall sind die Daten von guter Qualität, aber es kann auch in diesem Fall vorkommen, dass Linien in mehreren (konkurrierenden) Interpretationen Verwendung finden können, z.B. können parallele Linien bestimmte einfache Strassen darstellen oder aber, wie wir gesehen haben, Teil eines Autobahnsymbols sein. In diesem Fall ist es eine gute Heuristik, zu versuchen, zuerst die am stärksten strukturierten Gebilde zu erkennen, also z.B. Autobahnen vor einfachen Strassen. Die Linien, die zu einem Autobahnstück gehören, werden markiert und können daher nicht mehr ohne weiteres zu einer Erkennung von gewöhnlichen Strassen verwendet werden.

3   Von Strassenstücken zu Strassennetzen

Der oben erwähnten Möglichkeit, dass unter gewissen Bedingungen Fehler in der Interpretation auftreten, wird durch die Zuerkennung sogenannter Vertrauensfaktoren (certainty factors) an die erkannten Strassenstücke Rechnung getragen. Oft ist es aus einer zu wenig globalen Sicht unmöglich zu beurteilen, ob es sich bei einer durch Parallelen begrenzten Fläche nun um ein durch zwei (zufälligerweise parallel verlaufende) Strassen eingeschlossenes Gebiet (in Fig. 1b unten rechts), oder um die Strassenfläche selbst handelt. Ein anderes Problem bilden binäre Entscheidungen, ob aufgrund eines Schwellwerts eine Linie als Element einer höheren Struktur berücksichtigt oder zurückgewiesen werden soll. Beispiele dafür sind minimale Länge, minimaler paralleler Anteil einer Linie oder maximal zulässige Varianz der Distanzwerte. Damit die Möglichkeit besteht, auf diese Entscheidungen zurückzukommen, müssen auch in diesen Fällen Vertrauensfaktoren eingeführt werden, die dann beim Zusammenbau des Strassennetzes eine wichtige Rolle spielen.

Vertrauensfaktoren können z. B. abhängen von der Länge der erkannten Entität (je länger, desto sicherer) oder der Lage im Bild (Randlage ungünstig). Wichtig ist, dass sie auch nachträglich nach unten oder oben korrigiert werden können, je nach Kompatibilität mit in der Nähe gelegenen Entitäten. Die Handhabung unsicheren Wissens ('probabilistic reasoning') erfolgt z.B. mit den Methoden von Dempster-Shafer [6].

Es liegen nun also eine Menge von Hypothesen über Strassenstücke verschiedener Typen mit bekannten Positionen (Anfangspunkt, Endpunkt, Verlauf) und gewissen Vertrauensfaktoren vor. Bis dahin wurden alle Interpretationen isoliert voneinander, lediglich auf Zeichenregeln basierend, gemacht. Gemäss einem Strassennetz-Modell mit Strassenknoten und Strassenverbindungen (wir haben es wieder mit einem Graphen zu tun, die Elemente haben jedoch mehr Semantik) müssen sich elementarere Strassenstücke in einem globalen Kontext widerspruchsfrei verknüpfen lassen. Beim diesem Versuch ergeben sich Komplikationen u.a. an den Verbindungsstellen verschiedener Strassentypen sowie bei Lücken, die durch Eisenbahnlinien, Über- und Unterführungen, Tunnels, etc. verursacht werden. Die Behandlung dieser Sonderfälle erfordert die Formulierung einer Menge von speziellen Regeln. Dies ist nicht trivial, da gerade in der Nähe der Vereinigung von Strassenstücken dieselben mehr oder weniger deformiert sind und also ihre Identität nicht bis zur tatsächlichen 'Kreuzung' nachgewiesen werden kann. Hier muss Wissen über die Typen von Strassen, die zusammengeführt werden dürfen (Konsistenz), Zeichenregeln über die Ausgestaltung der Vereinigungszone etc. eingesetzt werden. Andere Regeln betreffen den Typ anderer Strassen, die ein Fussweg treffen, bzw. fortsetzen darf, wie eine Strasse allenfalls abrupt enden kann, etc.. Die Hypothesen werden also verifiziert und man versucht, die unabhängig voneinander erhaltenen lokalen Interpretationen (nur Zeichenregeln) konsistent zu machen. Das heisst, man lokalisiert Fehler oder Lücken in der Deutung und verwirft entweder eine Hypothese ganz oder gibt sie mit entsprechender Zusatz-Information zu einer Neubeurteilung an einen Prozess auf unterer Stufe zurück. Aufgrund der Existenz einer bestimmten Entität kann unter Umständen auch die Existenz einer andern Entität in der Nachbarschaft vorausgesagt werden. Dies veranlasst dann die Ablaufstrategie in der 'richtigen Region' (focus of interest) zu suchen.

Fig. 1b zeigt die vorläufige Erkennung von Stücken eines bestimmten strukturierten Linientyps wie sie in einem repräsentativen Kartenausschnitt einer Agglomeration (Fig. 1a) vorkommen. Fig. 1c zeigt die bereinigte Version von 1b, in dem alle parallelen Liniensegmente, deren Varianzwerte bezüglich Distanz einen Schwellwert überschreiten, eliminiert wurden. Die fiktiven Mittelachsen der strukturierten Linien (und somit der Strassenstücke) sind dicker dargestellt. Man erkennt leicht eine mehr oder weniger diagonal verlaufende Autobahn, die unten rechts in eine andere mündet. Daneben sind Segmente von parallelen Linien mit grossem, mittlerem und kleinem Abstand erkennbar, die für Strassen verschiedener Ordnung stehen, sowie von parallel-strichlierten Linien für Fahrwege.

Eine Strasse kann i.a. nicht einfach abrupt aufhören, sondern hat in der Nachbarschaft wahrscheinlich eine Fortsetzung durch ein gleichartiges Stück. Vielfach handelt es sich bei einem solchen Fall um einen durch eine Eisenbahn- oder Strassenüberführung hervorgerufenen Unterbruch. Diesen überbrückt man für die Beschreibung (Fig. 1d) und merkt sich dabei den Ort und die Art des Unterbruchs. Zwei Strassenstücke, die verbunden werden sollen, müssen folgende Bedingungen erfüllen: Erstens müssen sie gleichartig sein, zweitens die benachbarten Endpunkte genügend nahe beieinander haben, drittens in diesen Endpunkten Kollinearität aufweisen. Dazu gehört auch, Strassenstücke so zu verlängern (allenfalls auch aufzuteilen), dass sie von Strassenknoten zu Strassenknoten führen. Bei Strassenkreuzungen treffen sich die Mittelachsen nicht in einem einzigen Punkt. Bei der Skelettierung wird nämlich ein 4er-Knoten (4 inzidierende Linien) in zwei nahe beieinanderliegende 3er-Knoten aufgespaltet. Diese Konstellation lässt sich leicht erkennen, und mit entsprechenden korrigierenden Massnahmen zu einem einzigen 4er-Knoten reduzieren.

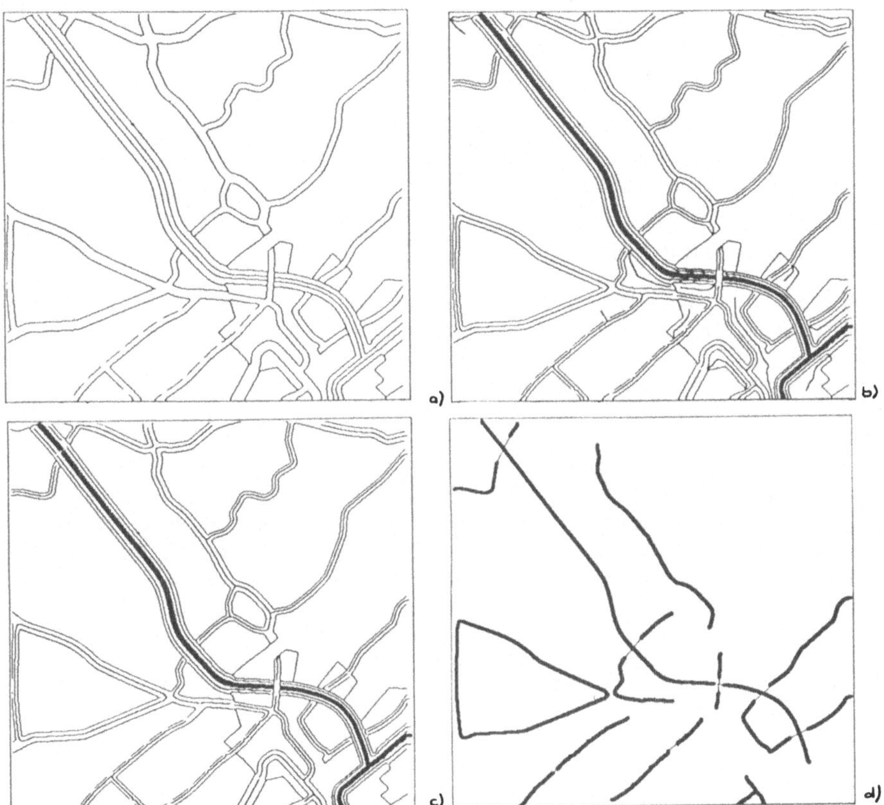

Figur 1: a) Ausschnitt aus der Schicht 'Situation', b) erkannte Strassentypen, c) bereinigtes Bild von b), d) durch Geradenstücke überbrückte benachbarte kollineare Liniensegmente

4 Implementation

Für die Low-Level-Bildverarbeitung ist in unserer Fachgruppe ein umfangreiches Software-System in Fortran und Pascal entwickelt worden, mit dem sich die Bildprimitive extrahieren lassen. Zur Zeit werden ergänzende und neue Routinen im Zusammenhang mit der Umstellung auf UNIX in C oder C++ geschrieben. Die ersten Gehversuche bis hin zu einem ersten Prototypen zur Erkennung von strukturierten Linien in einem Schema einer industriellen Anlage wurden mit PROLOG durchgeführt [4], u. a. weil sich die Prädikatenlogik für die Formulierung der Produktionsregeln gut verwenden lässt. Bildprimitiven wie Linien, Punkte und Knoten sowohl vom Vorder- wie auch vom Hintergrund wurden als PROLOG-Fakten abgelegt, die Zeichenregeln als PROLOG-Regeln. Ein Problem hat sich durch die hohe Belegung des Speicherplatzes innerhalb des PROLOG-Systems ergeben. Beim Backtracking werden nämlich die Fakten sequentiell durchsucht, was sehr zeitaufwendig werden kann.

Neuerdings erfolgt die Implementation in Common Lisp unter Berücksichtigung des objekt-orientierten Programmier-Paradigmas. Die Vorteile sind offensichtlich, insbesondere auch für Anwendungen in der Computer Vision: Es sind dies Modularität der einzelnen Programmteile (z.B. liessen sich die Intermediate-Level-Datenstrukturen zu einem späteren Zeitpunkt ohne weiteres dem Problem besser anpassen), Erweiterbarkeit (während der Entwicklung wurde es als sinnvoll erachtet, weitere Abstraktionsebenen oder Klassen von Objekten einzuführen), sowie Wiederverwendbarkeit. Es wurde versucht, einen Satz von generischen Operatoren bereitzustellen, mit denen Eigenschaften auf den punkt- und linienförmigen Entitäten oder Beziehungen zwischen diesen bestimmt werden können: z.B. die Tangentialrichtung an den Linienenden, ebenso wie Kollinearität, Parallelität, Anti-Parallelität, Orthogonalität, etc. zweier Linien, oder Inzidenz, Nähe, relative Position zweier Entitäten. Zu jedem Operator gibt es auch ein entsprechendes Prädikat, welches bestimmt, ob die angegeben Entitäten gewisse Eigenschaften haben oder nicht.

Die Regeln, die den Zusammenbau des Strassennetzes gemäss einem Modell betreffen, werden mit einer regelbasierten 'expert system shell' implementiert, welche sowohl Vorwärts- wie Rückwärtsableitungen unterstützt. Dies ergibt die benötigte Flexibilität, indem das in den Regeln vorhandene Wissen unabhängig vom verwendeten Inferenzmechanismus dargestellt werden kann. Die oben erwähnte Menge von Operatoren und Prädikaten lässt sich leicht in die Regeln einbinden.

5 Schluss und Ausblick

Das Verstehen graphischer Dokumente kann als ein eingeschränkter Spezialfall des Verstehens allgemeiner Szenen gesehen werden. Trotz dieser Vereinfachung weist ein System zum Verstehen graphischer Dokumente alle wichtigen Komponenten und Stufen eines allgemeinen Computer-Vision-Systems in nicht-trivialer Form auf und dient daher in idealer Weise als Vorstufe zum viel komplexeren Problem des Verstehens von allgemeinen Szenen.

Zum Überprüfen des generischen Charakters der Methoden haben wir bereits erste Versuche gemacht, anstelle der Strassenkarten Grundbuchpläne zu interpretieren. Im Unterschied zu Strassenkarten gibt es gerade Liniensegmente und eine Menge charakteristischer Strukturbegrenzungselemente wie Kreise, T-Junctions u.a.m.. Letztere eignen sich sehr gut als Ausgangspunkt für die Synthese. Ebenfalls soll geprüft werden, ob die aus Grauwertbildern extrahierten Liniensegmente mit Hilfe der allgemeinen Operatoren und Prädikate relativ einfach und schnell, zusammen mit dem Wissen über das Vorhandensein und die Art von linearen Strukturen, interpretiert werden können.

Die Handhabung von unsicherem, unvollständigem und unscharfem Wissen hat einen hohen Stellenwert. Jedenfalls wollen wir diesem interessanten Forschungsgebiet nach ersten eigenen Erfahrungen vermehrte Aufmerksamkeit schenken. Ebenso erscheint der Einbezug mehrerer Wissensquellen (multiple knowledge source fusion) von Wichtigkeit.

Literatur

1. Ade, F., Gerig, G., Ilg, M., Klein, F., Verstehen von Landkarten, Proceedings 8. DAGM Symposium, Paderborn 1986, pp. 1–11.
2. Ade, F., Ilg, M., Verstehen grafischer Dokumente, Techn. Rundschau TR 37, Sept. 1988, pp. 104–113.
3. Hanson, A.R., and Riseman, E.M., The VISIONS Image-Understanding System, from: Advances in Computer Vision, Vol. 1 (Ed. C. Brown), Lawrence Erlbaum Associates, Publishers, 1988, pp. 1–114.
4. Egeli, E., Modellgesteuerte Analyse komplexer Strukturen in Graphiken, Diss. ETH, Nr. 7987, 1986.
5. Ogniewicz, R., Kübler, O., Klein, F., und Kienholz, U., Lage- und skalierungsinvariante euklidische Skelette zur robusten Beschreibung und Erkennung binärer Formen, dieser Tagungsband.
6. Shafer, G., A Mathematical Theory of Evidence, Princeton, 1976.

Modellbasierte Objekterkennung aus Entfernungswerten
eines Laser-Radar-Systems

Thomas Knieriemen, Ulrich Krüll, Ewald von Puttkamer

Universität Kaiserslauten, Fachbereich Informatik

Tel.: 0631/205-2656   Telefax: 0631/205-3200

Zusammenfassung: Dieser Beitrag beschreibt ein wissensgesteuertes Verfahren mit dem typische Objekte einer Gebäudeumgebung wie Wände, Türen, Stühle usw. erkannt werden. Die ursprüngliche Sensorinformation wird von einem Laser-Radar-System geliefert. Dieses System ist Teil der Sensorik eines autonomen mobilen Roboters, der für den Einsatz innerhalb von Gebäuden konzipiert wurde. Aus den Rohdaten des Sensorsystems wird in einem ersten Abstraktionsschritt ein geometrisches Modell generiert. Die dabei erzeugten Karten bestehen aus mehreren Schichten, die je einen projizierten Aufnahmeschnitt der Umgebung darstellen. Die für die Identifikation wichtigen Daten der potentiellen Objekttypen werden in einer Wissensbasis bereitgestellt.

1. Einleitung

Ein Fahrzeug, das sich selbständig innerhalb einer zunächst unbekannten Umgebung frei bewegen kann, wird als autonomer mobiler Roboter (AMR) bezeichnet. Neben vielen anderen Komponenten wie Sensorik, Motorsteuerung oder Bahnregelung ist hierzu das Weiterverarbeiten von Sensordaten zu Umgebungskarten notwendig. In diesen Karten wird das Wissen des Roboters über seine Umgebung gespeichert und somit die Grundlage für Planungsaufgaben bereitgestellt.

Grundlage der globalen Wegplanung ist eine topologische Beschreibung der Umgebung, die wiedergibt, welche Plätze oder Räume durch welche Wege miteinander verbunden sind. Diese kann beispielsweise dadurch generiert werden, daß eine geometrische Freiraumkarte in einen Wegegraphen transformiert wird. Die Freiraumplätze werden in diesem Fall durch konvexe Polygone dargestellt und entsprechen den Knoten des Wegegraphen, während die Kanten eine Verbindung zwischen diesen repräsentieren /1/. Im Hinblick auf Aktionsplanungen in realer Umwelt bringt jedoch eine topologische Karte, mit der die tatsächliche Umgebungsstruktur adäquat wiedergeben wird, einige Vorteile. Die Knoten des entsprechenden Graphen sollen den Objekten und Raumbegrenzungen (Wände, Türen etc.) entsprechen, die Kanten die topologischen Beziehungen zwischen den Objekten wiedergeben /2/.

Um eine topologische Karte dieser Art zu generieren muß eine Objektidentifikation durchgeführt werden. Ein Ansatz zur Lösung dieses Problems unter den gegebenen Randbedingungen eines konkreten Projekts (MO-BOT-III, vgl. /3/) wird in diesem Beitrag vorgestellt.

2. Geometrische Modellbildung

Die Sensordatenverarbeitung zur Modellierung und Generierung von Umgebungskarten basiert im wesentlichen auf einem Laser-Radar-System /4/. Dieses System wurde so auf dem mobilen Roboter montiert, daß eine Rundum-Aufnahme der Umgebung möglich ist. Dazu wurden Laserentfernungsmesser (Meßrate: 720 Messungen pro Sekunde) auf 4 vertikalen Schwenkarmen montiert, die um +/-10 Grad geneigt werden können. Aus den Entfernungsmeßwerten und weiteren Parame-

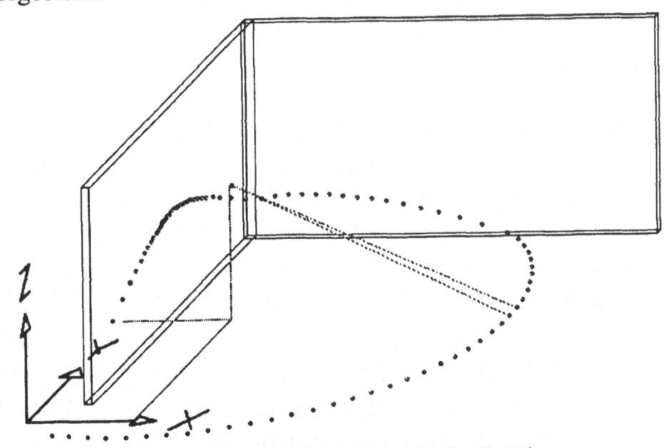

Grafik 1: Ergebnis einer Laser-Radar-Rotation

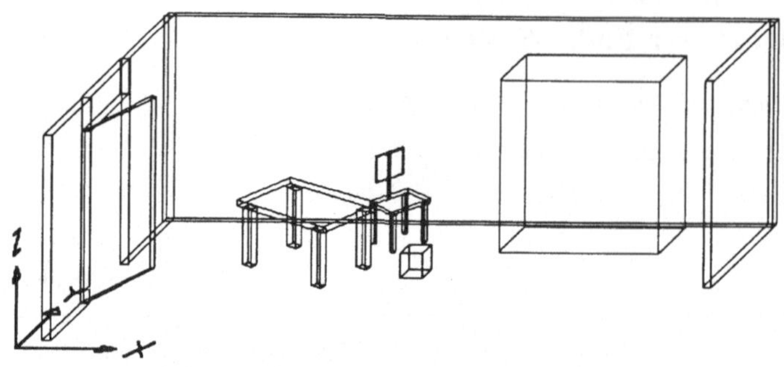

Grafik 2: Ausschnitt einer typischen Gebäudeumgebung

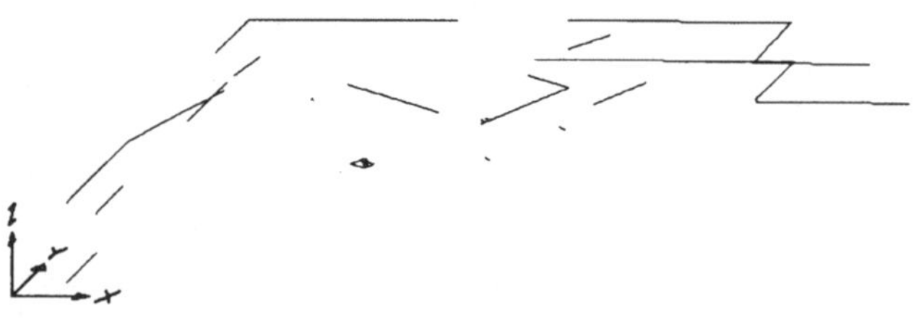

Grafik 3: Ergebnis der geometrischen Kartographie

tern können 3D-Meßpunkte berechnet werden. Grafik 1 zeigt das Ergebnis einer simulierten Laser-Radar-Rotation in einer einfachen Umgebung mit zwei Wänden.

Als Kompromiß zwischen der Forderung nach minimaler Rechenzeit einerseits und maximaler Information andererseits werden die 3D-Meßpunkte zu Linien in einer 2 1/2-dimensionalen Darstellung abstrahiert. Es werden zunächst Höhenintervalle definiert, die die Umgebung in dreidimensionale Schichten aufteilen. Innerhalb dieser Schichten werden alle Punkte projiziert und zweidimensionale Linien generiert. Grafik 2 zeigt den modellierten Ausschnitt einer Gebäudeumgebung mit typischen Objekten. Für diese Umgebung liefert die geometrische Kartographie eine Schichten-Karte (Grafik 3), in der mehrere Radaraufnahmen fusioniert sind. Diese Schichtenkarte ist relativ einfach zu verarbeiten und enthält zumindest einen Teil der dreidimensionalen Information, die durch die Sensoren erfaßt wurde. Natürlich hat diese Abstraktion bezüglich der Höheninformation Auswirkungen auf die Weiterverarbeitung der Daten zu topologischen Karten. Die Anzahl der Ebenen sollte so gewählt werden, daß sie der Raum- und Objektidentifikation dient. Von der Anzahl der Höhenintervalle ist aber auch der Kartierungsaufwand abhängig. Es ist daher wichtig, daß Anzahl und Lage der Intervalle variabel bleibt.

3. Objektklassifikation

Es erscheint zunächst schwierig, aus den zur Verfügung gestellten Daten Objekte zu erkennen, weil Art und Umfang der Daten erstens stark von der Position des Roboters abhängen und zweitens bedingt durch die Laserkonfiguration im Normalfall unvollständig sind. Andererseits ist es für einen mit der Materie Vertrauten kein Problem, aus den Linienzeichnungen der geometrischen Karten Objekte und Raumbegrenzungen zu erkennen. Man stützt sich dabei auf ein Grundwissen über die

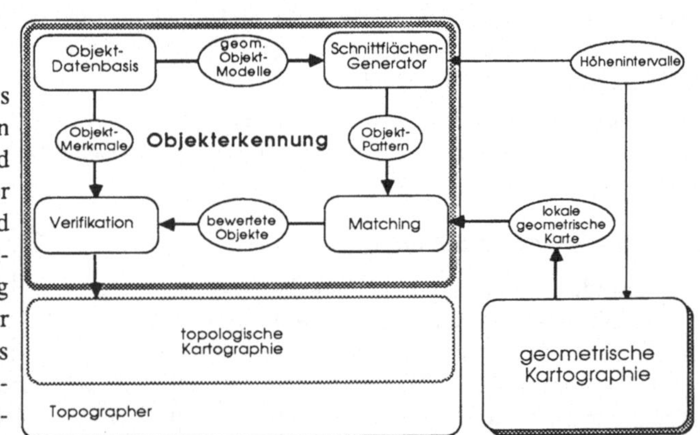

Grafik 4: Struktur und Komponenten der Objektklassifikation

Objekte (Form, Abmessungen etc.) und vergleicht dieses mit den Merkmalen der kartierten Linien (Länge, Lage etc.). Dieser Entscheidungsvorgang soll bei der Objekterkennung nachgebildet werden.

In Grafik 4 sind die grundlegenden Komponenten der Objekterkennung dargestellt. Es existiert eine Objekt-Datenbasis, in der dreidimensionale geometrische Modelle als Prototypen der Objektklassen und weitere Objektmerkmale gespeichert sind. Aus den geometrischen Modellen werden durch den Schnittflächengenerator

Objektpattern generiert. Diese Pattern sind Projektionen der Objektmodelle innerhalb der einzelnen Höhenintervalle. Die aus den geometrischen Objektmodellen extrahierten Pattern werden in der Matching-Komponente gegen die Daten der geometrischen Karte gematcht und für jedes Objektmodell bewertet. Abschließend werden in der Verifikationskomponente weitere Objektmerkmale verglichen und die Bewertung evtl. angepaßt. Das am besten bewertete Objekt wird in die topologische Karte eingetragen. Im folgenden werden die einzelnen Komponenten der Objektidentifikation näher betrachtet.

Objektdatenbasis: Für die Konzeption der Objektdatenbasis sind folgende zwei Fragen von zentraler Bedeutung: welches Wissen über die Einsatzumgebung muß gespeichert werden und wie wird diese Information intern repräsentiert? Die Antwort auf die erste Frage wird sehr stark durch die gewählte Sensorkonfiguration beeinflußt. Nimmt man beispielsweise das entfernungsmessende Sensorsystem von Mobot-III, so wären Attribute wie z.B. die Farbe eines Objekts nicht zu gebrauchen. Die folgende Liste gibt einen Eindruck über die gespeicherte Information, die zur Objekt- und Raumklassifikation im Projekt Mobot-III genutzt wird:
- ein Raum wird begrenzt durch eine geschlossene Kette von Wänden und Türen, enthält eine Menge von Objekten, hat Verbindungen zu anderen Räumen (Türen, Durchgänge,...) usw.;
- eine Wand hat die Form eines Quaders mit einer geringen Tiefe und einer Mindesthöhe von 200 cm; i.a. ist eine Wand unbeweglich und es existieren Verbindungen zu anderen Wänden / Türen;
- eine Raumtür ist eine spezielle Ausprägung einer Wand, die jedoch in einem definierten Freiheitsgrad beweglich und immer mit einer Wand verbunden ist; die Breite beträgt maximal 120 cm;
- ein Möbelstück läßt sich durch eine Menge von Quadern abstrahieren, wobei die Konstellation der einzelnen Quader genau festgelegt ist; es befindet sich innerhalb eines Raums, hat evtl. Kontakt zu anderen Objekten und ist beweglich in einem definierten Freiheitsgrad;
- ein (typischer) Stuhl ist eine Ausprägung eines Möbelstücks; besteht aus einer Sitzfläche, einer Rückenlehne und i.a. aus vier Stuhlbeinen; ist leicht beweglich; Abmessungen ca. 45 * 45 *100 cm.

Die aufgelisteten Attribute stellen teilweise Information über Objektklassen (z.B. ein Stuhl ist beweglich) und teilweise Information über bestimmte Instanzen dieser Klassen (z.B. in welchem Raum sich ein bestimmter Stuhl befindet) dar.

Für diese verbale Beschreibung muß eine geeignete Repräsentation gefunden und damit die zweite Frage beantwortet werden. Die gewählte Wissensrepräsentation sollte einerseits flexibel genug sein, um das ambivalente Wissen kodieren zu können und andererseits so strukturiert sein, daß sämtliche Informationen leicht zugänglich sind. Bei Expertensystemen, die ähnliche Anforderungen an die Wissensrepräsentation stellen, hat sich das Konzept der "Frames" als geeignet erwiesen. Mit der von M. Minsky /5/ entwickelten "Frame Theorie" wird ansatzweise versucht, eine deskriptive Datenbasis zu erzeugen, die Wissen über die reale Welt in einer strukturierten aber flexiblen Weise kodiert. Ein Ansatz dieser Art war deshalb zur Wissensrepräsentation naheliegend. Dabei wurden folgende Bestandteile des Framekonzepts zur Beschreibung der Objektdatenbasis integriert:
- jede Objektklasse wird durch einen examplarisches Frame beschrieben
- in den Slots können bestimmte Defaultwerte (ein Tisch ist etwa 120 cm breit) oder auch zwingend vorgeschriebene Werte (eine Wand ist unbeweglich) eingetragen werden
- über Constraints lassen sich Forderungen bezüglich der räumlichen Konstellation formulieren, wie z.B. ´eine Tür muß immer mit einer Wand verbunden sein
- eine Instanz einer Objektklasse erfüllt die in einem Frame formulierten Constraints und Slotwerte in einem Grade, der über einem bestimmten Schwellwert liegen muß

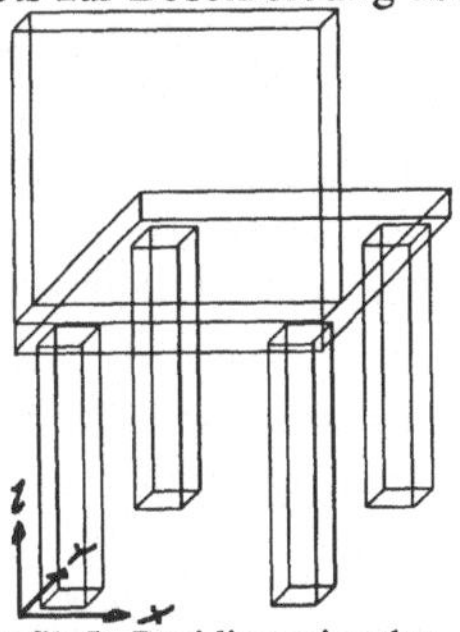

Grafik 5: Dreidimensionales Modell eines Stuhles

Jedes Frame hat einen spezifischen Slot (3D-Data), in dem eine dreidimensionale Darstellung des beschriebenen Objekts gespeichert ist. Damit wird eine Abstraktion für eine Menge von konkreten Objekten dargestellt. Jedes Modell setzt sich dabei aus einer Menge von Quadern zusammen, die jeweils durch 6 Attribute spezifiziert werden: die Abmessungen des Quaders in den 3 Dimensionen [sc_x, sc_y, sc_z] und die relative Lage zum Nullpunkt des Objektkoordinatensystems in den 3

Dimensionen [tr_x, tr_y, tr_z]. Denkbar als Erweiterung sind noch Rotationswinkel um die drei Achsen des Koordinatensystems sowie andere Volumenkörper als Grundkörper (z.B. Kugeln, Zylinder, Kegel, etc.).

Ein Beispiel für das Objektmodell eines Stuhls ist in Grafik 5 skizziert. Kein realer Stuhl wird genau dem in Grafik 5 dargestellten Prototyp entsprechen, aber auch ein Schreibtischstuhl mit nur einem Bein und Rollen oder ein Stuhl mit Armlehnen hätte zumindest die Sitzfläche und Rückenlehne mit diesem gemeinsam und sollte auf dieser Basis als Stuhl erkannt werden.

Schnittflächengenerator: Die Daten aus den geometrischen Karten (vgl. Grafik 3) sind nicht direkt mit den dreidimensionalen Objektmodellen vergleichbar. Deshalb müssen die 3D-Modelle zunächst in eine den Daten entsprechende Form gebracht werden. Dazu werden aus den oben beschriebenen geometrischen Modellen entsprechende Schablonen (im folgenden Pattern genannt) extrahiert. Diese Pattern sind die Projektionen des Modells innerhalb der einzelnen Höhenschichten (vgl. Grafik 6) analog zur Projektion der dreidimensionalen Sensordaten in die 2 1/2-dimensionale Darstellung. Jedes Pattern wird definiert durch die Abmessungen des Rechtecks [sc_x, sc_y], durch die relative Lage zum Nullpunkt des Objektkoordinatensystems [tr_x, tr_y] und durch die Angabe der jeweiligen Höhenschicht [level].

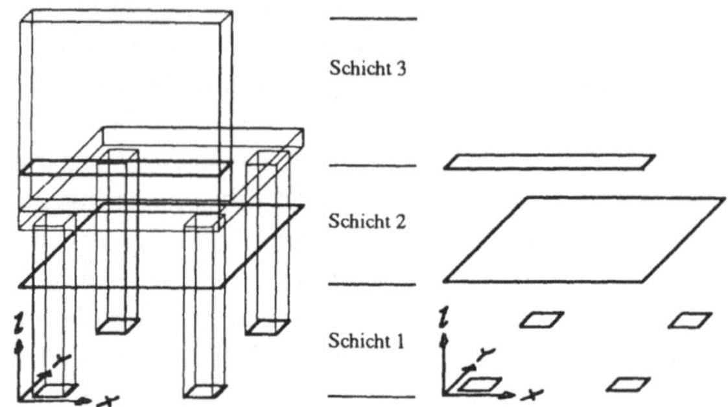

Grafik 6: Schnittebenen und Pattern eines Stuhls

Nachdem die Objektmodelle zu einer Menge von Pattern modifiziert worden sind, werden sie ebenfalls in der Datenbasis gespeichert. Zwar wäre es einfacher direkt die Suchpattern anstatt der dreidimensionalen Modelle zu spezifizieren, doch in diesem Fall müßte man sich frühzeitig (schon bei der Eingabe der Datenbasis) auf bestimmte Höhenintervalle festlegen. In der gewählten Form kann eine Datenbasis unabhängig von den durch die Radarkarten spezifizierten Intervallen verwendet werden. Es ist also möglich, mit der Lage und Anzahl der Intervalle zu experimentieren, um eine für die geometrische Kartographie (bzgl. Zeit und Speicherbedarf) und für die Objektidentifikation gleichermaßen günstige Konstellation zu finden.

Matching: Nachdem die Abstraktionsstufe der Datenbasis in etwa der Information in den geometrischen Karten entspricht, können die Pattern der Objektprototypen gegen die Segmente der geom. Karte gematcht werden. Jedes Segment wurde aus einer Menge von Meßpunkten in einer bestimmten Höhenschicht generiert, während jedes Pattern eine bestimmte Region eingrenzt, in der nach Meßpunkten gesucht werden muß. Der Matching-Prozess wird durch eine Reihe von Ungenauigkeiten beeinflußt:

- <u>Meßwerte:</u> das Objekt wird meist nicht von allen Seiten durch die Sensoren erfasst; das Objekt wird oft nicht in allen Höhenschichten erfasst; die Sensoren arbeiten mit Meßfehlern.

- <u>Kartographie:</u> Beim Zusammenfassen der Meßpunkte können irreale Segmente generiert werden.

- <u>Wissensbasis:</u> das in der Datenbasis gespeicherte Modell ist lediglich eine Abstraktion des realen Objekts und kann damit nicht exakt den Sensordaten entsprechen;

Diese Fehlermöglichkeiten werden im Matching-Prozess durch entsprechendeToleranzen berücksichtigt. Der eigentlichen Match-Algorithmus muß sämtliche Segmente der geometrischen Karte bearbeiten, d.h. jedem Segment muß ein entsprechendes Pattern und damit ein bestimmtes Objekt zugeordnet werden. Der Algorithmus besteht im Prinzip aus vier Schritten, die solange wiederholt werden, bis sämtliche Segmente der geometrischen Karte gematcht sind:

1.: Bestimme aus der geometrischen Karte ein geeignetes Startsegment;

2.: Speichere sämtliche Objektmodelle, die auf dieses Startsegment passen, in einer Agenda.

3.: Bestimme für jeden Agendaeintrag sämtliche Kartensegmente, die zu den Pattern des Objektmodells passen, und bewerte das jeweilige Ergebnis.

4.: Beende den Prozeß falls sämtliche Kartensegmente gematcht sind.

Der momentan implementierte Algorithmus wurde u.a. dadurch optimiert, daß Segmente und Pattern in ein Einheitskoordinatensystem transformiert werden und damit die Berechnung von Schnittpunkten zwischen Segment und Patternkante auf den viel einfacheren Test, ob die Segmentkoordinaten zwischen 0 und 1 liegen, reduziert wird.

Verifikation: Nach dem Matching-Prozeß enthält die Agenda mehrere Listen von Objektinstanzen, die jeweils von einem bestimmten Startsegment initiiert wurden. Für jede Instanz ist dabei festgehalten, welche Liniensegmente gematcht und wieviel Punkte insgesamt erreicht wurden. Unter diesen Instanzen muß nun die wahrscheinlichste ausgewählt und in die topologische Karte eingetragen werden.

Dazu werden für jede Instanz die entsprechenden Objektmerkmale, die als Constraints in der Objektdatenbasis gespeichert sind, überprüft. Beispielsweise werden damit Minimal/Maximal-Abmessungen und vorgeschriebene bzw. mögliche Relationen verifiziert. Die jeweilige Punktzahl wird entsprechend angepaßt und evtl. kann eine Instanz sogar aus der Agenda gelöscht werden (eine Tür z.B. kann niemals frei im Raum stehen, sondern muß immer an einer Wand aufgehängt sein).

Nachdem alle Spezifikationen überprüft und die Punkte der potentiellen Instanzen entsprechend korrigiert sind, wird diejenige Instanz in die topologische Karte übernommen, die am höchsten bewertet ist.

4. Bemerkungen

Der in diesem Beitrag beschriebene Ansatz zur Objekterkennung, wurde mit einer prozeduralen Programmiersprache implementiert und in einer Simulationsumgebung getestet. Grafik 7 zeigt das Ergebnis des Verfahrens für die in Grafik 2 dargestellte Umgebung. Bedingt durch die spezielle Konzeption der Sensorik stützt sich die Objekterkennung fast ausschließlich auf geometrische Daten, die von entfernungsmessenden Sensoren produziert wurden. Dabei sind die Sensordaten bzgl. ihrer Quantität stark eingeschränkt und von der aktuellen Aufnahmeposition abhängig. Im Vergleich zur Bildverarbeitung werden die Entfenungsdaten direkt von der Sensorik geliefert und damit eine Verarbeitung in Echtzeit unterstützt.

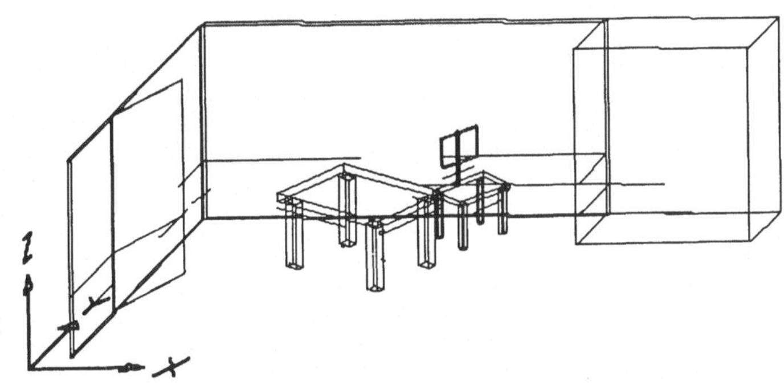

Grafik 7: Ergebnis dargestellt als Umgebungsszene

Die in der Datenbasis gespeicherten Modelle der Objekte stellen eine stark auf die Problemstellung hin optimierte Abstraktion dar. Eine mehr funktionale Beschreibung der Objekte (z.B. ein Stuhl ist ein Objekt, das man zum Sitzen benutzen kann), die mehr der menschlichen Sicht entspricht, wäre unter den gegebenen Randbedingungen nicht geeignet die Identifikation zu unterstützen.

Literaturverzeichnis

/1/ R. Chatila, J.-P. Laumond: Position Referencing and Consistent World Modeling for Mobile Robots; Intern. Conference on Robotics and Automation; St. Louis 1985
/2/ D. Marr, H.K. Nishihara: Representation and Recognition of the Spatial Organization of three-dimensional shapes; Proc. of the Royal Society B-200; 1978
/3/ R. Hinkel, T. Knieriemen, E.v. Puttkamer: MOBOT-III an Autonomous Mobile Robot for Indoor Applications; 19th Intern. Symposium and Exposition on Robots; Sydney 1988
/4/ R. Hinkel, T. Knieriemen: Environment Perception with a Laser Radar in a Fast Moving Robot; Symposium on Robot Control; Karlsruhe 1988
/5/ M. Minsky: A Framework for Representing Knowledge; in the Psychology of Computer Vision; ed. P.H. Winston, New York 1975

Effiziente Verfahrensentwicklung für die Bildauswertung durch objektorientierten, adaptierbaren Dialog und videoschnelle Verarbeitung im Funktionsverbund KIBAS - VISTA

Rainer Schönbein

Fraunhofer-Institut für Informations- und Datenverarbeitung (IITB)

Fraunhoferstraße 1, D-7500 Karlsruhe

Zusammenfassung

Das Bildauswertesystem KIBAS (konfigurierbares interaktives Bildauswertesystem) verfügt über eine grafische, objektorientierte Dialogkomponente. Bildschirmeinteilung und Systemverhalten sind vom Benutzer über ein Profil einstellbar. Die Verwendung einer vertrauten Begriffswelt und der durchgängige Gebrauch von leichtverständlichen Interaktionskonzepten ermöglichen eine einfache, komfortable Benutzung des Systems. Das System ist als Experimentalsystem zur Untersuchung des Benutzerverhaltens und der wechselnden Benutzeranforderungen über einen längeren Benutzungszeitraum hinweg konzipiert. Die Wahl unterschiedlicher Eingabemedien und -formen sowie die Veränderung der Anforderungen an die Antwortgestaltung des Systems sind dabei die Untersuchungsthemen. KIBAS verwendet einen UNIX-Arbeitsplatzrechner (Apollo); die Bildverarbeitungsalgorithmen stehen in Form von PASCAL-Programmen zur Verfügung. VISTA (Visuelles Interpretations-System für technische Anwendungen) ist ein spezieller Bildauswertungsrechner, der zur Lösung von Aufgaben aus den Bereichen Güteprüfung, Objekterkennung, Positionsbestimmung, Prozeßüberwachung und Prozeßsteuerung entwickelt wurde. VISTA ist ein busorientiertes Multiprozessorsystem. Es besteht aus vier Hauptkomponenten: der Video-Ein-/Ausgabe-Einheit für den Bildeinzug und die Bildwiedergabe, der ikonischen Verarbeitungsstufe zur Vorverarbeitung und Merkmalextraktion, der symbolorientierten Verarbeitungsstufe zur Umwandlung der Merkmale in Aussagen sowie der Steuerung zur Koordination aller ablaufenden Vorgänge. Die wichtigsten Verbindungen zwischen den Baugruppen des Systems werden durch den VMEbus und den Videobus geschaffen. Die Bildverarbeitungsalgorithmen wurden in Form spezieller Schaltungen realisiert. VISTA ist ein modulares System und deshalb ausbau- und anpassungsfähig an Aufgaben wachsenden Schwierigkeitsgrades. Der Funktionsverbund KIBAS - VISTA vereint die komfortable anpassungsfähige Benutzungsoberfläche des KIBAS mit der videoschnellen Bildverarbeitung des VISTA. Dieser Beitrag beschreibt den Funktionsverbund KIBAS - VISTA und stellt die Vorteile einer damit möglichen Verfahrensentwicklung für die Bildauswertung dar.

1.   Einleitung

Die Verfahrensentwicklung für die Bildauswertung ist gekennzeichnet einerseits durch die Anforderungen an die Programmentwicklung (Erstellung, Übersetzung, Test) mit den verschiedenen Möglichkeiten zur Bilddarstellung und -analyse und zum anderen der Forderung nach effizienten Verbindungsmöglichkeiten des zu erstellenden Teilschrittes mit vorgefertigten Verarbeitungsschritten. Wissenschaftler, Ingenieure verschiedener Fachrichtungen, Programmierer und technische Assistenten bilden den Personenkreis der Verfahrensentwickler, deren unterschiedliche Anforderungen von kommerziell verfügbaren Bildauswertesystemen nur unzureichend berücksichtigt werden (Faasch 87).

Der im folgenden vorgestellte Funktionsverbund KIBAS - VISTA besitzt eine grafische, objektorientierte Dialogkomponente, erlaubt die Berücksichtigung unterschiedlicher Anforderungen der verschiedenen Benutzerklassen (Schönbein & Roller 88) und ermöglicht damit eine komfortable, effiziente Benutzung des Systems. Bildschirmeinteilung und Systemverhalten können vom Benutzer über ein "Profil" beeinflußt werden. Als Interaktionselemente dienen ein Zeigeinstrument (Maus), Sprachein- und -ausgabesysteme, eine Tastatur und ein Farbmonitor zur Bilddarstellung und Dialogführung. Der Funktionsverbund KIBAS - VISTA unterstützt die Verfahrensentwicklung durch die Bereitstellung videoschnell ablauffähiger Verarbeitungsschritte (VISTA), verschiedener Verarbeitungs-, Darstellungs- und Dienstleistungsfunktionen (KIBAS), sowie durch eine Programmierumgebung, die die Integration neuer Verfahren in eine komfortable Benutzungsoberfläche mit geringem Programmieraufwand erlaubt. Dabei ermöglicht die adaptierbare, objektorientierte Benutzungsoberfläche des Systems eine Berücksichtigung der Benutzerkenntnisse und übernimmt einen Großteil der Eingabedatenprüfung.

2. KIBAS-Struktur und Ereignisverwaltung

Ein wesentlicher Aspekt der KIBAS-Struktur ist die Trennung zwischen Anwendungs- und Benutzerschnittstelle. Die Kommunikation zwischen Benutzer und Anwendungsroutinen wird durch das KIBAS-Basissystem gesteuert. Bei der Integration zusätzlicher KIBAS-Funktionen können die Bereiche Anwendungsprogrammierung, d.h. die Erstellung von Algorithmen zur Bildmanipulation, und Dialogprogrammierung, d.h. die Gestaltung der Bildschirmobjekte und der Kommunikation zwischen Benutzer und Anwendungsroutinen, unterschieden werden. Die Aktivierung von Anwendungsroutinen im KIBAS erfolgt ereignisgesteuert (Abb. 1). Die vom Benutzer initiierten Ereignisse werden vom Spracheingabeprozeß (SPE-

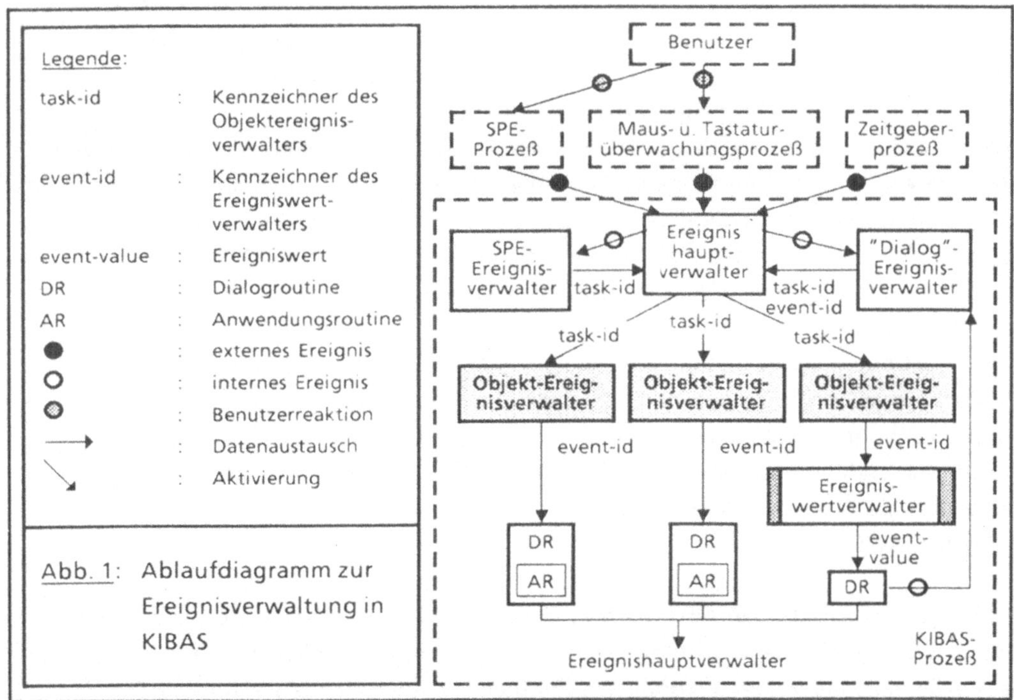

Abb. 1: Ablaufdiagramm zur Ereignisverwaltung in KIBAS

Prozeß) bzw. Maus- und Tastaturüberwachungsprozeß erfaßt und an den Ereignishauptverwalter weitergeleitet. Nach der Analyse der Ereignisse durch die "SPE-" und "Dialog-"Ereignisverwalter werden die entsprechenden Anwendungsroutinen über Objektereignisverwalter bzw. Ereigniswertverwalter aufgerufen. Der Aufruf der Anwendungsroutinen ist in Dialogroutinen eingebettet, die die Dialogkontrolle übernehmen. Anwendungsroutinen stehen als Programmcode auf der Arbeitsplatzstation oder auf dem angeschlossenen VISTA zur Verfügung. Die im nächsten Kapitel beschriebenen Interaktionskonzepte sind in den Dialogroutinen realisiert. Die Einbeziehung eines Zeitgeberprozesses erlaubt die zeitliche Modellierung des Dialoges.

3. Benutzungsoberfläche und Interaktionskonzepte

Die optische Darstellung des Systems und der Systemkomponenten sowie die Verwendung von allgemein bekannten "Metaphern" (z.B. Verkehrszeichen als Klassifikationssymbol) zur Unterstützung der Informationsdarstellung vermitteln dem Benutzer eine vertraute Begriffswelt. Dabei teilt KIBAS den Bildschirm in mehrere funktionale Bereiche ein (Abb. 2):

- Der Bildbereich (oben links) mit Verschiebebalken dient zur Anzeige der zu bearbeitenden Bilder. Der sichtbare Bildbereich stellt dabei stets nur einen Ausschnitt aus einem größeren internen Bildbereich dar. Mit Hilfe der Verschiebebalken können Teile von Bildern, die über den Bildbereich hinausragen, in den sichtbaren Bereich verschoben werden.

- Der Informationsbereich (unten links) stellt die Koordinaten und Größe des aktuellen Bildes (Quellbild für die nächste Bildmanipulationsfunktion) sowie die aktuelle Farbtabellenbelegung des Systems dar. Darunter befindet sich ein Meldungsbereich mit vorangestelltem Klassifikationssymbol zur Ausgabe von Fehlermeldungen, Handlungsanweisungen und vom Benutzer angeforderten Hilfemeldungen.

- Der Bildverwaltungsbereich (rechts) mit Verschiebebalken beinhaltet die "Bildalben" (visualisiert durch Albenetikette und darunter angegebenen Albennamen), in denen Bilder aus dem Bildbereich abgespeichert bzw. aus denen Bilder in den Bildbereich eingelesen werden. Des weiteren sind dort die in den Abbildungen 2 und 3 nicht dargestellten Verwaltungsfunktionen lokalisiert sowie die Funktionen zum Kamerabildeinzug und zur Bildausgabe auf einem Drucker.

- Die Funktionenleiste (zwischen Bildbereich und Bildverwaltungsbereich) enthält farblich unterschieden die Gruppen der Verarbeitungs-, der Darstellungs- und der Dienstleistungsfunktionen (s. Abschnitt 4).

- Die Kommandoeingabezeile (Abb. 3, unten) dient zur direkten Eingabe von Kommandos über die Tastatur und steht damit als Alternative zur Mausbedienung und Spracheingabe zur Verfügung.

Der Übergang von Abb. 2 zu Abb. 3 erfolgt durch die Ein- und Ausblendung von Objekten (Kommandoeingabezeile, Informationsbereich und Funktionenleiste) sowie durch die Auswahl der an den Mauszeiger gekoppelten "Einblend"-Funktionenleiste über das Profil.

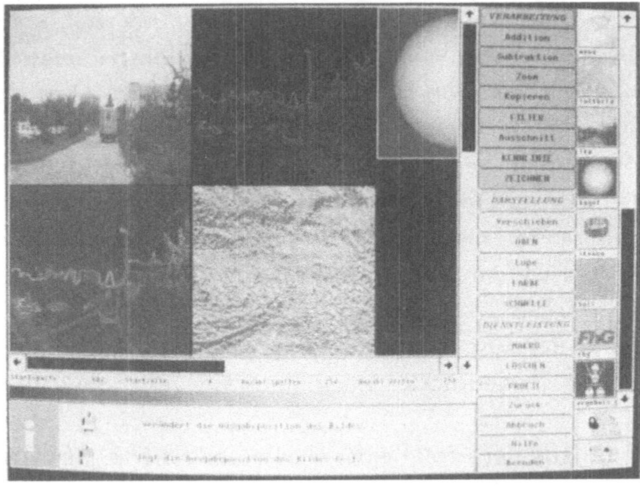

Abb. 2: KIBAS-Benutzungsoberfläche mit permanent
angezeigter Funktionenleiste und Informationsbereich

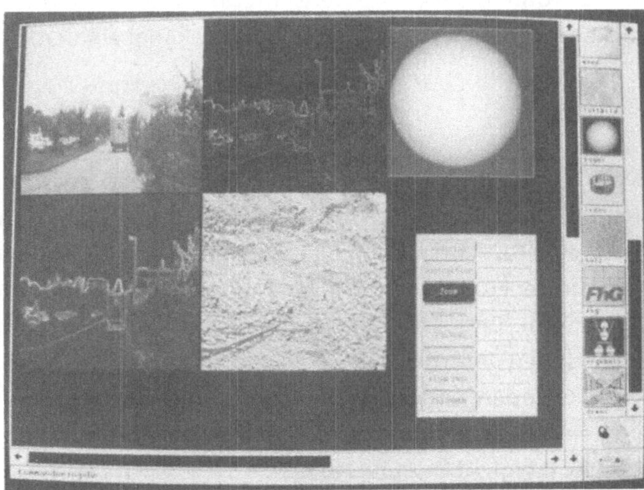

Abb. 3: KIBAS-Benutzungsoberfläche mit Kommandoeingabezeile und an den Mauszeiger gekoppel-
ter Einblend-Funktionenleiste

Die "direkte Manipulation" ermöglicht die unmittelbare Ansprechbarkeit von Objekten (Bilder und Funktionen) über Zeigeinstrumente, Funktionstasten oder Spracheingabe sowie die unmittelbare Visualisierung der Wirkung von Benutzeraktionen. Der "objektgebundene Kontext" erlaubt dem Benutzer die Unterbrechung und spätere Fortführung einer Aktion. Damit ist auch der Abbruch und die Zurücknahme einer Aktion möglich. Der Einsatz der Spracheingabe zur Manipulation von Objekten (Funktionsauswahl) und die akustische Rückmeldung über Sprachausgabe gestattet dem Benutzer die Verwendung der Interaktionsform "Zeigen und Sprechen". Die Kommunikation des Benutzers mit KIBAS basiert auf speziellen Interaktionskonzepten:

● Aktivierungskonzept. Durch Selektion eines Objektes mit der linken Maustaste werden Funktionen aufgerufen oder Metaobjekte (z.B. Bildalben) geöffnet.

- "Lokale Hilfe"-Konzept. Die Selektion eines Objektes mit der mittleren Maustaste bewirkt die Ausgabe eines Hilfefensters, in dem eine kurze Beschreibung des betreffenden Objektes dargestellt ist.
- Parametrierungskonzept. Die Einstellung von Funktions- und Interaktionsparametern kann nach Selektion der betreffenden Funktion mit der rechten Maustaste über ein eingeblendetes Parameterfenster vorgenommen werden.
- Bildrahmenkonzept. Die im Bildbereich ausgegebenen Bilder sind entsprechend ihrem Entstehungszeitpunkt durch verschiedenfarbige Rahmen markiert. Das zuletzt ausgegebene Bild (aktuelles Bild) ist rotumrandet, die davor ausgegebenen zwei Bilder grün bzw. blau. Früher entstandene Bilder werden mit einem schwarzen Rahmen markiert und sind nicht mehr direkt zugreifbar.
- Positionierungskonzept. Zur Angabe der Ausgabeposition eines darzustellenden Bildes wird ein weißer Rahmen benutzt, der mit gedrückter linker Maustaste verschoben werden kann.
- Dimensionierungskonzept. Zur Vorgabe der Größe eines Ergebnisbildes (z.B. bei Ausschnitt, Zoom) wird dem Benutzer ein gestrichelter, weißer Rahmen angeboten, dessen Größe durch Verschieben der Maus verändert werden kann.

Die Texte in den Ikonen der Funktionenleiste entsprechen genau den Kommandos, die der Benutzer eintasten oder sprechen kann, um die jeweilige Funktion aufzurufen. Über die verschiedenen Kommunikationskanäle können somit mehrere Interaktionen gleichzeitig durchgeführt werden. Der Benutzer kann den für ihn geeigneten Eingabekanal wählen.

4. Anwendungsfunktionen

Eines der wesentlichen Merkmale der KIBAS-Struktur ist die explizite Trennung zwischen Dialog- und Anwendungsfunktionalität. Diese Trennung ist für den Benutzer jedoch nicht erkennbar, für den Benutzer stellt sich eine KIBAS-Funktion als eine Einheit von Dialog und Anwendung dar. Die Dialogkomponente der Funktionen, d.h. die Auswahl und Aktivierung von Funktionen (Selektion), die Parametrierung sowie die Ergebnisdarstellung (Bildrahmenkonzept) wurde im vorangegangenen Abschnitt beschrieben. Die verschiedenen KIBAS-Funktionen lassen sich in vier Gruppen einteilen. Die Zuordnung der einzelnen Funktionen zu diesen Gruppen wird dem Benutzer durch die geometrische Anordnung und einheitliche Farbgebung deutlich gemacht (vgl. Abb. 2 und Abb. 3):

- Verarbeitungsfunktionen erzeugen ein neues Ergebnisbild, ohne das Quellbild zu verändern. Sie umfassen Funktionen zur Addition und Subtraktion von Bildern, zur Größenveränderung (Zoom), sowie zur Kopie- und Ausschnittbildung. Zur Konturverstärkung oder -abschwächung stehen eine Reihe von Filtern zur Verfügung. Die Funktionen "Kennlinie" und "Zeichnen" verfügen über eigene einblendbare Benutzungsoberflächen. Die Funktion "Kennlinie" erlaubt die interaktive Veränderung der Kennlinie oder die Benutzung vorgefertigter Abbildungsfunktionen zur Manipulation der Kennlinie.

Mit der Funktion "Zeichnen" ist die interaktive Markierung von Strukturen im Bild auf einer dem Bild überlagerten "Folie" möglich. Diese Folie bleibt erhalten und kann damit auch für andere Bilder übernommen werden. Zeichenfarbe, Pinselform und Zeichenmodus sind einstellbar.

- Darstellungsfunktionen verändern die Systemdarstellung, ohne ein neues Bild zu erzeugen. Mit der Funktion "Verschieben" läßt sich das aktuelle Bild in der Darstellungsfläche verschieben. "Oben" bietet Funktionen zur Manipulation der Bildhierarchie (vgl. Bildrahmenkonzept) an. Die Funktion "Lupe" stellt ein Vergrößerungsfenster (Lupe) zur Verfügung, welches mittels der Maus verschoben werden kann. Die interaktive Manipulation der Systemfarbtabelle oder die Auswahl vorgefertigter Farbtabellenmuster ermöglicht die Funktion "Farbe". Mit der Funktion "Schwelle" kann die Auswirkung von Schwellwertoperationen interaktiv erprobt werden.

- Dienstleistungsfunktionen stellen Funktionen zur Systembedienung dar. Die Funktion "Makro" erlaubt die Zusammenfassung vergangener und zukünftiger Aktionsfolgen zu einer Einheit. Diese Einheit kann mit einem Namen versehen und als Makro in die Benutzungsoberfläche integriert werden. Mit der Funktion "Löschen" können Bilder selektiv oder die komplette Bildfläche gelöscht werden. Die benutzerinitiierte Veränderung der Systemdarstellung und des Systemverhaltens ermöglicht das "Profil". "Zurück" hebt die Wirkung der zuletzt durchgeführten Funktion auf. Zum Abbruch einer aktiven Verarbeitungsfunktion dient die Funktion "Abbruch". "Hilfe" bietet eine globale Hilfestellung (zusätzlich verfügt jedes Objekt über lokale Hilfe, komplexere Einblendoberflächen wie in "Farbe", "Kennlinie", "Zeichnen" und "Makro" verfügen über zusätzliche Hilfen). Die Funktion "Beenden" dient zum Verlassen des Systems.

- Die in den Abb. 2 und 3 nicht dargestellten Verwaltungsfunktionen umfassen Funktionen zur Verwaltung von Bildalben, von in Alben enthaltenen Bildern und von weiteren Peripheriegeräten. Dabei werden Albenverwaltungsfunktionen zur Auflistung der Namen der Bilder innerhalb des aktiven Albums, zur Abspeicherung des aktuellen Bildes im Album, zum Aufbau eines neuen Albums, zur Albumbenennung, zur Albumetikettenbildung aus dem aktuellen Bild und zur Löschung eines Albums angeboten. Die Bildverwaltungsfunktionen beinhalten Funktionen zur Ausgabe eines Bildes im Bildbereich (Bilddarstellung), zur Namensgebung eines Bildes und zum Löschen eines Bildes im Album. Die Druckerverwaltungsfunktion ermöglicht den Ausdruck des aktuellen Bildes auf einem Laserdrucker. Mittels der Funktion "Kamera" kann ein Bild über eine an das VISTA angeschlossene Kamera eingelesen werden.

5. VISTA (Tatari & Paul 88)

Das Sichtsystem VISTA wurde für Anwendungen in der schnellen Oberflächenprüfung und der Bildfolgenauswertung konzipiert. Besondere Merkmale des Systems sind:
- Neben Fernsehbild-Kameras lassen sich auch Diodenzeilen oder Laser-Abtaster als bildgebende Sensoren an VISTA anschließen. Damit sind Aufnahmen mit sehr hoher Ortsauflösung möglich.

- Bildformate sind bis zu einer maximalen Zeilenlänge von 16 484 Bildpunkten frei wählbar und lassen sich damit den Dimensionen der aufzunehmenden Szene anpassen. Eine lückenlose Verarbeitung von "Endlosbildern" ist vorgesehen.
- VISTA kann mehrkanalige Bilder aufnehmen und verarbeiten. Mehrkanalige Bilder sind z.B. Farbbilder (RGB) oder Stereobilder. Damit können neben dem Grauwertverlauf auch Informationen über Farbe und Höhenrelief in der Szene gewonnen werden.
- Für die Bildauswertung werden im VISTA spezialisierte Verarbeitungsmodule eingesetzt. Diese Module arbeiten synchron zu einem Bildpunkttakt von 10,7 MHz. Zur Ausführung von Fließbandarbeiten können die Module hintereinandergeschaltet werden oder auch auf demselben Datenstrom parallel arbeiten. Die Schaltung der Datenwege zwischen den Modulen ist programmgesteuert und kann zur Laufzeit des Systems geändert werden.

Die Architektur von VISTA ist die eines busorientierten Systems mit spezialisierten Verarbeitungsmodulen. Für die Übertragung von Steuerinformationen dient der VMEbus, für die Übertragung von Bildern und Listen ein VIDEObus mit sechs Parallel-Kanälen zu 8 bit, einem Parallel-Kanal zu 16 bit und einem "Fließbandkanal", mit dem auf der Rückplatte des Systems lokale Verbindungen zwischen Modulen geschaffen werden können.

6.   Verbindung KIBAS - VISTA

Die Kopplung KIBAS - VISTA vereint die komfortable, anpassungsfähige Benutzungsoberfläche des KIBAS mit der videoschnellen Bildverarbeitung des VISTA. Mittels eines Buskopplers sind der Ein-/Ausgabebus der KIBAS-Arbeitsstation (Apollo/Domain DN 4000) und der VMEbus des VISTA-Systems verbunden. Zur Kommunikation der beiden Rechner wird ein Zwei-Tor-Speicher ("dual ported RAM") benutzt.

7.   Verfahrensentwicklung

Der KIBAS - VISTA Funktionsverbund unterstützt die Verfahrensentwicklung in verschiedenen Bereichen:
- KIBAS-Rahmenprogramme übernehmen die Darstellung und Verarbeitungsvorbereitung für das Quellbild sowie die Darstellung des Ergebnisbildes. Der Anwendungsprogrammierer kann sich somit auf seinen Algorithmus konzentrieren.
- Eventuell notwendige Vor- und Nachverarbeitungsschritte können in Echtzeit vom VISTA ausgeführt werden.
- Die komfortablen, grafikunterstützten Parametrierungsmöglichkeiten des KIBAS in Verbindung mit den anpaßbaren Handlungsanweisungen ermöglichen dabei ein schnelles Testen unterschiedlicher Parametersätze.
- Darstellungsfunktionen erlauben die Darstellung des Ergebnisbildes in verschiedenen Auflösungen oder Pseudofarben und erleichtern die Interpretation des Ergebnisbildes.

- KIBAS übernimmt die Verwaltung der Peripheriegeräte und entlastet den Benutzer von der Bilddateiverwaltung.
- Durch eine große Anzahl vordefinierter Dialogobjekte wird der Aufbau einer komfortablen Benutzungsoberfläche für das neue Verfahren in effizienter Form ermöglicht.

Neben der Komfort- und Geschwindigkeitssteigerung für die Entwicklung ermöglicht der Funktionsverbund eine anschauliche Verfahrensablauf- und Ergebnispräsentation für Kooperationspartner und Auftraggeber. (Bisher sind nur ikonische Bildverarbeitungsoperationen implementiert.)

8. Ausblick

KIBAS wurde auf einer Apollo/Domain-DN-4000-Farbgrafik-Arbeitsplatzstation entwickelt. Angeschlossen sind neben dem VISTA-System ein Spracheingabegerät der Firma Marconi (SR 128X) und ein Sprachausgabesystem von Speech Design (Audiocard 600).

Zur Zeit wird das System auf das Betriebssystem UNIX-BSD 4.3 mit dem Fensterverwaltungssystem X-Window und dem Dialogmanagementsystem Open Dialogue übertragen. Damit ist die Portierung auf unterschiedliche Arbeitsplatzrechner möglich. Neben der Überarbeitung der Kommunikationsprogramme zwischen KIBAS und VISTA im Hinblick auf höhere Übertragungsgeschwindigkeiten ist eine Vergrößerung des Funktionsumfanges des Funktionsverbundes vorgesehen. Die Integration eines Benutzermodelles zur automatischen, modellgestützten Anpassung des Systems an den individuellen Benutzer ist ein weiteres Ziel innerhalb des Projektes. Anhand einer systematischen Sammlung von Erfahrungen mit dem Funktionsverbund ist auch eine Weiterentwicklung in Richtung eines Ausbildungssystems vorgesehen.

Die diesem Beitrag zugrundeliegenden Arbeiten wurden vom Bundesministerium der Verteidigung (BMVg) im Rahmen eines Forschungsvorhabens gefördert.

Literatur

Faasch 87
H. Faasch: Konzeption und Implementation einer objektorientierten Experimentierumgebung für die Bildfolgenauswertung in ADA. Dissertation, Universität Hamburg, Fachbereich Informatik, 1987

Schönbein & Roller 88
R. Schönbein, W. Roller: Dialoggestaltung für die interaktive Bildauswertung. FhG-Berichte 3/88, Mitteilungen aus dem Fraunhofer-Institut für Informations- und Datenverarbeitung, IITB, 1988, 4-7

Tatari & Paul 88
S. Tatari, D. Paul: Verfahren und Geräte zur schnellen automatischen Prüfung texturierter Oberflächen. Technisches Messen tm, Heft 12/1988, 504-510

Ein paralleles Transputersystem zur digitalen Bildverarbeitung mit schneller Pipelinekopplung

B. Lang

Technische Informatik I
Technische Universität Hamburg-Harburg

Zusammenfassung

Dieser Beitrag stellt ein modulares, paralleles Bildverarbeitungssystem vor. Mit dem Einsatz von Transputern besitzt das System eine durch Links aufgebaute Verbindungsstruktur, die sich topologisch an unterschiedliche Algorithmen anpassen läßt. Es wird ein schneller Pipelinebus vorgestellt, der die Module des Systems zusätzlich verbindet und ein effizientes Laden und Entladen parallel angeordneter Transputer ermöglicht. Anhand zweier Beispiele wird die Algorithmenimplementierung und der jeweils angepaßte Lade- und Entladevorgang erläutert.

1 Einleitung

1.1 Algorithmische Möglichkeiten von Transputersystemen

Im Bereich der Parallelverarbeitung hat in den vergangenen Jahren der Transputer durch seine Möglichkeiten zum einfachen Aufbau vernetzter Systeme immer stärkere Bedeutung gewonnen. Durch den geringen Aufwand mithilfe von Link-Verbindungen vielfältige Verbindungstopologien zu realisieren, ist eine spezielle Anpassung paralleler Transputersysteme an unterschiedliche Algorithmen möglich. Weiterhin bietet insbesonders der Typ T800 durch 32-Bit Verarbeitung und interne Gleitkomma-Arithmetik schon als Einzelprozessor eine sehr hohe Rechenleistung.

Auch in der digitalen Bildverarbeitung hat aus obigen Gründen der Transputer bereits eine starke Beachtung gefunden und Systeme mit hohen Rechenleistungen ermöglicht [5,6]. Viele Algorithmen im Bereich der Bildverarbeitung weisen klar strukturierte Berechnungstopologien auf, die sich für eine Implementierung auf Transputersystemen sehr gut eignen.

1.2 Verbindungstopologien für Parallelrechner in der Bildverarbeitung

Die Effizienz der Abbildung eines Algorithmus auf eine Parallelprozessor-Struktur wird wesentlich durch die Verbindungstopologie des Verbindungsnetzwerks zwischen den parallelen Prozessoren bestimmt. Vielfältige Untersuchungen haben sich mit dem Entwurf und der Analyse solcher Netze beschäftigt. Die einfachste und flexibelste Lösung, die sich durch vollständiges Verbindung aller Prozessoren untereinander ergibt, scheitert schon bei mittleren Prozessorzahlen N durch ihren quadratischen Aufwand $O(N^2)$ bezüglich der Verbindungsleitungen. Zwei Netzwerke, deren Aufwand nur linear mit der Prozessoranzahl wächst, und auf welche sich ein großer Teil wichtiger Bildverarbeitungsalgorithmen abbilden läßt, verdienen dabei besondere Beachtung.

Im Bereich der Bildvorverarbeitung, wo Operationen auf einer lokalen Umgebung durchgeführt werden, eignen sich Netzwerke mit Nachbarschaftsverbindungen zwischen Prozessoren. Auf diese lassen sich lokale, lineare Bildfilterung [10] und morphologische Operationen [8], insbesonders auch iterative morphologische Algorithmen mit Nebenbedingung, direkt abbilden.

Die effektive Implementierung schneller, globaler Transformationen [1], zu denen die schnelle Fouriertransformation, der Viterbi-Algorithmus, translationsinvariante Transformationen und weitere lineare und nichtlineare [2] Transformationen zählen, ist auf dem 'Perfekt-Shuffle'-Netzwerk [9] möglich. Mit diesem 'Perfekt-Shuffle'-Netzwerk lassen sich weiterhin vielfältig nutzbare, mehrstufige, in allen Schichten homogene Verbindungsnetzwerke aufbauen, deren topologische und funktionale Äquivalenz zu weiteren bekannten mehrstufigen Netzen bewiesen ist [11,12].

1.3 Ladeanforderungen an ein Bildverarbeitungssystem

Neben den algorithmischen Anforderungen treten im Bereich der Bildverarbeitung hohe Anforderungen bezüglich des Ladens und Entladens von Massendaten auf. Sie lassen sich nur schwer mit den algorithmischen Anforderungen an Verbindungsstrukturen verknüpfen. Ein für algorithmische Zwecke entworfenes Verbindungsnetzwerk zwischen Prozessoren kann bezüglich der Ladeanforderungen denkbar ungeeignet sein.

Weiterhin benötigt ein paralleler Algorithmus, der auf mehreren Prozessoren implementiert wurde, eine bestimmte Aufteilung der Eingangsdaten, die beim Laden herzustellen ist. Auch liegen nach der Berechnung die Ergebnisdaten verteilt auf den Prozessoren vor und sind im allgemeinen gegenüber den Eingangsdaten permutiert. Ein bekanntes Beispiel für permutierte Eingangs- oder Ergebnisdaten findet man beim Bitreversal der schnellen Fouriertransformation.

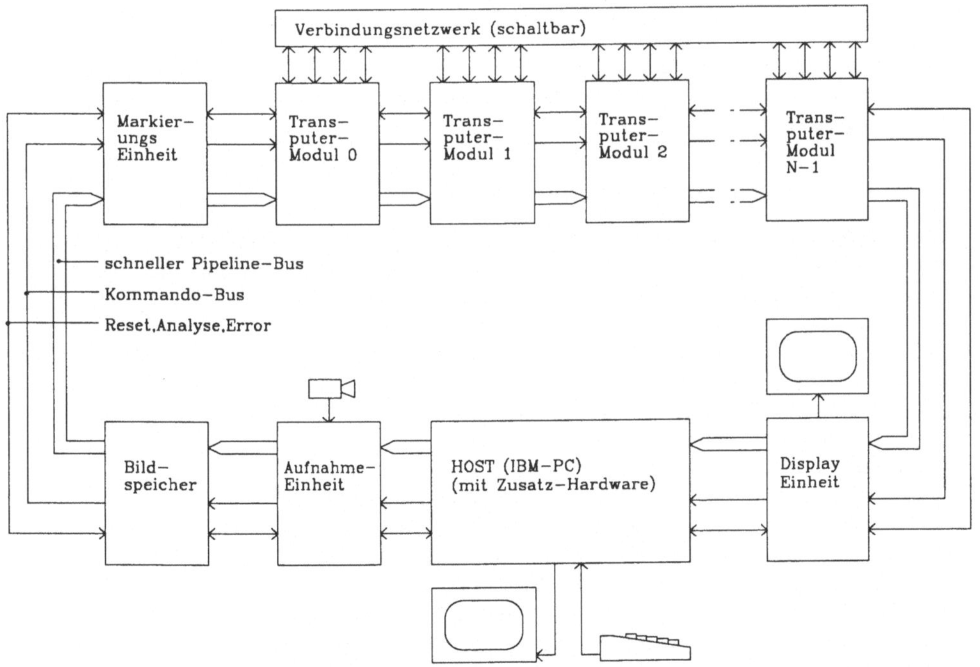

Bild 1: Paralleles Transputersystem mit schneller Pipelinekopplung: Systemübersicht

2 Systemübersicht

Zur Realisierung sowohl der algorithmischen als auch der Lade-Anforderungen wurde an der TUHH ein paralleles Transputersystem entwickelt (siehe Bild 1). Es ist modular aufgebaut, wodurch eine Konfiguration des Systemes für unterschiedliche Leistungsanforderungen möglich ist. Den für die Implementierung von Algorithmen relevanten Teil bilden Transputermodule; Bild 2 zeigt ihre Struktur. Weitere Module dienen der Host-Ankopplung, der Bildaufnahme,

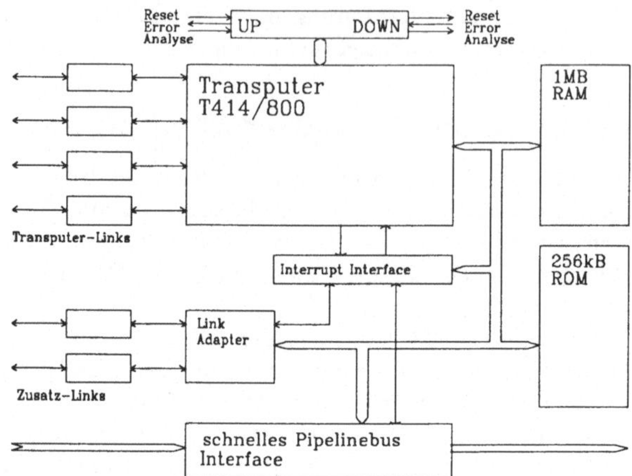

Bild 2: Transputermodul des paralleles Transputersystemes

-wiedergabe und -speicherung sowie der Verteilung der Daten auf parallele Prozessoren. Zur Kopplung der Module untereinander existieren 3 Verbindungsstrukturen im System, die sich in Topologie, Protokoll und Datenraten unterscheiden.

2.1 Verbindungsstrukturen des Systemes

Die erste Verbindungsstruktur ist mit Transputerlinks aufgebaut und verbindet N Transputermodule zu einem parallelen, verkoppelten Prozessorkern. Das in Bild 1 dargestellte System besitzt nur einen solchen Kern, allgemein kann ein System jedoch mehrere solcher Kerne beliebiger Größe enthalten. Das Link-Netzwerk bildet die algorithmische Verbindungsstruktur des Systemes. Es kann flexibel über Link-Switch Bausteine geschaltet werden, so daß eine beliebige Anpassung an Algorithmen-Topologien möglich ist. Die bisher auf einem Transputer verfügbaren 4 schnellen, seriellen Links stehen für die algorithmischen Belange voll zur Verfügung. Die mögliche Übertragungsbandbreite dieses Netzwerkes ist proportional zur Anzahl der verwendeten Links und damit variabel mit der Anzahl der angeschlossenen Prozessoren.

Die zweite Verbindungsstruktur bildet ein schneller, getakteter paralleler Pipelinebus mit einer an Standard-Videoraten angepaßten mittleren Datenrate von bis zu 10 $MByte/sek$. Auf diesem werden parallel zu den Bilddaten Kontrollinformationen übertragen, die den Fluß der Bilddaten im System steuern. Er wird im folgenden näher erläutert.

Die dritte Verbindungsstruktur ist wiederum durch Links aufgebaut und bildet den Kommandobus. Zu seiner Realisierung wurden zusätzliche, über Interrupt gesteuerte Linkadaptoren auf den Transputermodulen vorgesehen. Auf dieser Struktur ist ein komplexes Kommandoprotokoll zur Steuerung des Gesamtsystemes durch den Host implementiert, welches zum Host hin ein leistungsfähiges, komfortables Interface zur Systemsteuerung bietet und auf jedem Modul durch eine lokale Software-Betriebsumgebung unterstützt ist. Dieser Kontrollbus dient ausschließlich der Systemsteuerung und benötigt die geringste Bandbreite im System.

2.2 Ein schneller Pipelinebus mit Möglichkeiten zur Addressierung

Sollen mehrere parallele Prozessoren ein einzelnes Bild verarbeiten, so muß, wie in Abschnitt 1.3 angesprochen, vor der Berechnung jeder Prozessor die von ihm benötigten Daten erhalten. Dabei ergeben sich die zwei Hauptfälle, daß jeder Prozessor entweder das komplette Bild oder nur ein Teilbild benötigt. Der erste Fall erfordert das Senden eines einzelnen Datums an viele Empfänger, der zweite das Senden eines Datums an einen bestimmten einzelnen Empfänger.

Nach der Berechnung müssen die errechneten, auf die parallelen Prozessoren verteilten Ergebnisdaten wieder zu einem Gesamtergebnis, z.B. zu einem transformierten Bild, zusammengesetzt werden. Dazu ist ein synchronisiertes Auslesen der Daten aus den Prozessoren nötig.

Zur Realisierung solcher Lade- und Entlademodi wird auf dem schnellen Pipelinebus parallel zu jedem Datum ein Kontrollteil und eine Adresse übertragen. Diese Einheit aus Kontrollteil, Adresse und Datum soll im folgenden als Token bezeichnet werden (siehe Bild 3). Jedes an den

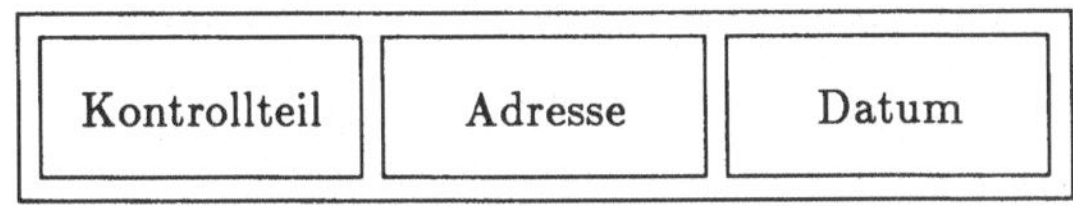

Bild 3: Token des schnellen Pipelinebuses

Pipelinebus angeschlossene Modul des Systemes besitzt ein Pipelinebus-Interface, welches den Kontrollteil der ankommenden Token liest und entsprechend dessen Definition die Adresse der Token interpretiert. Zur Auswertung der Tokenadresse befindet sich auf diesem Interface eine einstellbare Pipelinebus-Adresse. Mit ihr wird die Tokenadresse hardwaremäßig verglichen. Die oben angeführten Modi lassen sich mit der folgenden Definition des Kontrollteiles realisieren:

Single: Sind Token- und Pipelinebus-Adresse gleich, so wird das Datum des Tokens in das Modul eingelesen. Stimmen beide Adressen nicht überein, so wird das Token unverändert an das nächste Modul in der Pipeline weitergereicht.

All: Das Interface liest das Datum des Tokens in das Modul ein. Somit wird bei Gleichheit von Token- und Pipelinebus-Adresse das Token vom Bus genommen; bei Ungleichheit wird es hingegen zusätzlich an das nächste Modul weitergereicht.

Empty: Bei Adreßgleichheit wird das Interface durch diesen Tokentyp zur Ausgabe eines neuen Tokens stimuliert. Es liest ein Datum aus dem Modul aus und ergänzt es durch einen einstellbaren Kontroll- und Adreßteil. Bei Ungleichheit der Adressen wird das Token unverändert an das nachfolgende Modul weitergereicht. Ein Token vom Typ 'Empty' enthält kein gültiges Datum.

None: Mit diesem Tokentyp wird der leere Pipelinebus markiert, sein Adreß- und Datenteil enthält keine gültigen Daten.

2.3 Allgemeines Modulinterface für den schnellen Pipelinebus

Die Struktur eines Interfaces zur Interpretation obiger Tokendefinition zeigt Bild 4. Die Eingangssteuerung liest das ankommende Token (Token_UP), abhängig vom Tokentyp, in den Moduleingang oder in ein 'Bypass'-Register ein. Sind Moduleingang oder 'Bypass'-Register noch durch alte Daten belegt, so erfolgt eine Warteanforderung (Wait_UP) solange an das vorgeordnete Modul, bis das Einlesen durchgeführt werden konnte.

Die mit dem Eingang verkoppelte Ausgangssteuerung wählt über einen Ausgangsmultiplexer das auszugebende Token aus. Als Quelle stehen dabei das 'Bypass'-Register, der Modulausgang als Reaktion auf ein 'Empty'-Token, ein Warte-Register zur Behandlung einer Warteanforderung

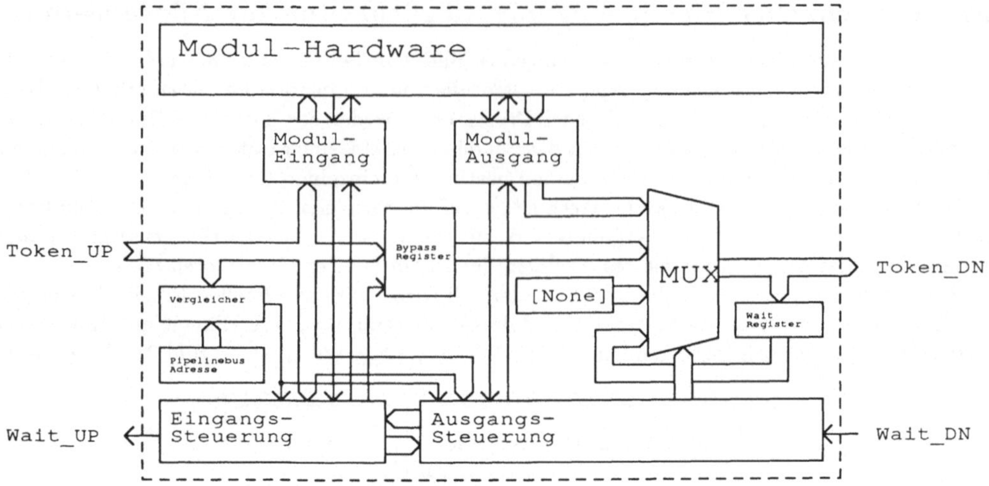

Bild 4: Allgemeines Modulinterface für den schnellen Pipelinebus

des nachgeordneten Modules (WaitDN) und ein 'None'-Token Generator zur Verfügung. Sowohl Ein- und Ausgangssteuerung arbeiten mit dem maximalen Bustakt von $10MHz$ und können pro Taktzyklus ein Token verarbeiten.

2.4 Markierungseinheit für den schnellen Pipelinebus

Die Markierung der Token nach obiger Definition wird durch eine Markierungseinheit, die jedem Kern von Transputermodulen vorgeschaltet ist, durchgeführt (siehe Bild 1). Bei der Verteilung eines Bildes auf N Transputermodule sendet die Bildquelle (z.B. der Bildspeicher) Daten zunächst an diese Markierungseinheit.

Bei segmentweiser Aufteilung des Bildes zwischen N nachfolgenden Transputermodulen läßt diese Einheit die Daten unverändert, ergänzt sie jedoch pixelweise mit dem Kontrollteil 'Single' und Adressen gemäß der gewünschten Verteilung auf die Transputermodule.

Will man ein komplettes Bild auf jeden der N Transputermodule eines Kernes laden, so markiert man alle Pixel des Bildes mit dem Kontrollteil 'All' und der Adresse des letzten Modules des Kernes.

Zum Entladen verteilter Ergebnisdaten generiert die Markierungseinheit Token vom Typ 'Empty' mit Adressen gemäß der vorliegenden Verteilung der Ergebnisse auf den Transputer-Modulen des Kernes. Die einzelnen Module ersetzen die an sie adressierten 'Empty'-Token durch ihre Ergebnisdaten mit einem neuen Kontroll- und Adreßteil (z.B. 'Single' und der Adresse des Bildspeichers).

3 Algorithmenimplementierung

Durch seine Modularität und Flexibilität ist das vorgestellte parallele Transputersystem für die Implementierung vielfältiger Algorithmen geeignet. Exemplarisch werden zwei Algorithmen behandelt, die hohe Anforderungen bezüglich des Rechenaufwandes und auch an das Laden und Entladen der Bilddaten stellen.

3.1 Implementierung der schnellen Fouriertransformation

Den algorithmischen Teil der Implementierung der schnellen Fouriertransformation (FFT) auf einem parallelen Prozessor mit 'Perfect-Shuffle'-Verbindungsnetzwerk stellt Stone in [9] ausführlich dar. Dort wird die Abarbeitung einer N-Punkte Transformation $(N = 2^n)$ auf $N/2$ Prozessoren behandelt.

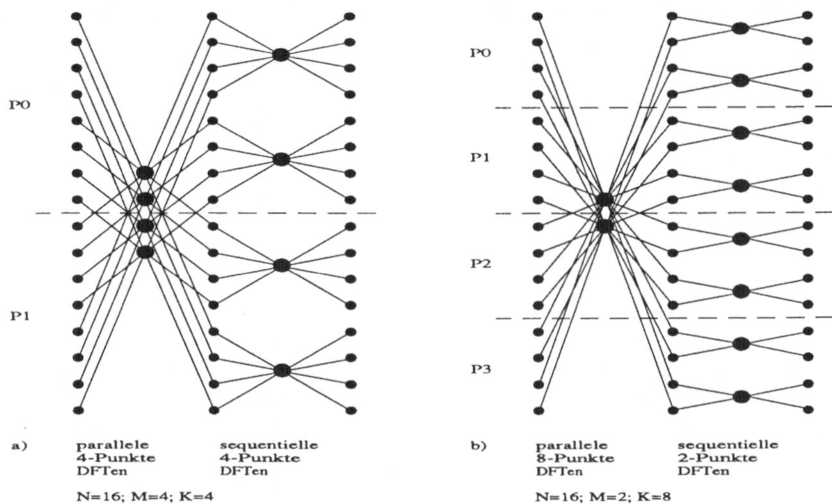

Bild 5: Topologie einer 16-Punkte DFT bei Abarbeitung auf a) 2 Prozessoren b) 4 Prozessoren

Hat man weniger Prozessoren zur Verfügung, läßt sich eine digitale eindimensionale N-Punkte Fouriertransformation (DFT) als 2-dimensionale $K \cdot M$-Punkte DFT formulieren [7] $(N = K \cdot M)$. Zur Berechnung der ersten Dimension muß man M DFTen von K Punkten, z.B. nach obigem Algorithmus von Stone, parallel auf $K/2$ Prozessoren durchführen. Anschließend werden zur Berechnung der zweiten Dimension jeweils 2 DFTen von M Punkten sequentiell auf jedem der $K/2$ parallelen Prozessoren ausgeführt. Eine bei Rabiner und Gold [7] hergeleitete Korrektur der Zwischenergebnisse nach Berechnung der ersten Dimension läßt sich in diese Berechnung ohne zusätzlichen Aufwand mit einarbeiten.

Zur Vorbereitung der algorithmischen Berechnungen muß bei obiger Parallelisierung jeweils ein Teilvektor der Länge $2M$ in einen Prozessor geladen werden. Dies erreicht man durch Markieren der $2M$ aufeinanderfolgenden Elemente eines Teilvektors mit Kontrollteil 'Single' und der Pipelinebus-Adresse des zugehörigen Prozessors. Die Adressen der Prozessoren für aufeinanderfolgende Teilvektoren werden beim zitierten Algorithmus linear aufsteigend vergeben.

Nach dieser von Stone vorgestellten Berechnung weisen die Ergebnisdaten nach der Transformation die bekannte 'Bitreverse'-Permutation auf. Zum Entladen der Daten generiert daher die Markierungseinheit 'Empty'-Token, deren Adreßteil der 'Bitreverse'-Permutation unterworfen ist. Dies bewirkt ein permutiertes stimulieren der Prozessoren und somit ein Auslesen der Daten in der gewünschten Reihenfolge.

In Bild 5 ist die Topologie einer 16-Punkte DFT bei der Implementierung auf 2 und auf 4 Prozessoren dargestellt.

Burkhardt und Barbosa beschreiben in [3] eine Klasse verallgemeinerter schneller Transformationen, zu denen auch die FFT zählt. Diese Transformationen, mit interessanten Applikationen im Bereich der Mustererkennung und Bildrestauration, lassen sich ebenfalls direkt auf das oben angesprochene 'Perfect-Shuffle'-Verbindungsnetzwerk abbilden und stellen ähnliche Ladeanforderungen an das System.

3.2 Implementierung der Bildrotation

Die Bildrotation ist ein rechenintensiver Teil schneller Algorithmen zur räumlichen Bewegungsschätzung ebener Bildvorlagen [4]. Ihre parallele Implementierung ermöglicht eine deutliche Leistungssteigerung gegenüber der sequentiellen Abarbeitung.

Die für die Rechenzeit der Auswertung wesentlichen ebenen Bildrotationen lassen sich jeweils durch die Gleichungen 1 bis 3 beschreiben. Darin bezeichnet $\mathbf{I_1(x)}$ die Grauwerte eines Eingangsbildes in Abhängigkeit von der Position $\mathbf{x} = (x_1, x_2)$ und $\mathbf{I_2(x)}$ ein um den Winkel ϕ rotiertes Bild. Die Matrix $\mathbf{A}$ ist als Drehmatix bekannt.

$$\mathbf{I_2(x)} = \mathbf{I_1(y)} \tag{1}$$

$$\mathbf{I_2(x)} = \mathbf{I_1(A \cdot x)} \tag{2}$$

$$\mathbf{A} = \begin{pmatrix} cos\phi & -sin\phi \\ sin\phi & cos\phi \end{pmatrix} \tag{3}$$

Zur Berechnung von $\mathbf{I_2(x)}$ ermittelt man für jeden Punkt $\mathbf{x}$ des gedrehten Bildes die zugehörige Koordinate $\mathbf{y} = \mathbf{A \cdot x}$ im Eingangsbild und bestimmt den Grauwert $\mathbf{I_1(y)}$. Bei digital vorliegenden Bildern, deren Grauwerte nur an diskreten Koordinatenpunkten bekannt sind, wird $\mathbf{I_1(y)}$ durch Interpolation ermittelt (z.B. nächster Nachbar, bilineare Interpolation).

Eine Möglichkeit der Parallelisierung ergibt sich, wenn man jedem von N Prozessoren das komplette Ausgangsbild $\mathbf{I_1}$ zur Verfügung stellt. Auf jedem Prozessor wird der Rotationsalgorithmus einschließlich Interpolation und Randbehandlung implementiert, ein Prozessor berechnet jedoch nur einen Teilbereich des gedrehten Bildes $\mathbf{I_2(x)}$.

Der Ladevorgang von $\mathbf{I_1}$ wird beim vorliegenden System über die schnelle Pipeline mit einer Markierung 'All' vorgenommen, damit erhält jeder der Prozessoren das gesamte Bild. Das Entladen geschieht mithilfe von 'Empty'-Token und Adressen gemäß der für die Berechnung vorgenommenen Aufteilung des Bildes.

Betrachtet man die Bildrotation als Bestandteil des von Diehl [4] vorgestellten, schnellen Algorithmus zur Bewegungsschätzung in Bildfolgen, so muß der Ladevorgang eines zu rotierenden Referenzbildes auf jeden der N Prozessoren einmal beim Start des Algorithmus erfolgen. Beim Laden der zu untersuchenden Bilder zur Bestimmung der Bewegungsparameter gegenüber dem Referenzbild kann eine vorgegebene Bildaufteilung durch Markieren mit 'Single'-Kennungen wieder direkt vorgenommen werden. Weiterhin entfällt, falls nur die Bewegungsparameter interessieren, das Entladen des rotierten Bildes.

4 Ausblick

Mit dem vorgestellten parallelen Transputersystem mit schneller Pipelinekopplung steht ein aus universellen Modulen bestehendes System für Bildverarbeitungs-Aufgaben zur Verfügung. Es ermöglicht die Entwicklung und effektive Implementierung unterschiedlicher paralleler Bildverarbeitungs-Algorithmen ohne teure Hochleistungsrechner. Anhand zweier beispielhafter Algorithmen wurden die Möglichkeiten der Implementierung aufgezeigt. Aufbauende Abeiten werden sich mit dem Ausbau des Systemes, der Entwicklung von Software-Hilfsmitteln und der Implementierung weiterer paralleler Algorithmen beschäftigen.

Literatur

[1] H.Burkhardt: Methoden der Digitalen Signalverarbeitung in der Bildverarbeitung und Mustererkennung. 8. DAGM-Symposium "Mustererkennung", Informatik Fachberichte Nr. 125, Springer-Verlag, 1986, S.43-55.

[2] H. Burkhardt: Transformationen zur lageinvarianten Merkmalerkennung. VDI-Fortschritt-Bericht, Reihe 10 (Angewandte Informatik), Nr. 7, VDI-Verlag Düsseldorf, Okt. 1979.

[3] H. Burkhardt, L.C. Barbosa: Contributions to the Application of the Viterbi-Algorithm. IBM Reasearch Report, RJ 3377(40413) 1/22/82, San Jose, Ca. und IEEE Trans. on Information Theory, Vol. IT-31, No. 5, 1985, S.626-643.

[4] N. Diehl: Methoden zur allgemeinen Bewegungsschätzung in Bildfolgen. VDI-Fortschritt-Bericht, Reihe 10 (Angewandte Informatik), Nr. 92, VDI-Verlag Düsseldorf, 1988.

[5] E. Hiltebrand: Arbeitsstation zur interaktiven Bearbeitung und Darstellung medizinischer Volumen-Bilddaten. 10. DAGM-Symposium "Mustererkennung", Informatik Fachberichte Nr. 180, Springer-Verlag, 1988, S.31-38.

[6] F.-D. Kübler: A cluster-oriented Architecture for the Mapping of Parallel Processors to High Performance Applications. Schweizer Informatik Gesellschaft, Conference on Economical Parallel Processing, Bern 1988.

[7] L.R. Rabiner, B. Gold: Theory and Application of Digital Signal Processing. Prentice-Hall, 1975.

[8] J. Serra: Image Analysis and Mathematical Morphology. Academic Press, 1982.

[9] H.S. Stone: Parallel Processing with the Perfect Shuffle. IEEE Trans. Comp., Vol. C-20, Febr. 1971, 153-161.

[10] F. Wahl: Digitale Bildsignalverarbeitung. Springer-Verlag, 1984.

[11] C.-L. Wu, T.-Y. Feng: On a Class of Multistage Interconnection Networks. IEEE Transactions on Computers, Vol. C-29, Aug. 1980, p.694-702.

[12] C.-L. Wu, T.-Y. Feng: The Reverse-Exchange Interconnection Network. IEEE Transactions on Computers, Vol. C-29, Sept. 1980, p.801-811.

Ein Bildsegmentierer für die echtzeitnahe Verarbeitung

C.Anderer, U.Thönnessen

Forschungsinstitut für Informationsverarbeitung und Mustererkennung (FGAN/FIM),
Eisenstockstr. 12, D-7505 Ettlingen 6

M.F.Carlsohn, A.Klonz

Philips GmbH, Systeme und Sondertechnik, Hans-Bredow Straße 20,
D-2800 Bremen 44

Zusammenfassung: Die echtzeitnahe Segmentation interessierender Objekte ist eine grundlegende Aufgabe der Bildverarbeitung. Es wird ein regionenbasiertes, zeilenorientiertes Segmentationsverfahren [1] [2] [3] vorgestellt, dessen Basis eine Binarisierung des Bildes mit einer Vielzahl von Grauwertschwellen ist. Die segmentierten Objekte werden ikonisch/symbolisch mit Hilfe von Merkmalen beschrieben. Durch einen mehrstufigen Zuordnungsprozeß werden die interessierenden Objekte ausgewählt und ähnliche Objektrepräsentanten aus verschiedenen Grauwertintervallen zusammengefaßt. Der Zuordnungsprozeß läßt sich an das Datenmaterial adaptieren und wird über Regelwerke gesteuert, deren Regeln sich durch eine einfache Sprache formulieren lassen.

Da das Verfahren durch seine Komplexität erheblichen Verarbeitungsaufwand erfordert, wurde es für den echtzeitnahen Einsatz parallelisiert und als spezielle Hardware realisiert.

Beschreibung des Verfahrens

Das Verfahren wurde für Bilddaten entwickelt, bei denen sich die interessierenden Objekte durch eine höhere/geringere Grauwertintensität vom Bildhintergrund abheben. Unter dieser Annahme wird das Bild mit einer einstellbaren Anzahl von Grauwertintervallen S_i systematisch binarisiert. Die Grauwertintervalle im Intensitätsbereich $[0, 255]$ sind wie folgt definiert:

$$I^o := \{I_1^o, \ldots, I_{20}^o\} \qquad \text{obere Schwellwerte}$$
$$I^u := \{I_1^u, \ldots, I_{20}^u\} \qquad \text{untere Schwellwerte}$$
$$\text{mit } I_i^u < I_i^o \quad \forall i, 1 \le i \le 20$$
$$\text{und } I_i^o \le I_{i+1}^o \wedge I_i^u \ge I_{i+1}^u \quad \forall i, 1 \le i \le 19$$

Die Grauwertintervalle $S_i := [I_i^u, I_i^o]$ bilden die monotone Folge:

$$S_1 \subseteq \ldots \subseteq S_{i-1} \subseteq S_i \subseteq S_{i+1} \subseteq \ldots \subseteq S_n,$$

d.h. das vorhergehende Grauwertintervall ist eine Teilmenge des nachfolgenden. Für die extrahierte(n) Binärfläche(n) des Objektes wird eine ikonisch/symbolische Beschreibung in Form eines Merkmalvektors angelegt.

Wählt man nun den Abstand der Binarisierungsschwellen klein genug, so ist in der Regel gewährleistet, daß alle interessierenden Objekte durch die Binarisierung separiert werden. Dies führt bei Objekten mit starkem Kontrast dazu, daß diese mehrfach als nahezu gleiche Fläche mit einem ähnlichen Merkmalvektor extrahiert werden. Diese Redundanz wird verfahrensmäßig dadurch verringert, daß Flächen eines Objektes mit nahezu "gleichem Aussehen" zusammengefaßt werden und abschließend nur ein Merkmalvektor als Ergebnis ausgegeben wird. Dies ist nur möglich über einen Zuordnungsprozeß der Flächen aus benachbarten Grauwertintervallen und der Überprüfung des Kontextes "ähnlich". Abb. 1-1 zeigt die synthetische 3-dimensionale Darstellung des Grauwertbildes. In Abb. 1-2 ist der prinzipielle Ablauf der Segmentation mit der Extraktion der Flächen dargestellt. Über der x/y-Ebene ist die Intensität aufgetragen. Die einzelnen Grauwertintervalle S_i sind angedeutet. Die Ergebnisse der Binarisierung sind als Flächen eingezeichnet.

Im ersten Schritt des Zuordnungsprozesses werden korrespondierende Flächen zwischen zwei benachbarten Grauwertintervallen S_{i-1} und S_i gesucht. Sind mehrere Flächen aus S_{i-1} in der

Abb. 1-1: 3-dimensionale Darstellung eines synthetischen Grauwertbildes

korrespondierenden Fläche von S_i enthalten, werden sie entsprechend der Relation "Enthalten in" in einer Liste verkettet.

Hieran anschließend werden die korrespondierenden Flächen auf ihre Ähnlichkeit überprüft (Abb. 1-3). Der Kontext "Ähnlichkeit" ist vom Benutzer über Regeln frei programmierbar und an die verschiedenen Aufgabenstellungen adaptierbar. Einander ähnliche Flächen werden als Paar gekennzeichnet. Der nächste Zuordnungsschritt erfolgt iterativ und faßt als ähnlich erkannte Paare korrespondierender Flächen zu größeren Tupeln zusammen. Dies geschieht unter Ausnutzung der angenommenen Transitivität der Ähnlichkeit. Die Iteration wird abgebrochen, wenn die Ähnlichkeit zwischen den korrespondierenden Flächen nicht mehr gegeben ist. Die Fläche des subsumierenden Grauwertintervalls ist der Repräsentant der Zusammenfassung.

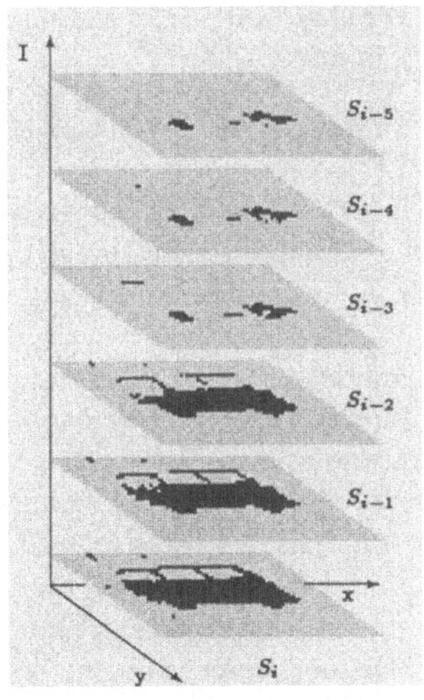

Abb. 1-2: Prinzip der Flächenextraktion

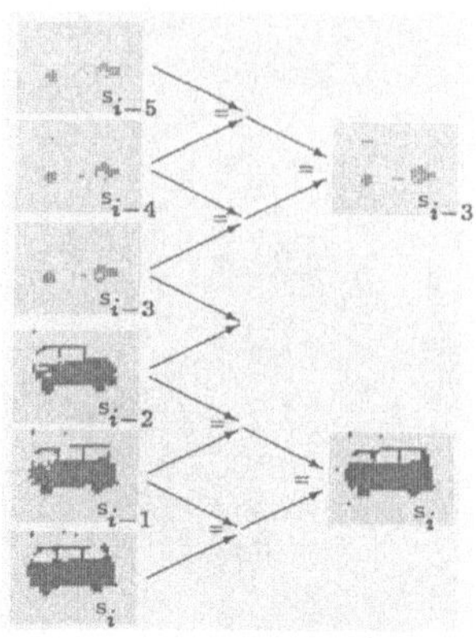

Abb. 1-3: Zuordnungsprinzip

Im letzten Schritt besteht die Möglichkeit, die Ergebnisse aus der Zusammenfassung zu bewerten. Die Repräsentanten werden entsprechend vorzugebender Kriterien ausgewählt (z.B. Flächen einer bestimmten Größe).

Die Regeln werden in einer einfachen Sprache, die für beide Regelwerke gleich ist, formuliert und von einem Compiler übersetzt. Es sind logische und arithmetische Operationen formulierbar. Weiterhin lassen sich Zuweisungen und bedingte Anweisungen (IF THEN ELSE Strukturen) einsetzen. Diese sind durch Sprunganweisungen und Label zu komplexeren Strukturen erweiterbar. Alle Merkmalkomponenten zweier korrespondierender Flächen sind verfügbar und auf funktionale Zusammenhänge überprüfbar. Ebenso lassen sich einfache Klassifikatoren in den Ablauf integrieren und die extrahierten Flächen vorklassifizieren. Diese Lösung bietet dem Benutzer größtmögliche Flexibilität und Transparenz für die Anwendung.

Systemarchitektur

Die notwendige Reduzierung der Verarbeitungszeit für ein echtzeitfähiges System ist nur durch eine Parallelisierung des Verfahrens zu erreichen. Dazu wurde der Datenstrom der Bilddaten auf parallele Binärdatenströme aufgefächert, die in identischen, modularen Funktionseinheiten weiterverarbeitet werden. Der parallelisierte Zuordnungsprozeß erfordert nur noch eine Kommunikationsstruktur zwischen direkt benachbarten Funktionseinheiten. Die SIMD-Architektur des Bildsegmentierers unterstützt die parallel ablaufenden Signalverarbeitungsprozesse des Mehrfachschwellenverfahrens.

Die Anzahl der im System verwendeten Prozessormodule bestimmt das Auflösungsvermögen des Bildsegmentierers und erlaubt mit einer modular erweiterbaren Parallelprozessorstruktur eine individuelle Anpassung der Systemkonfiguration an die Leistungsanforderungen des Anwenders. Die vorgestellte Version ist für 20 Grauwertintervalle ausgelegt.

Zuerst werden im Bildpunkttakt alle von der Intensität abhängigen Merkmale berechnet und zusammen mit den Grauwerten an dedizierte Hardware-Module übergeben. Auf diesen erfolgt die "intervallspezifische" Binarisierung und Merkmalextraktion schritthaltend mit dem Bildtakt eines 256x256 Pixel großen Bildes. Da die Berechnungen zeilensequentiell erfolgen, werden zunächst Teilmerkmale für die einzelnen Objektsegmente der aktuellen Zeile bestimmt und durch geeignete Etikettierung erst in einer folgenden Prozeßstufe für die gesamte Binärfläche akkumuliert. Bis dahin erfolgt die Verarbeitung synchron zum Zeilensprung des Bildes.

Die sich daran anschließende Verarbeitung der Merkmallisten erfolgt nun ebenfalls für jedes Grauwertintervall gleichzeitig, jedoch in jedem der 20 Prozessoreinheiten asynchron.

Das Universal-RISC-Prozessor-Modul CLIPPER (33 MHz) ist der Kern der Prozessoreinheit. Es besteht aus CPU/FPU, Befehls- und Daten-Cache-Management Unit (I-CAMMU und D-CAMMU). Die Prozessorperipherie gliedert sich in einen Merkmalspeicher (384 KB SRAM), einen Hauptspeicher für Befehls-Code und Daten (1 MB DRAM) und einen Kommunikationsspeicher (8 KB SRAM, dual port). Die Prozessor- Interkommunikation erfolgt über eine "Mailbox-Kette", die jedem Prozessor den Datenaustausch mit seinen beiden direkten Nachbarn erlaubt.

Eine Synchronisation der Prozessoreinheiten erfolgt nur in der Prozeßphase der Zuordnung und Ergebnisauswahl durch paarweisen Interrupt-Handshake. Dadurch wird eine Verteilung der Prozeßlast auf die 20 parallel arbeitenden Prozessoren möglich.

Der für die Parallelarchitekturen typische "Flaschenhals" beim Zugriff auf einen gemeinsamen Bus fehlt und es treten keine Warteschlangen auf.

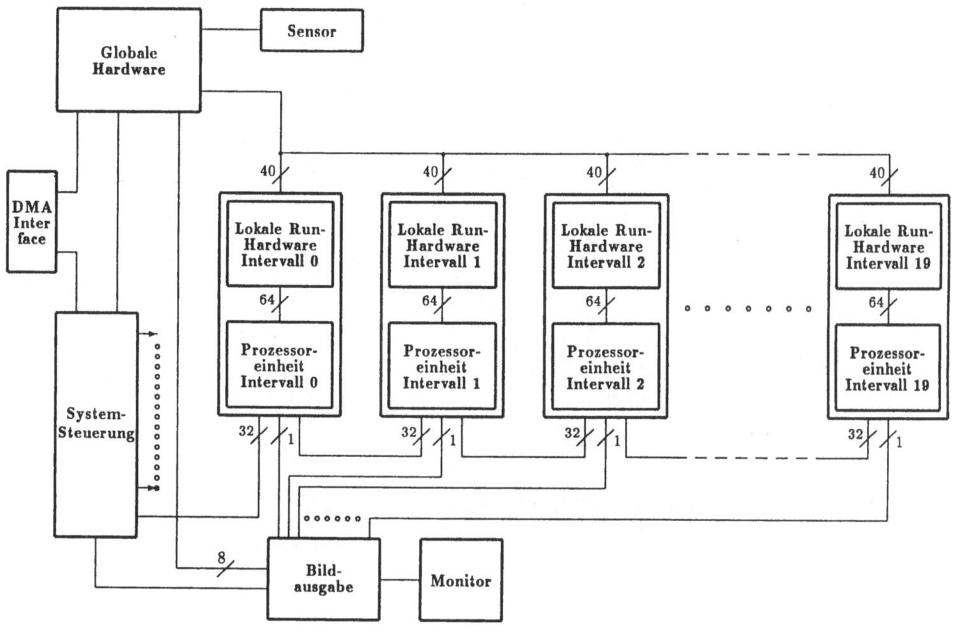

Abb. 1-4: Hardwarekonzeption

Ergebnisausgabe

Das System ist über die Anzahl der Binärschnitte modularisiert, so daß je nach Anwendung ein Betrieb mit 3 - 20 parallel bearbeiteten Grauwertintervallen möglich ist, ohne daß die Verarbeitungszeit wesentlich verlängert wird. Die Leistungsfähigkeit des Systems liegt im Durchschnitt bei ca. 3-4 Bildern/sec bei einer Bildgröße von 256x256 Bildpunkten und ca. 50 segmentierten Ergebnisobjekten/Bild. Sie ist naturgemäß von der Bildinformation abhängig. Zu einer VAX11/780 ergibt sich eine Leistungssteigerung um ca. einen Faktor 1000.

Gleichzeitig zur "ikonischen" Darstellung des Segmentierungsergebnisses auf einem Monitor wird auch eine "symbolische" Beschreibung der Objekte für weitere Verarbeitungsstufen bereitgestellt.

Das in Abb.1-5 dargestellte Gerät wurde von der Firma Philips, Systeme und Sondertechnik, in Bremen realisiert. Seit Anfang des Jahres steht es als leistungsfähige Hardware zur Verfügung.

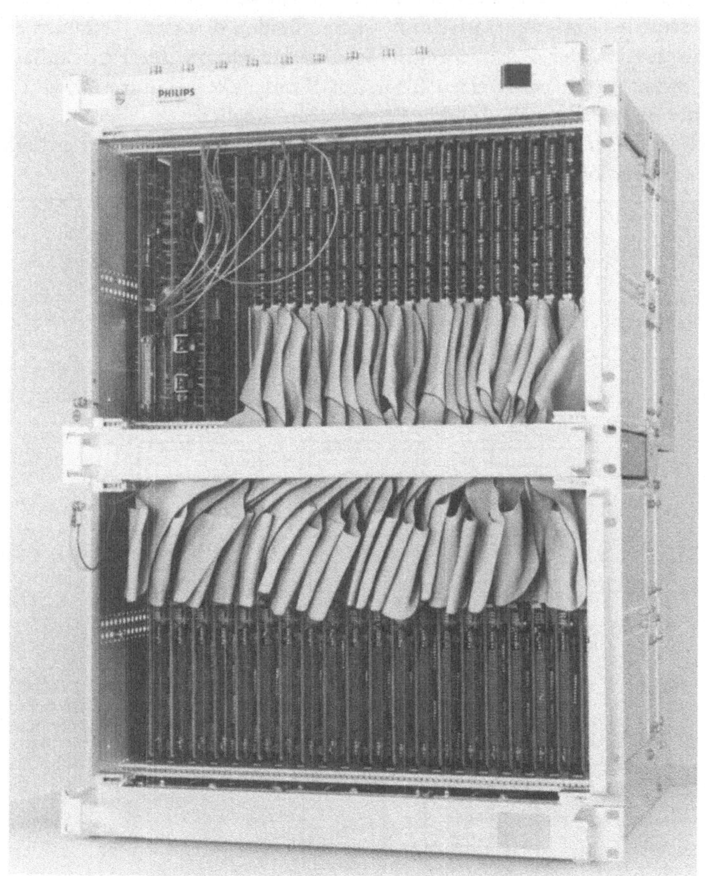

Abb. 1-5: Geräteansicht des Bildsegmentierers

Literaturverzeichnis

[1] H.Cipovic, D.Milivojevic, Z.Kajmakovic, *Planar Object Recognition by Computer Vision Methods*, RoViSec 3, SPIE No.449, 1984, pp. 9-16.

[2] F. Veillon, *One Pass Computation of Morphological and Geometrical Properties of Objects in Digital Pictures*, Signal Processing, Vol. 1, No. 3, Juli 1979, pp. 175-189.

[3] P.Vuylsteke, A.Oosterlinck, H.Van den Berghe, *Labeling and Simultaneous Feature Extraction in One Pass*, SPIE Vol.301, Design of Digital Image Processing Systems, 1981, pp. 173-180.

Integrierte Software-Werkzeuge
zur Erstellung und Benutzung von Bildverarbeitungssystemen

Mauer E., Behrens K.

Forschungsinstitut für Informationsverarbeitung und Mustererkennung
Eisenstockstraße 12   7505 Ettlingen 6

In diesem Artikel werden drei (im praktischen Einsatz stehende) komfortable und robuste Software-Werkzeuge vorgestellt, die, aufeinander abgestimmt, Unterstützung während unterschiedlicher Phasen der Software-Entwicklung in der digitalen Bildverarbeitung geben.

1. Motivation

Als Modell für die einzelnen Phasen bei der Entwicklung größerer Software-Systeme wird in der Regel der Software-Lebenszyklus (Abb. 1-1, [SOM87]) angegeben. Bei dem Umfang, den viele Software-Systeme mittlerweile annehmen, ist eine solche Entwicklung ohne unterstützende Werkzeuge kaum noch möglich. Für verschiedene Gebiete sind daher schon komplexe Entwicklungsumgebungen oder Komponenten einer solchen Umgebung im Handel erhältlich.

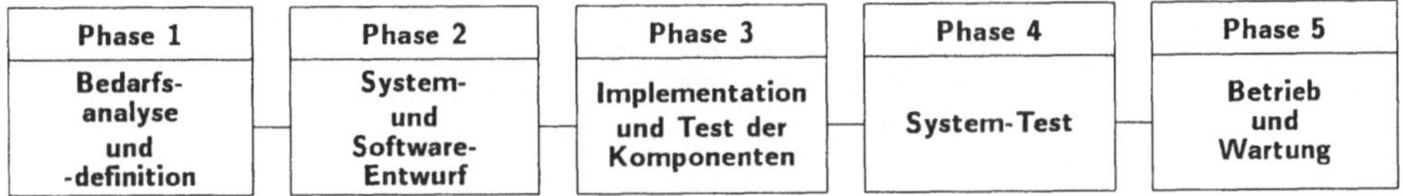

<u>Abb. 1-1:</u> Modell des Software-Lebenszyklus

Um die speziellen Belange der digitalen Bildverarbeitung berücksichtigen zu können und dem evolutionären Charakter der Forschung auf diesem Gebiet Rechnung zu tragen, wurden für die Phasen 3, 4 und 5 des Software-Lebenszyklus drei aufeinander abgestimmte Werkzeuge realisiert:

- das *Image Manipulation System (IMS)* zur Realisierung und zum Test einzelner Bildverarbeitungsoperationen (s.a. [SCH85]),

- der *Generator für interaktive Programmsteuerungen (GRIPS)* zum Test des Zusammenspiels der einzelnen Operationen in einem komplexeren System (s.a. [BEH87]) und

- das *Program Management System (PMS)* zur Verwaltung und Wartung von entwickelter Software und deren Zusammenfassung zu einem Gesamtsystem (s.a. [MAU88]).

In den folgenden Kapiteln werden diese drei Werkzeuge mit ihren Eigenschaften grob beschrieben und kurz erläutert. Weitere Details werden während der Poster-Präsentation gezeigt.

2. Entwicklung und Benutzung einzelner Bildverarbeitungsroutinen

Die Realisierung und Anwendung einzelner Bildverarbeitungsoperationen umfaßt neben der Implementierung des eigentlichen Algorithmus auch die Lösung stets wiederkehrender Probleme (z.B. Anbindung der zur Verfügung stehenden Hardware und Fremd-Software, Programmparameter- und Datenakquisition, Form der Programmbedienung, etc.). Das *IMS* reduziert einerseits den Aufwand bei der Programmentwicklung durch eine konsistente Anbindung unterschiedlicher Peripheriegeräte und durch die Emulation verschiedener Softwareschnittstellen. Andererseits wird die Programmerprobung mittels *IMS* durch dessen variable Programmbedienung, durch die intergrierten Prüfmechanismen und durch die speziellen Testhilfen spürbar vereinfacht.

2.1 Konzept von IMS

Für die Kommunikation der Verfahren mit der "Außenwelt" werden von *IMS* Datenstrukturen, sogenannte Parameterbeschreibungsblöcke (PBBs) zur Verfügung gestellt, aus denen die Parameterrampe für
den Operationsmodul zusammengesetzt werden kann. Momentan werden fünf Klassen von Parameterbeschreibungsblöcken unterschieden: PBBs für *boolesche Werte, ganzzahlige Werte, reelle Werte, Texte* und
für *die Struktur der zu verarbeitenden Bilddaten*. In Abb. 2-1 ist beispielhaft ein PBB eines ganzzahligen
Parameters **xxx** dargestellt. Dabei ist in der oberen Zeile die Bedeutung der jeweiligen Komponente, in
der mittleren Zeile der Name dieser Komponente und in der unteren Zeile der aktuelle Werteintrag aufgeführt.

Status	Default geprüft	Informationstext	Länge des Informations- textes	aktueller Parameter- wert	Untergrenze	Obergenze	Defaultwert
xxx.S	xxx.DEF	xxx.T	xxx.TE	xxx.I	xxx.U	xxx.O	xxx.D
TRUE	TRUE	"Schwellwert"	11	15	10	100	11

<u>Abb. 2-1:</u> Ausschnitt eines Parameterbeschreibungsblocks für einen ganzzahligen Parameter

Zwischen den Programmentwicklern und der "Außenwelt" wird mit den Parameterbeschreibungsblöcken
eine einheitliche Zwischenschicht eingefügt, aus der jeder Entwickler die für ihn sinnvollen PBBs aus-
wählt. Verschiedene logisch zusammenhängende Parameterbeschreibungsblöcke sind von vornherein zu einem Satz zusammengefaßt (z.B. die PBBs zur Beschreibung von Bildausschnittkoordinaten) und werden
auch "en bloc" angesprochen. In Abb. 2-2 ist die Zwischenschicht aller zur Verfügung gestellten Parameterbeschreibungsblöcke eingetragen und mit dicken Strichen gekennzeichnet. Der Entwickler wählt mittels
PBB-Schlüsselwörter die notwendigen PBBs für die zu entwickelnde Operation aus. Die PBB-Auswahl
wird in Abb. 2-2 durch das dünne Linienraster symbolisiert und soll den variablen und maskenhaften Aus-
wahlmechanismus verdeutlichen.

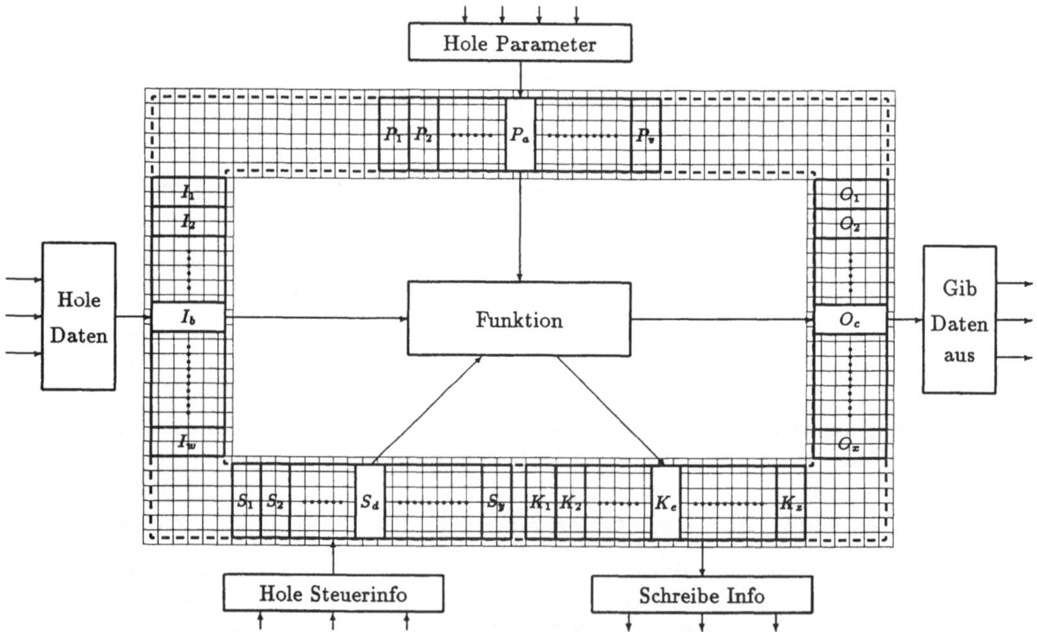

<u>Abb. 2-2:</u> Schema der IMS-Umgebung mit eingeblendeter PBB-Maske

Das Füllen der Parameterbeschreibungsblöcke mit aktuellen Werten erfolgt zur Laufzeit durch die Routine zur Parameterakquisition. Nach erfolgreicher Parameterakquisition können die Datenakquisition und
später die Ergebnisdatenausgabe automatisch in Abhängigkeit von den gewählten Parametereinstellungen

erfolgen. Dabei werden periphere Geräte über logische Namen angesprochen, und alle notwendigen Anpassungsaktivitäten werden automatisch durchgeführt. Außerdem werden für die synchrone und asynchrone Informationsausgabe und Datenvisualisierung spezielle Routinen zur Verfügung gestellt, die je nach Vorgabe von extern bedienbaren Steuerschaltern aktiviert oder deaktiviert werden können.

2.2 Entwicklung eines Programms mit IMS

Der erste Schritt zur Entwicklung eines Programms ist die Implementierung des Algorithmus als (Menge von) Unterprogramm(en) in einer beliebigen Sprache. Dabei können zur synchronen bzw. asynchronen Datenvisualisierung und zur Informationsausgabe die von IMS zur Verfügung gestellten Routinen verwendet werden.

Anschließend wird ein Rahmenprogramm erstellt, bei welchem geeignete vordeklarierte PBBs ausgewählt, deren Bedeutung für den Benutzer bekanntgegeben und die Defaultbelegungen festgelegt werden. Gegebenenfalls werden an gleicher Stelle alle Informationen für eine automatische Datentypkonvertierung (unterschiedliche externe und interne Datenrepräsentation) oder eine automatische Geometrieanpassung (unterschiedliche externe und interne Bilddatenstrukturen) eingetragen. Anschließend können (bei Bedarf wiederholt) die automatische Parameterakquisition und Datenakquisition aufgerufen werden. Das in der Regel kleine Rahmenprogramm endet mit der Aktivierung der Ergebnisausgabe und der damit verbundenen abschließenden Ablaufdokumentation. Zum Übersetzen, Binden und Testen der erstellten Quelltexte werden von IMS alle notwendigen Kommandoprozeduren zur Verfügung gestellt.

2.3 Benutzung eines mit IMS erstellten Programms

Bei der Definition der Benutzerschnittstelle von IMS wurde besonders auf Robustheit und auf benutzeradaptive Aktivierungsmodi geachtet. Robustheit wird durch erneutes Abfragen nicht verständlicher oder inkonsistenter Eingaben erreicht. Dabei werden sowohl Inkonsistenzen zwischen den einzelnen Eingaben als auch Inkonsistenzen zwischen den Eingaben und den Programmerwartungen überprüft. Je nach dem Kenntnisstand des Anwenders und dem jeweiligen Anwendungskontext eignen sich unterschiedliche Aktivierungsmodi:

- Geführter Modus mit sequentieller Anforderung aller Operationsparameter mit detaillierten Erläuterungen.

- Ungeführter Modus anhand einer einzelnen Kommandozeile, bei welcher der Benutzer über Schlüsselwörter die Defaultwerte der einzelnen Operationsparameter verändern kann.

- Durch Abrufen der Operationsparameter von bereits aktivierten Operationen mittels automatisch angelegten und editierbaren Parameterdateien (Pilotfiles).

Durch die Vereinheitlichung der Programmbedienung und des äußeren Programmrahmens wird dem einzelnen Anwender der Zugang zu unterschiedlichen Operationen erleichtert, die teilweise von unterschiedlichen Verfassern erstellt worden sind oder deren Erstellungsdatum bereits länger zurückliegt. Auch ist es so nicht notwendig, zu jeder einzelnen Operation ein Bedienungshandbuch zu erstellen. Statt dessen reichen in der Regel kurze operationsspezifische Funktionsbeschreibungen, die dann automatisch (siehe dazu auch Kapitel 4) in einer Help-Library oder einem Nutzerhandbuch zusammengefaßt werden können.

3. Zusammenfassung mehrerer Programmodule unter einem Rahmenprogramm

Um geeignete Operationssequenzen für die jeweiligen Aufgabenstellungen zu erproben, sind unterschiedliche Vorgehensweisen zur Kombination der implementierten Operationen möglich. Zum einen besteht die Möglichkeit, etwa mit Hilfe von Kommandofiles, mehrere Operationen nacheinander zu aktivieren und die Datenweitergabe zwischen diesen Operationen über externe Files durchzuführen, was ein erhebliches Maß an Ein-/Ausgabetätigkeit mit sich bringt. Zum anderen kann man in einem gemeinsamen Hauptprogramm die Unterprogrammaufrufe der einzelnen Operationen miteinander verknüpfen und so die Datenweitergabe intern durchführen. In diesem Fall läßt sich jedoch nicht jederzeit die Reihenfolge der Operationen ändern, sondern es muß bei Bedarf neu editiert, übersetzt und gebunden werden. Um diese Nachteile zu vermeiden, ist es sinnvoll, einen Auswahlmodul (*Dispatcher*) zu erzeugen, der sowohl eine vorgegebene als auch zur Laufzeit eine beliebige Reihenfolge der Operationsausführung ermöglicht (Abb. 3-1).

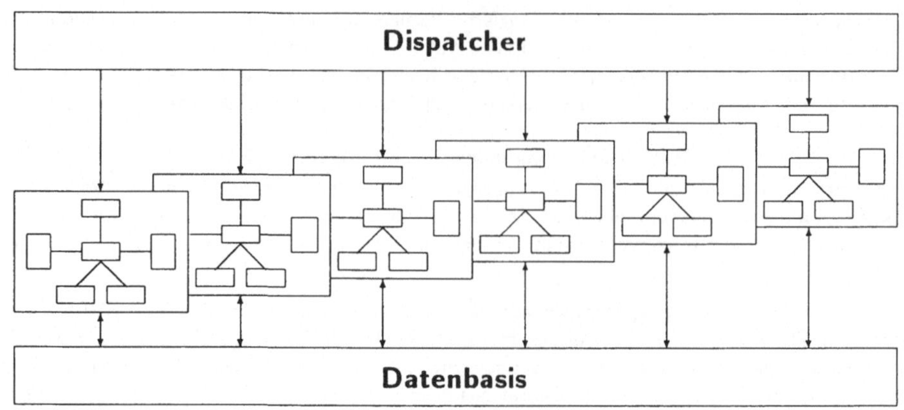

<u>Abb. 3-1</u>: Zusammenfassung von Bildverarbeitungsoperationen unter einem Dispatcher

Der *Generator für interaktive Programm-Steuerungen (GRIPS)* nimmt dem Benutzer die Erstellung eines solchen Dispatchers ab und erzeugt zusätzlich eine Menüsteuerung für die Operationsauswahl.

3.1 Konzept von GRIPS

Das von *GRIPS* gelieferte Ergebnis ist der Dispatcher, ein Programmodul, bei dem die einzelnen Operationen mit je einer, vom Benutzer wählbaren, Taste des Terminal-Keypads verknüpft sind. Dabei können unterschiedliche Abhängigkeitsmodi und besondere, vom Benutzer vorgebbare, Bedingungen zwischen den einzelnen Operationen berücksichtigt werden. Die Belegung der Tasten wird in einem Menü auf dem Bildschirm ausgegeben. Das Menü umfaßt auch Fenster für Textein- und -ausgaben und Fehlermeldungen. Für die Nutzung dieser Fenster werden von *GRIPS* Routinen zur Verfügung gestellt. Ferner existiert eine Routine zum Umschalten zwischen verschiedenen Menü-Ebenen, so daß durch das Erzeugen weiterer Menü-Ebenen quasi beliebig umfangreiche Programmpakete und -steuerungen erstellt werden können.

3.2 Anwendung von GRIPS

Die Erzeugung des Dispatchers erfolgt, wie in Abb. 3-2 dargestellt, in zwei Stufen. In der ersten Stufe wird mit Hilfe des speziell für diese Zwecke geschaffenen Editors *EDGRIPS* eine Datei erstellt, in der der Benutzer angibt, welche Operation unter welchen Bedingungen mit welcher Taste auf welcher Menü-Ebene verknüpft sein soll. Die zweite Stufe besteht aus dem eigentlichen Dispatcher-Generator, der das mittels *EDGRIPS* erzeugte File in einen PASCAL-Programmodul umsetzt, das den Dispatcher für das Programmpaket enthält. Dieser Modul wird dann mit den Einzelverfahren zusammengebunden und ergibt ein tastendruckgesteuertes Programm.

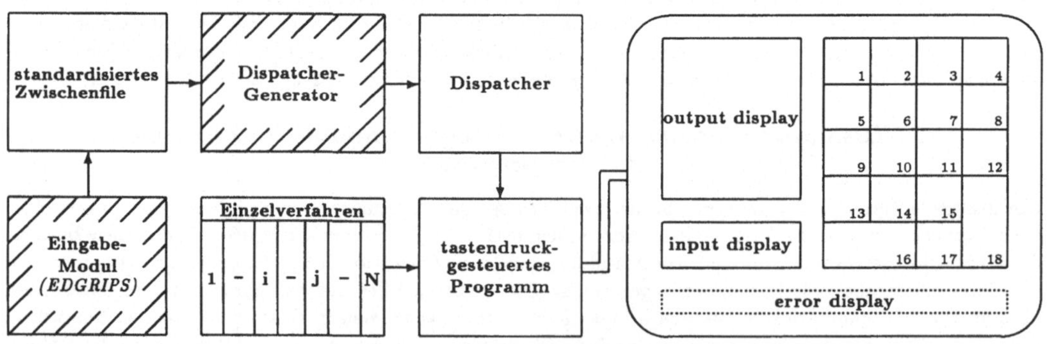

<u>Abb. 3-2</u>: Schema der GRIPS-Umgebung

Die Menübedienung kann anstelle durch Drücken der entsprechenden Tasten auch durch Anwahl der Funktion aus der Bildschirmgrafik mit Hilfe einer "Maus" erfolgen. Die automatische Generierung des Dispatchers entkoppelt den Benutzer von der Menüverwaltung und der Maussteuerung. Sie verhindert Programmierfehler und macht den Benutzer unabhängig von der zugrundeliegenden Systemsoftware. Das Hinzufügen neuer Operationen ist sehr einfach, da lediglich mit Hilfe von *EDGRIPS* die entsprechende Datei zu erweitern ist und der Dispatcher-Generator diese Änderung automatisch in den Dispatcher einträgt.

Durch Abspeichern von Tastendruckreihenfolgen lassen sich bewährte Operationskombinationen wiederholt aufrufen (z.B. im Batch-Betrieb). In einer weiteren Ausbaustufe ist beabsichtigt, daß über die Tastendrücke, die Reihenfolge und über den Erfolg der damit verbundenen Operationen eine Statistik geführt wird, so daß Vorschläge für nachfolgend zu aktivierende Operationen und Aussagen über ihre Erfolgswahrscheinlichkeiten gemacht werden können.

4. Verwaltung eines komplexen Bildverarbeitungssystems

Aus der Sicht des Anwenders besteht ein großes Interesse daran, (sowohl eigene als auch fremde) Programme, Module und Dokumentationen in einem größeren System zur Verfügung gestellt zu bekommen, ohne sich selbst um die Verwaltung eines solchen Systems kümmern zu müssen. Diese Systeme entwickeln sich, zumindest im Bereich der Forschung, durch Hinzunahme neu erstellter Software oder neuer Hardware ständig fort. In der Regel existiert zu jedem größeren System ein System-Manager, der für die Verwaltung und Erneuerung zuständig ist. Ein Teil der Arbeit des System-Managers läßt sich automatisieren, wie dies im *Program Management System (PMS)* realisiert und auf *IMS* und *GRIPS* angewandt wurde.

4.1 Konzept von PMS

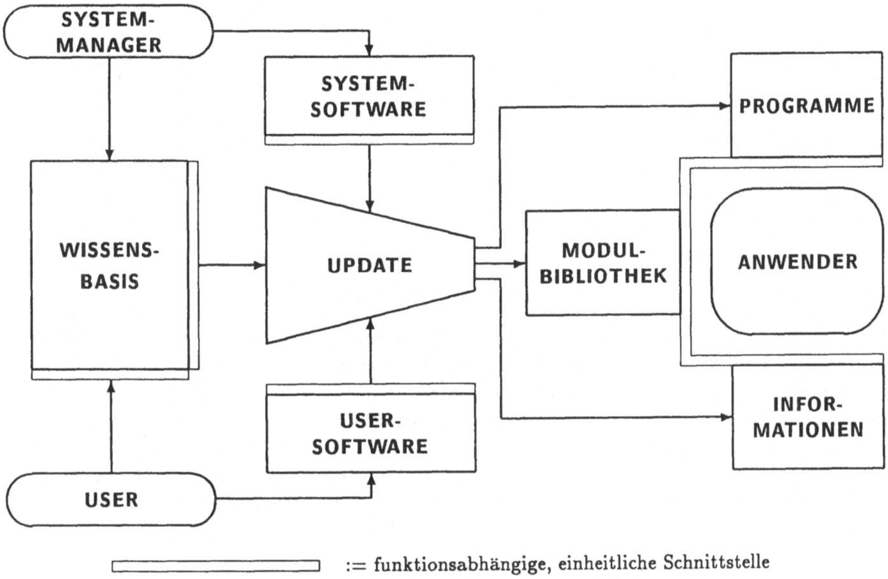

Abb. 4-1: Schema der PMS-Umgebung

Die wesentlichen Bestandteile von PMS sind in Abb. 4-1 wiedergegeben und haben folgende Aufgaben: Der System-Manager erstellt die System-Software (z.B. zur Bedienung neuer Hardware) und trägt die erforderlichen Informationen in die Wissensbasis ein. Der User erstellt die Anwender-Software und gibt diese über ein standardisiertes Eingabemodul ebenfalls der Wissensbasis bekannt. Der UPDATE-Modul beschafft sich die nötigen Informationen aus der Wissensbasis und kann so auf die Quelltexte von System- und User-Software zugreifen. Er erzeugt daraus die einzelnen Programme, faßt die Unterroutinen in Modulbibliotheken zusammen und bringt auch die zugreifbaren Informationen (Help-Library, System-Dokumentation, Operationsbeschreibungen) auf den neuesten Stand. Dieser Vorgang wird getriggert durch Veränderungen in der Wissensbasis. Wenn diese eintreten, wird der UPDATE-Modul aktiviert, und nach Abschluß dieses Vorgangs stehen alle Neuerungen sämtlichen Anwendern zur Verfügung. Dem Anwender

bleiben diese Aktivitäten verborgen. Er hat nur eine einheitliche Oberfläche vor sich, die ihm den Zugriff
auf die Programme, die Unterprogramme und die Informationen erlaubt.

4.2 Anwendung von PMS

Zur Bekanntgabe der Informationen über neue Software steht dem jeweiligen Software-Ersteller *(User)*
ein Programm zur Verfügung, das die Eintragungen in der Wissensbasis prüft und vornimmt. Ein in regelmäßigen Abständen automatisch aktiviertes Überwachungs-Programm überprüft, ob an der Wissensbasis Veränderungen stattgefunden haben. Ist dies der Fall, wird der UPDATE-Modul aktiviert und führt
eine Erneuerung des Systems (oder Teile davon) über Nacht durch. Dazu wird zunächst eine Sicherungs-
kopie des laufenden Systems angelegt. Im Anschluß daran wird aus der *System-Software* die Systemum-
gebung in dem Umfang erzeugt, wie sie vom *System-Manager* in der Wissensbasis bekanntgegeben wurde
(d.h. noch nicht freigegebene System-Software wird auch nicht berücksichtigt!). Danach werden die notwendigen Programm-Umgebungen (Include-Files, Environment-Files, etc.) auf eigenen Unterverzeichnis-
sen gesammelt. In einem dritten Schritt werden die Quelltexte der *User-Software* von den einzelnen Usern
geholt (wenn nötig auch von fremden Rechnernetzknoten), übersetzt und in einer *Modulbibliothek* zusammengefaßt. Durch Binden der Hauptprogramme mit dieser Bibliothek entsteht dann eine Menge einzelner
Programme. Für die Benutzung dieser Programme unter einer einheitlichen Benutzeroberfläche werden
Kommandoprozeduren erzeugt, in denen alle notwendigen Deklarationen von Symbolen und logischen Namen stattfinden, so daß durch Absetzen eines einzelnen Befehls der Anschluß an das System durchgeführt
werden kann. Nach Abschluß dieser Arbeiten wird das veränderte System auf alle dem System bekannten Rechner verteilt. Dadurch, daß alle diese Aktivitäten automatisch und nur nachts ablaufen, wird der
normale Rechenbetrieb tagsüber nicht gestört, wie das sonst bei der Wartung größerer Systeme durchaus
üblich ist.

Ferner wird zur Verbindung von *IMS* und *GRIPS* eine Umwandlung der *IMS*-Hauptprogramme in für
GRIPS verwendbare Funktionen durchgeführt. Dadurch und aufgrund der Eigenschaften von *GRIPS* lassen sich sehr leicht und schnell eine beliebige Kombination von *IMS*-Programmen zu einem aufgabenspezifischen Programmpaket zusammenfassen. Zusätzlich werden mittels *PMS* die zu jedem Programm
existierenden Help-Files gesammelt, durch einen automatischen Editor in eine für das Textverarbeitungs-
system LaTeX([LAM86]) verständliche Form gebracht und zu einem Operationshandbuch zusammengefaßt,
so daß zusammen mit dem Systemhandbuch jederzeit eine komplette, aktualisierte Version der *Informationen* über das System vorliegt.

5. Zusammenfassung

Es wurden drei aufeinander abgestimmte Software-Werkzeuge für die Realisierung und Verwaltung von
Bildverarbeitungs-Programmsystemen vorgestellt. Diese wurden auf DEC-Rechnern unter VMS entwickelt
und haben sich bereits seit mehreren Jahren im praktischen Einsatz bewährt. Sie werden institutsweit zur
Verfügung gestellt und für unterschiedliche Anwendungen in der Bildverarbeitung genutzt. Aus den mit
IMS und *GRIPS* erzeugten Einzelprogrammen hat sich unter Einsatz des *PMS* ein sehr komplexes, umfassendes und sich stetig weiterentwickelndes Bildverarbeitungssystem gebildet, in dem eigene und fremde,
neue und alte, einfache und komplexe Verfahren durch eine einheitliche Bedienoberfläche gleichermaßen
leicht zugänglich sind.

Literatur

[BEH87]   **Behrens K.** *GRIPS - Generator für interaktive Programm-Steuerungen*, Handbuch, FIM-
Kurzbericht Nr. 180, FIM/FGAN 1987

[LAM86]   **Lamport L.** LaTeX: *A Document Preperation System*, Addison-Wesley 1986

[MAU88]   **Mauer E., Schmied A.** *PMS - Program Management System*, FIM-Kurzbericht Nr. 207,
FIM/FGAN 1989

[SCH85]   **Behrens K., Mauer E., Schmied A.** *IMS - Image Manipulation System*, Handbuch, FIM-
Kurzbericht Nr. 147, FIM/FGAN 1985

[SOM87]   **Sommerville I.** *Software Engineering*, Addison-Wesley 1987

PICASYS - Ein Bildanalysesystem zur Identifikation
von Leiterplatinen

K.-U. Höffgen, M. Goerke, H. Noltemeier

Lehrstuhl für Informatik I der
Universität Würzburg
Am Hubland, 8700 Würzburg

Zusammenfassung : Das hier vorgestellte modellbasierte Bildanalyse-
system zur Identifizierung von Leiterplatinen wurde auf sehr einfacher
Hardware und für eine relativ geringe Bildauflösung entwickelt. Der
gewählte Modellbereich ist dabei nur beispielhaft; eine Anwendung des
Systems in anderen Bereichen sowie eine Erweiterung des Systems zur
Qualitätskontrolle ist durchaus möglich.

1. Einleitung

Im Bereich der Produktionssteuerung und Qualitätskontrolle wurden bereits in den vergangenen Jahren eine Reihe von modellbasierten Bild-
analysesystemen entwickelt.
Der dem System PICASYS (vgl. auch [4] und [6]) zugrundeliegende
Modellbereich ist eine Anzahl von Computersteckkarten (bestückte Leiterplatinen), und das Ziel ist eine automatische Identifizierung einer
einzelnen Karte auf einem Eingabebild, das eine Auflösung von 212 x
256 Pixeln mit jeweils 256 Grauwerten hat.

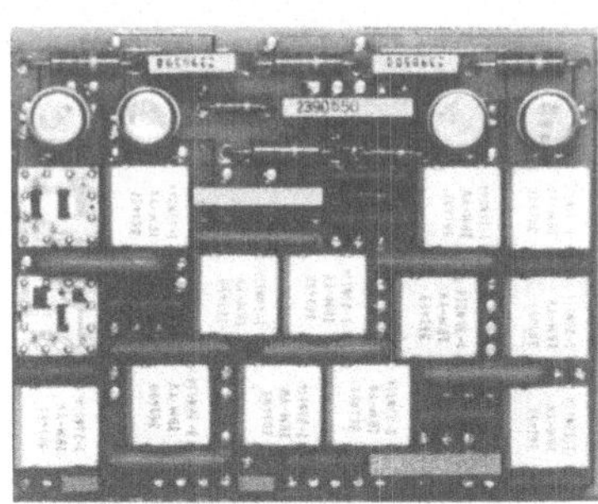 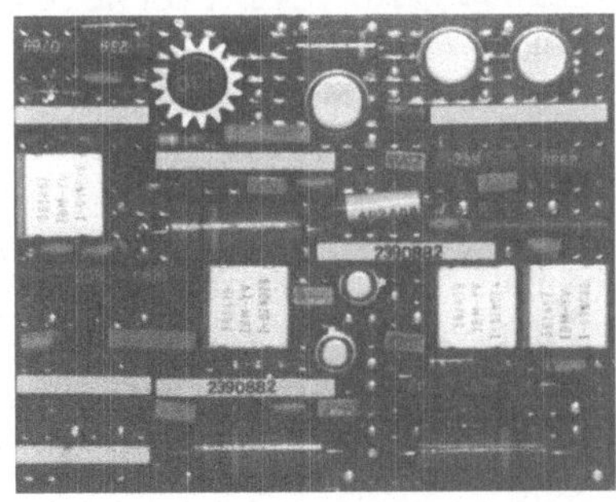

Abb.1: 2 der 9 Leiterplatinen aus dem Modellbereich von PICASYS

Es werden nun die einzelnen Komponenten des Systems PICASYS beschrieben: nach der Segmentierung wird ein Stringmatchingverfahren vorgestellt und anschließend die Kontrollkomponente des Systems beschrie-

ben. Die praktischen Erfahrungen mit dem System stehen am Schluß dieser Arbeit.

2. Segmentierung

In einem ersten Verarbeitungsschritt wird die Lage und Orientierung der Karte im Bild festgestellt. Die Randpunkte der Karte werden hierbei durch ein Schwellwertverfahren und die Orientierung der Karte selbst mit Hilfe der Hough-Transformation bestimmt. Sollte die Karte nicht achsenparallel ausgerichtet sein, so wird sie explizit im Bild gedreht.

Die Segmentierung der einzelnen Bauteile (Bausteine) auf der Karte, die in Klassen zusammengefaßt sind, erfolgt in zwei Stufen durch eine überwacht lernende Klassifizierungsstrategie [5]. In der ersten Stufe werden Pixel anhand ihrer Farbwerte und ihrer Dichte (= Anzahl der Pixel in einer Umgebung mit dem gleichen Farbwert) klassifiziert. Resultat hiervon ist ein Binärbild, das weitgehend aus den Bildpunkten einer gesuchten Bausteinklasse besteht. Nach einer Verbesserung dieses Binärbildes (Region growing, sowie Löschen isolierter Punkte) werden dann existierende Zusammenhangsgebiete durch kleinstmögliche Rechtecke eingerahmt. Hierbei kann man verschiedene, auf die jeweilige Bausteinklasse ausgerichtete Arten von Zusammenhangsgebieten betrachten (z.B. Punktmengen, die horizontal,vertikal und diagonal zusammenhängen und zusätzlich Lücken von einem Pixel in horizontaler Richtung aufweisen dürfen). Die Rechtecke werden anschließend anhand der Merkmale Größe und Form weiter klassifiziert, wobei hier die Merkmalsausprägungen durch das Modell der entsprechenden Bausteinklasse bestimmt sind.
Im Gegensatz dazu werden die Merkmalsausprägungen Farbe und Dichte der Bausteinklassen durch Testaufnahmen (Trainingsgebiete) initialisiert und können nach erfolgreicher Analyse des Bildes verbessert werden. Dieses ermöglicht insbesondere die Anpassung des Systems an veränderte Aufnahmebedingungen (Lichtverhältnisse).
Die Klassifizierung selbst konnte aufgrund der niedrigen Dimension des Merkmalsraumes durch ein Table-Lookup Verfahren realisiert werden.

3. Matching

Das Matchingverfahren kann ausschließlich strukturelle Informationen ausnützen und muß aufgrund der geringen Bildauflösung sehr robust sein. Unter diesen Voraussetzungen war das Stringmatchingverfahren von Chang et. al. [2] ein geeigneter Ausgangspunkt. Die Idee dabei ist, zunächst jedem Objekt im Bild einen repräsentativen Punkt der Pixelmatrix zuzuweisen und dann das Bild durch zwei Strings darzustellen, die die Reihenfolge der Objekte in x- bzw. y-Richtung beschreiben. Um nun gewisse Verschiebungen der repräsentativen Punkte relativ zueinander tolerieren zu können, wurde das Bild in jeder der beiden Richtungen in n Pixel breite Streifen unterteilt. Innerhalb dieser Streifen können dann die Objekte beliebig angeordnet sein, da die Reihenfolge

der Objekte nur noch auf die Streifen bezogen ausgewertet wird. Damit
bleibt das Problem, daß ein Objekt, das nahe an einer Streifengrenze
liegt, in den benachbarten Streifen verschoben wird. Diesen Objekten
wird jetzt die Möglichkeit gegeben, beiden Streifen anzugehören; d.h.
das Matching mit dem Modell entscheidet die exakte Zugehörigkeit dieser Objekte.

Konkret sieht damit das Matchingverfahren wie folgt aus: zunächst wird
jedes Eingabeobjekt lokal mit jedem Modellobjekt verglichen. Wenn die
beiden Objektgrößen bis auf eine gewisse Fehlertoleranz übereinstimmen, so wird von einem lokalen Matching dieser beiden Objekte gesprochen. Um nun die Menge der lokalen Matchpartner eines Eingabeobjektes
einzuschränken, wird zusätzlich eine globale Übereinstimmung bezüglich
der Lage des Objektes in Eingabebild bzw. Modell innerhalb gewisser
Fehlertoleranzen gefordert. Nach der Bildung der lokalen Matchmengen
wird in x- bzw. y-Richtung die Übereinstimmung bzgl. der Anordnung der
Objekte in Eingabebild und Modell überprüft. Dabei wird verlangt, daß
Objekte, die im Bild in einem Streifen liegen, nur mit solchen Modellobjekten matchen, die auch in einem Streifen liegen, oder daß die
Beziehung "liegt rechts von" (bzw. "liegt links von") im Eingabebild
auch im Modell erfüllt ist. Der Algorithmus durchläuft dabei in Depth-
First-Search die möglichen lokalen Matchpaare, wobei die Objekte, die
nahe an den Streifengrenzen liegen, beiden Streifen angehören können.
Als Ergebnis liefert der Algorithmus entweder ein erfolgreiches
Matching zwischen Bild und Modell, oder er bricht erfolglos ab.

4. Kontrolle

Die erste Aufgabe der Kontrolle beim Identifikationsprozeß ist die
Erstellung von Hypothesen, die dann im weiteren Verlauf verifiziert
werden.

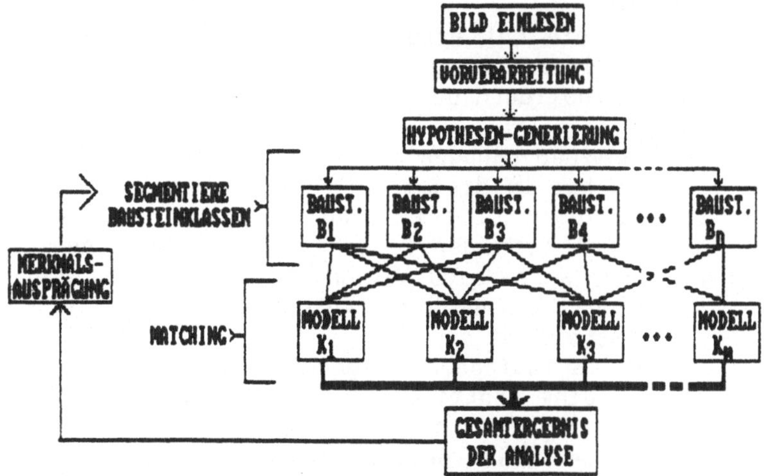

Abbildung 2 : Ablaufplan des Gesamtsystems

In PICASYS werden jeweils drei mögliche Modelle als gleichwertige
Hypothesen generiert, wobei dazu das Grauwerthistogramm ausgenutzt
wird: in einer Trainingsphase wurden die Grauwerthistogramme verschiedener Bilder auf signifikante Grauwerte untersucht und deren Größenbe-
reiche den einzelnen Modellen zugeordnet. Anschließend werden für jede
dieser 3 Hypothesen die Bausteinklassen aus einer Tabelle abgelesen,
die für das jeweilige Modell die besten Segmentierungsergebnisse erwarten lassen (diese Tabelle wurde wiederum in einer Trainingsphase
aufgestellt).
Nach der Segmentierung einer Bausteinklasse wird dann in x- und y-
Richtung ein Matching durchgeführt. Da obiges Matchingverfahren grundsätzlich keine Bewertungsmöglichkeit für die Übereinstimmung zwischen
Bild und Modell besitzt, wurde folgende Idee eingeführt: in den
Matchingprozeß wurden von Anfang an nur die Eingabeobjekte aufgenommen, deren lokale Matchmenge nicht leer war (die anderen hätten in
jedem Fall zu einem Abbruch des Verfahrens geführt). Ist das Matching
nun erfolgreich, so wird ein Matchergebnis berechnet, das die Anzahl
der Bausteine mit nichtleerer lokaler Matchmenge in Verhältnis setzt
zur Anzahl der segmentierten Bausteine bzw. der Anzahl der passenden
Modellbausteine. Anschließend wird dieses Ergebnis noch gewichtet, bevor daraus dann für jede Hypothese das Gesamtmatchingergebnis berechnet wird. Abschließend werden die Gesamtmatchingergebnisse der drei
Hypothesen verglichen und, sofern ein deutliches Maximum auftritt, das
entsprechende Modell als Ergebnis akzeptiert.

5. Experimentelle Ergebnisse

Das Gesamtsystem wurde so konzipiert, daß einerseits das Eingabebild
automatisch ausgewertet werden kann, um den Kartentyp anzugeben, und
andererseits die Bilddaten interaktiv verarbeitet werden können.
Hierzu steht dem Anwender eine umfangreiche menügesteuerte Benutzeroberfläche zur Verfügung.
Die Implementierung erfolgte auf einem IBM PC-AT in der Programmiersprache 'C'. Zur Digitalisierung wurde das Videocomputer-System MSX-
8280 von Phillips mit Farbbildern der Auflösung 212x256 und einem Byte
pro Pixel benutzt.
Bei bestmöglichen Aufnahmebedingungen füllt die Karte ungefähr 60% des
gesamten Bildes aus. Sie hat in diesem Fall eine Kantenlänge von 203 x
158 Pixeln. Auf Ihr befinden sich ungefähr 80 Bausteine, wobei der
größte ca. 700 Pixel groß ist, der kleinste jedoch nur ungefähr 40.
Aufgrund der geringen Hardware-Voraussetzungen, die an das System gestellt wurden, ist eine größtmögliche Ausnutzung des Informationsgehaltes der Bilder entscheidend.
Es wurden insgesamt 98 Testaufnahmen gemacht. Um die Robustheit des
Systems in Hinblick auf eine praktische Anwendung zu überprüfen, wurden die Bilder mit unterschiedlichen Aufnahmebedingungen digitalisiert. Insbesondere wurden die Beleuchtungsverhältnisse, Lage und
Größe der Karte im Bild verändert sowie unscharfe Aufnahmen digitalisiert.

Von den 98 Aufnahmen der Karten konnten 95 eindeutig identifiziert
werden, wobei der Modellraum aus neun verschiedenen Kartentypen bestand. Zwei der Analysen lieferten keine eindeutigen Aussagen, und
eine Aufnahme wurde falsch identifiziert.
Die Gesamtzeit einer Analyse, bestehend aus Segmentierung der Karte
und der Bausteinklassen sowie des Matchings, lag im Bereich von 38 bis
56 Sekunden. Der durchschnittliche Wert betrug 47 Sekunden.
Die Segmentierung der einzelnen Bausteinklassen sowie das Matching
sind jeweils unabhängig voneinander und somit parallel durchführbar.
Da jeder einzelne dieser Schritte auf einem PC ungefähr 8 Sekunden
dauert, läßt sich mit entsprechender Hardware die Gesamtverarbeitungszeit auf diesen Wert reduzieren; ein On-Line-Einsatz (d.h. eine mittlere Gesamtverarbeitungszeit von etwa 3 Sekunden) erscheint bei Verwendung schnellerer CPU's realisierbar.

6. Ausblick

Das in diesem Beitrag vorgestellte System demonstriert die Möglichkeiten des Einsatzes der Bildverarbeitung in der Automatisierung eines
Produktionsablaufs zur Herstellung von Leiterplatinen. Insbesondere
wurde gezeigt, daß das Erkennen von Objekten, die sich durch Strukturmerkmale voneinander unterscheiden, mit relativ einfacher Hardware
realisierbar ist. Die hierzu verwendeten Verfahren zeichnen sich durch
eine größtmögliche Ausnutzung des Informationsgehaltes der Bilder aus
und lassen sich damit auf eine Vielzahl von Anwendungsgebieten übertragen. Aufgrund der Robustheit des Systems, der reduzierbaren Ant-
wortzeiten und der geringen Qualitätsanforderungen an die digitalisierten Bilder ist eine industrielle Anwendung, etwa im Bereich der
Werkstückkontrolle, durchaus denkbar.

7. Literatur

[1]   Horst Bunke
      "Modellgesteuerte Bildanalyse" Stuttgart : Teubner, 1985
[2]   S.K.Chang, Q.Y.Shi, C.W.Yan
      "Iconic Indexing"   IEEE Trans. PAMI Vol.9, No.3, 1987
[3]   S.K.Chang, C.W.Yan, D.C.Dimitroff, T.Arndt
      "An Intelligent Image Database System"
      IEEE Trans. on Software Engineering Vol.14, No.5, 1988
[4]   Mark Goerke
      "Anwendung der Bildverarbeitung in der Produktionssteuerung
      Würzburg; Diplomarbeit am Lehrstuhl für Informatik I, 1989
[5]   Peter Haberäcker
      "Digitale Bildverarbeitung : Grundlagen und Anwendungen"
      München : Hanser, 1987 - 2.Auflage
[6]   Klaus-Uwe Höffgen
      "Strukturelle Matchingverfahren in der modellbasierten
       Bildanalyse"
      Würzburg; Diplomarbeit am Lehrstuhl für Informatik I, 1989
[7]   Heinrich Niemann / Horst Bunke
      "Künstliche Intelligenz in Bild und Sprachanalyse"
      Stuttgart : Teubner 1987

Anwendung eines schnellen,morphologischen Bildverarbeitungsrechners am Beispiel der Metaphasensuche

J. Hagelberg-Wölfing, B. Schleifenbaum

Wild Leitz GmbH, Wetzlar

Zusammenfassung

Als Metaphase bezeichnet man einen durch geeignete Präparation fixierten Zustand einer biologischen Zelle. Zur automatischen Suche dieser Metaphasen, die für eine Routineanwendung geeignet ist, wird eine schnelle Bildverarbeitungsrechnerarchitektur vorgestellt. Mit Methoden der mathematischen Morphologie für Grau- und Binärbilder und entsprechenden Hardwareprozessoren gelingt eine automatische Metaphasensuche, die den Forderungen des Anwenders nach Geschwindigkeit und Qualität der gefundenen Metaphasen bei vertretbarem Hardwareaufwand genügt.

1. Grauwertmorphologie

Durch Verallgemeinerung der Methoden der mathematischen Morphologie/SER82/ von Binärbildern auf Grauwertbilder ist es möglich, vorhandenes a-priori Wissen im Hinblick auf die zu lösende Bildanalyseaufgabe schon bei der Grauwertvorverarbeitung gezielt einzusetzen. Ein systematisches Vorgehen bei der Planung von morphologischen Algorithmen zur Segmentierung und Merkmalsextraktion ist in /SER86/ beschrieben.

Die Begriffe der mathematischen Morphologie sind gleichermaßen auf Binärbilder als auch auf Grauwertbilder anwendbar. Grauwertbilder werden hier als zweidimensionale, wertediskrete Helligkeitsfunktionen $h(x)$ einer diskreten, (hexagonal) gerasterten Ebene E mit den Schnitten

$$X_t(h) = \{ x : h(x) > t \}, \text{ für alle } t \in h$$

betrachtet.

Aus der Forderung, daß sich für morphologische Operationen an Grauwertbildern Funktionen mit den gleichen Schnitten ergeben, die man auch als Ergebnis binärer morphologischer Operationen auf Schnitten der Ausgangsfunktionen erhalten würde, ergibt sich z. B. als Verallgemeinerung des Mengendurchschnitts von Binärbildern die untere Einhüllende zweier Funktionen $f(x)$ und $g(x)$:

$$(f \cap g)(x) := \min(f(x), g(x))$$

Das Komplement eines Graubildes entspricht dem Negativbild. Weitere wichtige, elementare Begriffe der Morphologie auf Grauwertbilden sind:

Erosion $\quad h \ominus B := \min(h(y)),$	$B_x := \{z \in E : z = y + x, x \in E, y \in E\}$	Translation von B um x
$\qquad y \in B_x$	$\check{B} := \{x \in E : -x \in B\}$	Spiegelung
	$B \in E$	Nachbarschaft (Strukturelement)

Eine entsprechende Definition ergibt sich für die

Dilatation $\quad h \oplus B := \max(h(y))$

$\qquad y \in B_x.$

Auf ähnliche Art und Weise lassen sich auch die Abmagerung und Verdickung von Binärbildern auf Grauwertbilder übertragen. Für diese beiden Transformationen bleiben bei der Anwendung der entsprechenden zusammengesetzten Strukturelemente im hexagonalen Raster die topologischen Beziehungen im Grauwertgebirge erhalten.

Mit diesen elementaren Begriffen lassen sich weitere wichtige zusammengesetzte Transformationen angeben, wobei H die elementare hexagonale Umgebung des Ursprungs ist:

Ouverture $\quad h \ominus H^r \oplus H^r \qquad$ Elimination von hellen Objekten mit einem Radius < r.

Fermeture $\quad h \oplus H^r \ominus H^r \qquad$ Elimination von dunklen Objekten mit einem Radius < r.

Ouverture und Fermeture sind idempotent (Filtereigenschaft). Diese Transformationen bilden die Ausgangsbasis für eine Familie von sogenannten morphologischen Filtern /SER88/.

Top-Hat-Transformation /MEY77/

$$X_t(h - h \ominus H^r \oplus H^r)$$

Diese Transformation erzeugt ein Binärbild der Differenz von Originalbild h und einer Ouverture. Damit können helle Objekte auf dunklem Hintergrund mit einem Radius ‹ r segmentiert werden, ferner linienhafte Strukuren mit einer Breite ‹ 2r. Die Segmentierung gelingt dabei unabhängig von der lokalen Helligkeit des Hintergrundes. Ebenso ist die Segmentierung von dunklen Objekten auf hellem Hintergrund möglich

$$X_t(h \oplus H' \ominus H' - h)$$

Gradient $(h \oplus H) - (h \ominus H)$

Diese Transformation erzeugt ein richtungsunabhängiges Gradientenbild.

2.   Bildverarbeitungsrechner mit schnellen

   morphologischen Prozessoren.

Eine Rechnerarchitektur(Leitz MIAC, Modular Image Analysis Computer), auf der die oben angegebenen Algorithmen für Grau- und Binärbilder in Echtzeit ablauffähig sind, zeigt Bild 2.

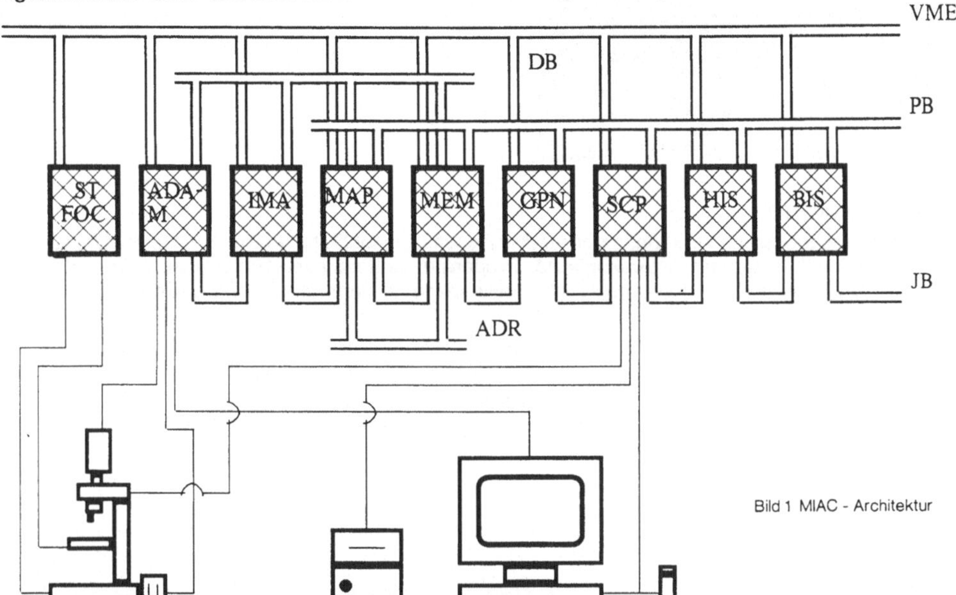

Bild 1 MIAC - Architektur

Für die Lösung von Routineaufgaben bei der Analyse von Mikroskopbildern kommt dazu ein Leitz-Medilux-Mikroskop mit rechnergesteuertem Scanningtisch für die x,y und z - Verstellung, rechnergesteuerter Objektivrevolver und rechnergesteuerte Lampenstromversorgung.

Die wichtigsten Merkmale der MIAC-Architektur sind:

- Parallelvideobus (PB), 2 * 8 Bit   -Bildpunktrate 13.5 MHz

- Pipelinevideobus (JB), 2 * 8 Bit   -Steckplatzkodierung

- Displayvideobus (DB), 1 * 8 Bit, 1 * 4 Bit (Overlay)   -Modulidentifikation

Folgende Hardwaremodule sind einsetzbar:

ADAM: TV-IO mit analoger Shadingkorrektur.

IMA: 4-Bit Grafikmodul mit 256 KByte DRAM

MAP/MEM: Modularer Bildspeicher, bis zu zwei 16-Bit Videotore, gleichzeitig

                Steuerrechnerzugriff (Grau- oder Binärmode)

GPN: Nichtlinearer (morphologischer) Grauwertprozessor, geeignet für Fließbandvearbeitung

BIS: Binärprozessormodul mit vier elementaren Binärprozesorstufen

HIS: Grauwerthistogrammberechnung und Runcode-Generierung in 20 ms

SCP: Steuerrechner mit MC 68000, 1 MByte DRAM und Bildpunktzähler

Die Steuerung des Gesamtsystems erfolgt durch ein UNIX - ähnliches Echtzeitbetriebssystem (Idris). Die Verwaltung der Bildverarbeitungshardware übernimmt ein hardwareunabhängiges Betriebssystem (IMOS, / HEI87/).

Zur einfachen experimentellen Algorithmenentwicklung steht eine Grafik- und Maus - unterstützte Programmierumgebung zur Verfügung (Bild 2).

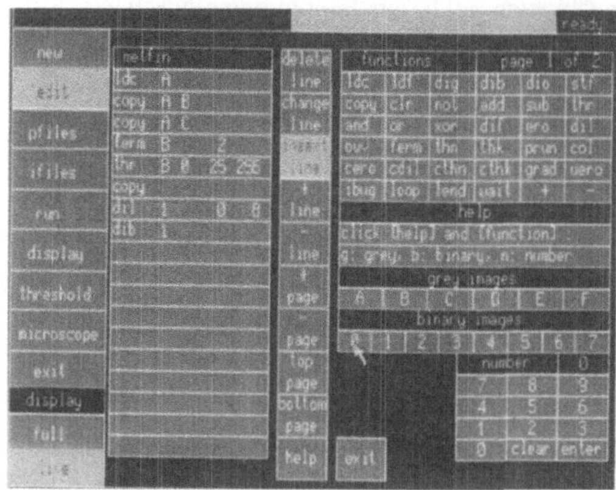

Bild 2 Interaktive Programmierumgebung

3. Schnelle Metaphasensuche

Ein solcher Bildverarbeitungsrechner kann in der medizinischen Diagnostik für Aufgaben eingesetzt werden, wo es auf schnelle Durchmusterung mikroskopischer Präparate nach seltenen Ereignissen ankommt ("rare event detection"). Die wohl am meisten untersuchte Anwendung ist dabei die Metaphasensuche /BER84, LÖR84/.

Als Metaphase bezeichnet man einen durch geeignete Präparation fixierten Zustand einer biologischen Zelle, in der die in den Chromosomen enthaltene genetische Information sichtbar gemacht ist (Bild 3). Nach der automatischen Metaphasensuche schließt sich meist ein interaktives Analyseprogramm an, mit dem eine genetische Diagnose (Karyotypie) erstellt oder eine Aberrationsanalyse als Mutationstest durchgeführt wird.

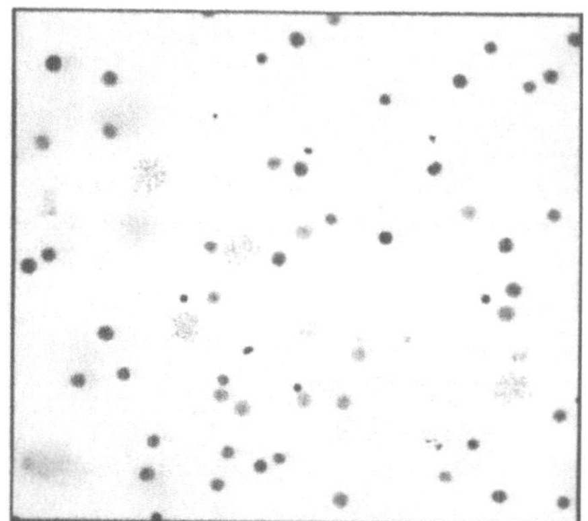

Bild 3 Originalbild mit Metaphasen

Für den Anwender muß aus ökonomischen Gründen neben der Zuverlässigkeit und einfachen Bedienung eine bedeutende Zeitersparniss gegenüber einer manuellen Durchmusterung der Präparate gewährleistet sein.

Der erste Schritt des Segmentierungsalgorithmus ist ein schneller morphologischer Test zur Entscheidung ob überhaupt Metaphasen bzw. Metaphasenkandidaten im Bildfeld sind. Die Chromosomen sind bei der gewählten Vergrößerung (20x) dunkle Bildbereiche mit r ‹ 3 Bildpunkte (Bild 3). In einer Metaphase sind davon bis zu 46 in einem Bereich mit r ‹ 20 vorhanden. Zur Segmentierung dient eine Top-Hat-Transformation (Bild 4b) mit einem Strukturelement entsprechender Größe. Die Schwelle (t) wird allerdings adaptiv ermittelt, um Chromosomen mit schwachem Kontrast und solche mit starkem Kontrast und dunklerem Hintergrund gleichermaßen segmentieren zu können (Bild 4c).

Eine anschließende binäre Dilatation läßt dann die Chromosomen verschmelzen (Bild 4d). Einen optimalen Datenflußgraphen ("Pipeline") /Hein87/ für einen Teil der morphologischen Bildverarbeitung bis zu diesem

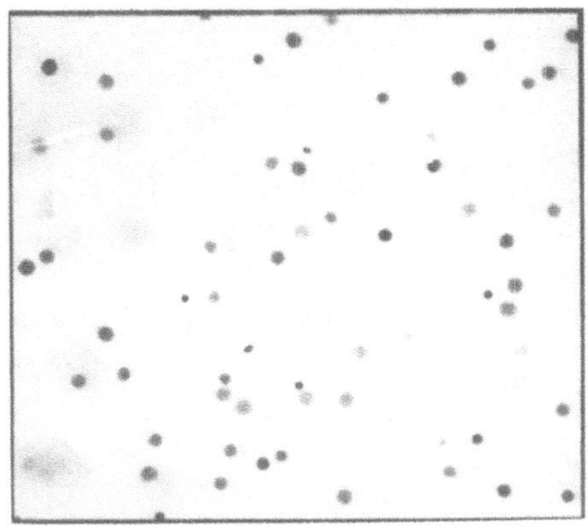

Bild 4a 2-mal dilatiertes Originalbild

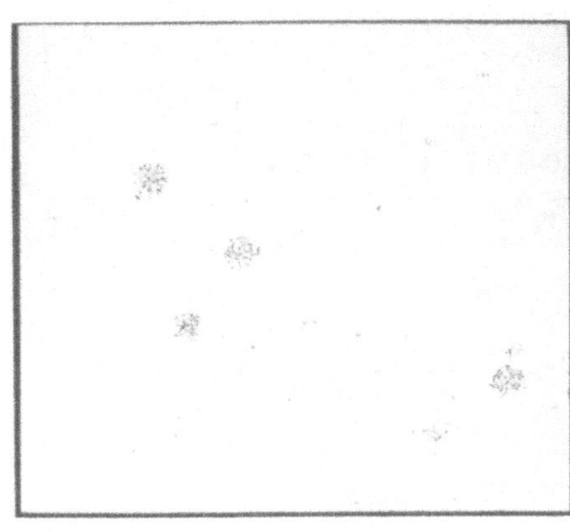

Bild 4b Top-Hat vor Schwellwertbildung

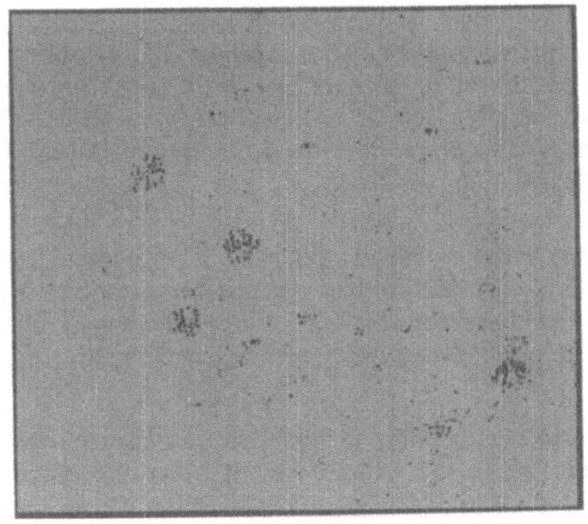

Bild 4c Top-Hat nach Schwellwertbildung

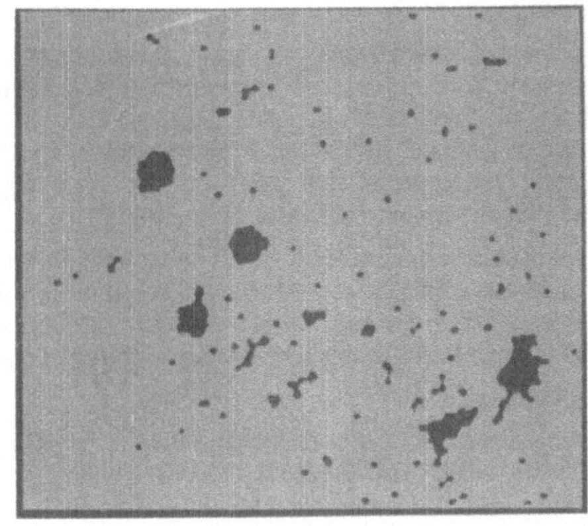

Bild 4d Verschmelzung von Chromosomen (binär)

Bild 4e Marker (binär)

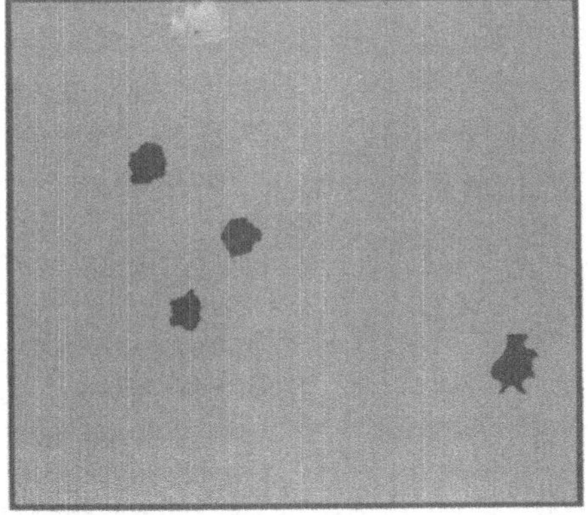

Bild 4f Masken (binär)

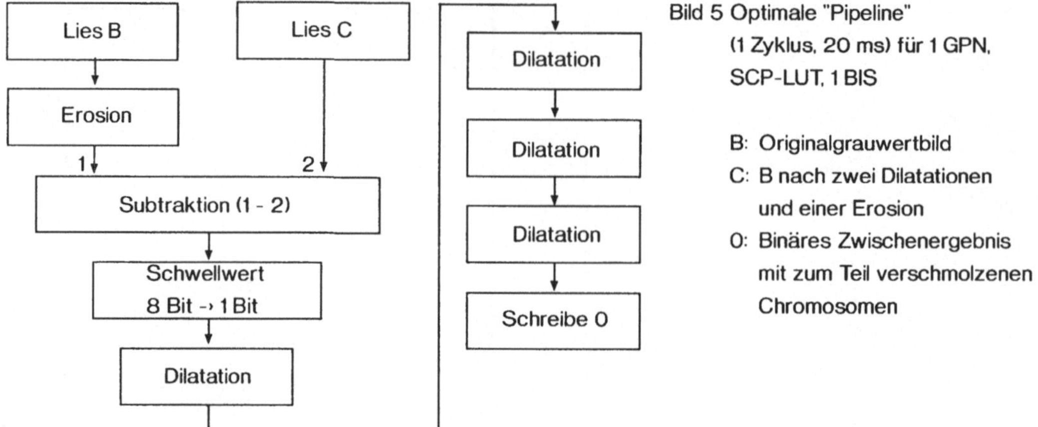

Punkt zeigt Bild 5.

Eine weitere Erosion eliminiert kleine Artefakte läßt aber die Metaphasenbereiche (Marker, Bild 4e) übrig. An dieser Stelle wird durch eine Bildpunktzählung entschieden, ob überhaupt Metaphasen im Bildfeld vorhanden sein können. Eine bedingte Dilatation mit Bild 4d rekonstruiert dann die potentiellen Metaphasenbereiche vollständig (Maske, Bild 4f). Eine logische Differenzbildung zwischen Bild 4c und einem Artefaktbild (geeigneter Schwellwert auf große dunkle Bereiche (Bild 4a)) dient zur Verhinderung einer Fehlsegmentation z.B. bei Präparationsartefakten.

Durch eine Individualanalyse dieser Masken (Fläche, Form) und der in diesen Masken enthaltenen Objekte (Chromosomen, Bild 4c) nach Anzahl und Fläche, wird die Güte der Metaphasenkandidaten ermittelt. Wenn der für jede Maske ermittelte Gütewert oberhalb einer vorgegebenen Schranke liegt, werden die x, y und z - Koordinaten und der Gütewert der Metaphase für eine spätere (interaktive) Analyse abgespeichert.

4. Ergebnisse

Die Suchzeiten hängen sehr stark von der Art der Präparation ab. Je nach Metaphasen und Artefaktdichte werden für 1000 Bildfelder von je 0.3 mm^2 8 - 20 min benötigt. Das entspricht einer Zeitersparnis gegenüber der manuellen Suche von etwa 2 - 3. Dabei ist die Qualität der automatischen Suche immer gleichbleibend und nicht von subjektiven Größen beeinflußt.

Literatur

/SER82/   Serra, J.   Image Analysis and Mathematical Morphology
Academic Press, London, 1982

/SER86/   Serra. J.   From Mathematical Morphology to artificial Intelligence
In: Proc. Eigth International Conference on Pattern Recognition
Paris, 1986. Vol. 2, S. 1336 - 1343

/SER88/   Serra, J.   Image Analysis and Mathematical Morphology
Vol. 2: Theoretical Advances, Academic Press, 1988

/Mey77/   Meyer, F.   Contrast features extraction
In: Quantitative Analysis of Microstructures in Materials Science,
Biology and Medicine, Dr. Riederer Verlag, Stuttgart, 1977,
S. 374 - 380

/HEI87/   Heinrich, K.   Leitz-Bildanalysegeräte als symbolische, programmierbare
Palic, J.   Datenflußrechner, Tagungsband DAGM 1987

/BER84/   v. d. Berg, H.T.C.M.   The Automation of Metaphase Finding and Chromosome
Analysis, Dissertation, Universität Leiden, 1984

/LÖR84/   Lörch, T.   Automatischer Metaphasenfinder
Frieben, M.   Tagungsband DAGM 1984,
Bille, J.   S. 63 - 69

ECHTZEITSPURERKENNUNG MIT EINEM SYSTOLISCHEN ARRAYPROZESSOR

F. Klefenz, R. Männer
Physikalisches Institut, Universität Heidelberg, D-6900 Heidelberg, F.R. Germany

1. Einleitung

In vielen Detektoren der Hochenergiephysik werden Bilder von Teilchenspuren (Abb. 1) generiert. Um den enormen Datenstrom zu reduzieren, müssen in diesen Bildern bestimmte Spurtypen in Echtzeit erkannt werden. Die interessanten Spuren entstehen in der Mitte des Detektors und verlaufen kreisförmig nach außen. Zusätzlich treten statistisch Rauschpunkte auf. Bei der OPAL-Kammer [1] beträgt die Aufnahmerate 1 Bild/25 µs. Mit dem hier beschriebenen systolischen Arrayprozessor können die Krümmungsradien und zugehörigen Startwinkel beliebig vieler gut definierter Spuren innerhalb dieser Zeit ausgelesen werden.

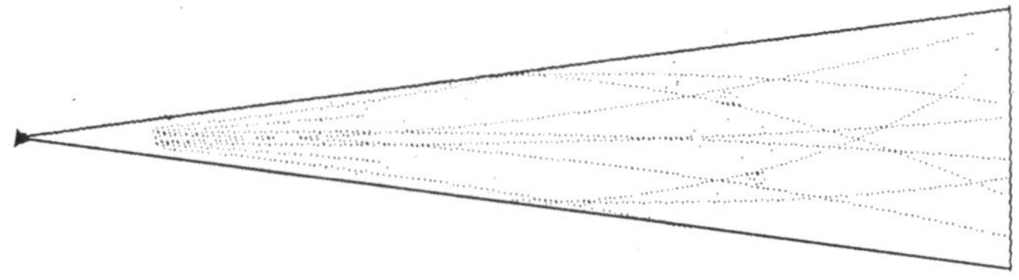

Abb.1: Sektorausschnitt aus der OPAL-Driftkammer

2. Prinzipielle Wirkungsweise

Die interessanten Teilchenspuren bestehen aus einer Reihe von Punkten, die auf Kreisbahnen durch den Ursprung liegen. Jeder Punkt ist durch seine Detektorkoordinaten r und ϕ gegeben und kann einer ganzen Schar von Bahnen mit unterschiedlichem Krümmungsradius $1/r_c$ und Startwinkel ϕ_s zugeordnet werden. Die Hough-Transformation [2] zwischen den Detektorkoordinaten und den Koordinaten in der $(1/r_c,\phi_s)$-Ebene ergibt den funktionalen Zusammenhang (Abb. 2)

$$\frac{r/2}{r_c} = \sin(\phi - \phi_s) \qquad (1)$$

Wird $1/r_c$ gegen ϕ_s aufgetragen, ergibt sich ein Sinus mit der Amplitude $2/r$ und der Phasenverschiebung ϕ. Für zwei oder mehrere Bahnpunkte schneiden sich solche Sinuskurven (Abb. 3). Die Überschneidungshäufigkeit in einem Punkt ist ein direktes Wahrscheinlichkeitsmaß für das Vorliegen einer Bahn mit den entsprechenden Parametern $1/r_c$ und ϕ_s, wobei die Wahrscheinlichkeit proportional zur Anzahl der beitragenden Bahnpunkte ist.

3. Simulation des Verfahrens

Zur Simulation des Verfahren wird die kontinuierliche Wahrscheinlichkeitsverteilung in der $(1/r_c,\phi_s)$-

Ebene durch ein zweidimensionales Histogramm angenähert. Dazu wird die $1/r_C$-Achse in 100 Bins geteilt (Intervall $5m < r_C < \infty$), die ϕ_S-Achse in 250 Bins (Intervall $-7{,}5° < \phi_S < 7{,}5°$). Für jeden Datenpunkt wird ein Sinus mit Amplitude $2/r$ generiert, der um ϕ phasenverschoben in dem $(1/r_C,\phi_S)$-Raster aufgetragen wird. Jedes vom Sinus überdeckte Rasterelement wird um 1 inkrementiert. Das resultierende zweidimensionale Histogramm gibt die Wahrscheinlichkeit für das Vorliegen einer bestimmten Bahn wieder.

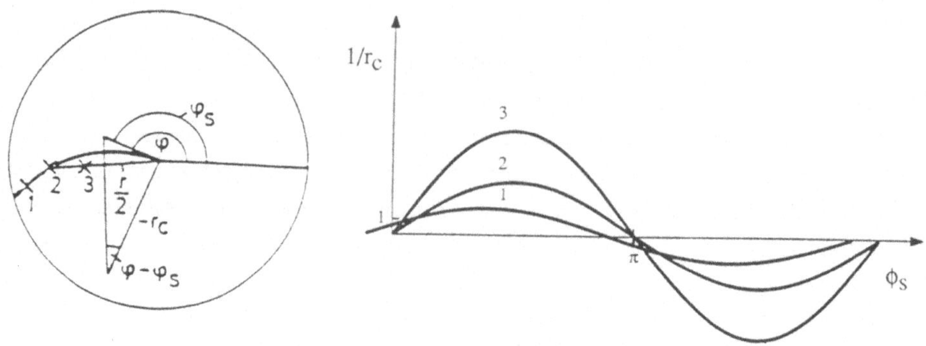

Abb. 2, 3: Beziehung zwischen den Koordinaten (r,ϕ) und $(1/r_C,\phi_S)$

Es wurden zunächst ideale Kreisbahnen mit unterschiedlichen r_C- und ϕ_S-Werten untersucht. Linksgekrümmte Bahnen werden in die untere Halbebene abgebildet, rechtsgekrümmte in die obere. Es ergibt sich eine diagonale Streifenstruktur mit einem scharf ausgeprägten Maximum (Abb. 4). Aufgrund der Diskretisierung werden allerdings nicht alle Bahnpunkte einer Kreisbahn im gleichen Bin akkumuliert. Bei idealen Bahnkurven ist das Maximum der Verteilung krümmungsradiusunabhängig (Abb. 4, 5) und die Häufigkeitsverteilung bis auf die Binningeffekte unabhängig von einer Verschiebung in ϕ_S.

Echte Bahnpunkte liegen jedoch nicht exakt auf Kreisbahnen. Dies liegt z.T. an der begrenzten Ortsauflösung des Detektors, z.T. an der Ablenkung der Teilchen durch Vielfachstreuung. Außerdem sind die Teilchenspuren in einen Störuntergrund eingebettet.

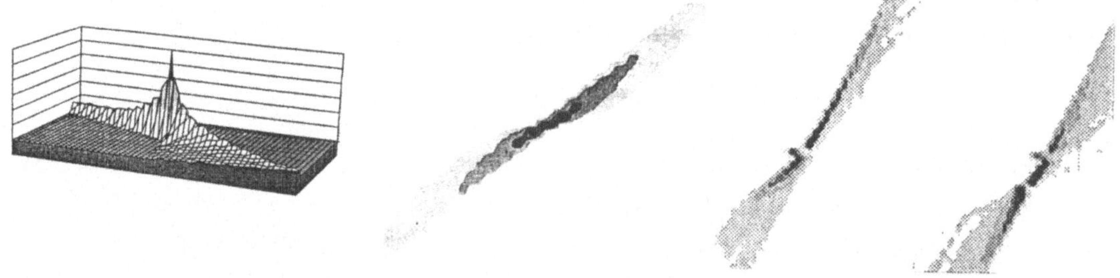

Abb. 4, 5, 6: Akkumulation der Bahnpunkte in der $(1/r_C,\phi_S)$-Ebene (a: ideal, b: Detektordaten)

Bei einer Simulation mit Detektordaten treten in den jeweiligen Sektoren ebenfalls charakteristische Streifen auf, die trotz Störpunkten eindeutig einzelnen Spuren zugeordnet werden können (Abb. 6). Das Verfahren ist sehr unempfindlich gegenüber Verrauschung, da die zu Rauschpunkten gehörenden Sinuskurven in der $(1/r_C,\phi_S)$-Ebene nur zu einem gleichmäßigen Untergrund führen. Abb. 7 zeigt künstlich verrauschte Bilder, die ohne Probleme korrekt verarbeitet werden können.

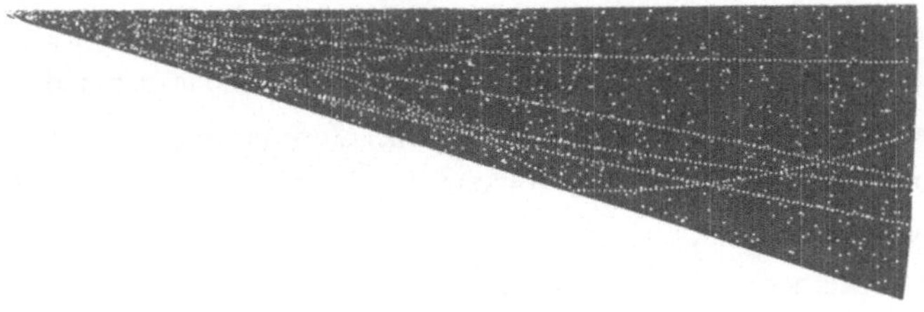

$$\begin{bmatrix} -1 & 0 & 1 \\ 0 & 2 & 0 \\ 1 & 0 & -1 \end{bmatrix}$$

Abb. 7: Künstlich verrauschte Sektordaten   Abb. 8: Faltungsoperator

4. Spurerkennung im Histogramm

Aus dem Histogramm muß auf die Anzahl der vorliegenden Spuren zurückgeschlossen werden. Eine einfache Peakdetektion über eine globale Schwellwertoperation scheidet aus, da stark gekrümmte Bahnen ein schwächeres Maximum als schwächer gekrümmte herausbilden. Daher wird ein Merkmalsklassifikator verwendet, der die Streifenmaxima detektiert. Der Merkmalsklassifikator wird durch eine 3×3-Maske (Abb. 8) realisiert, mit der das Eingangsbild gefaltet wird. Das Resultat einer solchen Faltung ergibt ein maximales Korrelationssignal an den Orten der Streifenmaxima. In einer konstanten Umgebung ergibt der Merkmalsklassifikator dagegen wegen der sich dort kompensierenden Diagonalgewichtungsterme kein Ausgangssignal. Nach Anwendung dieser Filteroperation erscheinen selbst stark gekrümmte Bahnen signifikant über dem Rauschniveau. Die Schwelle kann nun global gesetzt werden. Dies reduziert die Häufigkeitsverteilung auf ein Binärbild. Es besteht aus isolierten Clustern zusammenhängender Pixel, wobei jeder Cluster zu einer Spur in der $(1/r_C, \phi_S)$-Ebene gehört.

Die Zahl der vorhandenen Cluster läßt sich über die Eulerrelation [3] ermitteln, die im Rechteckraster folgende Form annimmt:

$$\text{Konnektivitätszahl} = n \cdot \begin{bmatrix} 1 & 0 \\ 0 & 0 \end{bmatrix} - m \cdot \begin{bmatrix} x & 1 \\ 1 & 0 \end{bmatrix}.$$

n gibt an, wie oft das erste, m, wie oft das zweite 2×2-Muster im Bild vorkommt. Die Konnektivitätszahl entspricht direkt der Zahl der Objekte, falls keine Einschlüsse auftreten. In unserem Fall treten Einschlüsse jedoch praktisch nicht auf; deshalb ist die Konnektivitätszahl hier gleich der Zahl der gefundenen Cluster, d.h. der Zahl der Spuren. Bei Anwendung dieses Verfahrens auf Detektordaten wird die Zahl der vorliegenden Spuren bei einer festen Schwelle von 260 mit ausreichender Genauigkeit korrekt ermittelt (Tabelle 1).

Tabelle 1: Klassifikationsergebnisse

Sektornummer	4	7	8	12	13	17	19	21	22	13 verrauscht (Abb. 7)
Vorliegende Spuren	1	2	2	1	4	2	2	1	1	4
Rekonstruierte Spuren	1	2	2	0	4	1	2	1	1	4

5. Implementierung

Das vorgestellte Verfahren erfordert, daß zu jedem Bahnpunkt ein Sinus amplituden- und phasenrichtig in der $(1/r_c, \phi_s)$-Ebene eingetragen wird. Aufgrund des Detektoraufbaus werden die Datenpunkte in radialer Richtung gleichzeitig ausgelesen, während - nach entsprechender elektronischer Vorverarbeitung - zu jedem Zeitpunkt t nur solche Datenpunkte anfallen, die zu einem $\phi = \text{const} \cdot t$ gehören. Dadurch wird die Hardwareimplementierung ganz erheblich vereinfacht, weil nun zu jedem Zeitpunkt t nur noch eine Schar von Sinuskurven unterschiedlicher Amplitude, aber gleicher Phasenverschiebung anfällt.

Bei einer direkten Implementierung des Verfahrens würde diese Schar von Sinuskurven mit konstanter Geschwindigkeit über ein Zählerarray hinweggeführt. Jeder Bahnpunkt würde die zugehörige Sinuskurve ansteuern und die davon überdeckten Zähler des Arrays würden inkrementiert. Wird das Zählarray als CCD [4] realisiert, so ist es jedoch wesentlich einfacher, eine ortsfeste Schar von Sinuskurven zu verwenden und die CCD-Elemente mit der entsprechenden Geschwindigkeit elektronisch darunter zu verschieben.

Bei einer vollelektronischen Implementierung wird die Sinusmaske lithographisch auf das CCD aufgebracht. Wird die entsprechende Sinuskurve angesteuert, so wird allen überdeckten CCD-Elementen über eine Konstantstromquelle eine Einheitsladung zugeführt. Nach einer synchronen Verschiebung aller CCD-Inhalte können die zum nächsten Winkelelement gehörenden Datenpunkte bearbeitet werden. Bei einer optisch-elektronischen Hybridlösung werden die Sinuskurven über ein Hologramm generiert, so daß die Ladungsakkumulation in den CCD-Elementen direkt optisch erfolgen kann. Zur Zeit werden Versuche gemacht, die jeweiligen Sinuskurven über eine Reihe von Lichtquellen anzusteuern, die entsprechend den gerade vorliegenden Datenpunkten in beliebiger Kombination eingeschaltet werden können.

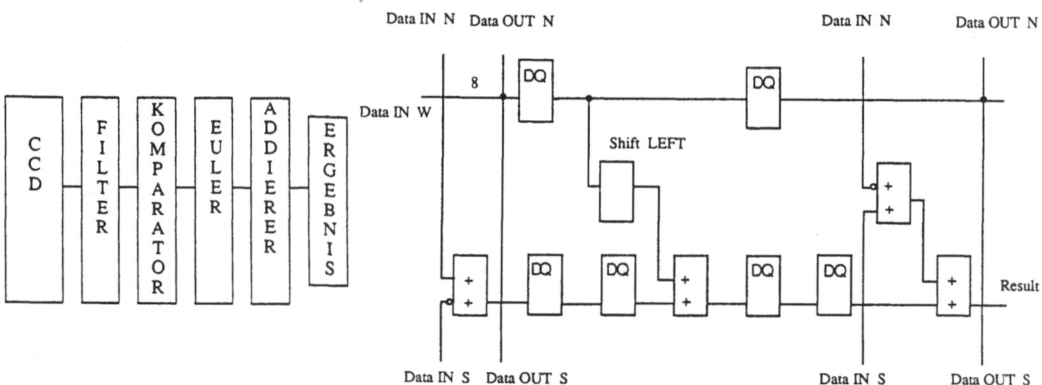

Abb. 9: CCD und Pipeline-Prozessor Abb. 10: Aufbau des Faltungs-Prozessors

Die Histogrammauswertung kann in einem Pipeline-Prozessor erfolgen. Dieser Prozessor besteht aus drei Stufen: der Faltung des Histogramms mit dem Merkmalsklassifikator, der Binärbilderstellung über eine Schwellenoperation und dem Zählen der verbleibenden Cluster durch Errechnen der Konnektivitätszahl (Abb. 9). In der ersten Stufe genügt die vorzeichenrichtige Addition der Inhalte von Nachbarzellen zu dem doppelten Inhalt der betrachteten Zelle (Abb. 10). Dies kann durch eine geeignete Verschaltung von Addierern und Verzögerungsgliedern realisiert werden; die Multiplikation mit 2 wird durch das Verschieben des Operanden um ein Bit erreicht. Zum Errechnen der Konnektivitätzahl werden nur 2×2-Matrizen mit binären Elementen (0, 1, x= don´t care) verwendet, so daß sich die zugehörigen Verknüpfungen über Gatter leicht realisieren lassen (Abb. 11). Diese Pipelinestufe generiert Signale, die zu den für die Konnektivitätszahl wesentlichen Größen m und n aufaddiert werden müssen. Dies geschieht über vier Stufen von table-lookup-ROMs (Abb. 12). Die einzelnen Bits werden in einem binären Baum schrittweise zu dem Ergebnis zusammengefaßt, von dem dann gegebenenfalls ein Detektor-Auslesesignal abgeleitet wird.

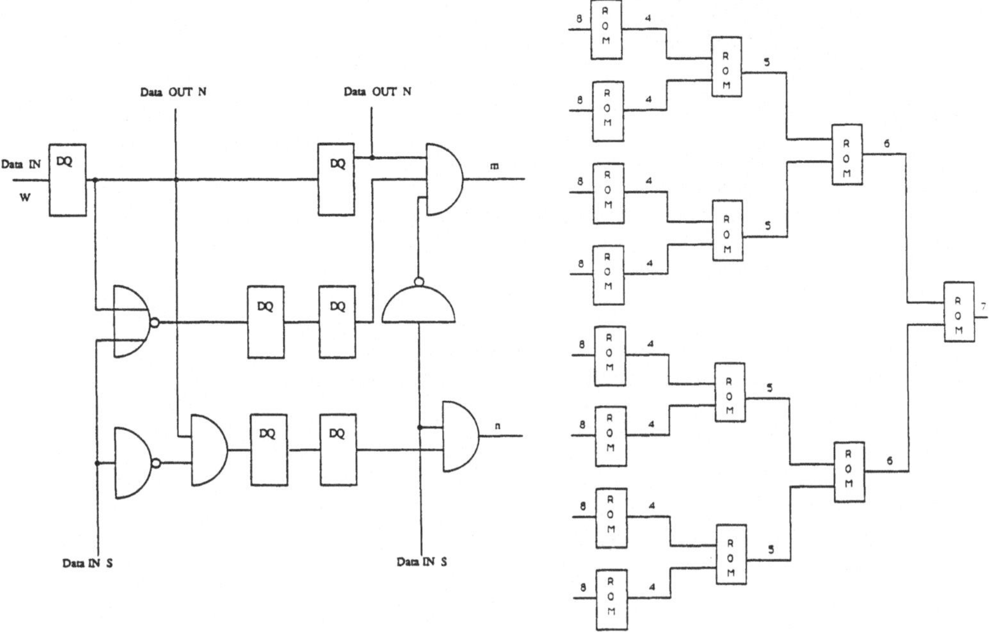

Abb. 11: Aufbau des Euler-Prozessors   Abb. 12: Aufbau des Parallel-Addierers

6. Schlußfolgerungen

Es wurde ein Verfahren vorgestellt, mit dem Kreisbögen beliebigen Durchmessers detektierbar sind. Die Resultate zeigen, daß das Verfahren hochauflösend und darüberhinaus unempfindlich gegen Störuntergrund ist. Mit der beschriebenen Hardwarerealisierung könnte eine beliebige Anzahl von Kreisbögen innerhalb von 25 µs detektiert und nach Krümmungsradius und Startwinkel klassifiziert werden. Ein entsprechender Prozessor soll bei großen Detektorsystemen in der Hochenergiephysik eingesetzt werden.

7. Referenzen

[1] Heuer R.D., Wagner A.: The OPAL Jet Chamber; *Nucl. Instr. Meth.*, Vol. A265 (1988) 11-19
[2] Davies E.R.: A Modified Hough Scheme for General Circle Location; *Patt. Recog. Lett.*, Vol. 7 (1988) 37-43
[3] Duff J.: *Cellular Logic Image Processing*; Academic Press, London (1986)
[4] Schroder D.K.: *Advanced MOS Devices*; Addison Wesley, Reading, MA (1987)

Ein Expertensystem zur frame-basierten Steuerung der Low- und Medium-Level-Bildverarbeitung

Gudrun Polensky[*], Tilo Messer[+]

* Dachauerstr. 123, 8000 München 2

\+ Bayerisches Forschungszentrum für wissensbasierte Systeme
Forschungsgruppe Kognitive Systeme
Orleansstr. 34, 8000 München 80

Zusammenfassung

Diese Arbeit beschreibt ein Expertensystem zur wissensbasierten Steuerung der Low- und Medium-Level-Bildverarbeitung, mit dessen Hilfe eine schnelle Anpassung an Problemstellungen in der Bildverarbeitung ermöglicht werden soll. Weiterhin soll dem Laien eine Möglichkeit geboten werden, Bildverarbeitungssysteme einfacher zu bedienen. Das Wissen über die Bildverarbeitungsoperationen wird mit Hilfe von Frames strukturiert, die eine objektorientierte Sicht der Wissensbasis erlauben. Ein auf der Frame-Struktur basierender Suchalgorithmus unterstützt den Zugriff auf die Wissensbasis. Weiterhin existiert eine Regelbasis, die Wissen über die Anwendbarkeit der Bildverarbeitungsoperationen und die Auswirkungen bei Parameteränderung darstellt. Der Regelinterpreter wird durch den Procedural- Attachment-Mechanismus in dem jeweiligen Operationsframe bzw. Parameterframe angestoßen. Diese Arbeit soll einen groben Überblick über den Aufbau der Wissensbasis und den Zugriffsmechanismus auf das Wissen vermitteln.

1. Einleitung

In der High-Level-Bildverarbeitung, der symbolischen Bildverarbeitung, werden schon seit längerem Methoden der Künstlichen Intelligenz (KI) eingesetzt. Aber auch auf niedrigerer Ebene, bei der Low- und Medium-Level-Bildverarbeitung, wird in letzter Zeit zunehmend auf die KI zurückgegriffen. Systeme, welche die wissensbasierte Steuerung der Low- und Medium-Level-Bildverarbeitung beinhalten, sind GOLDIE, XAMBA, BIKOS und XRAY. GOLDIE liefert einen Mechanismus zur zielgesteuerten Low- und Medium-Level-Bildverarbeitung [Kohl87]. Die Low- und Medium-Level-Bildverarbeitung wird durch ein Top-Down-Verfahren so gesteuert, daß die gewünschten Daten, die von Zielen der High-Level-Verarbeitung abhängen, erzeugt werden. XAMBA ist ein Programmgenerator, der wissensbasiert Bildverarbeitungsprogramme für eine gegebene Aufgabe generiert [Hess88]. BIKOS und XRAY sind zwei Expertensysteme für die automatische Konfigurierung zur visuellen Qualitätskontrolle von Werkstoffen, also einem eingeschränkten Aufgabenbereich in der Bildverarbeitung [Sysk86, Neum86, Stre88].

In dieser Arbeit wird ein System zur wissensbasierten Steuerung der Low- und Medium-Level-Bildverarbeitung vorgestellt [Pole89]. Es soll den nichterfahrenen Benutzer insbesonders bei der Auswahl geeigneter Operationen und Parametersätze für allgemeine Bildverarbeitungsaufgaben unterstützen. Ähnlich wie in den Systemen BIKOS und XRAY sind die Bildverarbeitungsoperationen und die dazugehörigen Parametersätze in Frames strukturiert. Die Anwendbarkeit von Operationen und die Auswirkungen bei Parameteränderungen werden innerhalb der Frames durch Regeln beschrieben.

Frames als Darstellungsform haben den Vorteil einer strukturierten aber dennoch vielseitigen Wissensrepräsentation. Sie wurden zuerst als eine Art Datenstruktur von Minsky in seiner Arbeit [Mins75] beschrieben. Bildverarbeitungsoperationen können hierarchisch in Instanzen (einzelne Bildverarbeitungsoperationen) und Klassen (Operationsklassen) verwaltet werden. Dadurch brauchen gemeinsame Charakteristika von gleichartigen Operationen nur einmal abgespeichert zu werden. Es kann eine hohe Ökonomie bei der Wissensakquisition, der Wissensmanipulation und der Wissensrepräsentation erlangt werden. Weiterhin bieten Frames die Möglichkeit, deklaratives und prozedurales Wissen darzustellen. Mit Hilfe einer deklarativen Repräsentation lassen sich Fakten darstellen, wie die Aufgabenbeschreibung, Parameterbeschreibung und Parameteranzahl der Operationen. Die prozedurale Repräsentation von Fakten ist zum Beispiel geeignet zur Ermittlung der Anwendbarkeit von Bildverarbeitungsoperationen mit Hilfe von Regeln, zur Berechnung von Bildeigenschaften aktueller Bilder, zur Parameteränderung und zur Berechnung ihrer Auswirkungen. Dadurch wird auch die Möglichkeit eines objektorientierten Aufbaus gegeben.

2. Aufbau der Wissensbasis

Frameartige Strukturen bilden die Grundlage für die Darstellung des Wissens über die Bildverarbeitung (Domänenwissen), die Bildinhalte und die Sprachschnittstelle.

2.1 Einteilung in Klassen und Instanzen :

Operationen und die dazugehörigen Parameter sind in Klassen (Operationsklassen, Parameterklassen) mit gleichen Eigenschaften eingeteilt. Die Instanzen (Operationen, Parameter) und auch Klassen selber ererben, wenn sie keine eigenen spezifischen Eigenschaften (Slot-Einträge) enthalten, die generellen Klasseneigenschaften übergeordneter Klassen.

Der hierarchische Aufbau kann durch einen Hierarchie-Graphen dargestellt werden. Ein Beispiel wird in Abbildung (Abb. 1) gegeben.

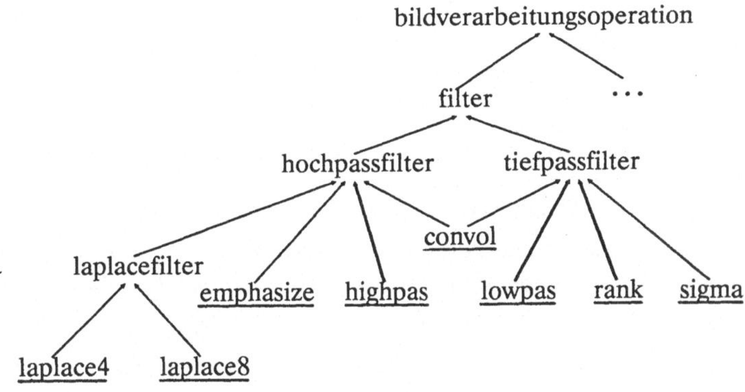

Abb. 1   Teil des Hierarchiegraphen (is_a-Hierarchie), Beispiel: Filter

2.2 Darstellung des Wissens :

Ein Frame besteht aus einer Reihe von vierstelligen Relationen [Schn87] der Form :

<Frametyp> (<Framename>, <Slot>, <Facette>, <Datenliste>).

Jeder Frametyp, der die Wissensbasis in verschiedene Teilgebiete einteilt, beinhaltet eine Reihe von Objekten und Klassen, die durch den Framenamen bezeichnet werden. Ein Objekt wird mit Hilfe von Slots (Attributeigenschaften) beschrieben. Ein Slot kann durch mehrere Facetten dargestellt werden, unter denen jeweils eine Datenliste steht.

Es existieren zwei Frametypen, die unterschiedliches Wissen repräsentieren :

1. *op_frame*:   beinhaltet das Wissen über Bildverarbeitungsoperationen
2. *par_frame*:   beinhaltet das Wissen über die Parameter

Der Vorteil dieser Einteilung in verschiedene Frametypen liegt in einer differenzierten Abarbeitung des unterschiedlichen Wissens über die Operationen und die Parameter. Jeder Frametyp repräsentiert seine eigene Hierarchie und besitzt eigene Attributeigenschaften (Slots). Weiterhin wird durch die Einteilung des Wissens in Teilgebiete der jeweilige Suchraum verkleinert, wodurch ein schnellerer Zugriff auf die Daten erreicht wird.

Die Operationen und die dazugehörigen Parameter werden durch folgende Slots repräsentiert :

1. Slots der Operationen-Frames: name, is_a (Verbindung zu übergeordneten Operationsklassen), beschreibung, auswirkung, alternativen (alternative Operationen),kosten, anwendbarkeit (Grad der Anwendbarkeit, siehe Kap. 2.4), parameteranzahl, parameter (Verbindung zu Parameterframes)

2. Slots der Parameter-Frames: name, is_a (Verbindung zu übergeordneten Parameterklassen), stelle, beschreibung, art (Art des Parameters: eingabe_objektparameter, ausgabe_objektparameter, eingabe_steuerparameter, ausgabe_steuerparameter)

a) Slots bei Parameterart eingabe_objektparameter: wert (Bildobjekt), grauwertkomponentenanzahl (Anzahl der benötigten Grauwertkomponenten (Grauwertbilder, die unter einem Bildobjekt zusammengefaßt sind, hier Eingabebilder)), ausdehnung_region (Regionengröße des Bildobjektes)

Folgende Slots sind für jede Grauwertkomponente (Grauwertbild) vorhanden :

grauwertkomponente : aktuelle Grauwertkomponente des Eingabebilds
vorverarbeitungsschritte : angewandte Vorverarbeitungsschritte, die Werte werden vom Benutzer abgefragt
kontrast : Kontrast des Eingabebildes, automatische Instantiierung durch Berechnung der Standardabweichung des Eingabebildes (hoch, mittel, gering)
helligkeit : Helligkeit des Eingabebildes, automatische Instantiierung durch Berechnung des Mittelwertes des Eingabebildes (Wertebereich: hell, mittel, dunkel)
komplexität : Komplexität des Eingabebildes, der Wert wird vom Benutzer abgefragt (hoch, mittel, gering)
objekt_hintergrund_grauwert : Unterschied der Grauwerte von Objekt und Hintergrund, der Wert wird vom Benutzer abgefragt (gemischt, verschieden)
anzahl_verschiedener_grauwerte : Anzahl verschiedener Grauwerte im Eingabebild, der Wert wird mit Hilfe des Grauwerthistogramms automatisch instantiiert (hoch, mittel, gering)
verrauscht : Verrauschungsgrad des Eingabebildes, der Wert wird vom Benutzer abgefragt (stark, mittel, gering)

b) Slots bei Parameterart eingabe-steuerparameter: wert (Wert des Parameters), vorgaenger_folge_wert (Wert des ursprünglichen und des gänderten Wertes bei Parameteränderung), parameteraenderung, auswirkung (Auswirkung bei Parameteränderung)

Die Facetten beschreiben die Art der folgenden Datenliste, mit deren Hilfe dann der eigentliche Wert ermittelt und überprüft werden kann. Es gibt folgende Facetten :

- *wert*: ermittelte Fakten, die am Ende einer Programmsitzung wieder gelöscht werden
- *value*: bekannte Werte.
- *if_needed*: Prozedur, mit deren Hilfe der eigendliche Wert bei Bedarf ermittelt oder berechnet werden kann (prozedurale Repräsentation)
- *default*: Klasseneigenschaften, die nicht auf alle zu der Klasse gehörenden Objekte zutreffen, aber auch Parametervorbesetzungen und nicht sicheres Wissen
- *range*: Prozedur zur Überprüfung und Ausgabe des Wertebereichs
- *number*: minimale und maximale Anzahl der eigentlichen Werte

2.3 Beispiele zum Frame-Aufbau :

In der folgenden Abbildung (Abb. 2) sind jeweils ein Beispiel für den Aufbau einer Operationsklasse und einer Instanz gegeben.

Operationsklasse : hochpassfilter		
Slot :	**Facette :**	**Daten :**
is_a	value	filter
name	value	Hochpassfilter
alternativen	if_needed range number	ermittle_aternativen no_range mindestens1
Operation : highpas		
Slot :	**Facette :**	**Daten :**
is_a	value	hochpassfilter
auswirkung	value	Hervorhebung von Grauwertkanten Informationsverlust
anwendbarkeit	if_needed range number	bearbeite_anwendbarkeit_regel(r_highpas) in_liste, [gut, geeignet, möglich, eventuell_möglich, schlecht] genau 1
kosten	value	1
parameteranzahl	value	6
parameter	value	highpas_eingabeobjekt, highpas_ausgabeobjekt, highpas_zeilen, highpas_spalten, highpas_grauwertkomponente_ein, highpas_grauwertkomponente_aus

Abb. 2   Frame-Aufbau für eine Operationsklasse und ein Operationsobjekt

Die Operation *highpas* gehört zur Klasse der Hochpaßfilter. Durch einen Grauwertvergleich jedes Pixels mit seiner Umgebung mit Hilfe einer Filtermatrix, wird eine Hervorhebung hochfrequenter Anteile im Bild bewirkt.

2.4 Regelbasis :

Zum Ermitteln der Anwendbarkeit von Bildverarbeitungsoperationen wird eine Regelbasis verwendet. Sie unterstützt die Auswahl der für eine bestimmten Aufgabe am besten geeigneten Operation und überprüft, ob eine Operation sinnvoll eingesetzt werden kann. Die Regel wird durch die if_needed-Prozedur 'bearbeite_anwendbarkeit-regel', die den Regelinterpreter aufruft, bearbeitet. Als Grad der Anwendbarkeit wird dabei einer der Werte gut, geeignet, möglich, eventuell_möglich, schlecht oder nicht_möglich ermittelt.

Das folgende Beispiel ist ein Ausschnitt der Regel *r_highpas* zum Ermitteln des Anwendbarkeitsgrades für den Hochpaßfilter *highpas* :

```
WENN    die Anzahl verschiedener Grauwerte im Eingabebeild gering
        das Eingabebild gering verrauscht
DANN    Anwendbarkeit der Operation highpas :  geeignet

WENN    die Anzahl verschiedener Grauwerte in Eingabebild hoch
DANN    Anwendbarkeit der Operation highpas :  schlecht
```

3. Zugriffsmechanismus auf die Wissensbasis

Auf die Objekte und ihre Daten kann mit Hilfe eines Suchalgorithmus, in dem auch die Vererbung realisiert ist, zugegriffen werden. Der Suchalgorithmus besteht aus den vier Teilen, die nacheinander ausgeführt werden :

- Suchen im Objekt nach einem existierenden Wert unter der Facette *wert*
- Suchen im Objekt und den übergeordneten Klassen (Breitensuche) nach einem *value*-Eintrag
- Suchen im Objekt und den übergeordneten Klassen (Breitensuche) nach einer *if_needed*-Prozedur
- Suchen im Objekt und den übergeordneten Klassen (Breitensuche) nach einem *default*-Eintrag

Wird ein Wert gefunden bricht die Suche ab, und der entsprechende Wert (value- , default-Wert, bzw. durch die if-needed-Prozedur ermittelte Wert) wird zurückgeliefert. Ein ähnlicher Algorithmus, die N-Vererbung, wird von P.H. Winston [Wins84] beschrieben. Bei der N-Vererbung wird zuerst ein Objekt und die übergeordneten Objektklassen über die is_a-Hierarchie nach einer Facette durchsucht, bevor die nächste Facette bearbeitet wird.

4. Implementierung und Ausblick

Das System wurde an der Technischen Universität München für die Bildverarbeitungssprache HORUS/PROLOG [Ecks88] entwickelt. Es ist in IF/PROLOG implementiert. In der jetzigen Version wird primär auf die Wissensrepräsentation eingegangen. Die anderen Komponenten eines Expertensystems, wie die Wissensakquisition und die Erklärungskomponente werden nur knapp behandelt. Weitergehende Entwicklungen sind unter anderem eine komfortablere Benutzerschnittstelle (z.B. menügesteuerte Oberfläche) und eine stärkere Automatisierung der Ablaufsteuerung. Wünschenswert wäre die Realisierung einer vollautomatischen Steuerung der Bildverarbeitung. Dies ist zur Zeit wegen der hohen Komplexität in der Bildverarbeitung nur für stark eingeschränkte Aufgabenbereiche vorstellbar.

5. Literatur :

[Ecks88]   W. Eckstein, *Das ganzheitliche Bildverarbeitungssystem HORUS*, Mustererkennung 1988, 10. DAGM-Symposium Zürich, IF 180, S. 53-59, (September 1988)

[Hess88]   R. Hesse, R. Klette, *Knowledge-based Program Synthesis for Computer Vision*, J. New Gener. Comput. Syst., 1:1, S. 63-85, (1988)

[Kohl87]   C.A. Kohl, A.R. Hanson, E.M. Riseman, *Goal-Directed Control of Low-level Processes for Image Interpretation*, Proc. of Image Understanding Workshop, Vol. 2, Los Angeles, S.538-551 (1987)

[Mins75]   M. Minsky, *A Framework for Representing Knowledge*, The Psychology of Computer Vision, P. Winston (Ed.), McGraw-Hill, New York (1975)

[Neum86]   B. Neumann, *Wissensbasierte Konfigurierung von Bildverarbeitungssystemen*, 8. DAGM-Symposium Paderborn, IF 125, S.206-218 (Sept./Okt. 1986)

[Pole89]   G. Polensky, *Darstellung des Wissens über Bildverarbeitungsoperationenen in einer Wissensbasis*, Diplomarbeit, Institut für Informatik, Techn. Universität München (1989)

[Schn87]   P. Schnupp, C.T. Nguyen Huu, *Expertensystem-Praktikum*, Springer-Verlag (1987)

[Stre88]   H. Strecker, K. Pfitzner, *XRAY - Ein prototypisches Konfigurierungs-Expertensystem für die automatische Röntgenprüfung*, KI, 2, S.4-8 (1988)

[Sysk86]   I. Syska, *Überlegungen zur automatische Konfigurierung von industriellen Bildverar-beitungssystemen*, Studienarbeit, Fachbereich Informatik, Universität Hamburg (1986)

[Wins84]   P.H. Winston, *Artificial Intelligence*, 2. Aufl., Addison-Wesley, Massachusetts (1984)

Digitale Kamera mit CCD-Flächensensor und programmierbarer Auflösung bis zu 2994 x 2320 Bildpunkten pro Farbkanal

Reimar Lenz
Lehrstuhl für Nachrichtentechnik
Technische Universität München, Arcisstraße 21, 8000 München 2

Zusammenfassung: Unter Ausnutzung eines am Lehrstuhl für Nachrichtentechnik der TU München entwickelten neuartigen Prinzips, Microscanning mit zweidimensionalen Piezo-controlled Aperture Displacement (PAD), wurde die derzeit höchstauflösende Digital-Farbkamera mit Flächensensor entworfen und aufgebaut. Hohe geometrische Genauigkeit, Farbtreue, Empfindlichkeit und die Elimination von Farbkonvergenzfehlern werden durch die Verwendung nur eines, auf kleine Detektorflächen hin optimierten Charge Coupled Device (CCD)-Flächensensors mit Farbstreifenmaske erzielt, ohne daß die Notwendigkeit für einen mechanischen Verschluß oder einen Farbfilter-Revolver besteht. Zeitliche und örtliche Auflösung sind programmierbar und in weiten Grenzen gegeneinander austauschbar, ihr Produkt bildet eine Konstante. Die höchste zeitliche Auflösung ist Fernsehechtzeit (50Hz Halbbildrate), die höchste örtliche Auflösung ist 2994(H) x 2320 (V) x 3Farbkänale (in < 8sec). Damit ist es erstmals möglich, das Bild natürlicher, ruhender Vorlagen direkt, ohne photographisches Zwischenmedium, mit einem Rechner in der Qualität eines 35mm-Farbdiapositivs aufzunehmen. Nahezu "quadratische Pixel" sind ebenso wie hexagonale Abtastraster programmierbar, wobei durch Bildpunkt-synchrone Analog/Digital-Wandlung eine stabile Bildgeometrie erzielt wird.

Einleitung und Stand der Technik: Eine Reihe von Anwendungen (Film- oder Schriftseitenabtastung, Textildesign, Photogrammetrie, industrielle Qualitätskontrolle, Farbtelefax, Werbegraphik etc.) bei denen das Bild einer Vorlage zur Weiterverarbeitung einem Digitalrechner zugeführt werden soll, bedürfen einer hohen örtlichen Auflösung. Dazu standen bislang drei prinzipiell verschiedene Techniken zur Verfügung:

1.) Trommelscanner: Sie erreichen mit bis zu 100 000 Bildpunkten pro Zeile (Spalte) die höchste Ortsauflösung. Die Bildvorlage wird auf eine rotierende Trommel aufgespannt und mit einem sich längs der Trommelachse langsam verschiebenden Photodetektor abgetastet (**zwei**dimensionale Relativverschiebung von Vorlage und **Punkt**sensor). Farbtauglichkeit kann durch Verwendungen dreier mit Filter versehenen Detektoren erzielt werden.

2.) CCD-Zeilenkameras: Die Auflösung liegt hier üblicherweise bei 1000 bis 4000 Bildpunkten pro Zeile, spezielle Industriesysteme erreichen durch Aneinanderreihung bis zu 30 000 Bildpunkte. Die zweidimensionale Bildgewinnung erfolgt durch **ein**dimensionale Relativverschiebung zwischen Vorlage und **Zeilen**sensor (1D-Sensor). Farbtauglichkeit kann erzielt werden durch Verwendungen von Filterrevolvern, farbiger Beleuchtung oder durch eine Farbstreifenmaske, die direkt auf dem Sensor aufgebracht ist. Durch letztere Maßnahme leidet allerdings die Auflösung in Zeilenrichtung.

3.) Kameras mit hochauflösenden CCD-Flächensensoren: Aufgrund technologischer Schwierigkeiten bei der Herstellung fehlerfreier, hochauflösender 2D-Sensoren sind erst Kameras mit einer Auflösung von bis zu 1320 x 1035 Bildpunkten verfügbar, angekündigt sind bereits 2048x2048 Bildpunkte. Da bei diesem Prinzip die Vorlage **nicht** relativ zum **Flächen**sensor verschoben wird, eignet es sich zu Momentaufnahmen bewegter oder sich verändernder Objekte, ist also im Gegensatz zu 1) u. 2) echtzeitfähig (allerdings nicht Fernseh-Echtzeit, sondern maximal 7Bilder/sec).

Farbtauglichkeit kann unter Verlust der Echtzeitfähigkeit erzielt werden durch Verwendung von Filterrevolvern oder farbiger Beleuchtung. Sie ist dadurch eingeschränkt, daß bei dem zur Anwendung kommenden Frame-Transfer-Sensor von Kodak die Dotierung des Siliziumsubstrats so gewählt ist, daß für den Blaukanal die Empfindlichkeit weniger als 1/10 als die des Rotkanals beträgt. Werden gleichzeitig Farbe und Echtzeit gefordert, besteht die Möglichkeit der Farbauftrennung durch Prismen bei Verwendung dreier, sehr genau zueinander zu justierenden Sensoren. Eine solcherart aufgebaute Kamera ist jedoch mit dem hochauflösenden Kodak-Sensor derzeit nicht auf dem Markt erhältlich, auch wäre die Auflösung für die nahezu verlustfreie Abtastung eines Farbdias noch nicht ausreichend.

Da aufgrund der großen Bildpunktezahl auf die bei Frame-Transfer-Sensoren übliche Speicherzone verzichtet wurde - sie hätte die Ausbeute bei der Sensorherstellung weiter vermindert - kommt zur Vermeidung vertikaler Bildverwischung ein mechanischer Verschluß (Shutter) zum Einsatz, der den Sensor während des Auslesens abschattet.

Die derzeit höchstauflösende Fernseh-echtzeitfähige Farbkamera verwendet drei Sensoren mit jeweils 750 x 576 Bildpunkten.

Das neue Prinzip:

Die hier vorgestellte Farbkamera bedient sich eines neuartigen Prinzips: Der Bildwandler besteht aus einem CCD-**Flächen**sensor relativ niedriger Auflösung mit Farbstreifenmaske, dessen im Vergleich zum Bildpunktabstand kleine, das Bild nahezu punktförmig abtastende Detektoren um Bruchteile seines Bildpunktabstandes **zwei**dimensional rechnergesteuert verschiebbar sind, siehe Bild 1.

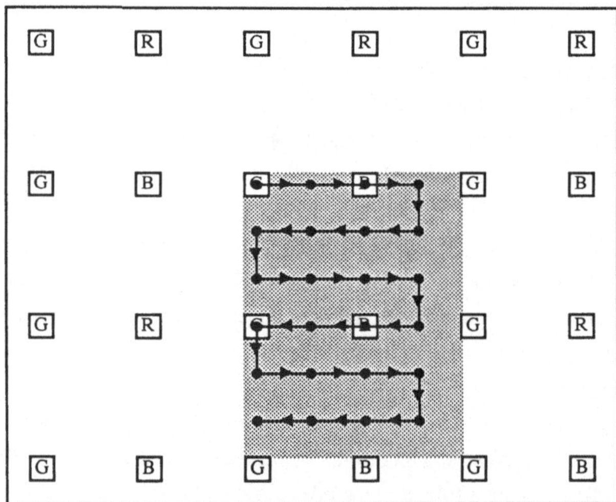

<u>Bild 1:</u> Ausschnitt des speziellen CCD-Flächensensors bestehend aus besonders kleinen, mit Farbfiltern (R,G,B) versehenen Detektorflächen. Grau unterlegt ist die elementare Farb-Auflösungszelle. Gezeigt wird das durch Microscanning mit Piezostellgliedern hervorgerufene Abtastraster einer Sensorzelle bei einer Auflösungserhöhung um den Faktor 24 gegenüber der Basisauflösung des Sensors (Faktor vier in x- und sechs in y-Richtung). Die höchstaufgelöste Bild mit 2994 x 2320 Bildpunkten für jeden Farbkanal ergibt sich aus insgesamt 192 Teilbildern.

Bei ortsfestem Sensor erhält man ein niedrig aufgelöstes Farbbild (mit 500 x 291 Bildpunkten, im weiteren Verlauf Teilbild genannt) in Fernsehechtzeit, das in PAL-Fernsehnorm ständig am analogen Ausgang der Kamera zur Verfügung steht. Durch zweidimensionales Microscanning (Aufnahme mehrerer, örtlich leicht versetzter Teilbilder) mit Piezostellgliedern kann nun unter Inkaufnahme einer geringeren zeitlichen Auflösung die örtliche Auflösung erheblich gesteigert werden. Die einzelnen, bereits im Kamerakopf digitalisierten Teilbilder werden nach Abspeicherung durch den Rechner zweidimensional zu einem resultierenden Bild höherer Auflösung verkämmt.

Merkmale	Trommel-Scanner	Zeilen-Scanner	CCD - TV Kameras	Spezial CCD Kameras	ProgRes 3000
Dimensionalität des Sensors	0	1	2	2	2
Dimensionalität und Ausmaß der Bewegung	2 / makro	1 / makro	0	0	2 / mikro
Auflösung	sehr hoch	hoch	niedrig	mittel	hoch
Zahl der Achsen mit variabler Auflösung	2	1	0	0	2
Fernseh - Echtzeit Fähigkeit	nein	nein	ja	nein	nein
Farbtauglich	ja	ja	ja	bedingt	ja
Lichtempfindlichkeit	sehr niedrig	niedrig	hoch	sehr hoch	hoch
Geometrische Genauigkeit	niedrig	mittel	sehr hoch	sehr hoch	hoch
Abtastraster	rechtwinklig	rechtwinklig	rechtwinklig	rechtwinklig	rechtwinklig, hexagonal
digitaler Ausgang	meist	meist	selten	ja	ja
analoger TV Echtzeitausgang	nein	nein	ja	nein	ja
Fokussieren	schwierig	schwierig	leicht	schwierig	leicht
Wahl des Bildausschnitts	schwierig	schwierig	leicht	leicht	leicht
Aussteuerungs-Kontrolle	zwei Durchläufe	zwei Durchläufe	einfach	einfach	sehr einfach
Erforderliche Puffergröße	klein	sehr groß	groß	sehr groß	mittel
Variable Auslesezeit	nein	bedingt	nein	nein	ja
Radiometrische Korrektur	sehr einfach	einfach	aufwendig	sehr aufwendig	mittel
Fehlerfreie Sensoren	immer	meist	teuer	sehr teuer	meist

<u>Bild 2:</u> Vergleich der Merkmale unterschiedlicher elektronischer Bildgewinnungsverfahren. Allen herkömmlichen Verfahren ist gemeinsam, daß die Summe aus der Dimensionalität des Sensors und der der Bewegung gleich zwei ist. Damit erhält man entweder eine niedrige Lichtempfindlichkeit oder eine feste, aufgrund technologischer Schwierigkeiten bei der Sensorherstellung noch relativ niedrige Auflösung. In der rechten Spalte sind die Eigenschaften der neuen Kamera aufgeführt (Erläuterung im Text. ProgRes 3000 steht für **P**rogrammable **R**esolution bis zu 3000 Bildpunkte pro Zeile).

Herkömmliche Verfahren im Vergleich zu Microscanning mit PAD:

Die mit dem neuen Prinzip erzielbaren Vorteile ergeben sich aus der Gegenüberstellung in Bild 2.

• Programmierbare Auflösung: Je nach Anwendung können zeitliche und örtliche Auflösung in weiten Grenzen gegeneinander ausgetauscht werden. Die niedrigste, in Fernsehechtzeit (40msec) erzielbare Basisauflösung entspricht der Anzahl der Bildpunkte eines Teilbildes. Für den speziellen in Anwendung kommenden Sensor sind dies 250 Bildpunkte pro Zeile für jeden der drei Farbkanäle, 290 Bildpunkte pro Spalte für den Grünkanal und je 145 für den Rot- und Blaukanal. Diese Aufteilung - die Hälfte der Bildpunkte für grün und jeweils ein Viertel für rot und blau - trägt der Tatsache Rechnung, daß grün für die vom menschlichen Beobachter empfundenen Bildschärfe die größte Bedeutung hat. Durch Ansteuern von Zwischenpositionen kann die Bildauflösung, für beide Achsen unabhängig, auf ganzzahlige Vielfache der Basisauflösung gesteigert werden. Die obere Grenze der örtlichen Auflösung ist für den auf kleine Detektorflächen hin optimierten Bildsensor bei 3000 x 2300 Bildpunkten erreicht. Sie werden aus insgesamt 192 Teilbildern mit einer Rohdatenmenge von 28MByte gewonnen. Die Dauer der Bildgewinnung ist proportional zur Anzahl der Teilbilder.

• Programmierbare Abtastraster: Die in beiden Achsen verschiedene Auflösung eines Teilbildes, dessen Bildpunkte zunächst nicht im quadratischen Raster angeordnet sind, kann durch unterschiedliche Überabtastfaktoren weitgehend kompensiert werden. Es ergeben sich dann im resultierenden Bild nahezu "quadratische Pixel". Ebenso kann die Bildvorlage durch geeignetes Microscanning in guter Näherung mit einem hexagonalen Raster, das der dichtesten Kugelpackung entspricht, abgetastet werden. Es entstehen neue Nachbarschaftsbeziehungen (6-er Nachbarschaft), die für die morphologische Bildverarbeitung von Bedeutung sein könnten.

• Hohe Lichtempfindlichkeit: Wegen der um etwa zwei Größenordnungen höheren Anzahl aktiv an der Bildgewinnung beteiligten Sensorelemente ergibt sich gegenüber Zeilenscannern eine entsprechend größere Lichtempfindlichkeit und damit besserer Signal/Rausch-Abstand. Die Kamera kann daher bereits bei normaler Raumbeleuchtung eingesetzt werden.

• Praktisch kein Aliasing: Bei der höchsten programmierbaren Auflösung wird die Nyquist-Ortsfrequenz, also die nach dem Abtasttheorem höchste korrekt aufnehmbare Frequenz, gerade unterdrückt. Die Detektorflächen der Sensorelemente wirken als wohldefinierter optischer Tiefpaß. Bei niedrigeren Auflösungen kann durch Bewegung des Sensors während der Lichtintegrationszeit für ein Teilbild ebenfalls eine definierte Tiefpaßwirkung erzielt und damit Aliasing vermieden werden.

• Hohe Geometriegenauigkeit: Da im Vergleich zu Zeilenkameras die maximale Verschiebung des Microscannings sehr gering ist, bleibt die ausgesprochen gute geometrische Genauigkeit von Flächensensoren (LENZ 88) selbst bei geringen Anforderung an die Relativgenauigkeit der Piezostellglieder nahezu voll erhalten.

• Keine Farbkonvergenzfehler: Im Gegensatz zu den Prismenblöcken in 3-Sensor-Kameras benötigt das Ein-Chip Prinzip mit PAD keinerlei Justage der Farbkonvergenz und unterliegt auch keiner thermischer Drift.

• Gleiche Abtastorte für alle Farbkanäle: Der Verlust an Ortsauflösung eines Farbstreifensensor im Vergleich zu einem äquivalenten Schwarz/Weiß-Sensor wird wettgemacht durch die Möglichkeit, die nebeneinander gelegenen, für rot, grün und blau empfindlichen Sensorelemente nacheinander auf den gleichen Abtastort im Bild zu verschieben. Auf gleiche Weise vermeidet man auch das oftmals störende, bei ruhenden Streifenmasken und stark strukturierten Vorlagen auftretende Farbmoiré.

• Echtzeit-Analogausgang: Im Gegensatz zu Zeilenscanner steht das Bildsignal mit Farb-Fernsehnorm ständig zur Verfügung. Es zeigt bereits das volle Bildformat und eignet sich daher gut zur Fokussierung, zur Wahl des Bildausschnitts und zur Farb- und Aussteuerungskontrolle.

• Hohe Ausbeute bei der Sensorherstellung: Aufgrund der relativ niedrigen Bildpunktzahl des Sensors ist ein großer Teil der Sensoren völlig fehlerfrei.

• Einfacher Rechneranschluß: Da die Farbkanäle im Zeitmultiplex und nicht parallel anliegen, werden nicht drei, sondern nur ein 8Bit breiter Datenkanal benötigt. Bedingt durch die geringe Anzahl von Bildpunkten eines Teilbildes muß im Rechnerinterface nur ein kleiner (schneller) Datenpuffer bereitgestellt werden. Bei langsamen Rechnern oder bei direkter Abspeicherung auf Platte dauert der Bildeinzug lediglich entsprechend länger.

• Radiometrische Korrektur jedes Bildpunktes praktikabel: Da jedes Sensorelement entsprechend der Anzahl der Teilbilder mehrfach verwendet wird, ist eine additive und multiplikative Korrektur selbst bei höchstauflösenden Bildern mit relativ geringem Speicherplatzbedarf durchführbar.

Um die Teilbilder korrekt verkämmen zu können, ist eine bildpunktsynchrone Analog/Digital-Wandlung mit dem Sensor-Auslesetakt erforderlich. Sie wird insbesondere für die Photogrammetrie gefordert (GRÜN 87) und wird bereits im Kameragehäuse in unmittelbarer Nähe des Sensors vorgenommen. Dadurch wird neben einer minimalen Signaldegradation durch Einstreuung von Störungen eine Unabhängigkeit von der Fernsehnorm erzielt. Weiterhin bleibt die geometrische Genauigkeit des Sensors weitgehend erhalten - der bei analogen Kameras auftretende horizontale Versatz einzelner Zeilen (Zeilenjitter) entfällt völlig und für Meßzwecke braucht der horizontale Skalierungsfaktor (LENZ 87) nicht geeicht zu werden, da ein Bildpunkt im Bildspeicher jeweils genau einem Bildpunkt auf dem Sensor entspricht.

Zusammenfassung:

Ein neuartiges Prinzip zur Farbbildgewinnung wurde vorgestellt und seine Funktionsfähigkeit durch praktische Realisierung bewiesen. Es kombiniert die Vorteile von Trommel- und Zeilenscannern (in Bezug auf die Flexibilität und Höhe der Auflösung) mit denen von Kameras mit Halbleiter-Flächensensoren (in Bezug auf Lichtempfindlichkeit, geometrische Genauigkeit, Geschindigkeit des Bildeinzugs und universelle Einsatzfähigkeit). Für den Rechneranschluß ist die Bildpunkt-synchrone Digitalisierung im Kamerakopf von besonderer Bedeutung, für den Bedienungskomfort die gleichzeitige Verfügbarkeit eines normgerechten Farbfernsehsignals in Echtzeit.

Literaturhinweise:

GRÜN, A., 1987: Towards Real-Time Photogrammetry, *Invited Paper to the 41. Photogrammetric Week*, Stuttgart, September 14-19

LENZ, R.K., 1987: Linsenfehlerkorrigierte Eichung von Halbleiterkameras mit Standardobjektiven für hochgenaue 3D-Messungen in Echtzeit, *Informatik-Fachberichte 149, Proc.9.DAGM-Symposium 1987*, Braunschweig, Sep.29 - Oct.1, Springer Berlin ISBN 3-540-18375-2, pp. 212-216

LENZ, R.K., 1988: Zur Genauigkeit der Videometrie mit CCD-Sensoren, *Informatik-Fachberichte 180, Proc. 10.DAGM-Symposium 1988*, Zürich, Sept. 27 - 29, Springer Berlin ISBN 3-540-50280-7, pp. 179-189

A Fast Generator for the Hierarchical Structure Code With Concurrent Implementation Techniques

Lutz Priese, Volker Rehrmann, Ursula Schwolle
Fachbereich Mathematik/Informatik
Universität-GH Paderborn
D-4790 Paderborn, FRG

Abstract

The sequential generator for the Hierarchical Structure Code (HSC) developed by G. Hartmann can be parallelized on a processor farm with an improved run time of a few seconds. To achieve a further improvement we suggest exploitation of its inherently asynchronous, parallel (i.e. concurrent) structure resulting in a concurrent OCCAM-HSC-generator.

Introduction to the HSC

A promising Hierarchical Structure Code (HSC) of G. Hartmann for picture recognition has been presented, see e.g. [Hartmann, 1987], where we refer to for more details.

In this context note that a hexagonal structure is chosen for the HSC. Seven neighbored pixels form one island of level 0, seven neighbored islands of level n form one island of level $n + 1$. For different levels k of resolution some symbolic form elements are detected to code the pixel information for all islands of the lowest levels. Such form elements for example can be bright/dark lines/areas and edges. To compute the connected coherent form elements for an island I of linking level $n + 1$ all previously computed (or detected) form elements of the seven sub-islands (of level n) of I are inspected. Figure 1 gives an example of a computed line form element on level k (resolution), $n + 1$ (linking) from the level (k, n) structures. The complete HSC data structure thus forms a forest (set of trees), where each tree presents some coherent form of the picture. The root contains some coarse data on the type of the form, and the information is getting more specific from son to son. On the leaves the original pixel information is stored.

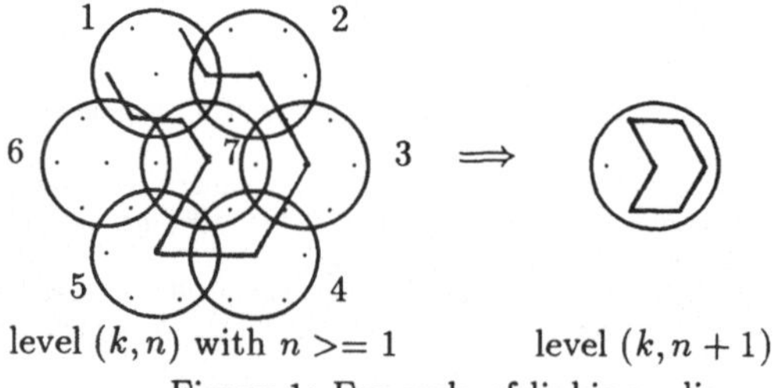

Figure 1: Example of linking a line

Picture recognition with the HSC is divided into two tasks:

- an HSC-generator computes the complete HSC-forest from the pixel information of a picture,

- AI-algorithms interpret the HSC-forest to 'recognize' the picture, i.e. answer specific questions about the content of the picture.

Previous HSC-Generators

For a typical 512×512 pixel picture of a real world environment a complete HSC-forest contains some $2 - 4$ Mbyte information and its computation involved extended search on all levels of resolution and linking. Thus, run time of an HSC-generator took several hours in early versions on a PDP-11. We reduced the run time of a sequential HSC-algorithm to $1 - 2$ minutes by using more adequate data structures and a 32-bit architecture.
For a further speed-up some parallel versions of an HSC-generator were needed.

As a first parallelizing paradigm we operated with a processor farm (compare [Packer, 1987], e.g.). For a parallel execution, a host processor divides the task into several canonical sub-tasks (here the data of 7 islands of level (k, n) for the computation of one island of level $(k, n + 1)$) and sends these tasks to different processors (which all use the same sequential algorithm).
We used trees (and graphs) of transputers as processor farms. Figure 2 presents the results.

# transputers	outdegree of host	run time (sec)
1	0	71.0
2	1	36.0
3	2	24.0
4	3	18.0
5	2	14.5
7	3	12.5

Figure 2: Run time of different processor farms for a complex picture

As the host transputer requires about 10 seconds for the mere sub-task handling this limit could not be minimized further.

An Asynchronous Parallel HSC-Generator

Elementary Processes

As the parallelizing technique of the previous section does not yield a speed-up proportional to the number of processors we looked to some inherent parallelism within the problem.

Therefore, regard a $(k, n+1)$-island and define the computation of all form-elements of this $(k, n+1)$-island from the form-elements of its seven (k, n)-sub-islands as a single process.

In principle, one might be inclined to run all these different processes asynchronously in parallel and hope for an automatical synchronization. Note that any $(k, n + 1)$-process can only be computed after the data of all form-elements of its involved sub-islands of level (k, n) are computed, i.e. its sub-processes have finished successfully. However, there is no need to compute all processes of level (k, n) first and start afterwards with all $(k, n + 1)$-processes. Note that any $(k, n + 1)$-process only needs local data from some (k, n)-processes no matter how far computation on a certain level has proceeded in some parts of the picture not involved at the moment. Thus, some $(k, n + 1)$-process may start even if not all its (k, n)-sub-processes have been completed. Also two $(k, n + 1)$-processes may need data from the same (k, n)-sub-process. If, in addition, $(k, n + 1)$-processes send some computed data to their (k, n)-sub-processes – as required in practice – the overall logic possesses many live-lock and dead-lock traps.

However, we have been able to avoid possible live- and dead-locks by using an appropriate data structure 'concurrent form'.

Data structure 'Concurrent Form'

A <u>head</u> is a 32-bit word that contains information on the type of the involved form, i.e. dark/bright line/area or edge (and some more types we have not mentioned there for simplicity), on the relative addresses of its involved sub-islands, and if it is a root (i.e. cannot be connected to other form-elements). In the case of a line the head also holds the relative addresses of the (at most) two sub-islands where the line may be continued to neighbored islands and whether such a continuation is possible. Obviously, one has to ensure by induction that this kind of information is <u>easily</u> computed from one level n to $n + 1$.

A 'concurrent form' is an $(s + 4)$-vector of 32-bit words:

1) head
2) father 1
3) father 2
4) s = # sub-structures
5) sub-structure 1

...

s+4) sub-structure s

Notice that any island of level (k, n) may be a sub-island of at most two islands of level $(k, n + 1)$, its father-islands.

<u>father i</u> contains information about its i-th father, namely the absolute address of the head of father i and the name of the transputer where it is stored. If father i is not yet computed or will not be existent, an adequate information is stored here.

<u># sub-structures</u> contains the number of sub-structures of one level below, which are involved to form the <u>head</u>. This number may exceed the number of involved sub-islands in general, as implied in figure 1.

<u>sub-structure i</u> gives the relative address of the sub-island where it belongs to and the absolute address (including transputer name) of the head of this sub-structure.

This data structure simplifies HSC-generating essentially. In order to compute a form-element F of level $n + 1$ only <u>one</u> sub-structure f_1 of level n has to be found. To compute a neighbored sub-structure f_2 fitting to f_1 one simply follows the pointer (if it does not point outside the island) from f_1 down to its sub-structure $f_{1,i}$ (of level $n - 1$) and from $f_{1,i}$ one goes up to the second father-element of level n of $f_{1,i}$.

No more search and computation is required to combine F. Only a simple gathering of fitting heads of one level below remains.

Some advantages are obvious: The HSC is just a forest, where each tree is composed of 'concurrent-forms'. The number of transputers involved is without any importance. All form-data of one (k, n)-process are stored on one transputer, the sub-structures and father-structure of any form of such a process may be stored on different transputers. Only gathering of information instead of searching is needed to compute a new concurrent-form. All concurrent-forms can be computed concurrently. As there are no pointers from level n to the same level, dead-locks and live-locks are easily avoided!

A Concurrent OCCAM-HSC-Generator

The number of the (k, n)-islands in a 512×512 pixel picture is about 120000. Thus, one should not try to implement each island as a single parallel OCCAM-process with its connections to its neighbored islands (of the same level), sub-islands (of one level less), and father-islands (of one level higher) as individual OCCAM-channels. The concept of individual processes and channels just defines the logic of the program but has to be simulated with different implementation techniques. If a pointer does not leave one transputer, it is implemented as a direct address of the own data base. If a pointer leaves one transputer a communication-manager fulfills its task. Note that using a pointer simply consists of asking for the data of another head-element or sending its own head-information. Thus, a communication-master will ask the communication-server (of the transputer where the pointer points to) these questions.

All processes on one transputer T operate concurrently but have to store their data in the storage of T. To do so, without conflicts, a further data-manager is required.

Let us call all processes that require data only from the storage of T the inner processes of T, all processes of T that need some information from other transputers the boundary processes of T. Thus, only boundary processes require the communication-master. Depending on the topology of the underlying transputer net, for any (x, y, k, n)-process it has to be computed whether it is an inner process or not. For standard topologies this task is trivial.

Thus the structure of a concurrent OCCAM-HSC-generator may be described as

```
PRI PAR
    Communication-server
    Communication-master
    Data manager
    PRI PAR
        PAR compute boundary processes
        PAR compute inner processes
```

Discussion

A preliminary version of our concurrent OCCAM-HSC-generator has been implemented and first tests have started. Even with very few transputers we have been able to reduce the run times as presented in the second section, see figure 2.

This is due to the fact that we no longer compute by searching but by gathering information. Because of the inherent simplicity of the communication between different transputers (send only addresses or single head-elements on request), we expect an almost linear speed-up with the numbers of transputers involved. Thus, we hope to achieve a 'real-time' HSC-generator with this technique.
Just two years ago our HSC-generator ran for several hours on a PDP-11, an existing version on transputers requires only a couple of seconds, and this approach should lead to a real-time behaviour. This demonstrates the power of parallelism. Our data structure 'concurrent-form' evolved from our knowledge of concurrent computations as modelled by Petri-nets, ACCS of Milner [Milner, 1983], or TCSP of Hoare [Hoare et al., 1984]. Further tests have to prove whether a linear speed-up is indeed possible – as we expect – using such concurrency techniques.

Literature

Brookes, S. D., Hoare, C. A. R., Roscoe, A. W.: A Theory of Communicating Sequential Processes. Journal of the ACM, Vol. 31, No. 3, July 1984, pp. 560-599

Hartmann, G.: Recognition of Hierarchically Encoded Images by Technical and Biological Systems. Biological Cybernetics 57, Springer, 1987, pp. 73-84

Milner, R.: Calculi for Synchrony and Asynchrony. Theoretical Computer Science No. 25, pp. 267-310, 1983

Packer, J.: Exploiting Concurrency; A Ray Tracing Example. INMOS Technical Note 7, Bristol 1987

SINAI
Ein objektorientiertes Bildverarbeitungssystem

Brigitte Wirtz und Christoph Maggioni

Siemens AG
Zentralbereich Forschung und Entwicklung
Informations und Wissensverarbeitung, Bildverarbeitung
Otto-Hahn-Ring 6, 8000 München 83

I. Einleitung

In diesem Beitrag wird das objektorientierte Bildverarbeitungssystem SINAI (Structured Interactive Analysis of Images) vorgestellt und ein Ausblick auf seine Weiterentwicklung in Richtung eines wissensbasierten Analysesystems gegeben.

Schwerpunkt der Arbeit in unserer Fachgruppe sind die Gebiete Bildanalyse und Bildinterpretation. Dabei interessieren uns insbesondere zum einen die Erkennung dreidimensionaler Objekte aus Grauwertbildern sowie ihre Modellierung und zum anderen der Entwurf eines integrierten Bildverarbeitungssystems, das sowohl traditionelle als auch wissensbasierte Methoden der Bildanalyse kombiniert.

Es gab für uns mehrere Gründe, ein Bildverarbeitungssystem zu entwickeln. Bei der Vielzahl der in unserem Arbeitsbereich entwickelten Algorithmen ist es nötig, eine "Bildverarbeitungsplattform" zu bilden, in der alle vorhandenen Bildverarbeitungsverfahren und Datenstrukturen gesammelt sind. Dies bietet zum einen die Möglichkeit, einen systematischen Pool jederzeit benutzbarer Algorithmen entstehen zu lassen, und ist zum anderen eine wichtige Voraussetzung zur Erweiterung in ein wissensbasiertes Analysesystem.

Ein großer Mangel der meisten Bildanalyseverfahren ist dabei ihre starke Abhängigkeit vom jeweiligen Anwendungsgebiet, für das Algorithmen und ihr Zusammenspiel jedesmal neu konfiguriert und optimiert werden müssen.

Es erscheint uns deshalb wünschenswert, ein System zu entwickeln, in dem nicht nur die Bildverarbeitungsverfahren vorliegen, sondern auch das Wissen darüber. Dazu gehören Wissen über ihre Funktionsweise, über ihre Anwendbarkeit, über Parametereinstellungen und über das Zusammenwirken mehrerer Verfahren.

Es gibt nun schon einige bekannte Bildanalysesysteme aus der Literatur: ACRONYM [5], 3DPO [4], 3-D-MOSAIC [8], [15], FORM [14] und VISTA [11]. Ein schöner Entwurf für ein framebasiertes System wird mit FORM in [14] vorgestellt. Weitere Vorschläge für Systeme und Verfahren finden sich in [1], [2], [3], [12] und [13], sowie ILA-IMP [6] und ImageCalc [9], zwei käufliche Systeme für die bei uns verwendeten Rechner. Bei einigen liegt dabei der Schwerpunkt der Arbeit neben einem groben Systementwurf auf den speziellen Erkennungs- oder Matching-Verfahren. Das System VIPEX [7] betont neben einem hierarchischen Operatorkonzept vor allem den Visualisierungsaspekt auf dem Anwendungsgebiet der Bildfolgenauswertung.

Unser Ziel ist hingegen schwerpunktmäßig ein Systemkonzept, das verfahrensunabhängig ist. Das Bildanalysesystem soll die einheitliche und wissengestützte Verwaltung von Algorithmen und Daten übernehmen und komplexe Bildanalyseprozesse interaktiv oder automatisch planen und konfigurieren.

Unser erstes Teilziel war dabei ein Basissystem, das es gestattet, die vorliegenden Bilddaten und Verfahren zu strukturieren und zu formalisieren [10]. Diese Strukturierung und ein modularer Aufbau ermöglichen die einfache Integration neuer Bildverarbeitungsverfahren und Systemkomponenten. Die leichte Bedienbarkeit wird durch eine komfortable, fehlertolerante Menüoberfläche gewährleistet.

II. Die grundlegenden Komponenten von SINAI

Im Folgenden sollen die wesentlichen Komponenten des Bildverarbeitungssystems SINAI vorgestellt werden. Es basiert auf je einem Konzept für Bilder, Fenster und Operatoren, sowie den dazwischen zulässigen Interaktionen. Die Datenstrukturen zu Bildern und Fenstern sind durch die Definition von objektorientierten Datentypen hierarchisch organisiert. Fenster sind die Ausgabemedien für Bilder und definieren die Sicht auf diese, während ein Operatorkonzept die Definition von strukturierten Bildverarbeitungsoperatoren auf Bildern ermöglicht.

II.1 Bilder - ein neuer Datenbegriff

Dem Entwurf eines Bildverarbeitungssystems vorausgehen muß eine Definition dessen, was man überhaupt unter dem umgangssprachlichen Begriff *Bild* verstehen will.

Inhalt eines Bildes kann eine Sammlung von Bildinformationen sein, die in unterschiedlichster Weise im Rechner abgelegt sein können. Bildinformationen können beispielsweise in Form von Binärbildern, Grauwertbildern, Linienstrukturen oder 3D-Objektbeschreibungen, wie z.B. Volumen- oder Flächenmodellen, vorliegen.

Wir definieren daher allgemein Bilder als Beschreibungen von Bildinhalten. Diese Beschreibungen umfassen beliebige Abstraktionsstufen vom einfachen Rasterbild, wie es aus der Kamera kommt, bis hin zu einer interpretativen Beschreibung einer Szene.

Die anfallenden Bilddatenmengen werden dazu durch die Definition von Bilddatentypen strukturiert. Spezielle Bildtypen werden oft

gemeinsame Merkmale haben. So sind z.B. sowohl Binär- als auch Grauwertbilder beide Rasterbilder vom Typ *pixelbild*. Außerdem sind verschiedene semantische Beschreibungsformen für ein und dasselbe Bild(-detail) denkbar. Eine Hierarchisierung des Bildkonzepts mit Hilfe eines objektorientierten Datenmodells ermöglicht die Klassenbildung von Bildtypen und damit eine weitere Strukturierungsmöglichkeit.

Zu einem Bild gehören neben den offensichtlichen Bilddaten (z.B. Bildmatrix) alle Informationen, die wir zu diesem Bild bis zu diesem Verarbeitungszeitpunkt jemals erhalten haben. Zu diesem Wissen gehört z.B.: ungültige Bereiche im Bild, angewendete Verfahren, Entstehungsgeschichte (die Abfolge der Verfahren), je nach Bildtyp charakteristische Parameter und Zusammenhänge zwischen den Abstraktionsstufen (wie habe ich die eine aus der anderen Beschreibungsmöglichkeit erhalten, z.B. eine Gerade aus dem Verlauf einer bestimmten Grauwertkante).

Im Hinblick auf ein System zur wissensbasierten Bildanalyse ist ein Bild die Menge der gesamten darüber verfügbaren Information.

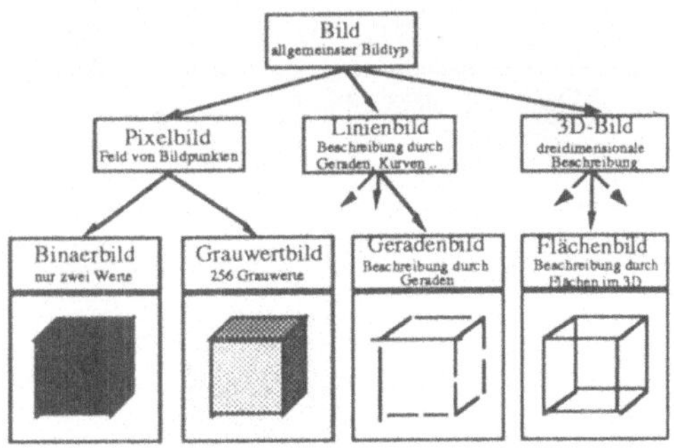

Abb. 1: Ausschnitt aus der Hierarchie der Bildatentypen

II.2 Fenster - flexible Ausgabemedien

Für einen Menschen ist es sicherlich einfach, sich Bildinformationen, die in Form eines Rasterbildes vorliegen, als optisches Bild vorzustellen. Schwieriger wird es hingegen, wenn die Bildinformationen beispielsweise durch Listen von im Bild gefundenen Umrißlinien von Bildobjekten gegeben sind. Des weiteren gibt es verschiedene Möglichkeiten, ein und dieselbe Bildinformation visuell aufzubereiten.

Aus diesen Gründen wurde ein Konzept entwickelt, das logische Sichten auf Bildinhalte erlaubt. Dieses Konzept ist das Konzept der *Fenster*, welche die Visualisierung von Bildinformationen bestimmen.

Fenster sind logische Ausgabegeräte, die durch ihre Bauweise (Programmierung) und ihre Einstellungen (Parametereinstellungen) die Art der Darstellung von Daten (Bildern) bestimmen. Verschiedene Darstellungsarten wurden dazu durch die Definition hierarchisch organisierter Fenstertypen realisiert.

Jeder Fenstertyp besitzt dazu ein Prädikat, das angibt, ob ein gewisser Bildtyp in einem Fenster dieses Typs darstellbar ist.

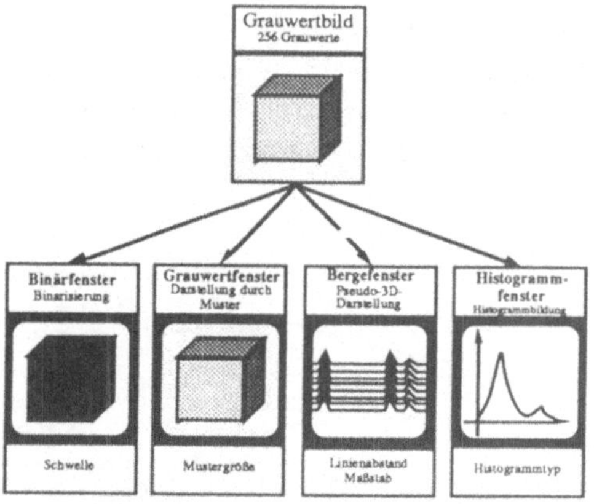

Abb. 2: Darstellung eines Grauwertbildes in verschiedenen Fenstern

II.3 Operatoren

Zur Integration von Bildverarbeitungsverfahren in SINAI wurde ein Operatorkonzept entworfen, das den Begriff des Bildverarbeitungsoperators exakt definiert und es ermöglicht, neben Bildern und Fenstern auch diese normiert anzugeben und zu benutzen.

Ein *Operator* ist hier eine Funktion von drei geordneten Argumenten, nämlich einer Liste von Eingabebildern, einer Liste von Ausgabebildern und einer Liste von Parametern.

Zu jedem Operator gehört ein Prädikat, das seine formal richtige, d.h. syntaktisch korrekte Anwendbarkeit bestimmt. Hierzu wird die Konsistenz zwischen der formalen Definition eines Operators und einer speziellen Instantiierung mit gegebenen Argumenten getestet. Dazu gehören z. B. die Überprüfung der Stelligkeit der Argumente (richtige Anzahl von Ein- Ausgabebildern, passende Parameter) sowie der Typen der Ein- und Ausgabebilder.

Dieses Konzept wird in SINAI in zwei Fällen benutzt. Zum einen wird überprüft, ob ein Benutzer einen Operator formal richtig anwendet. Zum anderen kann zu einem interessierenden Bild die Menge aller darauf anwendbaren Operatoren angezeigt werden.

III. Einbettung in ein System

Auf der Basis der oben vorgestellten Datenstrukturen wurde mit SINAI ein prototypisches Bildverarbeitungssytem implementiert. Das System verwaltet alle vorliegenden Datentypen (Bilder, Fenster, Operatoren) und alle damit erzeugten Objektinstanzen. Es stellt Standard-Betriebssystemfunktionen zum Umgang mit den Datenstrukturen zur Verfügung, wie z.B. das Anlegen, Löschen und Sichern von Bildern und Fenstern.

Darüberhinaus gibt es die Möglichkeit, über alle vorhandenen Objekte ausführliche Informationen zu erhalten. So wird zu einem Operator die Beschreibung seiner Funktionsweise sowie die Bedeutung der benötigten Parameter angezeigt, zu einem Linienbild die Anzahl aller Linien und deren Lage.

Mit Hilfe von speziellen Funktionen können Beziehungen zwischen Datenobjekten hergestellt werden. Dazu gehören das Binden eines Bildes an ein Fenster zur Visualisierung oder das Anwenden eines Bildverarbeitungsoperators auf ein oder mehrere Bilder.

Die konsistente Einbettung von neuen Datenstrukturen und Funktionalitäten geschieht durch die Definition von standardisierten Protokollfunktionen zu dem neuen Objekt. Damit ist eine leichte Erweiterbarkeit und Wartbarkeit des Systems gewährleistet. Es ist so auf einfache Weise verschiedenen Benutzern zugänglich und kann damit als *Pool* der bisher entwickelten Algorithmen, Bildtypen und Modelle dienen.

Auf den Systemkern wurde eine komfortable Benutzeroberfläche aufgesetzt. Sie ist objekt- und menüorientiert. Die Menüs bieten sinnvolle Voreinstellungen (*Defaulter*) und verhindern Fehleingaben durch zugehörige Testfunktionen. So ist es auch für mit Maschine und Implementation weniger vertraute Benutzer möglich, Erfahrungen mit Bildverarbeitungsalgorithmen zu sammeln.

SINAI wurde auf einer SYMBOLICS 3645 unter dem Betriebssystem GENERA 7.2 in SYMBOLICS-COMMON-LISP implementiert.

IV. Ausblick - was bringt die Zukunft ?

Das Bildverarbeitungssystem SINAI wurde bis jetzt zur Entwicklung eines Klassifikationsverfahrens von Schriftzeichen auf Werkstücken, zur Entwicklung von Segmentierungsalgorithmen sowie von Algorithmen und Modellen zur 3D-Objekterkennung benutzt.

Da insbesondere die Algorithmen zur 3D-Analyse sich immer weiter vom gängigen Rasterbild entfernen, muß die Fähigkeit des Systems erhöht werden, die Abhängigkeiten zwischen den gesammelten Daten zu verwalten und zu benutzen. Für Rückkoppelungsprozesse oder fokussierende Analysen ist dies eine wichtige Voraussetzung. Wir planen dazu eine framebasierte Systemkomponente in SINAI zu integrieren, mit *Slots* wie z.B. *ist-entstanden-aus*, *ist-Attribut-zu*, und *Dämonen*, die ein noch nicht existierendes benötigtes Attribut mit Hilfe eines Bildverarbeitungsoperators berechnen.

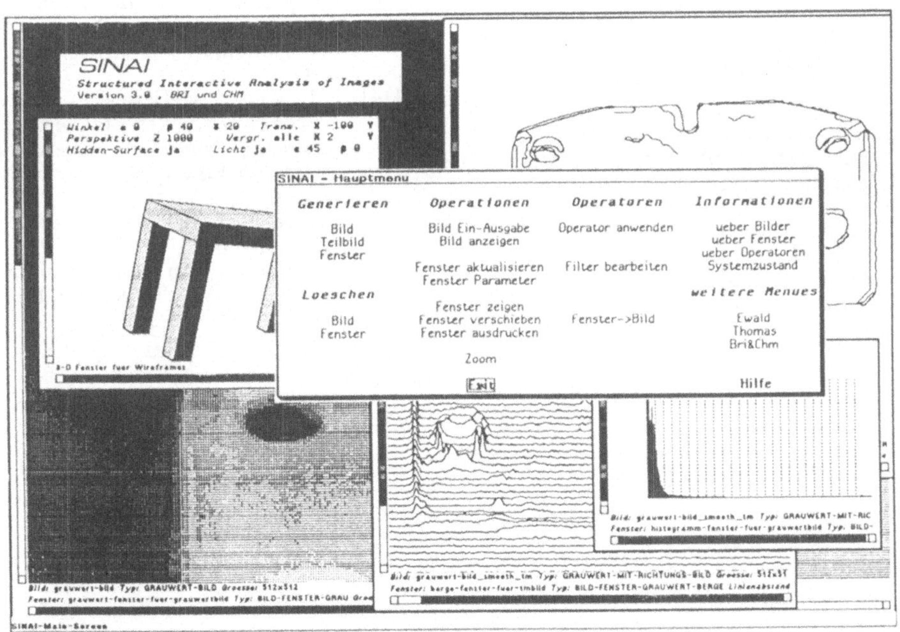

Abb. 3: Bildschirmansicht mit verschiedenen Fenstern und SINAI-Hauptmenü

Ein weiterer Schritt ist die Entwicklung einer Komponente zur Überprüfung der semantisch richtigen Anwendbarkeit von Operatoren. Dort soll z.B. überprüft werden, ob es in einer gegebenen Situation inhaltlich sinnvoll ist, ein bestimmtes Verfahren zu benutzen. Hierbei ist zu beachten, daß die Möglichkeit bestehen muß, durch Verknüpfung von mehreren Operatoren einen neuen Operator zu definieren. Teile seiner Syntax und Semantik sollen aus der Syntax und Semantik der Einzeloperatoren automatisch abgeleitet werden. Dies stellt einen Schritt in Richtung eines wissensbasierten Planungssystems dar, das wir mit Hilfe des KI-Tools JOSHUA entwickeln wollen, welches eine Spracherweiterung von Symbolics Common Lisp für die regelbasierte Programmierung ist.

Literatur

[1]   M. Barry, D. Cyrluk, D. Kapur, J. Mundy, V.-D.Nguyen: *A Multi-Level Geometric Reasoning System for Vision*, Artificial Intelligence 37, 1988, pp. 291 - 332

[2]   P. J. Besl, R. C. Jain: *Three-Dimensional Object Recognition*, Computing Surveys, Vol. 17, No. 1, March 1985, pp. 75 - 145

[3]   P. Bhaskaran, A. L. Pai: *An Intelligent Vision System for 3-D Object Recognition*, Proc. Fifth Annual Int. Phoenix Conference on Computers and Communications 1986, pp. 497 - 503

[4]   R. C. Bolles: *3DPO - A Three-Dimensional Part Orientation System*, Proc. IJCAI, Karlsruhe 1983, pp. 1116 - 1120

[5]   R. A. Brooks: *Symbolic Reasoning among 3-D Models and 2-D Images*, Artificial Intelligence, Vol. 17, No. 1-3, August 1981, pp. 285 - 348

[6]   D. Goldfarb: *ILA-IMP International Lisp Associates Image Processing Toolkit*, Dokumentation, Cambridge, MA, Februar 1988

[7]   V. Haarslev, R. Möller: *Eine graphische Umgebung zur experimentellen Bildverarbeitung*, Proc. 10. DAGM-Symposium Mustererkennung, Zürich 1988, pp. 319 - 325

[8]   M. Herman: *Representation and Incremental Construction of a Three-Dimensional Scene Model*, in A.Rosenfeld (ed.) *Techniques for 3-D Machine Perception*, North Holland 1986, pp. 149 -183

[9]   SRI International: *ImageCalc*, Menlo Park, California, 1988

[10] Ch. Maggioni, B. Wirtz: *SINAI Structured Interactive Analysis of Images*, Siemens AG, München, Interner Bericht Nr. ZFE F2 INF1-03/89 (1988)

[11] D. Paul, W. Hättich, W. Nill, S. Tatari, G. Winkler: *VISTA: Visual Interpretation System for Technical Applications - Architecture and Use*, IEEE PAMI, Vol. 10, No. 3, May 1988, pp. 399 - 407

[12] A. A. G. Requicha: *Representations for Rigid Solids: Theory, Methods, and Systems*, ACM Computing Surveys, Vol.12, No. 4, Dec. 1980, pp. 437 - 464

[13] S. Towers, R. Baldock: *Application of a Knowledge-Based System to the Interpretation of Ultrasound Images*, Proc. 9th. Int. Conf. on Pattern Recognition, Rome, Italy 1988, pp 107-110

[14] E. L. Walker, M. Herman, T. Kanade: *A Framework on Representing and Reasoning about Three-Dimensional Objects for Vision*, AI Magazine 9 (2), 1988, pp. 47 - 58

[15] E. L. Walker, M. Herman: *Geometric Reasoning for Constructing 3-D Scene Descriptions from Images*, Artificial Intelligence, Vol. 37, No. 1-3, Dec. 1988, pp. 275 - 290

Medical Imaging and Computer Vision:
An integrated approach for diagnosis and planning

Guido Gerig† , Walter Kuoni† , Ron Kikinis‡ and Olaf Kübler†

†Institute for Communication Technology , Image Science Division ,
ETH-Zürich , CH-8092 Zürich , Switzerland
‡Department of Radiology , Brigham and Women's Hospital ,
Boston MA 02115 , USA

Abstract

Recent advances in magnetic resonance imaging (MRI) show substantial improvement of image quality and acquisition speed. The acquisition of volume and/or multiecho data, flow measurements,and greater sensitivity offer new possibilities for diagnosis, therapy and operation planning.

In order to parallel the rapid progress in data acquisition, it is very important to provide efficient analysis tools which are appropriate for the complex multiple information inherent in the data. Computer Vision methods specifically adapted to the multidimensional and multispectral MR data can crucially improve the analysis by extracting, analyzing and visualizing the structural and functional properties of biological tissues. We have developed segmentation algorithms, that include edge-preserving smoothing and extraction of homogeneous regions. A multistage analysis scheme is proposed which performs the 3D segmentation of the brain (gray and white matter) and the ventricular system in multispectral MR volume data with only minimal user interaction. The quality of the segmentation demonstrates that the integration of MR-acquisition, multidimensional, and multispectral segmentation and 3D visualization represents a powerful new tool for medical diagnosis and research. The information from a large number of slices and multiple measurements is combined into representations which clarify the perception of 3D relationships and can highlight the locations and types of structural abnormalities. Based on a reliable segmentation, a further quantitative analysis of anatomical structures is accessible.

The segmentation of the human brain is illustrated by means of slice representations and 3D reconstructions of intermediate and final results. A discussion of the medical relevance of the proposed scheme to study the central nervous system shows its high potential for diagnosis and preoperative planning.

keywords: Medical image processing, 3D segmentation, 3D object recognition, 3D visualization, multispectral magnetic resonance volume data

1 Introduction

There is growing interest in the development of powerful methods for the automated analysis of digital medical images. Fast imaging techniques in computer tomography (CT) and magnetic resonance (MR) allow the acquisition of image series, sets of slices, or true volume data in clinical practice. We are therefore confronted with very large amounts of data (30 to 256 slices, 256^2 or 512^2 pixels each).

As MRI is capable of both providing structural as well as functional information of biological tissue it becomes a valuable and important tool that allows noninvasive examinations without disturbing the physiology of the organ system under study. Multiecho acquisition allows a better discrimination of different tissue types and anatomically functional units because specific characteristics are enhanced in multiple spectral channels.

The acquisition of volume data results in measurements of voxels in 3D space, and thereby volume measurements of the real 3D situation. For visual inspection, the data volume is usually cut into slices representing sectional information of 3D objects. The human visual system is extremely effective in the interpretation of our nmatural environment, but the act of 3D imagination out of sectional information is a highly spezialized human vision task and requires intensive training. It is possible for trained physicists to inspect 2D or simple 3D phenomena visually by analyzing sets of slices. But in complex situations, and if quantitative as well as qualitative results are desired, a visual analysis, or pure interactive processes, by far exceed practical time limits and must be supported by automated procedures.

This problem seems <u>well suited for computer vision</u>, which provides methods for analyzing complex 3D data and for transforming them into representations which are appropriate to our visual perception and cognitive processes. We need an integrated system that includes efficient, user friendly interactive manipulation techniques, multidimensional image processing methods, fully quantitative analysis capabilities, and new display facilities (color, 3D reconstruction etc.). This integrated system only allows a full exploitation of soft tissue structures and anatomical objects (a detailed discussion can be found in [1]). Fundamental operations are: segmentation of objects and tissue characteristics from the raw volume data, and visualization of the resulting surfaces using 3D display techniques. The visualization is mainly a computational problem and realized on several systems either as a voxel-based or surface-based algorithm. The segmentation of images into meaningful parts, however, is an important research issue which needs improved concepts particularly appropriate to the processing of medical images and to clinical requirements.

This paper focuses on 3D segmentation of multispectral and multidimensional MR image data. We have developed new segmentation processes especially adapted to the specific nature of MR image data, including data smoothing while preserving discontinuities, extraction of homogeneous regions, and morphological postprocessing using spatial context information.

2 Segmentation of multispectral and multidimensional data

The wish for interactive surgical planning, simulation of surgical steps, and especially the investigation of MR image data, clearly shows that the full exploitation of multidimensional and/or multispectral data significantly depends on the development of new segmentation techniques. It has been proposed to use the zero-crossings of DOG-(or LOG-) filtered images ([2], [3]) to isolate the brain out of MR images of the human head. Although the results look nice and seem to represent the brain surface, they must be analyzed very careful. They either represent a brain-like structure only representing a rough sketch of the surface, or serve solely as a display medium without the possibility to extract quantitative information. A further problem is the need for tedious interactive corrections in addition to the image processing.

2.1 Segmentation concept

The segmentation of MRI data into anatomically and functionally distinct regions requires the application of a multistage processing scheme. We choose a set of methods to segment MR-Data from the human head, resulting in a processing scheme with only minimal user interaction:

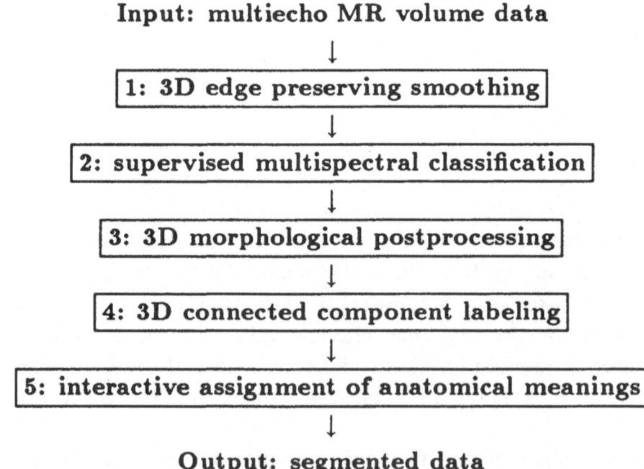

In the proposed segmentation scheme, human interaction is only necessary in stages 2 and 5. A supervised classification scheme requires the definition of training areas of distinct categories to calculate the discrimination functions in feature space. It seems possible to automate the selection of training areas in larger studies by introducing knowledge about the brain (brain model). Connected component labeling results in a set of labeled 3D regions. The brain can easily be assigned to the structure containing the largest number of voxels, whereas the ventricular system must be selected out of a larger number of liquid structures. Again, an automatic selection would be possible by utilizing the knowledge that the ventricular system is surrounded by brain tissue.

2.2 Smoothing while preserving discontinuities

Most important in clinical practice is the acquisition speed, that is why MR image data show a considerable noise component, which impeeds a successful application of segmentation algorithms. Conventional smoothing algorithms, e.g. low-pass filtering, show two main disadvantages: Edges and fine details are both blurred and shifted from their true locations. Recently Perona and Malik ([4]) presented an interesting approach that solves the problem of smoothing while preserving edges. Their so-called *anisotropic diffusion* results in a piecewise smoothing without blurring the discontinuities and/or altering their positions. This is a desirable result because it is reasonable to model MR image data as a piecewise constant image structure. The image is iteratively processed using only simple local operations, the process can be formulated using a lattice network where the pixels (voxels in 3D) represent nodes. Arc elements are chosen to have a conductance depending nonlinearily on the absolute differences (gradient) between node measurements. We have extended the algorithm to three dimensions by introducing voxel elements from neighboring slices:

$$I_{t+1}(x,y,z) \;=\; \frac{1}{C} \sum_{r=-1}^{+1} \sum_{c=-1}^{+1} \sum_{d=-1}^{+1} c(r,c,d)\, I_t(x+r,y+c,z+d) \tag{1}$$

$$c(r,c,d) \;=\; exp\left(-\left(\left|\nabla I(r,c,d)\right| K^{-1}\right)^2\right)$$

$$\nabla I(r,c,d) \;=\; (I(x,y,z) - I(x+r,y+c,z+d))$$

$$C \;=\; \sum c(r,c,d)$$

The result depends on the single parameter **K**, which controls the preservation or diffusion of discontinuities, and on the number of iterations. First experiments show the nice performance of the algorithm and its ability to enhance edges while propagating the information within smooth patches. The network-like structure of the algorithm favours its future implementation on parallel hardware.

2.3 Multispectral tissue classification

MR images are intrinsically multispectral, and therefore appropriate to multispectral pattern classification. The acquisition of multi-echo measurements results in a greater sensitivity and discrimination power, as different parameter settings enhance specific characteristics of soft tissues and liquid containing structures. It is the basic idea of multivariate processing to combine information from different echoes for the purpose of characterizing and identifying different tissue types ([5],[6]).

We can apply our software system that has been developed for remote sensing applications [7] with only minimal changes to the processing of multispectral MR data. It contains a comfortable, userfriendly interface to interactively select training areas, several methods to decorrelate and modify training data in feature space, and 7 parametric and nonparametric statistical classifiers. Most important is the extremely efficient classification of image data after the training stage by defining the decorrelated, classified feature space as a look-up-table; the classification of one voxel in a 3D data set only requires one array access, independent of the type of the classifier.

When applying multispectral statistical analysis to image data, several basic assumptions must be fulfilled: First, image data must be radiometrically homogeneous over their full 2- or 3-dimensional extension. Second, they further must be considered as piecewise constant. Lastly, different tissue categories should represent characteristic signatures in the multidimensional feature space. Our experiences clearly show that the data quality of recent data satisfies these conditions.

2.4 Morphological postprocessing

A pixel-based classification assigns the most probable category to each voxel without taking into account local context (a first inclusion of local neighborhood information is achieved while performing edge preserving smoothing!). Categories do not necessarily give unique descriptions of anatomical units (e.g. there are many different types of liquid-type regions). An important measure for segmentation is the connectivity and the spatial extension of segmented voxel-groups. We apply a 3D version of the erosion/dilation technique proposed by Serra ([8]) to separate volumetric regions of a certain category according to their spatial extension. An erosion followed by dilation with a spherical structuring element deletes structures thinner than the predefined diameter.

3 Application: Segmentation of brain from MR data

The proposed segmentation scheme is applied to carry out a segmentation of multispectral MR volume data. Input data are 36 axial MR-slices of the human head with an axial distance of $3mm$. Two echoes are measured in each slice ($TR = 3000ms$, $TE = 30\ and\ 80ms$), the resolution within the slices amounts to $1mm$. The noncubic voxels are transformed into cubic data using linear interpolation.

The two volume data sets specified by the two different echoes ($30\ and\ 80ms$) are preprocessed using the iterative anisotropic smoothing procedure. Noise could be significantly reduced after 3 iterations while preserving the fine detailed structure (see figure 1). The parameter **K** (exponential conduction function in equation 1) is set to 5.0. The evaluation of intensity profiles shows the data to be quite homogeneous within the field of view, a radiometric correction is therefore unnecessary.

Our analysis focuses on the three categories LIQUID, WHITE MATTER and GRAY MATTER. Training areas are chosen interactively in different slices of the 3D data set. The statistics of the trained categories (figure 2) clearly show that liquid is discriminated from brain tissue in echo 2, whereas white and gray matter can only be separated using the first echo. We have selected a nonparametric k_n-nearest-neighbor classifier, the classification result is shown in figure 2.

We make use of the rough a priori assumption that a typical brain surface contains structures with diameters larger than $5mm$. An erosion/dilatation of the brain classification result with a spherical structuring element of 2 pixel radius eliminates structures thinner than 5mm without changing the brain surface (the visual examination of the images before and after the postprocessing shows that only minimal changes at some sharp corners occur) (see figure 3).

Even after the local postprocessing step, we cannot directly obtain unique descriptions of anatomical units (e.g. there are many different types of liquid regions). A further important segmentation step is the grouping of local features to global structures. Connected component labeling results in sets of spatially connected voxel groups of certain categories, we define the connectivity to be the immediate 6 voxels in 3D space. In a last interactive stage, the largest group of brain category (white and gray matter) is selected as brain, whereas the ventricular system is interactively assigned after visual evaluation of liquid-type groups.

The segmentation defines object surfaces in 3D space. They can be directly displayed using our voxel-based prebuffer technique implemented on an efficient workstation ([2],[9]). Reconstruction results of brain and ventricular system are illustrated in figure 4. The liquid-filled eyeballs are additionally displayed to facilitate spatial orientation.

4 Medical relevance

The automatic segmentation of MR volume data into distinct anatomical parts is a prerequisite for the visualization of 3D phenomena and for a quantitative analysis. Using MRI in the analysis of the human brain is of special interest because it allows the noninvasive imaging (nonionizing radiation) of diseases to be detected as local morphological changes of soft tissue characteristics. Important is the diagnosis and location of structural abnormalities. The 3D visualization of anatomical and functional units gives a representation of complex spatial shapes and relations which are known, until now, only from preserved parts of the anatomy. Thus it can contribute very important information to a rough initial orientation and then for more detailed studies. Moreover, a 3D visualization combines information from a large number of slices into one representation. There are several research issues which deal with the analysis of images of the human brain:

- The extraction and visualization of the **ventricular system** of the brain can give valuable contributions to the diagnosis of different hydrocephalus types (e.g. occlusive vs. non-occlusive).

- It has been recently shown [10] that particular regions in the brain reflect the early morphological neuropathologic changes of **Alzheimer type dementia**. It is therefore necessary to analyze the characteristics of tissue atrophy within local regions of interest. The automatic segmentation of the brain represents a first step towards its detailed description.

- MRI represents a new diagnostic tool that allows to persecute the morphological changes caused by local inflammations of white matter, which is of special importance for **multiple sclerosis (MS)** cases. In a future project we will detect and describe temporal changes by comparing MR data acquired at different dates.

5 Conclusions

Recent advances in MRI have significantly improved data quality and acquisition speed. Volume data of high quality and resolution opens completely new possibilities for diagnosis and operation planning, as fine morphological changes in tissue characteristics can be detected using appropriate parameter settings or evaluation of multispectral measurements. The multi-echo data sets contain dense information which can not be completely explored by the usual visual examination of 2D slices. Computer vision methods and reconstruction procedures can help to analyze and visualize anatomically and functionally distinct structures in a way which comes very near to the surgeon's reality.

We have clearly shown that, despite of the large amount of data, the relatively low data quality, and the complexity of imaged structures, it is possible to apply efficient procedures to segment and visualize anatomical structures. In the proposed segmentation scheme, only minimal user interaction is required. We furthermore discussed that for operational use, it will be possible to integrate model knowledge to obtain a fully automatic segmentation procedure. The quality of image segmentation is evaluated by medical experts. It seems to represent the tissue categories in an accurate way without intrɔducing remarkable missclassifications. The 3D reconstructions of brain and ventricular system were accepted as realistic views comparable to what is known from the anatomy.

The 3D display of segmented objects allows us to study and qualify complex shapes and interrelations of different structures and provides specific reference to the position of abnormalities with respect to the surrounding anatomical structures. The segmentation of the brain surface and tissue categories opens all the possibilities to a subsequent analysis of predefined local areas in the brain, and to complete qualitative and quantitative examinations. In our future research, we plan to develop a model-based stage that integrates model knowledge to obtain a detailed description of segmented objects.

Finally, we discussed the medical relevance and the importance of segmentation, visualization and structural description to medical research and clinical practice. The two-pronged approach of medical imaging and Computer Vision seems to have high potential for the study of the central nervous system.

Acknowledgements

The research reported in this article is supported by the Swiss National Foundation grant NFP 18. The authors wish to thank Prof. Dr. Jolesz, Department of Radiology, Brigham and Women's Hospital, for providing the MR data. Siemens AG, Medical Systems is gratefully acknowledged for supplying the experimental computer hardware.

References

[1] R. A. Robb, *Multidimensional Biomedical Image Display and Analysis in the Biotechnology Resource at the Mayo Clinic*, Machine Vision and Applications 1 (1988), pp. 75-96

[2] J. Ylä-Jääski, F. Klein and O. Kübler, *Fast Direct Display of Volume Data for Medical Diagnosis*, Technical Report 101, Institute für Kommunikationstechnik, Division for Computer Vision, Zürich, Switzerland, 1987

[3] H. Bomans, M. Riemer, U. Tiede, K.H. Höhne, *3-D Segmentation von Kernspin-Tomogrammen*, in Mustererkennung 1987, 9. DAGM Symposium, Braunschweig, Informatik Fachberichte 149, Springer Verlag, September/October 1986, pp. 231-235

[4] P. Perona and J. Malik, *Scale Space and Edge Detection using Anisotropic Diffusion*, Proceedings of IEEE Workshop on Computer Vision, Miami, November 1987, pp. 16-22

[5] M.B. Merickel et.al., *Multispectral Pattern Recognition of MR Imagery for the Noninvasive Analysis of Atherosclerosis*, Proc. of 9th Int. Conf. on Pattern Recognition, Rome, Italy, Nov. 1988, pp. 1192-1197

[6] M.W. Vannier et.al., *Validation of Magnetic Resonance Imaging (MRI) multispectral tissue classification*, Proc. of 9th Int. Conf. on Pattern Recognition, Rome, Italy, Nov. 1988, pp. 1182-1186

[7] E. Egeli, G. Gerig, F. Klein and O. Kübler, *A Hardware and Software Optimized Program System for Interactive Image Processing*, Proc. SPIE 435 (1983), pp. 134-138

[8] J. Serra, *Image Analysis in Mathematical Morphology*, Academic Press, 1982

[9] O. Kübler, J. Ylä-Jääski and E. Hiltebrand, *3-D Segmentation and Real Time Display of Medical Volume Images*, Proc. of the Intern. Symposium on Computer Assisted Radiology, CAR'87, in Berlin. Eds.: H.U. Lembke, M.L. Rhodes, C.C. Jaffee, R. Felix. Springer-Verlag Berlin, Heidelberg, 1987, pp. 637-641

[10] T. Sandor, M. Albert, J. Stafford, S. Harpley, *Use of Computerized CT Analysis to Discriminate Between Alzheimer Patients and Normal Control Subjects*, American Society of Neuroradiology, AJNR 9 (1988), pp. 1181-1187

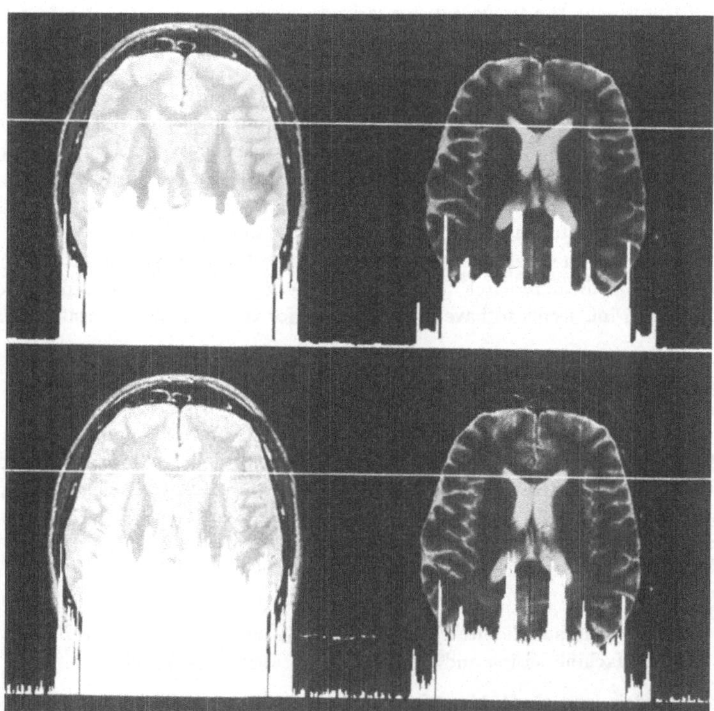

Figure 1: **Edge** preserving smoothing (overlay of intensity profile): top: smoothed images, bottom: original 1st and 2nd echo

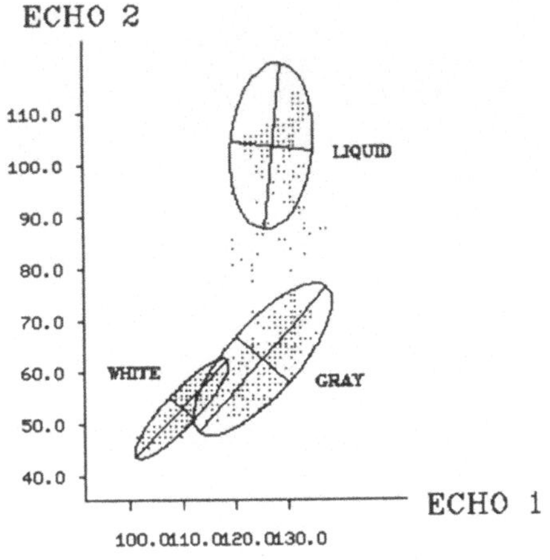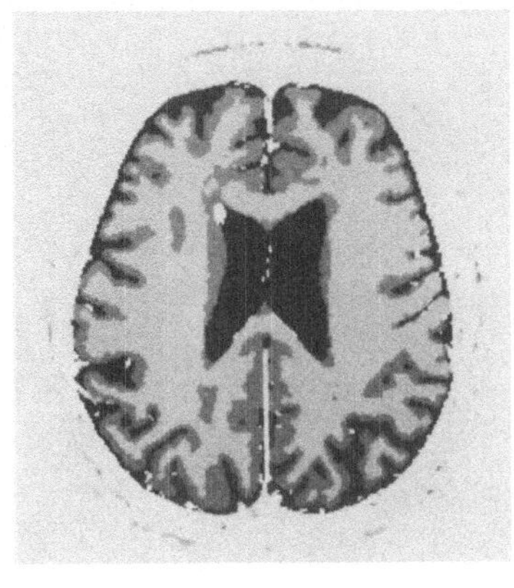

Figure 2: Multivariate classification: left: 2D feature space , right: classification result (white-, gray-matter, liquid)

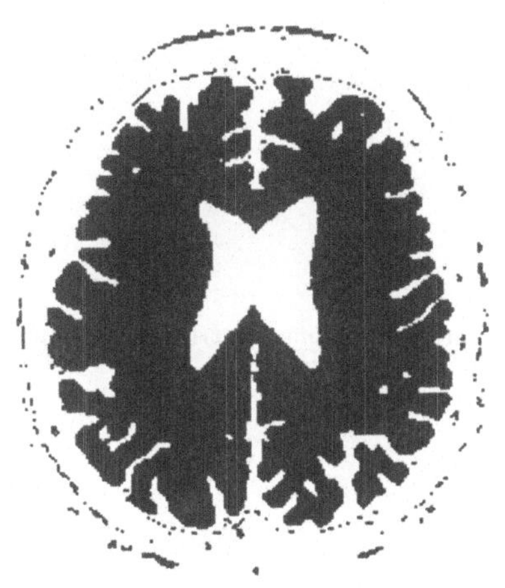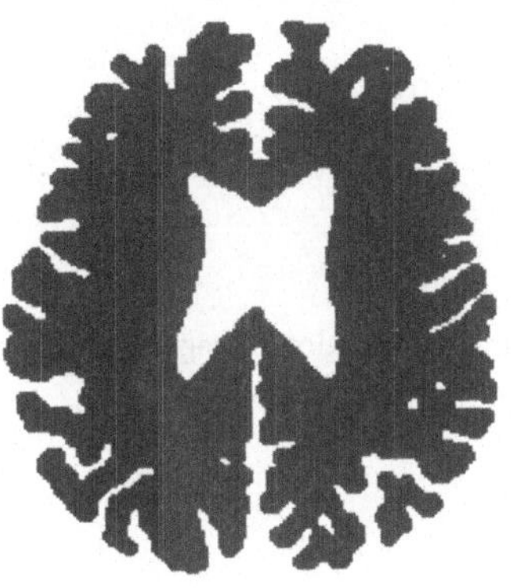

Figure 3: Morphological postprocessing: left: original image , right: after erosion/dilation

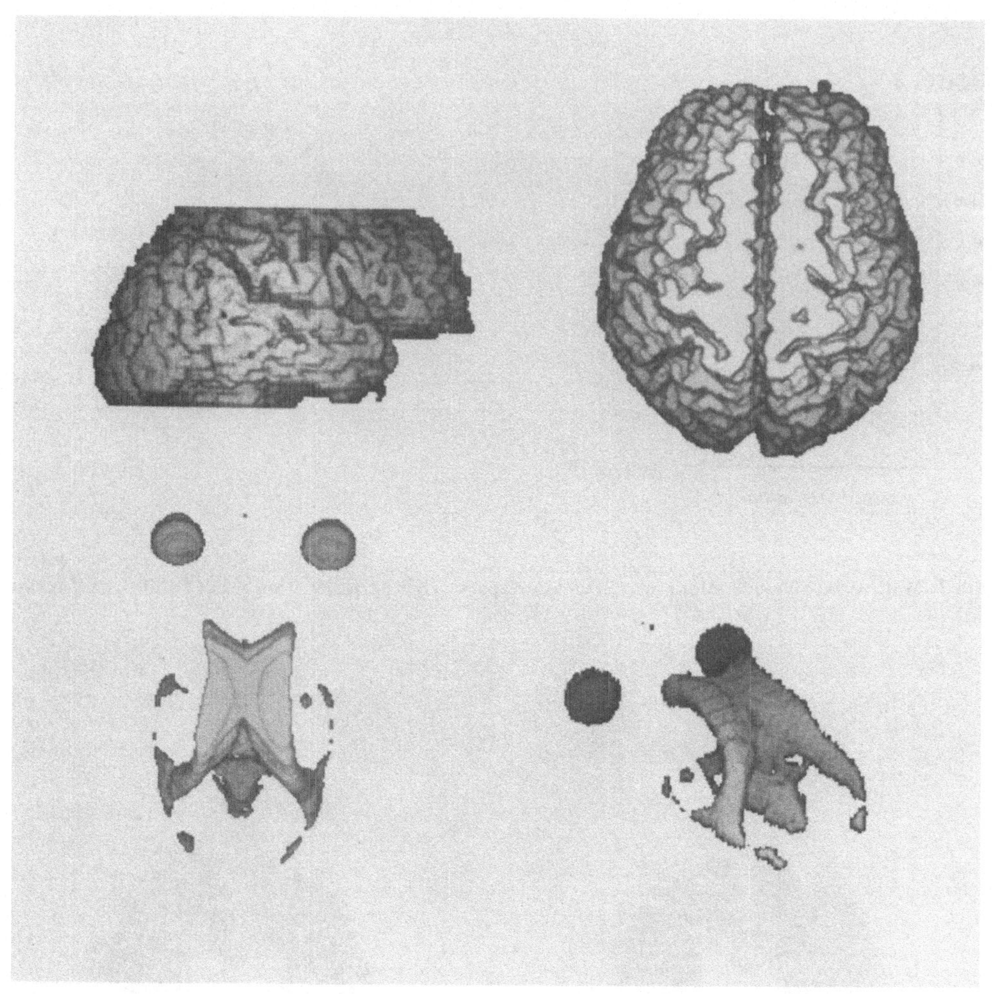

Figure 4: Visualization of segmentation results: top: brain surface, bottom: ventricular system and eyes

3D-Visualisierung von schwer segmentierbaren tomographischen Volumendaten

G. Wiebecke, M.Bomans, U.Tiede, K.H. Höhne
Institut für Mathematik und Datenverarbeitung
in der Medizin (IMDM)
Universitäts-Krankenhaus Eppendorf
D-2000 Hamburg 20

Zusammenfassung

Bei der 3D-Darstellung von Volumendaten muß entschieden werden, welche im Volumen enthaltenden Oberflächen visualisiert werden sollen. Oft ist jedoch eine Entscheidung, welche Volumenelemente eine Oberfläche bilden, nicht möglich. Bei der Darstellung von Knochen aus CT Daten liefert die binäre Segmentation zusammen mit der Grauwert-Gradienten-Schattierung gute Ergebnisse, jedoch treten hier Segmentationsprobleme bei dünnen Knochen auf. Generell schwierig ist die Segmentation bei Kernspintomogrammen. Man versucht deshalb, eine binäre Segmentation zu vermeiden, indem man Übergangsbereiche zwischen den Objekten definiert, diesen gewisse Opazitäten zuordnet und die Objekte semitransparent darstellt. Es zeigt sich dabei, daß im CT die Fehler z.B. bei dünnen Knochen durch die semitransparente Darstellung nicht behoben werden. Im MR können mit dieser Methode jedoch gute Ergebnisse erzielt werden, wenn eine binäre Segmentation keine richtigen Oberflächen findet. Reicht die Größe der Objekte für eine Schattierung nicht mehr aus, so kann in speziellen Fällen, wie z.B. bei Gefäßen, das Maximum der Grauwerte visualisiert werden.

Einleitung

Die 3D-Darstellung von Oberflächen anatomischer Objekte aus tomographischen Bildsequenzen gewinnt immer mehr an Bedeutung, sowohl bei klinischen Anwendungen in der Chirurgie, Orthopädie und Diagnostik, als auch für die medizinische Ausbildung. Eine Voraussetzung für die 3D-Oberflächendarstellung ist allerdings, daß die darzustellenden Objekte durch eine binäre Segmentation gewonnen werden können. Dieses gelingt jedoch bisher nur bei kompakten Knochen der Computertomographie problemlos. Dünne Objekte oder Brüche sind auch hier oft nicht eindeutig zu identifizieren. Noch schwieriger ist die Situation in der Kernspintomographie. Um 3D-Volumina trotzdem visualisieren zu können, kann man Verfahren anwenden, bei denen keine binäre Entscheidung darüber, welche Bildelemente eine Oberfläche bilden, nötig ist („weiche" Segmentation). Dies kann z.B. dadurch geschehen, daß man den ursprünglichen Grauwerten Durchsichtigkeitswerte zuordnet und eine semitransparente Darstellung erzeugt. Es kann in bestimmten Fällen jedoch auch die Darstellung anderer Parameter wie z.B. die des Maximums der Intensität im Grauwertvolumen herangezogen werden. Die letztliche Entscheidung darüber, was als Oberfläche erkennbar wird, trifft bei solchen Verfahren das menschliche Auge.

Ziel dieser Arbeit ist es, solche Visualisierungsansätze zu vergleichen und ihre Eigenschaften für die Darstellung verschiedenartiger Objekte aus Computertomographie und Kernspintomographie zu bewerten.

Methoden und Ergebnisse

Binäre Segmentation

Bei den klassischen Verfahren der 3D-Oberflächendarstellung werden die Oberflächen als 3D-Konturen durch Segmentation bestimmt. Zur Segmentation von Knochen im CT (Abb. 1) reicht im allgemeinen eine Schwellwertsegmentation aus (Abb. 3a). Bei Kernspintomogrammen (Abb. 2) ist dies nicht möglich. Wir benutzen hier zur Segmentation den Marr-Hildreth Operator, der allerdings bei den meisten Objekten keine exakten Konturen liefert [1].

Schattierung segmentierter Oberflächen

Ist es gelungen, eine 3D-Oberfläche zu segmentieren, so können die Oberflächennormalen berechnet werden. Eine Schattierung der Oberflächen mit Hilfe von klassischen Computergraphikmethoden ist dann möglich. Von den verschiedenen Methoden, die Normalen zu berechnen, wird für diese Untersuchung die Grauwert-Gradienten-Schattierung benutzt [4]. Die sich dem Betrachter darstellende Bildfunktion I ergibt sich als

$$I(x,y) = S(x,y,s),$$

wobei s der Abstand der dargestellten Oberfläche von der Bildebene ist und $S(x,y,z)$ der sich aus der Oberflächennormalen an der Stelle (x,y,z) ergebene Schattierungswert ist. In den Abbildungen dieses Artikels wird die Phong-Schattierung [2] benutzt. Abb. 3a zeigt die 3D-Darstellung eines Schädels, die mit dieser Methode gewonnen wurde. Im Bereich kompakter Knochen ist die Darstellung sehr gut, sogar Knochennähte werden sichtbar. Im Bereich der Augenhöhle und der Nase sind jedoch Segmentations- und Schattierungsartefakte zu erkennen. Diese rühren daher, daß bei kleinen oder dünnen Objekten eine einwandfreie Bestimmung der Oberflächen und der Oberflächennormale nicht mehr möglich ist[6].

Integral-Schattierung

Eine sehr einfache Methode zur Darstellung schwersegmentierbarer Objekte wie z.B. Herz oder Gehirn ist die Integration von Grauwerten, wie sie z.B. bei der Röntgenabbildung erfolgt. Sie hat jedoch nur dann Vorteile, wenn man sich zunächst mit einem Segmentationsverfahren dem Objekt grob annähert, und anschließend lokal um diese ungenaue Oberfläche herum weitere und feinere Details durch Integration visualisiert. Die dargestellte Intensität $I(x,y)$ ergibt sich aus der Gleichung:

$$I(x,y) = \sum_{z=s}^{s+n-1} g(x,y,z),$$

wobei $g(x,y,z)$ der originale Grauwert an der Stelle (x,y,z) ist, s ist die Tiefe des gefundenen Objektes. Die Integrationstiefe wird durch n angegeben. Ein Schattierungseffekt tritt auf, weil das Integral maximal ist, wenn die Oberfläche senkrecht zur Blickrichtung steht, und abnimmt, wenn die Oberfläche sich neigt. Abb. 4 zeigt eine Gehirnoberfläche, die mit dieser Methode erzeugt wurde. Man kann sehr viele Einzelheiten des Gehirns, wie z.B. die verschiedenen Sulci, deutlich erkennen und zuordnen.

Transparente Grauwert-Gradienten-Schattierung

Aufwendiger sind Verfahren, bei denen den Objekten nach Maßgabe ihrer Grauwerte eine Opazität zugeordnet und jedes Voxel schattiert wird. Bei dem von Levoy [5] beschriebenen Verfahren werden zusätzlich die Opazitäten noch durch den Grauwertgradienten gewichtet, um Oberflächen zu betonen. Die Schattierung erfolgt auch hier mit der Grauwert-Gradienten-Schattierung, mit dem Unterschied, daß man im transparenten Fall viele durchscheinende Oberflächen hat. Levoy benutzt dabei ein rekursives Verfahren, das auf einem Projektionsstrahl vom Betrachter aus gesehen von hinten nach vorn den Schattierungswert errechnet.

Unsere Implementation ist der von Levoy sehr ähnlich, nur geht die Berechnung vom Betrachter aus durch das Volumen und bricht dann ab, wenn die Restopazität für die letzten Volumenelemente keinen wesentlichen Beitrag mehr auf das Endergebnis leisten kann. Es ergeben sich die folgende Gleichungen:

$$I(x,y) \; = \; F(x,y,0,1).$$

Dabei ist $F(x,y,0,1)$ der Rekursionsanfang für die Tiefe 0 und einer Restopazität von 1. Der Schattierungswert an der Stelle (x,y,z) mit einer Restopazität d wird mit folgender Rekursionsgleichung berechnet:

$$F(x,y,z,d) \; = \; \begin{cases} d & d \leq t, z \geq zmax \\ d \cdot U(x,y,z) \cdot S(x,y,z) + & \\ F(x,y,z+1,d \cdot (1 - U(x,y,z))) & sonst \end{cases}$$

wobei $U(x,y,z)$ der Undurchsichtigkeitswert an der Stelle (x,y,z) ist und S der sich aus der Oberflächennormale ergebende Schattierungswert ist. Eine Eigenschaft der transparenten Darstellung ist die freie Wählbarkeit der Opazitätsparameter. Dies ermöglicht es, Objekte hervorzuheben oder aber auch zu unterdrücken. Der Freiheitsgrad in der Wahl der Opazitäten stellt aber auch einen Nachteil dar, da eine optimale Darstellung nur nach mehreren Versuchen gelingt und man trotzdem immer noch nicht sicher sein kann, ob das gezeigte Bild die Wahrheit widerspiegelt. Im Gegensatz zu der Methode von Heyers et al.[3] wird in diesem Verfahren die Schattierung ausschließlich auf Grund der Reflexion von Licht an festen Körpern berechnet und nicht auch durch gleichmäßige Emission, wie sie z.B. an Gasen gegeben ist.

In Abb. 3b ist der Schädel aus Abb. 3a mit der transparenten Grauwert-Gradienten-Schattierung dargestellt. Für die kompakten Bereiche ist das Ergebnis vergleichbar mit der Grauwert-Gradienten-Schattierung. Im Bereich der dünnen Knochen sind die Artefakte aus Abb. 3a zwar scheinbar reduziert, sie sind jedoch nur von anderen Strukturen verhüllt. Der visuelle Eindruck wird verbessert, Details wie das Nasenseptum verschwimmen jedoch.

Dieser Befund führt zu dem Schluß, daß bei Objekten, die im Prinzip einfach und genau zu segmentieren sind (wie z.B. Knochen im CT), bei denen aber die Segmentation wegen der kleinen Ausdehnung der Objekte Probleme macht (Partialvolumeneffekt), eine transparente Darstellung keine besseren Ergebnisse bringt.

Anders ist die Situation offensichtlich bei Kernspintomogrammen. Hier rühren die Fehler der Segmentation daher, daß prinzipiell eine Zuordnung von Regionen gleicher Grauwerteigenschaften zu einem Organ nicht möglich ist, so daß eine „harte" Segmentation selten zu befriedigenden Ergebnissen führt. Abb. 5 zeigt transparente Darstellungen eines Herzens, bei dem eine binäre Segmentation mit den uns zur Verfügung stehenden Mittel (z.B. Schwellwert, Marr-Hildreth-Operator) nicht gelang. Durch Veränderung der Opazitätsparameter wurde in Abb. 5a mehr das sehr helle Blutvolumen abgebildet, während in Abb. 5b der Herzmuskel besser zu sehen ist. Abb. 7 zeigt ein Knie, wobei die Gefäße transparent dargestellt sind. Dies zeigt, daß nach einer sinnvollen Einstellung der Opazitätsparameter auch relativ kleine Objekte dargestellt werden können.

Maximum-Darstellung

Noch feinere Strukturen wie z.B. intrakranielle Blutgefäße kann man mit der transparenten Grauwert-Gradienten-Schattierung nicht mehr darstellen, da sie nur noch aus sehr wenigen Volumenelementen bestehen und somit eine Bestimmung von Oberflächennormalen zu unsinnigen Ergebnissen führt. Man kann aber Kernspintomogramme so erzeugen, daß Gefäße sehr viel heller als alle anderen Objekte dargestellt werden. Dies kann man für die Gefäßdarstellung ausnutzen, indem man das Maximum auf einem Projektionsstrahl sucht und den so gefundenen Grauwert visualisiert.

$$I(x,y) \;=\; \max_{z \in \{0..zmax-1\}} g(x,y,z).$$

Abb. 6 zeigt ein so erzeugtes Bild. Das Verfahren hat die Eigenschaft, daß neben den Gefäßen auch andere Strukturen abgebildet werden. Da die Gefäße aber die hellsten Objekte sind, wird ihre Darstellung davon nicht beeinträchtigt. Ein Nachteil dieser Methode besteht darin, daß keine Aussage mehr über die Lage der Objekte im Raum gemacht werden kann. Eine Möglichkeit, dieses Problem zu umgehen, besteht darin, das Maximum transparent in ein aufgeschnittenes Objekt zu projizieren. Eine andere Möglichkeit, die räumlich Information zu erhalten, ist die Rotation des 3D-Objektes.

Schlußfolgerung

Als Ergebnis aus den Untersuchungen lassen sich folgende Aussagen zusammenfassen. Lassen sich die darzustellenden Objekte einfach segmentieren (z.B. Knochen bei CT) und sind die Objekte kompakt, so ergibt die binäre Segmentation zusammen mit der Grauwert-Gradienten-Schattierung die besten Ergebnisse. Schwer segmentierbare Details werden auch bei transparenter Darstellung nicht besser visualisiert.
Bei schwer segmentierbaren Objekten wie sie in der Kernspintomographie vorkommen sind „weiche" Segmentationsverfahren zur Visualisierung von Objekten nicht zu umgehen. Sie haben jedoch entscheidende Nachteile:

- da die Zuordnung von Opazitäten zu Objekten durch frei wählbare Parameter erfolgt, kann eine wahrheitsgemäße Abbildung nicht sichergestellt werden.

- eine quantitative Vermessung von Objekten, wie sie besonders in der Operations- und Strahlentherapieplanung benötigt wird, ist nicht möglich.

Für die reine Visualisierung von Bildvolumina stellen die gezeigten Verfahren jedoch einen Fortschritt dar. Optisch besonders gut erfaßbare Darstellungen können erreicht werden, wenn es

gelingt, die für die verschiedenen Objekte am besten geeigneten Segmentation- und Darstellungs-
verfahren zu kombinieren (Abb. 5,6,7).

Danksagung

Wir danken Siemens, Erlangen für die freundliche Überlassung der Bildsequenzen, aus denen wir
die Abbildungen 3-7 rekonstruiert haben. Herrn Riemer, Herrn Pommert und Herrn Schiers danken wir für die hilfreichen Kommentare zu früheren Versionen.

Literaturverzeichnis

[1] Bomans, M.; Riemer, M.; Tiede, U.; Höhne, K.H.: 3-D Segmentation von Kernspintomogrammen. In Paulus, E. (Hrsg.): Mustererkennung 1987 (Proc. 9. DAGM-Symposium)
Informatik Fachberichte 149, Springer, Berlin (1987) 231-235.

[2] Bui-Tuong, P.: Illumination for Computer Generated Images. Communication of the ACM
18 (1975), 311-317.

[3] Heyers, V.; Dengler, J.; Meinzer, H.P.: 3D-Visualisierung von Grauwertvoxelräumen. In
Bunke, H. et al. (Hrsg.): Mustererkennung 1988 (Proc. 10. DAGM-Symposium). Informatik Fachberichte 180, Springer, Berlin (1988) 39-45.

[4] Höhne, K.H.; Bomans, M.; Pommert, A.; Riemer, M.; Schiers, C.; Tiede, U.; Wiebecke, G.:
3D-Visualization of Tomographic Volume Data Using the Generalized Voxel Model. The
Visual Computer (1989) (im Druck).

[5] Levoy, M.: Display of Surfaces from Volume Data. Computer Graphics and Applications 8,
3 (1988) 29-37.

[6] Pommert, A.; Tiede, U.; Wiebecke, G.; Höhne, K.H.: Image Quality in Voxel-Based Surface
Shading. In Lemke, H.U. et al. (Hrsg.): Computer Assisted Radiology (Proc. CAR'89),
Springer, Berlin (1989) 737-741.

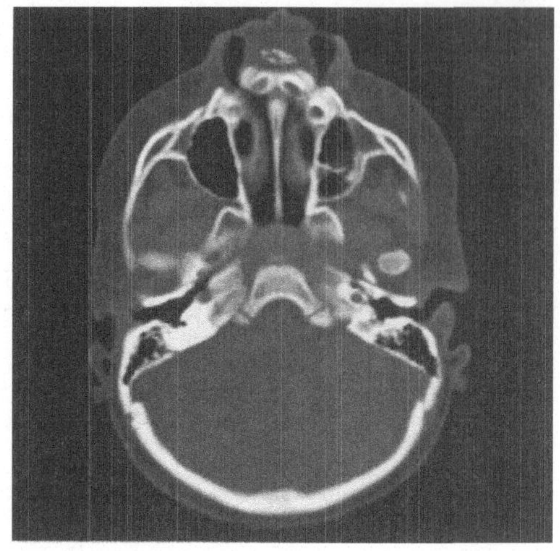

Abb. 1: Computer-Tomogram eines Kopfes.

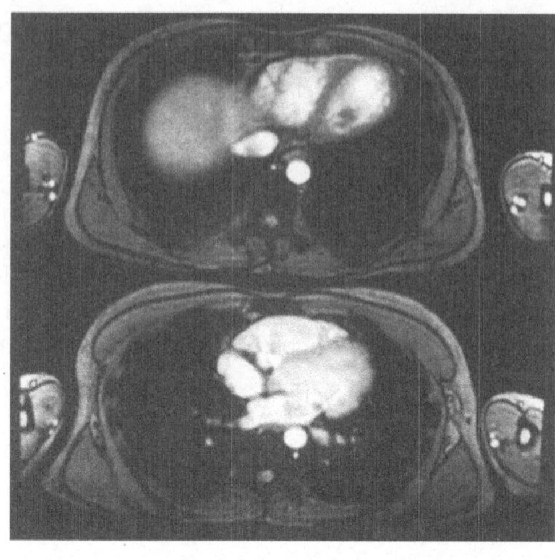

Abb. 2: Kernspintomogramme eines Herzens.

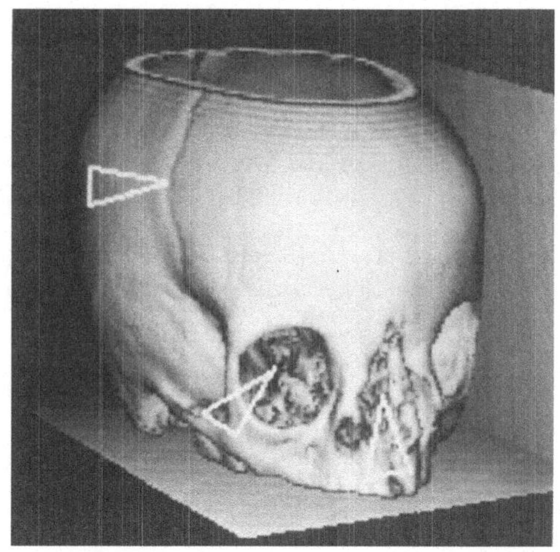

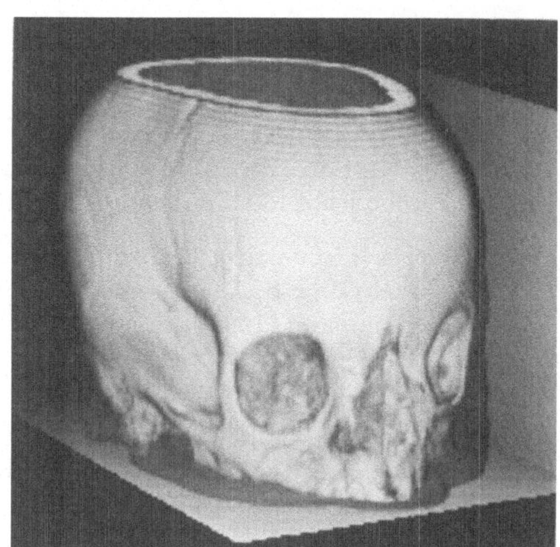

Abb. 3: Schädel eines 3 Jahre alten Kindes mit einer Gesichtsspalte. a) Schwellwertsegmentation mit Grauwert-Gradienten-Schattierung. b) transparente Grauwert-Gradienten-Schattierung. Die Pfeile in Abb.3a zeigen auf eine Knochennaht, in die Augenhöhle und auf das Nasenseptum.

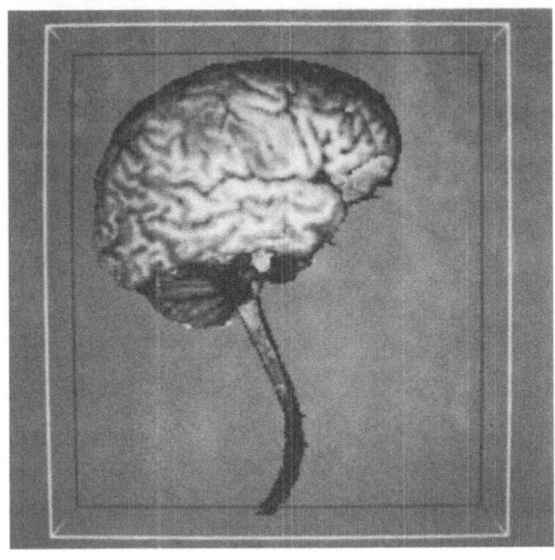

Abb. 4: transparente Darstellung eines Gehirns (Integral-Schattierung)

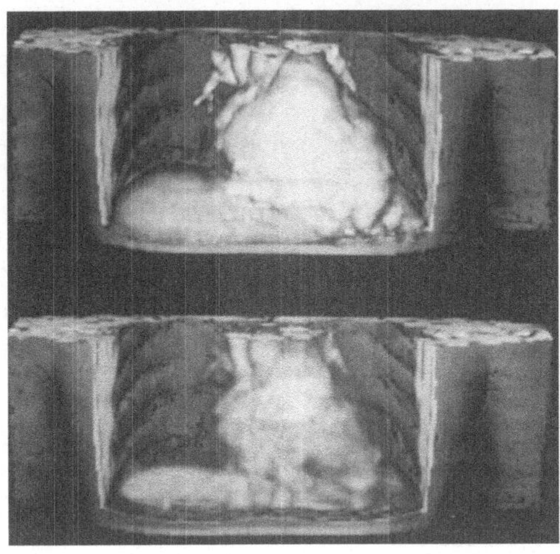

Abb. 5: Darstellung eines Herzens nach Extraktion der Hautoberfläche (transparente Grauwert-Gradienten-Schattierung). Die beiden Bilder unterscheiden sich in der Wahl der Opazitätsfunktion.

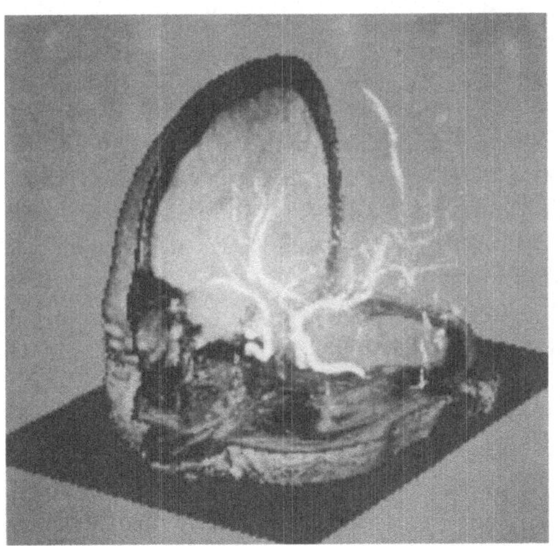

Abb. 6: Transparente Maximumdarstellung der intrakraniellen Gefäße.

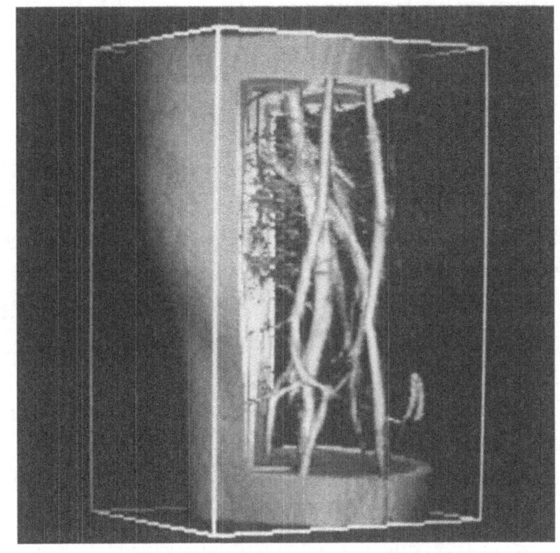

Abb. 7: Darstellung der Kniegefäße. (Haut: Grauwert-Gradienten-Schattierung, Gefäße: transparente Grauwert-Gradienten-Schattierung)

Ein Mehrgitterverfahren zur Korrespondenzfindung bei der 3D-Rekonstruktion von Elektronenmikroskop-Kippserien

Joachim Dengler, Michael Cop

Abteilung Medizinische und Biologische Informatik
(Leiter: Prof. Dr. C. O. Köhler)
Deutsches Krebsforschungszentrum, Heidelberg

Kurzfassung

Im Folgenden wird eine Mehrgittermethode zur 3D-Rekonstruktion von Transmissionselektronen-mikroskop-Kippserien vorgestellt. Das Verfahren dient zur Rekonstruktion von Kippserien, bei denen nur eine begrenzte Anzahl an Projektionen aus bekannten Winkeln vorhanden sind. Das Wesentliche des Verfahrens ist, daß die Verschiebungsparameter aus den Bilddaten selbst bestimmt werden, und somit keine zusätzlichen Goldpartikel notwendig sind. Dabei werden sowohl die Translations-, Rotations- und Skalierungsparameter bestimmt.

Im Zentrum des Verfahrens steht eine Kontrollstrategie von grob nach fein. Sie basiert auf der Beobachtung, daß durch ein entsprechendes Zuordnungsverfahren die Verschiebungsparameter auf einer groben Ebene gut bestimmt werden können. Diese Parameter dienen dazu, auf der nächst feineren Ebene eine bessere Rekonstruktion zu erzeugen, die wiederum Ausgang für eine verfeinerte Alignmentbestimmung ist. Dies wird bis zur feinsten Auflösungsebene wiederholt.

1. Einleitung

Die Standardtechnik zur 3D-Rekonstruktion aus Elektronenmikroskop-Kippserien (EM) ist die Methode der gefilterten Rückprojektion [De Rosier et al. 68; Crowther et al. 75; Rademacher et al. 78]. Da für diese Methode allerdings vollständige Information aus allen Raumrichtungen und eine präzise Ausrichtung (Alignment) der Einzelprojektionen Voraussetzung ist, stellen sich bei Verwendung der gefilterten Rückprojektion zwei Probleme:

Das erste Teilproblem ist der sogenannte "Missing Cone". Da aus apparativen Gründen bei EM-Aufnahmen nur ein maximaler Kippwinkel von 60 Grad möglich ist, fehlt Information in dem Bereich 60-90 Grad.

Zur Lösung dieses Problems wurden verschiedene Ansätze vorgeschlagen. Es kann durch apri-ori Wissen der Objektsymmetrie gelöst werden [Provencher u. Vogel 88], es konnte aber ebenfalls gezeigt werden, daß ein Winkelbereich von -60 bis +60 Grad in vielen Fällen auch für unsymmetri-sche Objekte ausreichend ist [Skoglund et al. 86]. Trotzdem ist im Bereich fehlender Projektionen nur eine begrenzte Raumauflösung vorhanden.

Das zweite grundlegende Problem ist die Ausrichtung der verschiedenen Projektionen auf einen gemeinsamen Ursprung, insbesondere bei unsymmetrischen Objekten. Im Gegensatz zur Compu-tertomographie kann die Ausrichtung der Projektionen nicht mechanisch vorgenommen werden, da die Objekte in der Größenordnung von 1nm sind. Zur Bestimmung der Ausrichtung fand die globale Kreuzkorrelationsfunktion zwischen aufeinanderfolgenden Projektionen Anwendung (siehe Abb. 1) [Hoppe 77]. Es hat sich aber gezeigt, daß sie besonders bei unsymmetrischen Objek-ten komplexer Struktur zu ungenauen Resultaten führt, da das Maximum der Korrelation nicht nur von der Verschiebung, sondern auch von der Struktur des Objekts abhängt. In der Praxis ist es daher üblich, kolloidale Goldkörner als Referenzpunkte auf den Objektträger aufzubringen [Olins et al. 83]. Die Verschiebungsparameter können dann durch eine Least Square Methode der Koordinaten der Goldkörner bestimmt werden.

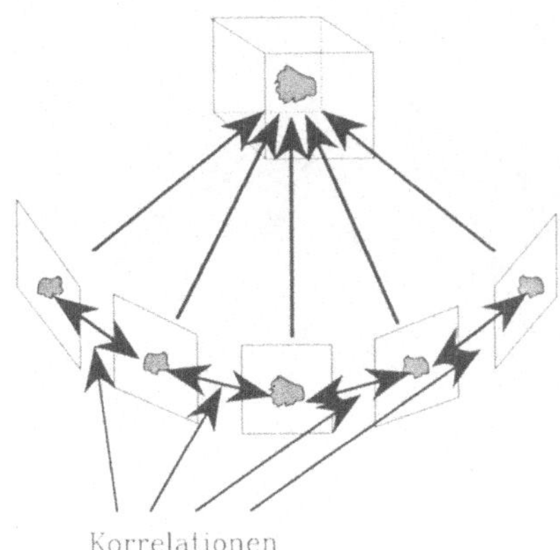

Abb.1: Traditionelle Rekonstruktionsverfahren verwenden zur Bestimmung der Verschiebungsparameter die maximale Korrelation zwischen Nachbarprojektionen. Danach wird das Objekt z.B. durch gefilterte Rückprojektion rekonstruiert.

2. Ein "Bootstrap"-Ansatz zur Lösung des Zuordnungsproblems

Der hier vorgestellte Ansatz löst das Korrespondenzproblem von EM-Kippserien ohne Goldkörner oder besondere Forderungen an die Symmetrie des Objekts.

Zwei Tatsachen weisen auf die Lösung des Problems hin:

1. Die Verschiebung wird durch globale Ähnlichkeitstransformationen der Projektionen erreicht, so daß für jede Projektion nicht mehr als 4 Parameter bestimmt werden müssen.

2. Die Ausrichtung wird durch die grobe Struktur der Projektionen des Objekts wiedergegeben.

Dadurch ist es möglich, die Verschiebungsparameter mit einer recht guten Präzision bei einer relativ groben Auflösung der Projektionen zu bestimmen. Die so bestimmten Parameter sollen strukturunabhängig und nur eine Funktion der tatsächlichen Verschiebung, Rotation oder Skalierung sein.

Diese Bedingungen schließen den direkten Vergleich der Projektionen miteinander aus, da dabei der Einfluß der Objektstruktur nicht vollständig von dem der Objektverschiebung separiert werden kann. Ein weiteres Problem bei dem paarweisen Vergleich der Projektionen ist die Fehlerfortpflanzung, die vor allem im Falle vieler Projektionen zum Tragen kommt [Saxton 84].

Der vorgeschlagene Mehrgitter-Bootstrap-Ansatz basiert auf der spektralen Zerlegung des Problems und einer iterativen Kontrollstrategie von grob nach fein. Daher wird zuerst eine Gauß-Pyramide von allen Projektionen erzeugt [Burt 84].

In jedem Stadium des iterativen Prozesses wird angenommen, daß eine vorläufige Rekonstruktion existiert. Von diesem vorläufigen 3D-Modell und den gegebenen Kippwinkeln werden digitale Pseudoprojektionen generiert. Die Verschiebungsparameter werden dann durch eine Zuordnung zwischen diesen Pseudoprojektionen und den Originalprojektionen berechnet (siehe Abb. 2). Zur Einleitung des Bootstrap-Verfahrens ist eine grobe Ausrichtung der Objekte ausreichend, was entweder per Hand oder durch Bestimmung des Schwerpunkts jeder Projektion bezüglich der Grauwerte vorgenommen werden kann. In diesem Stadium wird eine erste Rekonstruktion vorgenommen.

Die direkte Konsequenz aus dieser groben Zuordnung ist eine sehr geringe Auflösung des rekonstruierten Objekts, doch ist zu beachten, daß die erste Rekonstruktion mit Projektionen sehr geringer Auflösung gemacht wird. Es wird hier eine ausreichend grobe Ebene der Gauß-Pyramide gewählt. Die Seitenlänge der Projektionen beträgt nicht mehr als 5-20 Pixel. Dies gewährleistet, daß der relative Alignmentfehler nicht mehr als ein Pixel beträgt.

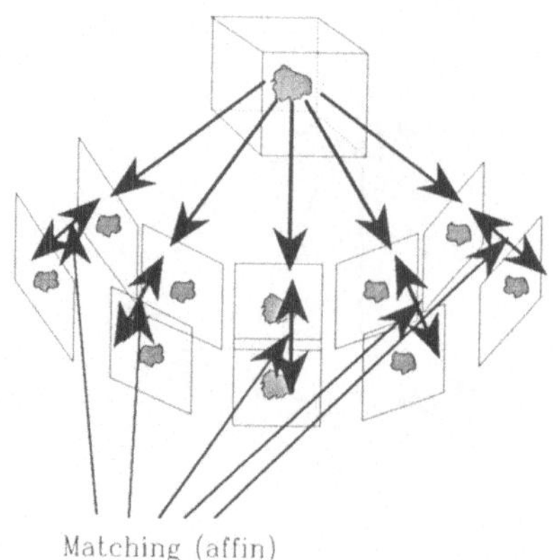

Abb. 2: Das Bootstrap-Verfahren generiert aus einer vorläufigen Rekonstruktion des Objekts Pseudoprojektionen, mit deren Hilfe dann durch Vergleich mit den Originalprojektionen die affinen Verschiebungsparameter bestimmt werden. So wird das Modell iterativ verbessert.

Bei dieser groben Auflösung sind die Berechnungen sowohl für die Rekonstruktion als auch für die folgende Zuordnung sehr schnell. Mit jeder gröberen Ebene beschleunigt sich die Rekonstruktion um einen Faktor 8 und die Zuordnung um den Faktor 4. Mit der vollen Auflösung wird erst gerechnet, wenn die Zuordnung schon optimal ist.

Das eigentliche 3D-Rekonstruktionsverfahren, das für alle Auflösungsebenen dasselbe ist, ist eine Kombination aus gefilterter Rückprojektion und einer algebraischen Rekonstruktionstechnik. Es wird ausführlich in [Dengler 89] beschrieben. Ausgehend von diesem rekonstruierten Objekt werden für dieselben Winkel wie bei den Originalprojektionen Pseudoprojektionen berechnet. Diese haben die gewünschte Eigenschaft, daß sie die Verschiebungen unabhängig von der Struktur des Objekts wiedergeben. Einige Verzerrungen, die durch die schlecht ausgerichtete Rekonstruktion entstehen, sind auf der groben Ebene vorhanden. Sie sind aber klein und können vernachlässigt werden, da das Ergebnis noch iterativ verbessert wird.

Das Zuordnungsverfahren wird solange wiederholt, bis die Änderung der Verschiebungsparameter unter einen Schwellwert fällt (etwa 0.2 Pixel). Die Verschiebungsparameter werden dann zur nächst feineren Ebene übertragen, d.h. auf die halbe Gittermaschenweite. Im Gegensatz zu den sonst üblichen Zuordnungsverfahren, die im Fourierraum arbeiten, wird in dem hier vorgestellten Verfahren ein ortsvariantes Verschiebungsvektorfeld mit Translations-, Rotations- und Skalierungsparameter bestimmt.

3. Das Zuordnungsverfahren

In den üblichen Standardverfahren wird die Zuordnung auf der Basis von Korrelationen berechnet. Bei reinen Translationsverschiebungen konnte mit traditionellen Korrelationsverfahren auch eine gute Zuordnung erzielt werden [Frank et al. 86]. Korrelationsansätze können allerdings nicht direkt Rotationsverschiebungen und Skalierungsunterschiede zwischen zwei Projektionen behandeln.

Die hier vorgestellte Methode kann jede Art der Verschiebung durch Translation, Rotation oder Skalierung in einem gewissen Bereich auffinden. Daher müssen die Orientierung der Kippachse und die Vergrößerungsfaktoren im voraus nicht genau bestimmt sein. Sie werden als Resultat des Verfahrens mitgeliefert. Außerdem werden mögliche Unterschiede in den absoluten Grauwertniveaus oder den Kontrasten durch das Schätzverfahren mit berücksichtigt.

Das Verfahren bestimmt die Verschiebung zwischen jeder Projektion und der entsprechenden Pseudoprojektion von einem vorläufigen Modell. Es nimmt für jede Projektion ein (im Raum) veränderliches Verschiebungsvektorfeld (VVF) an, das durch die vier Parameter der Translation, Rotation und Skalierung kontrolliert wird. Das VVF wird durch die Minimierung der Grauwertunterschiede in einem einzigen Minimierungsvorgang geschätzt. Bei der Annahme eines global konstanten VVF wäre dieses Verfahren konzeptionell identisch zur Maximierung des globalen Korrelationssignals [Frank 80]. Die Annahme eines veränderlichen VVF jedoch führt im Korrelationsmodell zu Problemen, die nicht auf einfache Art gelöst werden können. Das zu minimierende Funktional hat die Form (siehe auch [Nagel et al.86]):

$$\int \{I_2(x,y) - I_1(x-u,y-v)\}^2 \, dx\,dy$$
$$\approx \int \{I_2(x,y) - I_1(x,y) + (u(x,y),v(x,y)) \cdot \nabla I(x,y))\}^2 \, dx\,dy \qquad (1)$$

wobei das im Raum veränderliche VVF $U(x,y) = (u(x,y),v(x,y))^T$ als eine Ähnlichkeitstransformation zu verstehen ist, die sowohl Translation, Rotation und Skalierung enthält:

$$\begin{pmatrix} u \\ v \end{pmatrix} = \begin{pmatrix} A \\ B \end{pmatrix} + \begin{pmatrix} C & D \\ -D & C \end{pmatrix} \cdot \begin{pmatrix} x \\ y \end{pmatrix} \qquad (2)$$

Zur Bestimmung der Parameter $A-D$ muß das Funktional (1) in seine diskrete Form transformiert und bezüglich der Parameter $A-D$ minimiert werden. Um einen möglichst genauen Schätzwert zu erhalten, muß berücksichtigt werden, daß die Bestimmung des Gradienten bei einem diskreten Bild ein schlecht gestelltes Problem ist [Torre et al. 86], das durch adäquate Regularisierungsmethoden gelöst werden muß. Dies hat zwingend zur Folge, daß der Gradient des Bildpaares GR sowie die Grauwertunterschiede $I_2 - I_1$ von einem Regularisierungsparameter σ abhängen.

Die Wahl von σ bestimmt die Größenordnung der meßbaren lokalen Verschiebung. Daher sollte σ einen Wert entsprechend der größten zu erwartenden Verschiebung zwischen zwei Aufnahmen haben. Der meßbare Bereich bei einem gegebenen σ wird erheblich verbessert, wenn der Gradient aus dem Mittelwert von $I_1(x,y)$ und $I_2(x,y)$ bestimmt wird [Enkelmann 85]. Dies entspricht der Vorstellung, daß beide Bilder durch $u/2$ zueinander transformiert werden. In der tatsächlichen Transformation hingegen dient $I_2(x,y)$ als feste Referenz, und $I_1(x,y)$ wird mit u transformiert.

Die korrekten Schätzer für die Ähnlichkeits-Parameter werden durch Minimierung von (3) gefunden:

$$\sum_{x,y} \left\{ I_{2\sigma}(x,y) - I_{1\sigma}(x,y) + \left\{ \begin{pmatrix} A \\ B \end{pmatrix} + \begin{pmatrix} C & D \\ -D & C \end{pmatrix} \cdot \begin{pmatrix} x \\ y \end{pmatrix} \right\} \cdot GR_\sigma(x,y) \right\}^2 \to \min \qquad (3)$$

Das Ergebnis ist ein System aus vier linearen Gleichungen, das durch Standardverfahren gelöst werden kann.

Auf der gröbsten Ebene der Pyramide hat es sich gezeigt, daß eine Schätzung der Translationskomponenten A und B ausreichend ist.

Um eventuell vorhandene absolute Grauwertunterschiede in den Aufnahmen auszugleichen, werden die Aufnahmen mit einem breitbandigen Hochpaß vor der Zuordnung gefiltert. σ ist hier anders gewählt und unabhängig von dem obigen:

$$I(x,y) \longleftarrow I(x,y) - I_\sigma(x,y)(\sigma \gg 1) \qquad (4)$$

Die lokale Schätzung der Verschiebung hängt zum großen Teil von der lokalen Struktur des Bildes ab, d.h. seinen lokalen Gradienten. Der mittlere Gradient eines mittelwertfreien Bildes kann durch Kontrastverstärkungsverfahren vergrößert werden. Eines der einfachsten und effizientesten Verfahren ist der Pseudologarithmus [Burt 84a; Schmidt 88]

$$I(x,y) \longleftarrow \text{sign}\,(I(x,y)) \cdot \ln\,(1 + \alpha \cdot |I(x,y)|) \qquad (5)$$

Dies ergibt eine ähnliche Wichtung für Strukturen mit geringen Kontrasten, sowie für solche mit hohen Kontrasten. Dadurch werden alle möglichen Strukturen für die Berechnung benutzt. Ein weiterer Vorteil dieser Transformation ist die Tatsache, daß die Berechnung weitgehend unabhängig vom absoluten Kontrast wird. Die beste Wahl für die heuristische Konstante α ist nahe bei 1 für Aufnahmen, deren Grauwerte im Bereich 0 bis 255 liegen. Die Wahl von $\alpha \gg 1$ erzeugt ein Binärbild. Dieser "Binäre Grenzfall" wurde ebenfalls erfolgreich bei der Zuordnung von Bildsequenzen angewandt [Dengler et at. 88]. Es hat den Vorteil, daß es sowohl von den absoluten Grauwertniveaus, als auch von den absoluten Kontrasten unabhängig ist. Allerdings muß man einen Verlust an Auflösung in Kauf nehmen. Die Wahl $\alpha \gg 1$ ist nur eine lineare Skalenänderung, die keine Auswirkungen auf das Zuordnungsverfahren hat.

Zum Schluß wird der gesamte Satz an Projektionsaufnahmen mit den Parametern A bis D transformiert. Für die notwendige Interpolation hat sich die bilineare als ausreichend erwiesen. Eine einzige Transformation ist meist nicht ausreichend für eine verläßliche Verschiebungsbestimmung. Dies ist vor allem durch die Annahme eines lokal linearen Grauwertmodells bedingt. Eine Reihe von aufeinanderfolgenden Transformationen konvergiert allerdings sehr schnell, da sich die lineare Näherung bei geringerer Verschiebung verbessert. Daher wird die Transformation iterativ angewandt, bis die gesamte Änderung unter einen vorgegebenen Schwellwert fällt. Zwischen aufeinanderfolgenden Verschiebungsmessungen wird die 3D-Rekonstruktion mit dem zuletzt berechneten Verschiebungsfeld verbessert, was wiederum zu einer besseren Schätzung der Verschiebung führt. Ein Gütekriterium des aktuellen Alignments ist das Residuum der Ähnlichkeitstransformation. Es berechnet sich aus:

$$\max_{i \in \text{Ecken}} \{\|(A + Cx_i + Dy_i), (B - Dx_i + Cy_i)\|\} \tag{6}$$

Die Ecken sind die 4 Eckkoordinaten einer Projektion. Die die Konvexität erhaltenden Eigenschaften der Transformation stellen sicher, daß kein Pixel der Aufnahme mehr als dieses Residuum (in Pixel-Einheiten) transformiert wird. Die Iterationen auf jeder Ebene werden abgebrochen, wenn dieses Residuum unter 0.2 fällt.

Bei der Anwendung aufeinanderfolgender Ähnlichkeitstransformationen ist es wichtig, daß aufgrund der notwendigen Interpolation und dem daraus resultierenden Fehler die aufeinanderfolgenden Transformationen nicht auf ein bereits transformiertes Bild angewendet werden. Die die Ähnlichkeit erhaltenden Gruppeneigenschaften werden genutzt, um die Gesamtparameter A'' bis D'' der Gesamttransformation aus den Parametern A' bis D' und A bis D der Einzeltransformationen zu berechnen:

$$\begin{aligned}
A'' &= A' + C' \cdot A + D' \cdot B \\
B'' &= B' - D' \cdot A + C' \cdot B \\
C'' &= C' \cdot C - D' \cdot D \\
D'' &= C' \cdot D + D' \cdot C
\end{aligned} \tag{7}$$

Mit einem gefundenen Satz von Ähnlichkeitsparametern einer Auflösungsebene werden dann die Projektionen der nächst feineren Ebene transformiert. Um mit der Geometrie konsistent zu bleiben, müssen dazu die Werte A und B aufgrund der feineren Auflösung dieser Pyramidenebene verdoppelt werden. Hier wird dann eine weitere Rekonstruktion ausgeführt, ihre Pseudoprojektionen werden wieder den zugehörigen Originalprojektionen zugeordnet und so eine weitere Verfeinerung des Alignments erreicht. Diese Kontrollstrategie wird bis zur feinsten Ebene beibehalten.

4. Ergebnis und Ausblick

Der komplette Algorithmus wurde mit einem digital generierten 3D-Phantomobjekt getestet. Es handelt sich hierbei um eine Spirale, die zur Brechung der Symmetrie gekippt ist. Die Spirale ist in Abb.3 von zwei Seiten mit einem neu entwickelten Grauwert Ray-Tracing-Verfahrens [Heyers et al. 88] dargestellt. 25 Pseudoprojektionen wurden in einem Winkel von -60 bis $+60$ Grad mit

5 Grad Winkelauflösung und vertikaler Kippachse erzeugt. Diese Projektionen wurden wahllos in jeder Richtung um ±3 Pixel verschoben. Es wurden nur die Translationskomponenten angewandt, um keine Abtastfehler einzuführen. Abb.4 zeigt die Projektionen in Originalauflösung. Von diesen Projektionen wurde eine erste Rekonstruktion gemacht, und die Pseudoprojektionen mit den (groben) Originalen zur Deckung gebracht.

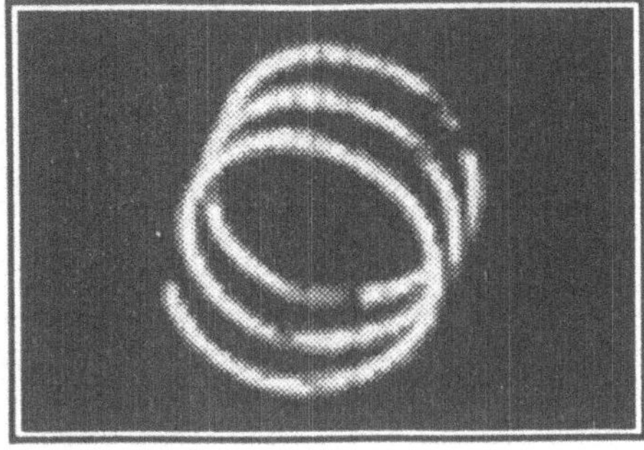

Abb.3: Original 3D-Spirale mit Grauwert-Ray-Tracing Verfahren dargestellt, Ansichten von vorne und von der Seite.

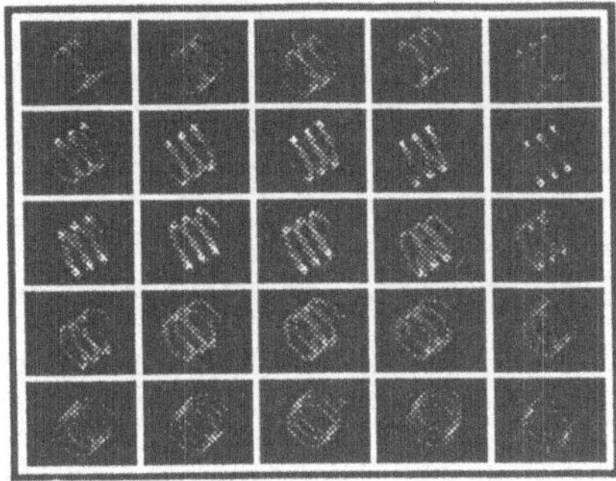

Abb.4: 25 Projektionen im Bereich -60 bis +60 Grad. Aus ihnen wird die Gauß-Pyramide erzeugt.

Es sollte erwähnt werden, daß das Alignment bei der hier beschriebenen Methode nicht die absolute 3D-Position liefert, doch ist diese sowieso willkürlich. Die Auswirkung davon ist, daß die absoluten Verschiebungen der Projektionen nur bis auf eine additive Konstante in y-Richtungbestimmt werden können. Die x-Komponente ist mit einer Funktion der Art

$$a \cdot \sin \Phi + b \cdot \cos \Phi \tag{8}$$

überlagert, wobei Φ der Kippwinkel ist. Da diese Funktion linear in a und b ist, können die aktuellen Residuen aus dem Zuordnungsverfahren mit dieser Funktion angepaßt werden. Das verbleibende Residuum ist der Fehler aus dem Zuordnungsverfahren. Die residualen Fehler aus beiden Anteilen sind in Abb.5 für alle Projektionen gezeigt. Hier wurden die residualen Verschiebungen der Schwerpunkte der transformierten Projektionen gewählt, da sie am besten die Objektstruktur bezüglich des Alignments wiederspiegeln. Man kann deutlich sehen, daß der Fehler der Zuordnung für einen Schwellwert von $e = 0.28$ 0.15 Pixel beträgt. Dies wird für das Rekonstruktionsproblem als ausreichend betrachtet.

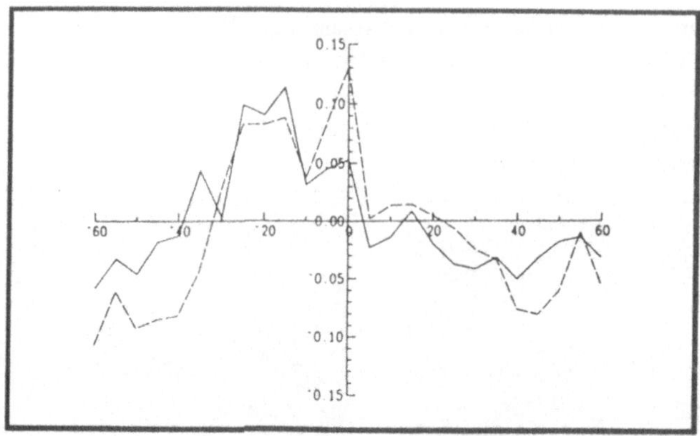

Abb.5: Verschiebungsfehler aller Projektionen (x- und y-Komponenten) gegen Projektionswinkel.

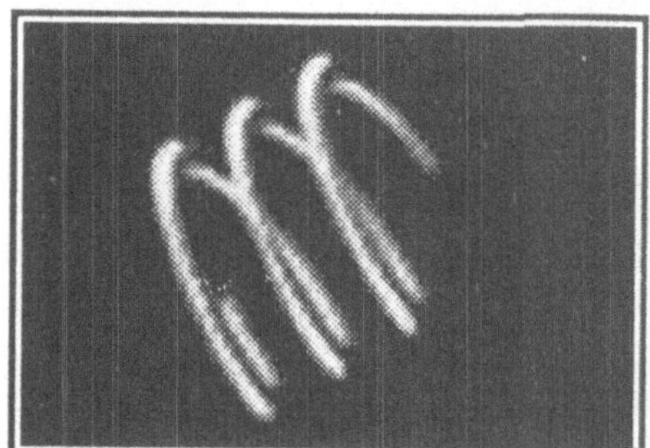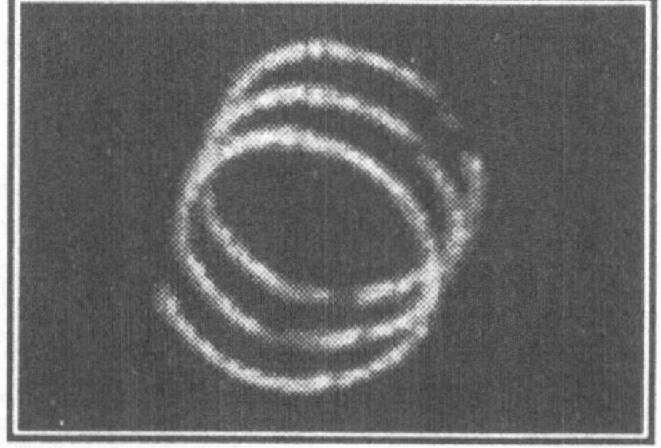

Abb.6:Rekonstruierte 3D-Spirale mit Grauwert-Ray-Tracing Verfahren dargestellt, Ansichten von vorne und von der Seite.

Die endgültige Rekonstruktion ist in Abb.6 mit dem schon erwähnten Ray-Tracing- Verfahren visualisiert. Die Rekonstruktion unterscheidet sich nur kaum vom Original. Aufgrund des "Missing-Cone" ist die Ansicht aus einem Winkel von 90 Grad stärker verschmiert, doch ist dieser Effekt bei weitem kleiner als bei der konventionellen gefilterten Rückprojektion.

In einem nächsten Schritt wurde das Verfahren auf reelle Daten angewandt. Es stand dafür eine recht verrauschte Kippserie einer 50S-Untereinheit eines Ribosoms zur Verfügung, die uns freundlicherweise von R. Vogel (EMBL Heidelberg) überlassen wurde (siehe Abb.7). Die Serie bestand aus nur 11 Aufnahmen in 8 Grad Schritten, in einem Bereich von -40 bis +40 Grad. Jede Originalaufnahme hatte die Größe 46 * 46 Pixel. Nach einer einfachen Vorverarbeitung, in der der Untergrund von den Aufnahmen abgezogen wurde, wurde der Algorithmus angewandt. Die Pyramide bestand dabei insgesamt nur aus drei Ebenen, so daß eine erste Rekonstruktion bei einer Größe von 11*11 Pixel gemacht wurde. In der Regel waren zwei bis drei Iterationsschritte pro Ebene ausreichend, um ein Residuum von höchstens 0.2 zu erreichen. Die erhaltenen Rekonstruktionen sind im Bereich -40 bis +40 Grad von sehr guter Qualität (siehe Abb.8), im Bereich des Missing-Cone aber leicht verschwommen (Abb. 9).

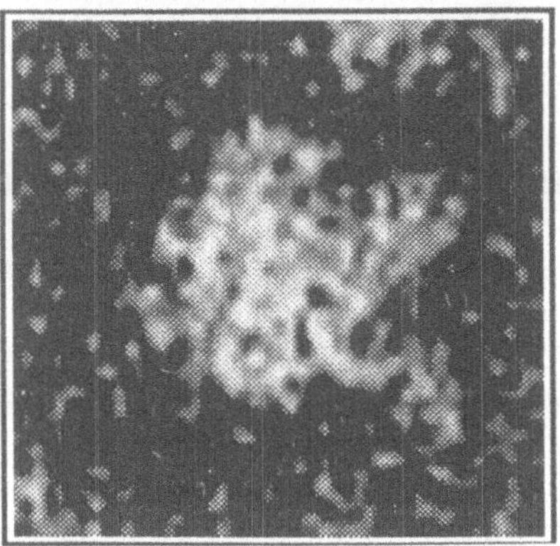 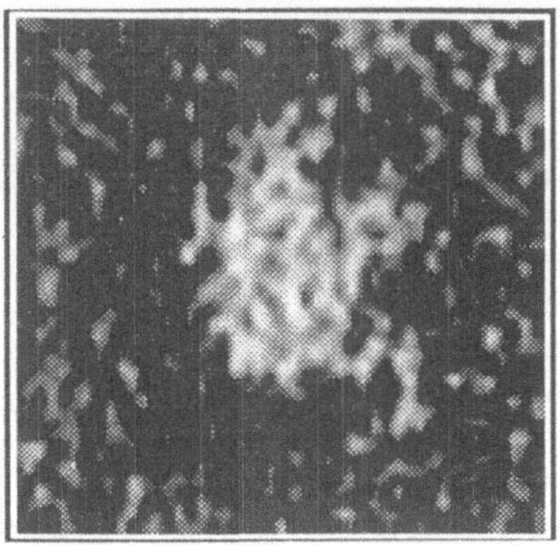

Abb.7: Projektionen einer 50S Untereinheit eines Ribosoms in 0 Grad und -40 Grad Richtung.

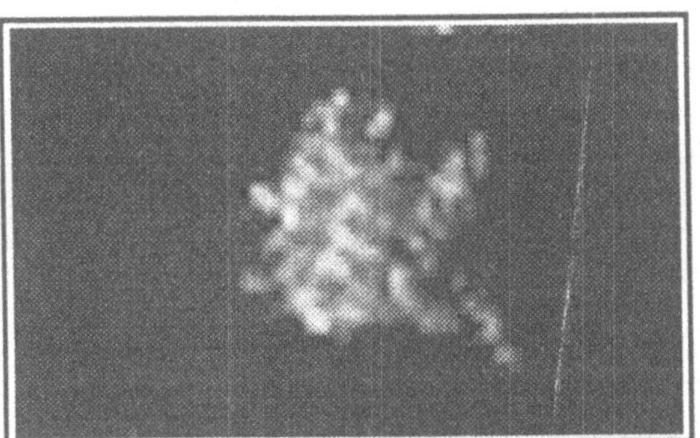 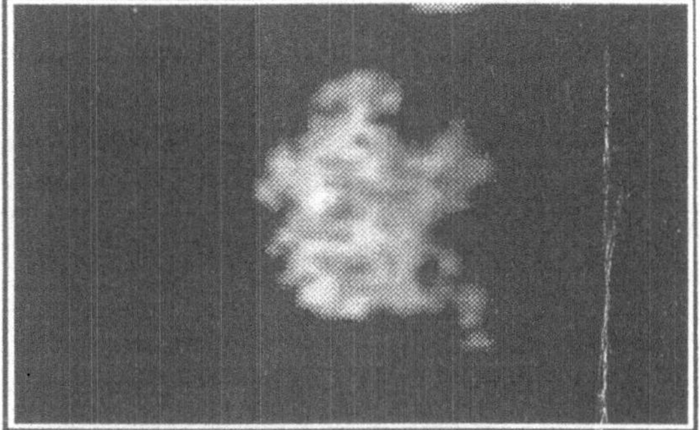

Abb.8: Rekonstruktion eines Ribosoms in 0 Grad und -40 Grad Richtung.

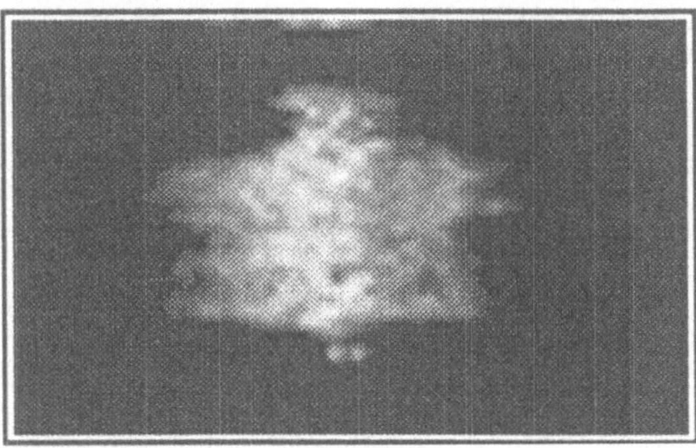

Abb.9: Ansicht des 3D-Objekts aus -90 Grad.

Weitere Rekonstruktionen wurden von Viren und Virus ähnlichen Objekten vorgenommen. Die bisher angewandten Rekonstruktionsverfahren behandelten nur eine einzige Kippachse. Der nächste Schritt unserer Arbeit ist die Erweiterung auf Projektionen in beliebiger Raumrichtung. Der Hauptunterschied besteht dabei in einer anderen Filtermaske für die gefilterte Rückprojektion [Harauz et al. 84], sowie programmtechnischen Einzelheiten der Behandlung von 3D-Objekten.

5. Literatur

[**Burt 84**] P.J. Burt, The Pyramid as a Structure for Efficient Computation, in Multiresolution Image Processing und Analysis, Ed. A. Rosenfeld (Springer, Berlin, 1984) 6

[**Burt 84 a**] P.J. Burt, Multi-Resolution Flow-Through Motion Analysis, RCA Technical Report, PRRL-84-TR-009 (1984)

[**Crowther et al. 75**] R.A. Crowther und A.Klug, Structural Analysis of Macromolecular Assemblies by Image Reconstruction from Electron Micrographs,
Annu. Rev. Biochem. 44 (1975) 161

[**Dengler et al. 88**] J.Dengler und M.Schmidt, The Dynamic Pyramid, A Model for Motion Analysis with Controlled Continuity, Int. J. of Patt. Rec. and Art. Intell.2 (1988) 275

[**Dengler 89**] J. Dengler, A Multi-Resolution Approach to the 3D-Reconstruction from an Electron Microscope Tilt Series Solving the Alignment Problem without Gold Particles, Ultramicroscopy (1989) in Druck

[**De Rosier et al. 68**] D.J. De Rosier und A. Klug, Reconstruction of Three Dimensional Structure from Electron Micrographs, Nature 217 (1968)130

[**Enkelmann 85**] W. Enkelmann, Mehrgitterverfahren zur Ermittlung von Verschiebungsvektorfeldern in Bildfolgen, Dissertation, Hamburg University (1985)

[**Frank 80**] J. Frank, The Role of Correlation Techniques in Computer Image Processing, in: Topics in Current Physics 13, Ed. P.W. Hawkes (Springer, Berlin, 1980) 187

[**Frank et al. 86**] J. Frank und M. Radermacher, Three-Dimensional Reconstruction of Nonperiodic Macromolecular Assemblies from Electron Micrographs, in: Advanced Techniques in Biological Electron Microscopy, Ed. J. Koehler (Springer, Berlin, 1986) 1

[**Harauz et al. 84**] G. Harauz, F.P. Ottensmeyer, Direct Three-Dimensional Reconstruction for Macromolecular Complexes from Electron Micrographs, Ultramicroscopy 12 (1984) 309

[**Heyers et al. 88**] V. Heyers, J. Dengler und H.-P. Meinzer, 3D-Visualisierung von Grauwertvoxelräumen, iebiger in: Informatik Fachberichte 180, Eds. H.Bunke, O.Kübler, P.Stucki (Springer, Heidelberg, 1988) 39

[**Hoppe 74**] W.Hoppe, Zur elektronenmikroskopischen dreidimensionalen Rekonstruktion eines Objektes, Naturwissenschaften 61, (1974) 534

[**Nagel et al. 86**] H.H. Nagel und W. Enkelmann, An Investigation of Soothness Constraints for the Estimation of Displacement Vector Fields from Image Sequences,
IEEE Trans. PAMI 8 (1986) 565

[**Olins et al. 83**] D.E. Olins, A.L. Olins, H.A. Levy, R.C. Durfee, S.M. Margle, E.P. Tinnel, S.D. Dover, Electron Microscope Tomography: Transcription in Three Dimensions,
Science 220 (1983) 498

[**Provencher et al. 88**] S.W. Provencher und R. Vogel, Three-Dimensional Reconstruction from Electron Micrographs of Disordered Specimens, Ultramicroscopy 25 (1988) 209

[**Radermacher et al. 78**] M. Radermacher und W.Hoppe: 3-D Reconstruction from Conically Tilted Projections, Proc. 9th Int. Congr. Electron Microsc., Microscopical Society of Canada, Toronto, Vol. 1 (1978) 218

[**Saxton 84**] W.O.Saxton, W.Baumeister, M.Hahn, Three-Dimensional Reconstruction of Imperfect Two-Dimensional Crystals, Ultramicroscopy 13 (1984) 57

[**Schmidt 88**] M. Schmidt, Mehrgitterverfahren zur 3D Rekonstruktion aus 2D Ansichten, Dissertation, Heidelberg University (1988)

[**Skoglund et al. 86**] U.Skoglund und B.Daneholt, Electron Microscope Tomography, Trends in Biochemical Sciences 11, No. 12 (1986) 499

[**Torre et al. 86**] V. Torre und T. Poggio, On Edge Detection, IEEE Trans. PAMI 8 (1986) 147

Morphologische Größenverteilungen zur Strukturanalyse medizinischer Bilder

Sabine Behrens, Joachim Dengler

Abteilung Medizinische und Biologische Informatik
Leiter: Prof.Dr. C.O. Köhler
Deutsches Krebsforschungszentrum, Heidelberg

1. Einleitung

Die Analyse vieler medizinischer Bilder geschieht bisher vorwiegend visuell bzw. manuell. Daraus resultierende Beschreibungen sind meist qualitativer Art und somit nicht reproduzierbar. Erwünscht ist eine verständliche, quantitative und reproduzierbare Beschreibung der interessierenden Bildstrukturen, um Bilder oder Objekte in den Bildern sowohl mit dem "Normalfall", als auch untereinander vergleichen zu können. Das Problem zahlreicher quantitativer Kenngrößen der Bildanalyse ist, daß sie für den Arzt schwer interpretierbar sind. Ein Merkmal, das die genannten Voraussetzungen erfüllt, ist die Größe einer Struktur. Sie ist intuitiv wahrnehmbar, jedem verständlich und auch quantitativer Art.

Eine Größenverteilung gibt die Wahrscheinlichkeit für die Größe einer bestimmten Struktur in einem Gebiet an. Sie wird mit einer Transformationsfamilie ermittelt. Anschaulich kann die Familie der Transformationen für eine Größenverteilung mit Sieben unterschiedlicher Maschenweite verglichen werden.

In diesem Beitrag wird die bisher nur auf binären Bildern angewandte globale Größenverteilung auf Grauwertbilder verallgemeinert. Eine regionale Verteilung in einem maskierten Ausschnitt wird beschrieben, und weiterhin die Größenverteilung als ein lokales Merkmal vorgestellt. An Hand einiger Beispiele soll gezeigt werden, daß dieses allgemeine Konzept der Größenverteilungen geeignet ist, um Textur, Form und Konfiguration zu charakterisieren.

2. Morphologische Größenverteilung

Die grundlegenden Definitionen der Größenverteilung sind seit längerem bekannt (Delfiner 1972, Matheron 1975, Serra 1982). Delfiner machte sich bereits 1972 Gedanken zur Verallgemeinerung des Konzepts "Größe". Matheron legte 1975 die Kriterien für eine Familie von Transformationen zur Ermittlung einer Größenverteilung in drei Axiomen fest. Serra schließlich beschreibt dazu in seinem Buch zahlreiche Weiterführungen und Beispiele. Aus der Literatur sind bisher jedoch ausschließlich Anwendungen globaler Art auf binären Bildern bekannt.

Matheron's Axiome machen keine Einschränkung bezüglich der zu untersuchenden Menge X. Sie ist aufgrund des Mengencharakters in erster Linie als Binärbild gedacht, kann aber bei entsprechender Verallgemeinerung auch ein Grauwertbild sein. Die konsistente Erweiterung der auf Mengenoperationen beruhenden binären Morphologie auf Grauwertobjekte ergibt sich mit Hilfe des Umbra-Konzeptes, bei dem zweidimensionale Grauwertbilder als dreidimensionale Binärbilder interpretiert werden (Sternberg 1986).

2.1 Die drei Größen-Axiome von Matheron

Transformationen, die die folgenden Matheron'schen Axiome erfüllen, sind geeignet, eine Größenverteilung zu ermitteln.

1. Anti-Extensivität: $\psi_\lambda(X) \subset X \quad \forall \lambda \geq 0$

2. Erhaltung der Inklusion: $Y \subset X \implies \psi_\lambda(Y) \subset \psi_\lambda(X) \quad \forall \lambda \geq 0$

3. Stabilität: $\psi_\lambda[\psi_\mu(X)] = \psi_\mu[\psi_\lambda(X)] = \psi_{Sup(\mu,\lambda)}(X) \quad \forall \lambda,\mu \geq 0$

Aus der Stabilität läßt sich ableiten, daß sich bei wiederholter Anwendung von Ψ mit demselben Parameter λ das Ergebnis nicht mehr ändert. Daraus ergibt sich die Filtereigenschaft der Transformation und insbesondere die Idempotenz:

4. Idempotenz: $\psi_\lambda[\psi_\lambda(X)] = \psi_\lambda(X)$

Wenn man die Transformation $\Psi_\lambda(X)$ als Überrest beim Sieben von X mit Maschenweite λ betrachtet, lassen sich diese drei Axiome durch das Sieben von Partikeln mit Sieben unterschiedlicher Maschenweite veranschaulichen. Sie sind quasi eine Übersetzung der intuitiven Wahrnehmung des Merkmals Größe in ein logisches Konzept.

2.2 Morphologische Transformation

$B(x,y)$ sei ein Bild oder ein Bildauschnitt mit $1 \leq x \leq n$ und $1 \leq y \leq m$, M ein binäres strukturierendes Element.

$$M \; ERO \; B \;\; = \;\; min\{B(x - m_1, y - m_2)\} \;\; \forall \; (m_1, m_2)\epsilon M$$
$$M \; DILA \; B \;\; = \;\; max\{B(x + m_1, y + m_2)\} \;\; \forall \; (m_1, m_2)\epsilon M$$

Diese Definitionen für die beiden grundlegenden morphologischen Operationen Erosion und Dilation (siehe auch Serra 1982, Sternberg 1986, Haralick 1987) gelten sowohl für binäre als auch graue Bilder. Rechnet man ausschließlich mit binären Bildern, können min und max durch $\wedge$ und $\vee$ ersetzt werden.

Zwei morphologische Operationen, die die genannten Axiome erfüllen, sind das Opening, das helle Strukturen bearbeitet, bzw. das Closing entsprechend für dunkle Gebiete.

$$M \; OPENING \; B \;\; = \;\; M \; DILA(M \; ERO \; B)$$
$$M \; CLOSING \; B \;\; = \;\; M \; ERO(M \; DILA \; B)$$
$$(M \; OPENING \; B)^c \;\; = \;\; M \; CLOSING \; B^c$$

2.3 Morphologische Verteilung

Führt man das Opening oder Closing mit einer Familie strukturierender Elemente $FORM_i$ der Form $FORM$ und den Größe $s_1 \ldots s_n$ durch, wobei alle $FORM_i$ kompakt, konvex und offen bezüglich des nächstkleineren sind, wird mit den beiden folgenden morphologischen Transformationen die Größenverteilung der hellen (1) bzw. die der dunklen (2) Strukturen von B bestimmt.

$$F_{s_i}(B) \;\; = \;\; \sum_{x,y}(B - (FORM_i \; OPENING \; B)) \tag{1}$$
$$F_{s_i}(B) \;\; = \;\; \sum_{x,y}((FORM_i \; CLOSING \; B) - B) \tag{2}$$

Bei einem Opening von B mit dem strukturierenden Element $FORM_i$ der Größe s_i verschwinden alle hellen Flecken, die kleiner als die verwendete Maske sind. Übrig bleiben diejenigen, die größer oder gleich der Maske sind. Die Differenz zwischen dem Bild B und seinem Opening enthält folglich genau die Flecken, die in die für das Opening verwendete Maske passen. Für wachsende Maskengrößen s_i mit $i = 1 \ldots n$ wächst diese Differenz monoton, da gilt

$$M_1 \subset M_2 \Rightarrow (B - M_1 \; OPENING \; B) \leq (B - M_2 \; OPENING \; B)$$

Entsprechendes gilt für dunkle Flecken und Closing. Durch eine entsprechende Normierung von $F_{s_i}(B)$ erhält man eine monoton steigende Funktion zwischen 0 und 1.

$$F'_{l_i}(B) \;=\; \frac{F_{s_i}(B)}{\sum_{x,y}(B - min\{B\})} \tag{3}$$

$$F'_{s_i}(B) \;=\; \frac{F_{s_i}(B)}{\sum_{x,y}(max\{B\} - B)} \tag{4}$$

Dies bedeutet, daß $F'_{s_i}(B)$ als normierte Verteilungsfunktion interpretiert werden kann. $F'_{s_i}(B)$ gibt für B die kumulative Wahrscheinlichkeitsverteilung der Struktur, die durch die Maske $FORM_i$ mit der Form $FORM$ und der Größe s_i vorgegeben ist. Form und Größenbereich der Maske ergibt sich aus der einzelnen Anwendung. Für isotrope Masken z.B. Kreise, Quadrate, Sechsecke, Achtecke, ... erhält man so eine isotrope Größenverteilung. Mit einer Familie anisotroper strukturierender Elemente wie Linien, Rechtecke, ... ergibt sich entsprechend eine anisotrope Größenverteilung, mit der z.B. gerichtete Strukturen in Länge oder Breite untersucht werden können.

3. Globale - regionale - lokale Größenverteilung

Die Funktion $F'_{s_i}(B)$ ist wie bereits oben beschrieben, die Wahrscheinlichkeitsverteilung für die Maskenfamilie $FORM_i$ in dem durch B vorgegebenen Gebiet. Repräsentiert B ein Bild kann man von einer **globalen** Größenverteilung bezüglich B sprechen. Sämtliche bisherigen Untersuchungen (z.B. Serra 1982, Meyer 1986) wenden die Größenverteilung global an.

Stellt B ein maskiertes Bild dar, das durch Multiplikation des Originals mit einer binären Maske entsteht, erhält man eine **regionale** Verteilung in dem durch die Maske vorgegebenen Gebiet. Der Vorteil dabei ist, daß die zu charakterisierende Region eine beliebige Form haben kann und nicht zwingend rechteckig sein muß.

Legt man um jedes Pixel eines Bildes eine kleine Umgebung B, erhält man für jedes Pixel eine Größenverteilung bezüglich dieser Umgebung, also eine **lokale** Größenverteilung. Diese lokalen Verteilungen können parallel berechnet werden, wenn in Gleichung (3) bzw. (4) B das gesamte Bild ist, aber nicht über dieses, sondern nur über die jeweilige Umgebung um jedes Pixel mittelt, was einer Faltung mit einer Maske in der Größe der Umgebung entspricht (SMOOTH). Diese Faltungsmaske kann alle Pixel gleichgewichten, wie es bei der globalen Größenverteilung der Fall ist, sinnvoller wäre jedoch bei lokaler Betrachtung z.B. die Gewichtung mit einer Gaußglocke.

$$F_{s_i}(x,y) \;=\; SMOOTH \; (B - (FORM_i \; OPENING \; B)) \tag{5}$$

$$F_{s_i}(x,y) \;=\; SMOOTH \; ((FORM_i \; CLOSING \; B) - B) \tag{6}$$

F_{s_i} ist bei der parallelen Bestimmung aller lokalen Verteilungen ein Bild, das für jedes Pixel die Wahrscheinlichkeit für die Maske $FORM$ der Größe s_i angibt. Mit $i = 1 \ldots n$ erhält man ein dreidimensionales Array der Dimension (n,x,y).

4. Bestimmung von Merkmalen aus der lokalen Größenverteilung

Bei Anwendung der lokalen Größenverteilung $F_{s_i}(B)$ mit $i = 1 \ldots n$ nach Gleichung (5) oder (6) erhält man die n-fache Datenmenge des Originals. Da zwischen diesen Merkmalen viele Abhängigkeiten bestehen, ist es meist ausreichend, den Erwartungswert und die Varianz der Größenverteilung für jedes Pixel zu errechnen. Die Dichtefunktion zu $F_{s_i}(B)$ ist

$$f_{s_i}(B) \;=\; \frac{\triangle F_{s_i}(B)}{\triangle s_i} \;=\; \frac{F_{s_i}(B) - F_{s_{i-1}}(B)}{s_i - s_{i-1}}$$

Aus f_{s_i} lassen sich Erwartungswert und Varianz folgendermaßen berechnen:

$$\bar{s} = \frac{\sum s_i f_{s_i} \cdot (s_i - s_{i-1})}{\sum f_{s_i} \cdot (s_i - s_{i-1})} = \frac{\sum s_i (F_{s_i} - F_{s_{i-1}})}{\sum (F_{s_i} - F_{s_{i-1}})} = \frac{\sum s_i \, \triangle F_{s_i}}{F_{s_n} - F_{s_1}}$$

$$\sigma^2 = \frac{s_i^2 \, \triangle F_{s_i}}{F_{s_n} - F_{s_1}} - \bar{s}^2$$

$$\triangle F_{s_i} = \sum_{x,y}(FORM_i \; OPENING \; B) - (FORM_{i-1} \; OPENING \; B)$$

Die Differenz aufeinanderfolgender Openings mit verschieden großen strukturierenden Elementen liefert Flecken, die größer gleich der kleineren Maske und kleiner als die größere Maske sind. Die

Wahl des Parameters s_i hängt sowohl von der Form der gewählten Maskenfamilie als auch von der Anwendung ab. Eine allgemeine Möglichkeit ist die Ableitung nach der Flächendifferenz der beiden Masken. Bei kreis- bzw. linienförmigen Masken ist auch eine Normierung mit dem Radius bzw. der Länge sinnvoll.

$$s_i = Area(FORM_i) \quad bzw. \quad s_i = Radius(FORM_i) \quad bzw. \quad s_i = Laenge(FORM_i)$$

5. Anwendungen

Beispiel 1: Globale binäre Größenverteilung

Abb. 1 zeigt einen Ausschnitt aus einer Mammographie mit segmentierten Verkalkungsflecken (Holder, Dengler 1988). Die Größenverteilung wurde mit Gleichung (1) ermittelt. In dem Diagramm in Abb. 2 ist die relative Häufigkeit von Kreismasken gegen deren Durchmesser aufgetragen. Diese Art der Größenverteilung, also global auf binären Bildern ist die aus der Literatur bekannte (Serra 1982). Damit kann eine fleckenhafte Textur, wie sie in medizinischen Bildern häufig vorkommt, beschrieben werden.

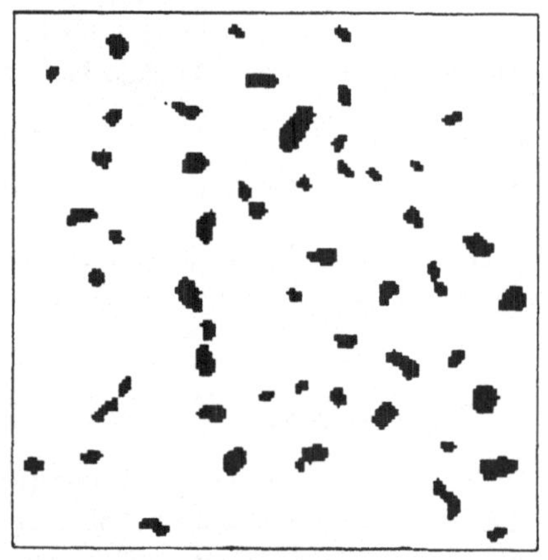

Abb. 1 Binarisierter Ausschnitt aus einer Mammographie

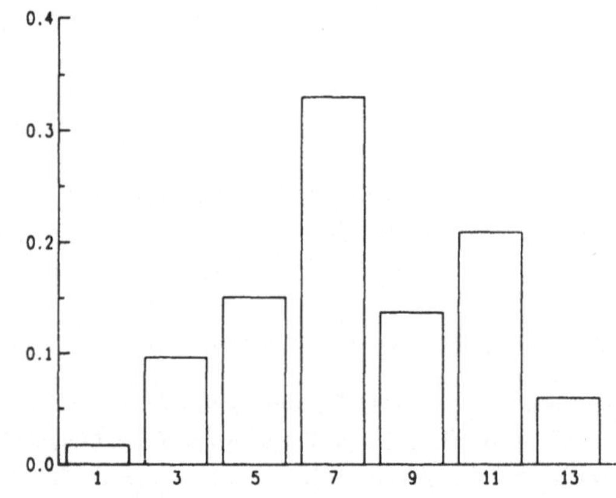

Abb. 2 binäre globale Größenverteilung von Abb. 5.1

Beispiel 2: Regionale Größenverteilung

Abb. 3 zeigt einen Ausschnitt aus einem CT-Schnitt mit Leber und Niere. In dem segmentierten binären Bild (Abb. 4) wurde die Größenverteilung sowohl mit der Opening- als auch mit der Closing-Operation durchgeführt. Dargestellt ist in Abb. 5 die relative Wahrscheinlichkeit für

Kreise über deren Durchmesser, wobei im 1. Quadrant die Kurve für die Opening-Operation dargestellt ist und im 4. Quadrant negativ die Verteilung bezüglich des Closings.

Der 1. Quadrant in Abbildung 5 zeigt, daß die Niere kleiner ist als die Leber, da in ihr keine so großen Kreise enthalten sind wie in der Leber. Außerdem läßt sie sich vorwiegend aus Kreisen einer Größenordnung rekonstruieren (10-15), wogegen die Leber sowohl aus großen wie auch aus kleinen Strukturen besteht, also "eckiger" ist. Aus dem 4. Quadrant kann man ablesen, daß die Leber nahezu konvex ist. Die Niere schließt ein konkaves Gebiet von etwa 13 Pixeln Durchmesser ein.

Berechnet man die Größenverteilungen nicht auf dem segmentierten, sondern auf dem maskierten Grauwertbild, erhält man Abb. 6. Bei großen Durchmessern hat sich lediglich die relative Häufigkeit geändert, die Form der Kurve ist erhalten geblieben. Dagegen ist die Wahrscheinlichkeit für kleine Kreise, die die Innenstruktur der Organe charakterisieren, deutlich gestiegen.

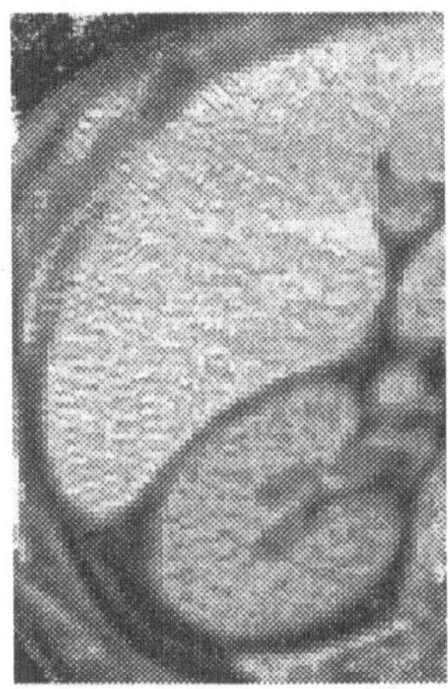

Abb. 3 CT-Schnitt von Leber und Niere

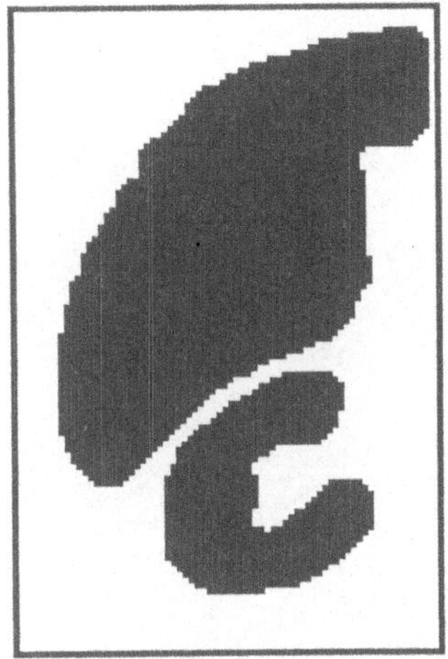

Abb. 4 Leber und Niere segmentiert

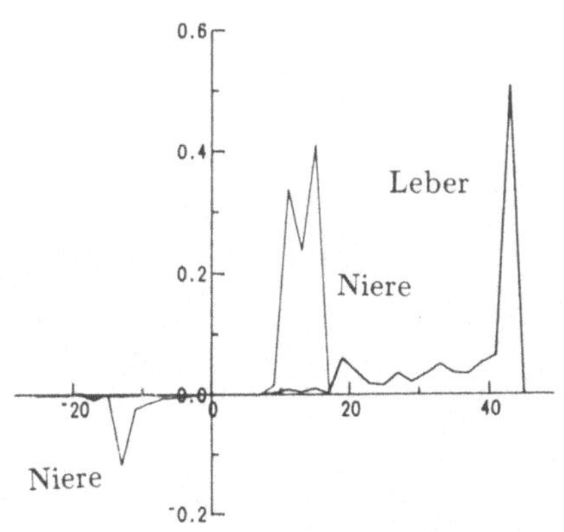

Abb. 5 regionale binäre Größenverteilung

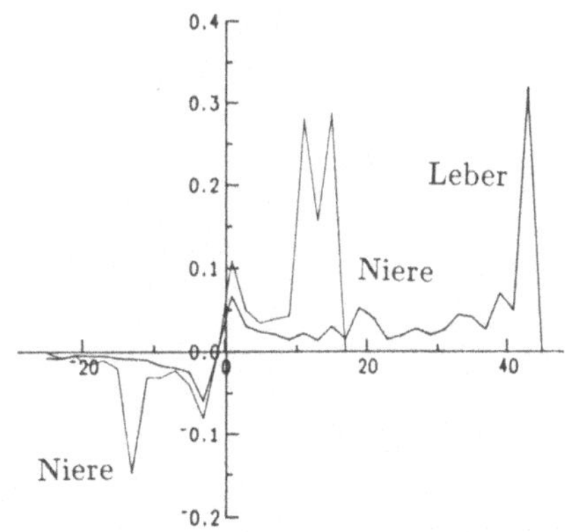

Abb. 6 regionale Grauwert-Größenverteilung

In Abb. 7 ist ein Ausschnitt aus einer Mammographie zu sehen, jedoch als Grauwertbild. Es wurde die Größenverteilung der drei markierten Regionen sowohl mit Gleichung (1) als auch (2) berechnet. Der mittlere Ausschnitt enthält im Gegensatz zu den beiden anderen zahlreiche Kalkflecken. Abbildung 9 zeigt von den drei Ausschnitten die Erwartungswerte der lokalen Größenverteilungen der hellen Flecken als Grauwertbild.

Mit diesem Bild läßt sich ein eventuell vorhandener Kalkherd in einer Mammographie lokalisieren, indem bei der mittleren zu erwartenden Kalkfleckgröße ein Schwellwert gelegt wird.

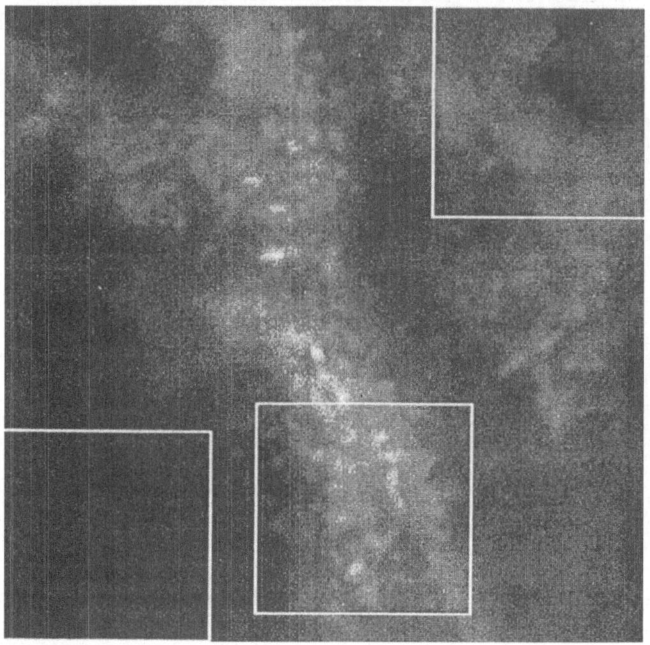

Abb. 7 Ausschnitt aus einer Mammographie

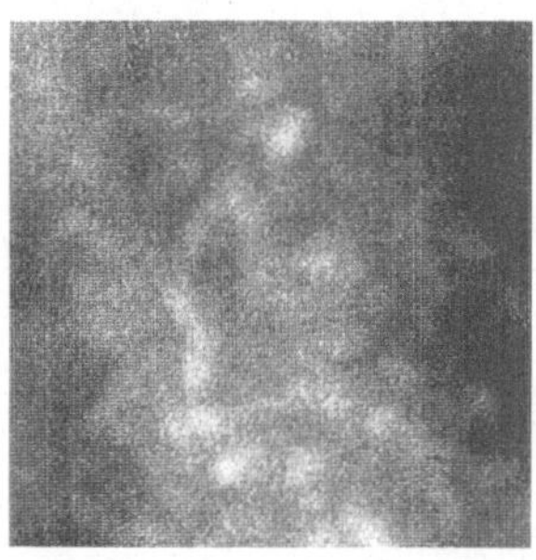  

Abb. 8 Auschnitte jeweils kontrastverstärkt

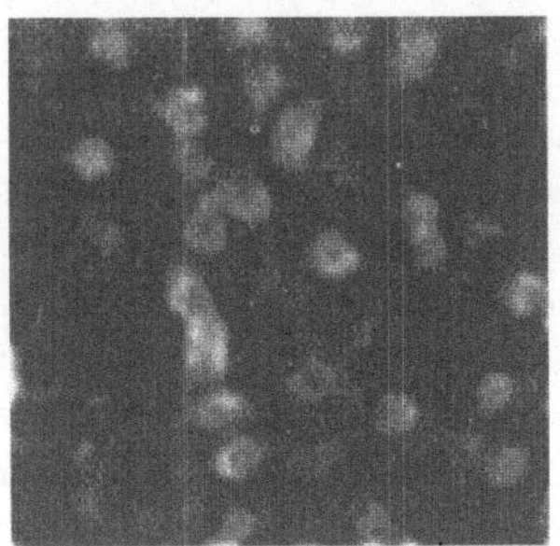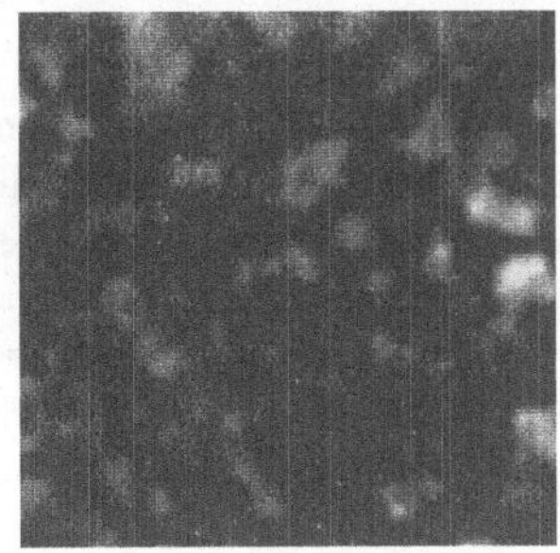

Abb. 9 Erwartungswerte der lokalen Größenverteilungen

In Abb. 10 ist die relative Häufigkeit der Erwartungswerte der lokalen Größenverteilung nach Gleichung (1) gegenüber der Fläche der Flecken aufgetragen. Der Ausschnitt mit Kalkflecken hat einen wesentlich größeren Anteil an großen Flecken (Fläche 17 bis 30), wogegen der Anteil der Flecken mit Fläche 5 bis 10 kleiner ist, dies entspricht einem kleineren Rauschanteil.

Abb. 11 zeigt die relative Häufigkeit der Erwartungswerte der lokalen Größenverteilung mit Closing. Sie beschreibt also die Verteilung des Hintergrundes bezüglich der Größe. Der Anteil der kleinen Flecken im Hintergrund ist bei dem Ausschnitt mit Kalkflecken bedeutend kleiner, d.h. der Anteil des Rauschens ist geringer, dagegen sind größere Flecken häufiger, die die Zwischenräume zwischen den eigentlichen Kalkflecken repräsentieren. Es läßt sich auch ablesen, daß die Kalkflecken größer sind als die Zwischenräume. Mit diesen beiden Diagrammen ist die Konfiguration einer Kalkgruppe gut beschrieben.

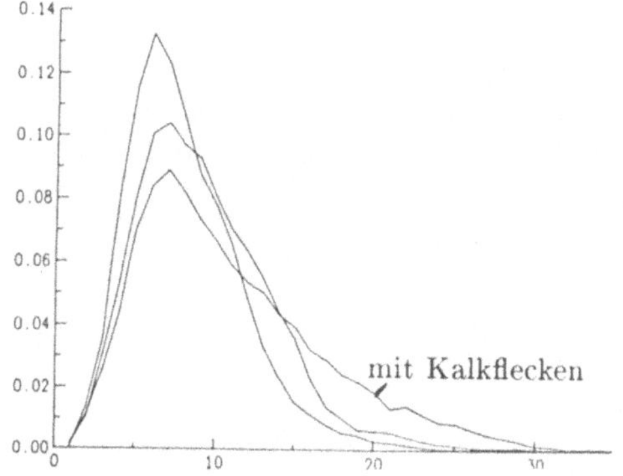

Abb. 10 Histogramm der Erwartungswerte (Opening)

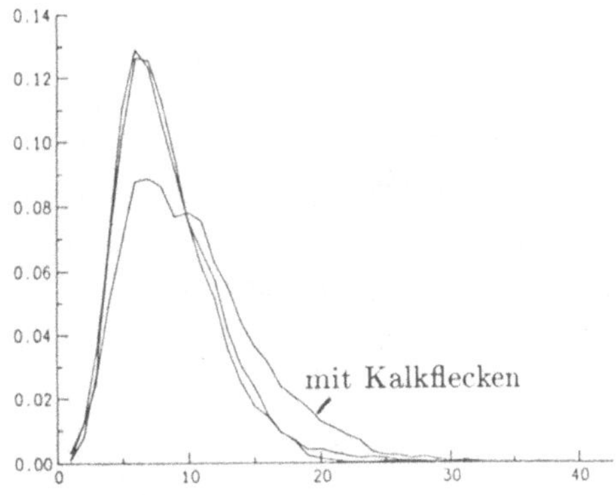

Abb. 11 Histogramm der Erwartungswerte (Closing)

6. Bewertung

Vorgestellt wurde die morphologische Größenverteilung als ein allgemeines Konzept zur Strukturanalyse von Binär- und Grauwertbildern. Aus der Anwendung auf graue Bilder ergeben sich spezifische Problemstellungen, die an entsprechender Stelle berücksichtigt wurden. Gerade die in der medizinischen Bildverarbeitung üblicherweise stark verrauschten Bilder machen ein binäre Behandlung sehr schwierig. Anders gesagt kann durch die Anwendung auf Grauwertbilder in vielen Fällen eine Vorsegmentierung ganz entfallen oder wird stark vereinfacht. Ein weiterer Vorteil der morphologischen Transformationen ist die separate Behandlung heller und dunkler Gebiete. Dadurch können in einer Konfiguration sowohl die Objekte als auch die Zwischenräume oder bei der Formanalyse konvexe Teile wie konkave Einschlüsse beschrieben werden. Die Beispiele beweisen, daß mit einem Konzept, nämlich der Größenverteilung, sowohl Textur wie auch Form und Konfiguration charakterisiert werden können.

Wir danken der Deutschen Forschungsgemeinschaft, die dieses Projekt im Rahmen des Schwerpunktes "Modelle und Strukturanalyse bei der Auswertung von Bild- und Sprachsignalen" seit einem Jahr fördert.

7. Literatur

Delfiner, P.
A Generalization of the Concept of Size.
Journal of Microscopy 95 (1972) 203-216

Haralick, R. M.; Sternberg, S. R.
Image Analysis Using Mathematical Morphology: Part I
IEEE Pattern Analysis and Machine Intelligence 9 (1987) 532-550

Holder, S.; Dengler, J.; Desaga, J. F.
Lokalisation von Mikrokalzifikationen in Mammographien.
In Bunke, H.; Kübler, O.; Stucki, P.(Hrsg.): Mustererkennung 1988, Proc. 10. DAGM-Symposium Zürich, 17-23, Informatikfachberichte 180, Springer, Berlin - Heidelberg - New York - London - Paris - Tokyo 1988

Matheron G.
Random Sets and Integral Geometry.
Wiley, New York 1975

Meyer, F.
Automatic Screening of Cytological Specimens.
Computer Vision, Graphics, and Image Processing 35 (1986) 356-369

Serra, J.
Image Analysis and Mathematical Morphology.
Academic Press London 1982

Sternberg, S. R.
Grayscale Morphology.
Computer Vision, Graphics, and Image Processing 35 (1986) 333-355

Object Location Based on Uncertain Models

Monika Sester and Wolfgang Förstner
Institut für Photogrammetrie – Universität Stuttgart
Keplerstraße 11, D-7000 Stuttgart 1

Abstract

The paper describes a concept for object location, when not only image features but also the model description is uncertain. It contains a method for probabilistic clustering, robust estimation and a measure for evaluating both, inaccurate and missing image features. The location of topographic control points in digitized aerial images demonstrates the feasibility of the procedure and the usefulness of the evaluation criteria.

1 Introduction

Object location is a central issue in Computer Vision. The task is to determine the pose, i.e. the position and the orientation of an object with respect to a reference frame, for which a model is known either derived from a CAD-system, (cf. GRIMSON/LOZANO-PEREZ 1984,1987, FAUGERAS/HEBERT 1987, HORAUD 1987), sensed from a prototype (FAN 1988), or, more challenging, described by a set of rules, which form a generic model of the object. In this case not only the pose i. e. few parameters are unknown, but also the individual structure of the object (FUA/HANSON 1987).

Main research issues are the computational complexity of the matching problem, which requires strong constraints or heuristics to lead to practially acceptable solutions and the problem of representing 3-dimensional shapes which are suitable for being derived from both CAD-systems and digital images automatically under broad conditions.

This paper deals with the problem of *uncertainty* encountered during object location from digital images. Uncertainty of the raw data, of the feature extraction, about the used thresholds, about the assumed model, due to wrong or missing correspondencies between model and image features altogether result in uncertain values for the pose parameters. It is the aim of the paper to show, that for this special task of object location the tools provided by mathematical statistics and probability theory are sufficient to describe the uncertainty of the pose parameters in a compact form. The quality of the pose estimation can be used as a means for *selfdiagnosis*, a prerequisite for any automated system to be used in practice.

In contrast to most other approaches, also the uncertainty of the model is taken into account. Moreover, the theory enables to test the validity of the model and to evaluate the effect of both imprecise and missing image features onto the result.

The motivation for this study resulted from a task at the photogrammetric department of the *Landesvermessungsamt Nordrhein-Westfalen*. For their orthophotoproduction they have built up a data base of more than 20000 topographic control points, mainly gable points of house roofs. They use their location in the image for determining the orientation of the aerial photos (scale appr. 1:12000). The X,Y and Z world coordinates of the gable points and a sketch of the roof in orthogonal projection are given. A first prototype program to locate the control points automatically reveiled the difficulty to transparently set thresholds in the image analysis and the matching procedure which required a theoretically more founded setup described in this paper.

The whole identification procedure consists of 4 steps (cf. Fig. 1).

1. *Interpretation* of the 2-D-sketches, resulting in a 3-D-description. It not only contains the coordinates in the world coordinate system, but also their uncertainty as the sketches are not fully in scale (cf. Fig. 1a → 1b).

2. *Projection* of the 3-D-model into the aerial image using the appropriate values for position and orientation of the camera from a flight plan leading to a 2-D-wire frame model in the image, again containing the uncertainty, now of both the sketch and the appropriate orientation parameters (cf. Fig. 1b → 1c).

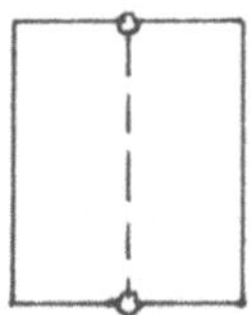 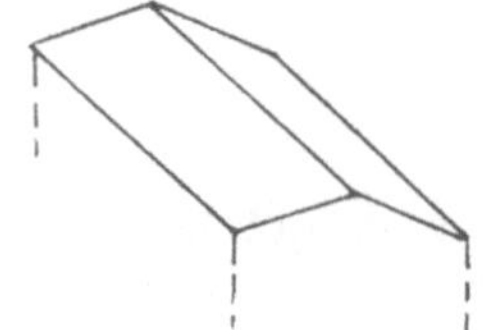 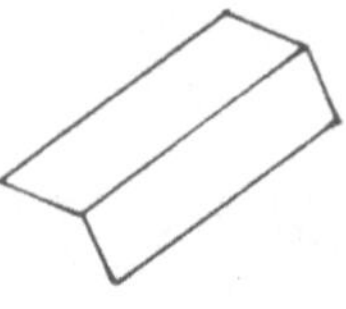

a. given sketch, orthogonal projection   b. interpreted sketch, 3D-model   c. projected model, 2D-wire model

 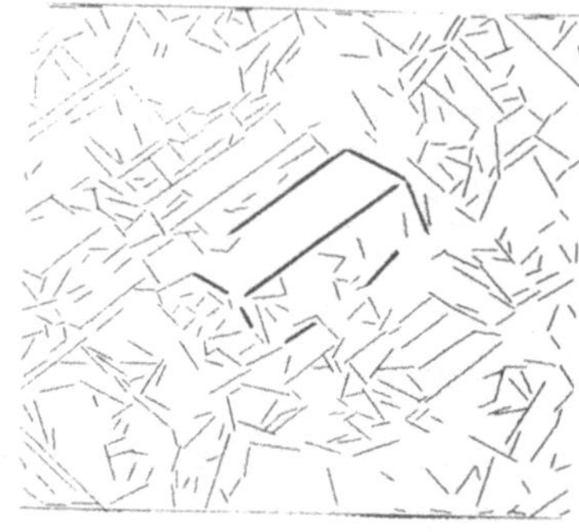

d. scanned image section, 240 x 240 pixels à 20 μ

e. extracted edges (thin), matched edges (thick)

Figure 1: Example for matching roof model to image edges

3. *Extraction* of straight line segments from a digitized subsection of the areal image, leading to a list of line segments, also here containing information about the geometric uncertainty due to the image analysis (cf. Fig. 1d → 1e).

4. *Matching* the image features to the model features taking the geometric relations and the uncertainty into account (cf. Fig. 1c ⇔ 1e).

Section 2 outlines the theoretical background of the approach. Sections 3 and 4 describe the chosen representation of model and image features. Section 5 contains the used matching procedure. An example of the procedure in section 6 demonstrates the feasibility of the approach and the usefulness of the chosen evaluation criteria.

2 Conceptual Background

The ultimate goal of our task is to determine the position of the control points, i. e. the gable points. A reasonable requirement is the location to be reliable in the sense that inaccuracies of model or image features on one hand and false matches on the other hand do not deteriorate the coordinates of the controlpoints too much. This notion of reliability has been developed by BAARDA 1967,1968 for the use in geodetic networks and seems to be appropriate here too, a review of the theory is given by FÖRSTNER 1987.

2.1 The Functional and the Stochastical Model

Assume a certain list of matched features $\mathbf{a}_o = \{a_k\}$ is hypothesized to be acceptable then we can determine the optimal transformation T with parameters $\mathbf{p}$ using the nonlinear model (stochastical variables are underlined)

$$E(\underline{\mathbf{a}}_o) = T(\mathbf{p}) = T(\mathbf{p}_t, \mathbf{p}_f, \mathbf{p}_l) \tag{1}$$

$$E(\underline{\mathbf{p}}_f) = \mathbf{p}_f \tag{2}$$

$$E(\underline{\mathbf{p}}_l) = \mathbf{p}_l \tag{3}$$

or after linearisation

$$\mathbf{x} + \mathbf{r} = \mathbf{A}\hat{\underline{y}} \quad or \quad \begin{pmatrix} \underline{\mathbf{x}}_o \\ \underline{\mathbf{x}}_f \\ \underline{\mathbf{x}}_l \end{pmatrix} + \begin{pmatrix} \mathbf{r}_o \\ \mathbf{r}_f \\ \mathbf{r}_l \end{pmatrix} = \begin{pmatrix} \mathbf{A}_t & \mathbf{A}_f & \mathbf{A}_l \\ 0 & \mathbf{I} & 0 \\ 0 & 0 & \mathbf{I} \end{pmatrix} \begin{pmatrix} \hat{\underline{y}}_t \\ \hat{\underline{y}}_f \\ \hat{\underline{y}}_l \end{pmatrix} \tag{4}$$

with

$\underline{\mathbf{x}}_o = (\mathbf{x}_{ok}^T)^T$	the vector of observations
$\underline{\mathbf{x}}_{ok} = \underline{\mathbf{a}}_k - T(\mathbf{p}^{(o)})$	the differences between image features and the predicted model features
$\mathbf{p}^{(0)} = (\mathbf{p}_t^{(0)T}, \mathbf{p}_f^{(0)T}, \mathbf{p}_l^{(0)T})^T$	approximate values for the transformation parameters for the translation $(\mathbf{p}_t)$, the form $(\mathbf{p}_f)$ and the location parameters $(\mathbf{p}_l)$
$\hat{\underline{y}} = (\hat{\underline{y}}_t^T, \hat{\underline{y}}_f^T, \hat{\underline{y}}_l^T)^T$	the estimated corrections to the approximations
$\mathbf{a}_k$	the parameters describing the image feature k
$\mathbf{r}$	the corrections to the observations $\mathbf{x}$
$\mathbf{A}_o = (\mathbf{A}_t, \mathbf{A}_f, \mathbf{A}_l)$	the partial derivations of T with respect to the unknown parameters

Eqs.1 to 4 contain the geometric relationship between the image features and the unknown parameters, in nonlinear and linearized form. They represent the projection of the object into the image. The unknown parameters describe the form of the object and the orientation of the image with respect to the object; the task is equivalent to locating the object with respect to the image coordinate system. The additional eqs. 2 and 3 contain the a priori knowledge about the form and the orientation parameters. The *functional model* thus is a statement about the expectation about the observed values, or more technically about the first moments of the stochastical variables associated with the uncertain input values.

The *stochastical model* describes the uncertainty of the observed image features $\mathbf{a}_k$ and of the approximate values $\mathbf{p}_f^{(0)}$ and $\mathbf{p}_l^{(0)}$. The uncertainty of the stochastical variables is represented by their second moments or their variances:

$$D(\mathbf{x}) = D(\underline{\mathbf{a}}_o, \underline{\mathbf{p}}_f, \underline{\mathbf{p}}_l)^T = diag\,(D(\underline{\mathbf{a}}_o), D(\underline{\mathbf{p}}_f), D(\underline{\mathbf{p}}_l)) = diag\,(\mathbf{C}_{oo}, \mathbf{C}_{ff}, \mathbf{C}_{ll}) = \mathbf{C}_{xx} \tag{5}$$

In case the covariance matrices $\mathbf{C}_{ff}$ and $\mathbf{C}_{ll}$ are diagonal the standard deviations σ_{p_f} and σ_{p_l} fully describe the uncertainty of the model and the approximate values. The matix $\mathbf{C}_{oo}$ in general will not be diagonal, describing the interdependencies between the geometric feature values derived from the image.

The model eq. 4 is general enough to represent a large class of object location tasks. We will specify the individual parts later.

2.2 Precision and Sensitivity of Estimated Parameters

The parameters can now be estimated from a best linear unbiased estimation procedure:

$$\hat{\mathbf{y}} = (\mathbf{A}^T \mathbf{C}_{xx}^{-1} \mathbf{A})^{-1}\, \mathbf{A}^T \mathbf{C}_{xx}^{-1}\, \mathbf{x} \tag{6}$$

The influence of random errors onto the result, i.e. the *precision* can easily be derived by error propagation. The covariance matrix

$$\mathbf{C}_{yy} = (\mathbf{A}^T \mathbf{C}_{xx}^{-1} \mathbf{A})^{-1} = \begin{pmatrix} \mathbf{C}_{tt} & \mathbf{C}_{tf} & \mathbf{C}_{tl} \\ \mathbf{C}_{ft} & \mathbf{C}_{ff} & \mathbf{C}_{fl} \\ \mathbf{C}_{lt} & \mathbf{C}_{lf} & \mathbf{C}_{ll} \end{pmatrix} \tag{7}$$

contains the variances of all unknown parameters. This specially holds for the translation parameters $\hat{\mathbf{p}}_t$ and the form parameters $\hat{\mathbf{p}}_f$.

A test on the residuals $\mathbf{r}_o$ reveals outliers in the image features, which may result from a wrong correspondence. A test on the residuals $\mathbf{r}_f$ may reveal errors in the assumed form of the object. For these tests an assumption about the distribution of the stochastical variables is required. A Gaussian distribution seems to be sufficient in all cases.

Even if the derived standard deviations are acceptable the result may be uncertain due to two reasons:

1. The ommittance of one of the image features say $\mathbf{x}_k$ may lead to a large change of the pose parameters

$\hat{\mathbf{y}}_p = (\hat{\mathbf{y}}_t, \hat{\mathbf{y}}_l)$. An upper bound for the influence (indicated by ∇) of image feature $\mathbf{x}_k$ onto an arbitrary function $f(\hat{\mathbf{y}}_p)$ of $\hat{\mathbf{y}}_p$ is given by

$$\nabla_k f(\hat{\mathbf{y}}_p) \leq \sigma_{f(y_p)} \, \mu_k \, t_k = \sigma_{f(y_p)} \, \bar{\delta}_k \tag{8}$$

where

- $\sigma_{f(\mathbf{y}_p)}$ is the standard deviation of $f(\hat{\mathbf{y}}_p)$
- μ_k describes the geometric sensitivity of the result with respect to the k-th feature and - for simplicity at the moment assuming no form parameters to be estimated - can be derived from

$$\mu_k = \lambda_{max}[\mathbf{C}_{\hat{x}_k \hat{x}_k} \, (\mathbf{I} - \mathbf{C}_{\hat{x}_k \hat{x}_k})^{-1}] \tag{9}$$

with $\mathbf{C}_{\hat{x}_k \hat{x}_k} = \mathbf{A}_k \mathbf{C}_{yy} \mathbf{A}_k^T$, where $\mathbf{A}_k$ is the k-th submatrix of $\mathbf{A}_o = (\mathbf{A}_t, \mathbf{A}_f, \mathbf{A}_l) = (\mathbf{A}_{o1}^T, \dots, \mathbf{A}_{ok}^T, \dots)^T$ (cf. FÖRSTNER 1983)

- t_k is the statistic for testing the significance of the residual $\mathbf{r}_k$ with respect to wrong correspondencies (namely the squareroot of a Fisher-teststatistic)
- $\bar{\delta}_k$ is an factor depending on μ_k and t_k, thus on the k-th feature only

Hence, even if σ_f and t_k are acceptable due to a weak geometry the k-th feature may heavily influence the result if μ_k is large. On the other hand even if σ_f and t_k are not acceptable the influence of $\mathbf{a}_k$ onto the result may be small, indicating that the feature may be superfluous. The upper bound $\sigma_f \, \mu_k \, t_k$ thus measures the *empirical sensitivity* of the result. The normalized and dimensionless *empirical sensitivity factor* $\bar{\delta}_k = \mu_k \, t_k$ therefore should be in the range up to 1 – 2.

2. The *theoretical sensitivity* does not depend on the data and measures the ability of the result to absorb wrong matches which are undetectable by a test on the residuals $\mathbf{r}_k$. The ability to detect (gross) errors depends not only on the geometry, thus on μ_k, but also on the used test, its degrees of freedom dof, the significance level $S = 1 - \alpha$ and the required minimum power β_0 for the test. We obtain the upper bound for the influence of an undetectable error to be

$$\nabla_{0k} f(\mathbf{y}_p) \leq \sigma_{f(y_p)} \, \mu_k \, \delta_0 = \sigma_{f(y_p)} \, \bar{\delta}_{0k} \tag{10}$$

with $\delta_0 = \delta_0(\alpha_0, \beta_0, dof)$, e. g. being $\delta_0 = 4.13$ for $dof = 1, \alpha = 0.001$ and $\beta_0 = 0.80$ (cf. BAARDA 1967, 1968, FÖRSTNER 1987).

The relevance of this measure is its ability to indicate *missing* image features, which are necessary for the determination of the pose parameters, as then the *theoretical sensitivity factors* $\bar{\delta}_{0k}$ of the *other* features may easily reach values greater than $4 - 5$. The relevance of the standard deviation of the pose estimates, measuring the influence of the random errors only, is reduced in this case. A similar reasoning can be applied to the approximate values specifying form and orientation. This leads to statements about the sensitivity of the result with respect to gross errors in the model or in the assumed orientation of the camera.

The upper bounds eqs. 8 and 10 contain information on the effect of random and gross errors, as well as missing data. As they are in the dimension of the resultant parameters, they are easily interpretable.

3 Representation of Uncertain Models

The basic idea of representing uncertain models is to describe the object in a parametric form and to specify the uncertainty of the parameters by their standard deviation. Though we only discuss the models used in a specific application the potential of the method will become clear (cf. SMITH ET AL. 1987).

In our case the model shows up in three different versions, which can be derived by geometric transformations each involving a well defined set of parameters:

a: The planar coordinates of the nodes in the sketch can be expressed by a few parameters, e.g. the width of the roof (cf. Fig. 2). The scale can directly be taken into account due to the given, and fixed coordinates of the two gable points, which therefore serve as an intuitive coordinate system. The imprecision of the sketches leads to reasonable standard deviations of appr. 10–15% of the lengths. The Z-values of the house depend on the slope of the roof which is unknown. Due to this lack of knowledge we allow each roof-plane to be of type horizontal, positive slope or negative slope unless geometric conditions form constraints on the choice. The interpretation of the sketches, which is necessary here, will not be discussed. We assume that one choice, among several has been made. Then the slopes are additional free parameters specifying the form. Their expected value is assumed to be approx. $\pm 45°$ or $0°$, the standard deviation to be $\sigma = 15°$.

The 3D-coordinates (X_i, Y_i, Z_i) of the model in full scale then depend on the parameters $\mathbf{p}_f$:

$$\begin{pmatrix} \mathbf{X} \\ \mathbf{Y} \\ \mathbf{Z} \end{pmatrix}_i = B_i(\mathbf{p}_f) = \begin{pmatrix} \mathbf{B}_{xi} \\ \mathbf{B}_{yi} \\ \mathbf{B}_{zi} \end{pmatrix} \mathbf{p}_f \tag{11}$$

For the simple roof in Fig. 2 we obtain with $\mathbf{X} = (X_i)$, $\mathbf{Y} = (Y_i)$ and $\mathbf{Z} = (Z_i)$ and the two parameters

i	B_{xi0}	B_{xi1}	B_{xi2}	B_{yi0}	B_{yi1}	B_{yi2}	B_{zi0}	B_{zi1}	B_{zi2}
1	0	0	0	0	0	0	0	0	0
2	0	0	0	1	0	0	0	0	0
3	0	-1	0	0	0	0	0	$-s$	$-w$
4	0	-1	0	1	0	0	0	$-s$	$-w$
5	0	+1	0	0	0	0	0	$-s$	$-w$
6	0	+1	0	1	0	0	0	$-s$	$-w$

Figure 2: Example for a parametric description of a sketch; w = width of the roof, l = length of the roof

$p_{f_1} = width = w$ and $p_{f_2} = slope$ and using $p_{f_0} = l$ the elements in the table of Fig. 2. $\mathbf{B}$ and $\mathbf{C}_{ff}$ symbolically represent the 3D-model.

b: The transformation T_l^w of this 3-D-model into the world coordinate system requires no new uncertain parameters.

c: The projection T_w^m of the 3-D-world model into the image depends on 6 parameters from which the translation $p_t = (X_0, Y_0)$ is assumed to be free (or with very large standard deviations, measured in pixels). The four other parameters p_l, namely scale and orientation, are assumed to have a precision of 5–10 % and $2^0 - 4^0$ resp. Thus the image coordinates of the projected model can be derived for each point i by

$$\mathbf{b}_i = \begin{pmatrix} x \\ y \end{pmatrix}_i = T_w^m(\mathbf{p}_t, \mathbf{p}_l)\, T_l^w\, B_i(\mathbf{p}_f) \tag{12}$$

For a point pair $\mathbf{a}_k = (\mathbf{b}_i, \mathbf{b}_j)$, forming an edge in the projected model this after linearization yields the right hand side of the first line in eq. 4.

Without additional knowledge the parameters then would be free. The a priori knowledge of the form and the orientation therefore is expressed by eqs. 2, 3 or the 2nd and 3rd line in eq. 4.

4 Representation of Straight Edges Extracted from Images

Any extraction scheme can be used for deriving straight edge elements (cf. e. g. BURNS ET AL. 1986, NEVATIA/BABU 1980). The uncertainty of the straight edges can usually be derived from the extraction process itself. We use a weighted least squares fit through the extracted edge elements, determined to subpixel position. The weights w_i are proportional to the squared value of the gradient. In a local coordinate system (u,v) (cf. Fig. 3), which lies near the principle axes of the edge elements the covariance matrix for

Figure 3: Representation of edge and local coordinate system (u,v)

the parameters a and m for the straight line $v = a + mu$ is diagonal and given by:

$$D\begin{pmatrix} \hat{a} \\ \hat{m} \end{pmatrix} = \sigma_0^2 \begin{pmatrix} 1/\sum w_i & 0 \\ 0 & 1/\sum u_i^2 w_i \end{pmatrix} \tag{13}$$

where the sums are taken over all edge elements. The uncertainty of the u-coordinates in this calculation can be neglected. The v-coordinates of the end points will be correlated due to the common factor $\hat{m}$. The u-coordinates of the edges have an accuracy which can be explained merely by rounding errors, thus can be treated as uncorrelated with standard deviation $\sigma_u = 1/\sqrt{12}$ pixels.

The extracted edge segment now is represented by three groups of observations:

1. The pair of v-coordinates with:

$$\mathbf{C}_{vv} = D\begin{pmatrix} \underline{v}_s \\ \underline{v}_e \end{pmatrix} = \begin{pmatrix} \sigma_{v_s}^2 & \sigma_{v_s v_e} \\ \sigma_{v_s v_e} & \sigma_{v_e}^2 \end{pmatrix} \quad or \quad \mathbf{W}_{vv} = \mathbf{W}\begin{pmatrix} \underline{v}_s \\ \underline{v}_e \end{pmatrix} = \begin{pmatrix} w_{v_s}^2 & w_{v_s v_e} \\ w_{v_s v_e} & w_{v_e}^2 \end{pmatrix} = \mathbf{C}_{vv}^{-1} \tag{14}$$

2./3. The u-coordinate system of starting point and end point resp.:

$$D(\underline{u}_s) = \sigma_{u_s}^2 \quad or \quad w_{u_s} = 1/\sigma_{u_s}^2 \quad\quad and \quad\quad D(\underline{u}_e) = \sigma_{u_e}^2 \quad or \quad w_{u_e} = 1/\sigma_{u_e}^2 \tag{15}$$

The weights will be modified in the robust estimation, allowing partial matches of image and model segments (4 cases). The complete weight matrix $\mathbf{W}^{(u,v)}$ of $(\underline{u}_s, \underline{v}_s, \underline{u}_e, \underline{v}_e)^T$ is transformed into the xy-coordinate system using the individual rotation matrix $\mathbf{R} = \begin{pmatrix} \mathbf{R}_\phi & \mathbf{0} \\ \mathbf{0} & \mathbf{R}_\phi \end{pmatrix}$ with $\mathbf{R}_\phi = \begin{pmatrix} \cos\phi & \sin\phi \\ -\sin\phi & \cos\phi \end{pmatrix}$

$$\mathbf{W}^{(x,y)} = \mathbf{R}\mathbf{W}^{(u,v)}\mathbf{R}^T \tag{16}$$

5 The Matching Procedure

The matching of the image edge segments to the model edges is performed in a two step procedure:

- The approximate position of the projected model in the image is determined via a *probabilistic clustering* technique, leading to a small set of candidate matches.
- The final correspondence is established via a *robust estimation*, which corresponds to the well known relaxation techniques.

5.1 Probabilistic Clustering for finding Candidates for Correspondencies

The approximate position of a reference point of the projected model is determined via clustering. In contrast to the usual procedure for clustering the varying uncertainty of the hypothesized position is taken into account when filling the accumulator. The final cluster is a discretized version of the probability density function (p.d.f) $p_{xy}(x,y)$ of the reference point.

For each match of image feature i with model feature j we get a range of expected positions for the reference point, as the image segments in general are shorter than the model segments. For a fixed position t, $t \in [0,1]$ the p.d.f. is given by

$$p_{xy}(x,y|i,j,t) = \frac{1}{2\,\pi\,\sigma_{ijt}^2} \exp\left\{ -\frac{1}{2}\frac{(x-x(t))^2 + (y-y(t))^2}{\sigma_{ijt}^2} \right\} \tag{17}$$

where the variance $\sigma_{ijt}^2 = (1/2 + (s_{ijt}/l_i)^2)\sigma_i^2 + \sigma_{s_{ijt}}^2$ depends on the distance s_{ijt} of the midpoint of the image segment from the reference point, its variance $\sigma_{s_{ijt}}^2$, the length l_i of the image segment and the uncertainty of its endpoints σ_i^2, neglecting the correlations in this step.

$P_0 = (x(0), y(0))$ and $P_1(x(1), y(1))$ are the extreme positions the reference point may occupy (cf. Fig. 4a). The p.d.f. $p_t(t|i,j)$ is assumed to be an equal distribution $e(t|i,j)$ on the straight line segment between P_0 and P_1. The probability $P_{ij}(i,j)$, that i and j correspond is assumed to be $1/|M|$ where M is the set of all matches. Then the cluster is determined by

$$p_{xy}(x,y) = \sum_{ij \in M} \left[\int_{t=0}^{1} p_{xy}(x,y|i,j,t)\, p_t(t|i,j)\, dt \right] \quad P_{ij}(i,j) = \frac{1}{|M|} \sum_{ij \in M} \int_{t=0}^{1} p_{xy}(x,y|i,j,t)\, e(t|i,j)\, dt \tag{18}$$

$p_{xy}(x,y|i,j,t)$ is approximated by a separable binomial distribution and the integrals by sums over all positions on the grid points of the accumulator between the extrems P_0 and P_1. Examples for the integral in eq. 18 are shown in Fig. 4b (s_{ijt} is approximated by $s_{ij0.5}$ here).

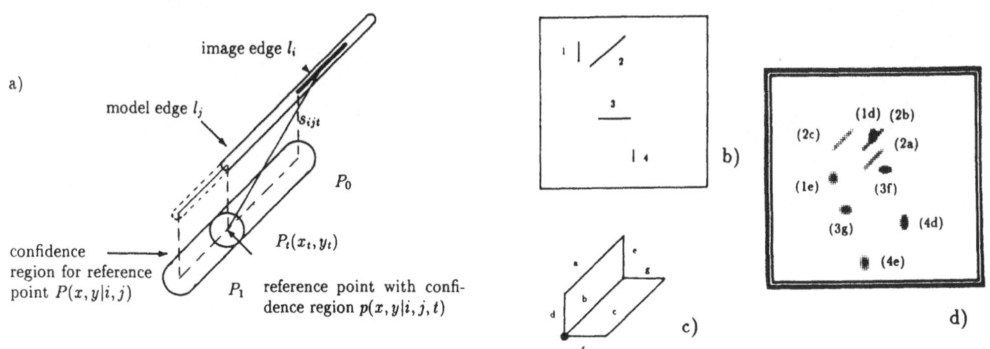

Figure 4: Principle of edge matching (a), example for distribution in accumulator for a few edges (d), selected edges (b), model (c)

With the approximate position of a reference point one easily can derive a set of candidate matches between image and model features. Specifically we obtain a set of straight image segments with a label indicating to which model segment they may correspond. The threshold for this conservative decision has to take the uncertainty of the model into account.

5.2 Robust Estimation and Final Evaluation

The list of preliminary correspondencies now is cleaned using the model eq. 4. As wrong correspondencies may be interpreted as outliers or with large deviations from the mean value, a robust estimation is performed (cf. HUBER, 1981). The principle is simply weighting down those observations which have large residuals or a large test statistic. In detail we use the following iteration scheme

$$\mathbf{W}_{vv}^{(\nu+1)} = \mathbf{W}_{vv}^{(0)} \, f(t^{(\nu)}) \qquad w_{u_s}^{(\nu+1)} = w_{u_s}^{(0)} \, f(t^{(\nu)}) \, f(t_{u_s}^{(\nu)}) \qquad w_{u_e}^{(\nu+1)} = w_{u_e}^{(0)} \, f(t^{(\nu)}) \, f(t_{u_e}^{(\nu)}) \tag{19}$$

For the first 3-6 iterations and the following 1 or 2 iterations the weight functions

$$f_1(t) = 1/\sqrt{1 + (t/c)^2} \quad and \quad f_2(t) = \exp\left\{-0.5(t/c)^2\right\} \tag{20}$$

resp. are used. t, t_{u_s} and t_{u_e} are the square-rooted Fisher-test statistics for the complete segments and the u-coordinates of the starting and the end point resp. c is a critical value, e. g. $c = 3$. The first weighting function leads to a mixture between L_1- and L_2-norm minimization and guarantees global convergency for linear problems, the second cancels the effect of large outliers onto the result.

As starting and end point are treated separately all possible 4 cases of an image segment matching a model segment can be realized. All correspondencies fitting to the final result, i. e. passing the test are assumed to be correct. The final result is then analyzed with respect to precision and sensitivity according to section 2.2. It is accepted if the image features, non detectable errors in the correspondencies and the assumptions contained in the model do have an effect onto the resultant position which does not exceed a prespecified threshold, e. g. a few pixels.

6   Example

From all the image segments in Fig. 1e the clustering process (cf. Fig. 5a) selected 17 segments as candidates (cf. Fig. 5b). The robust estimation selected the 8 segments shown in Fig. 5c. The precision of the reference point was $\sigma_p = 25.9 \ \mu m$. The check on the theoretical sensitivity reveals edge segment 7 to have the strongest possible effect onto the result with $\bar{\delta}_{07} = 4.9$ (cf. Table 1), i. e. with a maximum influence of $\nabla_{07}p = \sigma_p \cdot \bar{\delta}_{07} = 126 \ \mu m$, which is at the border of being acceptable. In case only the 4 segments in Fig. 5d would have been selected by the robust estimation, the precision of the reference point would

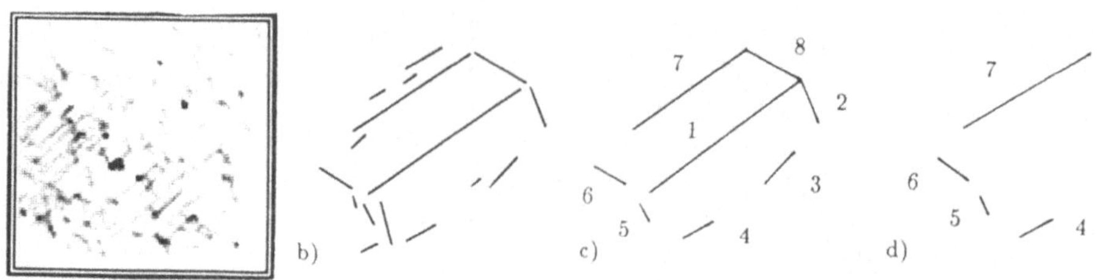

Figure 5: Cluster for approximate position (a), preliminary matches (b), matches selected by robust estimation (c), ficticiously reduced sample (d)

case	precision	sensitivity	x_1	x_2	x_3	x_4	x_5	x_6	x_7	x_8	width	slope
(c)	$\sigma_p = 25.9\mu m$	empirical	0.5	1.1	1.0	1.1	1.6	1.3	1.3	1.0	0.01	0.05
		theoretical	3.1	2.9	2.7	1.4	3.4	3.6	4.9	3.2	0.11	0.79
(d)	$\sigma_p = 81.5\mu m$	empirical	–	–	–	7.0	0.6	0.3	4.0	–	0.8	0.09
		theoretical	–	–	–	14.5	2.9	3.1	14.3	–	1.4	5.0

Table 1: Final evaluation of design for solutions (c) and (d) from Fig. 5. Sensitivity factors $\bar{\delta}_k$ and $\bar{\delta}_{0k}$

have been $\sigma_p = 81.5\ \mu m$. Also, due to the lack of stability errors in the matching may have influenced the result up to $\bar{\delta}_{04} = 14.5$ or up to 1.2 mm, which is appr. 60 pixels. Especially the high value of $\bar{\delta}_{04}$ indicates missing support for the model from the image. It is interesting to note, that also the apriori assumption about the slope of the roof has a significant influence onto the position of the reference point, namely up to 0.4 mm. As the test statistics all were accepted these values at the same time represent upper bounds on the empirical sensitivity of the result.

References

BAARDA, W. (1967) : Statistical Concepts in Geodesy, Publications in Geodesy, Vol. 2, No. 4, Netherlands Geodetic Commission, Delft, 1967.

BAARDA, W. (1968) : A testing Procedure for Use in Geodetic Networks, Publications in Geodesy, Vol. 2, No. 5, Netherlands Geodetic Commission, Delft, 1968.

BURNS, J.B., HANSON, A.R., RISEMAN, E.M. (1986) : Extracting Straight Lines, IEEE T-PAMI, Vol. 8., No. 4, 1986, pp. 425-455.

FAN, T.-J. (1988) : Describing and Recognizing 3D-objects using Surface Properties, Inst. for Robotics and Intelligent Systems, Report Nr. 237, School of Engineering Univ. of Southern California, L.A., Aug. 1988.

FAUGERAS, O.D., HEBERT, M. (1987) : The Representation, Recognition, and Positioning of 3-D Shapes from Range Data, in KANADE 1987, pp. 301-353.

FÖRSTNER, W. (1983) : Reliability and Discernability of Extended Gauß-Markov-Models, Deutsche Geod. Komm., A 98, pp.79-103.

FÖRSTNER, W. (1987) : Reliability Analysis of Parameter Estimation in Linear Models with Applications to Mensuration Problems in Computer Vision, Computer Vision, Graphics and Image Processing 40, 1987, pp. 273-310.

FUA, P., HANSON, A.J. (1987) : Resegmentation Using Generic Shape: Locating Cultural Objects, Pattern Recognition Letters 5, 1987, pp. 243-252.

GRIMSON, W.E.L., LOZANO-PEREZ, T. (1984) : Model-Based Recognition and Localization from Sparse Range or Tactile Data, Int. J. Robotics Res., Vol. 3, No. 3, 1984, pp.3-34.

GRIMSON, W.E.L., LOZANO-PEREZ, T. (1987) : Localizing Overlapping Parts by Searching the Interpretation Tree, IEEE T-PAMI, Vol. 9., No. 4, 1987, pp. 469-482.

HORAUD, R. (1987) : New Methods for Matching 3-D Objects with Single Perspective Views, IEEE T-PAMI, Vol. 9., No. 3, 1987, pp. 401-412.

HUBER (1981) : Robust Statistics, Wiley, NY, 1981.

KANADE, T. (1987), ED. : Three-Dimensional Machine Vision, Kluwer Academic Publishers, 1987.

NEVATIA, R., BABU, K.R. (1980) : Linear Feature Extraction and Description, Comp. Graphic and Image Proc., Vol. 13, 1980, pp.257-269.

SMITH, R., SELF, M., CHEESEMAN, P. (1987) : Estimating Uncertain Spatial Relationships in Robotics, Uncertainty in Artificial Intelligence, Vol. 2, 1987, North-Holland.

Object Recognition by Selective Focusing Using a Moore-Penrose Associative Memory

Wolfgang Pölzleitner [*] and Harry Wechsler [**]

[*]

Research Center Joanneum
Wastiangasse 6
A-8010 Graz, Austria

[**]

Computer Science
George Mason University
Fairfax, VA 22030

Abstract

This paper describes a rotation-, scale-, and distortion-invariant method of 2D object recognition based on the Moore-Penrose Distributed Associative Memory (DAM). Invariance to rotation and scale is achieved using conformal mappings. For distortion invariance the memory is adpated to the key stimulus before recall which leads to improved accuracy for both recall and classification. Based on known relationships between DAMs and regression analysis, a method is proposed to improve the selectivity of the association weights in an iterative way by discarding insignificant response vectors. Such selectivity allows the system to focus on the significant associations and to reduce cross-talk effects. The same formalism which allows to compute the significance of the association weights also provides for a reject option. Experimental results prove the feasibility and benefits of the new recognition method.

1 Introduction

In image recognition the indexing problem, i.e., the access to a memory by its contents rather than by addressing plays a central role. Recognition of an object in an image is an indexing problem. An object is presented to a memory which than outputs the address (or classification) of the object. A model for such a system is that of the Distributed Associative Memory (DAM) [1]. In this model pairs of key stimulus and response vectors are associated via a matrix operator. Retrieving information from memory is performed by multiplication of the unknown key vector with the memory matrix. The resulting output vector (the *recall* or *recollection*) is a particular stored data vector or an approximation thereof.

The central question is how to compute the matrix operator. The basic solution [1] is to use the Moore-Penrose generalized inverse (GI) to derive this operator. The DAM has several useful properties. First, when the key vector is incomplete, the recalled vector 'closest' resembles the corresponding response data vector in a minimum squared error (MSE) sense. This situation occurs e.g. when an object to be detected in an image is partly occluded by another object. Second, when the key vector is noisy the recalled vector is again optimal in the MSE sense.

When the DAM is used for image recognition one of the central problems is how to set up the input keys. It is well known that the performance of the memory rapidly deteriorates when the key-images are degraded versions of the data images. Linear transformations like rotation, scale and translation, if not properly accounted for, lead to faulty recalls [2,3]. Wechsler and Zimmerman [2] describe a DAM-based system which is invariant to rotation and scale changes.

2 The Moore-Penrose Associative Memory

This section presents a brief review of the Kohonen model of associative memory. Let x_k and y_k be the key (stimulus) vectors and data (response) vectors, respectively, with $x_k \in R^n$ and $y_k \in R^m$.

Typically, when dealing with 2D images, these vectors are derived by appending horizontal line scans. The pairs (x_k, y_k) are the vectors to be memorized and associated by matrix M as

$$y_k = M x_k \qquad \forall \, k = 1, \cdots, K \tag{1}$$

where K is the number of associations. The relation is characteristic to classical conditioning [4] and amounts to the recall of y_k by its key x_k and should hold for all k.

Let X and Y be defined as the $n \times K$ and $m \times K$ matrices $X = (x_1, \cdots, x_K)$ and $Y = (y_1, \cdots, y_K)$. Then Eq. (1) can be written as

$$Y = MX \tag{2}$$

Eq. (2) describes the general, *heteroassociative* case. For object recognition the *autoassociative* case is of special interest where $Y = X$.

A unique solution for Eq. (2) does not necessarily exist for any arbitrary group of associations that might be chosen. Usually, the number of associations K is smaller than m, the length of the vector to be associated, so the system of equations is underconstrained. The constraint used to solve for a unique matrix M is that of minimizing the square error, $\|MX - Y\|^2$, which results in the solution

$$M = Y X^+ \tag{3-a}$$

$$M = X X^+ \tag{3-b}$$

where X^+ is the Moore-Penrose generalized inverse (GI) of X; Eq. (3-a) is the solution for the heteroassociative memory and Eq. (3-b) is the solution for the autoassociative memory.

Note that in general X is not a square matrix and hence X^{-1} is undefined. When exact solutions for Eq. (2) exist, they are infinitely many, and $M = Y X^+$ is the solution with minimum norm $\|M\|^2$ for arbitrary Y. In this case the norm $\|MX - Y\|^2$ is zero. The condition for exact solutions is $X^+ X = I$ which only holds when the vectors x_k are linearly independent. When the x_k are linearly dependent Eq. (3-a) is an approximate solution [1]. The generalized inverse of X is found as $X^+ = (X^T X)^{-1} X^T$ where T is the transposition operator.

When noisy keys are presented to the memory, the recall operation for the autoassociative memory is

$$\hat{x}_k = X X^+ \tilde{x}_k \tag{4}$$

where $\tilde{x}_k$ is a noisy version of x_k, the noise is assumed to be an additive random variable with zero mean and variance σ^2, and $\hat{x}_k$ approximates x_k. The autoassociative memory can be shown to reduce the noise variance by a factor of K/n [1,3].

Object Recognition

Eq. (4) can be used to compute association weights, using $\hat{x} = X X^+ \tilde{x} = Xw$,

$$w = X^+ \tilde{x} \tag{5}$$

with $w \in R^K$, $w^T = [w_1, \cdots, w_K]$. The coefficients w_k indicate the strength by which the memory associates the input key $\tilde{x}$ with the stored data x_k. The higher the weight w_k, the stronger the association to the particular k-th data item. The weights can be used to implement a recognition procedure, e.g. by selecting the maximum w_k.

3   Relation to Multiple Linear Regression

Eqs. (5) and (4) are equivalent to the formalism used in the framework of multiple linear regression from statistics. In particular, Eq. (5) defines the computation of regression coefficients where x_k are used as the independent variables and y_k as the dependent variables [5]. Eq. (4) is the prediction equation that is used to predict the values of the dependent variable y for a new set of samples x.

The formalism of recall in an autoassociative memory can be viewed as predicting the data vectors from a given key vector. Let $\hat{y}$ be the prediction of y_k based on a collection of data vectors x_k,

$$\hat{y} = Y(X^T X)^{-1} X^T x \tag{6}$$

where x is an arbitrary (noisy) key-vector and $\hat{y}$ the prediction of y [5]. x in general is different from each of the x_k contained in X. In the autoassociative case we have

$$\hat{x} = X(X^T X)^{-1} X^T x \tag{7}$$

where $\hat{x}$ is the prediction of the 'noise-free x'. $\hat{y}$ can be written as the linear combination, i.e. regression, of the x_k, $\hat{y} = \sum_{k=1}^{K} w_k x_k$, with w_k being the regression coefficients given as $w = X^+ x$.

The major point of the above similarity is that we can carry over basic results from regression theory to refine the association process and reduce crosstalk effects. It is well known that the significance of a regression coefficient w_k can be tested by a statistical t-test. Given a regression problem as in Eq. (7) the estimated variance of the coefficient vector w is

$$\widehat{\mathrm{var}}(w) = \hat{\sigma}^2 (X^T X)^{-1} \tag{8}$$

where $\hat{\sigma}^2$ is an estimate of the association noise (or residual of the regression in this case) and for the autoassociative recall it is given by $\hat{\sigma}^2 = \|x - \hat{x}\|/(n - K)$ where n is the dimensionality of the input key and K is the number of associations.

The t-statistic t_k for a single coefficient w_k can be computed from

$$t_k = \frac{|w_k|}{\sqrt{\widehat{\mathrm{var}}(w_k)}} \tag{9}$$

The fundamental result from Eq. (9) is that using t_k an ordering of the coefficients w_k can be achieved. This ordering will be the basis for the *coefficient focusing* procedure described in the next section.

A method to reject an association can be found as follows. Let the association error E [3] be defined as $E = \epsilon(\|x - \hat{x}\|) = \mathrm{RSS}$, where $\hat{x}$ is the actual recall for an input x. E is the residual sum of squares (RSS) from the regression frame-work, $\| \cdot \|$ is the Euclidean norm, and $\epsilon(\cdot)$ is the expectation operator. The variation of x is $\mathrm{var}(x) = \|x - \bar{x}\|$, where $\bar{x}$ is the mean over all n elements of x. The above two entities are very useful to measure the goodness of an association by computing the coefficient of determination R^2 as

$$R^2 = \frac{\mathrm{var}(x) - \mathrm{RSS}}{\mathrm{var}(x)} \tag{10}$$

R^2 is 1 when the association is perfect and 0 when there is no correlation between the key x and the recall $\hat{x}$. This coefficient is important when a reject option should be implemented. When R^2 is small, the key-vector should be rejected as being not similar to any of the data vectors x_k.

4 Crosstalk and Coefficient Focusing

Assume an approximate key vector $\tilde{x}_k$ is presented to the memory M. When the memorized data vectors are dependent, then the vector recalled by Eq. (1) will be a combination of several x_i's. This phenomenon is known as *crosstalk*. Crosstalk not only degrades the association but also affects the weights w_k. The result is a diminished selectivity between the weights. Selectivity would be perfect, if only one $w_k = 1$ exists while all the others are zero. Several methods to increase the selectivity have been proposed [1], [3].

We describe an alternative method to increase the selectivity which is based on the idea that the poor selectivity (due to crosstalk *or* noise) is best handled by decreasing K. K can be decreased by deleting 'insignificant' response vectors from the memory, thus entirely eliminating their contribution to the error. The elimination of vectors from the memory is done in an iterative manner, by deleting one vector at a time. When a vector is deleted, the remaining vectors are used to recompute the association weights. This is done until only a few vectors remain in the memory. Since the weight of the 'best fitting' data vector increases while the weights of the remaining data vectors decrease this procedure is termed *coefficient focusing*. The algorithm is formally stated as follows.

Algorithm 1 DAM with Coefficient Focusing

Input: *The pairs of key vectors and data vectors x_k, y_k and the key x which is used for recall and identification.*

Output: *The sequence of weights $w^{(i)}$ ($i = 1, K$) which is found by successive elimination of the worst weight w_j at step i. Output is the sequence of recollections from the memory which finally contains only a single data-vector x_j.*

Method:

Step 1. *Initially set $i := 1$ and let the memory contain all K response vectors.*

Step 2. *At iteration i, compute the memory $M^{(i)}$ from Eq. (3-b) using the current number of $K - i + 1$ data vectors. Also compute t_k from Eq. (9).*

Step 3. *Discard the data vector x_j which has the largest value of p_j (or the smallest t_j). w_j is the i-th element in the sequence of deletions. Set $i := i + 1$. The remaining number of data vectors is now $K - i + 1$.*

Goto Step 2. until only one data vector remains in the memory.

Step 4. *Output the sequence of deleted data vectors. Output the identification of x as the data vector x_j which remained in the memory. Reject the object as unrecognized when R^2 is small, according to Eq. (10).*

The procedure described above can be speeded up considerably by intially deleting a predetermined number of data vectors that were found to be insignificant and start the focusing procedure from there. This is achieved by deleting all data vectors whose value of t_k are below a threshold. Alternatively, one can initially keep only a fixed (small) number of data vectors in the memory.

One could allow as well for the system to output the first l best candidates and to defer final selection to additional considerations based on the context or the semantics of the situation. Note that in Step 4 of the above algorithm a threshold for R^2 must be selected which defines rejection. It can be shown, however, that R^2 is exactly zero when only the constant vector x_0 remains in the memory. In practice this property makes the described method rather insensitive to the choice of the threshold for R^2.

5 Setting up the Key and Data Vectors

A fundamental problem for the application of the proposed DAM is that it performs best when the key-vectors are as orthogonal as possible, and that the key-vectors are not corrupted by 'positional' noise, i.e. small geometric shifts of pixels. Since both positional and intensity transformations are intrinsic among real images, much attention was payed in the past to extract invariant features

[6,7] from the images. A rotation-invariant recognition system uses conformal mapping to achieve (limited) rotation and scale invariance [2] and is adopted here. The complete set-up procedure for the DAM to yield the rotation-, scale-, and distortion-invariant behavior can be summarized as follows.

Algorithm 2 Loading an Invariant DAM

Input: *The 2D data images ξ_k and the key image ξ. ξ_k and $\xi \in R^{N \times N}$; A threshold T_p for the expected positional noise.*

Output: *The data vectors x_k and key vector x and the associative memory M, which is insensitive to rotation, scaling, or distortions in the original key vector x. x_k and $x \in R^n$.*

Method:

Step 1. Conformal Mapping.
Transform the images ξ_k and ξ into ξ_k' and ξ' by conformal mapping [2].

Step 2. Compensate Rotation and Scaling.
Shift ξ_k' vertically and horizontally into the best matching position to yield ξ_k'' such that the norm $\|\xi_k'' - \xi'\| = \min$.

Step 3. Compensate Distortions.
Rearrange the pixels in ξ_k'': At each position i,j in ξ_k'' find the pixel at i',j' in the neighborhood of i,j that is most similar to $\xi'(i,j)$ and place it at i,j:

$$\xi_k''(i,j) := \xi_k''(i',j') \tag{11}$$

such that $|\xi_k''(i',j') - \xi'(i,j)| = \min_{\alpha,\beta \in N_{i,j}} |\xi_k''(\alpha,\beta) - \xi'(i,j)|$, with the neighborhood $N_{i,j}$ of (i,j) defined as

$$N_{i,j} = \{\alpha,\beta \mid \alpha = i \pm k; \beta = j \pm l; \; k,l \in [-T_p \cdots + T_p]\} \tag{12}$$

Step 4. Set up data and key vectors.
The data vectors x_k and the key vector x which are input to the DAM are found by scanning the rows of the images ξ_k'' and ξ' and appending them to yield the one-dimensional vectors.

6　Experimental Results

The methods described in the previous sections were tested on images of wooden boards. The goal was to recognize resin galls, knots and holes, where the knots appear in three different subclasses: normal (i.e. bright knots), knots with partially dark surrounding, and dark knots. The resulting 6 classes should be discriminated. The prototype patterns are shown in Fig. 1. These objects were segmented by the method described in in [8]. In this approach, symbolic texture elements are found from the image, which are used to separate objects from background. Objects are then defined by connected component analysis of these texture elements. Elliptic approximation of the objects' border lines yields an approximate outline of the objects. The background pixels (the ones lying outside the ellipses) were uniformly set to a value corresponding to 80% of the maximum intensity before submitting the objects to the conformal mapping transform. These images are then subjected to Steps 2, 3 and 4 of Algorithm 2 to get the data keys that are stored in the memory matrix M. Note that these steps are already part of the recollection process for a selected key pattern and have to be repeated for each new key that is presented to the memory.

In addition to the 6 data vectors a 7th constant vector was included which is both used to permit the system to recognize the background (which is defined as constant intensity c_0) and to be insensitive to offsets between the data vectors and the key vector persented to the memory. Fig. 1 shows the result of computing the association weights from Eq. (5). It can be seen from part c) of Fig. 1 that the association weights w_3 and w_4 are relatively close. Part d) of Fig. 1 shows how these two weights change values after successive removal of the less significant weights. The sequence of removal was $w_5, w_2, w_4, w_0, w_6, w_4$ and the weight w_3 remains in the memory after the last iteration of Algorithm 1.

We conclude the description of the experiments by showing the results of recall when the input key vector is composed of a combination of 2 stored data vectors. The input image in Fig. 2 contains the upper half of data key x_6 and the lower half of data key x_1. Each of the objects occludes the other by 50%. As expected, the coefficients w_1 and w_6 are high initially as shown in Fig. 2.

7 Conclusion

We have described a method of using a DAM for image recognition purposes. This method benefits from two major components. First, the incorporation of a procedure to achieve insensitivity to geometric distortions, and second the focusing procedure which is used to decrease the crosstalk effects and enhance the selectivity of the weights. The performance of the method was tested on images of wooden boards and showed that it can in fact be used to recognize objects which are variable to a high degree. The reject option in our approach is relevant especially in cases where objects are imperfectly centered in the 'receptive field' of the DAM, or where unknown objects must be identified. We have been successful with the proposed rejection formula both in cases when no object was present in the key-image (i.e. it contained only background) and when the image contained unknown or unlearned objects.

References

[1] T. Kohonen, *Self-Organization and Associative Memory.* Springer-Verlag, 2nd ed., 1988.

[2] H. Wechsler and G. L. Zimmerman, "2-D Invariant Object Recognition Using Distributed Associative Memory," *IEEE Trans. Pattern Anal. Machine Intell.*, vol. 10, pp. 811–821, November 1988.

[3] K. Murakami and T. Aibara, "An Improvement on the Moore-Penrose Generalized Inverse Associative Memory," *IEEE Trans. Syst. Man Cybern.*, vol. 17, pp. 699–707, July/August 1987.

[4] D. O. Hebb, *The Organization of Behavior.* Wiley, 1949.

[5] S. Weisberg, *Applied Linear Regression.* Wiley, 2nd ed., 1985.

[6] L. Massone, G. Sandini, and V. Tabliasco, "Form-Invariant Topological Mapping Strategy for 2D Shape Recognition," *Computer Vision, Graphics, and Image Processing*, vol. 30, no. 1, pp. 169–188, 1985.

[7] R. A. Messner, "An Image Processing Architecture for Real Time Generation of Scale and Rotation Invariant Patterns," *Computer Vision, Graphics, and Image Processing*, vol. 31, no. 1, pp. 50–66, 1985.

[8] W. Pölzleitner, "A Hough Transform Method to Segment Images of Wooden Boards," in *8th Intl. Conf. Pattern Recognition (ICPR)*, (Paris), pp. 262–264, October 1986.

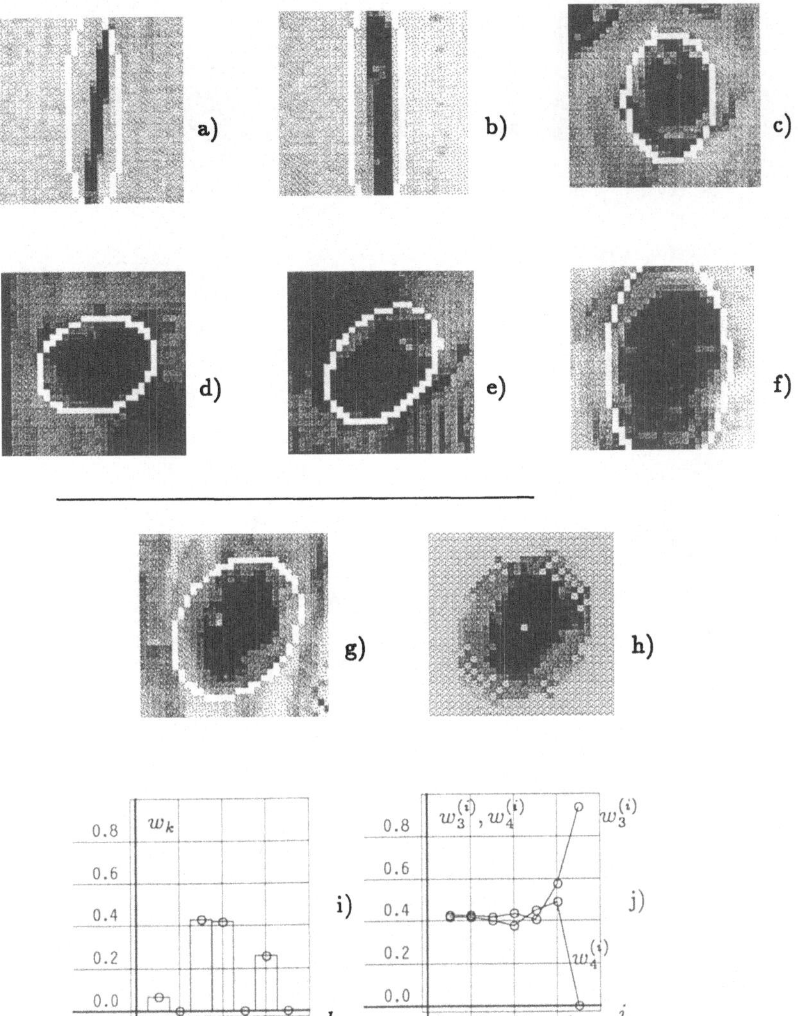

Figure 1: *Training objects* stored in the DAM and *recall* from the DAM when the key is a knot:
a) small resin gall, b) resin gall, c) small bright knot, d) dark knot, e) hole, f) big bright knot.
Recall: g) the input test pattern (the key image), h) the recalled image, i) the association weights,
j) the weights w_3 and w_4 at all 7 iterations; focusing converges to w_3.

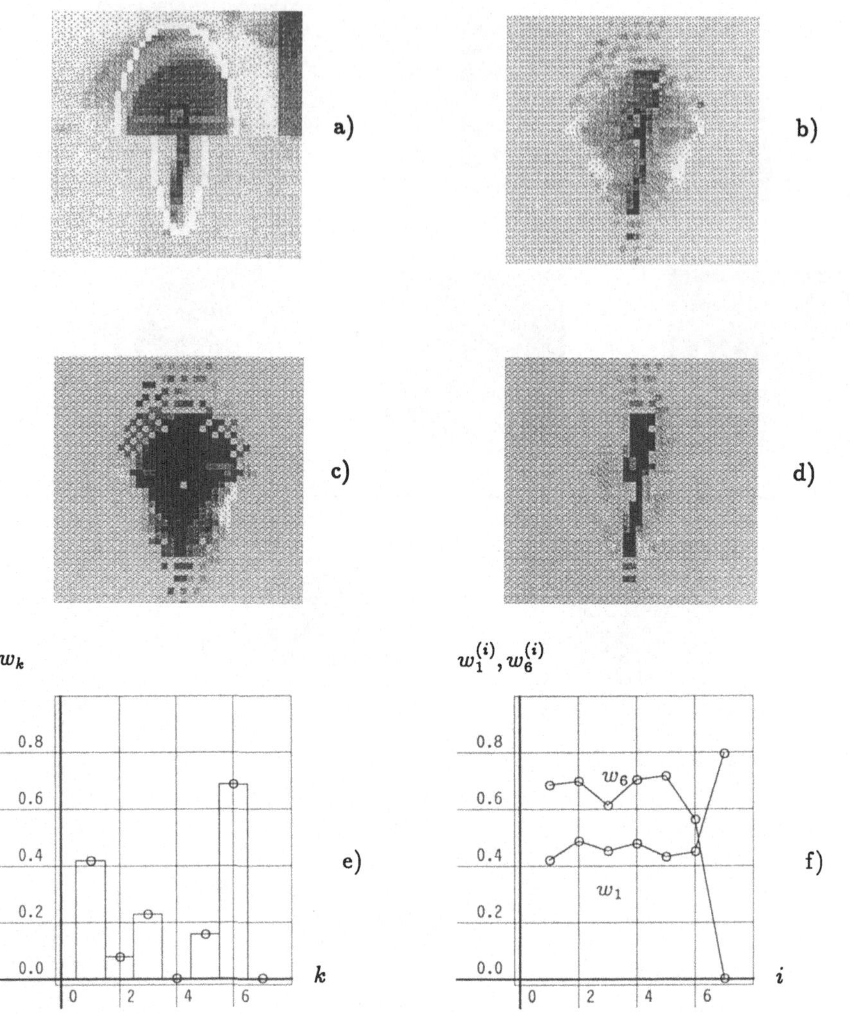

Figure 2: Recall from the DAM when the key is a combination of fractional training images. a) the input test pattern (the key image), b) the recalled image, c) the recalled image when focusing is forced to w_6, d) the recalled image after all iterations of the coefficient focusing procedure, e) the association weights, f) the weights w_1 and w_6 at all 7 iterations.

Ein ableitungsorientierter Ansatz
zur Detektion von Kanten in
multidimensionalen Bildfunktionen

Christian Drewniok und Leonie Dreschler-Fischer

Fachbereich Informatik der Universität Hamburg
Bodenstedtstraße 16
D-2000 Hamburg 50

Einleitung

Bis heute wurden zahlreiche Verfahren zur Kantendetektion in Grauwertbildern entwickelt. Hingegen finden sich nur sehr wenige Ansätze zur Detektion von Diskontinuitäten in multidimensionalen Bildfunktionen (z.B. Farbbildern). Die meisten dieser Ansätze stammen aus den siebziger Jahren und können dem heutigen Stand der Kunst nicht entsprechen. Erst in den letzten Jahren finden sich Hinweise auf einen fundierten ableitungsorientierten Ansatz zur Detektion von Farbkanten [*Di Zenzo 86, Novak + Shafer 87*].

Es soll gezeigt werden, wie sich — basierend auf einem formalen Ansatz — gradientenbasierte Detektionsverfahren für die Anwendung auf Bildfunktionen beliebiger Dimension verallgemeinern lassen. Die Unabhängigkeit des Ansatzes von der Anzahl der Dimensionen gilt dabei sowohl für den Werte- als auch für den Definitionsbereich der Bildfunktion. Die in den Ansatz eingehenden Voraussetzungen werden dargestellt. Schließlich wird die gute Eignung des Ansatzes anhand von Ergebnissen der Verarbeitung von RGB-Farbbildern demonstriert. Eine ausführliche Untersuchung der Problematik findet sich in [*Drewniok 88*].

Ein ableitungsorientierter Ansatz

Ableitungsorientierte Ansätze bilden eine wichtige Klasse von Verfahren zur Detektion von Kanten in Grauwertbildern. Kanten werden definiert als Extrema der ersten oder Nulldurchgänge der zweiten Ableitung der Bildfunktion in Richtung des Gradienten. Dabei wird ein Stufenkantenmodell zugrundegelegt: Die Gradientenrichtung sollte, als Richtung stärkster Änderung der Bildfunktion, senkrecht zur Orientierung der Stufenkante stehen; der Gradientenbetrag sollte mit der Stärke der Änderung — also dem Kontrast der Kante — korrespondieren (siehe z.B. [*Canny 83*]).

Der Gradient ist nur für skalare Funktionen definiert. Bei der Verallgemeinerung eines gradientenbasierten Ansatzes auf vektorwertige Bildfunktionen stellt sich somit die Frage: Wie läßt sich auch für multidimensionale Bildfunktionen Richtung und Stärke der lokal stärksten Änderung der Bildfunktion ermitteln?

Wir betrachten zunächst die m-komponentige Bildfunktion $\vec{C}$ sowie den durch den Winkel φ zur positiven x-Achse festgelegten Richtungsvektor $\vec{n}$:

$$\vec{C}(x,y) = \begin{pmatrix} C_1(x,y) \\ \vdots \\ C_m(x,y) \end{pmatrix} \qquad \vec{n} = \begin{pmatrix} \cos\varphi \\ \sin\varphi \end{pmatrix}$$

Die direktionale Ableitung der vektorwertigen Funktion $\vec{C}$ in Richtung $\vec{n}$ liefert einen Ableitungsvektor, dessen Komponenten die Richtungsableitungen der einzelnen Bildkomponenten sind. Diese lassen sich darstellen als Projektionen der jeweiligen Gradienten auf die Richtung $\vec{n}$. Durch Formulierung des Ergebnisses als Matrixprodukt erhalten wir schließlich:

$$\frac{\partial \vec{C}}{\partial \vec{n}} = \begin{pmatrix} \frac{\partial C_1}{\partial \vec{n}} \\ \vdots \\ \frac{\partial C_m}{\partial \vec{n}} \end{pmatrix} = \begin{pmatrix} \nabla C_1 \cdot \vec{n} \\ \vdots \\ \nabla C_m \cdot \vec{n} \end{pmatrix} = \begin{pmatrix} C_{1_x} & C_{1_y} \\ \vdots & \vdots \\ C_{m_x} & C_{m_y} \end{pmatrix} \cdot \vec{n} = J \cdot \vec{n}$$

Die durch die Richtungsableitungen der Komponenten nach x und y definierte Matrix J wird als *Jacobimatrix* (oder auch *Funktionalmatrix*) bezeichnet.

Durch Festlegen einer Berechnungsvorschrift läßt sich der Ableitungsvektor auf einen Skalar abbilden. Wir erhalten so ein skalares Maß für die Stärke der Änderung der Bildfunktion in Richtung $\vec{n}$. Als geeignetes Maß betrachten wir die Länge des Ableitungsvektors nach der euklidischen Norm (l^2 sei die quadrierte Länge des Vektors):

$$l^2 = \|J \cdot \vec{n}\|^2 = (J \cdot \vec{n})^T \cdot (J \cdot \vec{n}) = \vec{n}^T \cdot (J^T J) \cdot \vec{n}$$

Für die symmetrische 2×2-Matrix $J^T J$ seien die folgenden Koeffizienten definiert:

$$J^T J = \begin{pmatrix} a_{11} & a_{12} \\ a_{21} & a_{22} \end{pmatrix} \quad \text{mit}$$

$$a_{11} = C_{1_x}^2 + \cdots + C_{m_x}^2$$

$$a_{22} = C_{1_y}^2 + \cdots + C_{m_y}^2$$

$$a_{12} = C_{1_x}C_{1_y} + \cdots + C_{m_x}C_{m_y}$$

$$a_{21} = a_{12}$$

Um ein Analogon zum Gradienten zu erhalten, bestimmen wir nun diejenige Richtung $\vec{n}$, für die die Bildfunktion den größten Änderungsbetrag aufweist. Zu maximieren ist also l^2 in Abhängigkeit von $\vec{n}$. Der Ausdruck $\vec{n}^T \cdot (J^T J) \cdot \vec{n}$ entspricht gerade dem Rayleigh-Quotienten der symmetrischen Matrix $J^T J$. Die Extremwerte des Rayleigh-Quotienten einer Matrix stimmen mit den Eigenwerten dieser Matrix überein: Betrag und Richtung der stärksten Änderung der Bildfunktion entsprechen daher dem betragsmäßig größten Eigenwert sowie der Richtung des dazugehörenden Eigenvektors der Matrix $J^T J$. Die Schätzung von Richtung und Stärke der lokal stärksten Änderung einer multidimensionalen Bildfunktion läßt sich auf diese Weise zurückführen auf das Lösen eines Eigenwertproblems.

Reduziert sich die Bildfunktion auf eine einkomponentige — also skalare — Funktion, so stimmt dieser Ansatz exakt mit dem gradientenbasierten Ansatz für Grauwertbilder überein, den wir oben kurz skizziert haben. Bei der Herleitung des Ansatzes sind wir von einem zweidimensionalen Definitionsbereich der Bildfunktion ausgegangen; der beschriebene Lösungsansatz läßt sich jedoch auch in Bezug auf die Dimension des Definitionsbereiches verallgemeinern. Die Anwendung des Ansatzes auf eine m-komponentige Bildfunktion $\vec{C}$, deren Definitionsraum k Dimensionen umfaßt ($x_1, \ldots, x_k$), liefert für die Matrix $J^T J$ eine $k \times k$-Matrix mit den folgenden Komponenten a_{rs}:

$$a_{rs} = \sum_{i=1}^{m} C_{i_{x_r}} C_{i_{x_s}}$$

Während sich für das Eigenwertproblem im Falle einer 2×2-Matrix eine einfache analytische Lösung angeben läßt, wird man bei größeren Matrizen im allgemeinen auf aufwendigere, numerische Lösungsverfahren zur Eigenwertbestimmung zurückgreifen müssen.

Anwendbarkeit des Ansatzes

Zur Veranschaulichung des Ansatzes betrachten wir, als Beispiel für eine 3-komponentige Bildfunktion mit 2-dimensionalem Definitionsraum, ein RGB-Farbbild. Die (quadrierte) Größe der Änderung der Bildfunktion in eine bestimmte Richtung $\vec{n}$ entspricht, gemäß der verwendeten euklidischen Norm, der Summe der

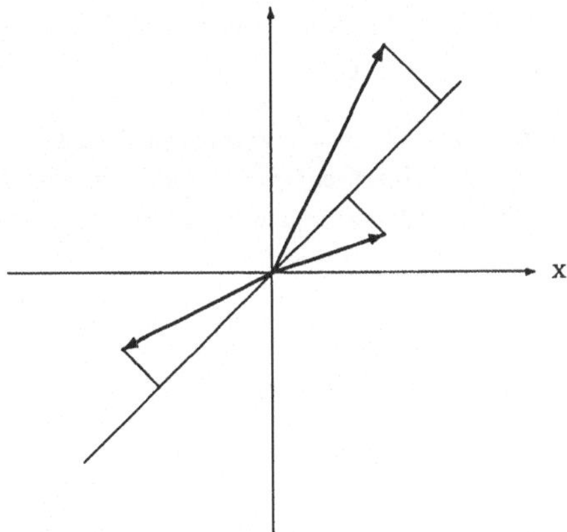

Abbildung 1: Projektion der Gradienten der verschiedenen Bildkomponenten auf eine Gerade

quadrierten Komponenten des Ableitungsvektors:

$$l^2 \;=\; \|J \cdot \vec{n}\|^2 \;=\; (\nabla R \cdot \vec{n})^2 + (\nabla G \cdot \vec{n})^2 + (\nabla B \cdot \vec{n})^2$$

Abbildung 1 soll dies verdeutlichen. Wir sehen dort die für die drei Bildkomponenten bestimmten Gradientenvektoren, sowie ihre Projektionen auf eine Gerade, deren Richtung durch $\vec{n}$ definiert ist. Die Richtung stärkster Änderung entspricht derjenigen Richtung, für die die Summe der quadrierten Projektionen maximal ist. Die so gefundene Lösung wollen wir im folgenden als Farbgradienten bezeichnen.

Farbkanten lassen sich nun — entsprechend der Kantendefinition im zugrundeliegenden gradientenbasierten Grauwertverfahren — definieren als Orte, die in Richtung des Farbgradienten ein lokales Maximum im Farbgradientenbetrag aufweisen. Das dargestellte Beispiel macht die in den Ansatz eingehende Voraussetzungen deutlich:

- Gute Korrelation bzw. Antikorrelation der in den einzelnen Bildkomponenten bestimmten Gradientenvektoren verstärkt den Gesamteffekt. Dies setzt die Abhängigkeit der Bildkomponenten voneinander voraus. Eine Anwendung des Ansatzes auf vektorwertige Bildfunktionen mit in der Regel voneinander unabhängigen Komponenten erscheint wenig sinnvoll.

- Die geschlossene Behandlung der Bildkomponenten setzt voraus, daß vergleichbar große Gradientenbeträge in den verschiedenen Komponenten vergleichbare Signifikanz für den Gesamteffekt besitzen. Insbesondere sollte die Varianz des in den einzelnen Komponenten auftretenden Rauschens vergleichbar sein [*Förstner 88*]. Gegebenenfalls muß deshalb der Anwendung des Ansatzes eine geeignete Wichtung der Bildkomponenten vorausgehen.

Alternative Berechnungsvorschriften

Die Abbildung des Ableitungsvektors auf einen skalaren Wert für die Stärke der Änderung der Bildfunktion mit Hilfe der euklidischen Norm ist zwar naheliegend, jedoch nicht zwingend. Denkbar wären hier verschiedene Berechnungsvorschriften. Aus der Vielzahl der Alternativen wollen wir zwei zum Vergleich heranziehen: die Verwendung der *Summe der Komponenten* des Ableitungsvektors sowie die *Wahl der betragsgrößten Komponente*.

Wie leicht zu zeigen ist, läßt sich der Ansatz bei Verwendung der Komponentensumme auf das gradientenbasierte Grauwertverfahren zurückführen:

$$l \;=\; \nabla C_1 \cdot \vec{n} + \cdots + \nabla C_m \cdot \vec{n}$$

$$= (\nabla C_1 + \cdots + \nabla C_m) \cdot \vec{n}$$
$$= \nabla(C_1 + \cdots + C_m) \cdot \vec{n}$$

l wird maximal, wenn $\vec{n}$ mit der Richtung des Grauwertgradienten der zuvor zu mittelnden Bildkomponenten übereinstimmt. Im Falle von RGB-Farbbildern entspricht dies der Berechnung des Gradienten auf dem entsprechenden Intensitätsbild.[1] Der damit verbundene Nachteil ist klar: Antikorrelierte Effekte in den einzelnen Bildkomponenten heben sich gegenseitig auf; eine Farbdiskontinuität, die sich nicht gleichzeitig durch eine Intensitätsdiskontinuität auszeichnet, kann mit Hilfe des Intensitätsansatzes nicht detektiert werden.

Die Maximumsnorm weist l die betragsmäßig größte Komponente des Ableitungsvektors zu:

$$l = \max \left(|\nabla R \cdot \vec{n}|, |\nabla G \cdot \vec{n}|, |\nabla B \cdot \vec{n}| \right)$$

l wird maximal, wenn $\vec{n}$ mit der Richtung des betragsmäßig stärksten Gradienten übereinstimmt. In einem praktischen Verfahren müssen also nur die Gradienten der einzelnen Bildkomponenten bestimmt und der betragsgrößte unter ihnen ausgewählt werden. Der ausgewählte, dominante Gradient wird in vielen Fällen ein guter Repräsentant für die Verhältnisse im betrachteten Bildausschnitt sein. Die Korrelation bzw. Nichtkorrelation der einzelnen Gradienten hat jedoch keinen Einfluß mehr auf das Endresultat! Ob nur ein starker Gradient vorhanden ist, zwei starke, aber nichtkorrelierte Gradienten vorliegen oder aber mehrere gut korrelierte Gradienten auf einen gemeinsamen Effekt hindeuten, spielt für das Ergebnis keine Rolle. Hieraus resultierende Detektionsfehler sollten sich in erster Linie in kontrastarmen bzw. stark verrauschten Bildbereichen bemerkbar machen.

Ergebnisse

Um die praktische Eignung zu demonstrieren, wurden mit Hilfe des Farbgradientenansatzes Kanten in RGB-Farbbildern detektiert. Das dabei verwendete Detektionsschema umfaßt folgende Schritte:

1. Ableitung (einschließlich Gauß-Glättung) der einzelnen Bildkomponenten nach x und y mittels zweidimensionaler Faltungsoperationen.[2]

2. Berechnung von Richtung und Betrag der lokal stärksten Änderung der Bildfunktion durch Bestimmung der Eigenvektoren mit maximalem Eigenwert unter Verwendung der Werte der einzelnen Richtungsableitungen.

3. Markierung lokaler Maxima im Änderungsbetrag durch bildpunktweisen Vergleich in Richtung der stärksten Änderung (gegebenenfalls ist hierzu eine Interpolation erforderlich).

Die mit dem Farbgradientenansatz erzielten Ergebnisse wurden sowohl mit den Ergebnissen des Intensitätsansatzes als auch mit den Ergebnissen bei Auswahl des betragsgrößten Gradienten verglichen. Die Bilder 2 bis 4 zeigen die binären Kantenbilder, die sich bei Verwendung der verschiedenen Detektionsansätze für das Apfel-Bild (Bild 1) ergeben. Die Kantenbilder enthalten alle lokalen Extrema ohne Rücksicht auf die Höhe der Änderungsbeträge (Varianz der zur Mittelung verwendeten Gaußfunktion $\sigma = 2$). Die Überlegenheit des Farbgradientenansatzes gegenüber dem Intensitätsansatz zeigt sich deutlich an den rechten Außenkonturen der drei kleineren Äpfel sowie im Schattenbereich in der unteren linken Bildecke. Während der Intensitätsansatz eine robuste Detektion dieser Kanten aufgrund der starken Textur im Hintergrund bzw. aufgrund des schwachen Grauwertkontrastes nicht gestattet, liefert der Farbgradientenansatz auch in

---

[1] Die Intensität entspricht dem arithmetischen Mittel der Farbkomponenten R, G und B.

[2] Die Form der Ableitungsoperatoren entspricht der ersten Richtungsableitung einer zweidimensionalen Gaußfunktion. Die Verwendung von Gaußfunktionen unterschiedlicher Varianz erlaubt die Detektion von Kanten in verschiedenen Auflösungsebenen.

diesen Bildbereichen gute Ergebnisse. Auch die Auswahl der betragsgrößten Gradienten liefert vergleichbar gute Resultate. Lediglich im Schattenbereich kann man — wie nach den theoretischen Überlegungen zu erwarten ist — dem Farbgradientenansatz eine Überlegenheit zusprechen.

Bild 5 und 6 zeigen die Endresultate von Intensitäts- und Farbgradientenansatz nach Anwendung eines Selektionsschemas, das die Festlegung verschiedener Auswahlparameter erlaubt (Mindestkontrast, Mindestlänge der Kantenzüge u.a.). Auf die Details des Selektionsschemas können wir an dieser Stelle nicht weiter eingehen (siehe hierzu [*Drewniok 88*]). Die damit gewonnenen Endresultate sollen lediglich demonstrieren, daß eine systematische Selektion signifikanter Merkmale aus der Menge aller detektierten Maxima möglich ist (siehe auch Bild 11 und 12). Die Bilder 7 bis 12 zeigen die Resultate des Farbgradientenansatzes für zwei weitere Szenen. Wie das Resultat für das Apfel-Bild belegen auch sie, daß sich mit Hilfe des vorgestellten Ansatzes auch für verrauschte Bilder sehr gute Ergebnisse erzielen lassen.

Zusammenfassung und Bewertung

Der dargestellte Ansatz erlaubt die leichte Übertragung aktueller ableitungsorientierter Grauwertverfahren zur Kantendetektion auf Bilder beliebiger Dimension. Denkbare Anwendungen sind, neben der Farbbildverarbeitung, beispielsweise die Detektion von Diskontinuitäten in mehrkanaligen medizinischen Bildern oder Bildkuben oder die Separation bewegter Objekte in dichten Verschiebungsvektorfeldern.

Im Rahmen des Projektes SISSY an der Universität Hamburg stellte sich für uns die Frage, ob und wie sich Farbinformation zur Unterstützung eines merkmalsbasierten Systems zur geometrischen Szenenrekonstruktion ausnutzen läßt [*Dreschler-Fischer 87*]. Die in unseren Experimenten mit dem vorgestellten Detektionsansatz erzielten und hier an verschiedenen Beispielen demonstrierten sehr guten Ergebnisse haben gezeigt, daß die Berücksichtigung der Farbinformation — im Vergleich mit den Ergebnissen eines entsprechenden Intensitätsverfahrens — sehr wohl zu robusteren und verläßlicheren Merkmalen führt. Die detektierten Merkmale bilden somit eine solidere Basis für die sich an die Merkmalsextraktion anschließende Korrespondenzanalyse und Szenenrekonstruktion.

Das Projekt SISSY wird von der DFG im Schwerpunktprogramm "Modelle und Strukturen bei der Analyse von Bild- und Sprachsignalen" unter dem Titel Dr176/2 gefördert. Wir möchten allen Mitarbeitern des SISSY-Projektes, insbesondere Carsten Schröder und Rainer Sprengel, sowie allen Mitgliedern des Arbeitsbereiches "Kognitive Systeme" für ihre Unterstützung danken. Gudrun Klinker danken wir für die hilfreichen Literaturhinweise.

Literatur

[Canny 83] *Finding Edges and Lines in Images*. John F. Canny. Technical Report TR-720, Artificial Intelligence Laboratory, MIT, Cambridge, MA, 1983.

[Di Zenzo 86] *A Note on the Gradient of a Multi-Image*. Silvano Di Zenzo. *Computer Vision, Graphics, and Image Processing* **33** (1986) 116–125.

[Dreschler-Fischer 87] *Das "Bootstrap-Problem" bei der geometrischen Szenenrekonstruktion — Eine Übersicht*. Leonie S. Dreschler-Fischer. Proc. 11th German Workshop on Artificial Intelligence GWAI-87, Geseke, 28. September – 2. Oktober 1987, K. Morik (Hrsg.), *Informatik-Fachberichte* **152**, Springer Verlag, Berlin – Heidelberg – New York – Tokyo 1987, 1–15.

[Drewniok 88] *Untersuchungen zur Detektion von Kanten in digitalen Farbbildern*. Christian Drewniok. Bericht FBI-HH-B-138/88, Fachbereich Informatik, Universität-Hamburg, 1988.

[Förstner 88] *Private Korrespondenz*. Wolfgang Förstner. Heidelberg, November 1988.

[Novak + Shafer 87] *Color Edge Detection*. Carol L. Novak, Steven A. Shafer. in: Proc. DARPA-Image Understanding Workshop, Los Angeles, CA, Februar 1987, Morgan Kaufmann Publishers Inc., Los Altos, CA 1987, Section 2 of the PI-Report *Image Understanding Research at CMU* by Takeo Kanade, 35–37.

Bild 1: Apfel-Bild, Intensität

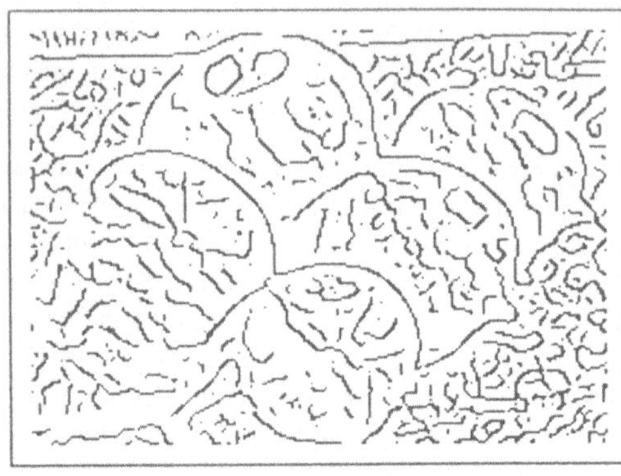

Bild 2: Intensitätsansatz, alle Maxima

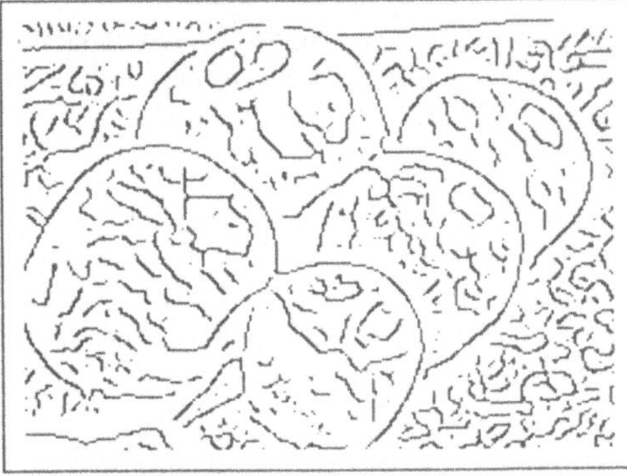

Bild 3: Farbgradientenansatz, alle Maxima

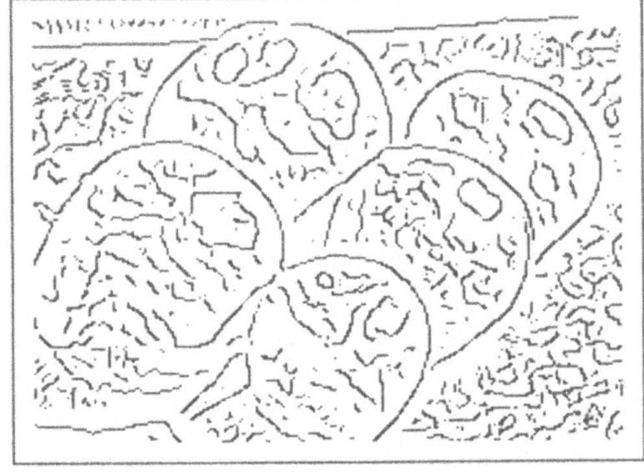

Bild 4: Betragsgrößter Gradient, alle Maxima

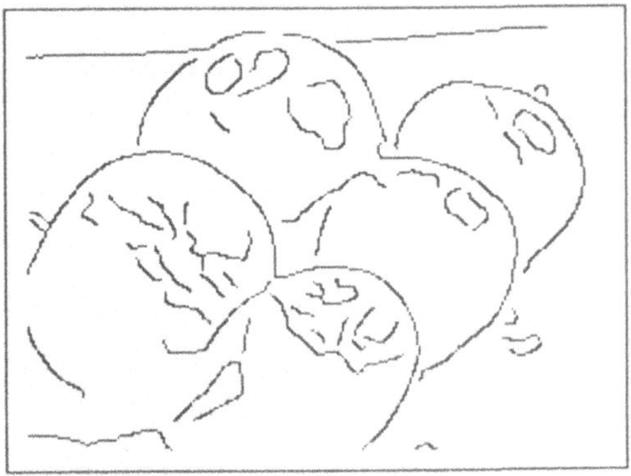

Bild 5: Farbgradientenansatz, Endergebnis

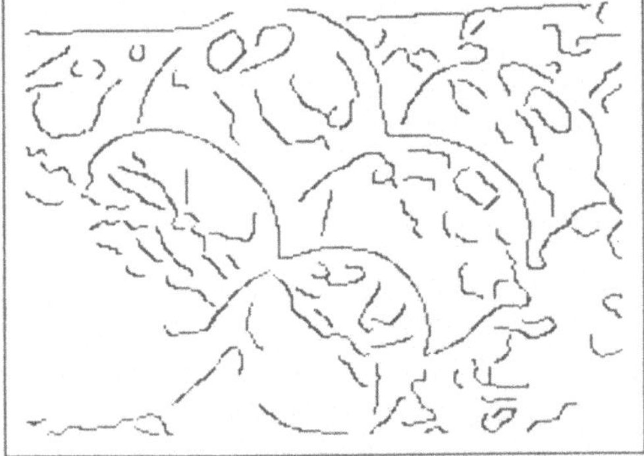

Bild 6: Intensitätsansatz, Endergebnis

Bild 7: Block-Bild, Intensität

Bild 8: Auto-Bild, Intensität

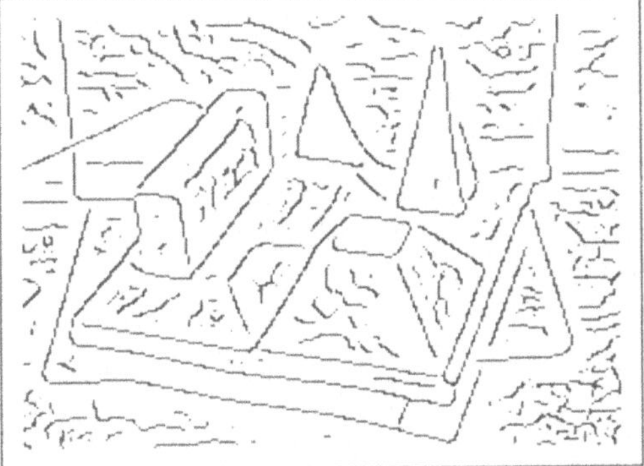

Bild 9: Farbgradientenansatz, alle Maxima

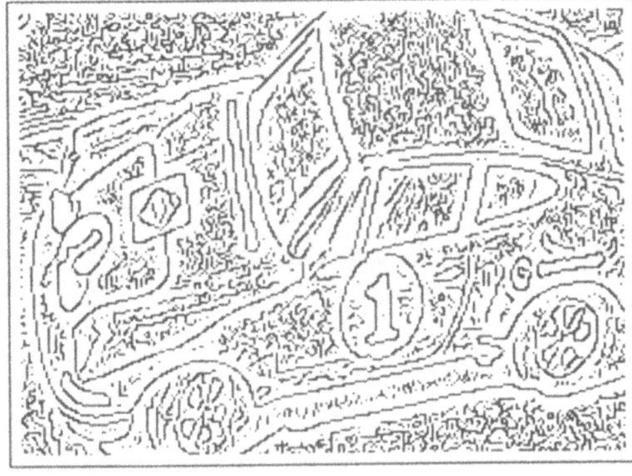

Bild 10: Farbgradientenansatz, alle Maxima

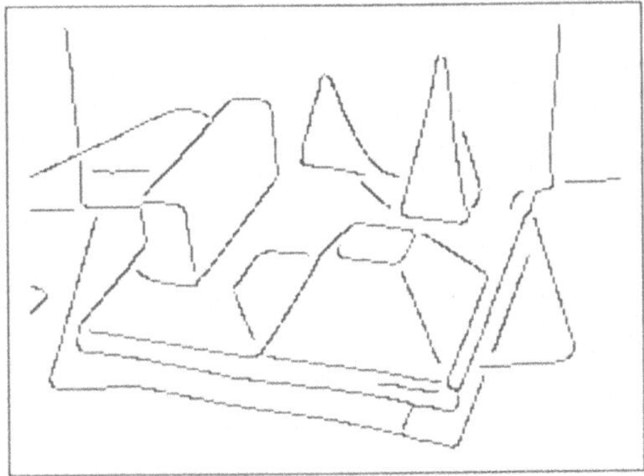

Bild 11: Farbgradientenansatz, Endergebnis

Bild 12: Farbgradientenansatz, Endergebnis

Erkennung handgeschriebener Ziffern
mit Hilfe neuronaler Netze

M. Schwarz, B.J. Hosticka und P. Richert

Fraunhofer Institut für Mikroelektronische Schaltungen und Systeme,
Finkenstraße 61, D-4100 Duisburg 1

Kurzfassung

Vorgestellt wird ein System zur Erkennung handgeschriebener Ziffern, das für die Klassifikation ein hierarchisches System von Merkmalen, wie Linienenden, Verzweigungspunkte, Geradenelemente und Bogenelemente verwendet. Für die Merkmalextraktion und Klassifikation werden Methoden neuronaler Netze berücksichtigt, so wird z.B. für die Klassifikation ein adaptives System - eine selbstorganisierende Merkmalkarte - eingesetzt.

1. Struktur des Bilderkennungssystems

Für die Klassifikation der Bildobjekte werden Merkmale aus den folgenden Klassen herangezogen.

- Linienenden und Verzweigungspunkte

- Geradenelemente   (dominant bei { 1, 4, 7 })

- Kreise und Bogenelemente   (dominant bei { 0, 2, 3, 5, 6, 8, 9 })

Neben der Unterscheidung der Teilobjekte nach ihrer Art, findet eine Unterscheidung nach ihrer Lage im Bildfeld statt (Bild 1).

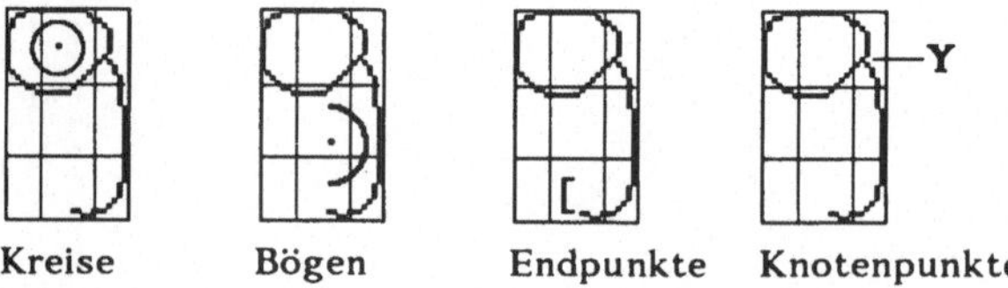

Bild 1: Beschreibung eines Bildobjektes durch Art und Lage seiner Teilobjekte.

Der Merkmalextraktion vorangestellt ist die Medianfilterung und Skelettierung des Eingangsbildes. Zusammen mit der Klassifikation, mit einer selbstorganisierenden Merkmalkarte [1], ergibt sich die in Bild 2 dargestellte Gesamtstruktur des Erkennungssystems.

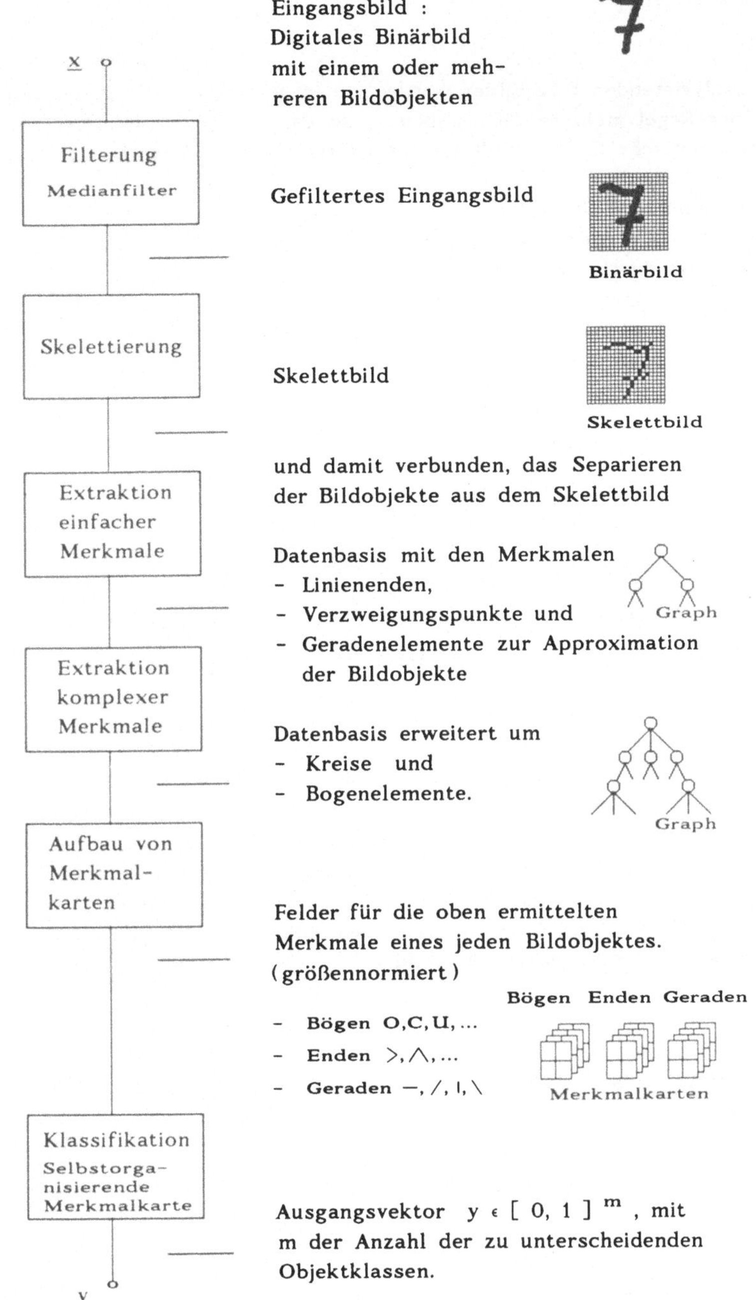

Bild 2 : Übersicht über die Gesamtstruktur des Bilderkennungssystems.

2. Vorverarbeitung

Die zu analysierenden Binärbilder werden zunächst digitalisiert. Dabei umfaßt ein Eingangsbild in der Regel mehrere Ziffernobjekte, so daß für die Erkennung der einzelnen Ziffern deren automatische Trennung durch das Erkennungssystem nötig ist.

Zur Entfernung von Störanteilen in Form einzelner Bildpunkte oder linienhafter Störungen (Bild 4), wird eine Medianfilterung vorgenommen. Hiezu wird ein Operatorfeld [2] (Bild 3) verwendet, das durch ein Modellneuron [3] mit der Funktion ($y = f (\sum_i w_i x_i)$) realisiert werden kann.

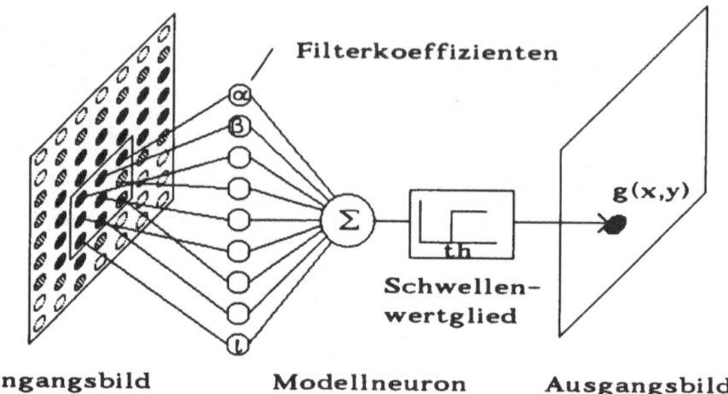

Bild 3: Mittels eines Modellneurons aufgebautes Medianfilter.

Die auf die Filterung folgende Skelettierung dient der Verdünnung der Linien der Bildobjekte auf die Breite von einem Bildpunkt. Dies ist zulässig, da Ziffernobjekte aus linienhaften Teilobjekten aufgebaut sind, deren Linienstärke für die Ziffernerkennung irrelevant ist.

Die Skelettierung wird durch sukzessives Löschen der Randpunkte der Bildobjekte vorgenommen, wobei zu beachten ist, daß ein Randpunkt nicht gelöscht werden darf, wenn er der Endpunkt eines nur einen Bildpunkt breiten Linienstückes oder der letzte verbleibende Punkt eines Bildobjektes ist oder wenn das Entfernen des Randpunktes die Konnektivität des Bildobjektes zerstören würde (Bild 5).

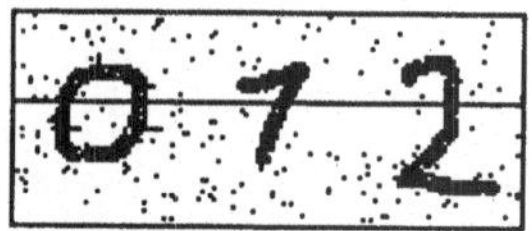

Bild 4 a): Gestörtes Eingangsbild.

b): Gefiltertes Eingangsbild.

Bild 5 a): Iterationsschritt 1

b): Skelettbild (Iterationsschritt 4)

3. Merkmalextraktion

Für die rechnerinterne Repräsentation des Skelettbildes wird eine hierarchische Datenstruktur in Form eines Graphen aufgebaut (Bild 6), der mit zunehmender Tiefe eine immer detailliertere Beschreibung der Ziffernobjekte erlaubt.

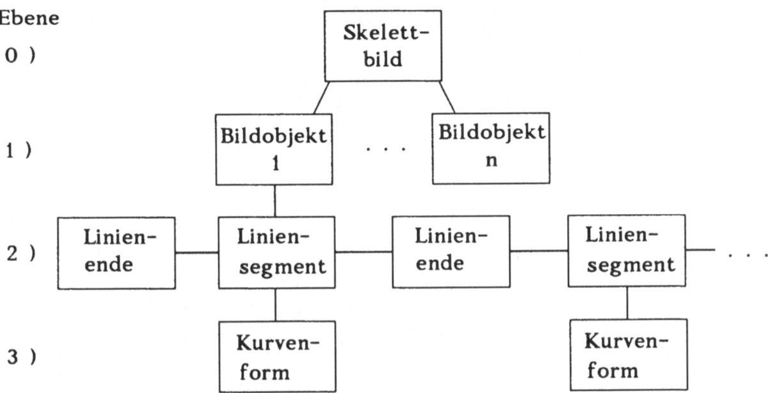

Bild 6 : Ebenen der Datenstruktur zur Beschreibung eines Skelettbildes.

Die erste Ebene (Ebene 1) der Datenstruktur beschreibt das Ergebnis der Objekttrennung. Hier werden alle durch Linienenden oder Verzweigungspunkte begrenzten Kurvenstücke aus dem Skelettbild herausgelöst und zu einem Bildobjekt zusammengefaßt, wenn die sie umschreibenden Rechtecke sich berühren oder überlappen (Bild 7 a und b).

In der zweiten Ebene des Graphen werden die einzelnen Bildobjekte durch Knoten für Linienenden, Verzweigungspunkte und die sie verbindenden Liniensegmente beschrieben (Bild 7 c). Diese Knoten sind entsprechend der Objekttopologie miteinander verbunden.

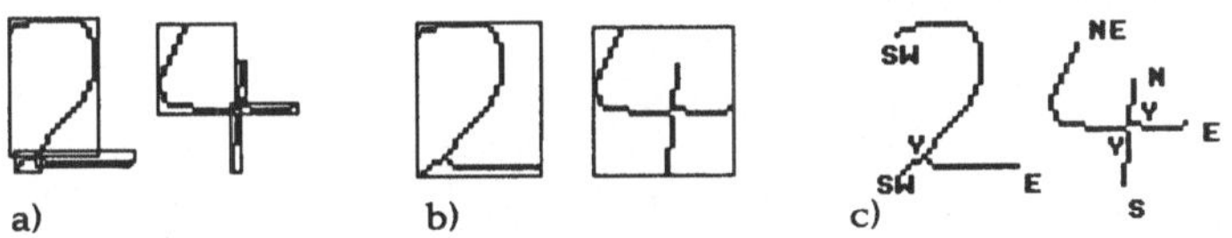

Bild 7: a) Herauslösen der Liniensegmente,
        b) Zusammenfassung zu Bildobjekten und
        c) Extraktion von End- und Knotenpunkten.

Für den Aufbau der 3. Ebene, die der Beschreibung der Kurvenform der Liniensegmente dient, werden die Liniensegmente, deren Form bisher nur durch den Kettencode beschrieben wird, zunächst durch Polygone approximiert. Für diesen Datenreduktionsschritt wird ein von der Medianfilterung [2] abgeleitetes Verfahren benutzt.

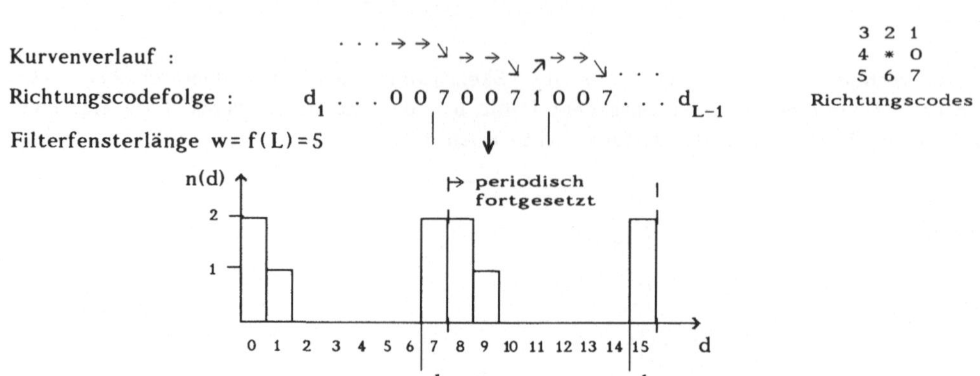

Bild 8: Periodisch fortgesetztes Histogramm der Richtungscodes.

Aus dem periodisch fortgesetzten Histogramm, der im Filterfenster gesammelten Richtungscodes, ist der Anfang d_a des Histogrammabschnittes zu ermitteln, in dem sich von Null verschiedene Einträge häufen (Clustererkennung) (hier d_a = 7). Für die Medianfilterung ist nur das Intervall $[d_a, d_e]$, mit $d_e = d_a + 8$, zu betrachten.

Sei nun s(m) die Summe der Histogrammeinträge (beginnend bei d_a),

$$s(m) = \sum_{d = d_a}^{m} n(d) \ , \quad m < d_e \quad \text{und} \quad th \quad \text{die Schwelle} \quad th = (w-1)/2 \ .$$

Hiermit ergibt sich der mediangefilterte Richtungscode zu

$$\tilde{d} = (\min_i m_i) \bmod 8 \ , \quad \text{mit} \quad m_i \in M, \quad M = \{ m \mid s(m) > th \} .$$

Diese Berechnung ist für jedes Element des Kettencodes vorzunehmen.

Die Geradenelemente zur Approximation des Skelettbildes werden so gewählt, daß sie die Punkte des Skelettbildes miteinander verbinden, an denen der gefilterte Kettencode eine Richtungsänderung aufweist (Bild 9).

Bild 9: Approximation des Skelettbildes durch Geradenelemente.

Für die Extraktion komplexer Merkmale [4] - Bogenelemente und Kreisbögen - ist es zunächst nötig, eine Größennormierung der Bildobjekte mittels einer affinen Abbildung vorzunehmen.

$$(x', y') = f(x,y) = \left(\frac{B}{\text{Objektbreite}} x , \frac{H}{\text{Objekthöhe}} y \right)$$

Hieran schließt sich die Übertragung der die Bildobjekte approximierenden Geradenelemente in Akkumulatorfelder für horizontale (A_H) und vertikale (A_V) Anteile der Geradenelemente an. Die Extraktion der komplexeren Merkmale (C-, D-, N- und O-Bögen) wird nun mittels kreuzförmiger Operatorfelder vorgenommen, die auf jedes Element der Felder A_H

und A_V angewendet werden (siehe Bild 10).

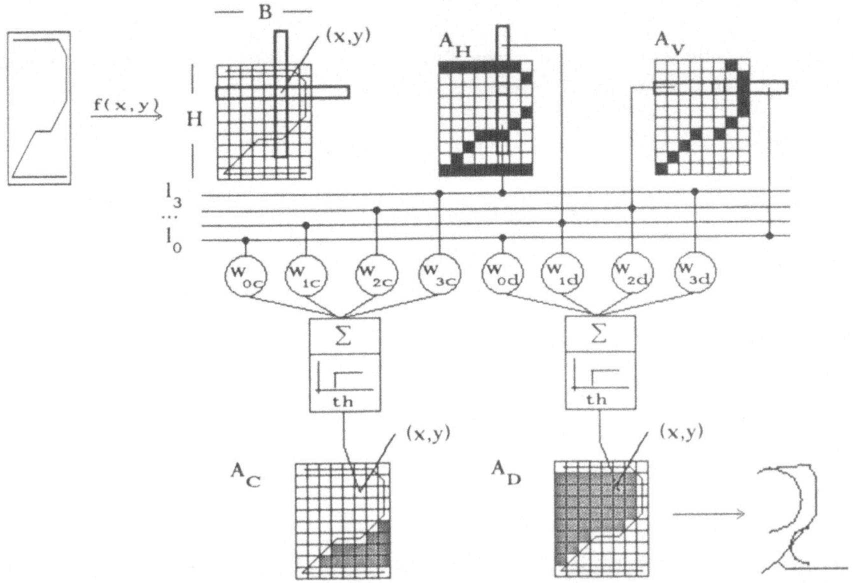

Bild 10: Bestimmung komplexer Merkmale am Beispiel von C- und D- Bögen.

Die Aktivitäten in den Ausgangsfeldern A_k, für die Merkmale $k \in \{ C, U, D, N, O \}$, ergeben sich aus der gewichteten Aufsummation der Aktivitäten l_i, $i \in \{ H, V \}$, die das Vorhandensein horizontaler bzw. vertikaler Anteile von Geradenelementen im Bereich der zugehörigen Operatorfeldarme beschreiben (siehe Bild 10, dort dargestellt für das Feldelement (x,y)).

Die Gewichtungen w_{ik} für die fünf komplexen Merkmaltypen ergeben sich aus einem Ungleichungssystem, das die Überschreitung der Schwelle th fordert, wenn an der Stelle (x,y) das gesuchte Merkmal vorliegt. Andererseits darf die Schwelle nicht überschritten werden, wenn bei der Suche nach einem C-förmigen Bogen z.B. ein D-förmiger Bogen am Korrelator anliegt.

4. Klassifikation

Für die Klassifikation wird ein Merkmalvektor $\underline{x}$ aufgebaut, der es ermöglicht, Art- und Lagemerkmale der Teilobjekte eines Bildobjektes zu unterscheiden. Dazu werden zunächst Akkumulatorfelder a(x,y,z) für die verschiedenen Teilobjektarten z aufgebaut; (x,y) beschreibt dabei die Lage eines Teilobjektes. Bild 11 zeigt die Aktivitäten in den Akkumulatorfeldern, hier dargestellt für eine Auflösung von 2×2 Feldelementen.

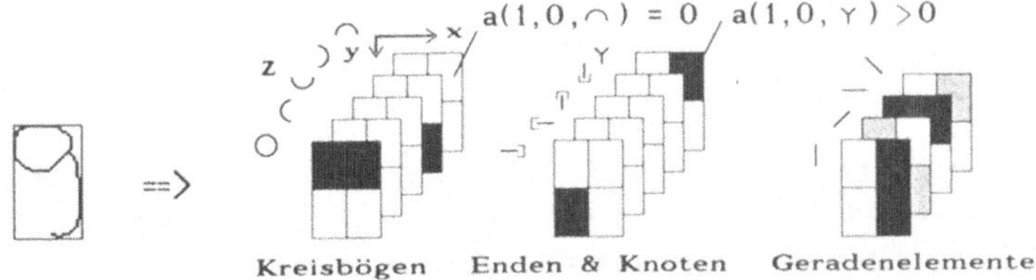

Bild 11: Beschreibung eines Bildobjektes durch das Aktivitätsmuster a(x,y,z).

Die Aktivitäten a(x,y,z) in den Akkumulatorfeldern ergeben sich aus der, mit der Wichtigkeit s eines Teilobjektes o, gewichteten Aufsummation aller Teilobjekte eines Bildobjektes, getrennt nach Lage (x,y) und Typ z des Teilobjektes. Als Maß für die Wichtigkeit s(o) eines Teilobjektes o, wird bei Bogenelementen deren Bogenlänge und bei End- und Knotenpunkten ein konstanter Wert verwendet. Es gilt damit :

$$a(\,x,\,y,\,z\,) \;=\; \sum_{o\,\epsilon\,O} \alpha(\,x,y,z,\,x(o),y(o),z(o)\,)\cdot s(o)\,,$$

mit O der Menge aller Teilobjekte eines Bildobjektes und

$$\alpha(\,x,y,z,\,x(o),y(o),z(o)\,) \;=\; \begin{cases} 1 & \text{wenn Lage und Typ von Akkumulator-} \\ & \text{torfeldelement } (x,y,z) \text{ und Teilobjekt } o \\ & (\,x(o),\,y(o),\,z(o)\,)\ \text{übereinstimmen,} \\ 0 & \text{sonst.} \end{cases}$$

Der zu klassifizierende Merkmalvektor $\underline{x}$ ergibt sich damit zu

$$\underline{x}^T \;=\; \underline{a}^T \;=\; \Big(\; a(0,0,\text{-}),\ \ldots,\ a(0,1,\text{-}),\ \ldots,\ a(0,0,/),\ \ldots,\ a(0,0,\cap),\ \ldots\;\Big)\;.$$

Bei der Klassifikation mit einer selbstorganisierenden Merkmalkarte [1] wird als Abstandsmaß des Merkmalvektors $\underline{x}$ von einem Referenzvektor $\underline{x}_{ref}$ der euklidsche Abstand $d(\,\underline{x},\,\underline{x}_{ref}\,) = \|\,\underline{x} - \underline{x}_{ref}\,\|$ verwendet (siehe Bild 12).

Aufgabe der Merkmalkarte ist die nichtlineare Abbildung des Merkmalvektors $\underline{x}$ in eine m-dimensionale Karte (hier m = 2), also die Dimensionsreduktion :

$$\varphi:\ \mathbb{R}^n \to G,\qquad G \subset \mathbb{R}^m,\quad m \ll n\ (\ n{=}56\ ,\ m{=}2\ )$$

so, daß

$$d(\,\underline{x}\,,\,\underline{w}(\underline{y})) \;=\; \min_{\xi}\ d(\,\underline{x}\,,\,\underline{w}(\underline{\xi})),\quad \text{mit}\quad \underline{x},\underline{w}\,\epsilon\,\mathbb{R}^n,\quad \underline{y},\underline{\xi}\,\epsilon\,G\;.$$

Man wählt also $\underline{y}(\underline{x})$ so, daß der Abstand von $\underline{x}$ zu einer Fläche $F \subset \mathbb{R}^n$, die durch $\underline{w}(\underline{y})$ festgelegt ist, minimal wird.

Da F hier durch eine Reihe von Stützstellen $\underline{w}_i$, $i\,\epsilon\,\{1,\ldots,\ M\}$, approximiert wird, gilt: Wähle die am besten zu $\underline{x}$ passende Stützstelle $i(\underline{x})$ so, daß

$$d(\,\underline{x}\,,\,\underline{w}_i) \;=\; \min_{j}\ d(\,\underline{x}\,,\,\underline{w}_j)\;.$$

Die Stützstellen $\underline{w}_i$ werden so eingestellt, daß der Erwartungswert der Summe aus dem Abstandsquadrat der Eingangsmuster $\underline{x}$ zur jeweils bestpassenden Stützstelle $\underline{w}_i$ und einem zusätzlichen Fehlerterm, der den Nachbarschaftserhalt der Abbildung erzwingt, minimiert wird.

$$E\{\,e\,\} \;=\; E\Big\{\; d^2(\underline{x},\underline{w}_i) \;+\; \sum_{j\,\epsilon\,N_i\backslash\{i\}} a(\,i\,,\,j\,)\cdot d^2(\underline{x},\underline{w}_j)\;\Big\}$$

Der zweite Fehlerenergieterm bewirkt, daß im $\mathbb{R}^n$ benachbarte Stützstellen $\underline{w}_j$ benachbarten Punkten in G entsprechen, was dadurch erreicht wird, daß a(i,j) für in G benachbarte Punkte i, j groß (a(i,i) = 1) und für weiter entfernte Punkte klein ist ($0 < a(i,j) < 1$ für $i \neq j$).

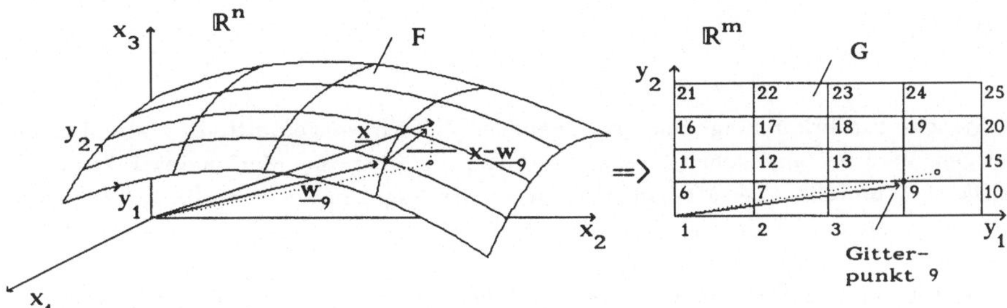

Bild 12: Abbildung von Merkmalvektoren auf G mit Hilfe einer durch
M = 25 Stützstellen approximierten Fläche F.

Für die Einstellung der Stützstellenvektoren $\underline{w}_i$ wird eine Gradientensuche verwendet.

——————— Adaptionsregel für die Merkmalkarte ———————

$$\underline{w}_j(k+1) = \underline{w}_j(k) + \mu(k)\, a(i,j)\, (\underline{x}(k) - \underline{w}_j(k)), \qquad j \in N_i(k)$$

mit

- i der am besten zu $\underline{x}(k)$ passenden Stützstelle $\underline{w}_i(k)$, also diejenige, für die gilt

$$\| \underline{x} - \underline{w}_i \| = \min_j \| \underline{x} - \underline{w}_j \|, \quad j \in \{1, \ldots, M\}.$$

- $N_i(k)$ einem Gebiet, das zunächst alle M Gitterpunkte aus G einschließt, sich aber im Verlaufe des Trainings immer weiter zusammenzieht, bis es am Trainingsende nur noch i und dessen nächste Nachbarn einschließt ($k \rightarrow \infty$).

- $\mu(k)$ der Lerngeschwindigkeit, eine mit der Zeit k monoton bis auf Null fallende reellwertige Funktion z.B. $\mu(k) = \mu_m$ für $k < k_{max}$ und 0 sonst.

- $a(i,j)$ einem Gewichtsfaktor, der umso größer ist, je dichter die Gitterpunkte i und j in der Merkmalkarte (in G) benachbart sind, hierüber wird der Nachbarschaftserhalt erzwungen.

Durch den Selbstorganisationsprozeß bilden sich in G Gebiete für die verschiedenen Objektklassen aus, die so angeordnet sind, daß die Gebiete ähnlicher Objekte in G benachbart liegen (siehe Bild 13).

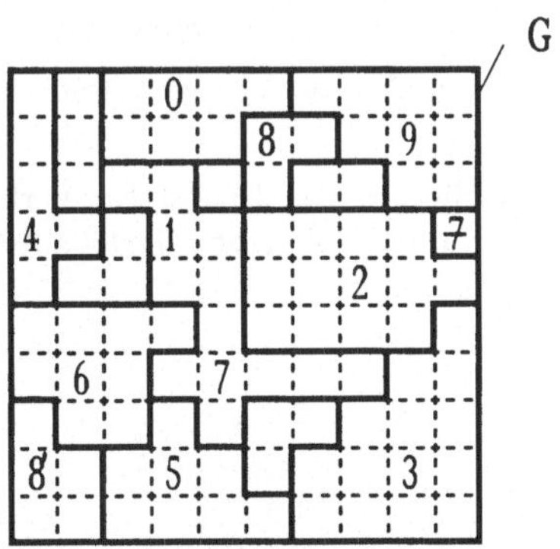

Bild 13: Merkmalkarte mit den Klassenbereichen der Ziffern 0 bis 9.

5. Zusammenfassung

Es wurde ein System zur Erkennung handgeschriebener Ziffern vorgestellt, das alle für die Erkennung nötigen Verarbeitungsschritte von der Filterung über die Merkmalextraktion bis hin zur Klassifikation umfaßt. Das Medianfilter ist in der Lage, neben Einzelbitfehlern auch linienhafte Störungen zu entfernen, wie sie z.B. durch eine fehlerhafte Zeile in einem CCD-Bildaufnahmeelement entstehen können. Da Schriftzeichen im wesentlichen aus linien- und punktförmigen Teilobjekten aufgebaut sind, wird im Anschluß an die Filterung eine Skelettierung der Bildobjekte vorgenommen, ohne daß hierdurch für die Klassifikation wesentliche Informationen verlorengehen.

Die Verarbeitungschritte Objekttrennung, Extraktion einfacher und komplexer Merkmale dienen der Übertragung des Skelettbildes in eine hierarchische Datenstruktur, einen Graphen, der die Topologie der Bildobjekte und auch die Kurvenform der einzelnen Linienelemente eines Zeichens beschreibt. Die Beschreibung der Kurvenform wird mittels einer Reihe von Bogenelementen vorgenommen, die den Kurvenverlauf approximieren.

Mit Hilfe der vorgestellten Merkmalkarte lassen sich auch stark nichtlinear verzerrte Eingangsmuster erkennen, vorausgesetzt die Trainingsfolge war genügend lang (1000 bis 5000 Trainingszyklen) und umfaßte auch Alternativmuster innerhalb einer Merkmalklasse, z.B. für die Ziffer 4 ein oben geschlossenes und ein oben offenes Exemplar. Aus der Auswertung der Merkmalkarte erhält man Hinweise auf Möglichkeiten zur Erhöhung der Erkennungssicherheit. So liegen die Gebiete der Ziffern 1 und 7 in der Regel benachbart, was es sinnvoll erscheinen läßt, zur besseren Unterscheidbarkeit beispielsweise einen Querstrich in der Ziffer 7 vorzuschreiben.

Alle Verfahrensschritte wurden in Form von Software in der Programmiersprache C realisiert, wozu Programmodule für Filterung, Merkmalextraktion und Klassifikation enwickelt wurden.

6. Literaturverzeichnis

[1] T. Kohonen, Self-Organization and Associative Memory, Springer, Berlin, 1984.

[2] William K. Pratt, Digital Image Processing, J. Wiley & Sons Inc. New York, 1978.

[3] R.P.Lippmann, An Introduction to Computing with Neural Nets, IEEE ASSP Magazine, April 1987.

[4] K. Fukushima, Neocognitron: A Hierarchical Neural Network Capable of Visual Pattern Recognition, Neural Networks, Vol 1, pp. 119-130, 1988.

Automatisches Erlernen struktureller Modelle für ein wissensbasiertes Werkstückerkennungssystem

W. Hättich*, H. Wandres*, P.-B. Krause**

* Fraunhofer-Institut für Informations- und Datenverarbeitung (IITB), Fraunhoferstraße 1, D-7500 Karlsruhe 1

**AEG Aktiengesellschaft Marine-Informationstechnik, Behringstraße 120, D-2000 Hamburg 50

1 Einleitung

Der Anwendungsbereich industrieller Handhabungssysteme kann deutlich erweitert werden, wenn ein solches System mit einem Sensor ausgerüstet ist, mit dessen Hilfe vorliegende Werkstücke erkannt und ihre räumliche Lagen bestimmt werden. Für die Werkstückerkennung und Lagebestimmung sind wissensbasierte Erkennungssysteme geeignet, in denen das Wissen über die Werkstücke und über Störeinflüsse in Form eines Modelles explizit verfügbar sind. Bei einem strukturellen Modell wird das Objekt in einfache Bestandteile bzw. Beschreibungselemente zerlegt und durch Angabe von Relationen zwischen den Bestandteilen beschrieben /Chin & Dyer 86/, /Wallace 88/. Durch diese Zerlegung des Objektes in einzelne Bestandteile wird es möglich, ein Objekt aufgrund nur weniger sichtbarer Bestandteile zu erkennen, wie es bei überlappenden Werkstücken erforderlich ist.

Erkennungssysteme mit strukturellen Modellen sind sehr leistungsfähig und flexibel, wenn das Modell gut ist. Die Erstellung eines guten Modelles stellt zur Zeit einen Engpaß bei der praktischen Anwendung dar, da die Modellerstellung wegen der notwendigen Detailkenntnisse über das Erkennungssystem nur von einem Spezialisten durchgeführt werden kann. Der Einsatz von CAD-Modellen für die Werkstückerkennung ist nicht möglich, weil diese die für die Erkennung notwendigen Angaben über die Reproduzierbarkeit und Trennungswirksamkeit von Objektbestandteilen nicht enthalten. Außerdem ist die Optimierung des Modelles wegen der Vielfalt der Möglichkeiten, ein Objekt in Bestandteile zu zerlegen, trotz guter interaktiver Werkzeuge ein zeitraubender Vorgang. Es ist deshalb wünschenswert, den Modellerstellungsprozeß zu automatisieren.

Die Modellerstellung für ein wissensbasiertes Werkstückerkennungssystem umfaßt drei Teilaufgaben:

1. Die Auswahl von Beschreibungselementen, die trennungswirksam und störunempfindlich sind.
2. Die Auswahl von Filterparametern zur Gewinnung der gewünschten Beschreibungselemente.
3. Die Auswahl einer Erkennungsstrategie, die festlegt, welche Beschreibungselemente in welcher Reihenfolge zur Erkennung benutzt werden sollen.

Mit der automatischen Modellerstellung für ein Werkstückerkennungssystem beschäftigen sich nur wenige Autoren. Sie behandeln überwiegend nur jeweils eine der drei genannten Teilaufgaben. Die Bestimmung von optimalen Filterparametern geschieht bei /Korn 86/, /Asada & Brady 86/ und /Ender 87/. Die Auswahl von Beschreibungselementen wird in /Bolles & Cain 82/, /Ayache & Faugeras 86/ und /Ettinger 87/ beschrieben, wobei eine feste Erkennungsstrategie zugrunde gelegt wird. /Mudge et al. 87/ haben die Erkennungsstrategie optimiert und dadurch im Vergleich zu einer heuristisch eingestellten Strategie bessere Erkennungsergebnisse erzielt. Allerdings wurde das Modell auf maximal zwei Beschreibungselemente begrenzt.

In der vorliegenden Arbeit wird die gemeinsame, aufeinander abgestimmte Lösung aller drei Teilaufgaben beschrieben. Die automatische Modellerstellung erfolgt nach der Methode "Lernen durch Zeigen". Im Gegensatz zu einer früheren Arbeit /Hättich 88/ wird anstelle einer Ballungsanalyse das Erkennungssystem selbst zur Modellerstellung eingesetzt.

2 Systemstruktur

Die Struktur des Erkennungssystems mit den Komponenten zur automatischen Modellerstellung ist in Abb. 1 dargestellt. In der Darstellung sind Operationen durch Rechtecke und Datenobjekte durch Kreise kodiert.

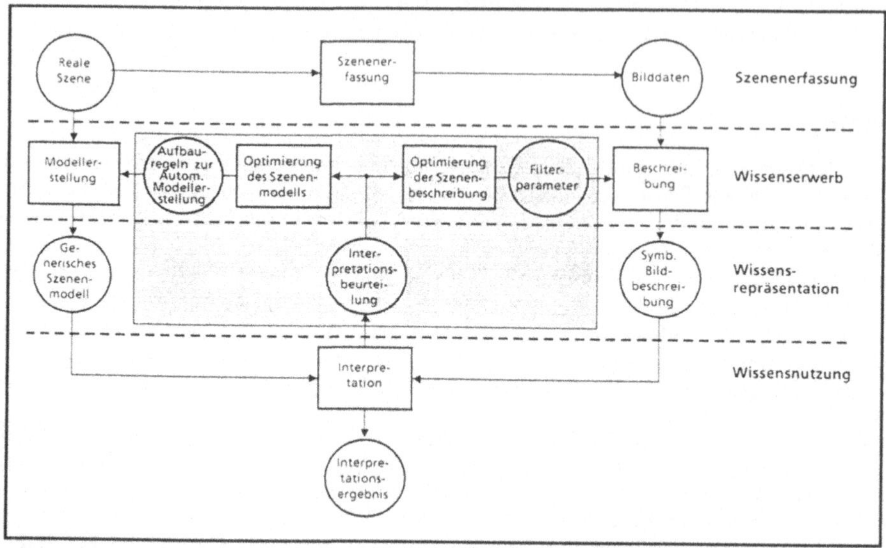

Abb. 1: Systemstruktur

In der Verarbeitungskomponente "Szenenerfassung" wird die reale Szene, die aus einer räumlichen Anordnung von Werkstücken besteht, in den Bildbereich abgebildet. Als Ergebnis dieser Abbildung erhält man Bilddaten in Form eines oder mehrerer Bilder der Szene. Mit Hilfe der Verarbeitungskomponente "Beschreibung" werden die Bilddaten in eine symbolische Beschreibung überführt, die nach Möglichkeit nur noch die für die Erkennungsaufgabe bedeutsamen Beschreibungselemente enthält. In der Verarbeitungskomponente "Modellerstellung" wird das generische Szenenmodell erzeugt. Es repräsentiert jede in der Szene mögliche Anordnung von Werkstücken. Durch die Verarbeitungskomponente "Interpretation" wird eine Zuordnung zwischen der symbolischen Beschreibung des Bildes einer Szene und dem generischen Szenenmodell hergestellt. Dies geschieht dadurch, das jenes Objektmodell ermittelt wird, das am besten zur symbolischen Bildbeschreibung der Szene paßt. Als Ergebnis des Vergleichs erhält man das Interpretationsergebnis, das Auskunft gibt, welche Objekte in der Szene enthalten sind, wie sie angeordnet sind und wo sie sich befinden.

Die in Abb. 1 gezeigte Systemstruktur ist in Einklang mit dem allgemeinen Strukturvorschlag für ein wissensbasiertes Werkstückerkennungssystems nach /Besl & Jain 85/, sie wurde jedoch zur Formulierung des Modellerstellungsprozesses erweitert. In der Abbildung ist die Systemerweiterung durch einen graugetönten Hintergrund hervorgehoben. Die Erweiterung besteht aus zwei Rückkopplungsschleifen, die dazu dienen, sowohl das generische Szenenmodell als auch die symbolische Bildbeschreibung hinsichtlich der Güte des Interpretationsergebnisses zu optimieren. In einem iterativen Prozeß werden in der Verarbeitungskomponente "Optimierung des Szenenmodells" die Aufbauregeln zur Erzeugung des generischen Szenenmodelles und in der Verarbeitungskomponente "Optimierung der Szenenbeschreibung" die Filterparameter zur Extraktion einer symbolischen Bildbeschreibung fortlaufend modifiziert, bis die Bewertung der Szeneninterpretation hinsichtlich der Stör- und Erkennungssicherheit optimal wird.

Zur Bewertung der Störsicherheit wird geprüft, welche Beschreibungselemente bei mehreren unterschiedlichen Bildaufnahmen zuverlässig und positionsgenau auftreten, d.h. gut reproduzierbar sind. Zur Bewertung der Erkennungssicherheit wird geprüft, welche Anordnung von Beschreibungselementen im Modell eine eindeutige Erkennung und Lagebestimmung bei Verwendung der kleinstmöglichen Anzahl von Beschreibungselementen ermöglicht.

3 Verfahrensbeschreibung

Für die oben angeführten drei Teilaufgaben der automatischen Modellerstellung werden nachfolgend die Verfahrensschritte erläutert.

3.1 Auswahl von Beschreibungselementen

Zur Auswahl der Beschreibungselemente in der rechten Rückkopplungschleife wird die Reproduzierbarkeit als Qualitätskriterium verwendet. Dieses Kriterium gewährleistet, daß die Beschreibungselemente bei verschiedenen Bildern unabhängig von der Lage der im Bild dargestellten Werkstücke möglichst zuverlässig und positionsgenau extrahiert werden und deshalb auch zur Modellbeschreibung geeignet sind.

Für die Bewertung der Reproduzierbarkeit wird ein struktureller Erkenner /Freytag et al. 86/ eingesetzt. Dies erscheint auf den ersten Blick nicht möglich, weil der Erkenner selbst bereits ein Modell benötigt, welches aber gerade gelernt werden soll. Es kann aber die Bildbeschreibung eines Referenzbildes als Ausgangsmodell für den Erkenner dienen, um die restlichen Referenzbilder anhand ihrer Bildbeschreibung zu interpretieren. Bei der Erkennung wird geprüft, ob sich ein bestimmtes Beschreibungselement in den verschiedenen Referenzbildern in den Positionen wiederfinden läßt, die aufgrund der bekannten Drehlage bei der Aufnahme erwartet werden. Alle Beschreibungselemente, die sich in den erwarteten Positionen befinden, werden einander zugeordnet und als zusammengehörig erkannt. Ein Maß für die Güte der Reproduzierbarkeit von Beschreibungselementen ist die Erkennungsgüte, die insbesondere davon abhängt, wie viele Beschreibungselemente sich in den erwarteten Positionen befinden und wie genau die erwarteten Positionen eingehalten werden.

Die Gewinnung gut reproduzierbarer Beschreibungselemente wird durch folgende automatisierbare Vorgehensweise erreicht:

Schritt 1:   Aufnahme von Referenzbildern des zu erkennenden Objektes in NL verschiedenen, aber bekannten Drehlagen.

Schritt 2:   Extraktion der symbolischen Bildbeschreibung für alle NL Referenzbilder von den Objekten in den verschiedenen Drehlagen.

Schritt 3:   Bewertung der Reproduzierbarkeit.

 Schritt 3.1:   Auswahl eines Beschreibungselementes in einem ersten Referenzbild.

 Schritt 3.2:   Erkennung der entsprechenden Beschreibungselemente in den restlichen Referenzbildern unter Verwendung des a priori-Wissens über die Lage der Objekte und Registrierung der Güten bei der Erkennung.

 Schritt 3.3:   Wiederholung der Schritte 3.1 und 3.2 mit allen Beschreibungselementen des in Schritt 3.1 ausgewählten Referenzbildes.

Schritt 4:   Auswahl der Beschreibungselemente mit den besten Erkennungsgüten.

Schritt 5:   Bestimmung von repräsentativen Modellbeschreibungselementen durch Mittelung der in Schritt 3.2 einander zugeordneten Beschreibungselemente der Bildbeschreibung.

Die Wirkungsweise des Verfahrens zur Auswahl von Beschreibungselementen wird anhand von Abb. 2 erläutert. Abb. 2a zeigt das Referenzbild eines Stanzteiles. Abb. 2b zeigt die Beschreibung dieses Stanzteiles durch Geradenstücke, die bei Anwendung einer Reihe von Filteroperationen mit einem vorgegebenen Parametersatz gewonnen wurden. In Abb. 2c ist die Rangfolgekurve der Erkennungsgüten dargestellt, die in Schritt 4 erhalten wurden. Nach einer anfänglich geringen Abnahme der Erkennungsgüte erfolgt ein Knick mit starkem Abfall der Erkennungsgüte. Alle Beschreibungselemente bis zum Knick in der Rangfolgekurve sind gut reproduzierbar und sind deshalb als Modellbeschreibungselemente geeignet. Die auf diesem Wege ausgewählten Beschreibungselemente sind in Abb. 2d mit entsprechender Rangfolgenummer zu sehen. Die Auswahl beschreibt den gut reproduzierbaren Teil der Kontur des Werkstückes; die für die Kreise unangepaßten Geradenstücke aus Abb. 2b sind durch das Verfahren automatisch eliminiert worden (vgl. Abb. 2d).

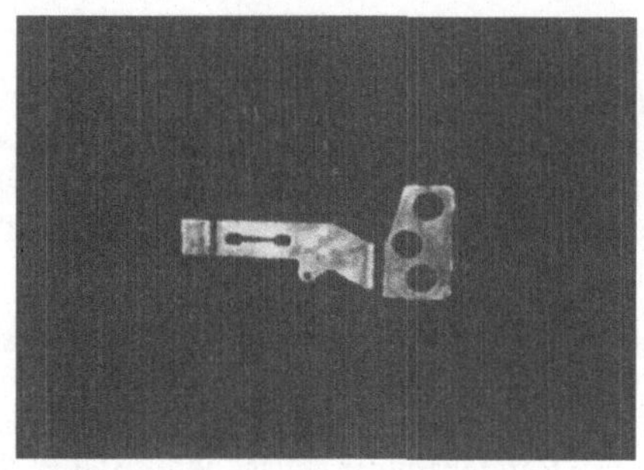

Abb. 2a: Referenzbild

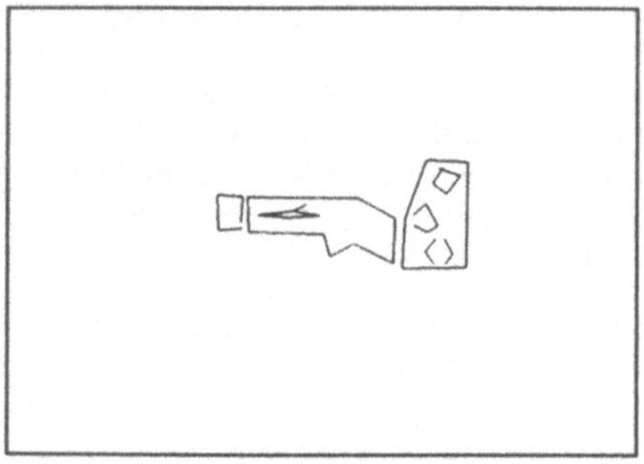

Abb. 2b: Bildbeschreibung

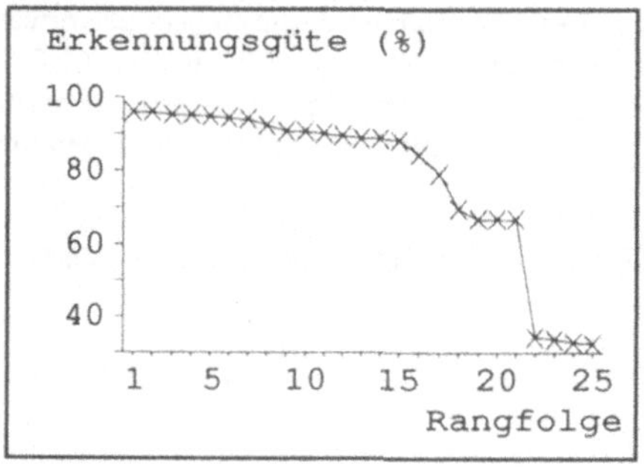

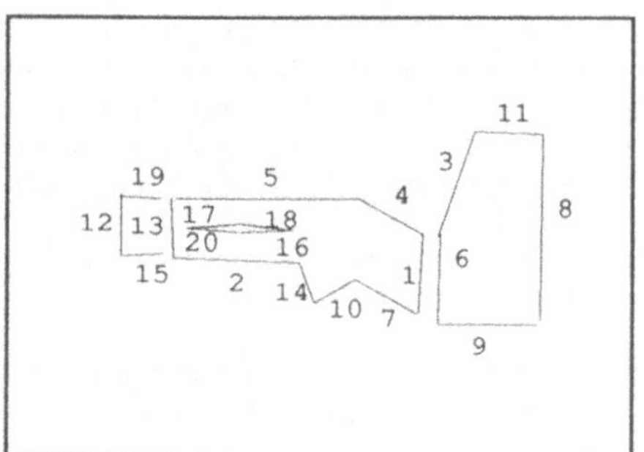

Abb. 2c: Rangfolgekurve

Abb. 2d: Ausgewählte Bildbeschreibungs-
elemente

3.2 Auswahl von Filterparametern

Das im vorangegangenen Abschnitt 3.1 beschriebene Verfahren zur Auswahl der Beschreibungselemente liefert eine Rangfolge für die zuvor extrahierten Beschreibungselemente bezüglich ihrer Reproduzierbarkeit. Die Reproduzierbarkeit hängt aber von den Parametern der Filter zur Extraktion der Beschreibungselemente ab. Es ist nicht gewährleistet, daß die verwendeten Parameter optimal für die Extraktion reproduzierbarer Beschreibungselemente sind. Die Aufgabe besteht nun darin, die Filterparameter zu finden und auszuwählen, bei deren Einstellung die Beschreibungselemente mit den besten Reproduzierbarkeitswerten erzeugt werden.

Zur Auswahl der optimalen Filterparameter werden die Schritte 1 bis 5 nach Abschnitt 3.1 bei verschiedenen Filtereinstellungen wiederholt, wobei für jede Filtereinstellung der Mittelwert der

Erkennungsgüten über alle ausgewählten Beschreibungselemente bestimmt wird. Der Filterparametersatz, der den größten Mittelwert der Erkennungsgüten liefert, wird als optimaler Parametersatz ausgewählt.

Die Erkennungsgüten, die nach Abschnitt 3.1 bei Variation eines Filterparameters erhalten wurden, zeigt Abb. 3a. Der im Beispiel variierte Parameter ist die Mittelungsbreite des Filters zur Segmentation der Konturlinien, der an die Form des Objektes angepaßt werden muß. Bei zu kleiner Mittelungsbreite ist das Filterergebnis störanfällig, bei zu großer Mittelungsbreite zu unempfindlich. Die Kurven in der Darstellung wurden bei zu kleiner, zu großer und bei optimaler Mittelungsbreite erhalten. In Abb. 3b ist die mittlere Erkennungsgüte bei Variation des Filterparameters "Mittelungsbreite" aufgetragen. Die Kurve zeigt ein deutlich ausgeprägtes Maximum. Das Maximum wird bei dem Parameterwert erreicht, der in Abb. 3a der optimalen Rangfolgekurve entspricht.

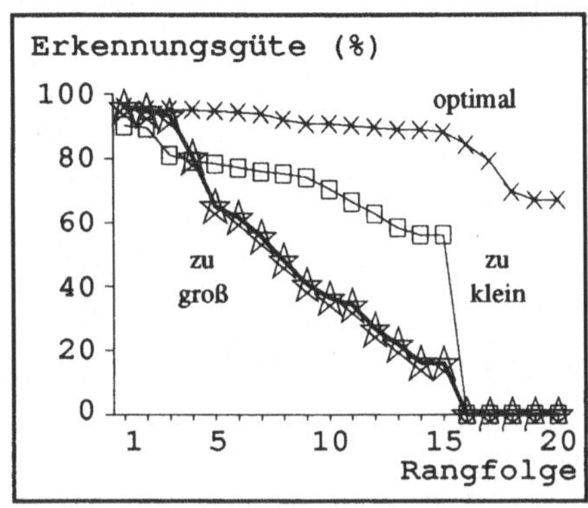

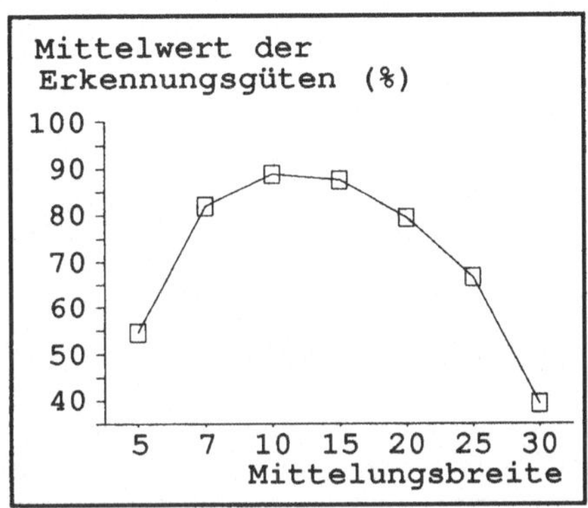

Abb. 3a: Rangfolgekurven bei Variation eines Filterparameters

Abb. 3b: Mittlere Erkennungsgüte bei Variation eines Filterparameters

3.3 Auswahl der Erkennungsstrategie

Zur Auswahl der Erkennungsstrategie geht man von Teilmodellen aus, die nur aus einem Beschreibungselement bestehen, und vergrößert diese, bis es gelingt, eindeutige Zuordnungen zwischen dem schrittweise erweiterten Modell und der Bildbeschreibung zu erhalten. Dies geschieht nach folgendem automatischen Verfahren:

Schritt 1: Aufbau von NE Teilmodellen, von denen jedes aus nur einem der NE besten Beschreibungselemente besteht, die nach Abschnitt 3.1 und 3.2 ermittelt wurden.

Schritt 2: Erkennungsexperimente mit Teilmodellen anhand der Bildbeschreibungen der Referenzbilder, die zur Ermittlung der gut reproduzierbaren Beschreibungselemente verwendet wurden.

Schritt 3: Abspeichern der Teilmodelle, mit denen nur richtige Erkennungen mit ausreichender Erkennungssicherheit erhalten wurden.

Schritt 4: Erweiterung der anderen Teilmodelle, die nicht nur korrekte Erkennungen liefern, um ein weiteres Beschreibungselement. Dabei werden nur solche Erweiterungen zugelassen, die zu Teilmodellen aus neuen, noch nicht vorhandenen Kombinationen von Modellbeschreibungselementen führen.

Schritt 5: Wiederholung der Schritte 2 bis 4, bis keine weiteren Erkennungen mehr möglich sind.

In Abb. 4 sind eine Auswahl aus 63 erhaltenen Teilmodellen zusammengestellt, mit denen an den Referenzbildern zur Modellerstellung nur richtige Erkennungen erhalten wurden. Die zu den Teilmodellen gehörenden Beschreibungselemente sind durch dicke Geradenstücke gekennzeichnet. Dem Auswahlverfahren standen die 15 besten Modellbeschreibungselemente in Abb. 2b zur Verfügung.

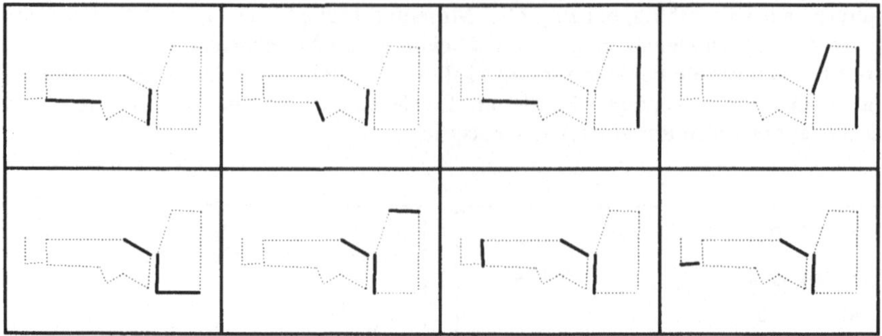

Abb. 4: Auswahl automatisch erstellter Teilmodelle

4 Ergebnisse und Ausblick

Zu jeder Teilaufgabe der automatischen Modellerstellung wurden umfangreiche Experimente durchgeführt. Es wurden Modelle für 10 verschiedene Werkstücke erstellt, die durch Geradenstücke, Ecken und Kreise beschrieben wurden. Zur automatischen Modellerstellung wurden dem Lernsystem von jedem Objekt 12 Referenzbilder angeboten, in denen die Objekte in verschiedenen, definierten Drehlagen abgebildet waren. Für jedes der angebotenen Referenzbilder wurden Bildbeschreibungen extrahiert, wobei jeweils 12 verschiedene Filterparametersätze erprobt wurden. Aus diesen verschiedenen Bildbeschreibungen wurden die Beschreibungselemente und Filterparametersätze automatisch bestimmt, die die besten Werte für die Reproduzierbarkeit ergaben. Unter Verwendung der ausgewählten, gut reproduzierbaren Beschreibungselemente wurden solche Modelle erstellt, die mit möglichst geringer Anzahl von Beschreibungselementen eine eindeutige Erkennung und Lagebestimmung erlauben.

Zur Überprüfung der Leistungsfähigkeit der automatisch erstellten Modelle wurden Erkennungsversuche mit automatisch und interaktiv erstellten Modellen durchgeführt. Von jedem Werkstück wurden dem Erkennungssystem 8 Testbilder dargeboten, in denen jeweils bis zu 4 teilweise stark überlappende Werkstücke abgebildet waren.

Als Ergebnis der Experimente läßt sich zusammenfassen:

- Die automatisch ausgewählten Beschreibungselemente und Filterparametersätze werden auch von erfahrenen Modellerstellern als gut empfunden.

- Die kombinatorische Vielfalt bereitet bei der verwendeten Optimierungsstrategie für die Modellerstellung keine Probleme. Zum Auffinden der Werkstückmodelle, die bei den Referenzszenen eine eindeutige Erkennung ermöglichen, wurden bei keinem Werkstück mehr als 200 Modelle konstruiert.

- Die automatisch erstellten Modelle sind leistungsfähig und praktisch einsatzfähig. In jeder Testszene sind mindestens zwei Werkstücke korrekt erkannt worden. Von insgesamt 174 zu erkennenden Werkstücken sind 154 richtig erkannt worden. Neben 20 stark verdeckten Werkstücken die nicht erkannt wurden, sind nur zwei Fehlinterpretationen aufgetreten. Eine ähnlich gute Erkennungsleistung wird bei interaktiver Modellerstellung nur mit viel Erfahrung erreicht.

Abb. 5 zeigt zwei Beispiele für Erkennungsergebnisse, die mit automatisch erstellten Modellen erzielt wurden. Links sind die Beschreibungselemente dargestellt, die bei den verschiedenen Werkstückmodellen verwendet wurden. Das Stanzteil wird durch Geradenstücke beschrieben (Abb. 5a), das Schlüsselblech durch Ecken (Abb. 5b). Rechts sind die Beschreibungselemente in die Testbilder eingeblendet, die zur Erkennung verwendet wurden. Die eingeblendeten Beschreibungselemente zeigen, daß alle drei Stanzteile (Abb. 5c) und alle vier Schlüsselbleche (Abb. 5d) korrekt erkannt wurden.

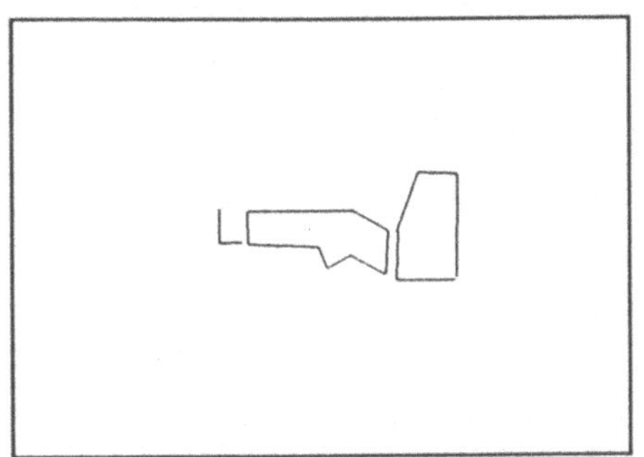

Abb. 5a: Modellbeschreibung durch Geradenstücke

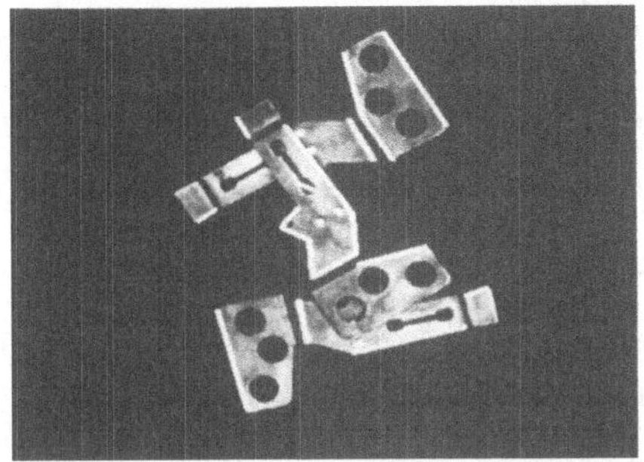

Abb. 5b: Erkannte Stanzteile

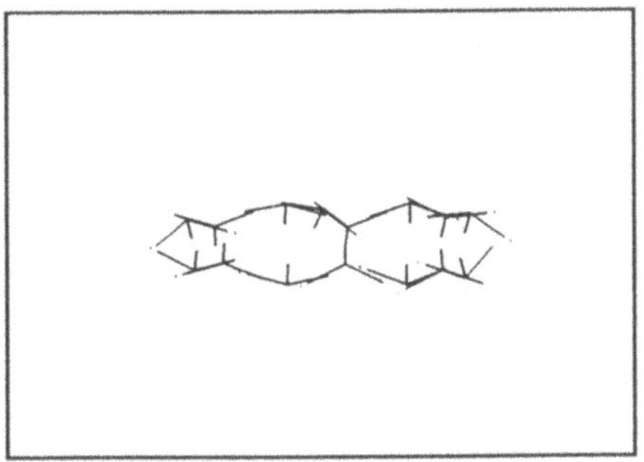

Abb. 5c: Modellbeschreibung durch Ecken

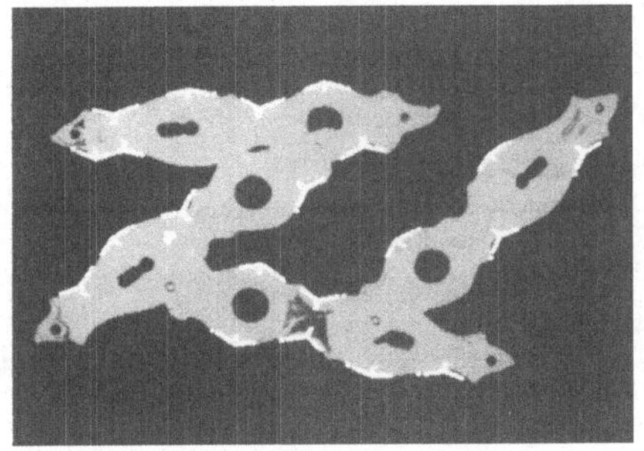

Abb. 5d: Erkannte Schlüsselbleche

Als Fortführung der Arbeiten ist eine Systemerweiterung zur Erstellung von 3-D Modellen geplant. Bei der Erweiterung auf 3-D Modelle müssen nur die Komponenten, die geometrische Operationen ausführen, modifiziert werden. Die Erweiterungsfähigkeit des Systems zur Erstellung von 3-D Modellen wurde an einfachen Beispielen bereits nachgewiesen /Stein 88/.

Die dem Beitrag zugrunde liegenden Arbeiten wurden mit Mitteln des BMFT und Partnerfirmen des Verbundprojektes "Multisensorielles System zur Deutung industrieller Szenen" durchgeführt. Die Verantwortung für den Inhalt liegt bei den Autoren.

Literatur

Asada & Brady 1986
The Curvature Primal Sketch.
Asada,H.; Brady,M.
IEEE Trans. Pattern Analysis and Machine Intelligence PAMI-8 , 1986, 2-14

Ayache & Faugeras 1986
HYPER: A New Approach for the Recognition and Positioning of Two-Dimensional Objects.
Ayache, N.; Faugeras, O.D.
IEEE Transactions on Pattern Analysis and Machine Intelligence, Vol. PAMI-8, No. 1, January 1986, 44-54

Besl & Jain 1985
Three-Dimensional Object Recognition.
Besl,P.J.; Jain,R.C.
Computing Surveys, Vol. 17, No. 1, March 1985, 75-145

Bolles & Cain 1982
Recognizing and Locating Partially Visible Objects: The Local-Feature-Focus Method.
Bolles,R.C.; Cain,R.A.
International Journal of Robotic Research, Vol. 1, Nr. 3, 1982, 57-82.

Chin & Dyer 1986
Model-Based Recognition in Robot Vision.
Chin,R.T.; Dyer,Ch.R.
Computing Surveys, Vol. 18, No, 1, March 1986, 67-108

Ender 1987
Ein Beitrag zur automatischen wissensbasierten Konfiguration von Bildinterpretationssystemen.
Ender, M.
Diss. Februar 1987, Fakultät für Maschinenwesen der Universität Hannover, in: VDI Fortschritt-Berichte, Reihe 10: Informatik/Kommunikationstechnik Nr. 68, VDI-Verlag, Düsseldorf, 1987

Ettinger 1987
Hierarchical Object Recognition Using Libraries of Parameterizised Model Sub-Parts.
Ettinger, G.J.
Technical Report AI-TR 963, Massachusetts Institute of Technology (MIT), Artificial Intelligence Laboratory, 545 Technology Square, Cambridge, MA, VA 22209

Freytag et al. 1986
Development Tools for a Model Directed Workpiece Recognition System.
Freytag,R.; Hättich,W.; Wandres,H.
Pattern Recognition, Vol. 19, No. 4, 1986, 267-278

Hättich 1988
Automatische Modellerstellung für ein wissensbasiertes Werkstückerkennungssystem.
Hättich, W.
FhG-Berichte 3-88, München, 1988, 15-17

Korn 1986
Invariante Formbeschreibung in verschiedenen Auflösungsebenen.
Korn, A.
Informatik-Fachberichte 125, 8. DAGM-Symposium Mustererkennung, Paderborn, Okt. 1986, G. Hartmann (Hrsg.), Springer-Verlag, Berlin, Heidelberg, New York, Tokyo, 1986, 89-93

Mudge et al. 1987
Automatic Generation of Salient Features for the Recognition of Partially Occluded Parts.
Mudge, T.N.; Turney, J.L.; Volz, R.A.
Robotica (1987), Vol. 5, 1987, 117-127

Stein 1988
Automatische Erstellung von geometrischen Modellen für ein Werkstückerkennungssystem.
Stein, K.
Diplomarbeit, Fachhochschule Furtwangen, Fachbereich Allgemeine Informatik, 1988

Wallace 88
A Comparison of Approaches to High-Level Image Interpretation.
Wallace, A.M.
Pattern Recognition, Vol. 21, No. 3, 1988, 241-259

Visiontool
Ein System zur Erkennung von Werkstücken

Martin Eichenberger
Institut für Informationssysteme
ETH-Zentrum, CH-8092 Zürich, Schweiz

Visiontool erlaubt nicht nur, Werkstücke zu erkennen, sondern macht den Erkennungsvorgang einsehbar: die verwendeten Merkmale, ihre Uebereinstimmung, die daraus gebildeten Hypothesen sowie die Erkennungsgüte sind schrittweise verfolgbar.

Damit unser Erkennungssystem zufällig plazierte, sich überlappende, gleiche oder verschiedene Objekte identifzieren und ihre Lage in der Szene bestimmen kann, sind zwei Voraussetzungen notwendig:

1) die mit der Videokamera digitalisierten Objekte werden nur silhouettenhaft erfasst;
2) die Objekte müssen dem System gezeigt werden, dann speichert es selbständig die Objektcharakteristiken.

Im folgenden sollen kurz die einzelnen Schritte des System Visiontool geschildert werden:

1. Binärisierung: Die Videokamera digitalisert die auf einem Leuchtpult liegenden Objekte zu einem Grauwertbild, das durch Binärisierung in ein Schwarzweissbild gewandelt wird.

2. Konturextraktion: Mit einem Konturextraktionsalgorithmus erhält man dann die Schattenumrisse der betrachteten Objekte in Form von Punktelisten.

3. Segmentierung: Um nicht mit einer Vielzahl von Konturpunkten arbeiten zu müssen, wird in einer ersten Abstraktion die Kontur in Geraden-, Kreis- und Kreisbogenstücke zerlegt. Die Konturlinie wird dazu dargestellt durch den Verlauf des Tangentenwinkel in Abhängigkeit zu der Bogenlänge. Die Linearisierung dieses Verlaufs erlaubt die Bestimmung der Segmente und eliminert gleichzeitig das Quantisierungsrauschen, das die Binärisierung verursachte. Die zur Bestimmung der Tangentenwinkel benutzte Sekantenmethode rundet scharfe Ecken ab; auch am realen Objekt treten Abrundungen scharfer Ecken auf: unsere Methode rekonstruiert in beiden Fällen die "fehlenden Ecken".

4. Merkmalsbildung:

Aus den binärisierten Silhouetten aller Modellobjekte und dem zu vergleichenden Szenenbild werden die äusseren und inneren Konturpunktelisten (Umrisse und Aussparungen) aufgebaut. Ein Objekt erscheint weniger als Folge von Geraden und Bogen, sondern als eine Kombination von Segmenten und den stärker charakterisierenden Ecken (Linien-Linien-, Linien-Bogen-, Bogen-Linien und Bogen-Bogen-Ecken) und Parallelen und konzentrischen Kreisbogen.

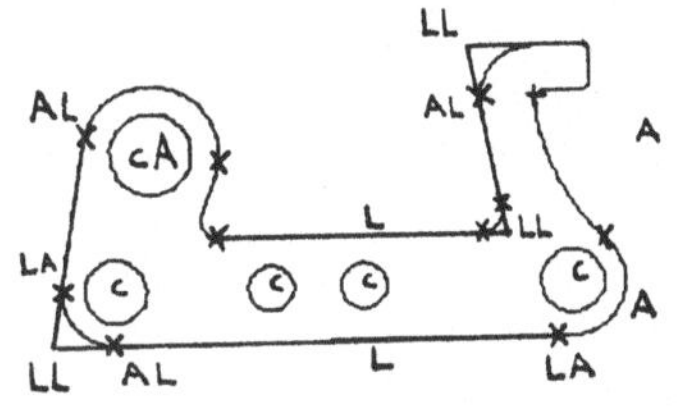

In einzelnen Fällen konstruiert Visiontool auch künstliche, nicht vorhandene Ecken, falls ein Konturstück nicht stabil segmentiert werden kann (z.B. ändernder Kurvenradius, Bruchflächen) **(Fig.1.1)**.

Solche künstlichen Ecken eignen sich zur Beschreibung besser, weil ihre Parameter "stark" sind: Eine durch zwei Geraden gebildete Ecke bleibt als Ecke mit Oeffnungswinkel, Richtung der Winkelhalbierenden und Ortskoordinaten unverändert, auch wenn die Länge der beiden sie bildenden Geraden durch Abdeckung verkürzt ist. Längen und Bogenöffnungswinkel hingegen bezeichnen wir als "schwache" Parameter, weil sie leicht abgedeckt werden können.

Schliesslich führen wir in der Merkmalsliste noch die zur Identifikation besonders wertvolle Beziehung zwischen Eckpunkten und Kreismittelpunkten ein, die wir "Struktur" nennen. Es ergeben sich dabei nicht nur verschiedene Relationen zwischen Eckpunkten und Mittelpunkten, sondern auch eine Orientierung über die Verbindungslinie dieser Punkte.

Um die durch verschiedene Lage im Bildraster bedingte Streuung der Merkmalsparameter beurteilen
zu können, wird das gleiche Modellobjekt in fünf bis zehn verschiedenen, nicht überlappenden Stellungen
der Videokamera zur Digitalisierung vorgelegt. Es werden nun die Varianzen der Parameter
korrespondierender Merkmale bestimmt. Merkmale mit zu grosser Streuung werden eliminiert. Diese
Varianzen werden später beim Modell-Szenenvergleich zur Festlegung der Uebereinstimmungstoleranz
verwendet. Dadurch ist der Benutzer nicht auf die Eingabe fester Schranken angewiesen.

5. Automatische Analyse der Merkmalsgewichtung

Um Visiontool industriell einsetzen zu können, ist es vor allem wichtig, minimale
Identifizierungszeiten zu erreichen. Dies soll durch eine zweckmässige Reihenfolge der Merkmale, nach
denen in der Szene gesucht wird sowie durch Berücksichtigung von Spiegel- und Rotationssymmetrie der
Modellobjekte ermöglicht werden. Die Zeit, die man dafür braucht, ist für eine bestimmte Objektmenge nur
einmal aufzuwenden, ist unabhängig vom Suchvorgang und fällt daher nicht ins Gewicht.

In welcher Reihenfolge müssen somit die Modellmerkmale benutzt werden, um einen optimalen
Modell-Szenenvergleich zu ermöglichen? Ihre Merkmalsparameter sollten:

1) durch eine eventuelle partielle Abdeckung wenig verändert werden und
2) eine zuverlässige Transformation ermöglichen.

Die Bestimmung der Transformation ist sehr wichtig, weil bei unserem Suchalgorithmus in jedem weiteren
Schritt geprüft wird, ob das nächste Modellmerkmal in der Szene mit der gleichen Transformation
gefunden werden kann.

Am aussagekräftigsten ist die "Struktur", da sie als komplexes Merkmal beim Modell-Szenenvergleich
besonders selektionnierend wirkt und eine sehr zuverlässige Transformationsbestimmung ermöglicht. Als
weitere Merkmale folgen konzentrische Bogen und anschliessend die aus Gerade und Kreisbogen
gebildeten vier Eckentypen. Merkmale geringer Güte sind Kreise, da sie keine Richtungsbestimmung
ermöglichen, ebenso Kreisbogen und Geraden, weil ihre Länge durch Abdeckung verkürzt wird.

Innerhalb der einzelnen Merkmalsgruppen wird danach gewertet, ob ein bestimmtes Merkmal noch in
anderen Objekten und wie häufig es gesamthaft vorkommt. Falls nämlich in einer Szene ein Merkmal
gefunden wird, das nur in einem Modellobjekt vorkommt, so muss die Szene dieses Objekt enthalten. Je
häufiger ein Merkmal im gleichen oder auch in verschiedenen Modellobjekten auftritt, desto vieldeutiger
wird seine Zuordnung; deshalb hat es für den Modell-Szenenvergleich eine tiefere Priorität.

Das Wesentliche unseres Suchalgorithmus besteht nun im folgendem: Wird ein Objektmerkmal in der
Szene gefunden, so bildet man eine "Hypothese", dass sich das Modellobjekt in der Szene befindet. Diese
Hypothese ist nur richtig, wenn das erwähnte Merkmal nur an diesem Objekt und nur einmalig vorkommt.
Findet es sich am gleichen Objekt mehrmals oder auch an andern, so müssen entsprechend weitere
Hypothesen gebildet werden. Die Besonderheit unseres Suchalgorithmus liegt darin, dass er nicht die erste
Hypothese zu Ende prüft, sondern schrittweise vergleichend sämtliche anderen möglichen Hypothesen
ebenfalls evaluiert. Es ist daher nötig, dass in der anfangs geschilderten "vertikalen" Merkmalsliste zu
jedem Merkmal "horizontal" das Vorkommen ähnlicher Merkmale am gleichen oder an andern Objekten
aufgeführt wird. Aehnliche Merkmale am gleichen Objekt können auch durch Rotations- oder
Spiegelsymmetrien entstehen, was automatisch überprüft wird. Die Zahl der horizontal aufgeführten
Merkmale und damit auch der zu überprüfenden Hypothesen kann dadurch verkleinert werden.
Weil auch spiegelbildliche Objekte erkannt werden sollen, verdoppelt sich die Anzahl der ähnlichen
Merkmale derjenigen Objekte, die keine Symmetrien aufweisen.

6. Der Modell-Szenenvergleich

Unsere Methode, die ähnliche Modellmerkmale an gleichen oder verschiedenen Modellobjekten
einbezieht, erzeugt im ersten Schritt eine Reihe von Hypothesen. Diese werden in weiteren Schritten in
bezug auf ihre Wahrscheinlichkeit miteinander verglichen. Es wird immer nur die wahrscheinlichste weiter
verfolgt, so dass der Suchweg zwischen horizontalen und vertikalen Abschnitten wechselt und zielstrebig
die richtige Lösung findet.

6.1 Grundalgorithmus:

Das Fortschreiten im Hypothesensuchbaum, wird durch eine **heuristische Bewertungsfunktion** gesteuert. Jeder Startpunkt erhält einen Uebereinstimmungswert zwischen 0 und 1, der sich aus der Merkmalsübereinstimmung ergibt. Erreicht eine Hypothse den höchsten Wert, wird das nächste zur Hypothese gehörende Modellmerkmal im Szenenbild gesucht. Findet es sich, so wird der Knotenwert um eine Konstante (optimal erwies sich 0.5) plus den neuen Uebereinstimmungswert erhöht. Findet es sich nicht, so wird der Knotenwert um dieselbe Konstante vermindert.
Es wird immer der Ast mit dem höchsten Endwert weiterverfolgt, d.h. vom entsprechenden Modellobjekt das nächste Merkmal gewählt und gesucht, ob es in der Szene gefunden werden kann.

```
FOR  ALL Modellmerkmale m DO
   FOR ALL Szenenmerkmale s with m = s DO
      CreateHypothesis h0: [(m,s)]
      FOR ALL smj ∈ m.similarfeatures DO
         IF smj = s THEN
            Bilde Hypothese hj: [(smj,s)]
         END; (* IF *)
      END;  (* FOR ALL *)
      WHILE Hypothesen vorhanden DO
         WHILE nicht alle Modellmerkmale des Modells der
               bestbewerteten Hypothese h untersucht
         DO
            Expandiere Hypothese h
         END (* WHILE *)
         Verifiziere beste Hypothese h
         IF Verifikation bestanden THEN
            Lösche alle zur Lösung beitragenden
            Szenenmerkmale aus der Szene
            Lösche alle Hypothesen
         ELSE
            Lösche beste Hypothese
         END;  (* IF *)
      END;  (* WHILE *)
   END;  (* FOR ALL *)
END;  (* FOR ALL *)
```

Jedes Mal, wenn eine Hypothese durch eine neue Korrespondenz bestätigt wird, wird der Rotationswinkel und der Translationsvektor neu berechnet. Zur Transformationsbestimmung wird ein gewichteter Mittelwert der Transformationen der bekannten Korrespondenzen berechnet. Grosses Gewicht zur Rotationsbestimmung haben lange Geraden und lange Verbindungen zwischen Kreiszentren. Zur Translationsbestimmung werden vor allem Kreise und lange Kreisbogen beigezogen.

6.2 Bestimmung der Transformation:

Bei der Expansion einer Hypothese wird ein weiteres Modellmerkmal mit der aktuellen Transformation der Hypothese in der Szene gesucht, welches mit dem Modellmerkmal der lageunabhängigen Parametern übereinstimmen und überprüft dann, ob das gefundene Szenenmerkmal in seiner Lage und Orientierung mit der bisherigen Korrespondenzliste der Hypothese keinen Widerspruch darstellt.

Je nach Vollständigkeitsgrad der Transformation unterscheidet man drei Hauptfälle:

a) *Transformation als ganzes bekannt :*

Aufgrund der bisher bekannten Korrespondenzen zwischen Modell- und Szenenmerkmalen konnte eine robuste Transformation berechnet werden. Zum Vergleich der lageabhängigen Parameter können die Modellparameter(Ort und Orientierung) in die Szene transformiert und dann dort mit den Szenenparametern auf Uebereinstimmmung getestet werden.

b) *Translation nicht bekannt :*

Bisher wurden nur Geraden-Geraden-Korrespondenzen gefunden, bei denen alle Szenengeraden nicht vollständig sichtbar sind (**Fig 1.2**). Die Orientierung ist zwar genau bestimmt, da jedoch das Zentrum einer Geraden ihr Mittelpunkt ist, kann die Translation nur sehr ungenau festgelegt werden. In diesem Fall wird die Orientierung und die relative Lage zu allen Merkmalen der Korrespondenzliste im Modell und in der Szene überprüft. Ist das gesuchte Modellmerkmal wieder eine Gerade, so wird versucht, diese Gerade mit einer anderen Geraden aus der Korrespondenzliste zu schneiden, mit der ein geeigneter Schnittwinkel gebildet werden kann. Dieser Schnittpunkt stellt eine robuste Punkt-Punkt-Korrespondenz dar, mit welcher sich nun auch die Transformation zuverlässig bestimmen lässt. Ist das neue Merkmalspaar von einem andern Typ, so ist die Translation ohnehin gegeben.

c) *Rotation nicht bekannt :*

In der Korrespondenzliste befinden sich nur Punkt-Punkt-Korrespondenzen, das heisst, es wurden erst Merkmale vom Typ Kreis, Bogen oder konzentrische Bogen gefunden . Es wird die relative Lage des Kandidatenpaares (Merkmal des Modells und der Szene) wie in Fall b) zu den bereits gefundenen überprüft. Lässt sich auch aus der neuen Korrespondenz die Rotation nicht bestimmen, so wird eine Verbindungslinie (aL: artificial line) zu einem andern Merkmalszentrum der Korrespondenzliste gebildet und dann anhand dieser die Drehung bestimmt (**Fig. 1.3**).

Wenn also aus der neu gefundenen Merkmalskorrespondenz eine nicht bekannte Teiltransformation immer noch nicht bestimmt werden kann, wird versucht, sie mit der Bildung eines virtuellen Merkmals zu berechnen. Eine künstlich eingeführte Verbindungslinie wird nur gebildet, falls sie eine gewisse Mindestlänge überschreitet. Ebenso wird ein künstlich gebildeter Schnittpunkt nur erzeugt, wenn die sich schneidenen Geraden einen deutlichen Oeffnungswinkel bilden. Diese beiden Massnahmen dienen einer robusten Bestimmung der noch fehlenden Teiltransformationen.

Die Berechnung der Transformation einer Hypothese sieht folgendermassen aus: die Spiegelung der Modellabbildung wird als erstes sogleich bei der Hypothesengenerierung festgelegt. Die anderen Teiltransformationen werden aus den Merkmalen der Korrespondenzliste durch Mittelung gewonnen. Es handelt sich dabei um ein gewichtetes Mittel, wobei die Gewichtung von der Güte der Merkmalsübereinstimmung, vom Merkmalstyp und von der Stabilität des entsprechenden Modellmerkmals abhängt. Nach jeder neu gefundenen Korrespondenz wird diese Transformation neu berechnet. Konnte eine Hypothese h verifiziert werden, so werden alle Szenenmerkmale sj der Korrespondenzliste der Hypothese h aus der Szene gelöscht.

7. Abbildungen

Abschliessend soll noch kurz über die Möglichkeiten der **Systemanalyse** gesprochen werden. Die Abbildungen lassen erkennen, aufgrund welcher Merkmale das System die einzelnen Objekte identifiziert hat. Zudem kann der Erkennungsvorgang schrittweise, filmartig verfolgt werden. Es kann auch analysiert werden, wie sich die verschiedenen Merkmale einzeln oder kombiniert zur Identifizierung von Objekten in der Szene eignen.

Modellobjekte   **Szenenbild**

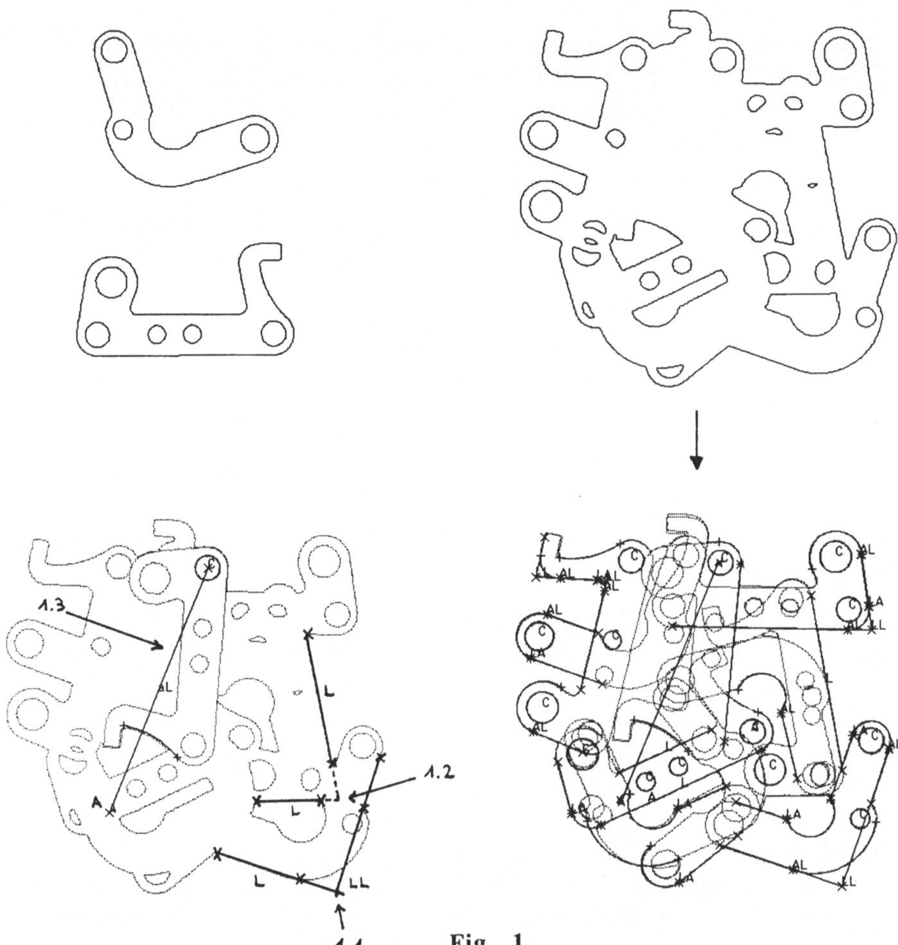

Fig. 1.

8. Literatur

Ayache, O.D. Faugeras, "HYPER: A New Approach for the Recognition and Positioning of Two-Dimensional Objects", IEEE Trans. Pattern Anal Mach. Intell. PAMI-8, No. 1, p. 44-54, January 1986

M. Eichenberger, G. Wong, "Two Contrasting Methods of Object Recognition: Geometric Features and Gabor Function Filter Values", Proc. of 6th Scandinavian Conference on Image Analysis, Oulu, June 19-22, 1989, Vol.1, p. 514-521.

W. Hättich, "Recognition of Overlapping Workpieces by Model Directed Construction of Object Contours", *Digital Systems for Industrial Automattion*, Vol. 1, Nr. 2-3, p. 223-239, 1982

F. Knoll und R. C. Jain, "Recognizing Partially Visible Objectw Using Feature Indexed Hypotheses", *IEEE Journal of Robotics and Automation*, Vol. RA-2, No. 1, (1986), p 3

Erkennung von chinesischer Druckschrift anhand der ''Schwarzsprungverteilung in angepaßten Teilbildern'': Verfahrensbeschreibung und Merkmalsanalyse

Jun Guo, Irmfried Hartmann, Richard Suchenwirth, Zhang Zheng

Technische Universität Berlin, Institut für Regelungstechnik und Systemdynamik,

Interdisziplinäres Forschungsprojekt 1074 TECHIS

Abstract— *''Schwarzsprungverteilung in angepaßten Teilbildern'' ist ein vom stroke density feature abgeleitetes neues Merkmalsextraktionsverfahren zur Erkennung von chinesischen und alphabetischen Druckschriften. Die Klassifikation erfolgt dynamisch anhand der city-block-Distanz. In einem Versuch über 19.241 Zeichenvorkommen aus 1544 Klassen (Gesamtzeichensatz 3755 Klassen) konnte eine Erkennungsrate von 98,21 % erreicht werden. Weiter wird ein neues Verfahren zur Effizienzmessung von mehrdimensionalen Merkmalen und somit zur Merkmalsanalyse vorgestellt.*

1. Einführung

Seit den ersten Versuchen zur optischen Erkennung von chinesischer Druckschrift [1] sind vielfältige Forschungsarbeiten auf diesem Gebiet geleistet worden, die verwendeten Verfahren haben allerdings die Praxisreife noch nicht erreicht. Die größten Probleme der chinesischen Schrift liegen im großen Umfang·des Zeichensatzes von mindestens 3.000 bis weit über 10.000 Zeichen und der meist erheblichen Komplexität der Zeichen, die sich voneinander teilweise nur in geringen Details unterscheiden.

An der Technischen Universität Berlin arbeitet eine interdisziplinäre Forschungsgruppe ''Teilautomatische Erkennung von chinesischer Schrift'' (TECHIS) an der Entwicklung und Bewertung von effizienten Verfahren zur optischen Erkennung von chinesischer Schrift. Der Arbeitszeichensatz umfaßt derzeit 3755 Zeichen und wird in der gegenwärtigen zweiten Projektphase auf 6763 Zeichen erweitert (Standardcodeplan GB 2312-80 der VR China). Fehlerkennungen sollen durch Ausnützung des Textzusammenhangs, wie Zeichen-, Übergangs- und Worthäufigkeiten, korrigiert werden. Die Eingabe erfolgt mittels CCD-Kamera oder Scanner. In der Vorverarbeitungsphase wird das eingelesene Bild in Zeilen und Zeichen segmentiert. Die Einzelzeichen werden in einer 64×64-Binärmatrix dargestellt, wobei Größennormierung und Glättung wahlweise erfolgen können.

In der ersten Projektphase wurde eine Reihe von effizienten Merkmalsextraktionsverfahren [2-3] untersucht und verbessert. Erkennungsversuche mit diesen Verfahren wurden über eine größere Textmenge durchgeführt, wodurch die Verfahren direkt verglichen werden konnten [4]. Ein neues Merkmalsextraktionsverfahren, Schwarzsprungverteilung in angepaßten Teilbildern (SSV-AT), wurde ebenfalls entwickelt.

2. Schwarzsprungverteilung

Das Merkmalsverfahren SSV-AT leitet sich aus dem bereits publizierten Verfahren 4-SDF (stroke density function in four directions) [5] ab. Es wird die Anzahl der Schwarzsprünge in vier Richtungen verwendet, wobei je Richtung eine geeignete Zerlegung in acht Teilbildern erfolgt. Jedes Teilbild liefert einen Merkmalswert, so daß ein Vektor von 4×8=32 Elementen entsteht. Diese Merkmalswerte geben die lokale Komplexität des Zeichens wieder. Um die Robustheit des 4-SDF-Verfahrens gegenüber verschiedenen typischen Störungen zu erhöhen, enthält nach dem SSV- AT-Verfahren jedes Teilbild die gleiche Anzahl schwarzer Pixel, während sie sich in der Größe durchaus unterscheiden können.

Gegeben sei ein quadratisches Binärbild $B(i,j)$ mit s Zeilen und s Spalten. Die Pixel können jeweils die Werte 0 (:weiß) oder 1 (:schwarz) annehmen.

Durch die folgende Transformation $\underline{T}_\mu$ wird das Bild $B(i,j)$ in einen Vektor $\underline{V}_\mu$ mit den Komponenten $V_\mu(k), k = 1, ..., s^2$ umgewandelt.

$$\underline{B} \in \Re^{s \times s} \xrightarrow{\underline{T}_\mu} \underline{V}_\mu \in \Re^N \qquad \text{mit} \qquad N := s^2, \mu = 1, ..., 4$$

Gegenwärtig werden vier Umordnungsvorschriften (horizontale, vertikale, und zwei diagonale Richtungen) zur Berechnung des Vektors $\underline{V}_\mu$ verwendet. Die Transformationen $\underline{T}_\mu$ wurden explizit in [4] dargestellt. Es sei $m_{00}(\underline{V}_\mu)$ die Anzahl der schwarzen Pixel in $\underline{V}_\mu$ während die Anzahl der Schwarzsprünge in $\underline{V}_\mu$ durch

$$N_j(\underline{V}_\mu) = \sum_{k=2}^{N} \sigma[V_\mu(k) - V_\mu(k-1)]$$

mit der Sprungfunktion $\sigma[x]$ gegeben ist. Weiterhin gelte für n die Bedingung $1 \leq n \leq m_{00}(\underline{V}_\mu)$, so daß man die Zerlegung

$$\underline{V}_\mu = [\underline{T}'_{\mu 1} \cdots \underline{T}'_{\mu n}]'$$

erhält. Der Merkmalswert $x_\mu(j)$ des j-ten Teilvektors ($j = 1, ..., n$) in der μ-ten Richtung ($\mu = 1, ..., 4$) gibt das Verhältnis der Schwarzsprünge im Teilbild $\underline{T}_{\mu j}$ zur Gesamtzahl der Schwarzsprünge in dem umgeordneten Bild $\underline{V}_\mu$ an, d.h.

$$x_\mu(j) = \frac{N_j(\underline{T}_{\mu j})}{N_j(\underline{V}_\mu)}$$

Die Division durch die Gesamtzahl der Schwarzsprünge bewirkt, daß der Wertebereich des Merkmals unabhängig von der Komplexität des Zeichens im Intervall $[0,1]$ liegt.

Die Gesamtzahl der berechneten Merkmale ist μn und für die Merkmale gilt

$$\sum_{j=1}^{n+1} x_\mu(j) = 1 \qquad \mu = 1, ..., 4$$

3. Klassifikation

Bei der Zeichenerkennung über große Zeichensätze wird oft eine hierarchische Klassifikation eingesetzt, bei der dem zu erkennenden Zeichen zunächst mit relativ schnellen Verfahren eine Vorklasse zugeordnet wird. Aus den Zeichen in der Vorklasse (Kandidaten) wird dann mit anderen Merkmalsverfahren das endgültige Ergebnis bestimmt.

Im Projekt TECHIS gehen wir von einer dynamischen Vorklassifikation relativ zum eingegebenen Zeichen aus. Verschiedenen Vorkommen des gleichen Zeichens können dabei unterschiedliche Vorklassen entsprechen - solange nur das korrekte Zeichen in der Vorklasse enthalten ist. Als Abstandsmaß zwischen dem Merkmalsvektor $\underline{X}$ des zu erkennenden Zeichens und dem Referenzvektor $\underline{X}_i$ der i-ten Zeichenklasse wurde die l_1-Distanz gewählt($N = \mu n$):

$$\|\underline{X}_i - \underline{X}\| = \sum_{j=1}^{N} |x_i(j) - x(j)|$$

Wird das Zeichen $\underline{X}$ der Zeichenklasse zugeordnet, zu der der Abstand am kleinsten ist, liefert dieses einstufige Klassifikationsverfahren bereits ein endgültiges Erkennungsergebnis, weitere Einzelheiten in [4].

4. Versuchsergebnisse

Für alle 3.755 Zeichen in Level 1 des chinesischen Codeplans GB 2312-80 wurde je ein Referenzzeichen auf eine Größe von 64×64 Pixeln normalisiert, binarisiert und in einer Referenzzeichendatei abgespeichert.

Um die Erkennungseigenschaften zu untersuchen, wurde eine Stichprobe von 15 Seiten aus einer chinesischen Zeitschrift (*Xiandaihua*), die insgesamt 19.241 Vorkommen chinesischer Zeichen enthielt, mit Kamera eingelesen. Die Zeichen wurden ebenfalls größennormiert und binarisiert und in einer Datei pro Vorlagenseite abgespeichert. Die Klassifikation erfolgte einstufig ohne Plausibilitätsprüfungen. Die Erkennungsraten für einzelne Seiten lagen zwischen 96,06 % und 99,05 %, wobei eine Gesamterkennungsrate von 98,21 % erreicht wurde.

In einem weiteren Versuch wurden 3.413 verschiedene Zeichen aus den chinesischen Normblättern GB 2312-80 mit einem Flachbettscanner mit einer Auflösung von 400 *dpi* erneut aufgenommen. Es handelte sich hier also um die gleichen Vorlagen, die jedoch mit unterschiedlichem Gerät erfaßt wurden. 3.396 Zeichen wurden korrekt erkannt, was einer "Wiedererkennungsrate" von 99,5 % entspricht.

5. Merkmalsanalyse

Wenn in Erkennungsverfahren mehrere Merkmale verwendet werden, ist die Effizienz der einzelnen Merkmale zu untersuchen. Hier können verschiedene Probleme auftreten: die Streuung der Vorkommen des gleichen Zeichens kann den Abstand zwischen verschiedenen Zeichenklassen übersteigen, oder die Korrelation zwischen zwei Merkmalen x_i und x_j kann groß sein, so daß beide zusammen nicht mehr Information über die Zeichenklasse vermitteln als jedes allein (z.B. Schwarzsprünge und Weißsprünge). Werden weniger Merkmale verwendet, so benötigt die Erkennung weniger Zeit, Speicherplatz und evtl. auch weniger Hardware; andererseits wächst die Erkennungsrate durch eine Erhöhung der Merkmalszahl, und nach Überschreiten einer optimalen Merkmalszahl sinkt die Erkennungsrate wieder ab.

Beim Entwurf eines Erkennungssystems ist die Klassentrennbarkeit ein nützliches Maß. Es ist eine Bewertung der Klassifikationseffizienz eines Merkmals oder einer Gruppe von Merkmalen vorzunehmen. Eine Kombination von Merkmalen nach verschiedenen Verfahren ist geeignet, wenn diese aufgrund von Untersuchungen eine höhere Effektivität liefert.

Unter den Annahmen einer vernachlässigbaren Korrelation der Merkmale untereinander und einer GAUSSschen bedingten Verteilungsdichte gilt näherungsweise

$$p(x|\omega_i) = \frac{1}{\sqrt{2\pi}\sigma_i} e^{-\frac{(x-\mu_i)^2}{2\sigma_i^2}}$$

wobei μ_i und σ_i der bedingte mittlere Erwartungswert bzw. die Standardabweichung der i-ten Klasse sind. Diese Parameter können rekursiv geschätzt werden [4].

Anhand der bedingten Verteilungen von Merkmalen, z.B. in Bild 1 für das erste, kann die Klassentrennbarkeit grob abgeschätzt werden [4].

Praktische Erkennungssysteme benötigen in der Regel mehrere Merkmale, um gute Ergebnisse zu erzielen. Offensichtlich gestaltet sich die Effizienzanalyse der Klassentrennung hier schwieriger. In [6] wurde eine Reihe von Kriterien zur Merkmalsbewertung für N-dimensionale Merkmalsräume erörtert. In einem Ansatz, die Klassentrennung für N-dimensionale Merkmalsräume darzustellen, wird die Verteilung von Mustern im N-dimensionalen Raum mittels orthogonaler Transformationen auf einen zweidimensionalen Beobachtungsraum abgebildet [7].

Hier wurde ein Verfahren entwickelt, bei dem man die Abstandsverteilung zwischen den einzelnen Zeichenklassen auswertet. Aus diesen Diagrammen lassen sich die Merkmalsgüte und geeignete Schwellwerte ablesen.

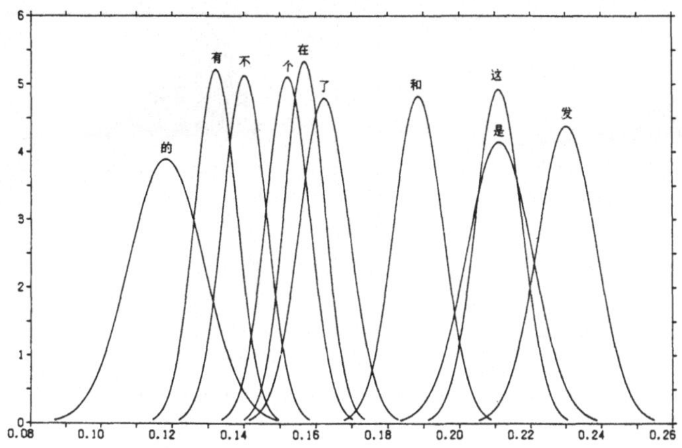

Bild 1 Verteilung des 1. Merkmals
"Schwarzsprünge in horizontaler Richtung" bei 10 chinesischen Zeichen

Es sei angenommen, daß die Merkmalsvektoren jeder Klasse ein bestimmtes Gebiet im Merkmalsraum einnehmen.

Sei $X_{il} = \{x_{il}(1), x_{il}(2), ..., x_{il}(N)\}$ das l-te Muster der i-ten Klasse im N-dimensionalen Merkmalsraum, $i = 1, 2, ..., K; l = 1, 2, ... L_i$.

Eine Metrik $\delta(X_{ik}, X_{jl})$ zur Messung des Abstands zwischen dem k-ten Muster der i-ten Klasse und dem l-ten Muster der j-ten Klasse für verschiedene Merkmalskombinationen kann auf verschiedene Weisen definiert werden [4], z.B. als

$$\delta_C(X_{ik}, X_{jl}) = \sum_{n=1}^{N} \left| x_{ik}(n) - x_{jl}(n) \right| f_c(n)$$

wobei f_c anzeigt, welche Merkmale kombiniert werden. So bedeutet

$$\{f_c(1), ..., f_c(N)\} = \{\ 00000000000000000001010000000000\ \},$$

daß das 20. und das 22. der 32 SSV-AT-Merkmale verwendet werden. Beispiele für Abstandsverteilungen δ_C für 3755 häufige chinesische Zeichen (Level 1 des GB 2312-80) und einige Kombinationen der Einzelmerkmale aus SSV-AT werden in Abb. 2 gezeigt.

Im oberen Teil der Diagramme (h_1) ist die Abstandsverteilung zwischen allen Paaren von 3755 Zeichenklassen und je einem Muster pro Klasse für eine gegebene Merkmalskombination dargestellt. Im unteren Teil der Diagramme (h_2) ist die Verteilung der Abstände von 10 Klassen für je 100 Muster dargestellt. Der Überlappungsbereich der beiden Kurven zeigt an, inwieweit Substitutionen (Falscherkennungen) möglich sind. Solche Diagramme wurden für eine größere Zahl von Kombinationen aus den SSV-AT-Merkmalen erstellt. Wenn ein euklidisches Abstandsmaß verwendet wird, ist die Verteilungsfunktion eine χ^2-Verteilung, deren Parameter geschätzt werden können. Wie aus Bild 2 zu erkennen, ist die Fehlerwahrscheinlichkeit

$$e = \int_{\theta}^{+\infty} h_2 d\delta + \int_0^{\theta} h_1 d\delta$$

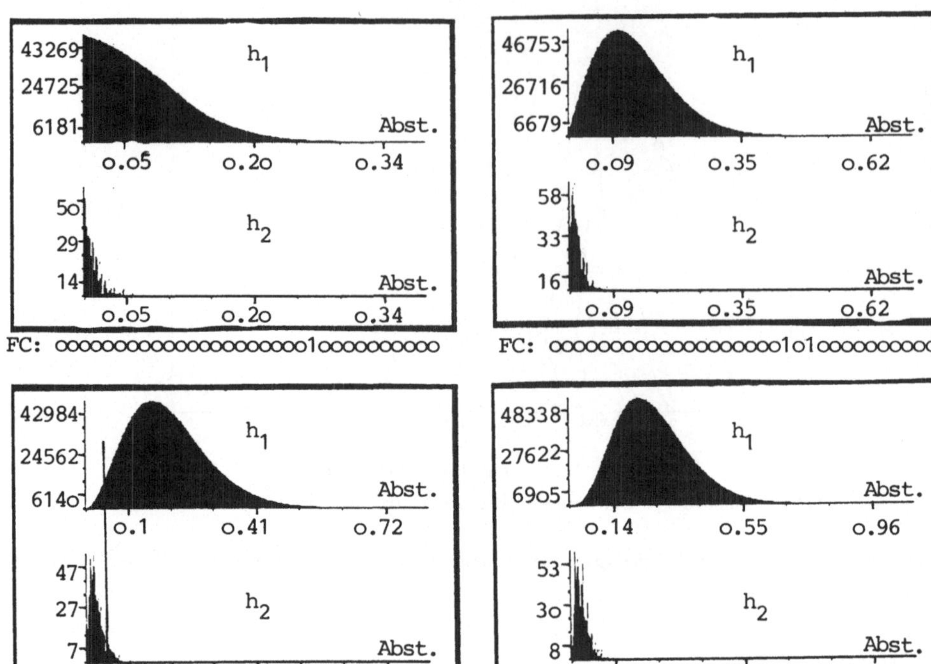

Bild 2 Abstandsverteilungen δ_C

6. Schluß

Das hier beschriebene Merkmalsverfahren SSV-AT hat sich als geeignet für die Erkennung von Zeichen aus einem großen Zeichensatz (hier 3.755 Zeichen) erwiesen, während es sich gleichzeitig gegen typische Störungen (Schmutzpunkte, Größenunterschiede, Verschiebungen) und Unterschiede zwischen verschiedenen Eingabegeräten sehr robust zeigte. Schon mit einem Lernmuster je Klasse konnte eine hohe Erkennungsrate von 98.21% erreicht werden. Durch mehrmaliges Lernen (Berechnung des rekursiven Durchschnitts mehrerer Vorkommen) konnte die Erkennungsrate gesteigert werden. Durch Effizienzanalyse der verwendeten Merkmale läßt sich die zur Erkennung benötigte Merkmalszahl von 32 bis auf ca. 10 reduzieren, ohne die Erkennungssicherheit erheblich zu beeinträchtigen.

7. References

[1] R. Casey, G. Nagy, "Recognition of printed Chinese characters", *IEEE Trans. Elect. Comput.* *15*, 91-101 (1966)

[2] M. Umeda, "A multifont printed Chinese character reader", *Trans. IECE Japan*, vol. J64-D, No. 8, pp. 908-915, 1984

[3] N. Hagita, I. Masuda, "Classification of handprinted Chinese characters by global and local stroke density", Trans. IECE Japan, vol. PRL 80-23, 1980

[4] R. Suchenwirth, J. Guo, I. Hartmann, G. Hincha, M. Krause, Z. Zhang, "Optical Recognition of Chinese Characters". *(Advances in Control Systems and Signal Processing, vol. 8)*. Wiesbaden: Vieweg 1989

[5] N. Hagita, M. Umeda, I. Masuda, "Classification of hand-printed Chinese characters by stroke analysis and two other features". (In Japanese, English abstract). *Papers of IECEJ Technical Group on Pattern Recognition and Learning*, PRL 79-27 (1979)

[6] J. Kittler, *Handbook of Pattern Recognition and Image Processing*, 61-70, (1986)

[7] Toshihiko Okada, Shingo Tomita, "An Optimal Orthogonal System for Discriminant Analysis", *Pattern Reconition*, Vol 18, No.2, 139-144, (1985)

Zur Erkennung von Bildstrukturen durch Analyse der Richtungen des Grauwertgradienten

Axel Korn

Fraunhofer-Institut für Informations- und Datenverarbeitung (IITB)

Fraunhoferstraße 1, D-7500 Karlsruhe

Zusammenfassung

In der vorliegenden Arbeit wird ein Verfahren zur Detektion von Bildbereichen mit örtlich sich langsam ändernden Grauwerten (Schatten,Schattierung) beschrieben sowie ein Verfahren zum Wiederauffinden gelernter Bildstrukturen (z.B. Fahrzeuge,Texturen) durch Auswertung eindimensionaler Histogramme der Richtungen des Grauwertgradienten (Winkelauflösung: 2 Grad). Das Erkennungsverfahren basiert auf einer Kreuzkorrelation zweier Winkelhistogramme. Nach jeder Korrelation werden die Winkelwerte des (gelernten) Referenzhistogramms um 1 oder einen beliebigen anderen Wert erhöht (modulo 360) und mit der verschobenen Funktion erneut korreliert. Bei dieser Vorgehensweise werden mögliche Drehungen des Vergleichs- gegen das Referenzmuster quantitativ erfaßt. Invarianz gegenüber Kontrastumkehrung ergibt sich aus der Tatsache, daß hierbei zu jedem Winkelwert der ursprünglichen Verteilung 180 Grad addiert wird. Die Größeninvarianz ergibt sich durch Normierung auf die Gesamtzahl der Bildpunkte in dem ausgewerteten Bildfenster. Invarianz gegenüber der Addition oder Multiplikation der Bildfunktion mit einer Konstanten ergibt sich aus der Vorschrift zur Berechnung des Gradientenwinkels. Die bisherigen Anwendungen auf Bildfolgen, Luftbilder und Texturmuster zeigen. daß gelernte Muster trotz erheblicher Größenänderungen in jeder Drehlage zuverlässig lokalisiert werden können.

1.   Einleitung

In den folgenden Abschnitten wird beschrieben, wie die Richtungen des Grauwertgradienten für die Lösung zweier unterschiedlicher Problemstellungen herangezogen werden können: Zur Detektion von Schattierungen und Schatten sowie zum Wiederauffinden gelernter Bildmuster in natürlichen Grauwertbildern. Die Winkelinformation wird mit Hilfe eines im IITB entwickelten Verfahrens gewonnen [1], [2], bei welchem der normierte Grauwertgradient für verschiedenene Skalierungen berechnet wird. Mit Hilfe zweier Masken wird der Gradientenwinkel mit 8 bit Auflösung berechnet, wobei wir später diese 256 Werte den 180 Winkelwerten 2,4,...,360 Grad zuordnen, also mit 2 Grad Winkelauflösung rechnen. In der vorliegenden Arbeit werden Faltungskerne mit 5 x 7 bzw. 11 x 13 Elemente benutzt. Bei diesen Maskengrößen wird das Digitalisierungsrauschen stark vermindert. Die größere der beiden Masken dient zusätzlich zur Reduktion störender Texturen innerhalb von Schatten-/ Schattierungsbereichen, wobei die Größe der Abschwächung von den betreffenden Ortsfrequenzen abhängt. Den weiteren Überlegungen wird die Häufigkeitsverteilung von Winkelwerten in verschiedenen Bildfenstern zugrundegelegt, deren Größe und Position einstellbar ist. Für Schatten und Schattierungen wird eine Vorzugsrichtung der Gradientenrichtung in einem Bildbereich erwartet, der deutlich größer als der Faltungskern sein sollte. Das entsprechende Winkelhistogramm hat in diesem Fall nur ein signifikantes Maximum. Für die zweite Problemstellung, d.h. das Wiederauffinden gelernter Bildmuster, bietet sich das

Winkelhistogramm als eine Art Fingerabdruck einer bestimmten Bildstruktur an. Angestrebt wird ein Erkennungsergebnis, welches invariant gegenüber möglichen Drehungen, Größen- und Kontraständerungen des gelernten Musters ist. Die Kreuzkorrelation im Winkelbereich erfüllt diese Invarianzforderung, wenn das Referenzmuster, d.h. das eindimensionale Histogramm "Häufigkeit der Winkel vs. Winkel" in geeigneter Weise bezüglich der Abszisse verschoben wird. Der Einsatz der beschriebenen Verfahren bietet sich u. a. zur Lösung folgender Aufgaben an: Wiederauffinden gespeicherter Bildmuster (z.B. Fahrzeuge, Kreuzungen, Texturen etc.) in Bildfolgen für die Zielverfolgung, zur Fehlererkennung bei der automatischen Oberflächenprüfung sowie zur Texturunterscheidung bei der Bildsegmentierung.

2.   Detektion von Bildbereichen mit örtlich sich langsam ändernden Grauwerten

In [1], [2] wird ausführlich auf die Problematik der Beschreibung von Grauwertänderungen mit Hilfe von Gradientenfiltern variabler Größe eingegangen. Es wurde eine Filterbank vorgeschlagen, in welcher über ein Auswahlkriterium automatisch die jeweils beste Maskengröße bestimmt werden kann. Aus Aufwandsgründen sollte man sich jedoch auf wenige Maskengrößen beschränken, welche mit Hilfe von a priori Wissen bez. der zu bearbeitenden Szene festgelegt werden können. Der Grundgedanke für das im folgenden beschriebene Verfahren besteht darin, durch eine relativ große Filtermaske Bildrauschen und Texturen soweit zu eliminieren, daß zusammenhängende Flächen mit möglichst homogenen Gradientenrichtungen entstehen. Diese Flächen sollten wesentlich größer sein als der verwendete Faltungskern. Örtlich sich langsam ändernde Grauwertverläufe werden aufgrund ihrer niedrigen Ortsfrequenz kaum beeinflußt und lassen sich in dem bandpaßgefilterten Bild relativ einfach durch die einheitliche Gradientenrichtung innerhalb des ansteigenden bzw. abfallenden Bereiches der Grauwertfläche detektieren. Schatten und Schattierungen führen zu einem signifikanten Maximum im Histogramm der Gradientenrichtungen. Dieses Maximum kann in Verbindung mit einem gewissen Toleranzbereich dazu benutzt werden, zusammenhängende Flächen im Bildbereich zu gewinnen und als mögliche Schatten- oder Schattierungsbereiche zu charakterisieren., wobei eine solche Interpretation i. a. Wissen über den Kontext erfordert. Es wird davon ausgegangen, daß diese Vorgehensweise bestehende Verfahren ergänzen kann (siehe z.B. [3]) oder Hinweise für die menschliche Formwahrnehmung aus Schattierung liefert (siehe z.B. [4]).

Zur Gradientenberechnung bei den im folgenden gezeigten Beispielen wurden 11 x 13 Masken verwendet, das entspricht einer Standardabweichung von σ = 2.0 für das Gradientenfilter (normierte Ableitung einer Gaußfunktion). Für das Beispiel des PKW in Abb. 1a) veranschaulichen Abb. 1b) und c) die beschriebene Vorgehensweise. In b) sind die helligkeitskodierten Gradientenrichtungen (siehe Beschreibung weiter unten) dargestellt. Die Auswertung des Histogramms für den markierten Bildausschnitt in b) ergibt, daß 95 % der betreffenden Bildpunkte Winkelwerte im Bereich 88 ± 16 Grad haben. Diese Winkelwerte sind in c) weiß dargestellt. Als größter zusammenhängernder Bereich ergibt sich der Schatten vor dem PKW. Wie aufgrund der Beleuchtungsbedingungen zu erwarten war, ergeben sich zusätzliche Bereiche, z.B. auf der Kühleroberfläche, welche Hinweise auf das Vorhandensein langsam veränderlicher Grauwertänderungen liefern. Die Konvention für den Gradientenwinkel geht aus Abb. 2c) hervor. Die Gradientenrichtung ist so festgelegt, daß der Vektor stets von dunkel nach hell zeigt, wobei der Winkel im Uhrzeigersinn zunimmt. Wie bereits im 1. Abschnitt beschrieben wurde, wird der gesamte Winkelbereich auf das Intervall [1,180] abgebildet, was einer Winkelauflösung von 2 Grad entspricht. Für die bildhafte Darstellung der Winkel (Abb. 1b, d) werden diesen 180 Werten Grauwerte mit denselben Zahlenwerten zugeordnet.

Abb. 1d) zeigt in der oberen Hälfte zwei einfache Objekte mit gekrümmten Oberflächen, deren Schattierung deutlich erkennbar ist. Die helligkeitskodierten Richtungen (σ = 2.0) sind in der unteren Hälfte von d) dargestellt. Das Glanzlicht auf der Oberfläche des kegelförmigen Objektes ist deutlich als Grenze zwischen zwei Gradientenrichtungen erkennbar, da diese sich um etwa 180 Grad unterscheiden. 94% aller Punkte in der oberen (dunkleren) Kegelhälfte liegen im Winkelbereich 130 ± 16 Grad.

3.   Lokalisierung gelernter Bildmuster in Grauwertbildern bei einer beliebigen Drehlage durch (eindimensionale) Korrelation von Richtungshistogrammen

Im Unterschied zu der im 2. Abschnitt beschriebenen Vorgehensweise werden im folgenden nicht signifikante Maxima des Winkelhistogramms ausgewertet, sondern das vollständige Histogramm der Gradientenrichtungen eines vorgegebenen Grauwertmusters als Modell oder Referenzmuster verwendet. Die größte Ähnlichkeit mit einem solchen Referenzmuster wird durch eine Kreuzkorrelation mit den Winkelhistogrammen anderer Gauwertmuster (Vergleichsmuster in einem Suchbild) ermittelt. Zunächst wird das Histogramm der Gradientenrichtungen in einem interessierenden Bildbereich berechnet. Man erhält eine eindimensionale Funktion: Häufigkeit von Winkelwerten vs. Winkel (Grad). Zur Rauschverminderung wird das Histogramm durch Faltung mit einer eindimensionalen Gaußfunktion, deren Breite einstellbar ist, geglättet. In Abb. 2b) ist ein solches geglättetes Histogramm für die Gradientenrichtungen (σ = 1.0) des Kastenwagens in a) dargestellt. Die Glättung erfolgte mit einer 1 x 9 Maske (σ = 2.0). Die vier Maxima des Histogramms bei 92, 180. 270, und 360 Grad repräsentieren die vertikal und horizontal orientierten Vorzugsrichtungen des Fahrzeugs. Abb.2a) ist ein Ausschnitt aus dem 1. Bild einer Bildserie, in welcher sich der Kastenwagen weg von der Kamera bewegt. Das geglättete Histogramm der Gradientenrichtungen von a) wurde als Referenzmuster verwendet, um das Fahrzeug im 300. Bild dieser Bildfolge automatisch zu lokalisieren. Trotz einer Größenänderung von 3:1 des Fahrzeugs und einer (künstlich) durchgeführten Drehung von 45 Grad befindet sich der Kastenwagen weitgehend innerhalb des Fenster mit dem größten Kreuzkorrelationskoeffizienten r = 0.869. Zusätzlich erhält man das Ergebnis, daß der höchste Kreuzkorrelationskoeffizient mit dem um 46 Grad verschobenen Referenzhistogramm erzielt wurde.

Drehungen des Vergleichs- gegenüber dem Referenzmuster lassen sich sehr einfach dadurch berücksichtigen, daß nach jeder Korrelation die Winkelwerte auf der Abzisse eines der beiden zu korrelierenden Histogramme jeweils um 1 oder einen beliebigen anderen Wert modulo 360 Grad erhöht werden und mit diesem Histogramm erneut eine Kreuzkorrelation durchgeführt wird. Mit Hilfe dieses Schrittes werden mögliche Drehungen der o.g. Muster quantitativ erfaßt durch Bestimmung des maximalen Wertes des Kreuzkorrelationskoeffizienten in Abhängigkeit von der oben durchgeführten Translation, welche dem Drehwinkel entspricht.

Weitere Anwendungen des beschriebenen Korrelationsverfahrens zeigten die Leistungsfähigkeit des Verfahrens bei der Texturklassifikation, welche u. a. bei der Luftbildauswertung oder bei industriellen Prüfaufgaben (siehe z.B. [5]) von großer Relevanz ist. Der Einfluß von Glättungsoperationen auf das Korrelationsergebnis sowie eine genaue Analyse der Klassifikationsergebnisse sind Gegenstand weiterer Untersuchungen. Zusätzliche Gesichtspunkte ergeben sich durch das geplante Einbeziehen von Histogrammen der Gradientenbeträge. Hauptzweck der relativ knappen Darstellung innerhalb dieses Beitrags war, mit Hilfe einiger repräsentativer Beispiele zu verdeutlichen, daß mit einfachen Methoden, die leicht in Form von Spezialprozessoren zu realisieren sind, eine Vielzahl von praxisrelevanten Aufgaben lösbar erscheinen.

Diese Arbeit wurde vom Bundesminister der Verteidigung gefördert.

Literatur

[1]   A. Korn: Combination of different spatial frequency filters for modeling edges and surfaces in gray-value pictures. SPIE vol. 595, 1985, pp. 22-30

[2]   A. Korn: Toward a symbolic representation of intensity changes in images. IEEE Trans. Pattern Analysis and Machine Intelligence PAMI-10, 1988, pp. 610-625

[3]   A.P. Pentland: Local shading analysis. IEEE Trans. Pattern Analysis and Machine Intelligence PAMI-6, 1984, pp. 170-187

[4]   V.S. Ramachandran: Formwahrnehmung aus Schattierung. Spektrum der Wissenschaft Nr.10, Oktober 1988, S. 94-103

[5]   M.S. Tatari: Auswahl von Texturanalyseverfahren zur automatischen industriellen Sichtprüfung, Dissertation an der Fakultät für Elektrotechnik der Univ. Karlsruhe, 1986, Fortschrittsberichte VDI Nr. 67, VDI-Verlag, Düsseldorf

a) Original

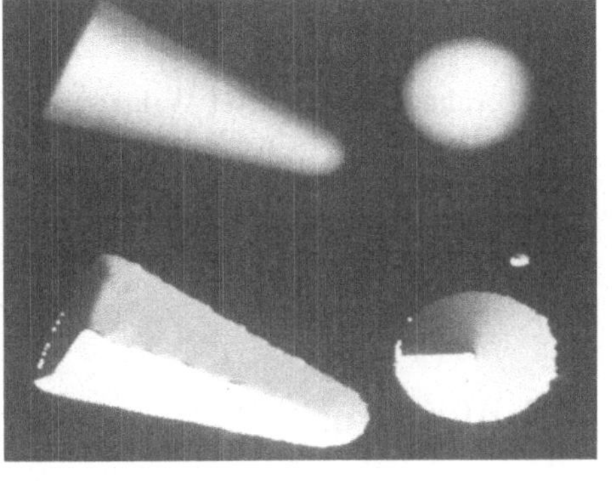

b) Gradientenrichtung ($\sigma = 2.0$)

c) Weiß: 88 ± 16 Grad Bereiche

d) Original und Gradientenrichtung

Abb. 1: Charakterisierung von Schatten- und Schattierungsregionen im Winkelbereich. Faltungskerne waren 11 x 13 Masken (siehe Text).

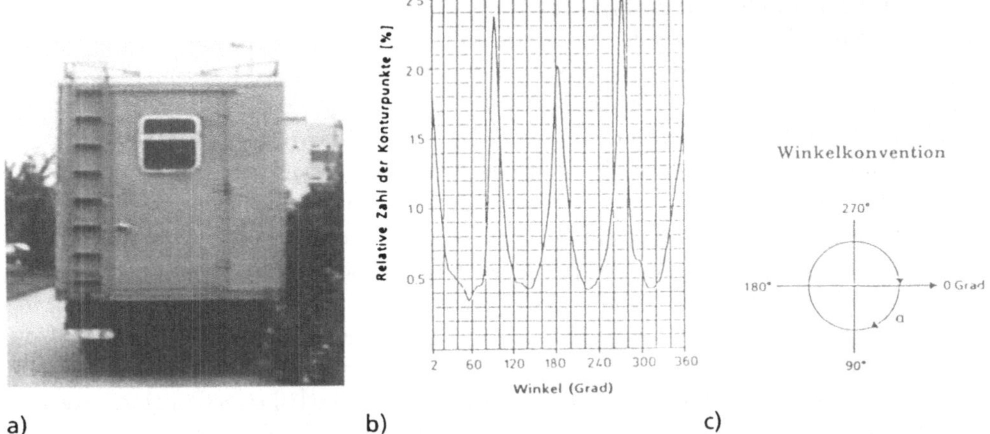

Abb. 2: Größen- und drehinvariante Detektion. In a) Grauwerte des Referenzbildes, b) Winkelhistogramm (s. Text), c) Bezugssystem.

Entropie-Varianz-Analyse,
ein Verfahren zur Bewertung von
Merkmalsdichteverteilungen

G. Sträßle, D. Feßmann
Universität Stuttgart, Institut für Physikalische Elektronik,
Pfaffenwaldring 47, D 7000 Stuttgart 80

Zusammenfassung

Anhand modellierter Wahrscheinlichkeitsdichtefunktionen wird der Funktionalzusammenhang zwischen Varianz und Entropie diskutiert. Dies führt zur Definition eines informationstheoretischen Ähnlichkeitsmaßes HV, das die Abweichungen von einer Normalverteilung monoton beschreibt und sehr einfach zu berechnen ist. Am Beispiel eines statistischen Klassifikationsverfahrens mit *linearem* Polynomansatz wird der Einsatz des HV-Maßes bei der Selektion von Merkmalskomponenten demonstriert.

1 Einleitung

Der erfolgreiche Einsatz statistischer Mustererkennungsverfahren hängt neben der erzielbaren Klassifikationsleistung R_c von der Komplexität und der Anzahl der benötigten Rechenoperationen O_c des Klassifikationsprozesses ab. Bestimmt wird O_c durch die Dimension des ein Objekt repräsentierenden Meßvektors D_f (Merkmalsvektors) sowie durch die Dimension des Klassifikationsansatzes D_p. Denkt man an Echtzeitanforderungen und an mögliche VLSI-Implementierungen, so stellt sich die Aufgabe O_c in Abhängigkeit von R_c zu minimieren. Da im folgenden ausschließlich *lineare* Klassifikationsansätze mit $D_p = 1$ betrachtet werden (z.B. Linearer Ploynomklassifikator [Schür77]), beschränkt sich die Optimierung auf D_f. Diese wird durch Merkmalsselektion durchgeführt. Da jedoch i.a. a priori nicht bekannt ist, welcher minimaler Satz von Meßgrößen zu einer optimalen Trennung im Merkmalsraum bezüglich einer gegebenen Klasseneinteilung mit $\bigcup_i \omega_i = \Omega$ führt, wird in der Designphase ein hochdimensionaler Merkmalsvektor berechnet. Es wird dann davon ausgegangen, daß dieser ein gewisses Maß an redundanter Information beinhaltet. Ziel von Merkmalsselektionsverfahren ist es nun problemspezifisch eine Dimensionsreduzierung des Meßvektors mit Hilfe von Bewertungs- bzw. Abstandsmaßen an einer gekennzeichneten repräsentativen Lernstichprobe vorzunehmen. Häufig trifft man auf die u.a. in [Niem83] und [Ycung74] aufgeführten Abstandsmaße *Bhattacharyya-*, *Mahalanobis-*, *Matusita-Abstand* und die *Divergenz* sowie auf das in [Blanz88] diskutierte informationstheoretische Beurteilungsmaß Q, das im folgenden stellvertretend für alle nichtparametrischen Bewertungsmaße betrachtet werden soll. Bewertet werden die Überlappungen der klassenspezifischen Wahrscheinlichkeitsdichtefunktionen $p(f|\omega)$ mit $Q = 0$ für vollständige Überlappung und $Q = 1$ für vollständige Trennung. Wählt man bei gegebenen

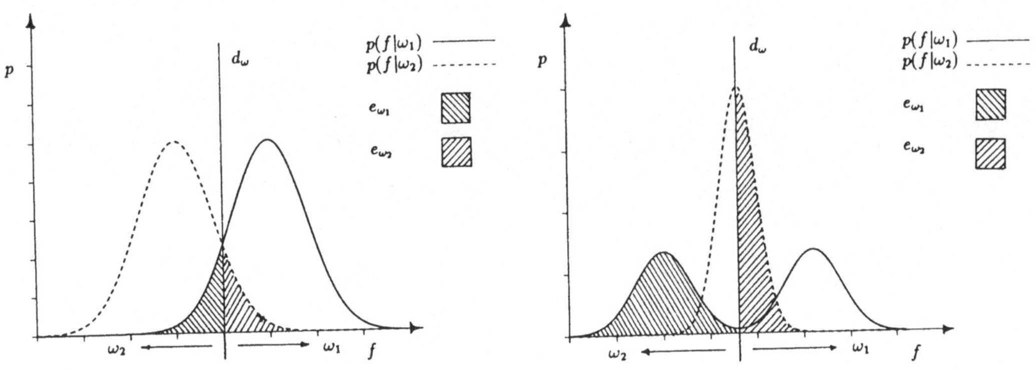

Abbildung 1: a) Übereinstimmung, b) Widerspruch von Überlappungsgrad der Dichtefunktionen $p(f|\omega_1)$, $p(f|\omega_2)$ und Klassifikationsfehler e_{ω_1}, e_{ω_2} bei linearer Entscheidungsfunktion d_ω

Dichteverteilungen $p(f|\omega_1)$ und $p(f|\omega_2)$ nach Abb. 1b einen linearen Klassifikationsansatz so wird deutlich, daß im Gegensatz zu Abb. 1a nicht vom Überlappungsgrad auf den Klassifikationserfolg bzw. den Klassifikationsfehler e_ω geschlossen werden kann. Es ist daher wünschenswert nicht nur den Überlappungsgrad der Dichteverteilungen zu bewerten, sondern auch deren Form. Das im folgenden Abschnitt präsentierte Entropie-Varianz-Maß HV besitzt nun die Eigenschaft, Abweichungen der Dichtefunktionen $p(f|\omega)$ von der zugrunde liegenden Normalverteilung $N(\mu,\sigma)$ monoton und quantitativ zu erfassen. HV nimmt den Wert 0 an für $p(\omega_i) = N(\mu,\sigma)$ und wird > 0 sonst. Angewandt auf Merkmalsselektionsverfahren erweitert es somit die Aussagekraft des Bewertungsmaßes Q für lineare Klassifikationsansätze, da HV die multimodale Charakteristik von $p(\omega_2)$ in Abb. 1b als signifikante Abweichung von einer Normalverteilung erfaßt.

2 Entropie-Varianz-Maß HV

2.1 Definitionen und Funktionalzusammenhänge

Ist $p(x)$ eine beliebige Wahrscheinlichkeitsdichtefunktion, mit

$$\int_{-\infty}^{+\infty} p(x)dx = 1, \tag{1}$$

so ist deren *Varianz V* definiert als

$$V[p(x)] = \int_{-\infty}^{+\infty} x^2 p(x)dx \tag{2}$$

und deren *Entropie H* als

$$H[p(x)] = -\int_{-\infty}^{+\infty} p(x)\ln[p(x)]dx. \tag{3}$$

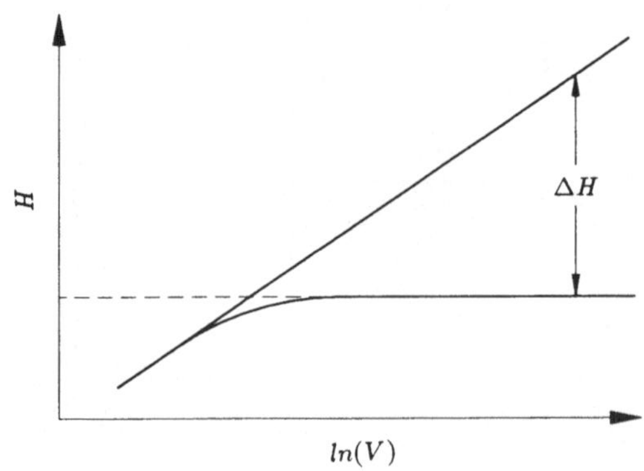

Abbildung 2: Graphische Darstellung des HV-Abstandes ΔH

Nach Shannon [Shan48] existiert für normalverteilte Wahrscheinlichkeitsdichtefunktionen $p(x) = N(\mu,\sigma)$ ein Funktionalzusammenhang von Varianz und Entropie:

$$V[N(0,\sigma)] = \sigma^2 \tag{4}$$

$$H[N(0,\sigma)] = \frac{1}{2} + \ln(\sqrt{2\pi\sigma^2}) \tag{5}$$

$$H(V) = \frac{1}{2}(1 + \ln(2\pi V)) \tag{6}$$

Da unter allen Wahrscheinlichkeitsdichten die Normalverteilung bei gegebener Varianz die größte Entropie besitzt ([Röhl67], [Guias77]) und somit alle von der Normalverteilung abweichende $p(x)$ eine kleinere Entropie annehmen, kann nun folgendes Distanzmaß HV definiert werden:

$$HV = \Delta H = H[N(0, V[p(x)])] - H[p(x)]. \tag{7}$$

Nach Bestimmung bzw. Messung von $V[p(x)]$ und $H[p(x)]$ beschreibt HV quantitativ die Abweichung von der zugrunde liegenden Normalverteilung. Im HV-Diagramm Abb. 2 ist der Verlauf von $H[p(x)]$ und $H[N(0, V[p(x)])]$ für variables $V[p(x)]$ mit gekennzeichnetem HV-Abstand dargestellt. $V[p(x)]$ wird variiert, indem für

$$p(x) = 0.5(N(\mu,\sigma) + N(-\mu,\sigma)) \tag{8}$$

und festem σ der Erwartungswert μ im Intervall $[0,\infty]$ variiert wird (vollständige Überlappung: $\mu = 0$, vollständige Trennung: $\mu \to \infty$).

2.2 Experimentelle Resultate

Gegeben sei ein Klassifikationsproblem mit 2 Klassen und einem Merkmalsvektor mit m Komponenten von denen hier nur 2 betrachtet werden sollen. Jede Klasse ist mit 2000 Ereignissen vertreten. In Abb. 3 ist der dazugehörende klassenspezifische Merkmalsraum mit seinen Randverteilungen abgebildet. Das Mustererkennungsproblem soll nun mit einem linearen Klassifikationsansatz gelöst werden. Eine der beiden Merkmalskomponenten

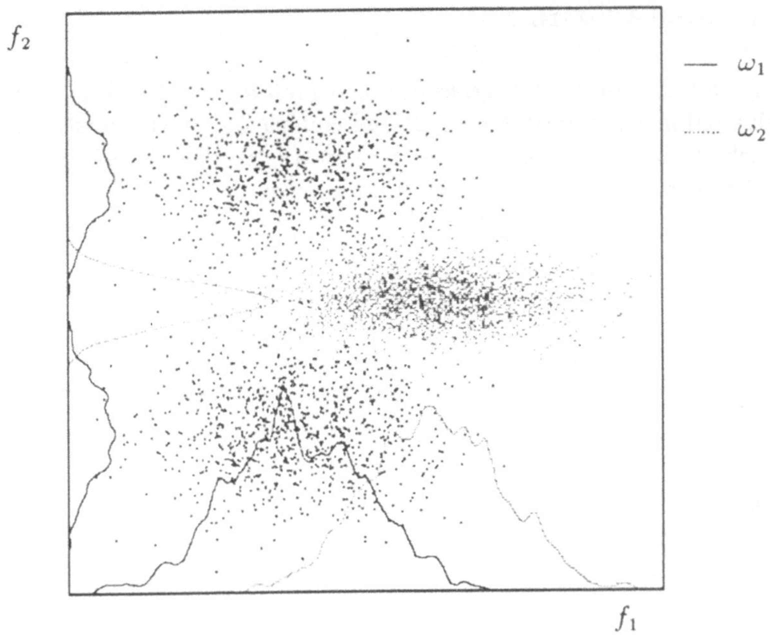

Abbildung 3: Merkmalsraum (f_1, f_2) mit Randverteilungen $p(f|\omega)$

$p(f|\omega)$	HV-Maß	Beurteilungsmaß Q	Klassif.-Leistg. R_c
$p(f_1|\omega_1)$	≈ 0.0	0.39	83.4%
$p(f_1|\omega_2)$	≈ 0.0		
$p(f_2|\omega_1)$	0.47	0.52	50.6%
$p(f_2|\omega_2)$	≈ 0.0		

Tabelle 1: Ergebnisse der Bewertungsmaße Q und HV bzgl. Klassifikationserfolg (linearer Ansatz)

soll mit Hilfe von Bewertungsmaßen unter der Nebenbedingung einer zu optimierenden Klassifikationsleistung deselektiert werden. Tabelle 1 stellt die Bewertung der Merkmalskomponenten f_1 und f_2 durch das Beurteilungsmaß Q ($Q = 1$: vollständige Trennung, $Q = 0$: vollständige Überlappung der $p(f|\omega_c)$) und durch das Ähnlichkeitsmaß HV nach Glg. (7) dar und stellt diese dem Klassifikationserfolg R_c gegenüber. Die Ergebnisse aus Tabelle 1 zeigen, daß das Bewertungsmaß Q die Merkmalskomponente f_2 aufgrund des geringeren Überlappungsgrades gegenüber f_1 besser bewertet. Eine Rückweisung von f_1 an Stelle von f_2 führt jedoch zu einer schlechteren Klassifikationsleistung. Die stark von einer Normalverteilung abweichende Form von $p(f_2|\omega_1)$ wird jedoch in Verbindung mit dem HV-Maß ΔH quantitativ erfaßt. Dies führt korrekterweise zur Rückweisung der Merkmalskomponente f_2.

3 Schlußbetrachtung

Es wurde eine informationstheoretisches Abstandsmaß HV vorgestellt, das Abweichungen von Normalverteilungen quantitativ erfaßt und einfach zu berechnen ist. Am Beispiel der Merkmalsselektion bei der Adaption statistischer Mustererkennungsverfahren konnte der erfolgreiche Einsatz dieses Maßes demonstriert werden.

Literatur

[Blanz88] Blanz, W.E., Non-Parametric Feature Selection for Multiple Class Processes, 9.th Conference on Pattern Recognition, Volume II, 1988.

[Guias77] Guiasu, S., Information Theory with Applications, McGraw-Hill New York, 1977.

[Niem83] Niemann, H., Klassifikation von Mustern, Springer Verlag Berlin, München, New York, Tokio, 1983.

[Röhl67] Röhler, R., Informationstheorie in der Optik, Wissenschaftliche Verlags-Gesellschaft, Stuttgart, 1967.

[Schür77] Schürmann, J., Polynomklassifikator für die Zeichenerkennung, Oldenbourg Verlag München, Wien, 1977.

[Shan48] Shannon, C.E., A Mathematical Theory of Communication, The Bell System Technical Journal, Vol XXVII, 1948.

[Young74] Young, T.Y., Calvert, T.W., Classification, Estimation and Pattern Recognition, American Elsevier Publishing Company Inc., New York, 1974.

A Simplex Design of Linear Hyperplane Decision Networks

PETER STROBACH

SIEMENS AG, Zentralabteilung Forschung und Entwicklung
ZFE F2, Forschung für Informatik und Software
Otto-Hahn-Ring 6, D-8000 München 83, West Germany

Abstract - A new type of two-layer aritficial neural network is presented. In contrast to its conventional counterpart, the new network is capable of forming *any desired decision region* in the input space. The new network consists of a conventional input (hidden) layer and an output layer that is, in essence, a *Boolean function* of the hidden layer output. Each node of the hidden layer forms a *decision hyperplane* and the set of hyperplanes constituted by the hidden layers perform a partition of the input space into a number of *cells*. These cells can be labeled according to their location (binary code) relative to all hyperplanes. The sole function of the output layer is then to *group* all these cells together appropriately, to form an *arbitrary decision region* in the input space. In conventional approaches, this is accomplished by a *linear combination* of the binary decisions (outputs) of the nodes of the hidden layer. This traditional approach is very limited concerning the possible shape of the decision region. A much more natural approach is to form the decision region as a Boolean function of the binary hidden layer "word" which has as many digits as there are nodes in the hidden layer. The training procedure of the new network can be split in two completely decoupled phases. First, the construction of the decision hyperplanes formed by the hidden layer can be posed as a linear programming problem which can be solved using the Simplex algorithm. Moreover, a fast algorithm for approximate solution of this linear programming problem based on an *elliptic approximation of the Simplex* can be devised. The second step in the design of the network is the appropriate construction of the Boolean function of the output layer. The key trick here is the adequate incorporation of *don't cares* in the Boolean function so that the decision regions formed by the output layer cover the entire input space but are still not overlapping.

I. Introduction

THE basic functioning of a multilayer neural network is that the first (hidden) layer uses linear hyperplanes to partition the input space into a number of cells. The function of the additional layers is then to group these cells into regions for classification, pattern recognition, or other purposes. Many approaches have attempt to combine more than two layers in a multilayer neural network from the believe that this might be the only way to form more complex decision regions. Little has been understood about the classification capabilities of such complex non-linear multilayer networks. The training of such networks basically relied on heuristic algorithms. This paper demonstrates that a two-layer neural network can form any desired decision region in the input space, when only the number of nodes in the hidden layer is sufficiently large and when the output layer is a Boolean function of the binary hidden layer output. Any layer beyond the second layer is redundant and the desired result is by no means improved from additional layers. An illumination of this striking statement is given in Section II of this paper. Section III discusses the design of the decision hyperplanes (nodes of the hidden layer) using a linear programming approach. A Simplex design of the linear hyperplanes is devised. In many problems of pattern recognition, it was found that the Simplex underlying the hyperplane design problem can be closely approximated by an inscribed sphere with the shape of a multi-dimensional ellipsis. This constitutes the basis of a very fast algorithm which provides an approximate solution to the given linear programming problem within only a single iteration and with an accuracy that is easily sufficient in most practical problems.

II. Classification Capabilities of Two-Layer Neural Networks

The classification capabilities of two-layer neural networks have been seriously studied by Makhoul, Schwartz and El-Jaroudi [1]. As a result, it was found that any desired decision region can be approximated with a desired accuracy, when only the number of nodes in the (single) hidden layer is sufficiently large. We shall explore this statement by investigating the partitioning of the input space by the first (hidden) layer. For this purpose, consider a two-layer network with two nodes at the input, four nodes in the hidden layer, and two output nodes, as displayed in Fig. 1. In the general case of a q-dimensional pattern (input) vector $[x_1, x_2, \ldots, x_q]^T$, the functioning of node j in the hidden layer is a *weighted linear combination* y_j of the components of the pattern vector according to

$$y_j = a_o^{(j)} + \sum_{i=1}^{q} a_i^{(j)} x_i \; ; \; 1 \leq j \leq n \; , \quad (1)$$

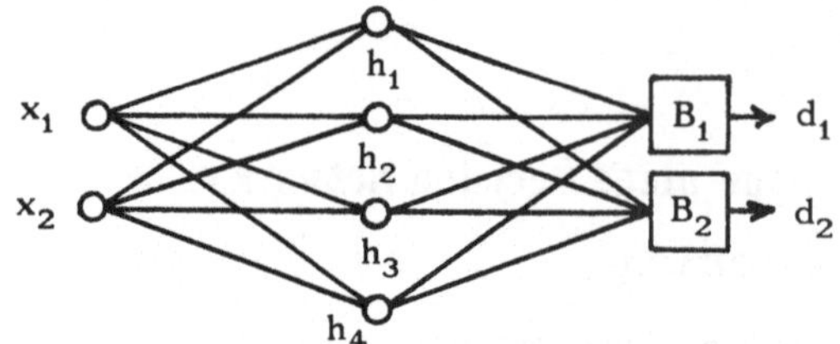

Fig. 1. Two-layer neural network with pattern vector $[x_1\ x_2]^T$, binary hidden layer outputs $\{h_1, h_2, h_3, h_4\}$ and an output layer with Boolean functions B_1 and B_2 which determine the decision regions d_1 and d_2.

where n is the number of nodes in the hidden layer and the real quantities $\{a_i^{(j)}, 1 \le i \le q, 1 \le j \le n\}$ are the *weights* (coefficient sets). The final output of a node, however, is a *binary* decision h_j, where $h_j = +$ for $y_j \ge 0$, and $h_j = -$ for $y_j < 0$. In a geometrical interpretation, the weights contributing to a node constitute a *linear hyperplane* that splits the input space in two *half-spaces*. The evaluation of (1) together with the binary decision h_j indicates whether a given pattern vector belongs to the "left" (-) or the "right" (+) half-space. Normally, such a hyperplane is designed such that it *seperates* two classes of pattern vectors or two point sets in the input space.

A. Partitioning of the Input Space.

When the hidden layer contains more than only a single node, the associated hyperplanes will *intersect* in the q-dimensional input space forming a q-dimensional *lattice* consisting of a number of *decision cells*. This statement is illustrated in Fig. 2 on the example of the two-layer network of Fig. 1. with four nodes in the hidden layer.

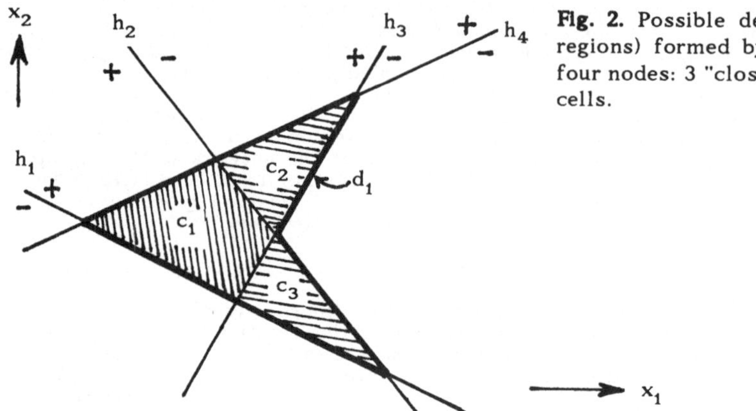

Fig. 2. Possible decision cells (decision regions) formed by a hidden layer with four nodes: 3 "closed" cells and 8 "open" cells.

Each cell in the lattice of Fig. 2 can be labeled with a binary word with n digits, where each digit indicates the *location* of the cell relative to the corresponding hyperplane. For example, the closed cell c_1 is completely determined by the 4-bit word (or "address") $c_1 = +++-$. One could now believe that there are $2^n = 2^4 = 16$ cells in this example, but a quick inspection of Fig. 2 reveals that one finds only a total number of 3 closed and 8 open cells = 11 cells. Consequently, there must be a number of 5 remaining addresses which are *physically impossible* and hence do not appear in the lattice. Such cells have been termed *"imaginary cells"* [1]. Next it is of interest to investigate, how many closed and open cells exist in a q-dimensional lattice formed by n nodes in the hidden layer. This problem was in fact already solved by L. Schäfli [2] in the last century. This was pointed out recently by Makhoul et al. [1]. Explicit formulas for the calculation of the number of open and closed cells as a function of q and n were reported in [1]. One important point in these results is that the number of open cells and the number of closed cells increases tremendously with increasing n and q. As a consequence, one finds that for many practical problems where (typically) q = 256 and n = 190, the input space is already segmented in a very large number of very small decision cells. By an appropriate *"grouping"* of a number of these cells, one can always approximate a desired decision region in the input space with an accuracy that is far more than sufficient in most practical applications.

B. The Boolean Output Layer.

This "grouping" of cells has to be performed by the nodes of the *output layer*. The number of nodes in the output layer is equal to the number of desired decision regions in the input space. In conventional neural network approaches, this grouping is performed again by a weighted linear combination of the already *binary* digits $h_1, h_2, \ldots, h_n$ of the hidden layer - a fairly *inappropriate* approach ! Clearly, from the considerations made so far, it is only a small step to the insight that a decision region d_k may better be formed as a *Boolean function*

$$d_k = B_k\big(h_1, h_2, \ldots, h_n\big) \tag{2}$$

of the hidden layer "word" $\{h_1, h_2, \ldots, h_n\}$. Consider again the example of Fig. 2. We may be interested in obtaining a decision region d_1 composed of the closed cells c_1, c_2, and c_3. Then d_1 may be expressed as a Boolean function of the hidden layer words for c_1, c_2, and c_3, hence: $d_1 = c_1 \vee c_2 \vee c_3$. More explicitly, we may write: $d_1 = h_1 h_2 h_3 \overline{h}_4 \vee h_1 \overline{h}_2 h_3 \overline{h}_4 \vee h_1 h_2 \overline{h}_3 \overline{h}_4$. This Boolean function can be easily *minimized* by appropriate exploitation of *don't cares*. One simple menthod to achieve this is the Karnaugh-diagram of Fig. 3.

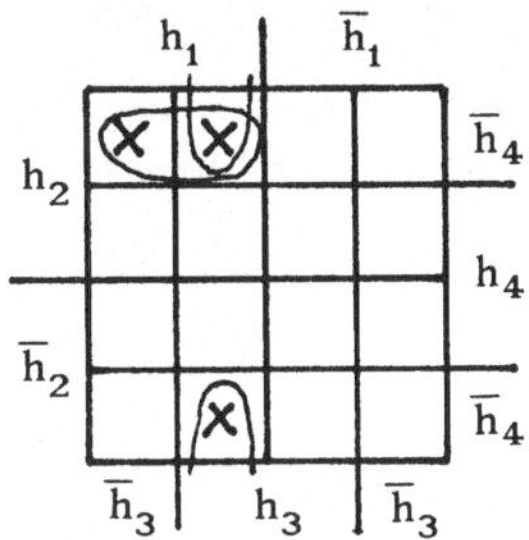

Fig. 3. Evaluation of Karnaugh-diagram for minimization of $d_1 = h_1 h_2 h_3 \overline{h}_4 \vee h_1 \overline{h}_2 h_3 \overline{h}_4 \vee h_1 h_2 \overline{h}_3 \overline{h}_4$ which gives the minimized result: $d_1 = h_1 h_2 \overline{h}_4 \vee h_1 h_3 \overline{h}_4$. Physically, the node B_1 is a *logical* network with n binary inputs and a single output.

It is emphasized that more sophisticated automatic methods have been developed for minimization of Boolean functions of a very large number of input variables. Next one may check that the number of cells in the lattice is orders of magnitude larger than the number of training patterns available in practical applications. So the training procedure "hits" only a tremendously small number of cells inside the desired decision region. This gives rise to the second problem, namely, that the Boolean function determining a decision region must be constructed from these *sparsely* distributed cells determined by the hidden layer words of a training sequence assuming a given hidden layer. This is in fact a difficult task and the search for an appropriate solution to this problem opens a new dimension in neural network design. Obviously, the complete decision region can be reconstructed from the sparsely distributed cells of a point set by some kind of a *"binary region growing"* where one introduces don't cares in the Boolean function of the output layer under the constraint that the cells thus-included in the decision region are not yet members of another decision region. The procedure can be repeated recursively until no more "free" cells are available in the lattice and all cells correspond to a distinct decision region. This procedure is time consuming. On the other hand, even with a training set consisting of an infinite number of training patterns, it is not guaranteed that all cells in the decision region are hit. Moreover, one may gain the interesting insight that in most cases there is a large number of different final solutions of the final decision region which all perform *identical* to the training set, but which may be more or less *robust outside* the training sequence. The optimal design of the Boolean output layer is therefore in a ongoing state of research.

III. A Simplex Design of the First Hidden Layer

We shall return to the early problem of optimizing the weights $\{a_i^{(j)}, 1 \le i \le q, 1 \le j \le n\}$ of the first hidden layer. Suppose we have given the problem of classification of m different patterns represented by m different training sets. In this case, one requires $n = (m-1)m/2$ nodes in the hidden layer for a *pairwise separation* of all m classes of patterns in the input space.

A. Linear Programming Formulation of Pairwise Separation Problem.

Let $\mathbf{X}^{(I)}, \mathbf{X}^{(II)} \in \mathbb{R}^{L \times q}$ be two point sets (training sets) of L points each in the q-dimensional input space. Then, we may define the set vectors $\{\mathbf{x}_i^{(I)}, 1 \le i \le q\}$, $\{\mathbf{x}_j^{(II)}, 1 \le j \le q\}$ where

$$\mathbf{X}^{(I)} = \left[\mathbf{x}_1^{(I)}, \ldots, \mathbf{x}_q^{(I)} \right] , \qquad (3a) \qquad\qquad \mathbf{X}^{(II)} = \left[\mathbf{x}_1^{(II)}, \ldots, \mathbf{x}_q^{(II)} \right] . \qquad (3b)$$

Let $\mathbf{a}$ be the coefficient vector of a linear hyperplane in the q-dimensional input space

$$\mathbf{a} = \left[a_0, a_1, \ldots, a_q \right]^T . \qquad (4)$$

The two point sets $\mathbf{X}^{(I)}$ and $\mathbf{X}^{(II)}$ might be evaluated according to their *distance* to the hyperplane defined by the coefficient set (4). This gives rise to two sets of inequality relations:

$$\left[-1 \;\; \mathbf{X}^{(I)} \right] \mathbf{a} \le \varepsilon , \qquad (5a) \qquad\qquad \left[-1 \;\; \mathbf{X}^{(II)} \right] \mathbf{a} \ge -\varepsilon , \qquad (5b)$$

or equivalently,

$$\left[-1 \;\; \mathbf{X}^{(I)} \right] \mathbf{a} \le \varepsilon , \qquad (6a) \qquad\qquad \left[1 \; - \mathbf{X}^{(II)} \right] \mathbf{a} \le \varepsilon , \qquad (6b)$$

where ε is a variable *margin vector* and $\mathbf{1}$ is the *eins vector*

$$\boldsymbol{\varepsilon} = \left[\varepsilon, \varepsilon, \ldots, \varepsilon \right]^T , \qquad (7) \qquad\qquad \mathbf{1} = \left[1, 1, \ldots, 1 \right]^T . \qquad (8)$$

Clearly, inequality (6a) constitutes, in any case, a region that contains the region covered by point set $\mathbf{X}^{(I)}$, if only the margin ε is sufficiently large. On the other hand, the counterpart situation is present in that inequality (6b) constitutes, in any case, a region that contains the region covered by point set $\mathbf{X}^{(II)}$, if again the margin ε is sufficiently large. This way, the margin ε can serve as a *distance measure* between the point sets and the separating hyperplane. In fact, we may check that if for the best choice of $\mathbf{a}$,

$\varepsilon < 0 \longrightarrow$ *Point sets are perfectly linearly separable.*

$\varepsilon \geq 0 \longrightarrow$ *Point sets are **not** linearly separable.*

Clearly, the hyperplane coefficient set $\mathbf{a}$ should be determined in a sense such that ε attains its **largest negative value**. In this case the coefficient set is the solution to the MAX–MIN problem:

$$\text{For all training patterns} \quad 1 \leq j \leq 2L: \quad min\left(|\,y_j\,|\right) \stackrel{!}{=} max \quad \longrightarrow \quad \mathbf{a} \ . \tag{9}$$

Hence, formulation (6a,b) can serve as a useful tool in setting up an optimization procedure for designing the hyperplane coefficient vector $\mathbf{a}$. For this purpose, we may rewrite (6a,b) in a more condensed form as

$$\begin{bmatrix} \mathbf{X}_s^{(I)} \\ -\mathbf{X}_s^{(II)} \end{bmatrix} \overline{\mathbf{a}} + \begin{bmatrix} -\mathbf{a}_O \\ \mathbf{a}_O \end{bmatrix} \leq \varepsilon \ , \qquad (10a) \qquad\qquad \varepsilon \longrightarrow min \ , \qquad (10b)$$

where $\overline{\mathbf{a}}$ is the *reduced* hyperplane coefficient vector and $\mathbf{a}_O$ is the *constant* vector

$$\overline{\mathbf{a}} = \begin{bmatrix} a_1, a_2, \ldots , a_p \end{bmatrix}^T \ , \qquad (11) \qquad\qquad \mathbf{a}_O = \begin{bmatrix} a_O, a_O, \ldots , a_O \end{bmatrix}^T ; \quad a_O > 0 \ . \quad (12)$$

$\overline{\mathbf{a}}$ should be determined such that ε attains its *largest negative value* (10b). This already constitutes an optimization problem of the form *"minimize a linear function (10b) with respect to a set of linear inequality constraints (10a)"*.The set of inequality constraints (10a) form a *Simplex* [3–5] in the q-dimensional parameter space. The minimum of ε is always located in an edge of the Simplex [3].
We shall give an meaningful interpretation of this optimization process. Formulation (10a,b) forces the hyperplane to be located exactly in the middle of the gap between the two point sets $\mathbf{X}^{(I)}$ and $\mathbf{X}^{(II)}$ since this location guarantees the *largest possible distance* of the closest members of the two point sets from the hyperplane, where the "distance" is defined as the weighted linear combination y_j of the coordinates of point (pattern vector) j according to (1). Next, one may transform this problem into a standard form in order to apply the *Simplex algorithm* [3–5] for its solution. After some algebra, one finds that (10a) can equivalently be written as

$$\begin{bmatrix} \mathbf{X}^{(I)} \\ -\mathbf{X}^{(II)} \end{bmatrix} \overline{\mathbf{a}} + \mathbf{g} \leq \begin{bmatrix} 2\,\mathbf{a}_O \\ 0 \end{bmatrix} \ , \qquad (13a) \qquad \mathbf{g} = \begin{bmatrix} \mathbf{a}_O \\ \mathbf{a}_O \end{bmatrix} - \varepsilon \ , \qquad (13b)$$

where

$$\mathbf{g} = \begin{bmatrix} g, g, \ldots g \end{bmatrix}^T \ . \tag{13c}$$

This is already a standard linear programming problem of the form
Minimize the linear function $f(\overline{\mathbf{a}}, g)$

$$f(\overline{\mathbf{a}}, g) \ = \ \varepsilon \ = \ a_O + \begin{bmatrix} 0, 0, \ldots , 0, -1 \end{bmatrix} \begin{bmatrix} \overline{\mathbf{a}} \\ g \end{bmatrix} \stackrel{!}{=} min \ , \tag{14a}$$

with respect to the set of linear inequality constraints

$$\begin{bmatrix} \mathbf{X}^{(I)} & 1 \\ -\mathbf{X}^{(II)} & 1 \end{bmatrix} \begin{bmatrix} \overline{\mathbf{a}} \\ g \end{bmatrix} \leq \begin{bmatrix} 2\,\mathbf{a}_O \\ 0 \end{bmatrix} \ , \quad (14b) \qquad\qquad \begin{bmatrix} \overline{\mathbf{a}} \\ g \end{bmatrix} \geq \mathbf{0}, \ (15a) \qquad \begin{bmatrix} 2\,\mathbf{a}_O \\ 0 \end{bmatrix} \geq \mathbf{0} \ , \ (15b)$$

Obviously, restriction (15b) is always satisfied since a_O was restricted to only positive values [recall (12)]. From (13b), we may also see that $\varepsilon = a_O - g$, and hence, $g \geq 0$ can always be guaranteed when only a_O is sufficiently large. When the relative location of the point sets in the q-dimensional input space is completely unknown, we set a_O to twice the dynamic range of the input space. The dynamic range in a practical recognition problem is always bounded. In many practical problems, however, it is sufficient to set $a_O = 1$ when there is a justified assumption that the point sets will be completely separable by the linear hyperplane $\mathbf{a}$.

B. Point Space Similarity Transformation.

We may encounter, at this state of the considerations that $\overline{a} \geq 0$ is not satisfied a priorily for arbitrary relative locations of the point sets. There exists, however, an adaptive similarity transformation which is invariant to our hyperplane design problem and which transforms point sets with arbitrary location into point sets where a subsequent design yields a hyperplane which strictly satisfies $\overline{a} \geq 0$. This transformation will next be described. For this purpose, one defines the center of gravity vector s of a point set under the assumption that each point attains the same weight:

$$s = \left[\sum_{i=1}^{L} x_1(i), \sum_{i=1}^{L} x_2(i), \dots, \sum_{i=1}^{L} x_q(i) \right]^T = \left[s_1, s_2, \dots, s_q \right]^T . \tag{16}$$

This way, we compute the vector $s^{(I)}$ from the point set $\mathbf{X}^{(I)}$, and $s^{(II)}$ from $\mathbf{X}^{(II)}$. The adaptive similarity transform is then defined such that

$$x_j^{(I)'} = MAX - x_j^{(I)}; \qquad x_j^{(II)'} = MAX - x_j^{(II)} \qquad if \quad \delta + s_j^{(I)} > s_j^{(II)}; \quad 1 \leq j \leq q, \tag{17a}$$

$$x_j^{(I)'} = x_j^{(I)}; \qquad x_j^{(II)'} = x_j^{(II)} \qquad otherwise, \tag{17b}$$

where δ is a distance and MAX is the absolute point set dynamic range. This transformation guarantees that $\delta + s^{(I)'} \leq s^{(II)'}$. When δ is sufficiently large then it is also clear that the separating hyperplane exhibits a strictly positive coefficient set $\overline{a} \geq 0$.

C. Elliptic Inscribed Sphere Approximation of Simplex.

In typical practical problems, the Simplex design of a coefficient set of a node of the first hidden layer requires in the order of 2000 - 5000 iterations - a rather time consuming and inefficient procedure ! On the other hand, the linear objective function ε typically decreases by only 0.1 percent of its total value in each iteration. This already indicates that the shape of the Simplex can be approximated quite closely by an *inscribed q-dimensional ellipsis* in most situations. An approximate solution of the linear programming problem (14a,b) is then found at the periphery of the ellipsis in only a *single* iteration. Such a procedure can be interpreted as the first (initial) step of the Karmarkar algorithm [6] with the only difference that Karmarkar used a *ball approximation* instead of an elliptic approximation.

IV. Conclusions

The paper presented a new type of two-layer neural network that is capable of forming any desired decision region in the input space when only the number of hidden nodes is sufficiently large. The training of the new network is performed in two steps. First, the nodes of the hidden layer are designed as pairwise seperating linear hyperplanes. Among several possibilities, the hyperplane design can be posed as a linear programming problem. A fast approximate algorithm for solving this linear programming problem employing an inscribed elliptic sphere approximation of the Simplex has been discussed. A nice feature of the linear programming approach is that the design delivers the value of the objective function ε, and the sign of ε indicates whether the two point sets are linearly separable or not. The hyperplanes form a q-dimensional lattice in the input space. The desired decision region is then approximated by an appropriate grouping of decision cells of the lattice. This is accomplished by the output layer, where each node was quite naturally defined as a Boolean function of the hidden layer word. It was shown that the Boolean function is never uniquely determined. The determination of the *optimal* Boolean function in terms of a maximum robustness of the decisions outside training is currently a topic of active research. Another topic of research is the derivation of fast approximative algorithms which solve the given linear programming problem.

References

[1] **J. Makhoul, R. Schwartz and A. El-Jaroudi**, "Classification capabilities of two-layer neural networks", in Proc. IEEE Int. Conf. on ASSP, paper S12.10, pp. 635–638, Glasgow, Scotland, May 1989.

[2] **L. Schäfli (1814–1895)**, *Gesammelte Mathematische Abhandlungen*, Vol. 1., Birkhäuser Verlag, Basel, 1950.

[3] **G.B. Dantzig**, *Linear Programming and Extensions*, Princeton, N.J., Princeton University Press, 1963.

[4] **C. Van De Panne**, *Methods for Linear and Quadratic Programming*, Studies in Mathematical and Managerial Economics, Henri Theil (Ed.), North Holland Publishing Comp., Amsterdam, New York, 1975.

[5] **G.B. Dantzing, A. Orden and P. Wolfe**, "Notes on linear programming, part I", Pacific Journal of Mathematics, Vol. 5, pp. 183–195, 1955.

[6] **N.K. Karmarkar**, "a new polynomial-time algorithm for linear programming", Combinatorica, Vol. 4, No. 4, pp. 373–395, 1984.

From optical flow of lines to 3D motion and structure *

Olivier Faugeras
Nourredine Deriche and Nassir Navab
INRIA Sophia Antipolis
2004 Route des Lucioles
06565 Valbonne Cedex
FRANCE
faugeras@inria.inria.fr

Abstract

We establish the motion equations for rigidly moving 3D lines and the structure equations that relate the optical flow of a line to its kinematic screw and 3D representation. We also show that if the time derivative of the optical flow is available, then five simple motion equations can be derived that relate the kinematic screw, the optical flow, and their time derivatives.

1 Introduction

In a previous paper, [3] we had investigated the problem of camera displacement estimation and recovery of structure from two views when point matches are available and three views when line matches are available. In this article, we want to extend this work to the estimation of motion and recovery of structure from motion as well as investigate the cooperation between motion and stereo.

2 What do we start from

The work described here builds on three previous pieces of work. The first one is a segment based trinocular stereo program that was developed by Ayache and Lustman [1]. The second one is a program for tracking line segments in sequences of images developed by Deriche [2,4]. The third one is the analysis developed in [3].

For our present purposes, it is sufficient to say that both program work on the same image description, i.e polygonal approximation of connected sets of edge pixels. The stereo program matches them between three images acquired simultaneously to avoid problems related to motion and reconstructs 3D line segments.

The token tracker program matches them between images taken at consecutive instants and builds a model of their 2D motion.

---

*This work was partially completed under Esprit P2502

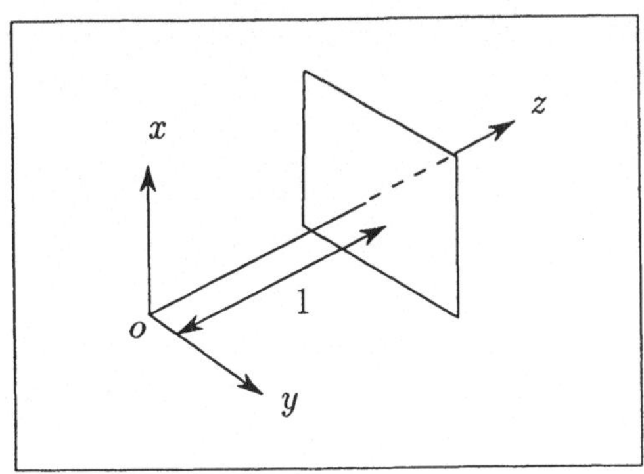

Figure 1: Pinhole model of a camera

3 Getting the equations

We model our camera with the standard pinhole model of figure 1 and assume that everything is referred to the camera intrinsic coordinate frame ($Cxyz$).

3.1 3D line representation

We represent a straight line D in 3D space with a direction vector $l = (\alpha, \beta, \gamma)^T$ and a point M of coordinates (x, y, z) which we leave unspecified at the moment. l is defined up to a scale factor.

3.2 2D line representation

A line d in the retina plane are represented by a vector $\mathbf{n} = (u, v, w)^T$ giving its equation:

$$ux + vy + w = 0 \tag{1}$$

The interpretation of $\mathbf{n}$ is that it is the normal to the plane defined by the 2D line and the optical center of the camera (see figure 3.

3.3 Relations between D and d

We now want to establish the relationships between the representation of a 3D line and its image in the retinal plane. The first relation is that the direction l of the line is perpendicular to the normal $\mathbf{n}$ to the plane it defines with the optical center.

$$l^T \mathbf{n} = 0 \tag{2}$$

We also need a point. If the line D is not parallel to the z-plane we can choose the point M to be the intersection M_z of D with the plane $z = 0$. Let us denote by p and q its x and y coordinates and by $\mathbf{r}$ the vector $(p, q, 0)^T$. Let us recall that the transformation between a point $\mathbf{M}$ in 3D space and its retinal image $\mathbf{m}$ is linear in projective coordinates

$$\mathbf{m} = \mathcal{P}.\mathbf{M} \tag{3}$$

in this equation $\mathbf{m}$ and $\mathbf{M}$ are projective representations of points m and M, i.e 3×1 and 4×1 vectors, respectively ; $\mathcal{P}$ is a 3×4 matrix which, with our choice of coordinates, has the very simple form :

$$\mathcal{P} = \begin{bmatrix} 1 & 0 & 0 & 0 \\ 0 & 1 & 0 & 0 \\ 0 & 0 & 1 & 0 \end{bmatrix} \tag{4}$$

Through relation (3) the point M_z is mapped on the retina on to point $\mathbf{r}$ of projective coordinates $(p, q, 0)$. Notice that this point is at infinity in the retina plane, it represents the direction of the image line. The relationship between $\mathbf{n}$ and $\mathbf{r}$ is:

$$\mathbf{r}^T \mathbf{n} = 0 \tag{5}$$

4   The motion equations for lines

We are dealing with the case where the Token Tracker algorithm of [2,4] tracks in the image of a camera a number of line segments whose supporting lines are represented by the vectors $\mathbf{n}$. From this tracking, we can estimate time derivatives $\dot{\mathbf{n}}$, $\ddot{\mathbf{n}}$ of those vectors. The vector $\dot{\mathbf{n}}$ is the optical flow of the corresponding line. If we have two or three cameras, this tracking process can be achieved for each image and the questions we ask and partially answer in the next sections are, how much information related to the 3D motion and structure of the rigid objects observed in the scene can be recovered from these quantities.

4.1   Recovering the direction of a 3D line from its optical flow and Ω.

In order to gain more insight into the problem, we now assume that the 3D line under consideration is attached to a rigid body whose motion is described by an instantaneous rotation Ω and linear velocity $\mathbf{V}$, its kinematic screw . We derive the differential equation satisfied by l. If we recall that the velocity $\dot{\mathbf{M}}$ of any point $\mathbf{M}$ attached to the rigid body is given by :

$$\dot{\mathbf{M}} = \mathbf{V} + \Omega \wedge \mathbf{M} \tag{6}$$

Since the direction l is a point at infinity, its differential equation is simply:

$$\dot{\mathbf{l}} = \Omega \wedge \mathbf{l} \tag{7}$$

We now express l as function only of the image measurements $\mathbf{n}$, $\dot{\mathbf{n}}$, and Ω. For this we can use equation (2) and take its time derivative :

$$\dot{\mathbf{n}}^T \mathbf{l} + \mathbf{n}^T \dot{\mathbf{l}} = 0 \tag{8}$$

Replacing $\dot{\mathbf{l}}$ in that equation by its value from equation (7), we obtain:

$$\dot{\mathbf{n}}^T \mathbf{l} + (\mathbf{n}, \Omega, \mathbf{l}) = 0 \tag{9}$$

where $(\mathbf{n}, \Omega, \mathbf{l})$ is the determinant of the three vectors $\mathbf{n}$, Ω, and l. Using standard properties of determinants, this equation can be rewritten as:

$$(\dot{\mathbf{n}} + \mathbf{n} \wedge \Omega)^T \mathbf{l} = 0 \tag{10}$$

Combining it with equation (2) shows that l is perpendicular to both vectors $\mathbf{n}$ and $\dot{\mathbf{n}} + \mathbf{n} \wedge \Omega$, therefore it is proportional to their cross-product; since l is a direction, i.e is defined up to a scale factor, we write:

$$l = \mathbf{n} \wedge (\mathbf{n} \wedge \Omega + \dot{\mathbf{n}}) \tag{11}$$

Equation (11) is fundamental. It allows to recover the spatial orientation of a 3D line from the knowledge of Ω and the 2D optical flow described by the image quantities computed by the Token Tracker : $\mathbf{n}$ and $\dot{\mathbf{n}}$.

4.2 Recovering a point of the 3D line from its optical flow , Ω and V

If the 3D line is not parallel to the plane $z = 0$ we can consider its intersection with that plane. Its coordinates p and q are given by the following equations:

$$\begin{cases} p = x - z\frac{\alpha}{\gamma} \\ q = y - z\frac{\beta}{\gamma} \end{cases} \tag{12}$$

which can be rewritten just as well as:

$$\begin{cases} p = \frac{1}{\gamma}\mathbf{M}^T\mathbf{u} \\ q = -\frac{1}{\gamma}\mathbf{M}^T\mathbf{v} \end{cases} \tag{13}$$

where $\mathbf{M} = (x, y, z)^T$, $\mathbf{u} = (\gamma, 0, -\alpha)^T = \mathbf{j} \wedge l$, and $\mathbf{v} = (0, -\gamma, \beta)^T = \mathbf{i} \wedge l$. Notice that $\mathbf{u} \wedge \mathbf{v} = -\gamma l$.

Taking time derivatives, replacing $\dot{\mathbf{M}}$ by $\mathbf{V} + \Omega \wedge \mathbf{M}$, and after some algebra, we find

$$\dot{\mathbf{r}} = \frac{1}{\gamma}\begin{bmatrix} \mathbf{u}^T \\ -\mathbf{v}^T \\ \mathbf{0}^T \end{bmatrix}(\mathbf{V} + \Omega \wedge \mathbf{r}) \tag{14}$$

This is the differential equation, or motion equation of M_z. Notice that its motion cannot described as a rigid planar motion. We now want to express $\mathbf{r}$ as function only of the image measurements $\mathbf{n}$, $\dot{\mathbf{n}}$, Ω, and $\mathbf{V}$. Taking the time derivative of equation (5) and replacing in it $\dot{\mathbf{r}}$ by its expression (equation (14)), we obtain:

$$\dot{\mathbf{n}}^T\mathbf{r} + \frac{1}{\gamma}\mathbf{n}^T\mathbf{B}(\mathbf{V} + \Omega \wedge \mathbf{r}) \tag{15}$$

where matrix $\mathbf{B} = \begin{bmatrix} \mathbf{u}^T \\ -\mathbf{v}^T \\ \mathbf{0}^T \end{bmatrix}$. Noticing that $\mathbf{n}^T\mathbf{B} = \gamma\mathbf{n}^T$, because of equation (2), we can write

$$\mathbf{r}^T(\dot{\mathbf{n}} + \mathbf{n} \wedge \Omega) + \mathbf{n}^T\mathbf{V} = 0 \tag{16}$$

Let now $\mathbf{h} = (v, -u, 0)^T$. Because of equation (5), $\mathbf{r}$ is proportional to $\mathbf{h}$. The constant of proportionality θ is obtained by replacing $\mathbf{r}$ with $\theta\mathbf{h}$ in equation (16):

$$\mathbf{r} = \theta\mathbf{h} = \frac{\mathbf{n}^T\mathbf{V}}{\Omega^T(\mathbf{n} \wedge \mathbf{h}) - \mathbf{h}^T\dot{\mathbf{n}}}\,\mathbf{h} \tag{17}$$

Equation (17) is just as fundamental as equation (11), and for the same reasons. In the case where the line D is parallel, or close to parallel to the plane $z = 0$, instead of considering the

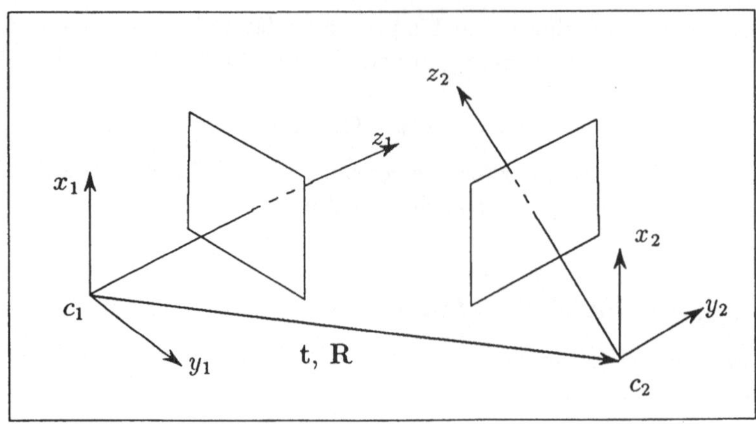

Figure 2: A stereo rig

point M_z, we may want to consider the point M_x, intersection of D with the plane $x = 0$. The vector $\mathbf{r}$ of interest is now $(0, p, q)^T$:

$$\begin{cases} p = y - \frac{\beta}{\alpha}x \\ q = z - \frac{\gamma}{\alpha}x \end{cases}$$

and equations similar to (17, 14) can be derived.

4.3 Recovering Ω and $\mathbf{V}$ from stereo and the token tracker

Suppose now that we have a second camera of known position and orientation with respect to the first one, as shown in figure 2. Events or measurements in the first camera will be sometime indexed by 1 while events in the second camera will be indexed by 2 or followed by '.

If a line segment represented by $\mathbf{n}_1$ is matched by the stereo program to a line segment represented by $\mathbf{n}_2$ in the second image at time t, we can reconstruct a 3D line represented by $\mathbf{l}, \mathbf{r}$. Considering equations (11) and (17) for the two cameras yields four equations in Ω and $\mathbf{V}$ which are actually linear equations. Therefore, if we assume that we have identified two segments located on the same rigid object, we can in general compute its rigid motion from the eight equations. Correspondingly, if the motion of the segment observed at time t is constant velocity (constant Ω and $\mathbf{V}$), then two observations of the same segment at two consecutive time instants will permit to compute Ω and $\mathbf{V}$.

Let us be more precise and analyze equations (11) and (17). From equation (11), we can verify that for a given direction of line, the rotation can only be recovered in the direction perpendicular to the line. Indeed, let us add $\lambda\mathbf{l}$ to Ω :

$$\mathbf{n} \wedge (\mathbf{n} \wedge (\Omega + \lambda\mathbf{l}) + \dot{\mathbf{n}}) = \mathbf{n} \wedge (\mathbf{n} \wedge \Omega + \dot{\mathbf{n}}) + \lambda\mathbf{n} \wedge (\mathbf{n} \wedge \mathbf{l}) \tag{18}$$

The last term is equal to $-\lambda(\|\mathbf{n}\|)^2\mathbf{l}$ since $\mathbf{n}$ is orthogonal to $\mathbf{l}$. Therefore, it is proportional to $\mathbf{l}$ and (11) is unchanged.

So the only information that can be recovered from (11) is the projection of Ω in a plane perpendicular to $\mathbf{l}$.

Assuming that we know from stereo which segment in the second image matches the one in the first image, a similar equation can be written in the coordinate system of this camera :

$$\mathbf{l}' = \mathbf{n}' \wedge (\mathbf{n}' \wedge \mathbf{\Omega}' + \dot{\mathbf{n}}') \tag{19}$$

If $\mathbf{R}$ is the rotation matrix from the first camera to the second, equation (19) can be written in the first coordinate system by applying $\mathbf{R}$ to it :

$$\mathbf{Rl}' = \mathbf{R}[\mathbf{n}' \wedge (\mathbf{n}' \wedge \mathbf{\Omega}' + \dot{\mathbf{n}}')] = \mathbf{Rn}' \wedge (\mathbf{Rn}' \wedge \mathbf{\Omega} + \mathbf{R}\dot{\mathbf{n}}') \tag{20}$$

Letting $\mathbf{m} = \mathbf{Rn}'$, the direction of the line is given by $\mathbf{n} \wedge \mathbf{m}$ in the first coordinate system. Both vectors $\mathbf{n}$ and $\mathbf{m}$ are orthogonal to the direction of the line so we can consider the projection $\mathbf{\Omega}_\perp$ of $\mathbf{\Omega}$ on the plane they define :

$$\mathbf{\Omega}_\perp = \alpha \mathbf{n} + \beta \mathbf{m} \tag{21}$$

Let us also project $\dot{\mathbf{n}}$ and $\dot{\mathbf{m}}$ on the triplet of vectors $(\mathbf{n}, \mathbf{m}, \mathbf{n} \wedge \mathbf{m})$

$$\begin{cases} \dot{\mathbf{n}} = u\mathbf{n} + v\mathbf{m} + w\mathbf{n} \wedge \mathbf{m} \\ \dot{\mathbf{m}} = r\mathbf{n} + s\mathbf{m} + t\mathbf{n} \wedge \mathbf{m} \end{cases} \tag{22}$$

Since only $\mathbf{\Omega}_\perp$ can be recovered, equation (11) can be written as :

$$\mathbf{l} = \mathbf{n} \wedge (\mathbf{n} \wedge \mathbf{\Omega}_\perp + \dot{\mathbf{n}}) \tag{23}$$

Replacing $\mathbf{\Omega}_\perp$ and $\dot{\mathbf{n}}$ by their expressions yields

$$\mathbf{l} = \mathbf{n} \wedge ((\beta + w)\mathbf{n} \wedge \mathbf{m} + u\mathbf{n} + v\mathbf{m}) \tag{24}$$

or

$$\mathbf{l} = (\beta + w)\mathbf{n} \wedge (\mathbf{n} \wedge \mathbf{m}) + v\mathbf{n} \wedge \mathbf{m} \tag{25}$$

Clearly, since $\mathbf{l}$ is parallel to $\mathbf{n} \wedge \mathbf{m}$ and $\mathbf{n} \wedge (\mathbf{n} \wedge \mathbf{m})$ is in the plane defined by $\mathbf{n}$ and $\mathbf{m}$, this equation implies :

$$\beta = -w = -\frac{\dot{\mathbf{n}}^T (\mathbf{n} \wedge \mathbf{m})}{\| \mathbf{n} \wedge \mathbf{m} \|^2} \tag{26}$$

In a similar way, we can use the equation :

$$\mathbf{Rl}' = \mathbf{m} \wedge (\mathbf{m} \wedge \mathbf{\Omega}_\perp + \dot{\mathbf{m}}) \tag{27}$$

to derive:

$$\alpha = t = \frac{\dot{\mathbf{m}}^T (\mathbf{n} \wedge \mathbf{m})}{\| \mathbf{n} \wedge \mathbf{m} \|^2} \tag{28}$$

This shows how to recover the projection of the angular velocity in the plane perpendicular to the direction of the line from the optical flows of its images in the two cameras.

Let us now look at the information that can be recovered from the intersection of the line with the plane $z = 0$. From the numerator of θ (17), it is clear that the only part of the translation velocity that can be recovered is the one perpendicular to the plane defined by the 2D line and the optical center of the camera. It is also instructive to look at the denominator of θ. The

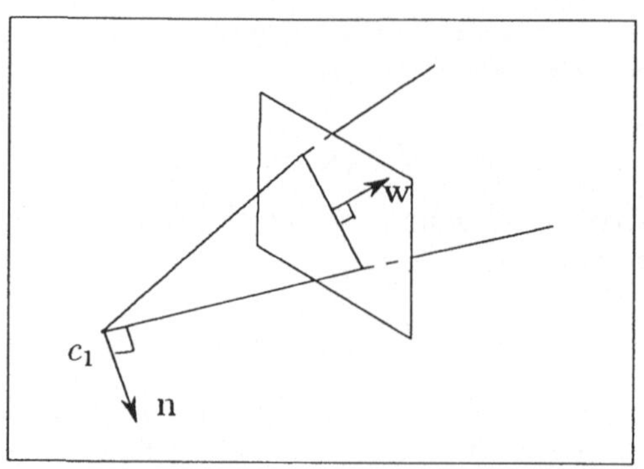

Figure 3: The vectors **n** and **w**

vector $\mathbf{w} = \mathbf{n} \wedge \mathbf{h}$ is parallel to the previous plane to the 2D line (see figure 3). Only rotations around such an axis can be recovered from the observation of p and q. Of course, since we are in fact using two cameras, we can reconstruct translation velocity in a plane perpendicular to the 3D line but not along it. We thus have:

$$\begin{cases} \mathbf{n}^T V = p(\mathbf{w}^T \Omega - \mathbf{h}_1^T \dot{\mathbf{n}}_1) \\ \mathbf{m}^T V = p'(\mathbf{w}'^T \Omega' - \mathbf{h}_2^T \dot{\mathbf{n}}_2) \end{cases} \qquad (29)$$

Let us look at the term $\mathbf{m}^T V$; the inner product $\mathbf{w}'^T \Omega'$ is equal to $(\mathbf{R}\mathbf{w}')^T \Omega$ and p' is computed as the intersection of the line with the plane $z' = 0$.

If we insist on having a situation similar to the one with 1 , i.e if we do not want to actually compute p and p', we can express the quantities $x = \mathbf{n}^T V$ and $y = \mathbf{m}^T V$ as functions of the measurements $\mathbf{n}$, $\dot{\mathbf{n}}$, $\mathbf{m}$, $\dot{\mathbf{m}}$ as follows.

We write

$$\begin{cases} \mathbf{r}_1 = x\mathbf{f}_1 \\ \mathbf{r}' = y\mathbf{f}' \end{cases} \qquad (30)$$

where, for example, $\mathbf{f}_1 = \frac{1}{\mathbf{w}_1^T \Omega - \mathbf{h}_1^T \dot{\mathbf{n}}_1} \mathbf{h}_1$ The second equation is to be taken in the second coordinate system. To express it in the first, we write:

$$\mathbf{r}_2 = \mathbf{R}\mathbf{r}' + \mathbf{t} = y\mathbf{f}_2 + \mathbf{t} \qquad (31)$$

We then simply notice that $\mathbf{r}_1 - \mathbf{r}_2$ is parallel to the line and therefore orthogonal to the vectors $\mathbf{u}$ and $\mathbf{v}$ defined previously since $\mathbf{u} \wedge \mathbf{v} = -\gamma \mathbf{l}$. Therefore, x and y are computed by solving the linear system of two equations:

$$\begin{cases} x\mathbf{f}_1^T \mathbf{u} - y\mathbf{f}_2^T \mathbf{u} - \mathbf{t}^T \mathbf{u} = 0 \\ x\mathbf{f}_1^T \mathbf{v} - y\mathbf{f}_2^T \mathbf{v} - \mathbf{t}^T \mathbf{v} = 0 \end{cases} \qquad (32)$$

4.4 Recovering the motion and structure of 3D lines from the Token Tracker or : what about the accelerations ?

So far we have seen how, from matches obtained from stereo and from computing optical flow in two images we were able to compute motion and structure without computing any more stereo matches. An intriguing question is, can we in fact do better and actually compute motion and structure from the line optical flow. We saw previously that it is impossible if we compute only the image velocities, but what about if we also compute the accelerations ? we now show that if we assume that the values of $\ddot{n}$ can be estimated from the images then, five polynomial equations in Ω, $\dot{\Omega}$, V, and $\dot{V}$ can be derived.

We take equation (11) and take its time derivative:

$$\dot{l} = (n^T \dot{\Omega})n - \|n\|^2 \dot{\Omega} + n \wedge \ddot{n} + (\dot{n}^T \Omega)n + (n^T \Omega)\dot{n} - 2(n^T \dot{n})\Omega \tag{33}$$

we then use equation (7) and replace l by its value given by equation (11) in it:

$$\dot{l} = (n^T \Omega)\Omega \wedge n + (\dot{n}^T \Omega)n - (n^T \Omega)\dot{n} \tag{34}$$

The final equation is

$$(n^T \dot{\Omega})n - \|n\|^2 \dot{\Omega} + 2(n^T \Omega)\dot{n} - 2(n^T \dot{n})\Omega - (n^T \Omega)\Omega \wedge n + n \wedge \ddot{n} = 0 \tag{35}$$

We notice that this equation is linear in $\dot{\Omega}$ and of degree 2 in Ω. It can be further simplified by assuming that $\|n\| = 1$ (which we can always do); this implies that $n^T \dot{n} = 0$, and the previous equation becomes:

$$(n^T \dot{\Omega})n - \dot{\Omega} + 2(n^T \Omega)\dot{n} - (n^T \Omega)\Omega \wedge n + n \wedge \ddot{n} = 0 \tag{36}$$

We now derive the last two equations from r. We use equations (14) and (17). Taking the time derivative of the second one, we write immediatly:

$$\dot{\theta}h + \theta\dot{h} = \frac{1}{\gamma}B(V + \theta\Omega \wedge h) \tag{37}$$

which establishes two equations relating Ω, $\dot{\Omega}$, V, and $\dot{V}$ to the measurements n, $\dot{n}$, and $\ddot{n}$.

If we assume constant velocities ($\dot{\Omega} = \dot{V} = 0$), a reasonable assumption for short time scales, we have five equations in the five unknowns which are the coordinates of Ω and the coordinates of V defined up to a scale factor. Notice that the last two equations (37) are linear in V. Thus, one line observed at one time instant is sufficient, in principle, to estimate Ω and the direction of V. Notice that equation (36) allows only to recover $\Omega_\perp$ and that once motion has been estimated, structure can be recovered from equations (11) and (17). This opens the door to some very interesting possibilities of cooperation between two token trackers and a stereo algorithm

5   Conclusions

We have derived the motion equations for a line, i.e the differential equations for its direction (7) and its point of intersection with a coordinate plane (14). From these equations, we have derived the motion and structure equations that relate the optical flow of a line to its kinematic screw and its 3D representation (equations (11) and (17)). We have further shown that if the time derivative of the optical flow was available, then five equations could be derived between the kinematic screw, the optical flow, and their time derivatives.
We are now in the process of experimenting with these equations and results of their application to robot navigation will be presented at the time of the conference.

Acknowledgments We have enjoyed many discussions with Thierry Vieville during the progress of this work.

References

[1] N. Ayache and F. Lustman. Fast and reliable passive trinocular stereovision. In *Proc. First International Conference on Computer Vision*, pages 422–427, IEEE, June 1987. London, U.K.

[2] O. D. Faugeras, R. Deriche, N. Ayache, F.Lustman, and E. Giuliano. Depth and motion analysis: the machine being developed within esprit project 940. In *Proceedings of the IAPR Workshop on Computer Vision (Special Hardware and Industrial Applications), Tokyo, Japan*, pages 35–44, October 1988.

[3] O. D. Faugeras, F. Lustman, and G. Toscani. Motion and structure from point and line matches. In *Proceedings of the First International Conference on Computer Vision, London*, pages 25–34, June 1987.

[4] G. Toscani, R. Deriche, and O.D. Faugeras. 3d motion estimation using a token tracker. In *Proceedings of the IAPR Workshop on Computer Vision (Special Hardware and Industrial Applications), Tokyo, Japan*, pages 257–261, October 1988.

Solids velocity estimation in two-phase turbulent flow
(as in circulating fluidized bed)

A. Borys
Technische Universität Hamburg-Harburg

1. Introduction

The purpose of this paper is to report about some specific problems arising in estimation of solids velocities in two-phase flow (as, for example, in a circulating fluidized bed [1]).

In [2] it has been deduced from measurements that the gas-solids flow in a circulating fluidized bed can be modelled by two phases: a dense phase and a lean phase. And of great importance is the knowledge of the local solids velocity distributions in both of these phases.

One of the tasks in [3] was to gain such distributions by processing on measured data.

2. Transit-time modelling

The measurements were carried out with an optic probe consisting of two spatially separated sensors. Hence it would be quite naturally to take over the time delay of arrival (TDOA) model used so widely in, e.g. sonar speed measurement applications [4], for the description of signals registered in our optical setup. This model has the following form:

$$\text{Sensor No. 1} \longrightarrow \quad x_1(t) = s(t) + n_1(t)$$
$$\text{Sensor No. 2} \longrightarrow \quad x_2(t) = s(t - D_o) + n_2(t) \tag{1}$$

where it is assumed that $s(t)$, $n_1(t)$ and $n_2(t)$ represent stationary, zero mean, band-limited signals uncorrelated with each other [5]. D_o is the unknown time-delay.

In one regard, however, the above model does not correspond at all with our problem of modelling the transit-time of solids in a circulating fluidized bed. In our case the transit-time is not constant and changes continually. However, this can be taken into account in the model of eq. (1) by introducing $D(t)$ instead of D_o, where $D(t)$ is assumed to possess the same properties as $s(t)$, $n_1(t)$, and $n_2(t)$. Note that the above assumption according to $D(t)$ is plausible in the case of a circulating fluidized bed working in the so-called "stationary state".

Now we rewrite the eqs. (1) for the case of $D(t)$

$$x_1(t) = s(t) + n_1(t)$$
$$x_2(t) = s\left[t - D(t)\right] + n_2(t) \tag{2}$$

Note that the model represented by eqs. (2) could be given a name a <u>stochastic-delay model</u>, and that the form of eqs. (2) is identical with the description of the so-called time-varying delay model, which is in very detail discussed by Stuller in his recently published paper [6]. Only difference between these two models lies in the fact that here $D(t)$ represents a stochastic process, and in [6] $D(t)$ is assumed to be some unknown deterministic function.

A question could now arise whether the model of eqs. (2) is already satisfactory for our purposes to model the flow process in a circulating fluidized bed. We can give a positive answer, in the sense of considering the first—order approximation model.

Let us now find the crosscorrelation function between the signals $x_1(t)$ and $x_2(t)$ given by eqs. (2), i.e.

$$R_{x_1x_2}(\tau) = E\left\{ x_1(t)x_2(t+\tau) \right\} \tag{3}$$

where $E\{\ \}$ denotes the expectation operator over the ensembles of the stochastic signals $s(t)$, $n_1(t)$, $n_2(t)$, and $D(t)$.

Substituting $x_1(t)$ and $x_2(t)$ from (2) into (3) gives

$$R_{x_1x_2}(\tau) = \mathop{E}_{D}\left\{ \mathop{E}_{s}[s(t)s(t-D(t+\tau)+\tau)|_D] \right\} + \\ + \mathop{E}_{D}\left\{ \mathop{E}_{n_1,s}[n_1(t)s(t+\tau-D(t+\tau))|_D] \right\} \tag{4}$$

On the other hand,

$$\mathop{E}_{s}\left\{ s(t)s(t-D(t+\tau)+\tau)|_D \right\} = R_{ss}[\tau-D(t+\tau)] \tag{5a}$$

and

$$\mathop{E}_{n_1,s}\left\{ n_1(t)s(t+\tau-D(t+\tau))|_D \right\} = R_{n_1s}[\tau-D(t+\tau)] = 0 \tag{5b}$$

Substituting (5a) and (5b) into (4) gives finally

$$R_{x_1x_2}(\tau) = \mathop{E}_{D}\left\{ R_{ss}(\tau-D) \right\} = \int_{-\infty}^{\infty} R_{ss}(\tau-D)f_D(D)\,dD \tag{6}$$

where $f_D(D)$ is the probability density function of D.

Transferring eq. (6) into the frequency-domain, we have

$$G_{x_1x_2}(f) = G_{ss}(f) \cdot G_{f_D}(f) \tag{7a}$$

or in other form

$$G_{f_D}(f) = \frac{G_{x_1x_2}(f)}{G_{ss}(f)} \tag{7b}$$

where $G_{f_D}(f)$ is the Fourier transform of $f_D(D)$.

Observation regarding eq. (7):

There is, under the assumption of knowledge of the power spectra $G_{x_1x_2}(f)$ and $G_{ss}(f)$, a possibility to determine the probability density function $f_D(D)$ globally. Replacing in (7b) $G_{x_1x_2}(f)$ and $G_{ss}(f)$ by their estimates, we have

$$\hat{G}_{f_D}(f) = \frac{\hat{G}_{x_1x_2}(f)}{\hat{G}_{x_1x_1}(f) - \hat{G}_{n_1n_1}(f)} \tag{8a}$$

and further

$$\hat{f}_D(f) = \mathfrak{F}^{-1}\left\{ \hat{G}_{f_D}(f) \right\} \tag{8b}$$

(Note that knowing $f_D(D)$ it is easily to find a solids-velocity distribution $f_V(v)$ from $f_V(v) = (L/v^2) \cdot f_D(L/v)$, where L is a distance between the sensors.)

Consider now the above method in more detail and assume for a moment that $D(t)$ equals some constant D_o. Then $G_{x_1x_2}(f)$ is equal to $G_{ss}(f) \cdot \exp(-j2\pi fD_o)$, and, from (7b), we have $G_{f_D}(f) = \exp(-j2\pi fD_o)$ or $f_D(D) = \delta(D - D_o)$, where $\delta(\cdot)$ means the Dirac-impulse. In practice, however, we would obtain in this case, from (8a), the following result:

$$\hat{G}_{f_D}(f) = \hat{G}_A(f)\, e^{-j2\pi fD_o}$$

$$\Updownarrow \tag{9}$$

$$\hat{f}_D(D) = \delta(D - D_o) * \int_{-\infty}^{\infty} \hat{G}_A(f)\, e^{j2\pi fD}\, df$$

where $\hat{G}_A(f) \neq 1$ being a function of $\hat{G}_{x_1x_2}(f)$, $\hat{G}_{x_1x_1}(f)$, and $\hat{G}_{n_1n_1}(f)$. The symbol "$*$" means the convolution operation.

Eq. (9) shows that the function $f_D(D) = \delta(D - D_o)$ is spread by the inverse Fourier transform of $\hat{G}_A(f)$. In consequence of this fact, and without *a priori* information $D = D_o$, we would not able to decide whether we have had really to do with the case of constant D or D representing a stochastic process. (Note that with the assumption of constant $D = D_o$, $\hat{D}_o = \max[\hat{f}_D(D)]$ can be used as an estimate of D_o.)

It is difficult to determine the value of errors caused by the not-very-good knowledge of the spectra $G_{x_1x_2}(f)$, $G_{x_1x_1}(f)$, and $G_{n_1n_1}(f)$ on the final results of the TDOA estimation. Chan *et al.* [7] have shown however that these errors can take on significant values, and, for example, disqualify in practice the theoretically optimum maximum-likelihood method (based on the knowledge of spectra) compared with others.

The disadvantages of the "global" approach presented above caused that in our concrete problem of finding the local distributions of solids velocity in a circulating bed, we decided to apply some other method. This method wil be discussed in the next section. However, before beginning with this material, we would like to call the reader's attention yet to the two facts:

1. It would be valuable to point out that there is the relation between the "global" method of eqs. (7) and (8) and the Wiener-filtering. One can easily show that

$$G_{f_D}(f) = \frac{H_W(f)}{1 - \dfrac{G_{n_1n_1}(f)}{G_{ss}(f) + G_{n_1n_1}(f)}} \tag{10}$$

where $H_W(f)$ is the transfer function of the Wiener-filter (which filters $x_1(t)$ to give the minimum mean square error (mmse) approximation to $x_2(t)$). Note that for

$$\frac{G_{n_1n_1}(f)}{G_{ss}(f) + G_{n_1n_1}(f)} \ll 1, \quad G_{f_D}(f) \cong H_W(f)$$

and $f_D(f)$ could be obtain as the impulse response of the corresponding Wiener-filter. Moreover, the Wiener-filtering corresponds directly to the generalized correlation method based on the so-called Roth weighting [8].

2. The transit-time model of eq.(2) corresponds to such a situation where the solids constitute only one phase. When one is interested in the refined structure, i.e. knowing separately what happens in the dense and lean phase, then one has to use the modified model:

$$x_1(t) = s_d(t) + s_\lambda(t) + n_1(t)$$
$$x_2(t) = s_d\big[t - D_d(t)\big] + s_\lambda\big[t - D_\lambda(t)\big] + n_2(t) \tag{11}$$

where the indices "d" and "λ" correspond to the dense and lean phase, respectively. On assuming that the signals $s_d(t)$, $s_\lambda(t)$, $D_d(t)$, and $D_\lambda(t)$ are uncorrelated with each other (although this can be questionable in a circulating bed), the equivalent of eq.(6) has the form

$$R_{x_1x_2}(\tau) = \int_{-\infty}^{\infty}\big[R_{s_ds_d}(\tau - D)f_{D_d}(D) + R_{s_\lambda s_\lambda}(\tau - D)f_{D_\lambda}(D)\big]dD \tag{12}$$

Note from (12) that only when

$$R_{s_ds_d}(\tau - D) = R_{s_\lambda s_\lambda}(\tau - D) = R_{ss}(\tau - D), \tag{13}$$

we can write

$$f_{D_d}(D) + f_{D_\lambda}(D) = f_D(D) \tag{14}$$

3. A method of segment-by-segment searching

The basic idea of this method can be briefly described as follows:
1. find a segment (pattern) in $x_1(t)$ possessing an element (or a number of elements) belonging to a dense (lean) phase,
2. detect (recognize) this segment in $x_2(t)$,
3. find the difference in the instants of occurrence of the pattern and its recognized version,
4. go to the next segment until all the registered samples of the signal $x_1(t)$ are searched.

An example of a solids-velocity distribution obtained by the use of the above method is shown in Fig. 1.

It should be noted here that the search-approach is presented in more detail in a report [3].

4. Summary

A method of "global" determination of the local solids-velocity distributions in a circulating bed has been criticized. As an alternative, a method of segment-by-segment searching has been presented. The first results obtained with the use of this method show its practical usefulness, and a further work to refine it is in progress.

Acknowledgement

This work was supported by the Deutsche Forschungsgemeinschaft (DFG).

References

[1] E.-U. Hartge, Y. Li, and J. Werther, "Analysis of the local structure of the two phase flow in a fast fluidized bed", in: Circulating Fluidized Bed Technology, Pergamon Press, Canada, 1986, pp. 153-160.
[2] E.-U. Hartge, D. Rensner, and J. Werther, "Solids concentration and velocity

patterns in circulating fluidized beds", Preprints of the 2nd Int. Conf. on Circulating Fluidized Beds (Compiegne,´Frankreich), 1988.

[3] N. Fliege, A. Borys, "Digitale Verarbeitung von Meßsignalen zur Erfassung der lokalen Struktur von Gas-Feststoff–Strömungen", Arbeitsbericht für die DFG (SFB 238), Technische Universität Hamburg-Harburg, 1988.

[4] Special Issue on Time Delay Estimation, IEEE Trans. Acoust., Speech, Signal Processing, vol. ASSP-29, June 1981.

[5] M. Azaria, and D. Hertz, "Time delay estimation by generalized cross-correlation methods", IEEE Trans. Acoust., Speech, Signal Processing, vol. ASSP-32, April 1984.

[6] J. Stuller, "Maximum-likelihood estimation of time-varying delay - Part I", IEEE Trans. Acoust., Speech, Signal Processing, vol. ASSP-35, March 1987.

[7] Y. Chan, J. Riley, and J. Plant, "A parameter estimation approach to time-delay estimation and signal detection", IEEE Trans. Acoust., Speech, Signal Processing, vol. ASSP-28, Feb. 1980.

[8] Ch. Knapp, and G. Carter, "The generalized correlation method for estimation of time delay", IEEE Trans. Acoust., Speech, Signal Processing, vol. ASSP-24, August 1976.

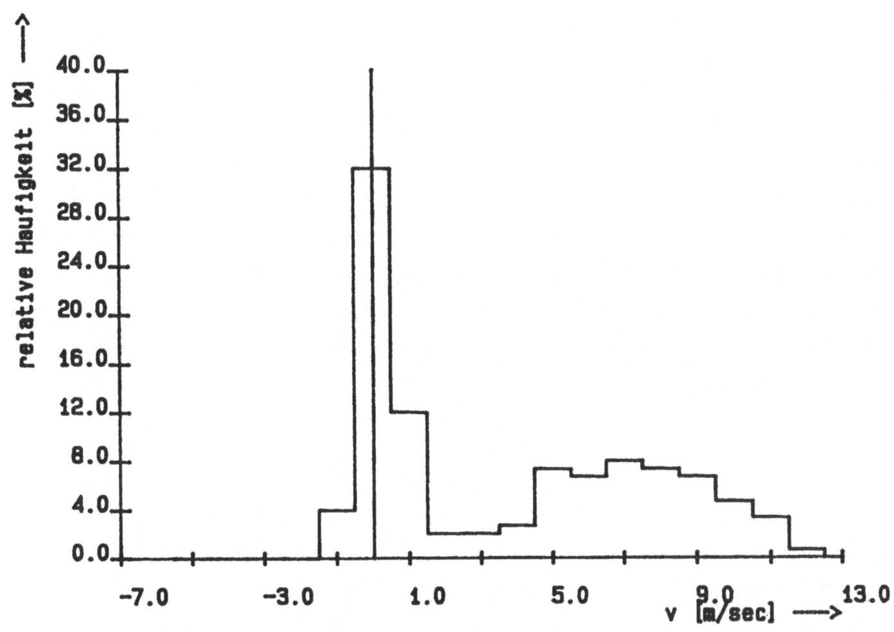

Fig.1. An example of a solids-velocity distribution.

535

Auswertung der Fokusintonation im gesprochenen Dialog

R. Bannert, J. Ph. Hoepelman & J. Machate
Fraunhofer Gesellschaft - IAO, Stuttgart

Einführung

Fokuserkennung und Fokusinterpretation bilden einen wichtigen Grundbaustein sprachlicher Kommunikation. Nur ein sprachverarbeitendes System, das in der Lage ist, Intonation zu erkennen, zu behandeln und selbst zu erzeugen, wird in Zukunft die benötigte Akzeptanz beim Benutzer finden. Im Rahmen des Schwerpunktprogramms "Modelle und Strukturanalyse bei der Auswertung von Bild- und Sprachsignalen" der Deutschen Forschungsgesellschaft wird am Fraunhofer Institut für Arbeitswirtschaft und Organisation ein Ansatz zur Auswertung der Fokusintonation im gesprochenen Dialog realisiert, der sich am Prinzip des Dialogtableau-Kalküls orientiert.

Fokuserkennung

Seit längerer Zeit herrscht in der Phonetik und der Perzeptionsforschung allgemein Übereinstimmung darüber, daß gesprochene Sprache durch die Prosodie strukturiert wird, was dem Hörer die Verarbeitung des ankommenden Sprachsignals erleichtert. Dies geschieht durch die zielgerichtete prosodische (tonale oder temporale) Aufteilung der Lautketten in Takte und prosodische Einheiten (Phrasen), deren Teile aber gleichzeitig tonal zusammengehalten werden. Prosodie umfaßt den Rhythmus und die Melodie (Intonation) der gesprochenen Sprache. Unter dem Sprachrhythmus versteht man die temporale Struktur von Segmenten und Silben. Die Intonation, die sich akustisch als Variation der Grundfrequenz (Fo) manifestiert, umfaßt im Deutschen erstens die Wortintonation, die sich im Wortakzent einer Silbe des Wortes oder Taktes äussert und zweitens die Satzintonation, die die größere Ebene des Satzes oder der Äußerung als Domäne hat und die den Satztyp, den Fokus, die Phrasierung und die Textverknüpfung signalisiert.

Der Fokus markiert das oder die bedeutungswichtigsten Wörter einer Äußerung. Fokussierte Wörter sind also die semantischen Stützpfeiler sprachlicher Kommunikation. Außer durch syntaktische Mittel, wie Wortstellung, wird der Fokus im Deutschen auch durch verschiedene phonetische Mittel signalisiert: in erster Linie tonal, aber auch temporal, spektral und durch höhere Intensität. Voraussetzung für die Erkennung des Intonationsfokus ist die Bestimmung und Auswertung des Fo-Verlaufs aus dem Signal. Die Information über die Fo-Werte entlang der Zeitachse ist die Eingabe für den Algorithmus zur Erkennung der Fokusintonation.

Algorithmus zur Erkennung von Wortakzenten

Bedingt durch die Sprachproduktion enthält das Sprachsignal stimmhafte und stimmlose Segmente. Die Grundfrequenz Fo ist also im akustischen Bereich nur an den Stellen der Äußerung zu finden, wo die Stimmlippen bei der Lautbildung schwingen. Auch bei fehlenden Teilen der Fo in einer Äußerung perzipiert der Hörer eine vollständige, ununterbrochene Melodie.

Eine analysierte Fo-Kurve enthält aber auch noch andere Eigenschaften, die als Folge der Lautproduktion der eigentlichen Fo-Kurve superponiert sind. Es handelt sich hier um die Erscheinung der sog. Mikrointonation, die durchaus als Störung der ursprünglichen, idealen Fo-Kurve zu sehen ist.

Generell ist festzustellen, daß jeder Laut durch seine spezifischen Eigenschaften die Frequenz der Stimmlippenschwingungen - und damit die Fo - in unterschiedlichem Maße beeinflußt. Diese Beeinträchtigungen der eigentlichen Fo zeigt sich besonders deutlich vor bzw. nach den Lücken in der Fo, die von stimmlosen Konsonanten stammen. Aber auch voll stimmhafte Konsonanten, vor allem stimmhafte Obstruenten wie z. B. [v, z, b] verändern lokal die Fo, indem die Fo in ihrem segmentalen Bereich stark absinkt, um gleich wieder stark zu steigen.

Weitere Fehler finden sich in den Werten der Fo-Analyse, die durch den Algorithmus der Fo-Analyse hervorgerufen werden. So kommt es vor, daß Fo-Werte überflüssigerweise angezeigt werden, z.B. mitten in stimmlosen Konsonanten, wo ja der Fo-Wert gleich Null ist, oder aber, daß Fo-Werte vereinzelt fehlen.

Bei der traditionellen Bearbeitung der Fo-Kurven versteht es der Phonetiker, diese genannten Fehler bzw. Abweichungen von der idealen, d.h. vollständigen und ungestörten Fo-Kurve dank seiner Kenntnisse, seines Wissens und seiner Erfahrung visuell zu berichtigen.

Deshalb erweist es sich als notwendiger Schritt, die Fo-Werte der Analyse automatisch zu berichtigen und zu ergänzen. Wir versuchen, es dem Phonetiker gleichzutun, indem wir von den Fo-Werten der Analyse die richtigen Werte bzw. Wertfolgen (d.h. Teile der Kurve) behalten wollen, die gestörten Teile der Kurve aber berichtigen bzw. ergänzen.

Nach dieser Berichtigung bzw. Nachbesserung der Fo-Werte einer Äußerung ergaben sich die Wortakzente dieser Äußerung als relativ deutliche und relativ große tonale Veränderungen in der Fo-Kurve der gesamten Äußerung. Diese tonalen Veränderungen können verschiedener Art sein: Ein Wortakzent zeigt sich als Gipfelakzent durch einen tonalen Anstieg, dem unmittelbar ein Fall folgt. Als Brückenakzent besteht der erste Akzent aus dem Anstieg, der zweite aus dem Fall. So gesehen ließe sich der Brückenakzent aus den beiden Hälften von zwei Gipfelakzenten betrachten.

Aus phonetisch-akustischer Sicht sind zur erfolgreichen automatischen Erkennung der Wortakzente allein mit Hilfe tonaler Information (Intonation) folgende Schritte notwendig, wobei zwei Hauptkomponenten zu unterscheiden sind: Restaurierung der aktuellen Fo-Kurve und Wortakzenterkennung.

Restaurierung der Fo-Kurve

1. Streichen falsch angezeigter Fo-Werte. Sie treten vereinzelt in größeren Lücken der Kurve auf (bedingt durch die Fo-Analyse).

2. Ergänzen fehlender Fo-Werte. Sie treten vereinzelt auf und sind ebenfalls durch die Fo-Analyse bedingt.

3. Glättung verzerrter Fo-Werte nach bzw. vor Lücken in der Kurve, die stimmlosen Konsonanten entsprechen. Diese falschen Fo-Werte sind eine Folge von Segmenteigenschaften.

4. Begradigung der Fo-Kurve über die lokale Störung (Fall-Anstieg) von stimmhaften Obstruenten.

5. Überbrückung der Lücken der stimmlosen Konsonanten. Die Vervollständigung der Kurve erfolgt, indem Ende und Beginn der Kurve um eine Lücke nicht linear (d.h. interpoliert), sondern mit einer Kurve höherer Ordnung verbunden werden.

Wortakzenterkennung

Nachdem die Fo-Werte berichtigt und ergänzt worden sind, beginnt die eigentliche Suche nach den Wortakzenten. Sie schlagen sich in den lokalen und relativ großen tonalen Veränderungen der Fo-Werte nieder, und zwar entweder als Anstieg oder als Fall der Fo-Kurve.

1. Feststellung des groben Kurvenverlaufs. Die Veränderung des Kurvenverlaufs nach oben bzw. unten wird festgestellt.

2. Bewertung des Kurvenverlaufs. Es wird festgestellt, ob in der Fo-Kurve ein sprachlich relevanter Verlauf vorliegt. Dabei sind zwei Bedingungen zu erfüllen:

    (a) die tonale Veränderung muß über eine genügend große Zeitspanne erfolgen.
    (b) die tonale Veränderung muß eine bestimmte Größe überschreiten

3. Markierung des entsprechenden Abschnitts auf der Zeitachse. Die tonale Erscheinung Wortakzent in der Fo-Kurve wird auf die Zeitachse projiziert und dort als Information für die Spracherkennungskomponente festgehalten.

Abb. 1 zeigt die Intonation (Fo) der Äußerung "Johánnes liebt Susánne", jeweils mit dem Wortakzent auf dem Subjekt bzw. Objekt. Zwei Intonationstypen sind dargestellt, links zwei Gipfelakzente, rechts der Brückenakzent, Zeitachse waagerecht, Frequenzachse senkrecht. Unter der Zeitachse erscheint jeweils die Markierung der beiden Akzente durch den Erkennungsalgorithmus.

Abb. 1: Fo-Kurven des Satzes "Johánnes liebt Susánne" und Markierung der Wortakzente.

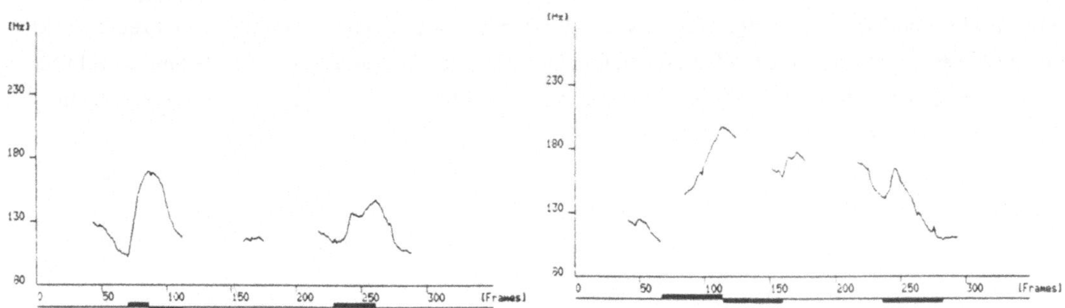

Das Ergebnis der Intonationserkennung, das sind die Sprachsegmente, über die sich der Fokus erstreckt, wird von der Spracherkennungskomponente auf die entsprechenden Worte projiziert und als Liste mit den markierten fokussierten Worten der Dialogkomponente übermittelt.

Dialogische Fokusbehandlung

Die Schnittstelle zwischen sprach- und fokuserkennender Komponente und Dialog-Komponente bildet ein Syntaxanalyse-Modul, das aus der Eingabe eine semantische Repräsentation erzeugt. Diese Repräsentation beschreibt eine operationelle Semantik, mit deren Hilfe es möglich ist im Rahmen der Tableaux Theorie entsprechende Dialogregeln zu gestalten. Eine Übersicht über die einzelnen Module des Systems bietet Abbildung 2.

Abb. 2: MAFID System-Übersicht

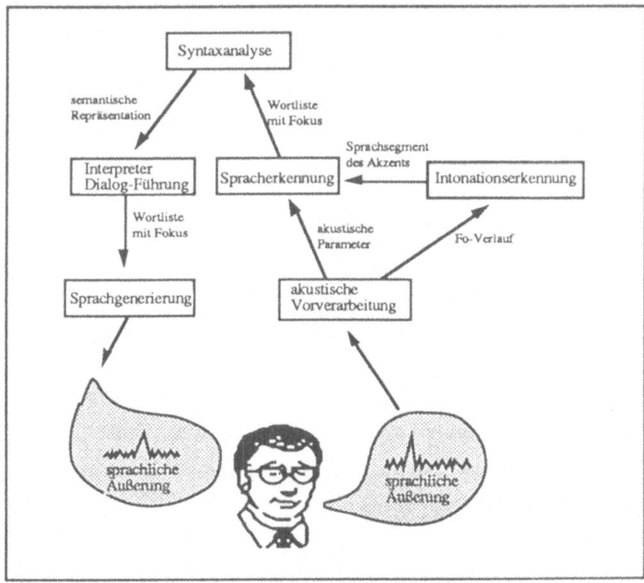

Die Forderung nach Flexibilität und Unterstützung der gerade für die Fokussierung bedeutsamen freien Wortstellung wurde in einer Syntaxanalyse-Komponente realisiert, die als Analyse-Ergebnis eine semantische Repräsentation liefert, in der von der Intonationserkennung gelieferte Wortakzente direkt integriert werden. Dazu wurde ein Fokusoperator eingeführt, der eine einheitliche Darstellung sowohl für Fokussierung mittels Intonation als auch durch sprachliche Mittel, wie z.B. Gradpartikel oder Negation, erhält. Im folgenden Beispiel wird die unterschiedliche Repräsentation der Frage "steht portix im ersten stock?" anhand zwei verschiedener Intonationsmuster veranschaulicht. Dabei kennzeichnet das Ausrufezeichen ein durch Intonation hervorgehobenes Wort.

Beispielsatz: steht portix im ersten stock?

(1)　　steht !portix im ersten stock?

(1')　　fokus(int, portix)
　　　　quest(X, [])
　　　　adjunct(X, in ($ref0))
　　　　phrase(X, steht, portix, [], [])

(2)　　steht portix im !ersten stock?

(2')　　fokus(int, erst($ref0))
　　　　quest(X, [])
　　　　adjunct(X, in($ref0))
　　　　phrase(X, steht, portix, [], [])

Als Anwendungsbeispiel wurde ein Informationssystem implementiert, dessen Wissensbasis über die Rechnerverteilung mittels gesprochener Sprache befragt oder korrigiert werden kann. ("portix" ist der Name eines der Rechner des Instituts).

Beschreibung eines semantischen Modells zur Behandlung von Fokus bei der Konstruktion und Abfrage einer Datenbasis

Als Paradigma zur Beschreibung eines semantischen Modells für die Auswertung der Fokusintonation diente die Theorie der Dialogspiele, in der es möglich ist, anhand von Satzoperatoren eine Zerlegung des Satzes in sprecher- und hörerrelevante Pflichten oder Rechte festzulegen. Ein unvollständiges Tableau, das eine Dialogsituation zwischen System und Benutzer beschreibt, zeigt Abbildung 3.

Beide Fragen des Benutzers können nicht positiv beantwortet werden, so daß die Dialogführungskomponente nach einer alternativen Antwort sucht. Bei der Suche nach einer alternativen Antwort müssen Fragen der thematischen Relevanz berücksichtigt werden, so daß der Benutzer nicht mit unsinnigen oder überflüssigen Systemantworten konfrontiert wird.

Abb. 3: Dialogmodell (informal)

<table>
<tr><td>System</td><td>Benutzer</td></tr>
<tr><td>portix steht im 2. stock
postfix steht im 1. stock</td><td></td></tr>
<tr><td>nein, !postfix steht im 1. stock</td><td>steht !portix im 1. stock?

steht portix im !1. stock?</td></tr>
<tr><td>nein, portix steht im !2. stock</td><td></td></tr>
</table>

Eine Regel zu Interpretation des Fokus bei ja/nein- Fragen bietet die folgende Regel, in der ein Dialog-zustand zwischen Benutzer und System dargestellt wird als ein Tripel der Form <DB, S, U>, mit DB als interner Datenbasis, S als Menge der Systemantworten und U als Menge der analysierten Benutzereingaben.

Abb. 4: Verwendung des Intonationsfokus in ja/nein Fragen

$$
\begin{aligned}
&\text{sei} \quad \alpha = [\,\text{fokus}(\text{int}, F), \text{quest}(X, [\,])|P\,] \\
&\qquad\qquad \text{die analysierte Benutzereingabe, dann folgt} \\
&\qquad\qquad \text{aus} \\
&\qquad \Phi = \; <DB, S, [\alpha|U]> \\
&\text{i) falls } P \in DB \\
&\qquad \Psi = \; <DB, [\,"ja"|S], [\alpha|U]> \\
&\text{ii) falls } \exists G(P' = P_{F/G} \in DB) \\
&\qquad \Psi = \; <DB, [\,"nein, [fokus(\text{int}, G)|P']"|S], [\alpha|U]> \\
&\text{iii) sonst} \\
&\qquad \Psi = \; <DB, [\,"weiß nicht"|S], [\alpha|U]>
\end{aligned}
$$

Weitere Regeln beschreiben die Verwendung der Fokusintonation als Widerspruchsmittel zur Korrektur oder auf der Systemseite als Mittel zur Hervorhebung relevanter Konstituenten bei der Beantwortung von W-Fragen. Ferner zeigten Gabbay und Moravczik (1978) und Hoepelman (1979), daß eine enge Beziehung zwischen Negation und Fokusintonation existiert, wobei letztere den Skopus der Negation bestimmt. So gibt es einen bedeutungsrelevanten Unterschied zwischen den folgenden Sätzen, die rein syntaktisch gesehen, die gleiche Struktur erhalten

(3)   !portix steht nicht im ersten stock.

(4)   portix !steht nicht im ersten stock.

Ausblick

Wir haben gezeigt, daß es möglich ist, ein prototypisches System zur integrierten Auswertung der Fokusintonation zu entwickeln und zu implementieren, in dem der Benutzer bei kontinuierlich gesprochener Eingabe eine die Fokusintonation berücksichtigende und anwendende generierte Ausgabe erhält.

Die Projektarbeit in den folgenden Jahren wird als Schwerpunkt zum einen den Algorithmus zur Wortakzenterkennung anhand unterschiedlicher prosodischer Kontexte verfeinern und empirisch absichern. Zum anderen sollen Konzepte der thematischen Relevanz in die Dialogführung integriert und die vorhandenen Regeln weiter ausgebaut werden. Als Schlußbemerkung sei noch erwähnt, daß die Entwicklung der Spracherkennungskomponente und der Sprachgenerierungskomponente nicht Bestandteil des Projekts ist, sondern auf hier am Institut entwickelte Modulen zurückgegriffen wird. Dabei wurde für die Spracherkennung das mit Markov-Modellen arbeitende System COSIMA verwendet, das bei kontinuierlich gesprochener Sprache, Sprecherunabängigkeit und Echtzeitverhalten leicht adaptierbar ist. Anliegen des Projekts jedoch ist die Entwicklung eines Algorithmus zur Wortakzenterkennung anhand des Fo-Verlaufs und dessen Integration in ein System mit Regeln zur Interpretation der Fokusintonation .

Literatur

Bannert, Robert. (1985 a)
"Towards a Model for German Prosody"
in: Folia Linguistica XIX, 321-341.

Bannert, Robert. (1985 b)
"Fokus, Kontrast und Phrasenintonation im Deutschen"
Zeitschrift für Dialektologie und Linguistik 52, 289-305.

Barth, E.M. & Krabbe, E.C.W. (1982)
From Axiom to Dialogue. A Philosophical Study of Logics and Argumentation
Berlin

Carlson, L.(1984)
"Focus and Dialogue Games"
in: Vaina, L. & Hintikka, J. (Eds), Cognitive Constraints on Communication,
Dordrecht

Gabbay, D.M. & J.M. Moravcsik (1978)
"Negation and Denial"
in: Günthner, F. & Ch. Rohrer (Eds), Studies in Formal Semantics. Intensionality,
Temporality, Negation, Amsterdam

Hoepelman, J.Ph. (1979)
"Negation and Denial in Montague Grammar"
in: Theoretical Linguistics, Vol. 6

Der IBM Spracherkennungsprototyp TANGORA

Anpassung an die deutsche Sprache

G. Walch, K. Mohr
U. Bandara, J. Kempf, E. Keppel, K. Wothke
IBM Wissenschaftliches Zentrum Heidelberg
Tiergartenstraße 15, D 6900 Heidelberg

Einleitung

TANGORA ist ein sprecherabhängiges Spracherkennungssystem für Satzerkennung bei isoliert gesprochenen Wörtern. Es ist ausgelegt für die Erkennung eines großen Vokabulars von etwa 20000 Wortformen in Echtzeit. Das System arbeitet mit rein statistischen Methoden ohne Anwendung von linguistischem Wissen. Die Erkennungsrate liegt bei der 20k Version für Englisch bei 95 bis 98%. Die deutsche Version kennt zur Zeit etwa 1300 Worte, das Vokabular wird im Laufe des Jahres auf etwa 10000 Worte erweitert.

TANGORA ist ein Forschungsprototyp basierend auf langjähriger Entwicklungsarbeit unserer Kollegen in IBM USA /JEL85/ /AAV86/. Der Name wurde in Erinnerung an den Weltrekordler im Maschinenschreiben Albert Tangora gewählt. Das System ist implementiert auf einem PC/AT und 4 Spezialkarten mit schnellen Signalprozessoren und schnellem Speicher.

Auf die allgemeine Problematik der Spracherkennung soll hier nur kurz eingegangen werden. Es genügt auf die große Variabilität des Sprachsignals für einen Sprecher abhängig von Sprechgeschwindigkeit, Stimmung, Konzentration hinzuweisen. Hinzu kommen die Variationen zwischen verschiedenen Sprechern auf Grund verschiedener Stimmlage, Intonation und regionaler Ausspracheunterschiede, die eine sprecherunabhängige Spracherkennung mit großem Wortschatz in Echtzeit mit der heutigen Technik als nicht realisierbar erscheinen lassen.

Im ersten Teil dieses Beitrags wird der prinzipielle Ablauf des Erkennungsprozesses bei TANGORA erläutert. Im zweiten Teil werden dann die zur Anpassung von TANGORA für die deutsche Sprache notwendigen Arbeiten beschrieben.

Erkennungsprozeß

Die dem System zu Grunde liegende statistische Theorie wird durch die folgenden Formeln ausgedrückt: /DOR87/

$$P(Ws|A) = \max_W P(W|A) \tag{1}$$

$$\max_W P(W|A) = \max_W \frac{P(A|W) \times P(W)}{P(A)} \tag{2}$$

$$Ws = \underset{W}{argmax}(P(A|W) \times P(W)) \tag{3}$$

Ziel ist es, diejenige Wortfolge Ws zu finden, die bei beobachteter akustischer Symbolfolge A die bedingte Wahrscheinlichkeit $P(W|A)$, also die Wahrscheinlichkeit der Wortfolge W bei gegebener Akustik A, maximiert (1).

Diese kann jedoch in (2) mit Hilfe der Bayes' Formel ausgedrückt werden als Produkt der Wahrscheinlichkeit $P(A \mid W)$, daß die artikulierte Wortfolge W die akustische Symbolfolge A produziert, und der a priori Wahrscheinlichkeit $P(W)$, daß gerade diese Wortfolge gesprochen wird, geteilt durch die a priori Wahrscheinlichkeit $P(A)$. Da $P(A)$ jedoch von W unabhängig ist, kann die Erkennungsaufgabe durch Ausdruck (3) beschrieben werden: es ist diejenige Wortfolge W gesucht, welche das Produkt $P(A \mid W) \times P(W)$ maximiert.

Wie im rechten Teil der Abbildung 1 zu sehen ist, gliedert sich die Realisierung des Erkennungsprozesses in 4 Stufen.

Signalverarbeitung und Vektorquantisierung

Das Sprachsignal wird zunächst (außerhalb des PC/AT) verstärkt und mit 20 kHz und 12 bit Auflösung digitalisiert (Datenrate 30 kB/sec). Die Daten werden auf einer der Signalprozessorkarten zwischengespeichert. Alle 10 ms wird dann über ein Fenster von 25 ms eine Fouriertransformation durchgeführt. Aus dem Fourierspektrum wird ein 20-elementiger Merkmalsvektor gewonnen, d.h. die Energien in 20 Frequenzbändern zwischen 200 und 8000 Hz, die auf Grund eines „Ohr-Modells" festgelegt sind, werden ermittelt. Dieser Vektor wird mit 200 sprecherspezifischen Prototypvektoren verglichen. Das Symbol des ähnlichsten Prototyps charakterisiert letztlich das Sprachsignal. Das akustische Signal wird somit in eine Symbolfolge(A) umgewandelt, die eine Datenrate von 100 B/sec ergibt. Die nachfolgenden Stufen der Erkennung arbeiten nur noch mit diesen komprimierten Daten. Die sprecherspezifischen Prototypvektoren werden in der Trainingsphase durch **k-Means Clustering** aus einer circa 5-minütigen Sprachprobe gewonnen.

Schnelles akustisches Modell: Fast Match

Die Aufgabe des **fast match** (FM) ist es, diejenigen Worte aus dem Vokabular zu finden, die einen großen Beitrag zum Term $P(A \mid W)$ liefern. Dafür ist jedes Wort des Vokabulars als eine Folge von phonetischen Symbolen (ähnlich einer Lautschrift) beschrieben, auch Aussprachevarianten sind zugelassen. Diese **phonetic baseforms** sind in einer Baumstruktur gespeichert.

Zu jedem phonetischen Symbol gibt es ein **Hidden Markov Model** (HMM) /RAB86/ mit 7 Zuständen und 13 Übergängen. Die Modellvorstellung ist, daß bei einem Übergang innerhalb des Markovmodells in einer Zeiteinheit (10 ms) genau ein akustisches Symbol produziert wird. Auch sogenannte Nullübergänge, die weder Zeit verbrauchen noch Symbole produzieren, sind möglich. Die Übergangs- und Ausgabewahrscheinlichkeite werden im Training ermittelt („Sprecherspezifische Daten" auf Seite 7). Das Markovmodell eines Wortes ist die Verkettung der Modelle seiner phonetischen Bestandteile.

Es sei hier betont, daß dieses System nicht auf der Erkennung einzelner Laute beruht, aus denen dann Worte synthetisiert werden, vielmehr sind die Einheiten der Erkennung ganze Wörter, aus denen ein Text gebildet wird. Beim **fast match** wird nun im Prinzip für jedes Wort des Vokabulars an Hand des Markovmodells die Wahrscheinlichkeit ermittelt, daß bei der Aussprache des Wortes die beobachtete Symbolfolge produziert wird. Da beim Abarbeiten der Baumstruktur die Wahrscheinlichkeit entlang eines Zweiges immer kleiner wird, kann bei Unterschreiten eines Grenzwertes die Berechnung abgebrochen werden. Die 100 bis 300 Worte mit größter Wahrscheinlichkeit sind dann die Kandidaten für die weitere Verarbeitung.

Sprachmodell

Im dritten Schritt der Verarbeitung wird für alternative Hypothesen über die Wortfolge (Pfade) der Term $P(W)$ ermittelt, und zwar für die im **fast match** gefundenen Wortkandidaten. Am Beginn des Erkennungsvorgangs (und am Beginn eines Satzes) wird die Hypothese (Satzanfang, $w1$) geprüft. Dafür werden a priori Wahrscheinlichkeiten der Worte $w1$ und der Folge

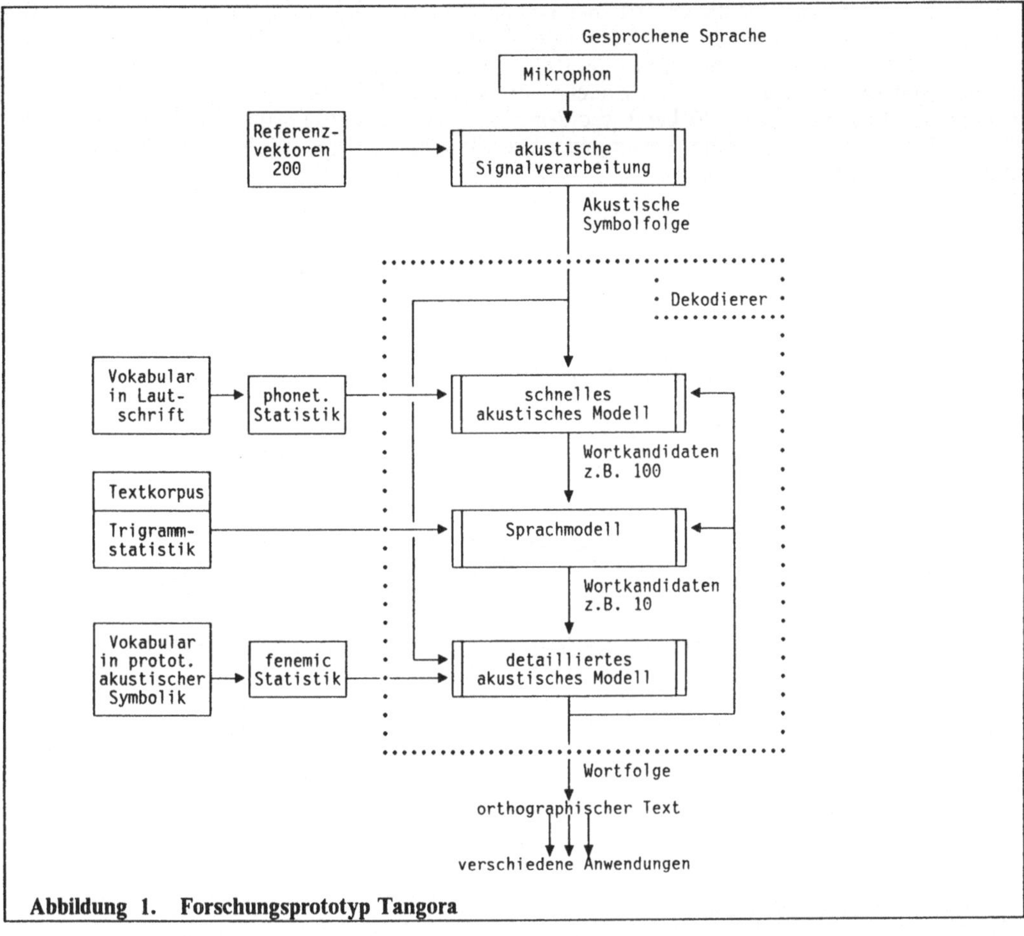

Abbildung 1.   Forschungsprototyp Tangora

(Satzanfang,$w1$) herangezogen. Die Kandidaten mit dem größten Wahrscheinlichkeitsprodukt aus FM und Sprachmodell (LM) bilden dann die Pfade für das nächste Wort.

Ab dem zweiten Wort eines Satzes sind nun die verschiedenen Kandidaten aus dem FM mit den verschiedenen Pfaden versuchsweise zu kombinieren, wobei außer den Wahrscheinlichkeiten $P(w2)$ und $P(w1,w2)$ auch die Trigramm-Wahrscheinlichkeiten $P(w0,w1,w2)$ gewichtet berücksichtigt werden. Die so gewonnenen Näherungswerte für $P(W)$ werden mit den akustischen Wahrscheinlichkeiten kombiniert, um damit die verschiedenen Pfade zu bewerten.

Die hier verwendeten Wahrscheinlichkeiten für Worte (Unigramm), Folgen von 2 Worten (Bigramm) und 3 Worten (Trigramm) wurden durch Analyse von großen Textsammlungen gewonnen.

Detailliertes akustisches Modell: Detailed Match

Die Wortkandidaten, die zu den wahrscheinlichsten Pfaden gehören, werden im 4. Schritt akustisch genauer überprüft. Im **detailed match** (DM) werden ähnlich wie im FM die Wortkandidaten an Hand von Markovmodellen und der beobachteten akustischen Symbolfolge bewertet. Im Unterschied zum FM sind die Einheiten, die durch die Modelle repräsentiert werden, nicht die bekannten Laute der deutschen Sprache, sondern künstliche Lauteinheiten (Feneme), die in ihrer mittleren Aussprachedauer (10 ms) einem Element der Symbolfolge entsprechen. Die Anzahl der Modelle entspricht der Anzahl der Prototypvektoren bzw. der verschiedenen Symbole. Jedes Modell ist für die Produktion eines bestimmten Symbols überwiegend (im statistischen Sinn) verantwortlich. Die Modelle der Worte sind wieder die Verkettung der Einzelmodelle der Feneme (etwa 100 bei 1 sec Sprechdauer). Sie werden aus der Aussprache aller Wörter durch mehrere Sprecher gewonnen. Die Modellparameter werden auch hier sprecherspezifisch trainiert.

Die Bewertung der Wortkandidaten im **detailed match** wird mit den Wahrscheinlichkeiten der bisherigen Pfade kombiniert. Daraus werden die wahrscheinlichsten aktuellen Pfade bestimmt. Den Enden dieser Pfade entspricht jeweils eine Position in der akustischen Symbolfolge. An diesen Positionen werden die Schritte 2 bis 4 (FM, LM, DM) wieder ausgeführt.

Anpassung der TANGORA für Deutsch

Im schematischen Aufbau der TANGORA in Abbildung 1 auf Seite 3 lassen sich von links nach rechts gesehen 3 Bereiche bezüglich der Sprecher- und Sprachabhängigkeit erkennen:

1. sprachspezifische Dateien
2. sprach- und sprecherspezifische Referenzmuster und Statistiken und
3. sprach- und sprecherunabhängige TANGORA Hard- und Software.

Sprach- und sprecherunabhängige TANGORA Hard- und Software

Die in den vorausgegangenen Kapitel beschriebene Spezialhardware und -software für Signalverarbeitung und Dekodierer kann auch für Deutsch verwendet werden. Zukünftige Weiterentwicklungen des englischen Systems können somit auch unmittelbar für das deutsche System übernommen werden. Die Anpassung der TANGORA an Deutsch betraf also im wesentlichen die sprachabhängigen Dateien, und die Grund-Schemata zur Gewinnung der sprecherspezifischen Daten für Deutsch.

Sprachspezifische Dateien

FM-Vokabular

Für das schnelle akustische Modell muß das gesamte Vokabular in Lautschrift transkribiert werden, Flexionsformen zählen dabei als eigenständige Worte. Für die Transkriptionen der deutschen Sprache verwenden wir zur Zeit 60 verschiedene Phone:
3 systemspezifische Phone: NULL-Phon, Stille, Satzgrenze;
25 Vokale;
30 Konsonanten;
2 „Offglides"(Ausklänge der Diphthonglaute au, eu, ei).

Die Schreibweise der Worte wurde maschinell mit Hilfe eines Regelapparates in eine IPA-Notation umgewandelt, die dann in das von uns verwandte Phonalphabet überführt wurde. Von Hand wurden danach allgemein verwendete Aussprachevarianten, wie z.B. das Verschlucken des E-Lauts in „en" Endungen, als zusätzliche Grundformen kodiert. Ein akustischer Vergleich mit Sprachaufnahmen mehrerer Sprecher führte zur Aufnahme weiterer Grundformen wie z.B. einer 2. Transkription für das Personalpronomen „es", das in unserem Umfeld auch sehr häufig mit langem E gesprochen wird. Dies führte bei 1300 Wortformen zu über 3000 Aussprachevarianten. Mit statistischen Methoden wurden schließlich in einem iterativen Verfahren die 2200 von unseren Testsprechern tatsächlich gesprochenen Grundformen ermittelt.

Dieses phonetische Vokabular wird in mehreren Teilbäumen zusammengefaßt. Dadurch ist es bei der Erkennung möglich, daß mehrere Prozessoren an unterschiedlichen Teilbäumen parallel arbeiten können.

Bei der Erstellung des FM-Vokabulars ergaben sich folgende Probleme:

Wortauswahl: Ein beschränktes Vokabular sollte unter Berücksichtigung folgender Aspekte gewonnen werden:

- hoher Abdeckungsgrad von Text auf Wortbasis,
- viele Sätze bildbar,
- Berücksichtigung aller Phone,
- genügend akustische Verwechselbarkeit, Homophone.

Mit den 1300 häufigsten Worten aus dem Wirtschaftsteil der Tageszeitung Mannheimer Morgen ließ sich ein Abdeckungsgrad von ca. 60% erzielen, jedoch sind nicht einmal 1 Prozent aller Sätze des Korpus komplett mit dem Vokabular bildbar.

Kodierung der Aussprachevarianten: Dabei stellte sich die Frage, ob einzelne Aussprachevarianten als eigenständige Grundformen oder durch spezielle Phone transkribiert werden sollen. Beispiele:

Die Aussprache einzelner Worte mit stimmhaften oder auch stimmlosem S. Diese Aussprachevarianten können nicht ignoriert werden, da der gleiche Sprecher sowohl stimmhaftes als auch stimmloses S, manchmal sogar bedeutungsunterscheidend, verwendet. Hier gibt es die Möglichkeit, entweder jedes solche Wort auf zwei Arten zu transkribieren, oder aber ein neues Phon einzuführen, das nur an den Stellen transkribiert wird, an denen sowohl stimmhaftes als auch stimmloses S vorkommen kann. Die erste Möglichkeit erhöht die Anzahl der Grundformen, und damit die benötigte Rechenzeit bei der Erkennung, die zweite Möglichkeit erhöht die Anzahl der Phone und vergrößert damit den Umfang des Trainingstextes.

Die unterschiedliche Aussprache des R-Lautes als bayrisches Zungen-R bzw. norddeutsches Zäpfchen-R benötigt weder alternative Phone noch alternative Aussprachegrundformen, da diese R-Varianten einerseits keine bedeutungsunterscheidende Funktion haben und andererseits die Statistik des für den R-Laut vorgesehenen HMMs sprecherspezifisch, und damit entweder zum Zungen-R oder zum Zäpfchen-R oder zu einer Mischung trainiert wird.

Durch die Transkription gängiger Aussprachevarianten eines Wortes, kann für ein anderes Wort die Situation entstehen, daß es akustisch nicht mehr eindeutig identifizierbar ist. Beispiel:

Die Kodierung von „Jahr" und „Ja". Durch Kodierung der, sehr gängigen, Aussprachevariante ohne den R-Laut für Jahr, entsteht für das Wort „Ja" eine Homophon-Situation, die es dem Benutzer unmöglich macht, das Wort „Ja" akustisch eindeutig zu diktieren.

Textkorpus und Trigrammstatistik

Das Sprachmodell hat die Aufgabe, die akustische Erkennung durch statistische Informationen über die Häufigkeit von Wortfolgen in einer bestimmten sprachlichen Domäne zu unterstützen. Bei Vorliegen von Homophonen, liefert es die einzige Grundlage für die korrekte Wortfindung. Das Sprachmodell ist um so sicherer, je mehr Wortfolgen beobachtet wurden, d. h. je größer der Textkorpus ist.

Die von uns ausgewählte Domäne ist der Wirtschaftsteil deutscher Zeitungen, da uns in diesem Bereich eine große Menge maschinenlesbaren deutschen Textes zur Verfügung steht. Zur Zeit enthält unser Textkorpus etwa 50 Millionen Wörter und wächst monatlich um ca. 1 Million.

Die uns zur Verfügung gestellten Texte müssen vor der Ermittlung der Wortfolgestatistiken quellenspezifisch bereinigt werden. Dazu gehört z.B. das Entfernen von Steuerzeichen, die Trennung von Überschrift, Text und Tabellen etc., das Ermitteln der Satzenden, und die Korrektur der Groß/Kleinschreibung am Satzanfang.

Danach werden für die Wörter aus dem gewählten Vokabular die Wortfolgestatistiken am bereinigten Korpus ermittelt. In diesem Schritt können Worte, die unterschiedlich geschrieben und auch ausgesprochen werden, in einer Klasse zusammengefaßt werden, um eine allgemeine Wortstatistik z.B. für die Klasse der Vornamen zu ermitteln. Das Vokabular des Sprachmodells muß jedoch in Kombination mit dem phonetischen Vokabular zu einer eindeutigen Identifikation der Schreibweise führen, d.h. die im Sprachmodell in einer Klasse zusammengefaßte Worte sollten nicht homophon sein.

Außer den Worten mit akustischer Entsprechung, gehören zum Vokabular des Sprachmodells auch die beiden Wörter:

Satzende ein künstliches Wort, dem keine Schreibweise entspricht. Damit wird die Erkennung am Beginn eines Diktats und die Entscheidung der Satzzeichenhomophone wie z.B. „." und „Punkt" verbessert.

Unbekanntes Wort Diese Wort repräsentiert alle nicht zum Vokabular gehörigen Worte. Über Ihre Aussprache liegt keinerlei Information vor.

DM-Vokabular

Das im DM verwendete Vokabular steht in einer 1:1 Beziehung mit dem Lautschrift-Vokabular des FM. Die darin enthaltenen Grundformen werden jedoch aus tatsächlichen Aussprachen mehrerer Sprecher gewonnen.

Dazu war es notwendig, das gesamte Vokabular in Sätzen zusammenzustellen, an Hand derer die Aussprache der Worte unter möglichst realistischen Bedingungen aufgenommen werden konnte. Die Aufnahme des Textes durch 9 Sprecher ergab für unser 1300 Worte umfassendes Vokabular etwa 500 MB Rohdaten. Nach der Signalverarbeitung dieser Daten wurden durch **k-Means Clustering** die 200 Systemprototypen ermittelt, die den Phon-Einheiten des DM, den sogenannten **Fenemen**, zu Grunde liegen. Insgesamt wird im DM mit 203 Phon-Einheiten gearbeitet, da auch hier die drei systemspezifischen Phone wie im FM hinzukommen.

Bei der Aufnahme des Textes wurden den Sprechern keine Vorgaben über bestimmte Aussprachevarianten gemacht. Vielmehr sollte an Hand der tatsächlichen Aussprachen mit Hilfe eines Viterbi-Algorithmus /RAB86/ eine Zuordnung zwischen phonetischen Grundformen des FM und zeitlichen Abschnitten des Sprachsignals ermittelt werden, um damit aus den vorbereiteten phonetischen Aussprachealternativen die auszuwählen, die tatsächlich gesprochen wurden.

Das DM-Vokabular wurde dann durch Mittelung der zeitlichen Abschnitte derjenigen Sprachaufnahmen erstellt, die jeweils einer Grundform zugeordnet wurden. Diese Mittelung kann im einfachsten Fall der Abschnitt mittlerer Länge sein. Bei Vorliegen trainierter DM-Statistik kann diese Mittelung wesentlich verbessert werden. Durch die Ableitung vom tatsächlich aufgenommen Sprachsignal ist im allgemeinen die Erkennungsgenauigkeit im DM höher als im FM, jedoch läßt sich auf Grund der größeren Verschiedenheiten der einzelnen Grundformen das Abarbeiten des Vokabulars im DM nicht durch eine Baumstruktur beschleunigen.

Zur Durchführung des Viterbi-Algorithmus wird neben dem Sprachsignal und den Strukturen der phonetischen HMMs auch eine trainierte Statistik benötigt. Die Strukturen der HMMs waren durch das FM-Vokabular und die vorgegebenen Sätze indirekt gegeben. Daran konnte ausgehend von einer Initialstatistik mit Hilfe der TANGORA-Trainingsprogramme die notwendige trainierte Statistik erzeugt werden. Das Problem, daß sich uns dabei stellte, war das Erzeugen von Initialwerten für die Emissionswahrscheinlichkeiten der HMMs unserer 60 deutschen Phone bezüglich der 200 Systemprototypen. Diese Initialwerte haben wir wie folgt ermittelt:

1. Aufnahme des englischen Trainingstextes von allen Referenzsprechern (siehe auch „Sprecherspezifische Daten" auf Seite 7).
2. Englisches Benutzertraining für alle Sprecher zusammen unter Verwendung der deutschen Systemprototypen für die Kodierung des Sprachsignals. Dadurch wird eine trainierte Statistik der englischen Phone bezüglich der deutschen Systemprototypen gebildet.
3. Jedem deutschen Phon wird mindestens ein englisches Phon zugeordnet.
4. Die Statistiken dieser englischen Phonmodelle für deutsche Systemprototypen werden mit einem geringen Anteil Gleichverteilung kombiniert um so die Initialstatistik für die deutschen Phonmodelle bezüglich der deutschen Systemprototypen zu gewinnen.

Mit einem Forward/Backward Algorithmus kann die Statistik der FM-HMMs trainiert werden. Die trainierten FM-HMMs liefern dann die Grundlage, um mit einem Viterbi-Algorithmus das Sprachsignal den jeweiligen Grundformen zuzuordnen und wie oben beschrieben, das DM-Vokabular aufzubauen.

Sprecherspezifische Daten

Zur Durchführung der Erkennung benötigt TANGORA folgende sprecherspezifische Daten, die in der schon mehrfach erwähnten Trainingsphase ermittelt werden:

- 200 Referenzvektoren (Benutzerprototypen)
- Übergangs- und Emissionswahrscheinlichkeiten für 60 Phon-HMMs
- Übergangs- und Emissionswahrscheinlichkeiten für 203 Feneme-HMMs

Ein Spracherkennungssystem für großen Wortschatz kann nicht verlangen, daß jeder neue Benutzer alle Worte des Systems diktiert, jedoch muß der Trainingstext alle Laute der Sprache in genügender Anzahl enthalten. Wir haben aus diesem Grund die von J. Sotscheck vorgestellten 100 Sätze /SOT84/ um 10 Sätze mit Nasalvokalen und anderen im Deutschen hauptsächlich in Fremdworten vorkommenden Lauten erweitert und als Trainingsskript für Deutsch verwendet.

In einer ca. 15-minütigen Sitzung diktiert der neue Benutzer die 110 vorgegebenen Sätze mit deutlichen Pausen zwischen den Worten in das System. Er wird dabei durch eine besondere Bildschirmdarstellung geführt, die ihm den noch ungewohnten Sprechrhythmus verdeutlicht. Die Sprechgeschwindigkeit ist dabei einstellbar.

Dabei wird das digitalisierte Sprachsignal in hoher Qualität (30 kB/sec) auf Platte aufgezeichnet. Die Rohdaten der 110 Sätze belegen dort in komprimierter Form ca. 10 MB Plattenplatz.

Referenzvektoren

Aus den ersten ca. 5 Minuten des Sprachsignals werden durch **k-Means Clustering** die 200 Benutzerprototypen ermittelt. Hierbei werden zuerst 200 zufällige Mittelpunkte im 20-dimensionalen Merkmalsraum angelegt, denen dann die ca. 30000 Vektoren nach einem Abstandskriterium zugeordnet werden. Zu den so gebildeten 200 Gruppen werden neue Mittelwerte errechnet, die in der nächsten Iteration die Rolle der Startwerte übernehmen. Die Iteration wird solange durchgeführt bis nur noch sehr wenige Vektoren (ca. 1%) ihre Gruppenzuordnung ändern. Die so gewonnenen 200 Zentren dienen dann als Benutzerprotoypen für die Kodierung des Sprachsignals dieses einen Sprechers.

DM-Statistik

Auch für das Training der Wahrscheinlichkeitsparameter der DM-Markovmodelle wird eine sinnvolle Initialstatistik benötigt. Die Übergangswahrscheinlichkeiten können für alle DM-HMMs gleich initiiert werden, da diese gerade so konstruiert wurden, daß sie genau einer Zeiteinheit entsprechen. Die Startwerte für die FM-Emissionswahrscheinlichkeiten werden durch paralleles Kodieren des Sprachsignals mit den Benutzerprototypen und nochmals mit den

Systemprototypen ermittelt. Danach wird für jeden Systemprototyp ausgezählt, wie häufig er mit welchem Benutzerprototyp zusammenfiel. In Kombination mit einer Gleichverteilung erhält man daraus die Initialwerte der Emissionswahrscheinlichkeiten. Mit Hilfe eines Forward/Backward Algorithmus /RAB86/ wird die DM-Statistik an den DM-Markovmodellen der 110 Sätze trainiert.

FM-Statistik

Hier wird die Initialstatistik wie folgt gewonnen: Die für alle Benutzer identischen Initialwerte der FM-Übergangswahrscheinlichkeiten sind die trainierten Übergangswahrscheinlichkeiten der Referenzsprecher. Für das Initiieren der Emissionswahrscheinlichkeiten wird das Produkt der FM-Emissionswahrscheinlichkeiten der Referenzsprecher mit der im vorigen Schritt ermittelten trainierten DM-Emissionswahrscheinlichkeiten des neuen Benutzers gebildet. Diese Statistik wird dann wieder mit einem Forward/Backward Algorithmus, diesmal an den FM-Markovmodellen der Sätze, trainiert.

Das gesamte Benutzertraining für einen neuen Benutzer dauert etwa 2 Stunden unter Verwendung einer der Signalprozessorkarten. Die Ergebnisse des Benutzertrainings werden auf der Festplatte gespeichert, wo sie etwa 360 kB Plattenplatz belegen.

Zusammenfassung

Die Anpassung des Systems TANGORA für Deutsch bestand nicht in der Modifikation der Systemsoftware, sondern in der Bereitstellung von sprachspezifischen Daten: Vokabular, Phonalphabet, phonetische Transkriptionen, DM-Vokabular, und Übergangs- und Emissionsstatistiken der Markovmodelle.

Die bisherigen Ergebnisse sind ermutigend. Die Erkennungsrate liegt bei über 95% für Texte, die mit dem noch sehr begrenzten Vokabular bildbar sind. Für durchschnittliche Sprecher kann selbst mit einem „sprecherunabhängigen" allgemeinen Benutzertraining eine Erkennungsrate von über 90% erzielt werden.

In den folgenden Monaten wird das Vokabular auf 10000 Worte erweitert werden. In der weiteren Forschungsarbeit soll untersucht werden, wo speziell im Deutschen Probleme auftreten und wie sie gelöst werden können. Ursache für solche Probleme können z. B. sein: hoher konsonantischer Anteil der deutschen Sprache, viele Flexionsformen und Komposita. Neben Arbeiten zur Verbesserung des Systems sind auch Untersuchungen über eine Benutzerschnittstelle zur Einbettung von Spracherkennung in Anwendungssysteme geplant.

Literatur

SOT84   J. Sotscheck: Sätze für die Sprachgütemessung und ihre phonologische Anpassung an die deutsche Sprache. Tagung der Dt. Arbeitsgemeinschaft für Akustik, DAGA 84, Darmstadt, 26.-30.3.1984

JEL85   F. Jelinek: The Development of an Experimental Discrete Dictation recognizer. Proc. IEEE, Vol. 73, No. 11, Nov. 1985, pp.1616-1624

AVE86   A. Averbuch et al.: An IBM PC Based Large-Vocabulary Isolated Utterance Speech Recognizer. Proc. ICASSP 1986, Tokio, Japan, Vol. 1, pp. 53-56

RAB86   L. R. Rabiner et B. H. Juang: IEEE ASSP Magazine, Januar 1986

DOR87   P. D'Orta et al.: A Speech Recognition System for the Italian Language. Proc. ICASSP 1987, Dallas, Texas, Vol. 2, pp. 841-843

Sprecherunabhängige Spracherkennung mit neuronalen Netzen

Peter Richert, Bedrich Hosticka und Markus Schwarz

Fraunhofer Institut für Mikroelektronische Schaltungen und Systeme
Finkenstraße 61, D-4100 Duisburg 1

Kurzfassung

Die Sprachkodierung mit linearer Prädiktionskodierung unter Verwendung partiell korrelierter Koeffizienten ermöglicht es, ein prinzipiell sprecherunabhängiges System zur Spracherkennung zu entwickeln. In diesem Beitrag werden der Aufbau eines derartigen Spracherkennungssystems und die experimentellen Ergebnisse vorgestellt. Das Problem der nichtlinearen zeitlichen Verzerrungen von Sprachsignalen wurde mit neuronalen Netzen gelöst. Aufbauend auf der Lautschrift erfolgt die Merkmalfindung und Klassifizierung der Sprache mit selbstorganisierenden Merkmalkarten. Die Leistungsfähigkeit des Systems wird an Hand der Erkennung von gesprochenen Ziffern für verschiedene Sprecher demonstriert.

1. Aufbau eines Sprachverarbeitungssystems

Das Sprachverarbeitungssystem wurde mit den Funktionsgruppen aus **Bild 1** aufgebaut. Die Sprachanalyse ermittelt aus der originalen Sprache die elektrischen Sprachparameter, die nach einer möglichen Speicherung und/oder Übertragung mit einer Sprachsynthese in synthetische Sprache zurückgewandelt wird.

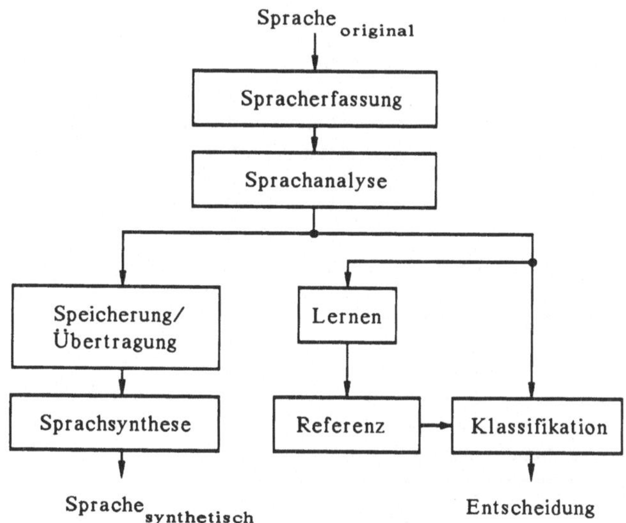

Bild 1: Aufbau eines Sprachverarbeitungssystems

Aus den Sprachparametern werden in einer Lernphase Referenzmuster erlernt und gespeichert, die in der Erkennungsphase einem Klassifikator zugeführt werden, um unbekannte Muster zu erkennen. Die Natur der Sprachdaten erfordert einen nichtlinearen Vergleich von unbekanntem Muster und Referenzwert.

2. Parametrisches Modell der Sprachsynthese

Das mechanische Röhrenmodell der parametrischen Spracherzeugung kann durch eine elektrische Ersatzschaltung ersetzt werden, wobei die Analogie zwischen akustischer und elektrischer Wellenausbreitung zu Vokaltraktmodellen mit Ketten- oder Lattice-Filtern führt [1]. Für kontinuierliche Sprache müssen die Modellparameter alle 2,5 ms bis 20 ms aktualisiert werden. Innerhalb dieser Intervalle kann das Filter als zeitinvariant angesehen werden. **Bild 2** zeigt den Aufbau eines Systems mit diesem Synthesemodell.

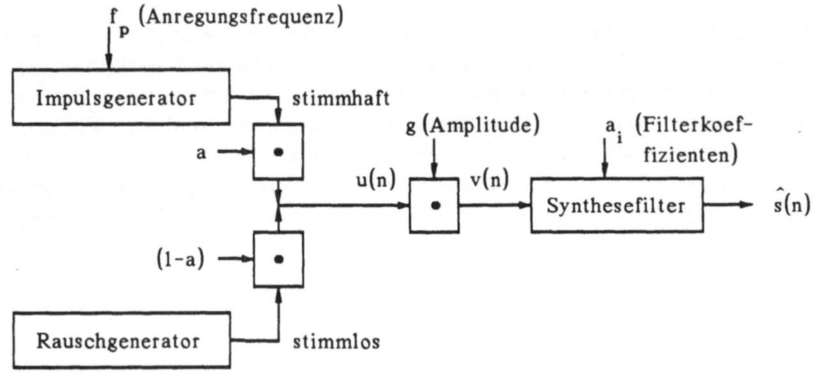

Bild 2: Parametrisches Modell der Sprachsynthese

3. Adaptive Sprachanalyse mit linearer Prädiktionskodierung (LPC)

Die Sprachanalyse ist das duale Verfahren der Sprachsynthese. Die Bestimmung der Sprachsignalparameter beruht darauf, das Sprachsignal durch ein zur Spracherzeugung inverses Filter (Analysefilter) zu geben und somit das Anregungssignal v(n) zurückzugewinnen. Aus den Parametern der Anregungsfunktion und den Filterkoeffizienten kann mit einem Synthese-Filter synthetische Sprache erzeugt werden.

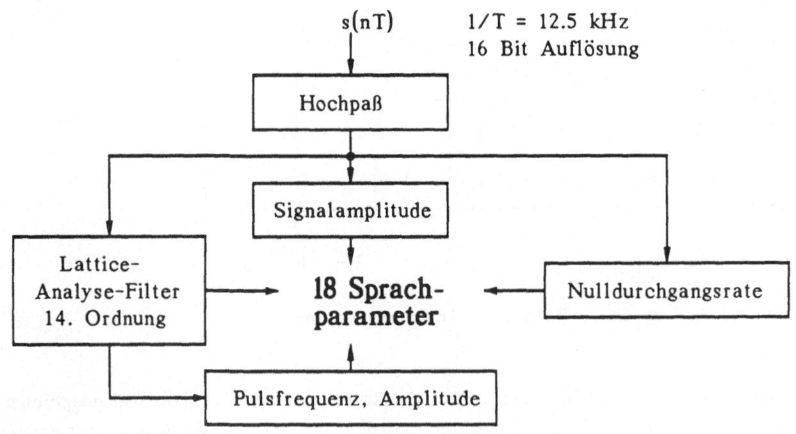

Bild 3: System zur Bestimmung aller Sprachparameter

Die lineare Prädiktionskodierung (LPC) [2] ist heute eines der wichtigsten Verfahren zur Sprachanalyse und -kodierung. **Bild 3** zeigt den Aufbau einer LPC-Sprachanalyse mit einem Lattice-Analy-
se-Filter. Zu Beginn werden die mit 40 dB/Dekade abfallenden hohen Frequenzanteile des abgetasteten Signals mit einem digitalen Hochpaßfilter 1. Ordnung angehoben, wodurch das Konvergenzver-
haltens des adaptiven Filters verbessert wird. Das sprecherabhängige Restfehlersignal der Analyse-
Filterung dient zur Bestimmung der Parameter der Anregungsfunktion. Aus dem originalen Signal
wird zusätzlich die Nulldurchgangsrate als Maß der Momentanfrequenz bestimmt.

Die partielle Autokorrelationsmethode [3] ist ein Verfahren zur rekursiven Berechnung der Korre-
lationskoeffizienten direkt aus dem Sprachsignal. Aus der Analogie zwischen akustischer und elektrischer Wellenausbreitung kann die Lattice-Struktur des im **Bild 4** dargestellten Analysefilters herge-
leitet werden.

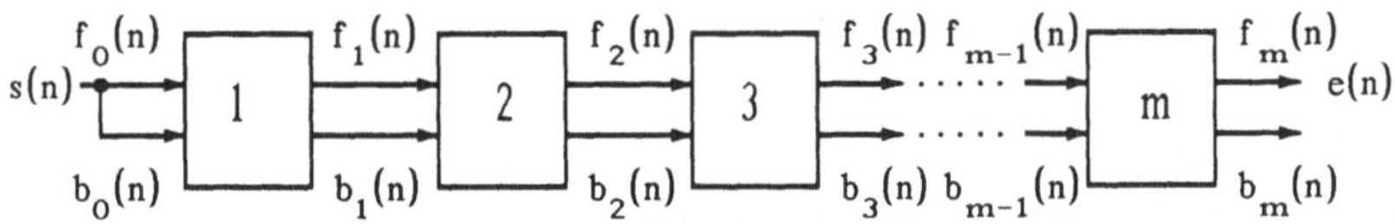

Bild 4 a: Blockdiagramm des Lattice-Analyse-Filters

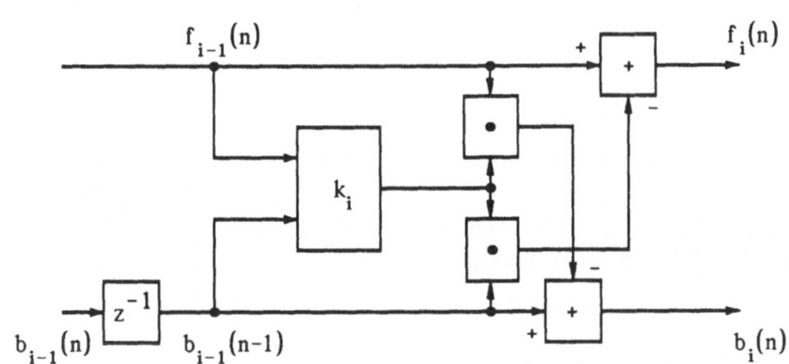

Bild 4 b: Stufe i des Lattice-Analyse-Filters

Die bei den Latticefiltern verwendeten Reflektionskoeffizienten haben einen direkten physikalen
Bezug zur akustischen Spracherzeugung. Sie sind das Verhältnis von zwei aufeinanderfolgenden
Querschnittsflächen des Vokaltraktes und sind somit weitgehend sprecherunabhängig. Die Bestimmung dieser Filterkoeffizienten läßt sich mit Hilfe des Orthogonalitätsprinzips durchführen. Für einen lokal stationären Prozeß kann für die Berechnung der Erwartungswerte ein Datenfenster der
Länge N verwendet werden, wie aus **Bild 5** zu ersehen ist. Der so bestimmte Reflektionskoeffizient
ist der Korrelationskoeffizient zwischen der vorwärts- und der rückwärtslaufenden Welle. Er wird
auch partieller Korrelationskoeffizient (PARCOR) genannt. Mit rekursiven Verfahren kann der
Rechenaufwand erheblich reduziert werden, indem nur der Prädiktionsfehler der aktuellen Zeit
berücksichtigt wird.

$\boxed{\begin{array}{l}
\textbf{(1) Initialisierung} \\
\quad f_0(n) = b_0(n) = s(n) \\[2mm]
\textbf{(2) Schleifensteuerung} \\
\quad \text{Inkrementieren von } i \text{ um } 1 \text{ von } i = 1 \text{ bis } i = m
\end{array}}$

(1) Initialisierung

$$f_0(n) = b_0(n) = s(n)$$

(2) Schleifensteuerung

Inkrementieren von i um 1 von $i = 1$ bis $i = m$

(3) Berechnung des Prädiktorkoeffizienten k_i

$$k_i = \frac{2 \cdot \sum\limits_{n=0}^{N-1} f_{i-1}(n) \cdot b_{i-1}(n-1)}{\sum\limits_{n=0}^{N-1} \left\{f_{i-1}(n)\right\}^2 + \sum\limits_{n=0}^{N-1} \left\{b_{i-1}(n-1)\right\}^2}$$

(4) Berechnung des Vorwärts- und Rückwärtsprädiktionsfehlers

$$f_i(n) = f_{i-1}(n) - k_i \cdot b_{i-1}(n-1)$$
$$b_i(n) = b_{i-1}(n-1) - k_i \cdot f_{i-1}(n)$$

(5) Wiederholen ab Punkt (2) bis Schleifenende

(6) Schlußberechnung

$$e(n) = f_m(n)$$

Bild 5: Algorithmus zur rekursiven Berechnung der Prädiktorkoeffizienten

4. Mustererkennung mit neuronalen Netzen

Wir wollen mit neuronalen Netzen einen unbekannten Mustervektor demjenigen Referenzvektor zuordnen, der entsprechend einem zu definierenden Kriterium am besten mit dem unbekannten Mustervektor übereinstimmt. Als Testmuster verwenden wir die Sprachparameter der zehn Ziffern 1 bis 10, deren Phonemzerlegung in **Tabelle 1** dargestellt ist.

Ziffer	Phoneme / Quasiphoneme	Phonem-Basis
eins	[ai], [n], [s]	[ai], [n], [s]
zwei	[t], [s], [v], [ai]	[t], [v]
drei	[d], [r], [ai]	[d], [r]
vier	[f], [i:], [r]	[f], [i:]
fünf	[f], [y], [n], [f]	[y]
sechs	[s], [ɛ], [k], [s]	[ɛ], [k]
sieben	[s], [i:], [b], [ə], [n]	[b], [ə]
acht	[a], [x], [t]	[a], [x]
neun	[n], [ɔy], [n]	[ɔy]
zehn	[t], [s], [ɛ], [n]	
	$\Sigma = 36$	$\Sigma = 17$

Tabelle 1: Phonemzerlegung der Ziffern

Aus den Eingangsdaten, die mit der Analyserate von 16 ms (200 Sprachwerte) in das neuronale Netz gelangen, wird in einer ersten Ebene jedem Datensatz ein Quasiphonem mit derselben Fensterlänge zugeordnet und in der folgenden Netzebene zu Phonemen zusammengefaßt. Durch die Einführung der Quasiphoneme als kleinstes Musterelement wird eine zeitliche Normalisierung der Sprache erreicht, wozu i.a. aufwendige Verfahren notwendig sind. Phoneme bestehen dann nicht aus einer konstanten Anzahl Quasiphoneme, sondern aus einer Mindestanzahl, wobei weitere, gleiche Quasiphoneme die zeitliche Länge des Phonems wiedergeben.

5. Merkmalfindung mit neuronalen Netzen

Die Anwendung neuronaler Netze zur Merkmalfindung wird mit dem Begriff »selbstorganisierende Merkmalkarten« (self organizing feature maps) bezeichnet [4]. **Bild 6** zeigt die Topologie einer hierarchischen Merkmalkarte.

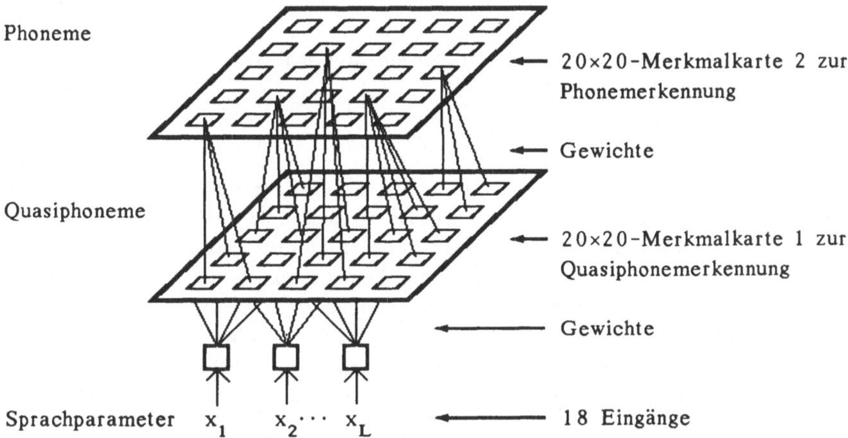

Bild 6: Hierarchisches System aus zwei Merkmalkarten zur Phonemerkennung

Der Algorithmus zur Adaption der Gewichte einer selbstorganisierenden Merkmalkarte ist in **Bild 7** dargestellt. Wir unterteilen die Lernphase der Merkmalkarte in drei Abschnitte. In der ca. 1.000 Zeitschritte umfaßenden Grob-Lernphase erfolgte die grobe Strukturierung der Karte anhand manuell ausgewählter Referenzvektoren eines Sprechers. Der Einflußbereich des Adaptionsalgorithmus umfaßte zu Beginn alle Neuronen und nahm zum Schluß auf ±1 Neuron ab. Der Lernfaktor α wurde linear von 0,5 auf 0,01 reduziert. In der folgenden ca. 10.000 Schritte umfaßenden Fein-Lernphase wurden die kompletten Sprachvektoren desselben Sprechers aller 10 Ziffern in zufälliger Reihenfolge zur Adaption verwendet. Der Adaptionsbereich wurde konstant auf ±1 Neuron gehalten, so daß nur noch ein lokales Lernen stattfand. Der Lernfaktor wurde mit der Zeit linear von 0,1 auf 0,001 verkleinert. Klassifikationen mit der so trainierten Karte ergaben für Sprache des trainierten Sprechers eine Fehlerrate von 3% und für Sprache weiterer drei unbekannter Sprecher eine Fehlerrate zwischen 25% und 38%.

(1) **Initialisierung für t = 0**

Die Gewichte aller Verbindungen der L Eingänge zu den Ausgängen der IxJ-Karte werden mit Pseudo-Zufallszahlen initialisiert. Die Größe der Umgebung U_0 wird zu max $\{I,J\}/2$ und der Lernfaktor α_s zu 0,5 gewählt.

(2) **Anlegen eines neuen Eingangsvektors $\underline{x}(t)$**

$t = t + \Delta t$

(3) **Bestimmen des Neurons mit minimaler Distanz**

Die Distanz d_j mit allen Referenzvektoren $\underline{r}_j$ wird berechnet und das Neuron N_c gesucht, für das d_j ein Minimum annimmt.

(4) **Adaption der Gewichte in Umgebung**

Die Gewichte in der Umgebung $U_c(t)$ werden an den Eingangsvektor $\underline{x}(t)$ adaptiert. Dabei ist $\alpha(t)$ der mit der Zeit kleiner werdende Lernfaktor.

(5) **Wiederholung ab Punkt (2), wenn $t \leq t_{ende}$**

Bild 7: Adaptionsalgorithmus einer selbstlernenden Merkmalkarte

In beiden Lernphasen wurden die Sprachdaten eines Sprechers in zufälliger Reihenfolge zur Adaption der Gewichte verwendet. Zumindest für die Grob-Lernphase ist dies eine unabdingbare Voraussetzung, da hierbei die Umgebung des besten Neurons und der Lernfaktor so groß sind, daß korrelierte Sprachdaten die Karte einseitig modifizieren würden. In der Fein-Lernphase wurde zwar die Umgebung auf das Minimum reduziert, aber der Lernfaktor ist zu Beginn dieser Phase noch hinreichend groß. Erst bei kleinen Lernfaktoren könnten korrelierte Daten benutzt werden. Genau das sind die Voraussetzungen für eine weitere, auf die nachfolgende Anwendung spezialisierte Lernphase. Dazu wurden die Sprachdaten von zwei Sprechern in ihrer natürlichen, also zeitlichen, Reihenfolge vom Netz adaptiert. Somit verstärken korrelierte Daten zwar lokale Bereiche der Karte, aber einzelne Laute ändern nicht die Ordnung der Karte, sondern nur deren Feinstruktur.

6. Ergebnisse der Simulationen

Als Ergebnis aller drei Lernphasen erhalten wir die im **Bild 8** dargestellte Merkmalkarte. Jedes angesprochene Neuron ist mit seinem entsprechenden Phonem gekennzeichnet. Bemerkenswert ist, daß sich zwischen verschiedenen Gebieten teilweise nicht angesprochene Neurone gebildet haben, die durch eine Linie verbunden bzw. mit einem Punkt markiert sind. Sie bilden somit eine selbstgebildete Trennungslinie zwischen verschiedenen Merkmalklassen, die sich aufgrund der minimalen Umgebung in der Lernphase ausgeprägt haben. Im Gegensatz dazu gibt es auch Klassen, die direkt aneinandergrenzen oder sogar vermischt sind. Eine direkte Nachbarschaft einzelner Phoneme, z.B. zur Pause, ist teilweise schon im Sprachsignal begründet, wohingegen sich eine ungewollte Nachbarschaft oft durch die lautliche Ähnlichkeit dieser Phoneme ergibt und somit im Sprachmodell zu suchen ist.

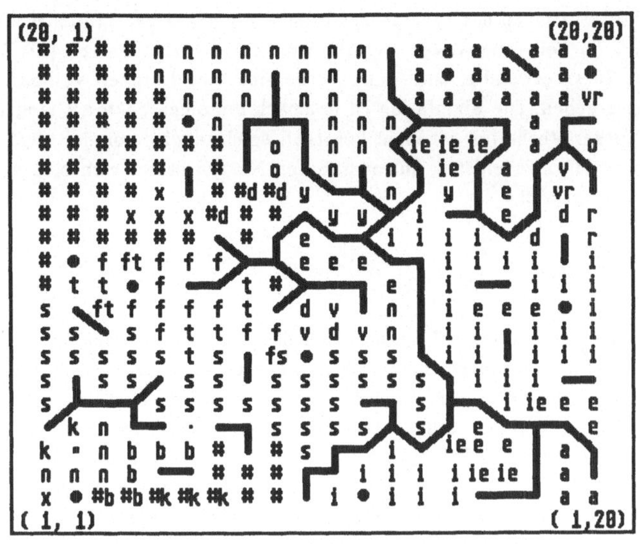

Bild 8: Merkmalkarte zur Phonemerkennung

Die Klassifizierung der gelernten Sprachdaten ergibt für den Sprecher 1 einen Erkennungsfehler von 0% von insgesamt 84 erkannten Phonemen. Für den Sprecher 2 liegt der Erkennungsfehler noch bei 10%, wobei hier von 80 erkannten Phonemen noch acht fehlende zu verzeichnen sind. **Tabelle 2** faßt die erreichten Ergebnisse zusammen.

	Sprecher 1				Sprecher 2			
Ziffer	Phoneme	richtig	fehlt	falsch	Phoneme	richtig	fehlt	falsch
1	2a i 2n 4ʃ *	10			2a i n ʃ	6		
2	tʃ 2v 4ai *	8			tʃ 4ai 2*	7	1	
3	d r 5ai *	8			* 2r 2ai 2*	7	1	
4	2f 4i r	7			2f 3i	5	1	
5	* 2f y 4n f *	10			* y 3n 3*	8	2	
6	2ʃ e 2k 3ʃ	8			* ʃ e * 2ʃ *	7	1	
7	3ʃ i b 2e n	8			2ʃ i b 2n *	7	1	
8	* 2a * 2x 2* t *	10			2a * 2x 4*	9	1	
9	* n 5ɔy 2n *	10			n 5ɔy 2n *	9		
10	tʃ e 3n	5			* tʃ 3e n *	7		
	0,0% ∑	84			10,0% ∑	72	8	

Tabelle 2: Ergebnis der Klassifizierung der zehn Ziffern

Gegenüber der Klassifizierung nach der Fein-Lernphase konnte der Erkennungsfehler von 3,6% für den Sprecher 1 bzw. 35,8% für den Sprecher 2 wesentlich reduziert werden. Besonders deutlich fällt die Steigerung der Erkennungssicherheit für den Sprecher 2 aus. Durch die zusätzliche Trainingsphase mit dessen Sprache konnten alle falsch klassifizierten Phoneme eleminiert und die Zahl der fehlenden Phoneme reduziert werden. Insbesondere der noch verbleibende Erkennungsfehler erfordert aber eine Fortsetzung dieser Lernphase bis zur letztendlichen Konvergenz der Gewichte.

Die Simulationszeit für 180 Iterationen der speziellen Lernphase beträgt ca. 9 Stunden CPU-Zeit auf einem VAX-Rechner 8550. Für ein einmaliges Vorsprechen der 20 Zahlwörter (10 Ziffern für 2 Sprecher) sind weniger als 20 s notwendig, und somit würde sich eine Lernzeit von ca. 1 Stunde bei einer Echtzeitverarbeitung ergeben. Die Simulation ist demnach um einen Faktor 9 langsamer als eine Echtzeitverarbeitung. Dieses Verhältnis ist nicht konstant, sondern wird von der Größe der Merkmalkarte bestimmt. Bei einer Hardwarelösung mit neuronalen Netzen ergibt sich eine von der Netzgröße unabhängige Verarbeitungszeit.

Zusammenfassung

Als wesentliches Ergebnis können wir festhalten, daß sich die Erkennungssicherheit bei Klassifizierung mit neuronalen Netzen durch weitergehende Trainingsphasen erhöhen läßt, daß aber andererseits eine weitgehende Sprecherunabhängigkeit bedingt durch das Sprachmodell nur mit einer Lernphase für verschiedene Sprecher erreicht werden kann. Die Sprecherunabhängigkeit des zugrundeliegenden Sprachmodells kann zu minimal 60% abgeschätzt werden. Es ist somit erforderlich, für eine sprecherunabhängige Klassifizierung die Anzahl der Neuronen sowie die der Sprecher soweit zu erhöhen, bis der Erkennungsfehler hinreichend klein wird. Für ein leistungsfähiges Spracherkennungssytem sind dazu aber Hardware-Realisierungen neuronaler Netze unumgänglich. Das vorgestellte System ist hierarchisch erweiterbar, wobei die Worterkennung aus den Phonemen unter Verwendung eines Wortlexikons die Erkennungssicherheit zusätzlich erhöhen würde.

Literatur

[1] S. Saito and U. Nakata, "Fundamentals of Speech Signal Processing", Academic Press Japan, Inc., ISBN 0-12-614880-5, 1985.

[2] J.D. Markel and A.H. Gray, jr., "Linear Prediction of Speech", Springer-Verlag Berlin Heidelberg New York, ISBN 3-540-07563-1, 1976.

[3] A.H. Gray, jr. and D.Y. Wong, "The Burg Algorithm for LPC Speech Analysis and Synthesis", IEEE Transaction on Accoustics, Speech, and Signal Processing, Vol. ASSP-28, No. 6, pp. 609 - 615, December 1980.

[4] T. Kohonen, "Self-Organizing Maps", IEEE International Symposium on Circuit and Systems, Presymposium Short Course Neural Networks, Helsinky University of Technology, Finnland, 1988.

Kontext-Disambiguierung in natürlichsprachlichen Anfragen an relationale Datenbanken

Jörg Noack[1]
Lehrstuhl für Angewandte Mathematik insbesondere Informatik
RWTH - Aachen, D-5100 Aachen

Zusammenfassung

In vielen natürlichsprachlichen Anfragesystemen werden lexikalische Mehrdeutigkeiten durch akzeptanzhemmende Klärungsdialoge behandelt. Die Arbeit beschreibt ein Verfahren zur Auflösung solcher Mehrdeutigkeiten mit Hilfe des Satzkontextes. Das zugrundeliegende Datenbankschema, das in einem Distanzgraphen kodiert ist, dient dabei als Wissensquelle. Das vorgestellte Disambiguierungsverfahren bildet eine von mehreren Analysestufen der transportablen, natürlichsprachlichen Datenbank-Schnittstelle NATHAN /4/.

1.Einleitung

Setzt man für natürlichsprachliche Anfragen an relationale Datenbanken voraus, daß sämtliche Wörter der Satzoberfläche eindeutig auf ein Konzept des Datenmodells abgebildet werden können, dann wird dadurch oftmals ein artifizieller Sprachgebrauch hervorgerufen, der weit vom Idealziel der Freiformulierbarkeit entfernt ist.

Die Grundidee, daß sich lexikalische Mehrdeutigkeiten durch die sprachliche Umgebung des Satzes auflösen lassen, wird bereits in /2/ beschrieben. Dort wird eine lokale Disambiguierung aufgrund von Kontext-Regeln, die zu jedem Datenbankschema aufzustellen sind, vorgeschlagen. Das in NATHAN eingesetzte Verfahren unterscheidet sich darin, daß der Begriff des *semantischen Zusammenhangs*, der für zwei Token gilt, durch eine Distanzfunktion formalisiert wird, daß eine Konfliktbehandlung nicht lokal gelöst werden muß und daß die Konstituentenzugehörigkeit einzelner Inhaltswörter beachtet wird.

2. Architektur von NATHAN

In NATHAN (**NAT**ürlichsprachliches **H**euristisches **AN**fragesystem) wird ein Teil des Vokabulars, welches dem Endbenutzer zur Verfügung steht, über einen Akquisitionsdialog von einem Diskursexperten erfragt und dann im Diskurslexikon festgehalten (Abb.1). In der Akquisitionsphase werden einzelne Wörter, hierbei handelt es sich um Nomina, Verben oder Adjektive, die zusammen die offenen Wortklassen des Lexikons bilden, mit sogenannten Datenbanktoken verknüpft:

---

[1] Diese Arbeit wurde vom MWF des Landes NRW gefördert.

Es sei D=(R,.) ein relationales Datenbankschema, wobei R = {R₁,...,Rₖ} aus einer Menge von Relationenschemata und jedes Ri mit $R_i = (S_i,.)$ für 1≤i≤k aus einer Menge Si von Attributen besteht. Ferner sei jedem Attribut A aus Ri ein Wertebereich dom(Ri.A), der auch Domain genannnt wird, zugeordnet. Dann ist durch T = R ∪ {R$_i$.A| R$_i$∈R, A∈S$_i$} ∪ {dom(R$_i$.A)| R$_i$∈R, A∈S$_i$} die Menge der zu D gehörigen <u>Datenbanktoken</u> gegeben.

Unter Datenbanktoken werden also die internen Bezeichner von Relationen(-schemata), Attributen oder Domains verstanden. Die sprachliche Modellierung in NATHAN setzt ein restringiertes Strukturmodell /1/ als Datenmodell des Diskurses voraus. Der Einfachheit halber gehen wir hier davon aus, daß der Diskursexperte einem Wort, das in einer Anfrage eine interne Relation oder ein internes Attribut bezeichnen soll, ein Attribut- bzw. Relationentoken zuordnet und daß Attributwerte mit den entsprechenden Domains (Domaintoken) verknüpft sind, zu denen sie gehören.

Lexikalische Mehrdeutigkeiten äußern sich darin, daß der Scanner zu einem Inhaltswort einer natürlichsprachlichen Anfrage mehrere Datenbanktoken im Lexikon vorfindet.

<u>Bsp.1:</u>   "Zeige den <u>Namen</u> des <u>Vermieters</u> in <u>Aachen</u>!"
    {[vermieter.name], {[vermieter.vnr]} {[d_vort],
    [wohngebiet.name]}   [d_wort]}

In Bsp.1 steht das Inhaltswort "Name" sowohl für das Attribut "vermieter.name" als auch für das Attribut "wohngebiet.name". Der Attributwert "Aachen" kann sowohl zu dem Domain "d_vort" als auch zu "d_wort" gehören.

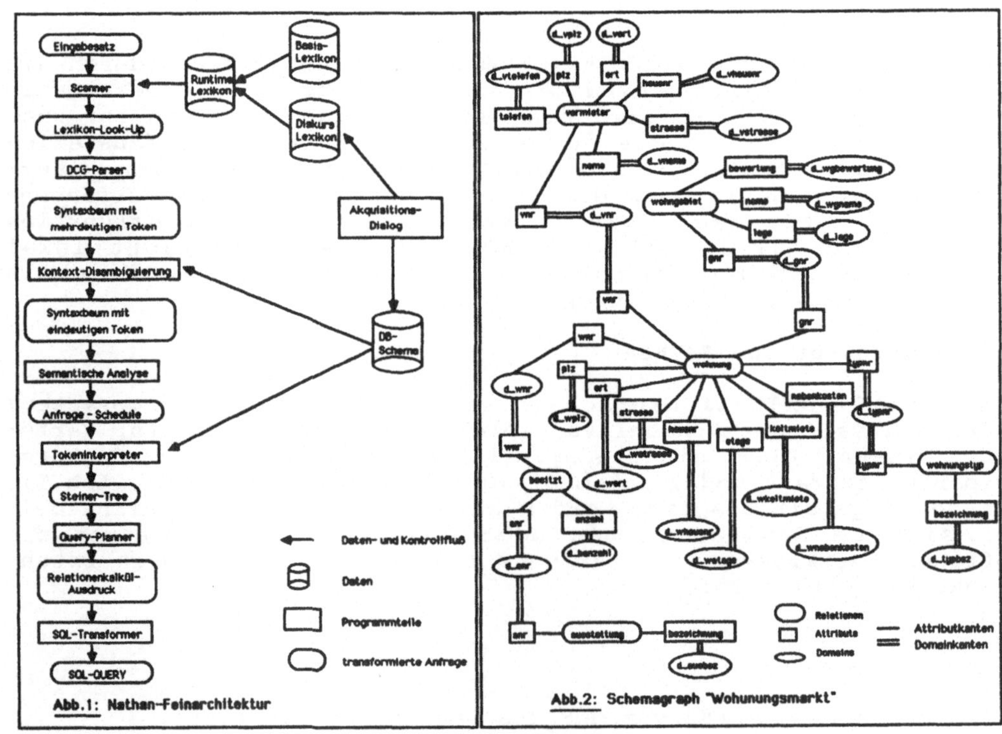

Zunächst wird das Datenbankschema $D=(R,.)$, welches in einer Anwendung benutzt werden soll, in einen Distanzgraphen $G_D=(V,E,d)$ überführt:

1. $V = R \cup \{R_i.A \mid R_i \in R, A \in S_i\} \cup \{dom(R_i.A) \mid R_i \in R, A \in S_i\}$
2. Es sei $e \in E$ mit $e = \{v_1,v_2\}$, falls entweder
   (i) $v_1=R_i.A$ und $v_2=R_i$, dann heißt e <u>Attributkante</u> oder
   (ii) $v_1=R_i.A$ und $v_2=dom(R_i.A)$, dann heißt e <u>Domainkante</u>.
3. $d: E \rightarrow \mathbb{N}$ mit

$$d(e):= \begin{cases} \mu & \text{falls } e \text{ Domainkante} \\ a & \text{falls } e \text{ Attributkante} \end{cases}$$

Ein solcher Distanzgraph heißt <u>Schemagraph</u>. Domainkanten werden wesentlich höher als Attributkanten bewertet. Es gilt also: $\mu \gg a$. Wir betrachten nur zusammenhängende Schemagraphen. Abb.2 zeigt ein Beispiel. Die Distanzfunktion d läßt sich auf beliebige Tokenpaare übertragen. Es sei T die Menge aller Token eines Datenbankschemas. Dann ist durch $d: T^2 \rightarrow \mathbb{N}$ mit $d(t1,t2):=k \cdot \mu + l \cdot a$ die <u>Distanz</u> zweier Token definiert, wenn auf dem minimalen Pfad $P_{t_1 t_2}$ im zugrundeliegen Schemagraphen k Domainkanten und l Attributkanten auftreten. Es ist leicht einzusehen, daß d nun die Eigenschaften einer Metrik hat. Im folgenden dient d der Bewertung des semantischen Zusammenhangs von Tokenpaaren.

3. Kontext-Disambiguierung

Das in NATHAN implementierte, heuristische Disambiguierungsverfahren versucht die durch den Syntaxbaum gegebene Konstituentenstruktur soweit wie möglich auszunutzen. Im Normalfall werden für jeden Nicht-Blattknoten die unterhalb verbundenen Teilbäume von links nach rechts durchlaufen. Sobald ein Teilbaum bearbeitet ist, wird eine Disambiguierung zwischen der Menge Tl_{i-1} von Tokenfolgen, welche als Resultat der Disambiguierung der ersten $i-1$ Teilbäume erzeugt wurde, und der Menge Tr_i von Tokenfolgen des i-ten Teilbaumes vorgenommen (Abb.3),

<u>Abb.3</u>: Abgleich des i-ten Teilbaumes mit seinen linken Bruderbäumen

indem die Abstände der übriggebliebenen Token der benachbarten Inhaltswörter paarweise verglichen werden, d.h. es sei

$$Tl_{i-1} = \begin{Bmatrix} [tl_{11}, \ldots, tl_{u1}], \\ \ldots \\ [tl_{1m}, \ldots, tl_{um}] \end{Bmatrix} \quad \text{und} \quad Tr_i = \begin{Bmatrix} [tr_{11}, \ldots, tr_{v1}], \\ \ldots \\ [tr_{1n}, \ldots, tr_{vn}] \end{Bmatrix},$$

dann ergibt sich

$$Tl_i = \begin{Bmatrix} \ldots \\ [tl_{11}, \ldots, tl_{u1}, tr_{1j}, \ldots, tr_{vj}] \\ \ldots \end{Bmatrix},$$

falls $d(tl_{ui}, tr_{1j})$ minimal ist für $1 \le i \le m$ und $1 \le j \le n$. Gilt jedoch
m=n=1, so erübrigt sich ein Vergleich und Tl_i ist ebenfalls ein-
elementig. Für den Fall, daß die Wurzel von Tr_i durch *res* markiert ist, d.h. der Syntaxteilbaum beschreibt einen Relativsatz,
muß von der bisherigen Strategie abgewichen werden. Die Notwendigkeit einer Ausnahmebehandlung ergibt sich dadurch, daß das
Bezugsnomen des Relativpronomens nicht das am weitesten rechts
stehende Inhaltswort des (i-1)-ten Teilbaumes zu sein braucht.
Mehrdeutigkeiten zwischen den Tokenfolgen von bereits bearbeiteten Teilbäumen und denen eines Relativsatzes werden durch Dis-
tanzvergleiche zwischen den aktuellen Token der potentiellen
Bezugsnomina und den aktuellen Token des im Relativsatz am weitesten links stehenden Inhaltswortes versucht aufzulösen, d.h.
es sei

$$
Tl_{i-1} = \begin{bmatrix} [tl_{11},\ldots,tl_{u11},\ldots,tl_{u_m1},\ldots], \\ \cdots \\ [tl1m,\ldots,tl_{u1m},\ldots,tl_{u_mm},\ldots] \end{bmatrix} ,
$$

wobei in den bereits verarbeiteten Teilbäumen u Inhaltswörter
vorkommen mögen und $u_1,\ldots,u_m$ mit $1 \le u_1 \le \ldots \le u_m \le u$ die Indizes der
potentiellen Referenznomina seien, und es sei

$$
Tr_i = \begin{bmatrix} [tr_{11}, \ldots, tr_{v1}], \\ \cdots \\ [tr_{1n}, \ldots, tr_{vn}] \end{bmatrix}
$$

die Menge der Tokenfolgen, welche aus der Disambiguierung des
Relativsatzes resultiert. Dann ergibt sich

$$
Tl_i = \begin{bmatrix} \cdots \\ [tl_{11}, \ldots,tl_{ui},tr_{1j},\ldots,tr_{vj}] \\ \cdots \end{bmatrix} ,
$$

falls $d(tl_{u_pi}, tr_{1j})$ minimal ist für $1 \le i \le m$, $1 \le j \le n$ und $1 \le u_p \le u_m$. Für
m=n=1 kann wiederum auf eine Distanzermittlung verzichtet werden.

<u>Bsp.2</u>: "Zeige die Wohnungen von Aachener Vermietern, die
         300 DM kosten"

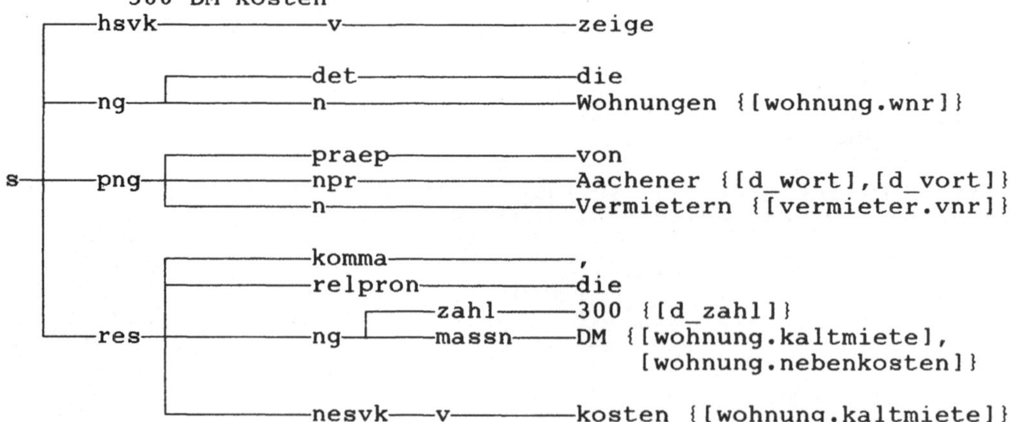

Zunächst sei der obige, vereinfachte Syntaxbaum mit annotierten
Tokenfolgenmengen gegeben. Die Syntaxanalyse in NATHAN basiert
auf einer umfangreichen *Definite Clause Grammar*, die aus dem ATN

des PLIDIS-Projektes /3/ abgeleitet wurde. Das Metatoken "d_zahl" repräsentiert sämtliche Domains (inkl. "d_wkaltmiete"), die für den Wert 300 mit Hilfe von Prädikaten herausgefiltert wurden. Eine Disambiguierung der Teilbäume mit den Wurzeln ng, png und res ergibt:

```
        ┌──────hsvk─...
        │            {[wohnung.wnr]}
        ├──────ng─ ...
  s─────┤            {[d_vort,vermieter.vnr]}
        ├──────png─...
        │              {[d_wkaltmiete,wohnung.kaltmiete,wohnung.kaltmiete]}
        └──────res─...
```

Die Tokenfolgenmengen der Knoten ng und png können zu

$$\{[\underline{wohnung.wnr},d\_vort,\underline{vermieter.vnr}]\}$$

verkettet werden. Als mögliche Referenztoken sind "wohnung.wnr" und "vermieter.vnr" zu beachten, da sie die Nomina "Vermieter" und "Wohnungen" des übergeordneten Satzes repräsentieren. Da

$$d(wohnung.wnr,d\_wkaltmiete) = \mu + 2a \text{ und}$$
$$d(vermieter.vnr,d\_wkaltmiete) = 3\mu + 2a$$

ergibt sich schließlich

```
     {[wohnung.wnr,d_vort,vermieter.vnr,
  s─── ... d_wkaltmiete,wohnung.kaltmiete,wohnung.kaltmiete]}.
```

Falls die aktuelle Tokenfolgenmenge am Ende des Disambiguierungsverfahrens aus mehr als einer Folge besteht, liegt eine "echte" Mehrdeutigkeit vor, die vom System nicht weiter behandelt werden kann. NATHAN präsentiert ein Menue, in dem der Endbenutzer aufgefordert wird, zwischen den einzelnen Folgen, die dort zusammen mit ihren externen Bezeichnern aufgelistet sind, auszuwählen.

Den Projekt-Diplomanden V. Hemmelrath, S. Özer und M. Itani gilt mein Dank für ihre Unterstützung bei der Implementierung.

4. Literatur

/1/ Wiederhold,G., El-Masri,R.: *The Structural Model for Database Design*, in P.P. Chen (ed.): *Entity-Relationship Approach to System Analysis and Design*, North-Holland, 1980, 237-257

/2/ Janas,J.M.: *Natürlichsprachliche Schnittstellen zu relationalen Datenbanken: Ein semantisch orientierter Ansatz*, Diss., Hochschule der Bundeswehr München, Neubiberg, 1982

/3/ Kolvenbach,M., Lötscher,A. und Lutz,H.D. (Hrsg.): *Künstliche Intelligenz und Natürliche Sprache*, Gunter Narr Verlag, Tübingen, 1979

/4/ Noack, J.: *NATHAN: Ein transportables Front-End zur Interpretation deutschsprachiger Anfragen an ein relationales Datenbanksystem*, Zwischenbericht, RWTH Aachen, 1988

Zum Stand der Normung in der Bildverarbeitung - Programmierschnittstelle und Bildaustauschformate

Detlef Krömker, Georg Rainer Hofmann
Fraunhofer-Arbeitsgruppe für Graphische Datenverarbeitung
Abt. Simulation und Animation

Wilhelminenstr. 7, D-6100 Darmstadt
Telex: 4 197 367 agd d   Telefax: +49 6151 1000 99

1. Allgemeines zur Normungsarbeit

"Die Normung in der Bundesrepublik Deutschland ist eine Aufgabe der Selbstverwaltung der Wirtschaft. Zentralorgan der Normung ist das DIN Deutsches Institut für Normung e.V.
Die Arbeitsergebnisse des DIN sind die DIN-Normen, die in ihrer Gesamtheit das Deutsche Normenwerk bilden." ([NORM], S.48).

"Die wachsende Bedeutung der Normen erfordert es, die bisherige Zusammenarbeit zwischen der Bundesregierung und dem DIN zu intensivieren. Dem DIN kommt hierbei in zunehmenden Maße die Aufgabe zu, die Bundesregierung durch Beratung zu unterstützen und durch Ausarbeitung von DIN-Normen allgemein anerkannte Regeln der Technik zu schaffen, die es ermöglichen, in Rechtsvorschriften auf Normen Bezug zu nehmen. Diese Möglichkeit der Verknüpfung von Rechtsvorschriften und technischen Normen entlastet die Bundesregierung davon, in jedem Einzelfall technische Regeln selbst erarbeiten zu müssen." ([NORM], S.48).

Das Verhältnis zwischen DIN und der Bundesregierung ist in einem Vertrag zwischen der Bundesrepublik Deutschland (vertreten durch den Bundesminister für Wirtschaft) und dem DIN (vertreten durch den Präsidenten) vom 5. Juni 1975 geregelt.

"Die fachliche Arbeit (im DIN) wird von ehrenamtlichen Mitarbeitern geleistet. Die ehrenamtlichen Mitarbeiter sind Fachleute aus den interessierten Kreisen (z.B. Anwender, Behörden, Berufsgenossenschaften, Berufs-, Fach- und Hochschulen, Handel, Handwerkswirtschaft, industrielle Hersteller, Prüfinstitute, Sachversicherer, selbstständige Sachverständige, Technische Überwacher, Verbraucher, Wissenschaft)." ([NORM], S.53).
Der DIN ist in Normenausschüsse (NA) gegliedert, die ihrerseits Arbeitsausschüsse (AA), Unterausschüsse (UA) und Arbeitskreise (AK) unterhalten.

Im Bereich der elektrotechnischen Normen unterhalten der DIN und der Verband Deutscher Elektrotechniker (VDE) gemeinsam die Deutsche Elektrotechnische Kommission (DKE).

Das DIN arbeitet in der internationalen Normung (Standardisierung) im Rahmen der ISO (International Organisation for Standardization), der IEC (International Electrotechnical Commission) und CEE (Commission International de Reglement en vue de l'Equipment Electrique), da vertreten durch die DKE; sowie in den europäischen Normungsgremien CEN/CENELEC (Comité Européen de Coordination des Normes / Electriques).

"Die weltweite Normung sollte möglichst allein in der ISO und IEC in enger gegenseitiger Zusammenarbeit durchgeführt werden, damt einer unübersichtlichen Vielfalt in den internationalen Normungsarbeiten vorgebeugt wird." ([NORM], S.36). Ein Ergebnis dieser Bemühung war 1987 die Gründung des ISO/IEC JTC1 (Joint Technical Committee 1), "Information Technology".

2. Zur Einordnung des "Imaging" in die internationale und nationale Normung

Das ISO/IEC JTC1 "Information Technology" ist in Unterkommittees (Subcommittees, SCs) gegliedert, die z.Zt. folgende Themen bearbeiten:
(Die für das Imaging - evtl. - relevanten SCs sind **fett** gedruckt.)

AG   Advisory group

SC1   Vocabulary

SC2   Character sets and information coding

SC6   Telecommunications and information exchange between systems

SC7   Software development and system documentation

SC11   Flexible magnetic media for digital data interchange

SC14   Representation of data elements

SC15   Labelling and file structure

SC17   Identification and credit cards

SC18   Text and office systems

SC20   Data cryptographic techniques

SC21   Information retrieval, transfer and management for open systems interconnection (OSI)

SC22   Languages

SC23   Optical digital data disks

SC24   Computer graphics

SC47B   Microprocessor systems

SC83   Information technology equipment

P&R   Planning and Requirements

Das jetzige ISO/IEC JTC1 ist aus dem ISO TC97 hervorgegangen. Die Numerierung der SCs im JTC1 repräsentieren eine historisch gewachsene Struktur - von daher sind die Lücken in der Numerierung zu erklären.

Mit der Neugründung des ISO/IEC JTC1 ging 1987 die Gründung des SC24 "Computer graphics" einher. Schon früh zeichneten sich für dieses SC neue Arbeitsbereiche ab, so z.B. "Windowing", "Imaging", usw. Auf der 1988er Plenarsitzung des SC24 in Tuscon (USA) wurde beschlossen, eine Study Group "Imaging" (Bildverarbeitung) einzusetzen. Diese Study Group hat einige Ziele und Aufgaben bis zur nächsten SC24 Plenarsitzung (Oktober 1989 in Recife, Brasilien) zu erfüllen:

-   "Scope", "Purpose" und "Justification" eines neuen Standards im Bereich Imaging festzulegen,

-   das Verhältnis eines solchen neuen Standards zu existierenden oder in der Arbeit befindlichen Normen festzustellen,

-   den Entwurf (Draft) und/oder die Inhaltsangabe (Outline) einer neuen Norm, falls möglich, zu erstellen,

-   einen Zeitplan er künftigen Normungsarbeit festzulegen.

Innerhalb der SC24 Study Group drängt insbesondere die amerikanische ANSI X3H3.8 Vertretung auf eine zügige Bearbeitung des neuen Themas. Sie verteilte daher schon im Februar 1989 den amerikanischen Entwurf eines "Programmers Interface Kernel (PIK), Version 3.0". ([PIK]).

Der DIN reagierte auf diese Aktion des ANSI mit der Gründung eines Arbeitskreises "Imaging" im Arbeitsausschuß AA24 des NI Normenausschuß Informationsverarbeitungssysteme. Basis der konstituierenden Sitzung des AK "Imaging" am 21. April 1989 war ein Fachgespräch "Bildverarbeitung (Imaging)", welches gemeinsam von der ITG FG "Mustererkennung", des GI FA.4.1 "Graphische Datenverarbeitung", der DAGM und des DIN-NI (Normenausschuß Informationsverarbeitungssysteme) am 20. und 21. April 1989 in Darmstadt veranstaltet worden ist.

Der DIN AK "Imaging" akzeptierte einen unter der Federführung von Gemmar und Hofele in der ITG erarbeiteten Vorschlag "Empfehlung für ein Ikonisches Kernsystem (IKS)" [IKS] als Grundlage einer deutschen Position; allerdings wissend, daß der IKS-Vorschlag nicht das ganze angestrebte Anwendungsspektrum abdeckt.

Das erste internationale ISO-Treffen fand am 18. und 19. Mai 1989 in Darmstadt statt, mit Vertretern aus den USA, aus Frankreich, aus der DDR, aus Grossbritannien, und den Vertretern des DIN. Japan hatte eine schriftliche Stellungnahme zu der Sitzung vorgelegt. Nach einem fachlichen Austausch (Referate der einzelnen Delegationen) wurde einstimmig beschlossen, dem ISO/IEC JTC1 zu empfehlen, ein "New Work Item on Imaging" einzurichten. Die hierfür notwendigen Unterlagen wurden in einem Rohentwurf erstellt und befinden sich zur Zeit in der weiteren Abstimmung innerhalb der beteiligten nationalen Organisationen.

Im folgenden sind die Grundlagen dieser Vorlagen im Original (Stand 21. Juli 1989) in Auszügen wiedergegeben.

3. Proposal for a New Work Item (Draft)

Title:
An Imaging Application Programming Interface with an Abstract Description of Image Data Formats and Image Processing Operations and with related Audit Trail and State Capture Facilities.

Scope:
The standard includes an abstract modelling of image data and image operators as well as audit trail and sophisticated state capture facilities to support advanced undo/redo strategies. The standard will be extensible in the sense of addition of new functions required for specific needs of future applications. It includes bindings to high-level languages and provides the basic building blocks and tools to build imaging applications within conventional, distributed, windowed or image-oriented computing environments. The basic building blocks encompass data objects, operator objects, peripheral objects and objects which establish the connection to various user interface systems.

Purpose and Justification:
The purpose of the standard is to allow applications developers to perform a rich set of imaging services without biasing the architecture used to implement/provide those services. The standard is intended to support operations on various classes of images from a wide range of imaging applications, from those which simply display images to those which require highly interactive human directed image processing. Where applicable, interfaces to existing standards will serve to handle non-imaging specific areas. The standard provides a means for the sharing and use of specialized display architectures and image processing hardware.

There is currently no standard API in the imaging community, de facto or otherwise. Image processing, manipulation, and analysis are widely performed operations. There is a growing use of imaging applications, yet there is no focused effort to develop a standard interface. Due to the lack of such a standard it is diffcult to produce and maintain applications which require imaging. It is becoming increasingly difficult to do so in contemporary environments using specialized imaging hardware or distributed and non-homogeneous models of computing.

Relevant Documents to be considered:
ANSI: X3H3.8 PIK Working Draft Version 4.0
DIN: Empfehlung für ein Ikonisches Kernsystem IKS (see English abstract)
Japanese Questionnaire on Imaging (May 1989)
NAG-SERC Image Processing Algorithms Library (report on UK Alvey Project No. MMI/127)
ISO NWI Requirements Log (will be provided by 16 October 1989)
Technical Outline from DIN, based on "Imaging" Rapporteur Meeting on 18-19 May 1989 (will be provided by 16 October 1989)

Requirements for Different Application Areas
Image processing is a technology which is employed in a number of different aplications and markets. These applications may include systems for: medical imaging, remote sensing, industrial vision, publishing, computer graphics art, computer animation, scientific visualization, mission planning, traffic control, and so forth.

4. The Overall Imaging System Environment

The system environment of imaging will deal with digital images. Thus, in the diagram below, the *digital image* is placed in the center. In the diagram, *three loops* may be identified. These loops are the interaction loop, the sense and display loop, and the imaging and visualization loop.

The *interaction loop* is for the direct manipulation and enhancement of the image by the user. The function of enhance is to transform imagery in a useful manner. For example, imagery can be modified to enhance the contrast, or to transform it geometrically, or to sharpen the details. Moreover, the user may use tools such as paint-boxes to alter and edit the digital images.

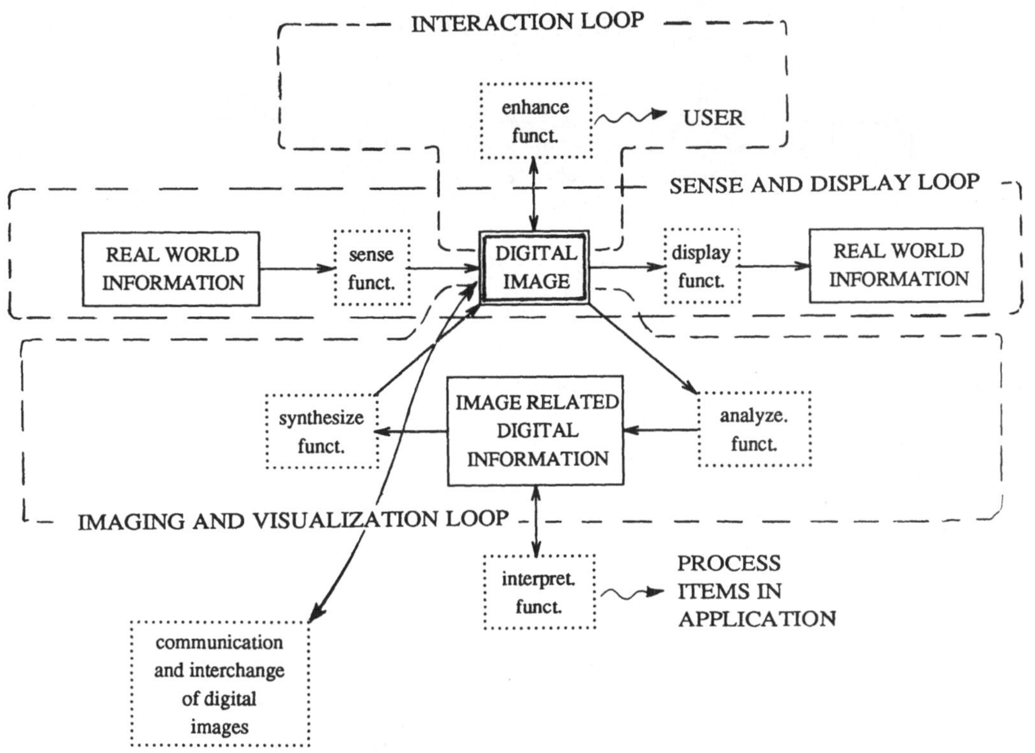

<u>Diagram: The Overall System Environment</u>

The *sense and display loop* consists symmetrically of sense and display for transacting the real world information into digital images and vice-versa. Sense includes such sub-functions as aim sensor, transduce radient scene energy, compensate sensor distortion, and reconstruct sensor image. Display is the complement of the sense function, it includes to combine image pieces, pre-compensate display distortion, decompose display image, and transduce pixel values.

The *imaging and visualization loop* covers those areas in Computer Graphics which deal very closely with image material. The analyze function transforms imagery into information. Analyses comes in many forms. It can be directed towards the image itself, or towards the real world represented by the information in the image. For example, the histogram analyses as well as object recognition, among others.
The synthesize function is the complement of the analyze function. It is mostly identified with the field of rendering images from their generic description in Computer Graphics.

From the point of view of applications which deal with image evaluation, a *basic image processing model is required*. Employing a fundamental definition and description scheme for structures, the atomic function processing element (SIO) utilizes the following description elements: The defining processing stage elements

- D*: Functional element,
- DN: Neighborhood element,
- DQ: Parameter element,
- DS: Control element,

along with the assocoiated input and output elements

- DI: Input element,

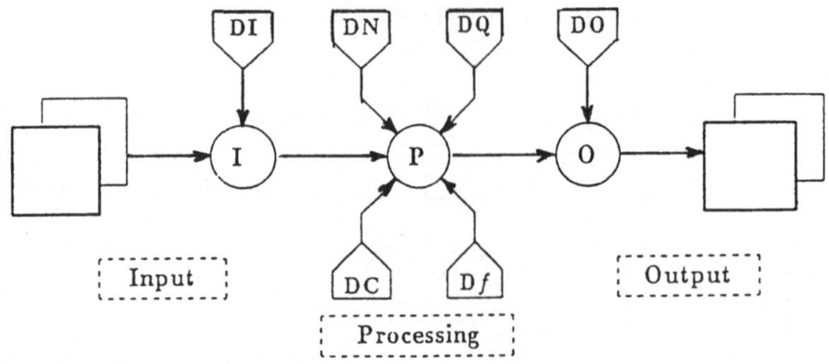

Diagram: Structural Iconic Operations (SIO); Data In (DI) is followed by Processor stage (P) and Data Out stage (DO).

- DO: Output element,

combine to produce a complete description model of the SIO.

The SIO is independent of its hardware implementation. Later on, several implementations may be found for SIOs. The atomic SIOs may be combined in different manners to form some lattices of imaging processors of higher complexity.

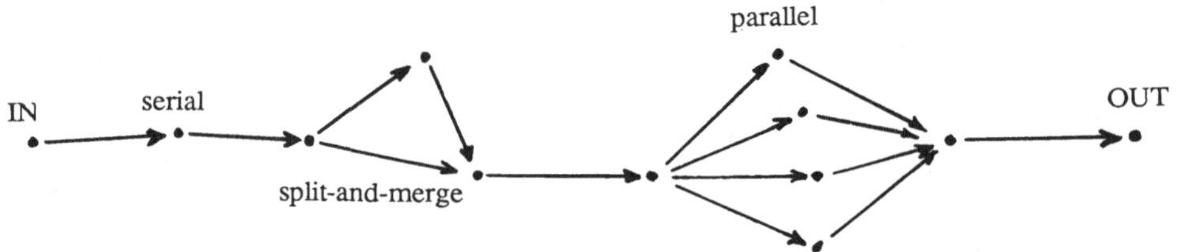

Diagram: Lattices of SIOs.

Such lattices may contain serial, parallel and split-and-merge facillities. Special data flow structures, such as pipelines, may be implemented.

5. Structure Layer Model of an Iconic Kernel System (IKS)

Selection of an object-oriented architecture upon which to base the structure of IKS provide the methodology for system representation which has proven advantageous in reaching the stated development goals. A primary feature of this system architectures is the natural provision for decomposition of the systems whole into an ensemble of objects. An object is a closed functional unit including private data and a set of methods that can access that data. An object is requested to perform one of its methods by sending it a message. Employing this technique provides a well-defined object interface and an unequivocal separation between object specification and object implementation (i.e. how an object implements its methods and how its private data is arranged), which leads to the desirable effect of information hiding and encapsulation. Additionally, this object-oriented design provides a hierarchical structure, with the resultant potential exploitation of common characteristics among classes of objects (inheritance).

Furthermore, employing an object-oriented architecture supports the representation of a comprehensive stock of fundamental iconic building-blocks in a problem-oriented format. Using this building-block approach, the organization of diverse data structures, operations, and devices is established. This allows the conceptual complexity of the system specification to be greatly reduced. In addition, the structure produced contains a high degree of system modularity and thereby supports a simple, structured system expansion without impacting the existing system. The possible architectures of IKS are based upon the following five object categories:

- Data Objects (DO) which provide data structure representations;
- Operator Objects (OO) which represent operations contained in the system and allow the integration of new operations;
- Peripheral objects (PO) which model, and serve as the pathway to, external devices;
- Communication objects (CO) which provide the connection of IKS with diverse use layers;
- Administration objects (AO) which establish system control.

Within individual object categories detailed refinement and specialization according to diverse points of view may be implemented. Using the representative object categories of data, operator, and peripheral, a description of the associated structural hierarchy is presented which can be easily extended if required (by future requirements).

Data objects provide data structures which are the objects of processing, along with the predescribed attributes and functions which facilitate their simple manipulation and autonomous administration. Within the category of data objects a differentiation between elementary, image-like, and SIO data objects class may be established.

Elementary data objects represent essentially general data structures, which reach beyond the domain of image processing. In the framework of IKS, they are used primarily for the data structure specification and representation which serve as an abstract description of symbols or symbol-like image information. In this way, a connection with symbolic image processing is provided.

Images, or image-like data objects are the central element of IKS, and are given by rasters or matrices of image points. They directly represent the geometric organization of the original image sensor data. Differentiation among diverse sub-classes of the image data class is based upon a consideration of spatial dimensions, as well as temporal, luminance, and chrominance attributes.

Operator objects represent image processing operations, and serve primarily as a mechanism facilitating integration of new (specialized) operations into IKS. Establishment of a hierarchical structure within this object class is based upon diverse criteria including operation type, operand count, definition- or value-set mapping, as well as mapping of single pixels (pixel-to-pixel mapping) or local windows (context-to-pixel mapping). Iconic image operations, described uniformly according to the operation model for SIO, are the central pillar of this element of IKS to become standardized.

Peripheral objects model external devices, and will provide a uniform connection to IKS of the spectrum of devices required by an image processing system. Peripheral objects provide both a logical and a functional description of the external device, and provide an encapsulation mechanism for device-specific features. As a result, they provide a logical communication interface between object and device. Hierarchy within this class is established based on requirements of individual device sub-classes.

The structures of IKS will be composed of machine-indepentent and machine-dependent system layers. Particular attention is given to the system interface specification which facilitates the systematic interconnection of diverse user layers and special hardware components to IKS. The generalized (machine-dependent) IKS-layer is the system kernel which, composed of IKS-specified objects, is completely machine and operating system independent. This layer is placed upon the system-specific IKS-layer which provides logical hardware and device-drivers, thereby adapting the operating system and hardware to the machine-independent level. The interface between the generalized layer the system-specific layer is given by way of a list of manufacturer-independent, machine-specific functions which should be as small as possible (procedural interface). As a result, IKS offers both a high degree of portability, which is the primary feature that suppliers of image processing system components are able to deliver new devices which mate directly with the defined procedural interface. Thus, system growth and expansion is easily accomodated within the defined, structured environment of IKS.

6. Issues Concerning a Digital Image Interchange Format

The whole imaging environment discussions have made clearence to a *local* manipulation of the imagery. Nevertheless, the communication of digital images becomes a more and more important item, since the (broadband) networks for the data transfer become available. Not only the television and video-telephone, but also the industrial and administrative applications of the imaging deal with the exchange of digital picture material via the networks.

For these purposes, a special format for the transfer of raster pictures has to be developed which is independent of differences in pixel resolution, color resolution, color primaries, number of colors per device and scan direction. For its use in the networks and on tapes, the file format has to initially have a character encoding to prevent any collision of pixel or image data with control sequences of terminal servers, network controllers and other hardware devices. Binary encodings are still to be developed; the format should be open for different data compressions and other data encodings according to different application profiles.

The file format FTCRP (which means *Format for the Transfer of Colored Raster Pictures*) will meet the lack of a file concept of very high functionality and integrative power in the computer graphics, especially in the so-called "electronic imaging" community. The authors of FTCRP are concerned about different image storage and interchange formats in both the electronic imaging and the printing industry.

FTCRP shall be definitively and truly designed and defined to be a superset of these formats' functionality and parameters.

Specified herein should be the function and a sketchy syntactical description of the file format; the format has to be open for different encodings and compressions (character, binary, run-length, etc.).
Further investigations shall deal with different encodings for the FTCRP.

The transfer of digital raster images and pictures is marked by great dissenses, which origin in the history of the main application areas

i   - printing industry and preprinting process,

ii   - computer graphics art,

iii   - image recognition and image understanding,

iv   - broadcasting and television,

v   - image processing and imaging.

These application areas differ in

-       their main **color models**: As application area i uses mostly YMCK or CMYK, ii and v are using RGB. As iii works with grey level images, iv has the Yuv, which is derived from the CIE1931 standard color space.

-       their **geometric resolution**: The printing industry and the preprinting processing use much more pixels per picture than most of the other areas.

-       their **environment**: As i works mostly with hardcopy and printing facilities, other applications use CRT monitors or both of the two display techniques.

-       their **geometric facilities**: Some application areas use the picture as it is, while area v uses cut-and-paste and other montage algorithms; area i uses to transform (rotate) color seperations against another to prevent aliasing-effects.

In all the mentioned applications areas, several standards have been elapsed to perform raster image transfer. Unfortunately, these standards are dependent of the application area and not suitable for a wider use, e.g. in upcoming digital networks, as ISDN and Broadband-ISDN. Moreover, the application dependent standards are "hard standards": Their definition is "down to the bit on the tape". They do not consider any flexibility anyhow.

FTCRP is going to be designed to compromise between the two extreme positions

insular solution ⟵⟶ hard standard

of standardization in order to achieve a broad acceptance. Therefore, FTCRP should follow the main ideas

-       The picture communication due to FTCRP may be an open communication: The raster picture may be completely specified by the sender of the FTCRP file; the receiver of the FTCRP file will "do his very best" to perform a reproduction and display of the transferred raster picture as it is possible for the receiver's hardware and software facilities.

-       The FTCRP concept and syntax should allow for a balancing of complexity. By transferring data (e.g. raster pictures) due to a common standard, there is a certain effort for the sender to code due to the standard, and a certain effort for the receiver to decode due to the standard. Since the functionality of FTCRP shall cover a very wide range of functionality, sender and receiver of the FTCRP file may have their own agreement whether they meet on a FTCRP level which is closer to the sender or which is closer to the receiver (if the two levels differ at all).

FTCRP should seperate clearly *concept*, *syntax* and *different codings*, as e.g. cleartext, character and binary encoding. Moreover, raster pictures should be transferred to different compressions, such as run-length and other space saving codings.

FTCRP is supposed to be able *to cover each and every parameter* which is relevant for a raster picture in each and every application in the computer graphics and imaging community. Of course, there will be a large set of parameters for specifying a raster picture with FTCRP.

However, we have to compromise between

efficiency $\longleftrightarrow$ true imaging.

Efficiency stands for fast preview, fast retrieval, fast display; true imaging means true color reproduction, true geometric reproduction, apply of sophisticated techniques, e.g. for hardcopy purposes.

Therefore, the parameter concept shall support *subsets of parameters* in terms of FTCRP generator user options and FTCRP interpreter user options. This is to allow for the user to find his own mode of transfer and reproduction of a FTCRP raster picture.

7. Ausblick

Während die Einrichtung eines "New Work Item" im Bereich des Application Programmers Interface als sehr wahrscheinlich gelten kann, bestehen hinsichtlich des Bilddaten-Austauschformates noch Differenzen. Zur Zeit liegt der Bereich nach Ansicht des DIN AK innerhalb der Zielsetzung der Normungsaktivitäten "Imaging", während der ANSI X3H3.8 das Thema ausgeklammert sehen möchte. Anerkannt wird jedoch auch beim ANSI die Notwendigkeit einer konsistenten Eingliederung eines Austauschformats in ein Application Programmers Interface. Für die Ausarbeitung eines Bildaustauschformats sind jedoch auch die weiteren vielfältigen, zum Teil parallel laufenden, Bestrebungen innerhalb der ISO und des CCITT (International Telegraph and Telephone Consultative Committee) zu berücksichtigen.

8. Referenzen

[IKS]:   P. Gemmar, G. Hofele, *Empfehlung für ein Ikonisches Kernsystem IKS*
ITG-Fachgruppe Mustererkennung; Ettlingen 1989.

[PIK]:   *Programmer's Imaging Kernel (PIK), Strawman 3*
ANSI X3H3.8 - Imaging Applications Programmer Interface Task Force Group; o.O. 1989.

[FTCRP]:   G.R. Hofmann (Ed.), *FTCRP- Implementation Manual and Report. Vol. 1: Concepts and Syntax; User Options (Version 1.1)*
Bericht-Nr. FAGD-89s003; Fraunhofer-Arbeitsgruppe für Graphische Datenverarbeitung; Darmstadt 1989.

G.R. Hofmann, W. Puchtler (Eds.), *FTCRP- Implementation Manual and Report. Vol. 2: Compressions and Codings (Version 1.1)*
Bericht-Nr. FAGD-89s004; Fraunhofer-Arbeitsgruppe für Graphische Datenverarbeitung; Darmstadt 1989.

[NORM]:   *Grundlagen der Normungsarbeit des DIN*
DIN-Normenheft 10, 5. Auflage; DIN Deutsches Institut für Normung e.V.; Berlin Köln 1987.

Autorenindex